ORNITHOLOGY

ORNITHOLOGY

Frank B. Gill

The Academy of Natural Sciences of Philadelphia
and
The University of Pennsylvania

Bird portraits by James E. Coe

Consultant for bird portraiture — Guy Tudor

W. H. Freeman and Company

New York

Cover image: Great Egret; photograph © Bryan Peterson/Stock Market.

Library of Congress Cataloging-in-Publication Data

Gill, Frank B.
Ornithology / Frank B. Gill.
p. cm.
Bibliography: p.
Includes index.
ISBN 0-7167-2065-5
1. Ornithology. I. Title.
QL673.G515 1989 89-16793
598—dc20 CIP

Printed in the United States of America

3 4 5 6 7 8 9 0 HL 9 9 8 7 6 5 4 3 2

To my grandfather, Frank Rockingham Downing, who was the first of many to share with me his knowledge and love of birds

Contents

Preface

Ornithology introduces the biology of birds from a contemporary ornithological perspective. The book is designed for undergraduate students, but I have always had in mind, as well, bird enthusiasts young and old, who simply want to know more about birds. Ornithology invites the participation of persons with great diversity of backgrounds and interests. I hope this book will be useful to similarly diverse readers, and, to this end, I have tried to share the excitement that birds and knowledge of them provide me.

Ornithology is divided into six major parts, which follow one another logically:

I Origins
II Form and Function
III Behavior and Communication
IV Behavior and the Environment
V Reproduction and Development
VI Populations

Throughout, I have emphasized the effects of evolution in birds, especially the integration of morphological, behavioral, and physiological adaptations. I give much attention to recent discoveries and to exciting prospects for study in the future. I have avoided theory for theory's sake and have stressed discovery rather than the mathematical models that may guide discovery.

Those who teach a one-semester course in ornithology will doubtless have to choose among the many available topics discussed in *Ornithology.* To allow for differences in the outlines of courses, I have packaged the chapters in short topical units. Also I have structured Appendix I, Birds of the World, in terms of the orders rather than the families, because I find this to be the best way to expose beginning students to the diversity of birds. Students easily master the orders of birds in one semester, but the memorization of families is often overwhelming. As a supplement to *Ornithology,* I recommend *The Birder's Handbook* by Paul Ehrlich and his colleagues (published by Simon and Schuster, Inc.), which provides students with information on the biology of the species of birds they learn to identify on field trips. Also available for lectures are slide sets covering the birds of the world, bird behavior, familiar eastern North American birds, eastern wood warblers, and others. These can be ordered directly from VIREO, c/o The Academy of Natural Sciences, 19th Street and the Parkway, Philadelphia, Pennsylvania 19103.

The participants in any field of endeavor develop their own jargon. My editors have ruthlessly slashed my unwitting use of esoteric shorthand, and together we have striven to define terms that are not in everyone's vocabulary when those terms first appear in the text. The English names of bird species are always capitalized to leave no doubt that, for example, a Brown Booby is *Sula leucogaster,* not just a booby that is brown. The illustrated taxonomy of the species in Appendix I, Birds of the World, conforms closely to that of John Morony and his colleagues (1975) and the sixth edition of the American Ornithologists' Union *Checklist for the Birds of North America.* James Clements (1974) was my standard for the common English names of bird species not found in North America. Scientific names appear in both Appendix I and Appendix II, an index of English names with cross references to species.

Many topics in avian biology — evolution of mating systems, communication, speciation, and the classification of birds, to name just a few — are entering an era of revolutionary reexamination. Rapid scientific development is always daunting to the author of a general work, but I have sought to warn of imminent change. Moreover, wrong notions abound in ornithology, as in all our attempts to understand the world; I look forward to spearing those wrong notions that I may have inadvertently sustained or promulgated, and I invite my readers to help slay a few dragons. Controversy, of course, also abounds in ornithology, surely a sign of intellectual well being. I have tried to present controversy impartially. Where I have taken sides, I hope it will turn out I have done so sensibly. More, I hope I have presented a summary, not of my own biases, but of ornithology by ornithologists.

Acknowledgments

Many colleagues generously reviewed manuscript and enhanced both the accuracy of the text and the effectiveness of the presentation. I sincerely thank them for their time and efforts: John Alcock, George Barrowclough, Bruce Beehler, Carl Bock, Jack Bradbury, Howard Brokaw, Donald Farner, John Farrand, Alan Feduccia, Alec Forbes-Watson, Mercedes Foster, Crawford Greenewalt, James King, Donald Kroodsma, Wesley Lanyon, David Ligon, Peter Marler, Douglas Mock, Douglass Morse, Gene Morton, Pete Myers, Ken Parkes, Robert Payne, Bob Ricklefs, Fred Sheldon, Charles Sibley, Peter Stettenheim, Robert Storer, Harrison Tordoff, Sandra Vehrencamp, Charles Walcott, Nathan Wheelwright, and Larry Wolf.

I owe special thanks to Guy Tudor for his supervision of the ornithological illustrations; to Jim Coe for his bird portraits; to Mary Dolack for her dedicated help in the production of the final manuscript, especially the bibliography; to editor Joseph Ewing for his substantial contributions to the introduction and for skillfully enhancing the manuscript in all respects; to Bob Ricklefs for his sustaining encouragement; and to my family, Frances, Diana, and Jim, for their loving patience.

ORNITHOLOGY

Introduction

With no other animal has our relationship been so constant, so varied, so enriched by symbol, myth, art, and science, and so contradictory as has our relationship with birds. Since earliest record, birds have served as symbols of peace and war, as subjects of art, as objects for study and for sport. Birds and their eggs range from the most exotic to the commonplace: the food on our table. Their command of our imagination is not surprising because they are astonishing creatures, most notably for their versatility, their diversity, their flight, and their song.

Birds are conspicuous and are found everywhere. We find Snowy Owls within the Arctic Circle and sandgrouse in the deserts of the Middle East, the White-winged Diuca-Finch at the highest elevations of the Peruvian Andes, Emperor Penguins diving hundreds of meters beneath Antarctic seas, parrots in the rainforest of Brazil, and ostriches in the arid plains of southern Africa.

These highly mobile creatures are travelers of the long distance and the short. Some birds, like the Nicobar Pigeon in Indonesia, move incessantly from island to island, while others are master navigators, traveling phenomenal distances. The Sooty Shearwater migrates from islands off Australia to the coasts of California and Oregon, the Arctic Tern migrates from New England to the Antarctic, and the Rufous Hummingbird migrates from Alaska to Mexico.

And birds please the eye. Little in nature is more extravagant than the Twelve-wired Bird-of-Paradise, more subtly beautiful than the Evening Grosbeak, more stylish than the Horned Sungem, or more improbable than the Javan Frogmouth.

All these qualities seem to have provoked wonder and a sense of mystery since the dawn of human existence. Indeed, in almost every primitive culture birds were divine messengers and agents: To understand their language was to understand the gods. To interpret the meaning of the flight of birds was to be able to foretell the future. Our words *augury* and *auspices* literally mean bird talk and bird view. By the time Greek lyric poetry was flourishing (5th and 4th centuries BC), the words for bird and omen were almost synonymous, and a person seldom undertook an act of consequence without benefit of augury and auspice. This practice prevails in Southeast Asia and the Western Pacific.

As symbols of ideology and inspiration, birds have figured largely in most cultures and in many religions. The dove was a symbol of motherhood in Mesopotamia, and was especially associated with Aphrodite, the Greek goddess of love. For the Phoenicians, Syrians, and Greeks the dove was the voice of oracles, and in Islam it is said to call the faithful to prayer. In Christianity it represents the Holy Spirit and is associated with the Virgin Mary. Bearing an olive branch in its bill, it continues to be a potent symbol of peace, most strikingly represented in Picasso's painting, The Dove of Peace. In contrast, the dove was a messenger of war in early Japanese culture.

The eagle appeared as a symbol in Western civilization as early as 3000 BC in the Summerian city of Lagash. In Greek mythology the eagle is the messenger of Zeus. At least since Roman times, the symbolic eagle in Europe was the Golden Eagle, and that species also was the war symbol of many North American Indians at the time of early English settlement. In 1782 when Charles Thomson, Secretary of the Continental Congress, designed an eagle as the symbol of the fledgling United States, he chose the Bald Eagle. Among the members of the Congress who opposed any eagle as a symbol for a republic was Benjamin Franklin, who opposed the Bald Eagle specifically because, he said, it was "a bird of bad moral character." He proposed instead the native wild turkey. As we know, Franklin lost the argument.

Less common than the eagle, but prevalent in myth and legend, the raven has had a long but checkered career. As Apollo's messenger the raven reported a nymph's infidelity, and, as a consequence, Apollo changed the bird's color from white to black. After 40 days Noah sent forth both a dove and a raven to discover whether the flood waters had receded. The faithless raven, according to some accounts, did not return and so earned Noah's curse and, once again, a color change from white to black. The belief in the raven's color change appears in a Greenland Eskimo legend in which the Snowy Owl, long the raven's best friend, poured sooty lamp oil over him in the heat of a disagreement.

In other legends the raven plays a more favorable role. North American Indian folklore described the raven's generosity in sharing its food with men stranded by flood waters. Norse sailors, like Hindu sailors half the world away, carried ravens, which they released to guide them to land. Two ravens were widely reported to have guided Alexander the Great through a duststorm on his long journey across the Egyptian desert to consult the oracle at the Temple of Ammon.

Centuries later, Konrad Lorenz shared the favorable view in his deeply moving account of a young raven that formed a lifelong attachment to him:

> He tried to be with me as much as he possibly could, accompanying me on all my walks, either flying from tree to tree, or, with a favoring breeze, sailing high above my head and following me in the same way as, with other motives, vultures follow a caravan. During the time I was [away] from home he searched for me everywhere, specially and very intelligently in those places where we had been together . . . .
>
> As accompanying a walking man means much troublesome wheeling and flying for a raven, I wanted to accelerate my movements by the use of a bicycle. It was difficult at first to convince Roah [the name Lorenz gave the bird in what he calls a feeble imitation of the raven's deep, unbirdlike call] that the bicycle was harmless, so I shut him in his cage with the bicycle, putting all the food on the saddle. He loved the bicycle for ever afterwards, the more so when he understood that it was much easier and more fun to follow me when I was cycling. I had only to wheel the machine out of the house to make him utter a joyous series of starting calls and send him flying along our usual route. He never understood why I could not fly, and to the end of his days tried to induce me to take wing.

Not only is our association with birds as old as human society, but it is characterized by the diversity of our interests in them. We can do no more here than give a few examples of the diversity and, by way of those examples, come finally to the rich and varied science of ornithology. We begin with avian flesh and eggs on our table.

The domesticated chicken existed in India before 3000 BC and was known in China by 1500 BC. It appeared in Mediterranean countries at the same time, though its early use there may have been more for religion than food. Large-scale breeding and raising of poultry for food by the Romans developed, but the practice on that scale disappeared after the fall of the Roman Empire and did not reappear in Europe until the 19th century. Mallard ducks and geese were domesticated in the Far East nearly 1000 years before Christ. Domestication of the turkey in Mexico appears to be very ancient, and the bird was imported into Europe by the middle of the 16th century. (There is some belief that Christopher Columbus carried turkeys to Europe in 1492.)

Domestic fowl were brought from England to the Jamestown Colony in 1609, and through the 18th century other forms were imported from Asia. Because of ever-growing interest in new breeds in the United States, the first American poultry exhibition was held in Boston in 1849. In 1873, the American Poultry Association (APA) was founded, the oldest association of livestock breeders and growers in the country. In 1905, the APA published the *American Standard of Perfection.* Now in its 1983 edition, the book is a wonderfully informative and entertaining illustrated guide to the ideal characteristics of more than 100 domestic fowl, ducks, geese, and turkeys, and is one piece of evidence that the chicken is certainly one of the most refined domestic animals.

Trapping small passerine birds, such as the skylark, began very early. Greek vase paintings of bird nets appeared by the 6th century BC. These birds were considered a great delicacy, as they still are in Italy and parts of the Far East. The invention of firearms enabled us to kill birds more easily, and the sport of hunting for its own sake developed with considerable style and ritual. In our own time, building a duck decoy has itself become an art.

Like birds themselves, birds' eggs have been prized as food for thousands of years. For both eggs and flesh there seems to be an interesting, and not very surprising, evolutionary connection between inconspicuousness and palatability. This connection provoked so much interest that a tasting panel was formed by the Department of Game and Tsetse Control in Zambia in the early 1980s. The results of eating and rating some 190 species generally bore out a correlation between declining palatability of flesh and increasing conspicuousness of the animal. A tasting panel in Cambridge, England, came to much the same conclusion about birds' eggs: the more conspicuous the egg shell, the less palatable the egg.

The pigeon has had a dual role as a carrier and as a prized food. There were ancient pigeon posts in Babylon, and the bird was used as a carrier in early Egyptian dynasties. Aelian, a Roman writer active in the early 3rd century AD, described the use of a carrier pigeon by an athlete named Themistocles to inform his father of his victory in the Olympic Games. Use of carrier pigeons as messengers was very well developed in Roman times and continued through the centuries until the invention of the radio, telegraph, and telephone, when the practice was largely abandoned except for sport and research.

Falconry, on the other hand, is blessed with a modest renaissance. The sport may have originated as long as 4000 years ago. It flourished in Europe in the Middle Ages, and the Crusaders brought back Islamic techniques that increased and refined European falconry. After a sharp decline of Peregrine Falcons and several small accipiters in Europe and North America in the 1960s, breeding and release programs arose, and now the ancient sport, with its historical tradition of studying and protecting birds of prey, is being revived.

Bird farming on a very large scale occurred in the years before World War I when more than three-quarters of a million ostriches were bred in South Africa each year. There was then a great demand for their feathers for ornamentation, a fashion that does not seem to have been popular in Western civilization before the late 19th century. Ostrich farming on a smaller scale continues today, with the skins harvested for use in leather goods. But use of feathers as ornamentation was widespread among North and South American natives, in Africa, and in the Western Pacific from the earliest known times. The elaborate feather capes of the Hawaiian kings and the feather mosaics of the Mayas and Aztecs were works of high art. Among native North Americans, particular uses of feathers as badges of rank and status were common. Feather clothing was also common for protection from weather, much as goose down is widely used today.

Perhaps the greatest influence of birds on art has been in music. The earliest piece of English secular music of which we know, "Sumer is Icumen in," is a canon for four voices and the words are those of the 13th-century lyric in which the cuckoo welcomes summer with its song. The cuckoo, nightingale, and quail are heard in Beethoven's Sixth Symphony. The 18th-century composer Boccherini wrote a string quartet called "The Aviary," perhaps the first complex composition in which a number of birds are imitated.

Birds as subject and as metaphor are found frequently in opera. Wagner wrote an aria about owls, ravens, jackdaws, and magpies for *Die Meistersinger.* In Puccini's *Madame Butterfly* a character sings of a robin, in *La Boheme,* another sings of swallows. In what is probably the most popular aria in the most popular opera of all time, the "Habañera" in Bizet's *Carmen,* the opening words are "Love is a rebel bird that no one is able to tame." Janacek's *The Cunning Little Vixen* is a 20th-century "nature opera" with animal characters that include numerous birds, and in Gershwin's *Porgy and Bess* there is a buzzard. These examples are only a few of the many that come to mind.

Composers Maurice Ravel and Bela Bartok were knowledgeable about bird species and their songs. Ravel made use of his knowledge in works for orchestra, voice, and, most strikingly, piano. Bartok's interest was so persistent that during a visit to the mountains of North Carolina, he took long walks and transcribed bird songs previously unknown to him. It is believed that much of the transcription found its way into the *Piano Concerto No. 3,* his final work.

In rock music, the best treatment of birds may be Jimmie Thomas's 1958 "Rockin' robin," a solid hit when first sung by Bobby Day and revived to equal acclaim by Michael Jackson and The Jackson Five. Swallows, chickadees, and crows urge the robin to "Go, bird, go," a raven teaches him the Charleston, and he turns out to be a better dancer than buzzards or orioles.

Conway Twitty was a contemporary of Elvis Presley and was one of the best early rock-and-roll singers. One of his most popular songs asks the question "Is a Bluebird Blue?" Unfortunately, Twitty is now less remembered for his own music than for his transformation into Conrad Birdie in the Broadway musical *Bye Bye Birdie.*

Another interesting confluence of the name of the musician—in this instance, the nickname—and the name of the music brought together one of the most memorable of American jazz musicians and one of the most memorable tunes: Charlie "Bird" Parker and "Ornithology."

Our enjoyment of bird sound in music extends beyond mere imitation. The phonograph recording of the singing of a real nightingale is played in performances of Respighi's *Pines of Rome,* and James Fassett, an American composer, has written a *Symphony of Birds* that consists entirely of the recorded songs and calls of real birds.

The role of birds in painting and sculpture is impressively large. Birds appear in paleolithic cave paintings in France and Spain as early as 14,000 BC and as neolithic cave paintings in Eastern Turkey 8000 years later. In Egyptian tombs at Thebes, very accurate bird paintings appear before 2000 BC. One depicts a man force feeding cranes with a funnel in exactly the same way geese are fed in the Perigord region of France today. In both painting and sculpture, the Egyptians' accuracy has seldom been matched. In *The Outermost House,* Henry Beston wrote of this unique power of ancient Egyptian artists "to reach, understand, and portray the very psyche of animals. . . . A hawk of stone carved in hardest granite on a temple wall will have the soul of all hawks in his eyes."

In Knossos, on Crete, a famous Minoan fresco of a partridge and a hoopoe survives from about 1800 BC. A few centuries later birds begin to abound in Greek vase painting, some depicted with a quite modern realism. In Roman frescoes and mosaics, birds are sometimes stylized, but many are not. Among the most vibrant and brilliantly colored are those in mosaics from Pompeii that are now in the National Museum in Naples.

In much of medieval art, birds became so highly stylized that it is often impossible to identify the species. A remarkable exception is an assemblage of species in a 13th-century illuminated manuscript of the Book of Revelations. The composition, in which the birds sit before a man preaching, is strikingly similar to Giotto's famous painting of St. Francis preaching to the birds, with one late-comer arriving after the sermon has begun. Hieronymus Bosch's *Garden of Delights* (about 1500), is filled with birds, some realistic, though sinister, and some monstrous hybrids.

Birds appear frequently in English and Dutch nature paintings and still lifes of the 17th century, but the Romantic painters of the 18th century showed little interest in birds despite great enthusiasm in Europe at the time for Japanese prints in which birds abound. The impressionists and postimpressionists also showed little interest in birds, though van Gogh's last work, painted on the day of his suicide, is of a flight of rooks across a somber sky. Among 20th-century painters, Matisse and Picasso showed recurring interest in birds. Some of the best of Picasso's work are the etchings of birds he did for a modern edition of Buffon's *Natural History.* In 20th-century sculpture, Brancusi's sleek birds in both chrome and stone are memorable.

Birds are ubiquitous in literature. In his comedy *The Birds,* Aristophanes mentioned 79 species and was well informed about their habits and appearance, as his Athenian audience must have been; otherwise they would have missed much of the playwright's wit. For its perfect matching of avian and human characteristics, the play has been described as an "ornithomorphic view of man." One of the earliest lyric poems in English describes a quarrel between an owl and a nightingale. The cuckoo, nightingale, and lark probably appear in poetry more often than other species. The nightingale and the lark appear together in the argument between Shakespeare's Romeo and Juliet about which of the birds they heard singing at dawn. Birds are prominent enough in Shakespeare's plays and poems to have led the scholar James Harting to write an entire book on the subject, *The Ornithology of Shakespeare,* first published in 1871.

Some lyric poets were excellent ornithologists, notably the 17th-century Englishmen Michael Drayton and Andrew Marvel, whose descriptions of birds are very precise. More recently, Shelley's skylark, Keats's nightingale, and Yeats's swan have become the best known birds in English literature.

Beginning as early as the 15th century, books with numerous bird illustrations began to appear. Pierre Belon wrote a *History of the Nature of Birds* with 160 woodcuts, which was published in Paris in 1555. Illustrated works began to appear in ever greater numbers in the 18th century. The Count de Buffon's *Natural History of Animals*, published in Paris over the last half of the 18th century, contained almost a thousand plates illustrating birds. Between 1731 and 1743, Mark Catesby, an Englishman, published many hand-colored plates of birds of the Carolinas, Florida, and the Bahamas. Catesby's birds are static and unlifelike, but the beauty of the colors is undeniable. At the end of the 18th century, another Englishman, Thomas Bewick, produced very fine bird engravings that were distinguished by detailed rendering of the background. He was followed by, and doubtless influenced, John James Audubon, the best-known painter of birds, who published between 1827 and 1838 his enormous four-volume work, *The Birds of America*. Also in the mid-nineteenth century, John Gould, aided by a number of other artists, produced dozens of grand books of bird illustrations. By the turn of the century, a great flourishing of bird illustration was under way and it continues to this day. It was closely associated with the rise of modern ornithology and of field guidebooks.

Among the finest illustrators of the early 20th century were Bruno Liljefors of Sweden, Archibald Thorburn of England, and Louis Agassiz Fuertes of the United States. Fuertes, with his unerring eye and his faultless sense of the salient characteristics of any bird, is believed by some to have made his birds more dazzlingly alive than any other painter.

With all the disparate appeal of birds, it is little wonder that some human beings have chosen to study them. Aristotle's 4th-century BC *History of Animals* is the first effort we know of in Western culture to account systematically for what we observe in nature, and the writing reflects the first organized scientific research. Birds figure prominently in all of Aristotle's work in natural history. Alexander of Myndos, in the 1st century AD, wrote a three-volume work on animals, two of which are about birds. Only fragments survive in quotation. Pliny the Elder (AD 23–79) produced an elaborate natural history encyclopedia in 37 volumes, all of which survive. He summarized the work of some 500 ancient authors and offered his own critical point of view. Aelian (AD 170–235), a Roman who wrote in Greek, devoted much attention to birds in his *On the Characteristics of Animals*. Until the Renaissance, our knowledge of the natural history of birds depended largely on these and other Greek and Roman writers. They told us much that was reliable, but they also left us with many wrong notions. The quotations from Alexander's work reflect close and accurate observation, but Aelian was steadfastly uncritical of his sources and perpetuated two remarkably wrong notions about the behavior of cranes: one, that they flew against the wind and swallowed a stone for ballast so as not to be swept off course; the other, that they posted sentinels at night, requiring them to stand on one foot while holding a stone in the other, thereby insuring that if the sentinel fell asleep, it would drop the stone and be awakened by the noise. More remarkable still was Aelian's notion that the Purple Gallinule would hang itself if it discovered an adulterous wife and was so modest it would faint at the sight of a bridegroom.

After Aristotle and Pliny, we have little in the way of systematic observation and cataloguing of information until the 1500s. We have mentioned Pierre Belon's 1555 *History of the Nature of Birds,* and in about 1600, Ulyssis Aldrovandus, working at the University of Bologna in Italy, produced a three-volume, 2600-page work entitled *Ornithologiae.*

The next step toward modern ornithology was the growth of field observation in the eighteenth century. Captain John Smith made a list of birds at the Jamestown Colony. In 1789, Gilbert White, an English clergyman, published a natural history of his parish, gathered over 40 years' time. His observations of birds were marvelously precise and beautifully expressed. In his *Notes on Virginia,* Thomas Jefferson commented on 77 species of birds and later instructed Meriwether Lewis to make what turned out to be a fine list of birds observed on the Lewis and Clark Expedition. William Bartram, a naturalist born in Philadelphia, described more than 200 species of birds from his observations during a trip through the Carolinas, Georgia, and Florida from 1773 to 1777. Then came Alexander Wilson, an emigrant Scotsman and school teacher, who (encouraged by Bartram's example) eventually produced his own nine-volume *American Ornithology,* the seminal modern work. In 1832, Thomas Nuttall, curator of the Harvard botanical garden, published a more manageable work, if not title, in *A Manual of the Ornithology of the United States and of Canada.*

From this now solid grounding arose the great museums of natural history and their collections, the ornithological societies, and the array of magazines and journals devoted to birds. At the beginning of this century, Frank Chapman of the American Museum of Natural History became a pivotal leader in modern ornithology. Lacking formal training, he nonetheless was a pioneer in the study of South American birds and possessed a remarkable ecological and evolutionary perspective. He assembled a formidable group of colleagues — Dean Amadon, James Chapin, Ernst Mayr, and Robert Cushman Murphy — whose efforts dominated the study of ornithology for half a century. Moreover, Chapman was a great popularizer. He championed the diorama mode of museum display and wrote the *Handbook of Birds of Eastern North America,* the first widely distributed pocket field guide to birds, published in 1895.

Contemporary ornithology has benefited from years of careful field observation by devoted amateurs as well as by professional ornithologists. Our knowledge of avian life histories and populations is more complete than that of most other classes of animals. Owing in part to this wealth of information and in part to the attributes of birds, birds have increasingly become the subject in primary biological studies. By the middle of the 1980s, as Ernst Mayr (1984) has pointed out, birds provided more textbook examples of biological phenomena than any other class of vertebrates.

Birds have been central to work on the formation of species and have been used in some of the most detailed molecular analyses of phylogeny. Perhaps the greatest contribution of bird studies has been to population and community ecology, but their contribution to evolutionary ecology and to the discovery of new connections between animal behavior and ecology are not far behind. Birds are particularly well suited for the study of mating systems and strategies. Birds are similarly useful in investigation of the roles of kinship in evolution and of altruism. The rules governing communication and physiological mechanisms that connect communication to behavior have been greatly elucidated by bird studies.

Because only humans, parrots, and songbirds can imitate sounds, birds are immensely important to the investigation of the interplay between inheritance and learning. Because birds in captivity continue partly to maintain their natural behavior, and to some degree because they are long-lived, they are useful in the study of effects of natural stimuli on physiology and behavior. The same characteristics also make it possible to study the environmental control of reproduction and the role of circadian and circannual rhythms.

Birds are useful in the study of hormone action on developing and adult brains, of the anatomy and development of brain neural circuitry, and of cell death and survival. Systems for song control have served for the study of sexual differences in the brain. Hormone action on receptors, regulation of gene expression, the molecular biochemistry of hormone action, the evolution of the neocortex, and the uses of brain maps are all better understood because of research on birds. Finally, birds show some of the capacity, common in lower vertebrates, for adult growth and regeneration of the brain. This could be important in determining why mammals lack the capacity.

Birds are, more obviously perhaps, ideal subjects for the study of adaptation to extreme conditions and unusual niches, of navigation, and of energetics of flight. And at the very origin of ethology, the work of Niko Tinbergen with gulls and of Konrad Lorenz with ducks and geese provide classical examples of the attempt to understand the evolution of behavior. Their studies earned them a Nobel prize. In cell biology and medicine, the discovery of B vitamins and their roles in nutrition came from studies of chickens, which readily reveal dietary deficiency. Albert Szent-Györgyi won a Nobel prize for the elucidation of the "Krebs" cycle from studies of pigeon breast muscle, as did Payton Ruos for studies of avian sarcoma that linked viruses to cancer for the first time.

Primary research has greatly affected our understanding and practice of conservation, especially research in ecology and evolution. Birds are, of course, sensitive indicators of environmental change, witness Rachel Carson's account in *Silent Spring* or failure of seabird reproduction, resulting from changes in food and climate, that warns of the El Niño phenomenon well before meteorologists announce it. We are only now beginning to understand that, besides urbanization and modern modes of transportation, the practices of agriculture and forestry, especially mechanized deforestation, have profound effects on habitats and populations. Because of their migration, birds more than other animals help us to understand the global nature of these effects. Sadly, study of birds is also helping us to understand the effects of our introduction into the environment of domestic animals, rats, nonnative species, and oil and chemical pollution.

A few stories of extinction in modern times are well known; those of the Passenger Pigeon and the Great Auk are dramatic. Less well known is an extraordinary story of a battle won by the birds. It occurred in Western Australia in 1932 and is known as the Emu War. At the time, it attracted much attention and was covered by the press. It seems that some 20,000 emus threatened wheat fields. Soldiers employing machine guns and artillery spent a month attacking the birds. The birds, in the words of Dominick Serventy of the Australian Wildlife Research Office, "apparently adopted guerilla tactics and split into small units. This made the use of military equipment uneconomic." After the soldiers withdrew, fences were built to separate the emus from the grain.

A satisfactory outcome indeed. Perhaps through basic research and thoughtful practice, we will come to a more harmonious relationship with those astonishing creatures that have charmed and fascinated us throughout our own existence.

An informal bibliography

Joseph Kastner's *A World of Watchers,* (Sierra Club, 1988) is a delightful ramble through the world of birdwatching and almost everything that can be connected to it. Mr. Kastner and Miriam T. Gross are the editors of *The Bird Illustrated from 1550–1900* (Abrams, 1988), a lovely set of annotated reproductions of paintings and drawings from the collection of the New York Public Library, where Ms Gross is a research librarian. Two books of wide-ranging interest are *A Book of Birds* by Mary Priestley (1937 but out of print) and Joseph Wood Krutch's *A Treasury of Bird Lore* (Doubleday, 1962). Peggy Munsterberg wrote a fine introduction to *The Penguin Book of Bird Poetry* (Penguin, 1984), and John R.T. Pollard's *Birds in Greek Life and Myth* (Thames and Hudson, 1977) is graceful and is the most complete account of its subject written entirely in English. Finally, *A Dictionary of Birds,* edited by Bruce Campbell and Elizabeth Lack (Buteo Books, 1985), is unfailingly entertaining and informative on almost every topic that has to do with birds.

part I

ORIGINS

chapter 1

Birds: Form and function

Some 150 million years ago, a small, bipedal reptile lived among the dinosaurs. Its stiff scales eventually became soft feathers. Its leaps and short glides led to graceful flight. Feathered insulation enhanced control of a high body temperature, increasing activity and endurance. Mastery of flight opened a world of ecological opportunities. A new group of vertebrates, the Class Aves, evolved.

Responding to the ecological opportunities, birds diversified in form and function. From a fundamental anatomy evolved huge, flightless ground birds such as ostriches; small, agile perching birds such as chickadees; nocturnal hunters such as owls; aquatic divers such as penguins; aerial masters such as albatrosses; and shoreline waders such as herons (Figure 1–1). In each major assemblage of birds appeared a host of variations, specialists eating particular prey or occupying particular habitats.

The variety of birds is the grand result of millions of years of evolutionary transformation and adaptation. Roughly 300 billion birds of over 9000 species now inhabit the earth. Yet this number is only a small fraction of the number of species that have ever lived since the age of dinosaurs.

Figure 1–1 Birds have evolved along major lines, each adapted to a particular mode of life. (From *Evolution of Vertebrates* by E.H. Colbert; Copyright © 1955 John Wiley & Sons, Inc.; reprinted by permission of John Wiley & Sons, Inc.)

The earliest bird, *Archaeopteryx lithographica,* known from fossils of the Jurassic period, 150 million years old, had feathers and could fly after a fashion. A major radiation of toothed birds (some volant, some not) followed in the Cretaceous period, roughly 100 million years ago. The basic kinds of modern birds, such as raptors, rails, ducks, woodpeckers, and songbirds, evolved in the subsequent epochs of the Tertiary period as climates changed, continents moved, and islands emerged. The distributions of species and the compositions of avifaunas changed as new taxa supplemented or replaced older ones.

What is a bird?

Birds, along with mammals, amphibians, reptiles, and bony fishes, are vertebrates, or animals with backbones. Despite their diversity of form, birds are an extremely well-defined group of vertebrates. They are distinguished from other vertebrates by feathers, which are unique modifications of the outer skin (Figure 1–2). No comparable structures exist in any other class of vertebrates. Feathers are essential for both temperature regulation and flight. They insulate the body and help maintain high body temperature. Lightweight and strong, the long feathers of the wing generate lift and thrust for flight.

All birds have bills, a distinctive attribute that facilitates instant recognition, even by very young children. Anatomically, the avian bill is a

Figure 1-2 Snowy Egret with display plumes. Feathers are the distinguishing feature of birds. (Courtesy A. Cruickshank/VIREO)

protuberant pair of toothless mandibles covered by a horny, epidermal sheath. Although its form varies greatly, the avian bill has no exact parallel among other vertebrates; only that of the duck-billed platypus of Australia comes close.

Birds are feathered flying machines. Their wings and their ability to fly are familiar attributes but are not diagnostic features; bats and insects also have wings and can fly. Nevertheless, the entire avian body is structured for flight.

Fused bones of the pelvis, feet, hands, and head contribute to the rigidity and strength of the body. Horizontal, backward-curved projections on ribs, called uncinate processes, overlap other ribs and so strengthen the walls of the thorax. The furcula, or wishbone, is uniquely avian, having evolved as a central element of the flight apparatus of the earliest known birds. The furcula prevents lateral compression of the chest during downstroke of the wings and also is an anchoring site for the pectoral flight muscles. The wing itself is a highly modified forelimb that, with a few remarkable exceptions, is nearly incapable of functions other than flight. Birds' bones are lightweight structures, being spongy, strutted, or hollow. Modern birds also lack teeth and the associated heavy maxillary bones of the jaw. The reduced anterior mass improves balance in flight and in precision landings.

Birds are endothermic. They maintain high body temperature as a result of metabolic heat production, as do mammals, over a wide range of ambient temperatures. Avian physiology therefore accommodates the extreme metabolic demands of flight and temperature regulation. The red fibers of avian flight muscles have an extraordinary capacity for sustained work and the ability to produce heat by shivering. Avian circulatory and respiratory systems are powerful and efficient in delivering fuel and removing waste, and metabolism for flight and high body temperature makes great demands on those systems.

The reproductive and sensory systems of birds are also unusual. Birds produce large, richly provisioned external eggs, the most elaborate reproductive cells of any animal. Nurturing the growth of the embryos and of the young after they hatch requires dedicated parental care, which is coupled to elaborate mating systems, nesting behavior, territoriality, brood parasitism, coloniality, and a suite of other social adaptations. Highly developed neural systems and acute senses mediate feats of navigation and communication. In particular, the extraordinary vocal abilities of birds have few parallels among animals. Only human vocal production is comparable.

Birds as reptiles

It has long been recognized that birds evolved from reptiles (Figure 1–3). Thomas H. Huxley, the great evolutionary biologist of the 19th century, asserted that birds were "merely glorified reptiles" and classified them together in the taxonomic category Sauropsida (Huxley 1867). Birds and reptiles share many characteristics. The skulls of both articulate with the first neck vertebra by means of a single ball-and-socket device, the occipital condyle (mammals have two condyles). Birds and reptiles have a simple middle ear that has only one ear bone, the stapes (mammals have three middle-ear bones). The lower jaws, or mandibles, of both birds and reptiles

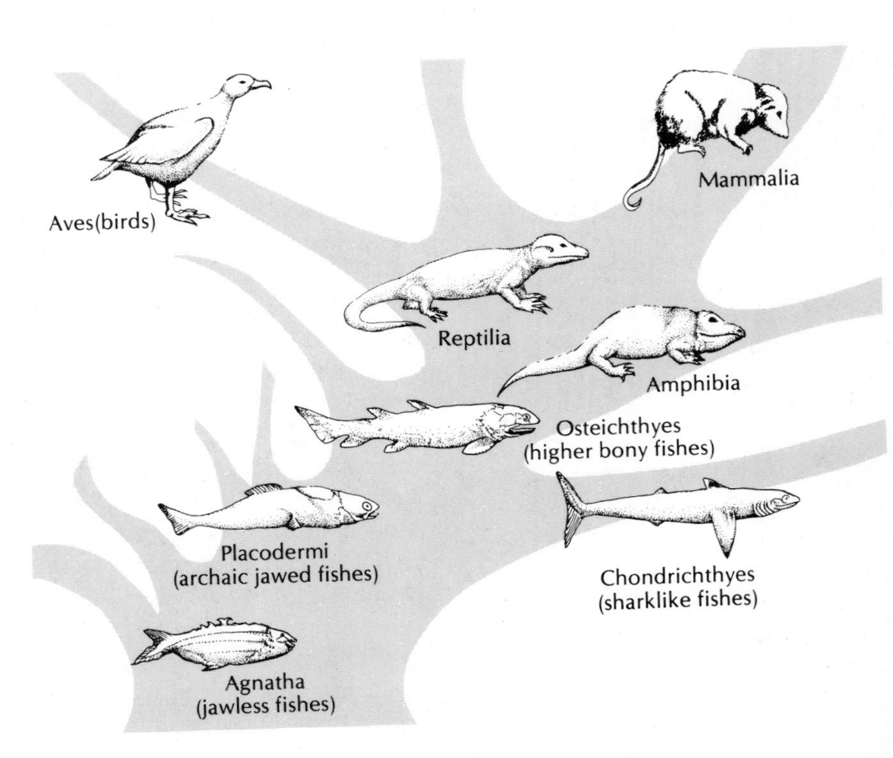

Figure 1–3 A simplified family tree of the vertebrates. Birds evolved from reptiles independently of mammals. (After Romer 1955)

Figure 1 – 4 Some reptilian features of the avian skull.

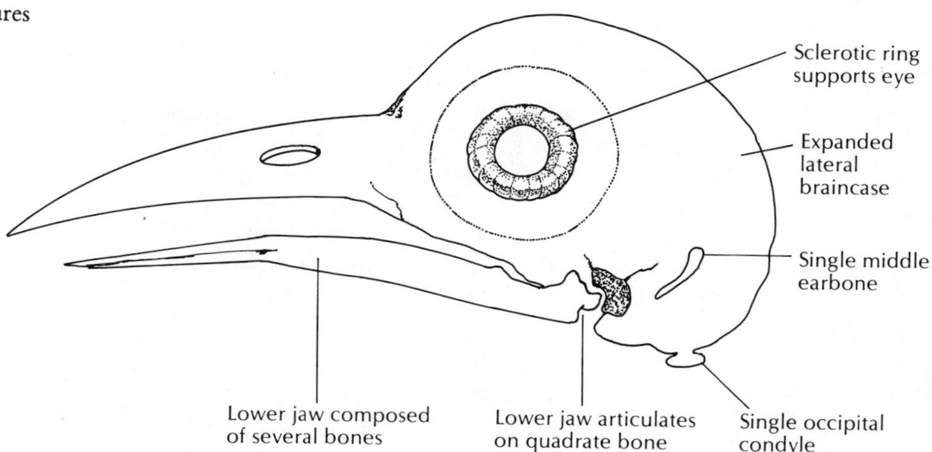

have five or six bones on each side (mammals have only one mandibular bone, the dentary) (Figure 1 – 4). Avian and reptilian ankles are sited in the tarsal bones, not between the tibia and tarsi as in mammals. The scales on the legs of birds are similar to the body scales of reptiles.

Both birds and reptiles lay a yolked (polylecithal), polar (telolecithal) egg in which the embryo develops on the surface by shallow (meroblastic) divisions of the cytoplasm on the surface of the egg. Female birds and some female reptiles have the XY (heterogametic) sex chromosome combination (in mammals males are the heterogametic sex). Birds and reptiles have nucleated red blood cells (the red blood cells of mammals lack nuclei).

Adaptive radiation of form and function

The evolution of birds from a reptilian ancestor and their subsequent diversification is the result of speciation, the irreversible splitting of one species into two with independent evolutionary futures and dissimilar habits. The appearance of a multiplicity of ecological types as a result of speciation is called adaptive radiation, which is characterized by variations in form and function that enable species to exploit their environment: Bill sizes and shapes change in relation to the types of food eaten; leg lengths change in relation to habits of perching or terrestrial locomotion; and wing shapes change in relation to types or patterns of flight and to aerodynamic efficiency.

Birds range in size from 2 grams (hummingbird) to 100,000 grams (ostrich). Mammals span a much greater range of sizes, and few of them are constrained by the limits on size dictated by flight. Indeed, ostriches, the largest living birds, are flightless. The lower limits of size among birds are set by loss of heat from the body's surface. As surface area increases relative to mass, so does the rate of heat loss and the need for energy to maintain a high body temperature.

Modern shorebirds, reflecting adaptive radiation, are of myriad ecological types. The ancestral shorebird has evolved into a host of forms, including aerial scavengers such as gulls, plunging divers such as terns, and pirates such as skuas, as well as a host of waders including sandpipers, plovers, turnstones, stilts, oyster-catchers, snipes, woodcocks, curlews, and godwits, each with characteristic bill length, shape, and curvature. As varied as the habitats they occupy, shorebirds also include aerial pratincoles, deep-water divers such as auks and possibly loons, and the grouselike seedsnipes of South American moorlands.

The varied diets of modern birds include green leaves, buds, fruits, nectar, seeds, invertebrates of all sizes, and vertebrates of many kinds including carrion (Morse 1975) (Table 1 – 1). Few birds are specialized herbivores; apparently, mammals have occupied most of the grazing and browsing niches. Fruits, seeds, and insects nourish the majority of birds, especially the passerine landbirds, whose adaptive radiation was coupled to that of flowering plants and associated insects (Regal 1977).

Corresponding to a diversity of diets is a diversity of bills. Illustrating the variety of bill forms that can evolve during the process of adaptive radiation are those of the Hawaiian honeycreepers, which have evolved from a flock of small finches that strayed out over the Pacific Ocean from Asia or North America millions of years ago. The finches made a landfall on one of the Hawaiian Islands, then flourished and spread throughout the archipelago. Isolated populations changed in genetic composition and appearance, at first imperceptibly and then conspicuously. Subtle changes in bill shapes and sizes led to a proliferation of bill types and feeding ecologies to match, from heavy grosbeak bills for cracking large legume seeds to long,

Table 1 – 1 Numbers of Families Feeding upon Different Food Categories*

Food Category	Primarily	Regularly	Sparingly	Total of Columns 2 – 4	Infrequently, If at All
Green plants, buds	2	13	13	28	140
Seeds	4	31	24	59	109
Fruit	19	34	21	74	94
Nectar	3	5	7	16	152
Insects	50	58	32	140	28
Other terrestrial invertebrates	2	10	23	35	133
Littoral and benthic invertebrates	6	10	13	29	139
Small vertebrates (<5 kg)	4	14	29	47	121
Large vertebrates (5 kg+)	0	0	3	3	165
Fish, crustacea, squid, etc. (nekton and plankton)	20	10	11	41	127
Carrion	1	4	6	11	157
General	—	—	—	56	112

* These data are derived from a detailed family-by-family analysis of the feeding habits of birds. (From D.H. Morse 1975)

sickle bills for sipping nectar from flowers or probing bark crevices for insects (Figure 1–5).

Various forms of locomotion further expand the ecological opportunities of birds. There are specialized flying birds, as well as specialized swimmers, runners, waders, climbers, and perchers. Shorebirds, as we have mentioned, include aerial, wading, and diving species. Birds soar through the sky, scurry and stride across the land, hop agilely from branch to branch, hitch up tree trunks, and swim powerfully to great depths in the sea. The

Figure 1–5 A classic example of adaptive radiation: Hawaiian honeycreepers have evolved bills that range from thin warblerlike bills to long sicklelike bills to heavy grosbeaklike bills. (From Raikow 1976; drawing by H. Douglas Pratt)

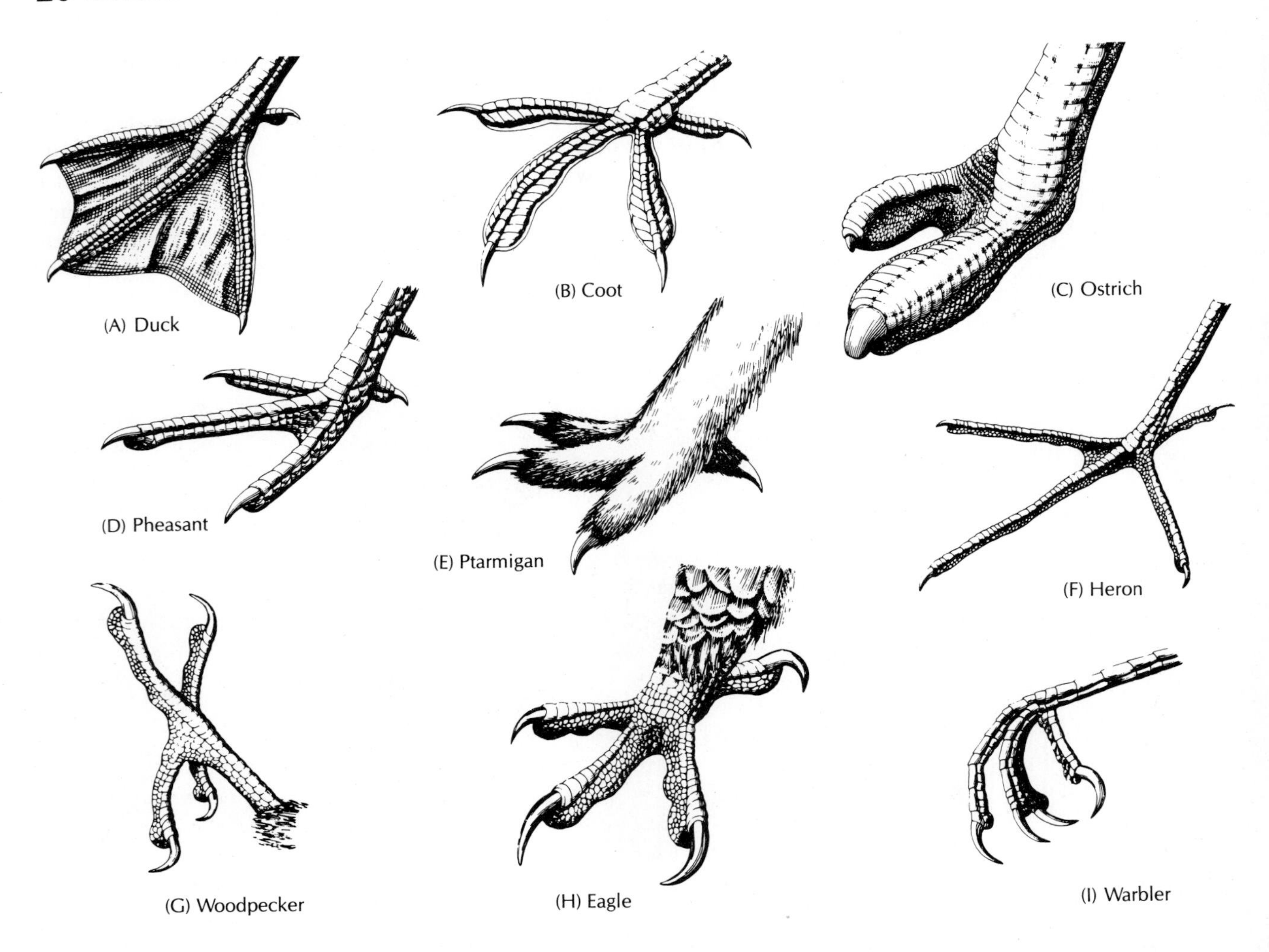

Figure 1–6 The feet of birds reveal their ecological habits: Waterbirds have (A) webbed or (B) lobed toes for swimming; terrestrial birds have toes specialized for (C) running, (D) scratching in dirt, (E) walking on snow, or (F) wading. Other landbirds have feet designed for (G) climbing, (H) holding prey, or (I) perching. (From Wilson 1980, with permission of Scientific American, Inc.)

combination of forelimbs adapted for flight and hindlimbs for bipedal locomotion gives birds a tremendous range of ecological options.

Wing shapes and modes of flight differ greatly, from the long, narrow wings of the albatross, adapted for soaring over the oceans, to the short, round wings of wrens, adapted for agile fluttering through dense vegetation. At another extreme are the adaptations of wing-propelled diving birds, such as penguins, which use their wings as flippers.

Like the structure of bills and wings, the anatomy of feet and legs tells us much about avian ecology (Figure 1–6). At one extreme are the tiny, weak feet and short legs of specialized aerial species such as tropicbirds and swifts. At the other extreme are the long, powerful legs of wading and cursorial birds such as storks and ostriches. The long toes of herons and especially of jacanas, which spread the bird's weight over a large surface area, facilitate walking on soft surfaces. Sandgrouse can scurry on soft desert sands, and ptarmigan can walk on snow by virtue of snowshoelike adaptations of their feet. Lobes on the toes of coots and webbing between the toes of flamingos and avocets reduce sinking into soft mud. Climbing birds such as woodpeckers have large, sharply curved claws, and nuthatches and other birds that climb downwards have prominent hind toes with a large claw.

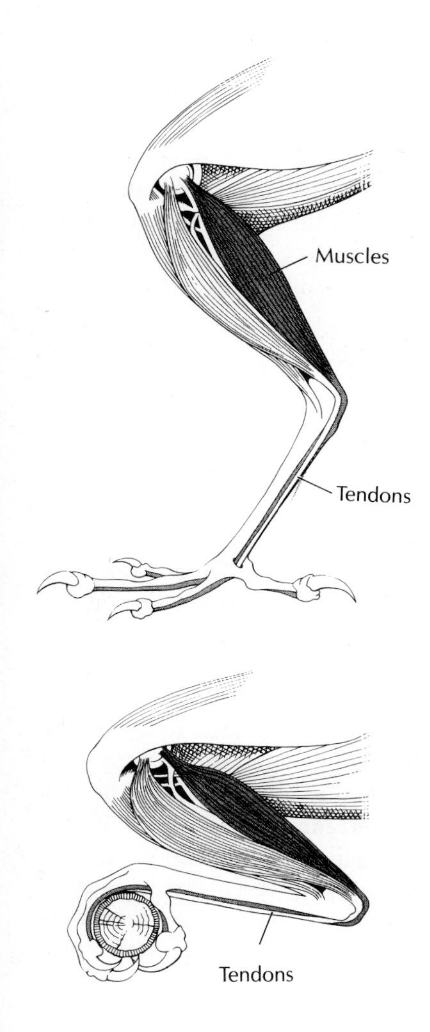

Figure 1–7 (above) When a perching bird squats, the leg tendons, which are located on the rear side of the ankle, automatically cause the toes to grip. (From Wilson 1980, with permission of Scientific American, Inc.)

The foot bones (three tarsals) are fused together and to the metatarsals, creating a long, strong single element, the tarsometatarsus, which enables birds to walk on their toes rather than on the whole foot. What appears at first glance to be a backward-bending knee joint is really the ankle joint. Legs as well as feet are adapted for various modes of locomotion. Whereas arboreal birds have short legs that aid balance on unstable twigs, terrestrial birds have long legs. Ostriches, which have long, strong legs and only two toes, are highly adapted for running by means of long strides. Other birds, such as sandpipers, scurry on short, thin legs. Short, heavy legs and big feet tend to evolve in large, flightless herbivorous birds such as moas and the dodo on islands without predators.

Arboreal, or tree-dwelling, species, which constitute the majority of birds, have feet designed for gripping branches tightly (see Chapter 2). Among the features of such feet is a long tendon that passes around the backside of the ankle joint. When a bird bends the joint to squat, the tendon automatically flexes, locking the toes around the branch (Figure 1–7). When a bird stands, the tension and the toe's grasp relax. The foot of the dominant group of perching birds, the songbirds, or Passeriformes, is perhaps the most advanced in this respect. The large, opposable single rear toe (hallux), which enhances the ability to grip a branch, is unusual among vertebrates. The tendons and muscles that flex the toes are arranged to facilitate perching at the expense of control of individual toe positions. A special system of ridges and pads between the tendons that flex the toes and the insides of the toe pads acts as a natural locking mechanism, which permits birds to sleep while perching.

A bird's center of gravity must remain over and between its feet, particularly when it perches, squats, or rises (Figure 1–8). Equal lengths of the two main leg bones (tibiotarsus and tarsometatarsus) of long-legged

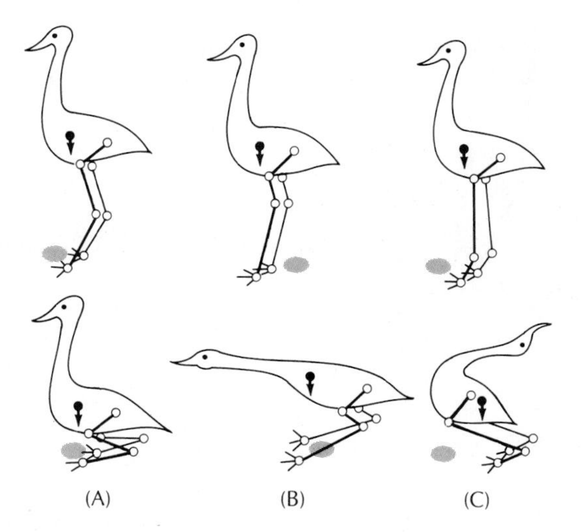

Figure 1–8 (A) Leg bones (tibiotarsus and tarsometatarsus) of equal lengths contribute to the balance of long-legged birds. When a bird crouches to incubate its egg, for example, leg bones of different lengths (B and C) would displace the center of gravity. (After Storer 1971a)

birds insure this relationship. Therefore, birds, unlike most swift quadrupeds, cannot increase running ability by evolving distal leg length (Storer 1971a).

Foot-propelled diving birds such as loons and grebes have sacrificed balance on land for swimming ability. They have powerful legs situated at the rear of a streamlined body. Articulating with their long, narrow pelvis are a short femur, a long tibiotarsus with an extension (cnemial crest) for muscle attachments, and a laterally compressed tarsometatarsus that reduces resistance in the water (Figure 1–9).

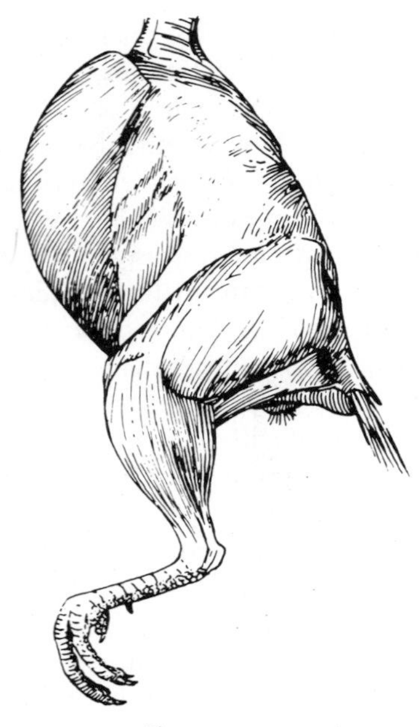

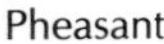

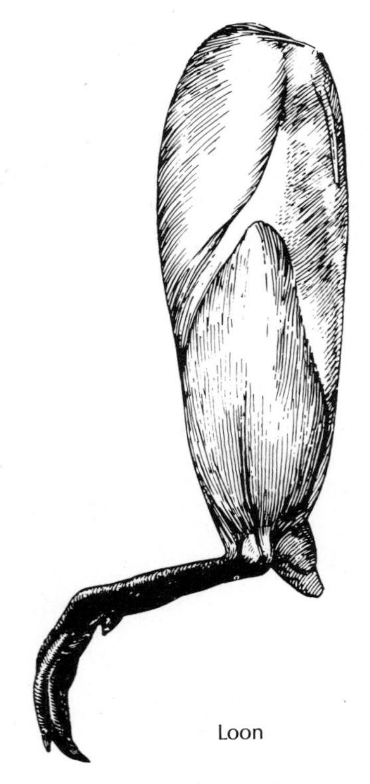

Figure 1–9 The body of a loon, which is streamlined for diving, compared with that of a pheasant. Note also the loon's highly developed hind limb musculature. (From Storer 1971a)

Summary

Characterized as vertebrates with feathers, birds have distinctive bills, are endothermic, produce large cleidoic (external) eggs, and have elaborate parental behavior, as well as extraordinary vocal abilities. The anatomy and physiology of most birds are adapted for flight. Birds share with reptiles many anatomical features that distinguish them from mammals, including a single occipital condyle on the back of the skull, a single middle-ear bone, and nucleated red blood cells.

The process of adaptive radiation is well illustrated today by the shorebirds, which include terrestrial waders, aerial plungers, and wing-propelled divers. The radiation of Hawaiian honeycreepers illustrates the way in which bill forms can evolve in relation to feeding habits.

Evolutionary adaptations of birds — anatomical, physiological, ecological, and behavioral — are the main theme of this text. To this point we have viewed the major features of bird evolution. In the next chapter we examine the early evolutionary history of birds, from the first known fossil bird, *Archaeopteryx lithographica,* through 150 million years of evolution of the established fundamentals of avian anatomy. In the chapters that follow we look at the details of avian adaptations: functional variations in feather structure, bill length, social behavior; regulation of body temperature; and many other aspects of the biology of birds.

Further readings

Perrins, C.M., and A.L.A. Middleton. 1985. The Encyclopedia of Birds. New York: Facts on File. *A comprehensive, colorful volume of the biology of birds for general audiences.*

Storer, R.W. 1971. Adaptive radiation of birds. Avian Biology 1:149–188. *An elegant summary of the diversity of birds.*

chapter 2

Evolutionary history

The similarities between birds and reptiles leave no doubt of their evolutionary relationship. Yet we are not content with that. We want to know as precisely as possible which reptiles gave rise to birds, and we want to understand the steps involved in this transformation. For this knowledge we must turn to the fossil record. Five specimens of a legendary fossil creature, *Archaeopteryx lithographica,* are the only evidence we have of the origin of birds. To paleontologists in quest of missing links those specimens are invaluable.

Archaeopteryx: The link between birds and reptiles

Archaeopteryx was a reptile with two incontrovertibly avian features: feathers and a furcula (wishbone). Long feathers, indistinguishable from modern feathers, extended the length of forelimb and tail. *Archaeopteryx* had an unusually large furcula and well-developed coracoid bones, completing a pectoral girdle that was intermediate between that of reptiles and that of modern flying birds. The furcula, as noted earlier, is a key adaptation of birds for powered, flapping flight. *Archaeopteryx* was an active, bipedal creature capable of running on the ground and, perhaps, leaping between large branches or fallen trees. It was at least semiarboreal and probably used its clawed fingers for help in climbing trees. It could fly by means of powered flapping flight.

The first evidence of *Archaeopteryx* and the origin of birds was an impression of a single feather found in a Bavarian quarry, in which limestone was mined for lithographic slabs; the impression was brought to the attention of Hermann von Meyer of Munich in 1861. A few months later a complete skeleton of a small reptilelike animal with feathers was also found and brought to von Meyer's attention (Figure 2-1). He named the fossil creature *Archaeopteryx* (*archios,* ancient; *pteryx,* wing) *lithographica.* The discovery of a second complete specimen of *Archaeopteryx* in another quarry near Eichstätt, Bavaria, followed in 1877. The first specimen is now in the British Museum; the second is in the Humboldt Museum für Naturkunde in East Berlin.

Figure 2-1 This fully articulated skeleton of *Archaeopteryx lithographica* was found in 1877 near Eichstätt, Bavaria. (Courtesy J. Ostrom)

These fossils were exciting, not only because of what they suggested about the evolution of birds from reptiles, but also because of their completeness and precise detail. Fossils are usually found as single, isolated bones, not fully associated skeletons. The specimen in the Humboldt Museum "may well be the most important natural history specimen in existence, perhaps comparable to the Rosetta Stone" (Feduccia 1980). It is fully articulated, revealing details of the wing bones, flight feathers, and pairs of feathers attached to each vertebra of its long tail.

Many years passed before three additional *Archaeopteryx* specimens were discovered. They are specimens of much poorer quality, so poor, in fact, that one was misidentified (ironically, by von Meyer) as a pterodactyl. The fifth specimen found near Eichstätt in 1951 was misidentified as a small dinosaur, *Compsognathus,* until 1973 when Franz X. Mayr of Eichstätt noticed the slight feather impressions.

These five fossils were preserved in fine limestone deposits that contain a record of creatures in central Europe during the age of dinosaurs —in the late Jurassic period, 150 million years ago. During this period, central Europe was tropical with palmlike plants. Great, warm, epicontinental seas and lagoons covered parts of the European continent. The coastal lagoons attracted pterodactyls, flying reptiles, some as small as sparrows and others as large as eagles, which flew with membranous wings. They occasionally perished in the lagoons where gentle fossilization in the fine calcareous sediments preserved features in fine anatomical detail. Also preserved there were the remains of the feathered reptile *Archaeopteryx.*

Archaeopteryx was a crow-sized reptile with a blunt snout instead of the bill of a modern bird and many small, well-developed reptilian teeth (Figure 2–2). The detailed osteology of its head matched that of reptiles. *Archaeopteryx* had a long, lizardlike tail with 18 to 21 separate caudal vertebrae (tail bones), each of which bore a pair of feathers. Its forearm was long, but the handbones, that is, metacarpals and digits, were not fused, and each of the long digits bore a sharply curved claw. Its sternum, if there was one, was unkeeled and probably cartilaginous, not the strong, ossified, well-developed structure of modern flying birds. The ribs of *Archaeopteryx,* as in most reptiles, lacked the horizontal reinforcements (uncinate processes) characteristic of most modern birds.

Because *Archaeopteryx* was a reptile with several of the characteristics of birds, it was a timely discovery of an animal that was intermediate between two higher taxonomic categories, a transition from ancestral to descendent stocks. It seemed that Darwin's prediction of intermediate evolutionary links in *On the Origin of Species by Natural Selection* (1859), published only two years previously, had been fulfilled. Some biologists think these specimens may still be the best fossil proof of the process of organic evolution (Ostrom 1976).

As the intermediate morphology of *Archaeopteryx* became known, the species quickly moved into the center of the debate between opponents and supporters of evolution by natural selection. Creationists, defending their views of the separate and unchanging appearances of birds and reptiles, insisted that Darwinists were misinterpreting the apparent intermediacy of *Archaeopteryx.* J. Andreas Wagner, an influential zoologist at Munich University, was adamantly opposed to the new Darwin theory and the idea that this fossil form might be a missing link. In contrast, Thomas H.

Figure 2–2 Skeletal features of (A) *Archaeopteryx* and (B) a Rock Dove. In modern birds (1) the brain case is expanded and head bones are fused; (2) the hand bones are also fused into several rigid elements; (3) the pelvic bones are fused into a single, sturdy structure; (4) the tail vertebrae are reduced and partially fused into a pygostyle; (5) the sternum is a large, keeled bony structure for the attachment of flight muscles; and (6) the rib cage is strengthened with horizontal uncinate processes. (From *Evolution of Vertebrates* by E.H. Colbert. Copyright © 1955 John Wiley & Sons, Inc.; reprinted by permission of John Wiley & Sons, Inc.)

Huxley, Darwin's eloquent champion, was convinced of the link between birds and birdlike reptiles and soon converted leading American paleontologists. Charles Marsh of Yale University was one of these. He wrote:

> The classes of Birds and Reptiles, as now living, are separated by a gulf so profound that a few years since it was cited by the opponents of evolution as the most important break in the animal series, and one which that doctrine could not bridge over. Since then, as Huxley has clearly shown, this gap has been virtually filled by the discovery of birdlike Reptiles and reptilian Birds. *Compsognathus* and *Archaeopteryx* of the Old World . . . are the stepping stones by which the evolutionist of today leads the doubting brother across the shallow remnant of the gulf, once thought impassable. (Marsh 1877; Feduccia 1980)

The discovery of *Archaeopteryx* thus contributed to the acceptance of Darwin's theory of evolution as well as to our understanding of the origin of birds. Still unanswered, however, were some key questions. What more does *Archaeopteryx* tell us about the ancestry of birds? Do its features reveal which reptiles were the ancestors of birds?

Origin of birds

There is little doubt that birds evolved from some line of Mesozoic reptiles. Which line is a matter of debate. One possibility is that birds evolved from a group of reptiles called thecodonts. Another possibility is that birds evolved from small theropod dinosaurs (Figure 2–3).

The thecodontian theory of the origin of birds looks to a large group of primitive reptiles, the pseudosuchians, that prevailed in the early years of the Mesozoic era. Among these were the lightly built thecodonts, a group that gave rise to various dinosaurs, to pterosaurs, to crocodiles, and perhaps also directly to birds. Some thecodonts, such as *Longisquama,* a thecodont with elongated scales, were quite like birds. Early crocodilians were quite different from the modern crocodiles; some were arboreal.

The early works of Maximillian Fürbringer (1888), Robert Broom (1913), and especially Gerhard Heilmann (1927) all favored the theory of thecodontian ancestry of birds. Recently, S. Tarsitano and Max Hecht (1980) and Larry D. Martin (1983) also found support for this theory. Fourteen characters seem to unite modern birds and crocodilians. This theory suggests that birds split very early — no later than the middle Triassic period — from the main stem of crocodilian evolution. If so, proponents of the dinosaur theory ask, why is there a gap of 90 million years in the fossil record between thecodonts and *Archaeopteryx?*

One of the attractions of the dinosaur theory of the origin of birds is that dinosaurs and *Archaeopteryx* are contemporaneous in the fossil record. Although we usually think first of the large, spectacular species, dinosaurs varied greatly in size and habits. One group, the theropod dinosaurs,

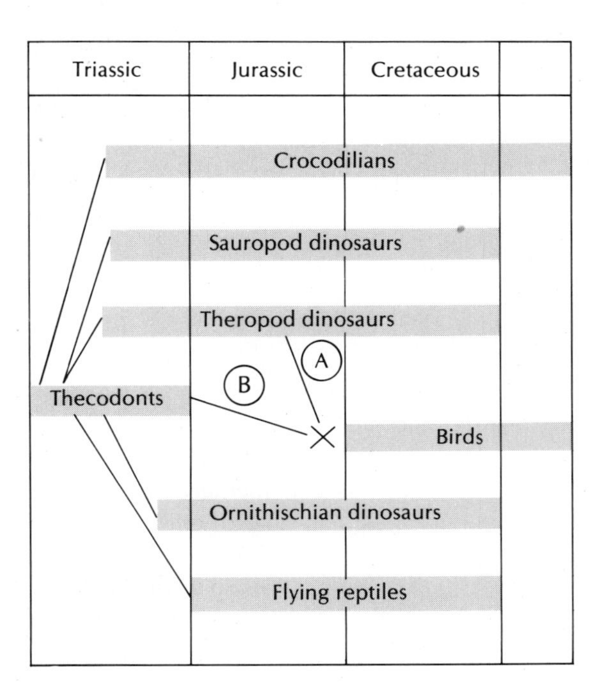

Figure 2–3 Alternative theories of the evolution of birds. (A) Some experts believe that birds evolved from small theropod dinosaurs. (B) Other experts believe that birds evolved directly from the thecodont ancestors of dinosaurs and crocodiles. (After Ostrom 1975)

(A)

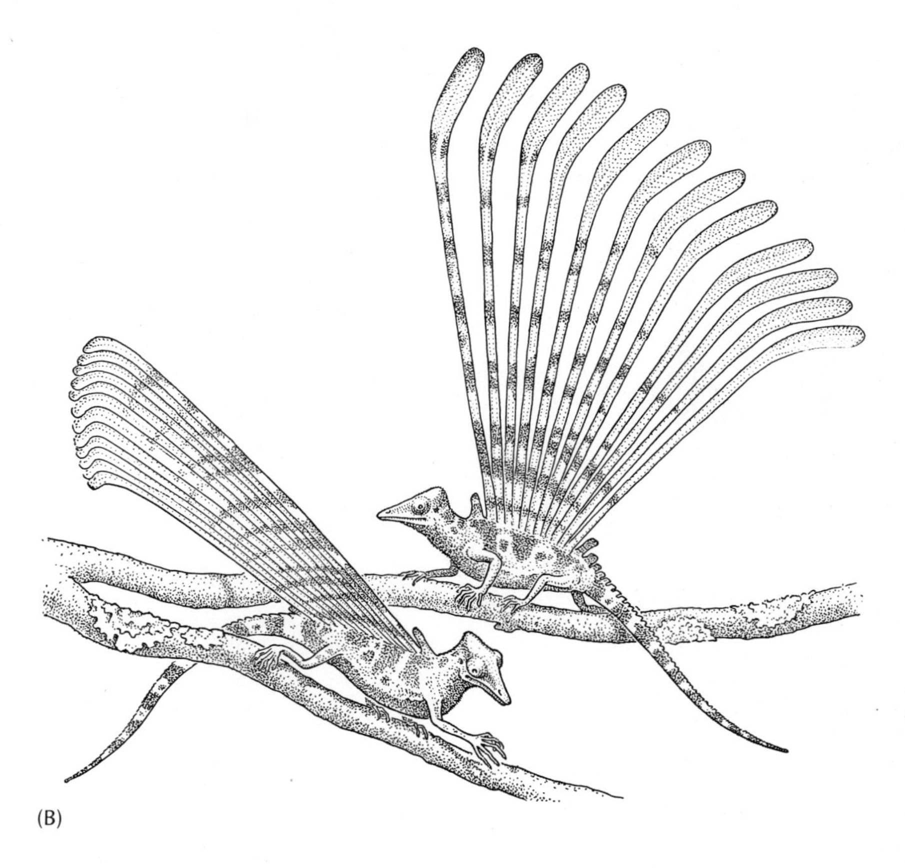

(B)

Figure 2–4 Two possible relatives of birds. (A) *Compsognathus* was a small theropod dinosaur that was preserved in the same limestone deposits as *Archaeopteryx.* (B) *Longisquama* was a lightly built, arboreal thecodont reptile with elongated scales. Whether a theropod or a thecodont reptile was the ancestor of birds remains uncertain. (A from Heilmann 1927, B from Bakker 1975, with permission of Scientific American, Inc.)

included not only large carnivores such as *Tyrannosaurus rex* but also many small ones, called coelurosaurs, that were close in size to modern iguanas. These were agile, lightly built, bipedal little dinosaurs with many small, sharp teeth. They probably chased small vertebrates and large insects. They may even have been endothermic.

The hypothesis that birds evolved from small dinosaurs goes back to the early debates that followed the discovery of fossil *Archaeopteryx.* T.H. Huxley (1868) was particularly impressed by the similarities between *Archaeopteryx* and *Compsognathus,* a small dinosaur preserved in the same limestone deposits (Figure 2–4). Recent studies by John Ostrom support Huxley's theory of a dinosaur ancestry of birds. *Archaeopteryx* and small dinosaurs share 23 of 42 specialized skeletal features, though some of these may not be comparable structures (Martin 1983). The principal features involved are structures of the hand, vertebrae, humerus and ulna, pectoral arch, hindlimb, and pelvis. Ostrom noted,

> It has been repeatedly observed that the *Archaeopteryx* specimens are very birdlike, but also possess a number of reptilian features. . . . the actual fact is that these specimens are not particularly like modern birds at all. If feather impressions had not been preserved in the London and Berlin specimens, they would never have been identified as birds. Instead, they would unquestionably have been labelled as coelurosaurian dinosaurs. The last three specimens (of *Archaeopteryx*) to be recognized were all misidentified at first and the Eichstätt specimen for twenty years was thought to be a small specimen of the dinosaur *Compsognathus.* (Ostrom 1975, p. 61)

Whatever the ancestor of *Archaeopteryx,* major features of modern birds remained to be evolved. In particular, the many steps to flight radically overhauled both body and physiology and opened the door for the diversification of birds.

Evolution of avian flight

The evolution of avian flight and how well *Archaeopteryx,* the putative ancestor of birds, could fly have long been debated among ornithologists. Quite likely, *Archaeopteryx* was a semiarboreal creature capable of at least gliding and weak flapping but not of long, sustained flights. Like the modern guans (Cracidae), it may have been a strong-running, terrestrial bird that could leap into trees, jump among large branches, and make short flights between trees.

Archaeopteryx probably did fly occasionally, but we will never know how well. Because it lacked both a bony keeled sternum and triosseal canal and therefore had neither a well-developed supracoracoideus muscle nor an associated pulley system for the tendon, *Archaeopteryx* probably could not launch itself into the air from the ground. Perhaps it flew only after jumping from a tree. Even so, these deficiencies do not altogether preclude powered flight (Olson and Feduccia 1979). *Archaeopteryx* may have had strong pectoral muscles. It had an unusually large furcula, which in modern birds is an important site of attachment of the flight muscles. Moreover, the acute angle of its scapula suggests that *Archaeopteryx* had dorsal elevator muscles that functioned like those of modern flying birds. The vanes of the pri-

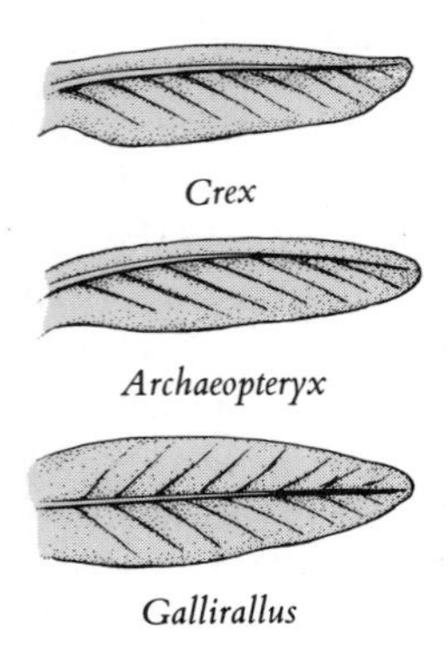

Figure 2–5 The vanes of the primaries of *Archaeopteryx* were asymmetrical as in modern flying birds, such as the rail, *Crex,* not symmetrical as in flightless birds, such as the rail, *Gallirallus.* The asymmetry has an aerodynamic function and presumably evolved in relation to flight of this primitive bird. (After Feduccia and Tordoff 1979)

maries of *Archaeopteryx* were asymmetrical, a characteristic that is shared by nearly all flying birds and is most pronounced in strong fliers (Feduccia and Tordoff 1979). In flightless birds, these vanes are symmetrical (Figure 2–5).

One wonders what caused the forelimbs of the reptilian ancestor of *Archaeopteryx* to evolve into protowings in the first place. Two theories exist—an arboreal theory and a cursorial theory. In brief, the arboreal theory proposes that parachuting and gliding from elevated perches were the first stages in the evolution of flight. The cursorial theory proposes that forelimbs first evolved in the direction of becoming wings because they heightened leaping ability in bipedal terrestrial reptiles. Both theories are concerned with the series of adaptive steps leading from terrestrial, bipedal, coelurosaurian dinosaurs to an *Archaeopteryx* capable of active, though primitive, gliding flight.

Most drawings and models of *Archaeopteryx* depict an arboreal reptile clambering around trees, grasping branches with clawed fingers. The arboreal theory proposes that extensions of the bones of the forelimb enhanced by elongated (flight) feathers enabled the ancestor of *Archaeopteryx* to parachute and glide between trees. The arboreal theory has been the favored theory for many years (Bock 1965; Feduccia 1980) (Figure 2–6). Recently, however, there has been a groundswell of support for the cursorial theory (Ostrom 1974, 1976a and b, 1979). Could a running bird leave the ground by leaping with outstretched forelimbs? Although it is doubtful that the creature could achieve adequate lift, extensions of the forelimbs would help control and extend its leaps and might lead to primitive flight, a process that has more in common with ballistics than with aerodynamics (Caple et al. 1983, 1984).

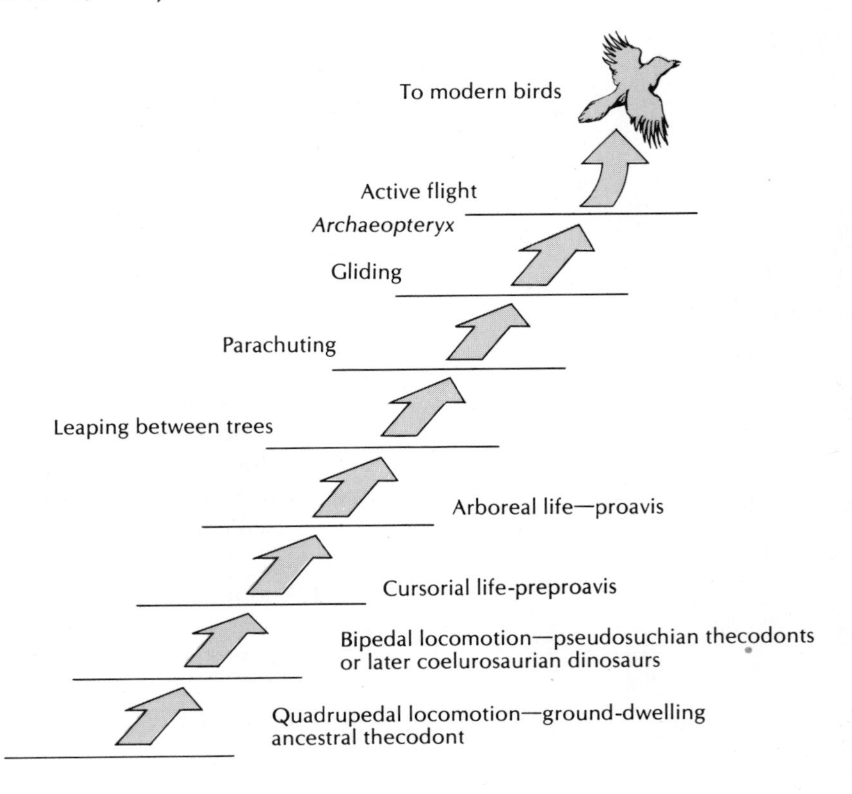

Figure 2–6 The arboreal theory of the evolution of avian flight suggests that, after evolving bipedal locomotion, the reptilian ancestors of birds became arboreal and leaped between trees. Active flight evolved from earlier stages of parachuting and gliding flight that enhanced the leaping abilities of the ancestors of *Archaeopteryx.* (After Feduccia 1980; adapted from Bock 1965)

The cursorial theory emphasizes the role of improved control of body orientation (pitch, roll, and yaw) during jumping rather than the role of lift and gliding. Small increases in length of the extended forelimbs increase potential control. Minor additions of lift of one to five percent, especially from three body radii — such as two wings and a tail — have a dramatic effect on the control of roll and pitch. Following this line of reasoning, the initial elongation of the "wings" and tail feathers would have been advantageous. (Proponents of the cursorial theory envision the ancestor of *Archaeopteryx* as a little dinosaur, similar to the bipedal theropods, that ran and jumped to catch insects in its jaws.) Elongation would lead to greater foraging volumes, enhanced maneuverability, and higher velocities of running and jumping. Protowings, increased arboreal habits, and gliding with feeble flapping might be the next evolutionary steps. The ballistic-cursorial theory thus provides a series of adaptive evolutionary steps independent of the aerodynamics of true flight. The flight capabilities of modern birds seem, in this light, a logical extension of the first small jumps by the ancestor of birds.

Evolution of feathers

Compared with the scales of reptiles and other inert epidermal structures made of keratin, feathers are filamentous, soft-textured, flexible, lightweight structures that resemble frayed scales. Given that birds evolved from reptiles, feathers probably evolved from scales of some kind (Waterman 1977). Birds have scales on their feet. Most avian scales are chemically allied to feathers, though some (reticulate scales on the soles of the feet) are closer in chemical composition to reptilian scales.

We are not certain what advantages promoted the evolution of feathers from scales. This question arouses great interest, debate, and speculation. The classical flight-origin hypothesis for the origin of feathers suggests that they first evolved in association with flight (Heilman 1927; Parkes 1966). It is conceivable that elongated, frayed scales on the trailing edges of the forelimbs enhanced either primitive gliding or parachuting. The much simpler membranes of bats, pterosaurs, and "flying" frogs were the evolutionary solution to gliding flight in these tetrapod vertebrates. As gliding abilities improved and steering requirements increased, the hypothesis continues, so did the elaboration of feathers on the wings and tail. There remains the question of why protofeathers evolved first on the forelimbs rather than other parts of the body. Furthermore, the flight-origin hypothesis does not adequately address the problem of what advantages the initial stages of scale elongation could have had.

Reptile experts tend to view the problem from the standpoint of the original scales rather than of fully evolved feathers. They favor physiological hypotheses. For example, feathers may have evolved as temperature regulation devices, either as insulation or as heat shields. The hypothesis that feathers evolved as a source of insulation stems from the observation that modern feathers, especially downy barbs, clearly have value as insulation. Although the arrangement of barbs and barbules in down cannot have been the initial evolutionary step (this is an advanced condition, *derived*

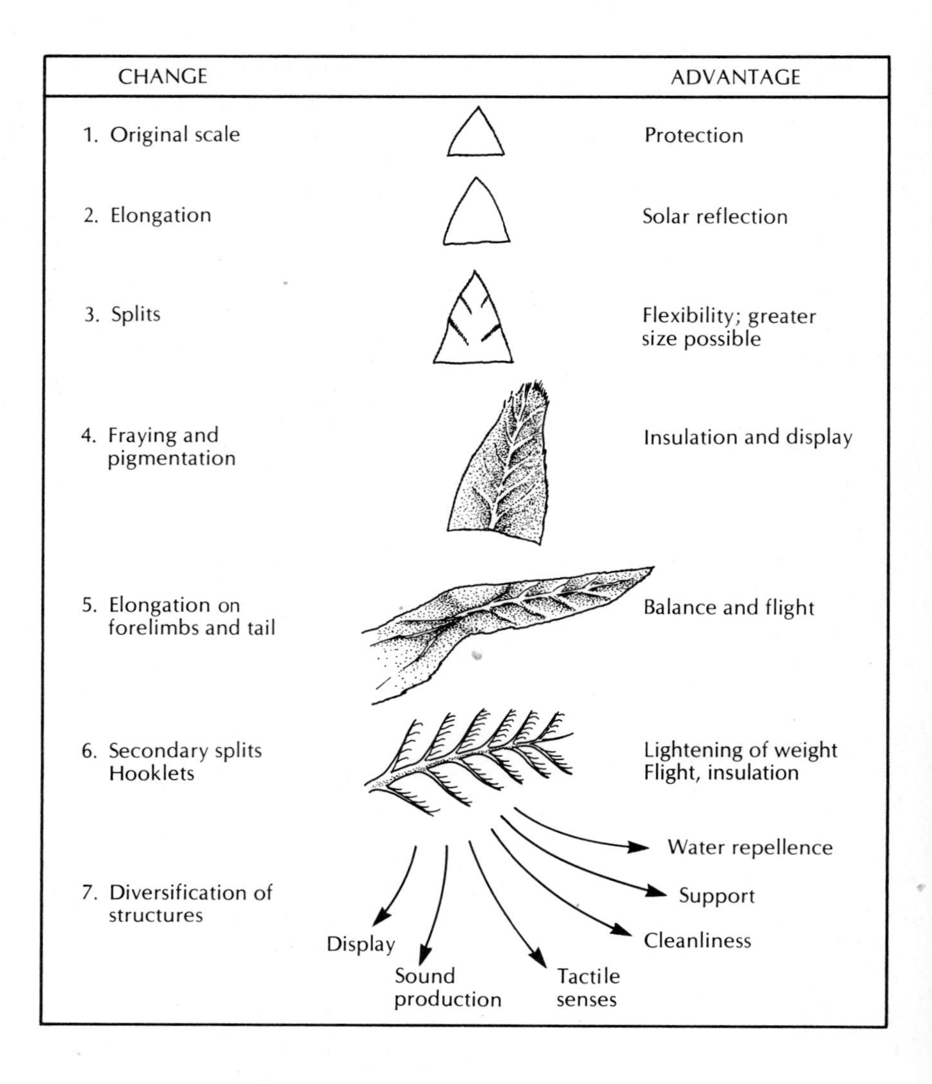

Figure 2–7 Hypothetical steps in the evolution of feathers from scales. Elongation and splitting of reptilian scales aided reflection of solar heat and permitted larger, flexible scales. Increased fraying and pigmentation of the large scales contributed to insulation and to displays. Elongation of feathers on the forelimbs and tail improved balance on extended leaps, and ultimately flight. Secondary splits that led to the evolution of barbules with interlocking hooklets gave rise to the modern light feather structure that aids flight and insulation. In addition, this versatile structure is easily modified for special purposes including sound production, tactile sensation, support, and water repellence.

from the vaned structure of feathers), any insulation contributed by primitive vaned feathers could have been favored by natural selection in scaled reptiles.

Philip Regal (1975) hypothesized that scales first evolved toward feathers because they permitted the ancestors of birds to be more active in hot, sunny habitats during midday. Supporting this view are four heat-transfer characteristics of modern reptile scales:

1. Long scales dissipate solar heat by improving microcirculation patterns of air and by casting shadows. These advantages favor large scales in modern lizards that live in hot, sunny climates.

2. The orientation of scale surface in relation to the sun affects the percentage of heat reflected away from the body; flexibility and control of scale position increase effectiveness.

3. Subdivision of scales adds flexibility to the structure in the face of otherwise cumbersome increases in scale size. Large, stiff scales would increase weight and decrease mobility.

4. Subdivisions, similar to those of modern feathers, enhance reflection.

Such functional hypotheses remain completely speculative. We know virtually nothing about the behavior and ecology of the reptiles that evolved into birds. Furthermore, the many functions performed by feathers must all be taken into account when trying to piece together their evolutionary history. The early advantages of featherlike scales probably complemented and reinforced each other once the process of transformation began. The advantages of the novel, flexible structure of the protofeather coupled with its usefulness in thermoregulation and primitive flight catalyzed the rapid evolution of definitive feather structure in the earliest birds (Figure 2–7).

Diversification and extinction

The evolutionary history of birds comprises a series of major phases: the transformation of reptile ancestors into a feathered flying bird, the temporary dominance of unsuccessful lineages of primitive toothed birds in the Cretaceous period, the early establishment of the attributes of modern birds, the diversification of form and function into modern orders of birds and additional lineages that became extinct, and the adaptive radiation of modern species during the past two million years. The currently accepted geologic time scale is shown in Table 2–1.

Table 2–1 Geological Time Scale

Era	Period	Epoch	Million Years before Present
	Quaternary	Recent	0.01
		Pleistocene	1.5–3.5
Cenozoic (age of birds and mammals)	Tertiary	Pliocene	7
		Miocene	26
		Oligocene	37–38
		Eocene	53–54
		Paleocene	65
	Cretaceous	Late	100
		Early	135
Mesozoic (age of reptiles)	Jurassic	Late	155
		Middle	170
		Early	180–190
	Triassic		230

(After Feduccia 1980)

Archaeopteryx is our only evidence of the transition from reptiles to birds. Whether it was a direct ancestor of modern birds is not certain. *Archaeopteryx* may have been a member of an early radiation of other toothed birds, the Sauriurae, known only from scattered ancient fossils of Cretaceous birds (Martin 1983). Among these were the genera *Gobipteryx* of Mongolia, *Alexornis* of Baja California, and *Enantiornis* of Argentina. Better known are the later, toothed Cretaceous seabirds, *Hesperornis, Ichthyornis,* and their relatives.

The hesperornithiformes comprised a variety of diving seabirds that superficially resembled modern loons. Their size ranged from that of a small chicken to that of a large penguin. The largest was *Hesperornis regalis,* one to two meters in length. All 13 known species were flightless with unkeeled sterna, unfused clavicles, and vestigial wing bones. They had large, powerful, lobed feet and inhabited the Cretaceous seas that covered the central portions of the North American continent. Flying above the same shallow seas were toothed, ternlike birds *(Ichthyornis),* of which six species are known from the fossil record. Both groups of toothed Cretaceous birds had well-formed, unserrated, reptilian teeth with constricted bases and expanded roots set into distinct sockets in their bony jaws. None of these lineages have survived. They disappeared along with dinosaurs in the cataclysmic period of mass extinction that marked the end of the Mesozoic era. Among the few survivors were the ancestors of toothless modern birds.

The modern orders of birds diverged from each other 60 million years ago, or early in the Tertiary period (Wilson et al. 1977). Specialized waterbirds, such as loons, auks, gulls, ducks, cranes, and petrels, invaded aquatic niches during the Eocene epoch, 55 to 60 million years ago. Primitive woodpeckers and their relatives also appeared during the early Eocene and became the predominant perching birds during the Miocene epoch. Relatives of rollers, kingfishers, and hornbills diversified in the Oligocene epoch. In the Miocene, the rapid evolution of flowering plants and insects opened new niches for insect-eating, fruit-eating, and nectar-feeding birds; this resulted in an explosive radiation of songbirds (Regal 1977). By the end of the Tertiary, about ten million years ago, birds had diversified into a broad range that included many modern genera.

The Tertiary produced some huge carnivorous birds that temporarily occupied some of the niches left vacant by bipedal dinosaurs. Diatrymas, lumbering predators over two meters tall, with powerful legs, clawed toes, massive horse-sized skulls, and tearing, eaglelike beaks, must have terrorized many lesser creatures before becoming extinct in the Eocene (Figure 2–8). In the Eocene, long-legged vulturelike birds *(Neocathartes)* lived in Wyoming beside shorebirds *(Presbyornis)* with ducklike heads. From the Oligocene to the Pliocene epochs, about 12 known species of phorusrhacids, predatory birds two to three meters tall with powerful, rapacious bills, ranged throughout South America and north to Florida.

Finally, in the Quaternary period (the last two million years) major glaciations and associated habitat changes fragmented the populations of modern birds. Some became extinct and some evolved into new species. The average longevity of a species in the Pleistocene epoch is estimated to have been half a million years, compared with three million years in the preceding Tertiary (Brodkorb 1971). Most of the modern species of birds appear

Figure 2–8 Large carnivorous birds flourished during the Tertiary period. These included diatrymas. (From Heilmann 1927)

to have evolved during the Pleistocene or later. In Pleistocene fossil deposits, paleontologists have found bones that are indistinguishable from those of over 900 modern species. We know nothing about the songs, behavior, plumage color, or nests of the birds represented by those fossils.

Also present during the Pleistocene was a multitude of vultures and raptors that exploited the carcasses of the increasingly diversified mammals. During the Ice Age, huge vulturelike teratorns dominated the skies. One teratorn *(Teratornis merriami)* with a four-meter wing span was abundant in southern California. Another *(T. incredibilis),* known from caves in Nevada, had a wing span of five to six meters, and yet another, recently discovered in Argentina, had an eight-meter wing span. These enormous birds symbolize past avian achievements.

Summary

Birds evolved from a small, bipedal reptile more than 150 million years ago in the Mesozoic era. *Archaeopteryx lithographica,* one of the most important fossils of all time, was a crow-sized, toothed, bipedal reptile with two incontrovertible avian features: feathers and a furcula. It could fly and clamber around trees. Known from five specimens preserved in fine limestone deposited in the late Jurassic period in central Europe, it represents an evolutionary link between birds and reptiles. Because of its timely discovery, *Archaeopteryx* fostered acceptance of Darwin's theory of evolution. Exactly which group of reptiles was the ancestor of *Archaeopteryx,* and hence which gave rise to birds, remains uncertain. Current debates focus on small theropod dinosaurs or on thecodont ancestors of the crocodiles as the most likely candidates.

Once established as a new group of vertebrates, birds diversified in both form and function. A variety of toothed birds established themselves in the Cretaceous period but then became extinct. The modern orders of birds diverged from each other 60 million years ago at the beginning of the Tertiary period. Waterbirds multiplied in variety in the Eocene epoch, as did landbirds in the Miocene epoch. The characteristics that distinguish modern species evolved during the last two million years.

Further readings

Feduccia, J.A. 1980. The Age of Birds. Cambridge: Harvard University Press.

Martin, L.D. 1983. The origin and early radiation of birds. *In* Perspectives in Ornithology (A.H. Brush and G.A. Clark, Jr., eds.), pp. 291–338. Oxford: Oxford University Press.

Olson, S.L. 1985. The fossil record of birds. Avian Biology 8:79–238.

chapter 3

Phylogeny

One strength of a comparative science such as ornithology is that it helps us to understand the evolution of diversity. Finding natural order in the evolutionary relationships of species, their phylogeny, is the first challenge that confronts us. Then follows the construction of a classification, or orderly arrangement, that reflects those relationships. A classification of the kinds of birds of the world helps ornithologists to communicate and serves as a tool in the continuing study of avian evolutionary relationships.

In this chapter we present an overview of the science of classification, which is based on similarities and differences caused by evolution, and identify the evolutionary processes that affect avian diversity. We review the history of formal classification, citing a long-standing controversy concerning the evolutionary relationships of flamingos as an example of the kinds of problems that confront ornithologists. We examine some of the morphological, behavioral, and genetic attributes of birds that provide clues to evolutionary history, as well as factors such as convergence — the independent evolution of similar adaptations in unrelated organisms — that can mislead scientists who attempt to determine which species have a common ancestor. Finally, we introduce some of the primary methodology of systematics, including cladistics—the study of evolutionary branching sequences—and biochemical genetics. A classification of the birds of the world appears in Appendix I.

Classification and phylogeny

The names given to a bird tend to vary with locale. The American Goldfinch, for example, is also called the yellow-bird, thistle-bird, wild canary, and beet-bird (Figure 3–1). However, it is essential that the names of birds be standardized if we are to have communication among ornithologists. The science of naming and classifying birds is called taxonomy, and the scientists who do this work call themselves taxonomists. A taxon (pl. *taxa*) is any group of animals that is recognized in a classification. The Class Aves is a taxon that includes all species of birds. The standard for communication is the system of nomenclature developed in 1758 by Carolus Linnaeus, a Swedish botanist. Linnaeus assigned two latinized names to each species: The first denotes the genus, a group of similar species; the second denotes the species. Thereby, the American Goldfinch is known formally as *Carduelis tristis,* which is a taxon that includes all populations of that species. This combination of names is unique; no other bird, indeed, no other animal, may have the same pair of names.

The genus name of a species reflects its evolutionary relationship to similar species. To reflect evolutionary relationships further, related genera are grouped under families, related families are grouped under orders, and related orders are grouped under classes (Table 3–1). Taxa that share a common evolutionary history are called a lineage. Details of how new species evolve are discussed in Chapter 25.

Figure 3–1 The American Goldfinch has many local names, such as wild canary, yellow-bird, thistle-bird, and beet-bird. (Courtesy A. Cruickshank/VIREO)

Table 3-1 Classification of Three Species of Woodpeckers

Taxon	Downy Woodpecker	Hairy Woodpecker	Northern Flicker
Class	Aves	Aves	Aves
Order	Piciformes	Piciformes	Piciformes
Family	Picidae	Picidae	Picidae
Genus	*Picoides*	*Picoides*	*Colaptes*
Species	*pubescens*	*villosus*	*auratus*

The diversity of birds is the result of three evolutionary processes: phyletic evolution, which is the gradual change of a single lineage through evolution; speciation, the splitting of one phyletic lineage into two or more; and extinction, the termination of a lineage. Some evolutionary changes result in groups having similar taxa; others encourage substantial diversity. Extinction of related lineages causes links to be missing from continuing lineages. Deciphering the historical patterns of speciation and phyletic evolution, a field of scholarly endeavor called systematics, is a difficult task. Systematists are scientists who study the evolutionary relationship of organisms. Most systematists work with large museum collections of fossils and preserved specimens to determine the relationships among species or groups of species. If key fossils are lacking, as they usually are, the systematist must assess existing taxa to hypothesize sequences of evolutionary events (Bock 1973).

Prevailing classifications of birds attempt to portray the evolutionary relationships of the various lineages. Theoretically, each taxon is monophyletic, that is, it contains birds related by evolutionary descent from a common ancestor. A hierarchical organization of taxa reflects relative closeness or distance of the evolutionary relationships among those taxa. Of the hypothetical taxa in Figure 3-2, five (A, B, C, D, and E) should be placed in

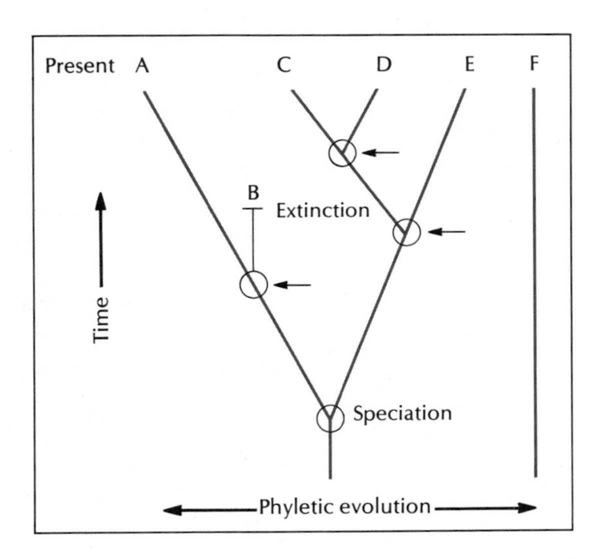

Figure 3-2 Diversification of evolutionary lineages includes speciation (circled nodes), the splitting of lineages; extinction, the loss of lineages; and phyletic evolution, the gradual change of a lineage with time. Clusters of similar, related taxa, such as C, D, and E, present in modern times, result from these changes. Taxon B became extinct. Taxon F is not related to the other taxa, which shared a common ancestor.

the same higher taxon, the same family perhaps. Taxon F probably is not related and thus belongs in a different family. If F had evolved from lineage C at some much earlier date, it might be assigned to the same order. Taxa C, D, and E would be grouped together in the same genus because they form a cluster of similar taxa. Because taxon B became extinct, the resulting gap between A and C might qualify A as a genus of its own. In *On the Origin of Species by Means of Natural Selection,* Charles Darwin reflected on the hierarchical structure of a classification:

> I believe that the arrangement of the groups within each class, in due subordination and relation to the other groups, must be strictly geneological in order to be natural, but that the amount of difference in the several branches or groups, though allied in the same degree in blood to their common progenitor, may differ greatly, being due to the different degrees of modification which they have undergone, and this is expressed by the forms being ranked under different genera, families, sections or orders (Darwin 1859, p. 420).

If we examine the natural world, we can see the possibility of constructing a hierarchy, or ranking, of differences. A cursory survey of, for example, some North American birds will distinguish woodpeckers from owls. Less obvious are the differences among Downy Woodpeckers, Red-bellied Woodpeckers, and Northern Flickers, or the differences among Great Horned Owls, Barred Owls, and Eastern Screech-Owls. A little experience, however, leads to mastery of this level of identification. Recognition of the subtle differences between Downy and Hairy Woodpeckers or between Eastern Screech-Owls and Whiskered Screech-Owls is more difficult and requires more expertise. Owls and woodpeckers are classified in different orders, reflecting the relative independence of their evolutionary lineages. All woodpeckers are classified in the same order and family (Table 3–1). The very similar and closely related Downy Woodpecker and Hairy Woodpecker are classified in the genus *Picoides,* but the less closely related Northern Flicker is classified in the genus *Colaptes,* along with the Andean Flicker and other flickers (Figure 3–3).

The birds of the world do not comprise just a few species of owls or woodpeckers, but over 9000 diverse species. A succession of efforts to classify them has culminated in a modern, comprehensive, 15-volume classification of all modern birds and their distributions: *Checklist of the Birds of the World.* James L. Peters started this monumental work in the 1920s, and it was completed, under the guidance of Ernst Mayr and R.A. Paynter, Jr., in 1986. Supplementing it are several condensed summaries of the birds of the world (Clements 1981; Howard and Moore 1980; and particularly, Morony et al. 1975).

The number of avian species recognized by ornithologists is subject to revision. A list of 19,000 species in the early 1900s shrank to 8600 by 1940, partly because better information revealed some "species" to be variations due to age or sex. More fundamental, the species concept itself broadened to incorporate geographical variations and reproductive continuities among populations once classified separately. Today, recent reassessments of hybridizing, isolated, and variant populations have drastically changed the official species list. Since 1970, 32 species of South American birds have

Figure 3–3 Three species of woodpeckers: (A) Downy Woodpecker, (B) Hairy Woodpecker, (C) Northern Flicker. The Downy Woodpecker and the Hairy Woodpecker are more closely related to each other than either is to the Northern Flicker.

been reassigned to subspecies or variant status. These deletions are balanced by the addition of 32 species, owing to taxonomic revisions and new discoveries (DeSchauensee 1982). Worldwide, taxonomic revisions and new discoveries in the past 20 years have increased the total number of bird species recognized by ornithologists from 8600 to about 9021. The complete classification of living birds is now a hierarchical arrangement of 30 orders, 174 families, 2044 genera, and 9021 species (Table 3–2). The rate of discovery of new species has dwindled to about two per year. Current taxonomic efforts focus increasingly on the assessment of evolutionary relationships, or phylogeny, and on the continued modification of the classification to reflect our understanding of phylogeny. A classification is more than an authoritative basis for orderly communication about birds. It is also a set of working hypotheses about the relationships, similarities, and differences among birds.

The process of naming and classifying birds is an ancient one. For contemporary taxonomists, deciphering sequences of branching and degrees of similarity among taxa and converting the best available information into a classification is a major challenge. Francis Willoughby and John

Table 3-2 Orders of Recent Birds, Class Aves

Order	Number of Taxa			Members
	Family	Genus	Species	
Tinamiformes	1	9	47	Tinamous
Rheiformes	1	2	2	Rheas, nandus
Struthioniformes	1	1	1	Ostrich
Dinornithiformes	1	1	3	Kiwis, moas (extinct)
Casuariiformes	2	2	5	Cassowaries, Emu
Podicipediformes	1	5	20	Grebes
Procellariiformes	4	23	104	Tube-nosed seabirds: petrels, shearwaters, albatrosses, storm petrels, and diving petrels
Sphenisciformes	1	6	18	Penguins
Pelecaniformes	6	9	62	Waterbirds with totipalmate feet: cormorants, pelicans, anhingas, boobies, gannets, frigatebirds, and tropicbirds
Anseriformes	2	45	150	Waterfowl: ducks, geese, swans, and screamers of South America
Phoenicopteriformes	1	3	6	Flamingos
Ciconiiformes	5	44	114	Long-legged wading birds: storks, herons, ibis, spoonbills, Hammerkop, and Whale-headed Stork
Falconiformes	5	81	288	Raptors: falcons, caracaras, hawks, eagles, vultures, kites, Osprey, and Secretary bird
Galliformes	4	78	268	Gallinaceous birds: grouse, quail, pheasants, chickens, curassows, guans, chachalacas, guineafowl, and moundbuilders; excludes Hoatzin (see Cuculiformes)
Gruiformes	11	81	209	Diverse terrestrial and marsh birds: rails, coots, sungrebes, cranes, Sunbittern, Kagu, Limpkin, seriamas, bustards, and roatelos of Madagascar
Charadriiformes	18	76	331	Shorebirds and their relatives: sandpipers, plovers, phalaropes, stilts, jacanas, painted snipes, pratincoles, gulls and terns, seedsnipes, sheathbills, skimmers, skuas, and auks
Gaviiformes	1	1	5	Loons
Pteroclidiformes	1	2	16	Sandgrouse

Order	Number of Taxa			Members
	Family	Genus	Species	
Columbiformes	1	42	303	Pigeons and doves
Psittaciformes	3	81	340	Parrots, macaws, lories, and cockatoos
Coliiformes	1	1	6	Mousebirds
Musophagiformes	1	5	18	Turacos, plaintain eaters
Cuculiformes	2	39	130	Cuckoos and Hoatzin
Strigiformes	2	30	146	Owls and barn owls
Caprimulgiformes	5	24	105	Nightjars, potoos, frogmouths, owlet-frogmouths, and Oilbird
Apodiformes	3	135	428	Swifts, crested swifts, and hummingbirds
Trogoniformes	1	8	37	Trogons and quetzals
Coraciiformes	10	44	200	Kingfishers and allies: todies, motmots, bee-eaters, rollers, Cuckoo-roller, Hoopoe, woodhoopoes, and hornbills
Piciformes	6	62	383	Woodpeckers and allies: wrynecks, piculets, barbets, toucans, honeyguides, jacamars, and puffbirds
Passeriformes	73	1104	5276	Perching birds, songbirds, passerines

(After Bock and Farrand 1980)

Ray's *Ornithologiae,* published in 1676, was the first formal classification of birds. The authors of this seminal work attempted to arrange all birds then known into a logical, hierarchical classification — the "cornerstone of modern systematic ornithology" (Zimmer 1926). Willoughby and Ray divided birds first into two obvious groups, Landfowl and Waterfowl, and then into additional subgroups. Nearly a century later, Linnaeus (1758) used this elementary classification in his *Systema Naturae,* which became the model for subsequent classifications. Important as they were, these early efforts tended to classify according to superficial similarities such as adaptations to aquatic versus terrestrial habitats, rather than according to any sound, overarching concept.

Darwin's theory of evolution by natural selection transformed the philosophical basis of ornithological systematics. His concept of evolutionary relationships provided a sound theoretical and rational basis for finding evidence of common ancestries. Thomas H. Huxley also helped lay the foundations of evolutionary systematics in birds with his study of the arrangement of the bones of the avian bony palate that form a partition between the nasal cavities and the mouth (Huxley 1867) (Figure 3–4). In

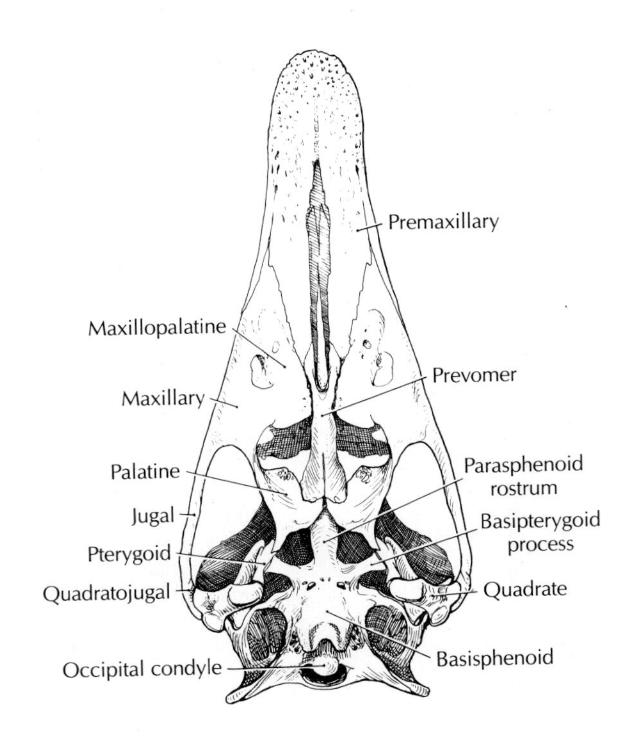

Figure 3–4 Bony palate of the Common Rhea, showing the complex arrangement of bones. Other orders of birds have different arrangements, which have been used to classify them. (From *Fundamentals of Ornithology* by J. Van Tyne and A.J. Berger; Copyright © 1976 John Wiley & Sons, Inc.; reprinted by permission of John Wiley & Sons, Inc.)

response to Darwin and Huxley, anatomists began to search for meaningful taxonomic characters, features that suggest common ancestry of groups of species. In addition to the bony palate, some of the most important features include the form of the nostrils; the structure of the leg muscles and tendons of the feet; the arrangement of toes; the form, size, arrangement, and number of scales (scutes) on the tarsus; and the presence or absence of the fifth secondary flight feather on the wings. Additional characters have been suggested that influence the assessment of relationships. The birth of a convincing classification took decades but was finally manifest in the work of Hans Friedrich Gadow (Gadow 1892, 1893). Based on the integrated assessment of 40 anatomical characters, Gadow's brilliant classification of birds was a milestone in ornithological systematics, one that still prevails. Modern classifications of the birds of the world—namely, those of Erwin Stresemann (1927–1934), Alexander Wetmore (1960), Robert Storer (1971b), and James Peters in the 15-volume *Checklist of the Birds of the World*—derive from Gadow's work. Despite a major effort in the first half of this century, ornithologists failed to discover phylogenetic clues that seriously challenged the prevailing arrangements of the higher categories of birds. Recently, Erwin Stresemann, a great German ornithologist, said,

> As far as the problem of the relationship of the orders of birds is concerned, so many distinguished investigators have labored in this field in vain, that little hope is left for spectacular breakthroughs. Science ends where comparative morphology, comparative physiology, and comparative ethology have failed us after nearly 200 years of effort. The rest is silence. (Stresemann 1959, pp. 277–278)

Flamingo relationships

Morphological comparisons yield different conclusions by taxonomists, and the different interpretations lead to vigorous debates. One such debate concerns the evolutionary relationships of flamingos. For years ornithologists have wondered whether flamingos are really ducks with long storklike legs or storks with ducklike, filter-feeding bills. Evidence for both conclusions seemed strong. Lacking firm resolution, taxonomists have given flamingos an order of their own (Phoenicopteriformes) between the Anseriformes and the Ciconiiformes (Storer 1971b); see Table 3–2. Gadow himself classified flamingos as a subgroup (Suborder Phoenicopteri) of the Ciconiiformes.

Adding to the long-lived uncertainty is the recent proposal by Storrs Olson and Alan Feduccia (1980a) that flamingos may not be related to either ducks or storks. They suggest instead that shorebirds, specifically stilts, are the modern relatives of flamingos. The past affiliations with ducks and storks, they assert, are based solely on "traditions and misconceptions". The musculature of flamingos is like that of the Banded Stilt of Australia, and both possess a unique leg muscle. Other characteristics, including the flamingo's distinctive life history, natal down, skeletal structure, and behavior, also resemble those of the Banded Stilt. Olson and Feduccia recommend that the taxonomic status of the flamingos be reduced to the family Phoenicopteridae, which should be placed immediately after the Recurvirostridae, putative ancestors of the flamingos in the Order Charadriiformes. The history of this changing classification is summarized in Table 3–3.

Taxonomic characters

What kinds of characters can be used to determine which taxa are related? Of greatest value to taxonomists are characters, called conservative, that do not change easily in the course of ecological adaptation. Yet there is no

Table 3–3 Different Classifications of Flamingos*

Taxon	Gadow 1893	Storer 1971b	Olson and Feduccia 1980a
Order	Ciconiiformes		
Suborder	**Phoenicopteri**		
Order	Anseriformes	Anseriformes	Anseriformes
Order		**Phoenicopteriformes**	
Order		Ciconiiformes	Ciconiiformes
Order	Charadriiformes	Charadriiformes	Charadriiformes
Family		Recurvirostridae	Recurvirostridae
Family			**Phoenicopteridae**

* These classifications reflect scientists' uncertainty of the evolutionary relationships of flamingos. Traditionally allied to storks (Ciconiiformes), either as a suborder (Phoenicopteri) or as a separate order (Phoenicopteriformes), Olson and Feduccia recently suggested that they are really shorebirds that merit only family status (Phoenicopteridae) in the order Charadriiformes.

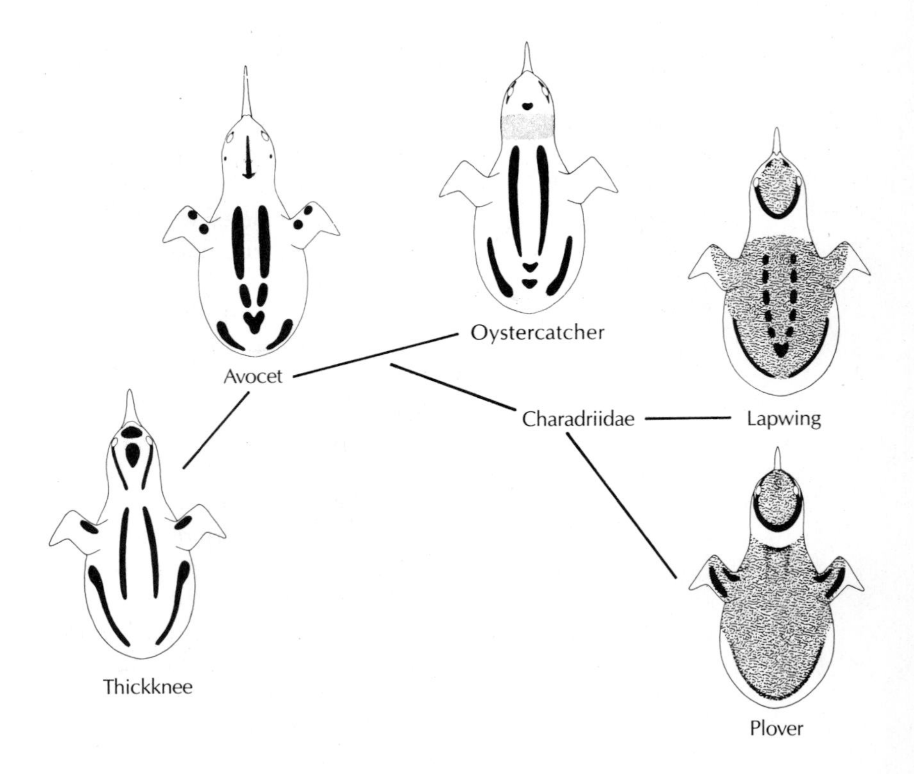

Figure 3–5 Plumage color patterns of downy young shorebirds provide clues to their evolutionary relationships. (Adapted from Jehl 1968)

single conservative attribute of birds that leads to a comprehensive solution. Choice of taxonomic characters for classification varies, as does their application from one group of birds to another. Behavior may yield clues to evolutionary relationships among some birds (Delacour and Mayr 1945, 1946; Storer 1963), and so may plumage patterns of downy young (Storer 1967; Jehl 1968) (Figure 3–5), calls and morphology of vocal apparatus (Ames 1971; Lanyon 1960, 1967, 1978), or genes and proteins. Unique characters may point to related groups of species, those with a common ancestor. Passerine birds, for example, have several unique characters. They have unique sperm (see Chapter 17), a specialized perching foot, and a preen gland (see Chapter 4) with a unique nipple structure. As mentioned, the passerine foot has a large hallux (rear-directed toe), uniquely arranged deep plantar tendons, and greatly simplified foot muscles that facilitate perching at the expense of more delicate toe movements (Raikow 1982) (Figure 3–6). These features indicate that members of the Passeriformes have evolved from a common ancestor; that is, they are monophyletic.

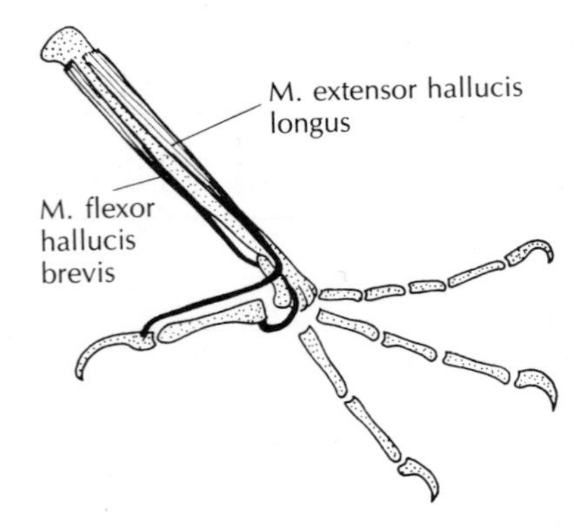

Figure 3–6 Simplified diagram of the intrinsic muscles of the passerine foot. (Adapted from Raikow 1982)

Convergence

Characters that change easily in the course of adaptation provide little information about ancestry and are sometimes misleading. Adaptation to similar ecological roles causes unrelated species of birds living in different places to become similar in details of appearance, morphology, and behavior. The meadowlarks of North American grasslands and the longclaws of

Africans grasslands are a classic case of convergence in color pattern. Both live in open grasslands and are about the same size and shape. Both have streaked brown backs, bright yellow underparts with a black V on the neck, and white outer tail feathers. But meadowlarks are classified as related to the Red-winged Blackbird and other members of the New World family Icteridae on the evidence of their conical bill shapes, distinctive jaw muscles, and lack of bristles at the base of the bill. Longclaws are classified as related to the pipits and wagtails in the family Motacillidae, based on their slender notched bills, configuration of jaw muscles, and presence of bristles at the base of the bill. A classic case of morphological convergence is that of the North Atlantic auklets, such as the Dovekie, and the diving-petrels of the South Atlantic (see Figure 5–15). Both are compact, black-and-white seabirds that use their wings to propel themselves underwater to capture marine crustacea. Dovekies are related to gulls and other members of the Charadriiformes; diving petrels are related to albatrosses and other members of the Procellariiformes, which have distinctive skeletons and tubular nostrils.

One could make a long list of instances of morphological convergence. Weavers of the Old World have counterparts among the blackbirds of the New World. Various tyrant-flycatchers share ecological and morphological attributes with shrikes, wheatears, tits, warblers, pipits, or thrushes (Keast 1972). Australian landbirds share so many morphological attributes with shrikes, flycatchers, and small insect-eating warblers that they were misclassified with superficially similar European and Asian birds. Genetic studies show them to be a major radiation of related Australian species that have filled a variety of niches on that continent (Sibley and Ahlquist 1983).

Convergence is unveiled by the study of fine anatomical details and the discovery of differences that reveal multiple evolutionary pathways to a particular functional solution (Bock 1965). Generally, the more complex the character, the less likely it is that anatomical details will be precisely the same. The details of avian foot structure reveal how unrelated birds evolved similar, but not identical, arrangements of the four toes (Bock and Miller 1959). Although most perching birds have anisodactyl feet (Figure 3–7), at least nine groups, including woodpeckers and their allies, most parrots, cuckoos, owls, the Osprey, turacos, mousebirds, Cuckoo Roller, and some swifts, have zygodactyl feet. Differing details in the orientation of the articulating surfaces (condyles) of cuckoo and of woodpecker toes indicate that these unrelated birds have evolved the zygodactylous foot arrangement in different ways. In woodpeckers, parrots, and cuckoos, the reversed position of the fourth toe is fixed, but in the other zygodactylous groups, the fourth toe can be used in a forward (anisodactyl) or in a reversed (zygodactyl) arrangement. The trogons appear zygodactylous, with two front and two rear toes, but their second toe, not their fourth, is directed backward (heterodactyl) to compensate for a weak hallux and to facilitate stable perching. As the figure also shows, the syndactyl foot characterizes the Coraciiformes; and in the pamprodactyl foot, the first and fourth toes can point forward or backward. Thus we see that convergent foot structures can be detected by careful study of the details of the toes themselves.

Keeping in mind that many of the current debates in avian systematics center on the possibility that similarities result not from ancestral relationship but from convergence, let us think for a moment of the trou-

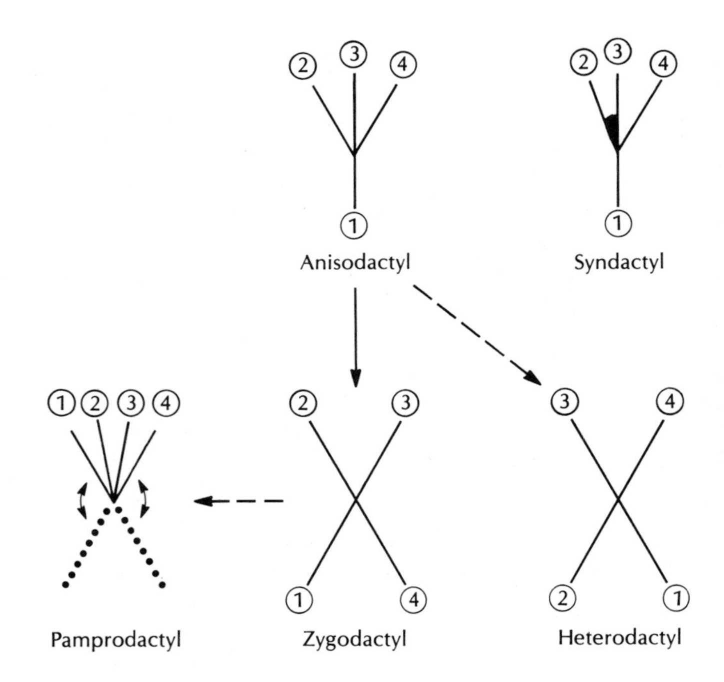

Figure 3–7 Toe arrangements of perching birds. Alternatives to the prevalent (anisodactyl) arrangement of three toes in front and the hallux (the first digit) pointing to the rear have evolved several times. The syndactyl foot, in which the bases of toes 2 and 3 are fused, characterizes the Coraciiformes. The zygodactyl arrangement, with two forward-pointing toes and two rear-pointing toes, has been achieved in different ways ten times during the evolution of birds. In trogons, toe 2, not toe 4, is rear-directed (heterodactyl). In the pamprodactyl foot, the positions of toes 1 and 4 are not fixed; all four toes may point to the front. Dashed arrows indicate uncertain derivations.

bled classification of flamingos. Could the similarities between flamingos and Banded Stilts reflect mere convergence? Both, after all, are adapted to an unusual environment, the shores of shallow, saline lakes.

Primitive and derived characters

Taxonomic characters must be *homologous* structures, that is, structures that are shared by two organisms and that "can be traced phylogenetically to the same feature in the immediate common ancestor of both organisms" (Bock 1973, p. 386). Particularly interesting for taxonomic studies are conservative, homologous characters that exist in both their original and their changed states. For example, penguins have evolved flipperlike wings from the typical avian wings of their petrel ancestors. In this case, the wings of petrels represent the ancestral, or primitive, character state whereas the flipperlike wings of penguins represent the advanced, or derived, character state. Where hooked bills have evolved from unhooked bills, the former represent the derived character state, and the latter represent the primitive character state.

If two species share a derived character state, we can hypothesize that they have a common ancestor with the same derived character state. The flipperlike wings of the various species of penguins presumably reflect a common ancestry. We can then draw simple hypothetical branching sequences, or cladograms, based on the characters of extant species and their hypothetical ancestors. Figure 3–8 shows three hypothetical birds that have different feet, crests, and bills. We assume that the cladogram with the fewest evolutionary changes, the most parsimonious, illustrates the most

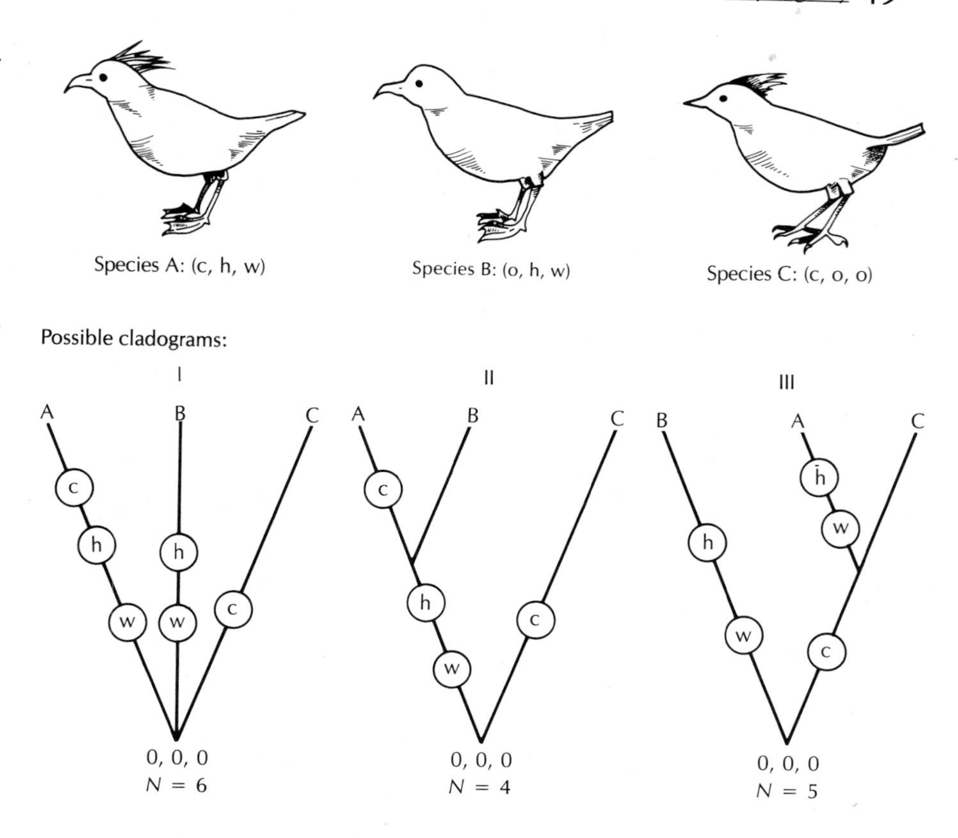

Figure 3–8 Possible cladograms for three hypothetical bird species, A, B, and C, that have three combinations of three derived characters: (c) crest, (h) hooked bill, and (w) webbed feet. Primitive character states (no crest, unhooked bill, or unwebbed feet) are denoted by 0. The changes from primitive (0) to derived character states (c, h, or w) are indicated for the evolution of each lineage (A, B, or C) from the common ancestor (0, 0, 0). The center cladogram ($N = 4$) is the most parsimonious, requiring fewest total changes (N) to account for the distribution of derived characters among the three species; it also has the advantage that it postulates no convergence between species A and B.

plausible phylogeny. In this case, cladogram II, which assumes a common ancestor for A and B that looked like B, would be the most likely one.

As in the hypothetical case in Figure 3–8, several cladograms can be made for any set of species. In constructing such cladograms, we assume that a particular change cannot take place independently in different lineages. In other words, we assume that convergence is not possible. We may also assume that once a derived character state evolves, it cannot be reversed; for instance, flightless birds do not reevolve the ability to fly, and hook-billed birds do not lose their hooks. There is, however, an exceptional case. Robert Raikow and his associates (1979) found that bowerbirds have reevolved a leg muscle, the *M. iliofemoralis externus,* which had been lost in the evolution of most passerine birds.

One example of an avian relationship based on primitive and derived character states is that proposed by Alan Feduccia (1977) in a study of the stapes (middle ear bone) of perching birds (see Chapter 8). The wood hoopoes and the Hoopoe share a unique, derived character state, an anvil stapes, which supports the traditional hypothesis that the two families are closely related (Figure 3–9). Other coraciiform birds, the kingfishers and their allies — motmots, todies, and bee-eaters — have a different-shaped stapes, a derived character state also found in trogons, a perplexing order of tropical birds without obvious close relatives.

Some songbirds, including New World flycatchers and their relatives (Tyranni), have a stapes unlike that of other passerines but very similar to the kingfisher stapes. Consequently, Feduccia first suggested that the suboscine passerines were related to kingfishers and not to other passerines,

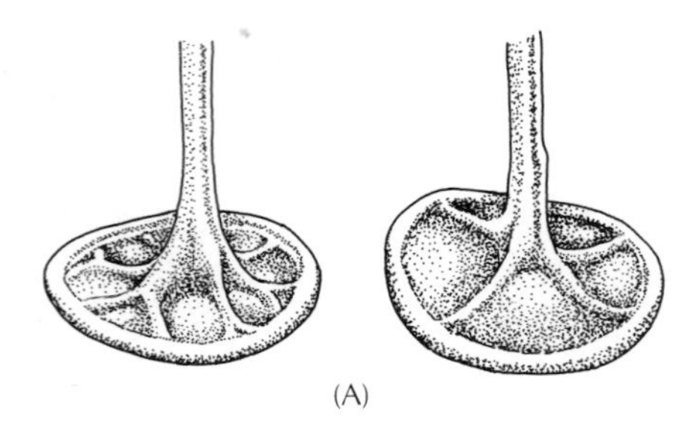

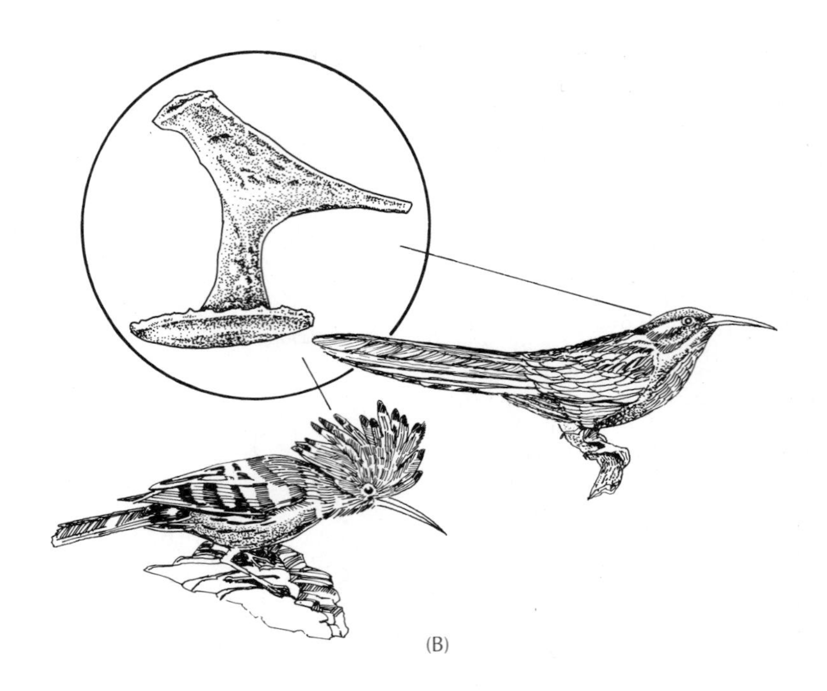

Figure 3–9 (A) Primitive and (B) derived anvil form of the middle-ear bone, or stapes. Both hoopoes and wood hoopoes share the derived character, which supports the hypothesis of close evolutionary relationship. (After Feduccia 1977)

but his own more recent studies of their sperm do not support his earlier radical view (Feduccia 1979). The similar stapes of these two groups must be convergent. As stated earlier, classifications based on single characters, such as the earbone, are weakened by the possibility of convergence. Additional characters must be considered to generate a meaningful classification.

Ratite relationships

The first major ornithological study of the derived states of many different characters was Joel Cracraft's work (1974a) on the relationships of ratites, the large, flightless species that include ostriches, rheas, emus, the extinct moas of New Zealand, and the extinct elephant birds of Madagascar. The tinamous of South America also belong to this assemblage, even though they can fly, because they possess the many other characters that unite the ratites as a monophyletic group, including a unique (paleognathous) palate structure (see Figure 3–4), an unusual segmented bill covering in the downy chick, similar configurations of the openings in the pelvis called the ilioischiatic fenestra, similar DNA nucleotide sequences, similar chromosome arrangements, and similar amino acid sequences of an eye lens protein. For his study, Cracraft distinguished between the primitive and derived states of ratite skeletal characters: Primitive character states were those that are also present in unrelated members of the Galliformes, and derived character states were those unique to ratites. Cladogram analyses of the characters suggested that tinamous separated from the main ratite lineage very early, that rheas are most closely related to ostriches, and that emus are more closely related to ostriches than are moas and kiwis.

Cracraft's conclusions reflect his hypotheses regarding primitive and derived character states. Anthony Bledsoe (1988), who reevaluated Cracraft's data, disagreed with some of his decisions and made allowance for

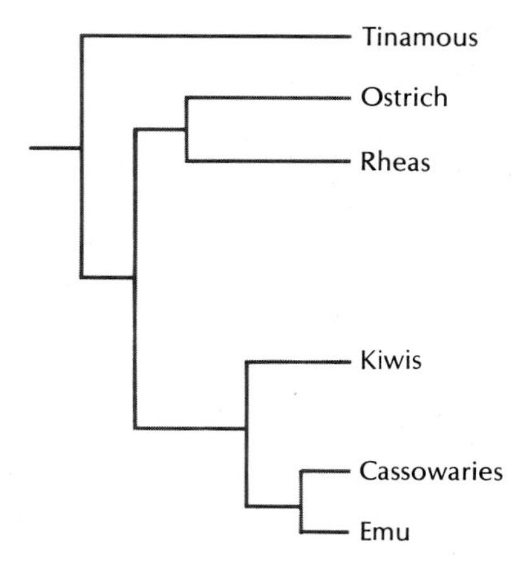

Figure 3–10 Proposed phylogeny of ratite birds. Tinamous, which are the only flying members of this group, represent the most primitive forms. The flightless forms purportedly dispersed throughout the southern continents prior to the breakup of Gondwanaland 75 to 80 million years ago. (After Sibley and Ahlquist 1981a)

evolutionary reversals. Bledsoe's alternative cladistic analysis suggests that emus and kiwis are more closely related to each other than either is related to the Ostrich (Figure 3–10), the opposite of Cracraft's proposed relationships. Bledsoe's results are consistent with biochemical evidence, which gives him the edge over Cracraft. Because presuppositions about character states can skew results, it is important to compare such findings with other types of evidence.

Biochemical systematics

Our rapidly increasing knowledge of DNA structure and of the genetic control of protein synthesis has enabled us to uncover new possibilities in the study of the evolutionary relationships of birds. We can test hypothetical relationships based on morphological similarity against other estimates of similarity based on comparative biochemistry. In general, biochemical studies tend to corroborate previous morphological evidence of relationships (Barrowclough and Corbin 1978; Barrowclough et al. 1981). Sometimes, however, biochemical analyses challenge traditional views, reveal overlooked cases of convergence, and suggest unsuspected relationships among taxa.

One of the primary techniques in biochemical systematics is protein electrophoresis, a method of separating proteins, usually different forms of the same enzymes, according to their mobility in a weak electric current. Proteins carry net electric charges, which reflect slight differences in amino acid composition and hence differences in the genes that control the amino acid composition of a particular protein. A survey of allele differences at 30 to 40 genetic loci gives a good estimate of the genetic divergence between species since the time that they had a common ancestor. Genetic divergence increases with taxonomic rank: Species in different genera, on average, are five times as different genetically as are species in the same genus; species in different families are seventeen times as different genetically as are species in the same family. Surveys of protein differences reveal clusters of closely related species. In the case of wood warblers, the analysis of protein similarities among nine species conformed to their division into three genera (*Vermivora, Dendroica,* and *Seiurus*) based on similar morphological characters (Barrowclough and Corbin 1978) (Figure 3–11).

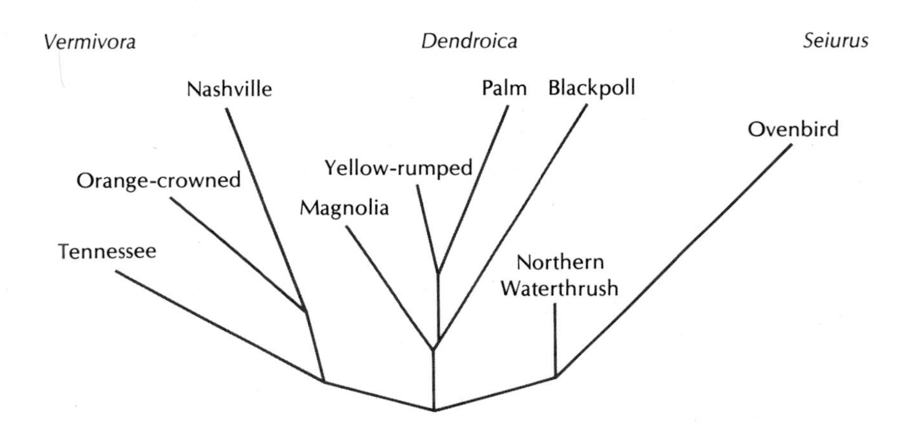

Figure 3–11 The genetic differences among bird species can be assessed by the electrophoretic separation of proteins. The results of this biochemical study of eastern wood warblers conformed closely to previous assessments of relationships based on morphological characters. Species are linked by branch lengths that are proportional to their genetic similarities. (After Barrowclough and Corbin 1978)

The pioneer electrophoretic studies of egg-white proteins by Charles Sibley and Jon Ahlquist produced startling information about the relationships of some odd birds, each of which was assigned to its own (monotypic) family. For example, the Hoatzin, an unusual bird of the Amazon's tributaries, is probably related to *Guira* cuckoos, not guans, as it has been classified (Sibley and Ahlquist 1973) (Figure 3–12). Another strange bird, the skulking Wren-thrush of highland bamboo habitats in Central America, was long thought to be an aberrant thrush (Turdidae), but study of its proteins revealed that it is really a wood warbler (Parulidae) (Sibley 1968). Analyses of the vocal apparatus (Ames 1975), hindlimb muscles (Raikow 1978), and life history (Hunt 1971) support Sibley and Ahlquist's protein-based hypothesis. Support from characters as disparate as protein structure, muscle arrangements, and behavior is strong support indeed.

DNA hybridization, a biochemical technique, is a promising way to determine avian relationships (Sibley and Ahlquist 1983; Gould 1985). Like protein electrophoresis, DNA hybridization measures the amount of genetic change that has taken place since the time that two groups diverged from their most recent common ancestor, but it does so more directly and more comprehensively. Instead of measuring divergence of the products of a sample of genes, as does protein electrophoresis, DNA hybridization measures divergence of the entire nuclear genome at the level of the nucleotides that constitute the genetic code. Fragments of single strands of avian DNA collected from two species form a double-stranded hybrid com-

Figure 3–12 Electrophoretic studies have suggested that the Hoatzin is related to *Guira* cuckoos, not to guans, as had been thought. (From original by G. M. Sutton, courtesy Academy of Natural Sciences, Philadelphia)

plex when allowed to associate under special laboratory conditions. The number of nucleotide base pairs the two birds have in common in their entire (nonrepeated) genome is then determined by measuring the thermal stability of the double-stranded complex; each one percent increase in the match between genetic materials requires a 1 °C increase in temperature to dissociate (separate). The DNA of a species matched against itself is very stable, but DNAs of distantly related species, such as a penguin and a warbler, readily dissociate at low temperatures. The fact that hybridized strands of thrasher and starling DNA dissociate at higher temperatures than those of thrashers and thrushes suggests that thrashers are more closely related to starlings than to thrushes (Figure 3 – 13).

In addition to confirming much of the classical classification of birds, which is based on morphology, the results of DNA hybridization support some new ideas (Sibley and Ahlquist 1985a, b). David Ligon (1967) suggested that New World vultures, including Turkey Vultures and Andean Condors, are more closely related to storks such as the Wood Stork than they are to raptors. Storks and vultures share a variety of skeletal and behavioral characters, including the unusual habit of defecating on their legs as a means of increasing heat loss through evaporation of liquid excreta. DNA analyses support Ligon's hypothesis. These same analyses offer no support, however, to Olson and Feduccia's proposal of a relation between flamingos and stilts. Instead, DNA links flamingos to the stork assemblage, as Gadow proposed nearly a century ago from his studies of morphology.

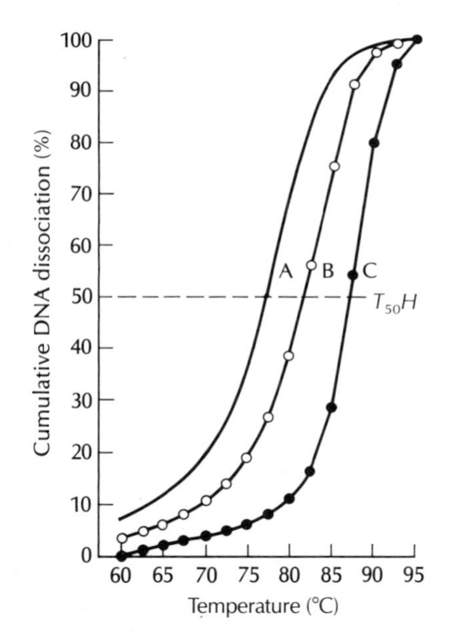

Figure 3 – 13 Results of DNA × DNA hybridization studies of the relationship of thrashers and starlings. Similar DNAs are more resistant to temperature stress than are dissimilar DNAs. Shown here are the cumulative thermal dissociation curves of DNA of the Long-billed Thrasher hybridized with DNA of (A) thrushes, (B) starlings, and (C) with itself. The temperatures at which half of the DNA dissociated ($T_{50}H$) are used for standardized comparison of the curves. (After Sibley and Ahlquist 1984)

DNA studies reveal the extraordinary influence of adaptive radiation and convergence among ecological equivalents, regardless of their ancestry. Charles Sibley and Jon Ahlquist (1980, 1985b) demonstrated that a major evolutionary assemblage of the relatives of modern crows diverged from the lineage of other passerines at the beginning of the Tertiary, 60 million years ago. Relatives of modern crows include birds-of-paradise, lyrebirds, bowerbirds, shrikes, drongos, monarch flycatchers, the nuthatchlike Australian sitellas, and perhaps the New World vireos. In addition, these DNA studies reveal that many Australian songbirds, including the brightly colored blue wrens, the warblerlike thornbills, and the honeyeaters, evolved as a separate branch from this assemblage. Some of these Australian songbirds had been previously classified with morphologically similar Asian and European forms. It is interesting to note that the adaptive radiation of Australian passerines apparently parallels that of marsupial mammals and eucalyptus plants.

DNA studies support the conclusion, based on skeletal characters, that tinamous diverged from other ratites before the other lineages separated from one another (Sibley and Ahlquist 1981a). The biochemical evidence, however, suggests that the evolutionary splitting of the other groups of large ratites occurred according to a different sequence. Ostriches of Africa and Asia apparently separated from the ancestors of South American rheas when the ancient supercontinent of Gondwanaland broke up 75 to 85 million years ago. The Australian ratites also split off at this time. Later, in the Eocene, the Australian ratites split again into two lineages, the moa – kiwi ancestor became isolated in New Zealand and the emu – cassowary ancestor became isolated in Australia – New Guinea (see Figure 3 – 10). Bledsoe's analysis of skeletal characters, mentioned earlier, also suggested this relationship of kiwis and emus.

Summary

The classification of birds, like the classifications of other organisms, is organized as a hierarchy of taxa, or groupings of species and populations, of the Class Aves into 30 orders, 174 families, 2044 genera, and 9021 species. The species is the primary unit of biological classification. Theoretically, each taxon is monophyletic, consisting of species more closely related to each other than to species in other taxa. The formal naming of species and their orderly arrangement in a hierarchical classification facilitates scientific inquiry and communication. The current ordinal classification of the birds of the world has changed little from the landmark work of Hans Friedrich Gadow in 1893, though the number of species recognized by ornithologists continues to change as a result of new discoveries and conceptual revisions.

The diversity of modern birds reflects historical patterns of speciation, extinction, and phyletic evolution. Conservative taxonomic characters — attributes that do not change easily in the course of adaptation — enable ornithologists to decipher which groups of species shared a common ancestor. Recognition of ancestral, or primitive, versus changed, or derived, character states aids in the reconstruction of the sequences of past evolutionary events. Convergence, the independent evolution of similar adaptations by unrelated species, can cause unrelated species to appear related. Cases of convergence can be revealed by detailed study of complex characters, such as the internal anatomy of the toes, and by biochemical evidence. Biochemical studies, which permit the assessment of genetic affinities of species, tend to confirm conclusions based on morphology but sometimes suggest unsuspected affinities and new patterns of adaptive radiation. The evolutionary relationships of certain birds, such as the flamingos, continue to puzzle ornithologists.

Further readings

Eldridge, N., and J. Cracraft. 1980. Phylogenetic Patterns and the Evolutionary Process. New York: Columbia University Press. *A provocative exposition of cladistic philosophy.*

Mayr, E. 1969. Principles of Systematic Zoology. New York: McGraw-Hill. *A summary of the basics of classical systematics.*

Sibley, C.G., and J.E. Ahlquist. 1983. Phylogeny and classification of birds based on the data of DNA–DNA hybridization. Current Ornithology 1:245–292. *An introduction to a powerful new biochemical methodology.*

Storer, R.W. 1971. Classification of birds. Avian Biology 1:1–18. *The standard classification of birds.*

Van Tyne, J., and A. Berger. 1976. Fundamentals of Ornithology. New York: John Wiley & Sons. *Contains an excellent review of taxonomic characters as well as of the families of birds.*

Voous, K.H. 1973. List of recent Holarctic bird species. Nonpasserines. Ibis 115:612–638. *A detailed annotated review of the classification of birds from the European viewpoint.*

part II

FORM AND FUNCTION

chapter 4

Feathers

Feathers, the most distinctive feature of avian anatomy, are an extraordinary evolutionary innovation. They are unique integumentary structures that are fundamental to many aspects of a bird's existence. They provide insulation essential for controlling body temperature, aerodynamic power necessary for flight, and colors useful for communication and camouflage. In addition to these primary functions, feathers perform a variety of other roles. Modified feathers are important in swimming, sound production, hearing, protection, cleanliness, water repellence, water transport, tactile sensation, and support (Stettenheim 1976).

Feathers are composed mainly of keratin, an inert substance consisting of insoluble microscopic filaments embedded in an amorphous proteinaceous matrix. Because keratin is resistant to attack by protein-digesting enzymes, it is an extremely long-lasting biological material. Although keratin in the generic sense is a standard constituent of other hard epidermal structures, such as claws, hair, fingernails, and scales, the ϕ-keratin in bird feathers is unique. It is not, as was once believed, the same as that found in reptile scales. The genes that control the production of avian ϕ-keratin have been isolated and partly sequenced and were found to be radically different from genes that control other keratin synthesis (Gregg et al. 1983, 1984). It appears that a genetic revolution of some kind led to the novel chemistry of feathers (Brush 1989).

In this chapter we shall consider feather structure and function, discussing first the basic feather structure, adaptive modifications of this structure, the major kinds of feathers in a bird's plumage, and the care of feathers by application of oily secretions of the preen gland. Feathers are replaced in seasonal molts, an important feature of every bird's annual cycle. We discuss the relationships of molts and plumages and the details of molt sequences in some feather tracts. We devote the last section to feather colors, both the basis of the colors themselves and the functions of plumage color pattern.

Feather structure

The detailed structure of bird feathers has fascinated biologists for centuries; it is an enormous topic (Lucas and Stetteheim 1972, pp. 235–276). We begin by reviewing the major feathers of a typical body, or contour, feather.

The primary features of a typical body feather are a long, central shaft and a broad flat vane on either side of this shaft (Figure 4–1). The hollow base of the shaft, the calamus, anchors the feather in a follicle below the surface of the skin. The rest of the shaft, the rachis, supports the vanes.

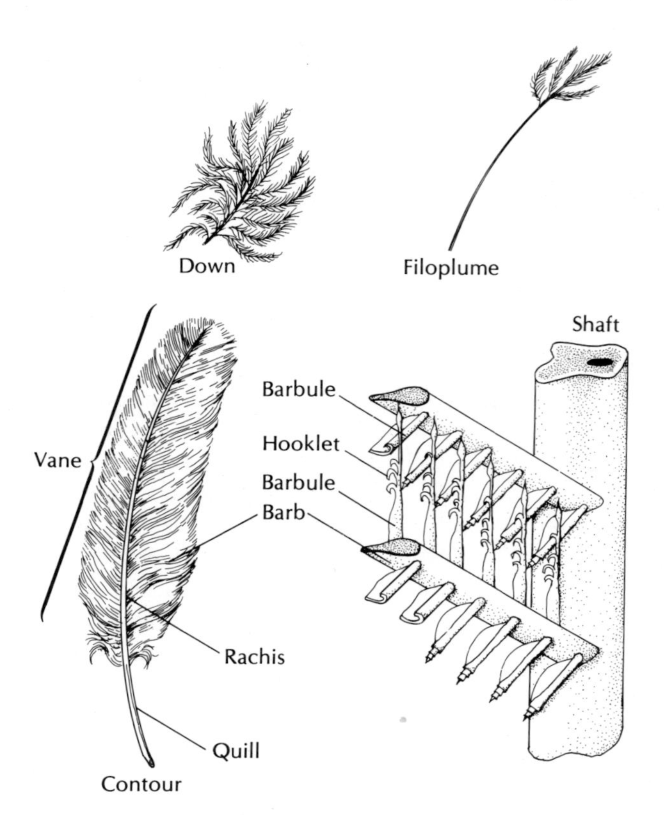

Figure 4–1 Structure of a contour feather.

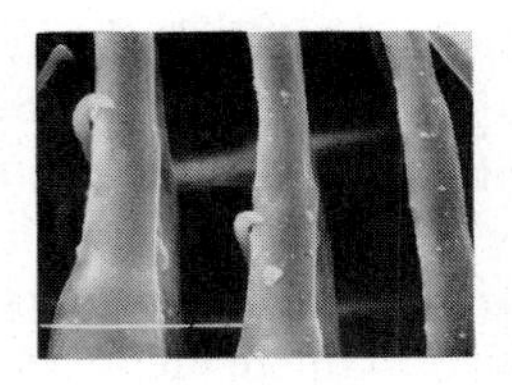

Figure 4–2 Scanning electron micrograph (2310×) of hooklets grasping adjacent barbules in feather of a Red Crossbill. (Courtesy B.A. Reaney)

Lateral branches off the rachis, called barbs, are the primary elements of vane architecture. Each barb consists of a tapered central axis, the ramus (pl., *rami*), with rows of smaller barbules, called vanules, projecting from both sides. A barbule consists of a series of single cells linked end to end; the cells may be simple or may bear projections called barbicels, some of which are elaborate. Barbs and barbules form an interlocking but flexible surface.

The vanes of a typical body feather grade from a hidden, fluffy basal portion, which provides insulation, to a visible, cohesive distal portion, which has a variety of functions. The barbules on the barbs at the base are long, thin, and flexible and do not have barbicels. With their similarly thin, flexible parent barbs, they create a downy or plumulaceous feather texture. In contrast, the outer pennaceous part of the vane is a firmly textured, tightly interlocking structure. Well-developed barbicels of various kinds are present on the pennaceous barbules that form this part of the vane. The cohesive structure of the outer vane is based on the interlocking arrangement of pennaceous barbules. Those on the distal side of the barb have hooklets that grasp thickened dorsal flanges on the next higher proximal barbules of the adjacent barb. The hooklets, which also slide laterally along the flange, are responsible for both cohesion and flexibility of the pennaceous vane (Figure 4–2).

The body feathers of birds typically include a secondary structure, or afterfeather, which emerges from the underside of the rachis where the first basal barbs of the vane branch off. With rare exception, the barb and barbule structure of afterfeathers is plumulaceous. The afterfeather's primary function is to enhance insulation. The afterfeathers of ptarmigan winter plumage are three fourths as long as the main feather. The afterfeathers of summer plumage are much shorter.

Feathers are subject to striking modifications. Extreme fusion of the developing barbs results in feathers that look like strips of plastic, as, for example, do the crown feathers of the Curl-crested Araçari, a small toucan, and the central tail feathers of the Red Bird-of-Paradise. The "plastic" feathers of the paradise birds function in courtship display, but why this araçari has such feathers is not known. The vane shapes of display feathers range from long and pointed, as are a rooster's hackles, to short and round, as are the head feathers of small birds. Rachises vary from thin and flexible, like those in the display tail feathers of some tropical hummingbirds, to heavy, stiff structures, like those in the bracing tail feathers of woodpeckers. Close spacing of large barbs with extra-long, curved barbicels makes water-repellent feathers in petrels, rails, and ducks. Flat, coiled barbules on the belly feathers of sandgrouse create a hairy mat that absorbs water for transport to young at nests remote from water sources.

Vaned feathers

Most conspicuous feathers are vaned feathers, which include the smaller contour feathers covering the body surface and the larger flight feathers of the wings and tail. The smooth overlapping arrangement of vaned feathers reduces air turbulence in flight. The tiny, flat contour feathers that cover a penguin's body create a smooth, almost scaly, surface that reduces friction during swimming.

Flight feathers (Figure 4–3) are large, stiff, almost completely pennaceous feathers without an afterfeather. They are adapted primarily to aerodynamic functions and have very little importance in insulation. Long, stiff flight feathers, called remiges (sing., *remex*), form the main horizontal surfaces of the wings. Rows of smaller feathers, called coverts, overlap the bases of the remiges and cover the gaps between them. The long shafts of the outer remiges, the primaries, attach to the hand bones. These feathers provide forward thrust on the downstroke of the wing during flight. Flight feather vanes are asymmetrical, presenting a narrow outer vane that cuts the air. Most birds have ten primaries; however, storks, flamingos, grebes, and rheas have eleven, ostriches have sixteen, and some passerines have nine.

Specialized barbules, called friction barbules, found on the inner vanes of the outer (primary) wing feathers, reduce slippage and separation of feathers during flight. Friction barbules have broad, lobed barbicels that rub against the barbs of overlying feathers. The longest friction barbules are found in the central part of the inner vane.

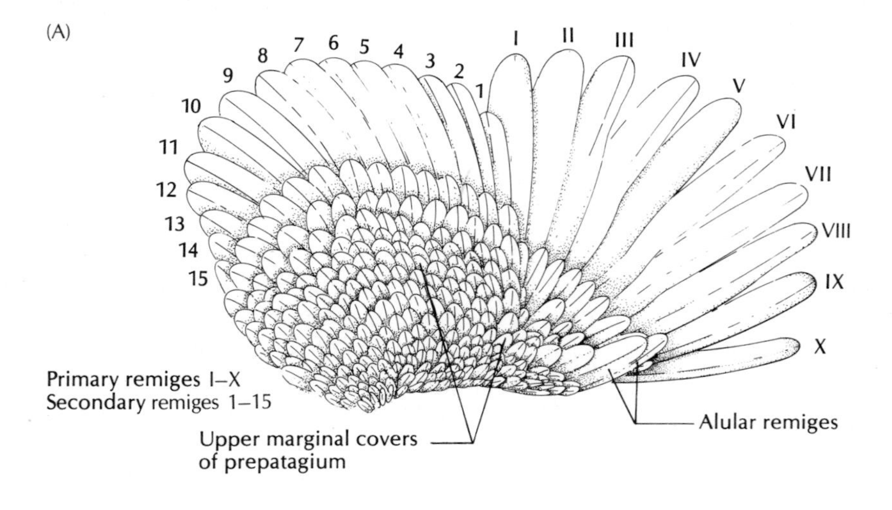

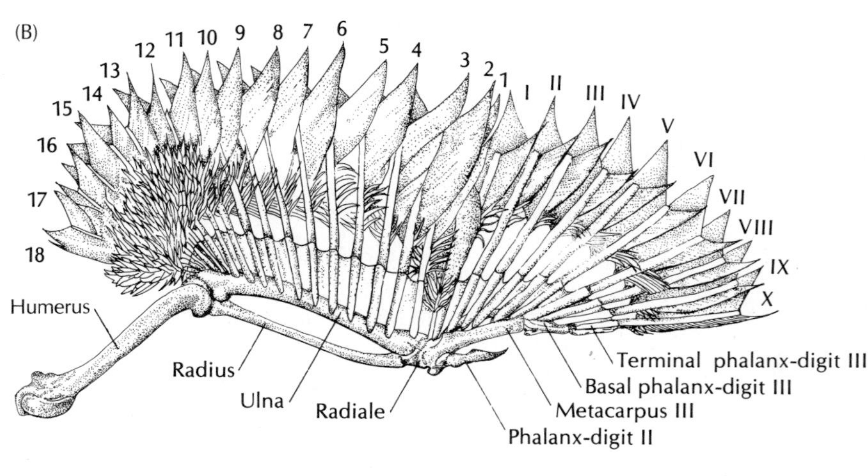

Figure 4–3 Dorsal view (A) of the extended left wing of a White Leghorn Chicken and (B) of the skeletal attachments of the primaries and secondaries of the same wing. (After Lucas and Stettenheim 1972)

The silent flight of an owl, which enables it to surprise prey, results in part from two special structural features that help to muffle feather sounds. The barbs on the leading edges of the owl's primaries are long, curved, well-separated structures that reduce air turbulence. Unusually long barbules help minimize the rubbing of overlapping feathers. Nightjars have a similar soft feather texture.

Because flight efficiency is directly linked to the structure of the primaries, major structural modifications are uncommon. The narrow outer primaries of the male American Woodcock, which produce trilling noises during courtship flights, are an exception. The primaries of flightless cassowaries consist only of strong spinelike extensions of the calamus. These 28-cm spines protect a cassowary's flanks from abrasive vegetation. During the breeding season, extensions of the second primary of male Pennant-winged Nightjars and Standard-wing Nightjars grow for temporary use in courtship (Figure 4–4). It is said that the nightjars discard the extensions by biting them off, but this remains to be proven.

The inner flight feathers of the wing, the secondaries, attach to the ulna and generate lift in flight. Ranging from 6 in hummingbirds to 19 in the Great Horned Owl, and 40 in albatrosses, the secondaries form much of the inner wing surface. They have also been modified for display purposes, for example, in the broad flaglike inner secondary that is essential for courtship in the Mandarin Duck. Quite a different kind of modification is found in the thickened, clublike feather shafts of the central secondaries of the Club-winged Manakin. The shafts make loud, castenetlike, snapping noises when the manakin claps its wings together.

Figure 4–4 The "standards" of the Standard-wing Nightjar are highly modified primaries, which are dropped shortly after courtship has been completed.

The flight feathers of the tail, the rectrices (sing., *rectrix*), attach to the fused caudal vertebrae, or pygostyle (Figure 4 – 5). There are usually 12 rectrices, which function primarily in steering and braking during flight. The elaborate tails of pheasants, lyrebirds, birds-of-paradise, and hummingbirds serve primarily in display and can be a handicap in flight. Some motmots, drongos, and kingfishers have racket-tipped rectrices with bare shafts and terminal vaned sections. The racket-tipped tail of the male Marvelous Spatuletail, a small hummingbird, includes only four rectrices, two of which are only thin, flexible rachises, 15 cm long, with large flags at the ends. The circular tail tips of a male King Bird-of-Paradise are tight whorls of rachises and inner vanes (Stettenheim 1976). Tail feathers are also modified for sound production in some snipes and the Lyre-tailed Honeyguide, and for bracing support in creepers, woodpeckers, woodcreepers, swifts, and penguins.

Down, bristles, and other kinds of feathers

Unlike firm-vaned feathers, down feathers are soft and fluffy. Birds such as ducks, whose natal down covers their entire bodies, are said to be ptilopaedic; psilopaedic chicks, such as a hatchling thrush, have only a few scattered feathers. The down feathers of adult birds, called definitive downs, also vary from thick continuous distributions underneath the contour feather coat to restricted distributions in association with the feather tracts or bare areas. A down feather typically lacks a rachis. Rather flexible, plumulaceous barbs and barbules extend directly and loosely from the basal calamus. Downy barbules tangle loosely with each other, trapping air in an insulating layer next to the skin. Down feathers provide, of course, excellent natural, lightweight thermal insulation (see Figure 4 – 1).

Semiplumes are intermediate in structure between down and contour feathers. They have a large rachis with loose plumulaceous vanes. Some are close to down in structure, whereas others more closely resemble contour feathers. Semiplumes are distinguished from down feathers by the length of their rachis, which is always longer than the longest barb. Semiplumes are found at the edges of the contour feather tracts but are usually hidden from view. They enhance thermal insulation and fill out the aerodynamic contours of body plumage.

Filoplumes are hairlike feathers distributed inconspicuously throughout the plumage in association with contour and flight feathers (see Figure 4 – 1). They monitor the movement and position of vaned feathers. They are most numerous near mechanically active or moveable feathers; flight feathers may have eight to twelve filoplumes per quill. Filoplumes consist of a fine shaft that thickens distally, ending in a terminal tuft of one to six short barbs with barbules. The barbules, similar to those of a down feather, are simple in structure. The slightest disturbance of a filoplume's enlarged tip is magnified and transmitted by the long, thin shaft to sensory corpuscles at its base, which then signal the muscles at the base of the vaned feather, causing them to adjust the feather's position. Filoplumes aid fine adjustment of the remiges during flight. Those in association with the breast feathers may also monitor air speed. As might be expected, filoplumes are absent in ostriches and other flightless ratites.

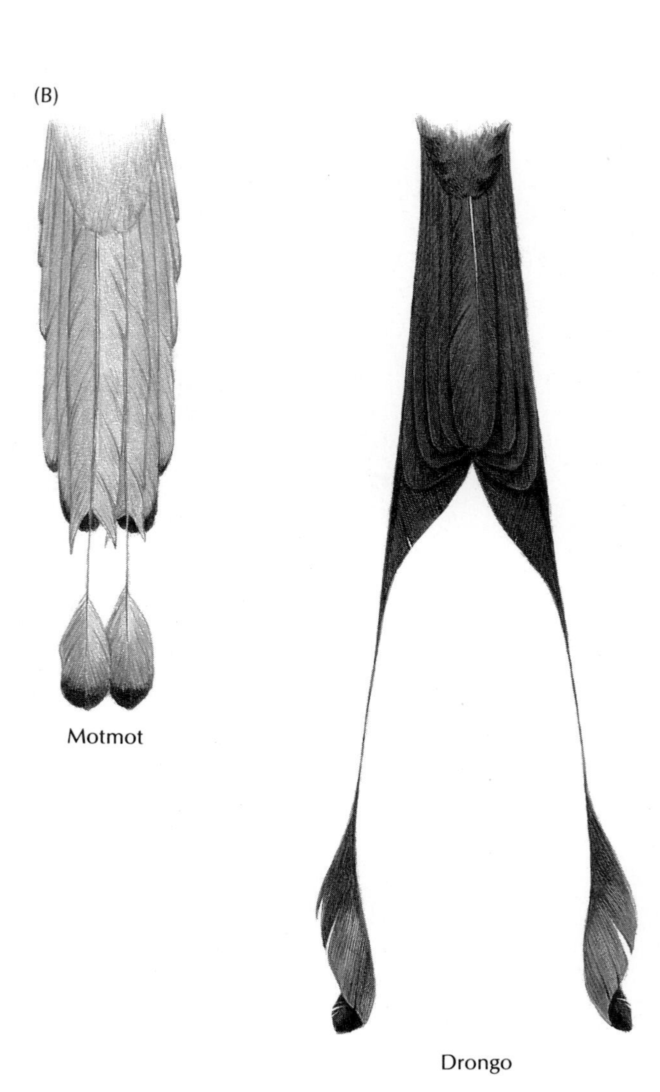

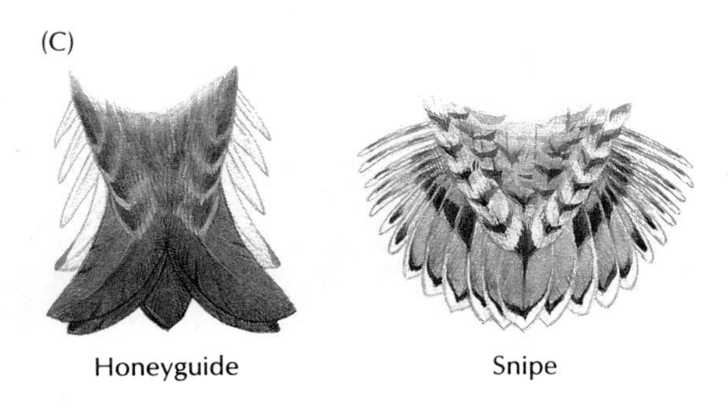

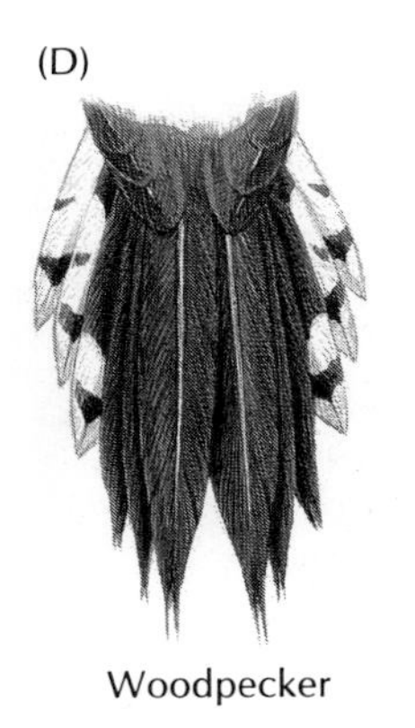

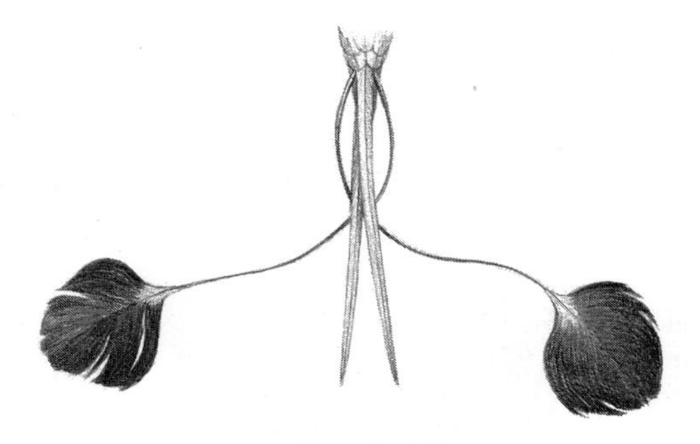

Figure 4–5 Tail feathers and their modifications: (A) unmodified tail of gull; (B) racket tails of a motmot, a drongo, and a hummingbird (Marvelous Spatule-tail); (C) sound-producing tails of a honeyguide (Lyre-tailed) and a snipe and (D) supporting tail of a woodpecker.

Bristles (Figure 4–6) are another specialized type of feather with both sensory and protective functions (Stettenheim 1973). Bristles are simplified feathers that consist only of a stiff, tapered rachis with a few basal barbs. Semibristles are similar but have more side branches. Except for those on the knees of Bristle-thighed Curlews and on the toes of some owls, bristles are found almost exclusively on birds' heads. The facial feathers of raptors tend to be simplified to bristles and semibristles, which are easier to keep clean than fully vaned feathers. This condition reaches an extreme in the carrion-eating vultures, which have bare heads with scattered bristles. The eyelashes of such birds as hornbills, rheas, and cuckoos consist of protective bristles, as do the nostril coverings of woodpeckers, jays, and crows. Most aerial insect-eating birds have bristles and semibristles around their mouths. The semibristles around the mouths of nightjars and owlet-

Figure 4–6 Bristles. (A) Whip-poor-will has well-developed bristles about the mouth. (B) Owlet-nightjar has elaborate bristles and semibristles around its bill. (C) An exception to the usual head locations of bristles are those on the knees of the Bristle-thighed Curlew.

nightjars are especially well developed, acting not only as insect nets, but possibly also as sensors of tactile information, in much the same way that a cat's whiskers do. Corresponding to their sensory functions, bristles, like filoplumes, have sensory corpuscles at their bases.

The feather coat

The feather coat of most birds consists of thousands of feathers (Wetmore 1936). A Tundra Swan has roughly 25,000 feathers, of which 20,000, or 80 percent, are on its head and neck. Songbirds typically have 2000 to 4000 feathers, of which 30 to 40 percent are on the head and neck. The lightness of a single feather belies the total weight of a bird's feather coat. In general, the feather coats of birds weigh two to three times as much as their bones. For example, the plumage of a Bald Eagle weighs about 700 grams, or 17 percent of its total mass (4082 grams) whereas its skeleton weighs only 272 grams, or 7 percent of its body mass, less than half that of the plumage (Brodkorb 1955).

Although feathers cover the entire body of a bird, they are not attached to the skin evenly or uniformly. Rather, feather attachments are grouped in dense concentrations called feather tracts, or pterylae, which are separated from each other by regions of skin with few or no feathers, called apteria. There are eight major feather tracts (Figure 4–7); these are subdivided into as many as 100 separate groupings, which can be used to distinguish avian taxa. The study of these arrangements is called pterylosis (Nitsch 1867). The functional significance of having feathers arranged by tracts and apteria has not yet been definitively established. Strategically placed feather groupings may reduce the cost of flight by decreasing total mass and the costs of feather production by economizing on the number of feathers produced. The apteria facilitate wing and leg movements and provide spaces for tucking these appendages beneath the feather coat. Apteria themselves may also facilitate heat loss (see Chapter 6).

The bases of feathers that are grouped tightly in feather tracts are linked by a complex network of tiny muscles that originate on one feather base and insert into an adjacent one. This elaborate muscle system enables a bird to control feather position for courtship displays or regulation of heat loss. If functional sets of feathers were not grouped closely, but were spaced evenly over the skin, controlling muscles would have to span greater distances. This would make them decidedly less effective unless their mass increased accordingly.

Feather care

Daily care of the feathers is essential. Birds may preen their feathers as often as once an hour while resting. They systematically rearrange their plumage with their bills, repositioning out-of-place feathers. They also draw the long flight feathers individually and firmly through the bill to restore the vane's integrity. Birds groom and delouse head and neck feathers by vigor-

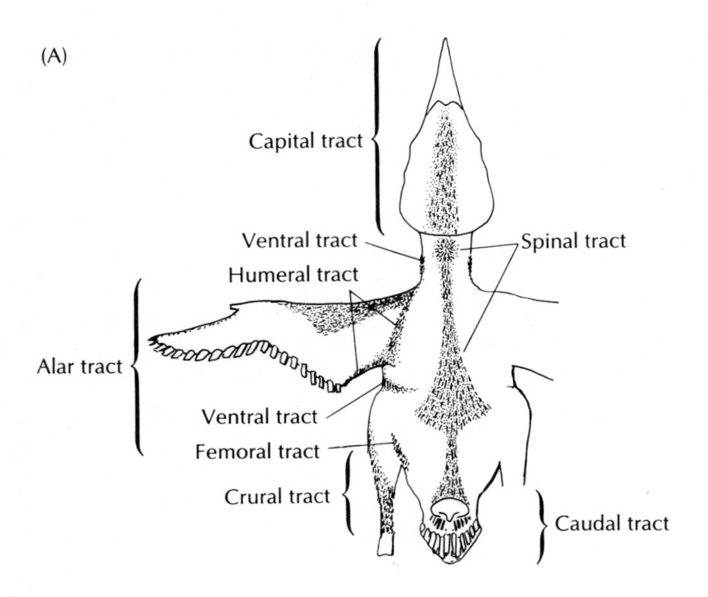

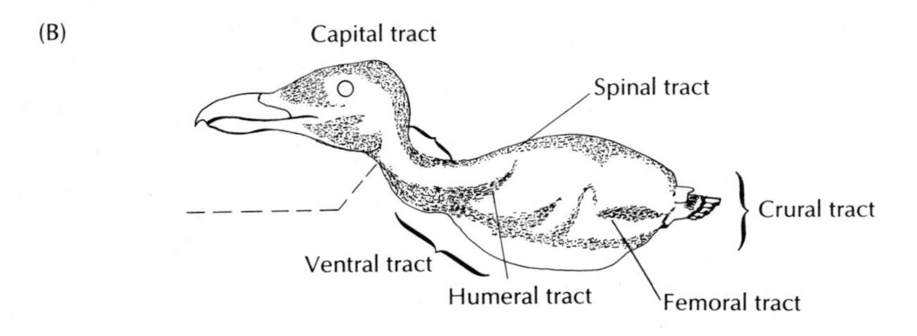

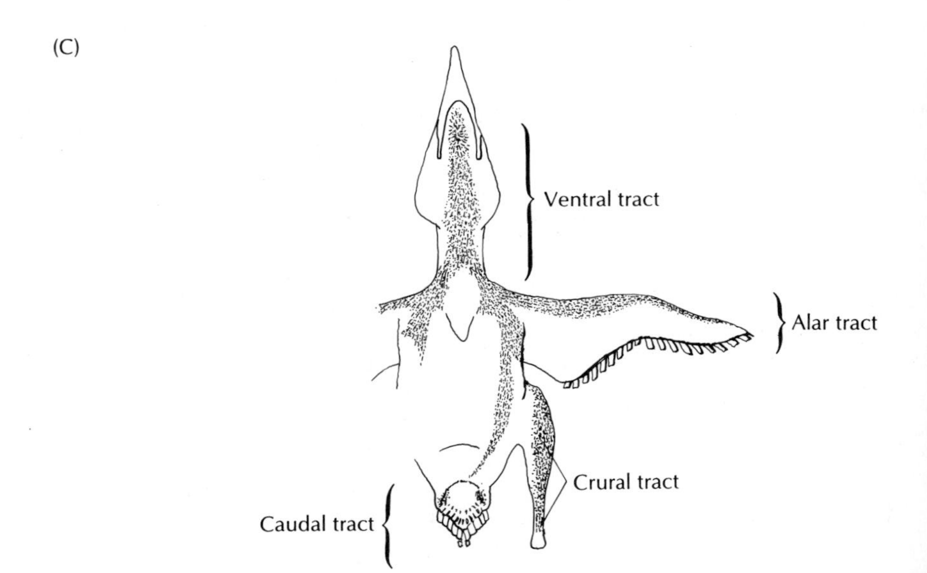

Figure 4–7 The eight major feather tracts, or pterylae, of birds. Bare or nearly featherless areas between the tracts are called apteria. Loggerhead Shrike: (A) Dorsal view. (B) Lateral view. (C) Ventral view. (After Van Tyne and Berge 1976)

Figure 4–8 Head scratching techniques. (A) Tennessee Warbler scratching directly with foot under the wing. (B) Golden-winged Warbler scratching indirectly with foot over the wing. (From Burtt and Hailman 1978)

ous scratching. Herons, nightjars, and barn owls have miniature combs on their middle toe claws that are used in grooming.

Most birds scratch their heads directly, reaching up under the wing with a foot, though some scratch indirectly, over the wing (Simmons 1957, 1964; Burtt and Hailman 1978) (Figure 4–8). The advantage of one method over the other is not apparent. Related birds tend to scratch alike; for example, sandpipers scratch directly whereas plovers and the related oystercatchers and stilts all scratch indirectly.

Feathers are inert and do not have an internal system of nourishment and maintenance. They would become brittle with age and exposure were it not for regular applications of the waxy secretions of the uropygial gland, or preen gland, located on the rump at the base of the tail. This gland, which is found in most birds, appears to have evolved as an essential accessory to feathers. Most preen glands are bilobed structures with a small tuft of downlike feathers encircling the glandular orifices of a well-differentiated papilla (Figure 4–9).

The preen gland secretes a rich oil of waxes, fatty acids, fat, and water, which, when applied externally with the bill, cleans feathers and preserves feather moistness and flexibility (Jacob and Ziswiler 1982). Unlike sebaceous secretions through the skin or the feather base, preen gland secretions do not harm the fluffy, insulative texture of the plumage. Regular applications of the secretion to the plumage sustain its functions as an insulating and waterproofing layer. Waterbirds typically have large preen glands, but whether the secretions of this organ are essential for keeping feathers dry and maintaining buoyancy is uncertain.

The waxy secretions of the preen gland also help to regulate the bacterial and fungal flora of feathers. Certain preen gland lipids protect feathers against fungi and bacteria that digest keratin (Baxter and Trotter 1969; Pugh and Evans 1970). Others may promote the growth of nonpathogenic fungi and discourage feather lice. Such chemical hygiene, researchers believe, is one of the most important functions of preen gland secretions. The foul-smelling preen gland secretions of hoopoes and wood hoopoes may also help repel mammalian predators.

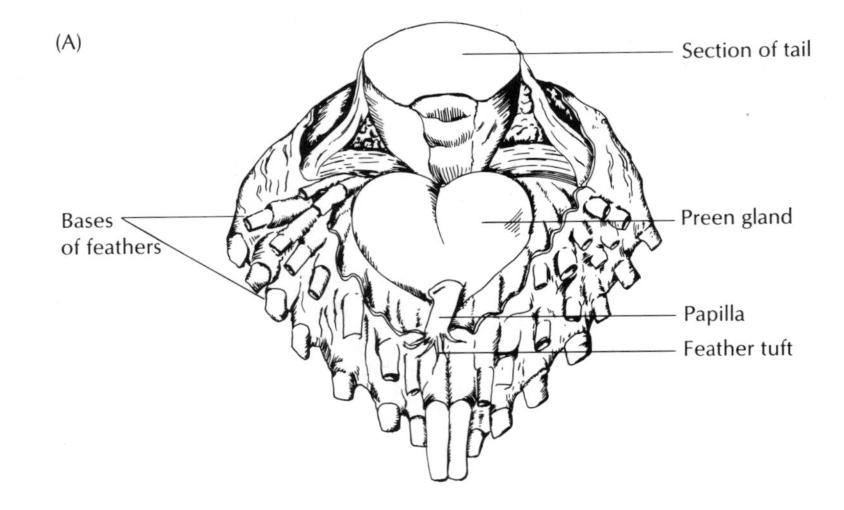

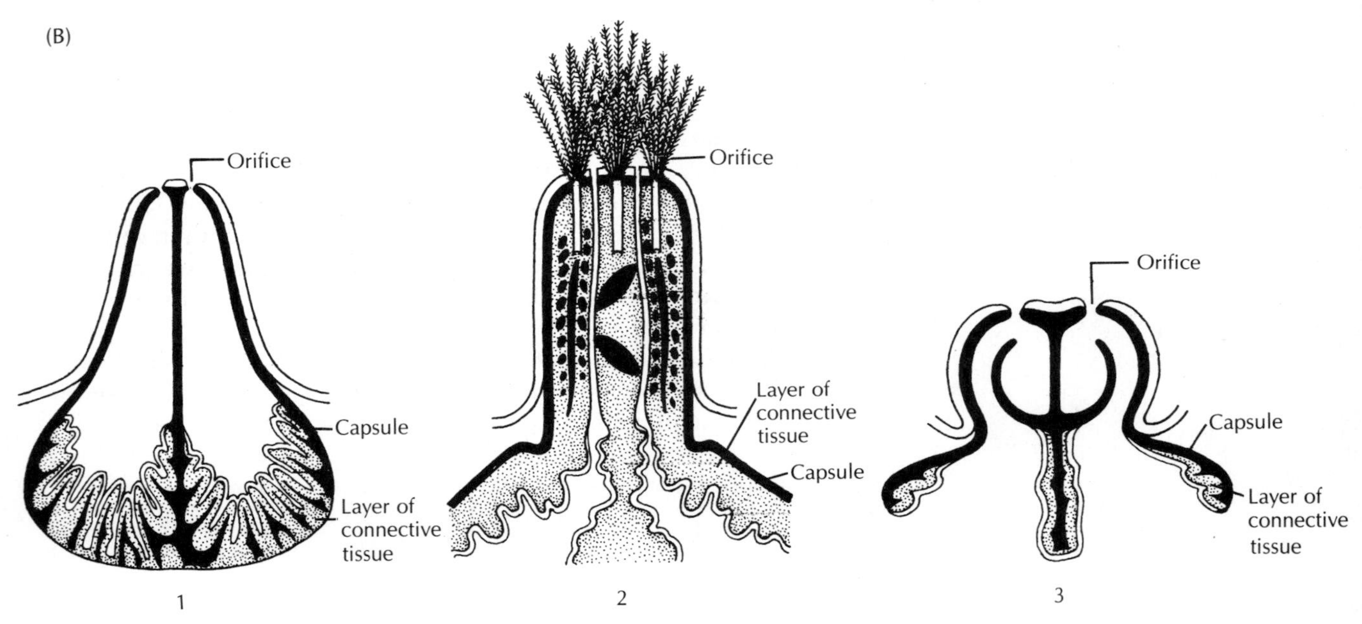

Figure 4–9 At the base of the tail on the lower back of most birds is the preen gland, which produces oily secretions that are essential for feather care. (A) Dorsal view of the gland and its environment on a White Leghorn chicken. (B) Details of papilla: (1) delicate type; (2) compact type; (3) unique passerine type. (A, after Lucas and Stettenheim 1972; B, adapted from Jacob and Ziswiler 1982)

A dustlike substance resembling talcum powder is present on the contour feathers of many birds. This is a waterproofing product of special feathers called powderdowns, which are dispersed throughout the feather coat among the ordinary downs. The powder consists of keratin particles 1 micrometer (μ) in diameter, which are sloughed continuously from the surface of the barbs. Powderdown feathers grow in dense, distinctly arranged patches on birds such as herons, Cuckoo Rollers, and Kagus.

Feather growth

Although a bird can change the position, visibility, and function of its plumage, it cannot change the structure of an individual feather. Feathers are inert structures. Once fully grown they cannot change color or form except through fading or abrasion. No nerves, muscles, or blood vessels lie beneath the outer surface of the exposed feather. The only mechanism for repair of damage is replacement of the whole feather. Except for cases of accidental loss of feathers and their immediate regrowth, feather replacement takes place regularly with age and with season.

New feathers grow from specialized pockets of epidermal and dermal cells called follicles, which periodically produce new feathers (Figure 4–10). The growth of a new contour feather starts in the follicle with the formation of a dermal papilla with a thickened apical cap. A thin layer of epidermal cells covering the papilla gradually thickens into the collar that develops into the feather. The dermal papilla develops into the feather pulp that supports the delicate epidermal cylinder. It also supplies nutrients to, and removes wastes from, the cells of the growing feather. The cells of the epidermal cylinder divide rapidly to form a tubelike structure, the main axis of which will form the rachis and the secondary barb ridges. Cells then differentiate into barbs and barbules.

The new feather grows rapidly, and toward the end of its growth the basal cells form a simple cylindrical calamus that anchors the mature feather in the follicle. The emerging new feather then pushes its predecessor out of the follicle.

The transformation of soft epidermal tissue into a hard durable structure is the last phase of feather growth. A major shift in cell function causes the new feather to fill with keratin, which constitutes 90 percent of the mature feather. As the process of keratinization ceases in the oldest, outermost feather cells, the sheath encasing the young feather cracks, allowing the keratinized feather tips to emerge and the barbs to unfold. The pulp, the core of living cells and blood vessels, is then resorbed by the follicle, rendering the completed new feather inert. The only evidence remaining of this early life-support system is a small hole at the end of the shaft, known as the inferior umbilicus.

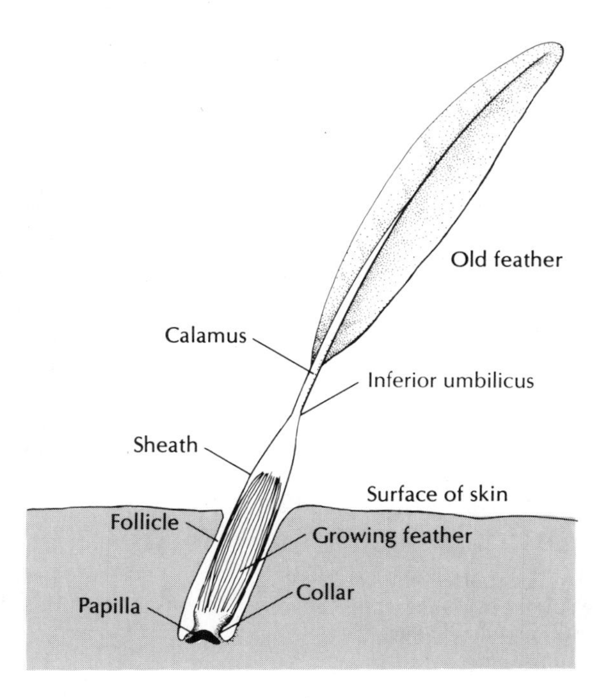

Figure 4–10 New feathers growing from a papilla and collar in the follicle push out the old feather. (After Watson 1963)

The follicle grips the feather at the calamus by a combination of muscular tightening and friction. Substantial force is required to pull a feather from this grip: 500 to 1000 grams for a single body feather of the average chicken. The tight grip of follicular muscles, which are controlled by the autonomic nervous system, may relax when a bird becomes frightened. The resultant loss of feathers is known as fright molt.

Molts and plumages

Every bird goes through a series of plumages in its lifetime. The first natal down may consist of a few scattered down feathers, as in the case of most hatchling landbirds, or it may be a dense, fuzzy covering like that of baby ducks and chickens. Such fragile feathers rarely last more than a week or two. They are soon replaced by a more substantial set of downy or pennaceous feathers.

Most birds have only one coat of natal down, which is pushed out of its follicles by incoming juvenal (pennaceous) feathers. Wisps of down may remain attached for a time to the new feathers. The baby bird's first set of wing and tail feathers appear at this time and grow rapidly in preparation for flight. As the young bird, now called a juvenile, approaches independence, it begins to exchange parts of its juvenal feathers for new plumage. In loons

Figure 4–11 The spotted plumage of a juvenile American Robin with residual tufts of down still attached to incoming head feathers.

and penguins, a second generation of down grows from the same follicles, pushing the old ones out. In hawks and waterfowl, a second coat of down grows from different follicles.

The replacement of juvenal plumage by feathers that closely resemble those of adults involves most of the body feathers, though not always those of the wings or tail. The young American Robin, for example, begins in midsummer to replace its spotted juvenal plumage with unspotted adult plumage (Figure 4–11). The first wing feathers remain. A few months later, its original flight feathers will propel the young robin on its first migratory flight. The bird will not molt again until it is just over one year old.

In the temperate zone the adult bird typically molts in the summer after breeding, replacing its entire plumage. It may keep its new set of feathers for 12 months, or it may replace some plumage before nesting the following spring, converting somber camouflage plumage to a brightly colored plumage for territorial display. Feathers of species that retain their plumage a full year may change in appearance because of wear. The European Starling, which is spotted in the winter, loses its spots as the feather tips wear off, and by spring it is sleek and glossy.

Close examination of the feathers of a male Scarlet Tanager reveals a series of feather generations, from which an observer can estimate the bird's age and the time of its last molt (Table 4–1). In its first month, a juvenile male Scarlet Tanager is olive-green with olive-brown wing feathers and streaked underparts. After it molts in July and August, it resembles its unstreaked, olive-green mother, except for black wing coverts that identify it as a male. In less than a year, the male will replace most of its olive-green body plumage with red-orange feathers. During this process, it has a mixture of old and new feathers: green, yellow, and orange. Even when fully red-orange in May, its olive-brown wings signal its status as a first-year individual; adult males have black wing feathers. In its second summer, the male tanager molts its entire plumage, replacing red-orange body feathers with olive-green winter plumage and replacing its olive-brown wing feathers with jet black feathers. The following spring it replaces winter plumage with a bright red breeding plumage; by this age it resembles all other adult males. After breeding, the adult tanager molts completely into camouflage plumage; and every year before breeding, it molts partially into brightly colored display plumage.

Table 4–1 Plumages and Molts of a Male Scarlet Tanager

	Year	Winter	Spring	Summer	Fall
Molt	1			Prejuvenal	First prebasic
Plumage				Juvenal	Basic 1
Color*				—	Green/Brown
Molt	2		Prealternate 1		Prebasic
Plumage		Basic 1	Alternate 1		Basic
Color		Green/brown	Red/brown		Green/Black
Molt	3+		Prealternate		Prebasic
Plumage		Basic	Alternate		Basic
Color		Green/black	Red/black		Green/Black

* Body/wings.

Phillip Humphrey and Kenneth Parkes (1959) recognized the need for a terminology of molts and plumages that is independent of seasonal aspects, which vary among species and regions. They proposed that molts be related to the incoming generation of feathers because feather loss is a passive result of the growth of new feathers. They also proposed (1) that the plumage that is renewed after breeding be considered the main component, or the "basic" plumage, of the annual cycle of plumages, and (2) that breeding adornments be considered temporary additions, or "alternate" plumages. Thus, the olive-green winter plumage of the Scarlet Tanager is its "basic" plumage, acquired via a "prebasic" molt that does not include replacement of the juvenal flight feathers. The following spring a "prealternate" molt produces the bright red "alternate" plumage.

The comparative study of molts and plumages reveals that some species, such as the American Robin and European Starling, undergo only a single annual molt, whereas others, such as the Scarlet Tanager, have a complex series of extra seasonal molts (Dwight 1907; Stresemann and Stresemann 1966). One complete molt a year was probably the primitive pattern from which more complex molt patterns evolved, and it continues to be the typical pattern. Gradual feather replacement imposes the least metabolic stress on an individual, and a yearly molt is sufficient to offset normal rates of feather wear. Multiple molts have proved advantageous for some birds as aids to seasonal display or as adaptations to severe feather wear or parasites. Some birds have more than one complete molt a year; others have both one complete molt and several partial molts.

Birds that live in environments that severely abrade feathers may molt more frequently than usual to maintain plumage. For example, in deserts, where wind and sand rapidly destroy feathers, some African larks molt completely twice a year. European larks, which suffer less abrasion, molt only once a year. Species, such as the marsh-dwelling Sharp-tailed Sparrow, that live in coarse grass habitats may also molt twice a year (Stresemann 1967). Shedding parasites is one apparent result of the double molt in this sparrow. It has fewer feather parasites than the Seaside Sparrow, which lives in the same marshes but molts only once a year (Post and Enders 1970). Another marsh and grassland species, the Bobolink (Figure 4–12), also molts completely twice a year. Male Bobolinks switch between streaked, brown winter plumage and a striking black-and-white alternate display plumage, whereas females replace their worn brown feathers with fresh ones of the same color.

A few birds molt three or four times a year, but the additional molts are only partial ones. Ruffs undergo the prebasic molt in the fall, the prealternate molt in the spring, which produces most of its breeding plumage, and then a third molt, which produces the "ruff" (see Chapter 16). To match their camouflage to the seasonal changes in the tundra, ptarmigan have three partial molts a year, and some populations of the Willow Ptarmigan have four.

Whereas geese have simple, annual molt and plumage cycles, many ducks of the north temperate regions have evolved more unusual sequences, in which the prealternate molt starts before the fall prebasic molt finishes. After they breed, drakes undergo a rapid prebasic molt that often includes simultaneous loss of all flight feathers, rendering them flightless and vulnerable for several weeks. The basic plumage that follows is a dull, hen-col-

Figure 4–12 Bobolinks molt completely twice a year. The male changes from a brown streaky (left) plumage like the female's in the winter to a bold black-and-white plumage (right) in the spring.

ored "eclipse" plumage that does not last long. Because the prealternate molt begins before the prebasic molt of the flight feathers is complete, the prealternate molt produces the drake's handsome breeding plumage by early winter, the season of courtship and pair formation.

Sequences of feather replacement

Molting in most birds follows a regular sequence within each feather tract. The usual sequence for the primary flight feathers, for example, is from the innermost primary outward to the last feather of the wing tip. In contrast with groups such as the ducks, which become flightless because they molt all their flight feathers at the same time, regular and symmetrical sequences of flight feather replacement seem geared to maintain flight ability. Replacement of the primaries and secondaries of the wings produces only small, temporary gaps in the wing surface and only a small reduction of flight power.

Like the flight feathers of the wing, tail feathers also typically molt from the innermost pair to the outermost pair, or centrifugally, with some exceptions. Large Asian partridges called snowcocks use their enormous tails, with 20 rectrices, for additional lift in sailing across steep ravines. To avoid loss of lift, snowcocks have a pattern of tail molt that differs from that of other partridges; it starts in the middle of each side of the tail and proceeds slowly in both directions. Another distinctive molt pattern is found in male Pennant-winged Nightjars and Standard-wing Nightjars. Their extended second primaries are a liability in flight, so they emerge out of normal sequence, after all other primaries are replaced.

We do not know what triggers the serial replacement of feathers (Payne 1972). Possibly, each follicle has a different sensitivity to the hormones that trigger molt. If so, the feather with the greatest sensitivity would drop first. Adjacent follicles may interact, directly or indirectly, in association with local expansion of blood vessels. Much remains to be learned about the control of molt sequences.

Feather colors

Feathers come in all shades, hues, and tints, and their colors are due either to biochrome pigments deposited in the feather or to special features of feather surfaces. Biochromes are naturally occurring chemical compounds that absorb the energy of certain wavelengths of light and reflect the energy of other wavelengths to produce the observed colors. Structural colors result from physical alteration of the components of incident light on the feather surface.

Pigments

There are three major categories of pigments in feathers: melanins, carotenoids, and porphyrins. The most prevalent are melanins, which produce earth tones, grays and blacks, browns, and buff colors. Carotenoids produce bright yellows, oranges, reds, and certain blues and greens. Porphyrins produce bright browns and greens, as well as the unique magenta found in turaco feathers.

With the exception of albinos, virtually all birds have some melanin pigment in their feathers. Melanin pigment is synthesized from tyrosine, an amino acid, by mobile pigment cells called melanoblasts, which creep about in the dermis layer of the skin. Melanoblasts manufacture and insert melanin granules into specific cells that are destined to become particular barbs and barbules. Periodic deposition into the primordial feather structures during development produces subtle color patterns such as barring or speckling. The shades of brown or gray depend on the density of melanin deposition.

Two kinds of melanin dominate in the barbs and barbules of bird feathers. Eumelanins are large, blackish, regularly shaped granules that produce dark brown, gray, and black. Phaeomelanins are smaller, irregularly shaped, reddish or light brown granules that produce tans, reddish browns, and some yellows. Color patterns often result from having mostly eumelanins in some places and mostly phaeomelanins elsewhere. In the plumage of the Gray Catbird, for example, the lead-gray color of most of the plumage results from eumelanin, and the rusty color of the undertail coverts comes from phaeomelanin.

Melanin performs many functions. The extra keratin associated with melanin makes the feather more resistant to wear (Burtt 1979). Dense melanin concentrations in the black wingtips of high-speed aerial species, such as gulls and gannets, reduce fraying of those feathers. Melanins help protect the feathers of desert species from sand abrasion. Melanins also absorb radiant energy, which aids thermoregulation (see Chapter 6). There is speculation that melanin granules promote drying of damp feathers by absorbing and concentrating radiant heat in the feather microstructure (Gill 1973; Wunderle 1981). If true, this could help to explain why birds of wet climates tend to be dark colored, a phenomenon known as Gloger's rule, in honor of Constantin Gloger who studied the relationship between climate and variation in birds.

Most carotenoid pigments, which are responsible for bright red, orange, and yellow colors, dissolve easily in lipids or organic solvents and are often stored in egg yolk, body fat, and the secretions of oil glands. These pigments also accumulate in droplets of lipid in the cells of growing feathers and are then left imbedded in the barbs and barbules when the natural fat solvents disappear during the last stages of keratinization.

The solubilities of carotenoid pigments enable chemists to extract them easily and to identify those responsible for particular plumage colors. Carotenoids take two chemically distinct forms, carotenes and xanthophylls. Xanthophylls have oxygen atoms attached to the carbon and the hydrogen molecules of their structure, whereas carotenes do not. Xanthophyll lutein is a common pigment that produces bright yellow feathers, such as those of breeding male American Goldfinches. Bright-red pigments include canthaxanthin, astaxanthin, and rhodoxanthin. The bright-red feathers of a male Scarlet Tanager contain canthaxanthin; the deep-red tips of a Cedar Waxwing's secondaries contain astaxanthin (Brush 1967). Bright-red pigments are usually chemically transformed yellow pigments. The red canthaxanthin of a Scarlet Tanager's summer plumage, for example, alternates with a related yellow pigment, isozeaxanthin, in this bird's yellow winter plumage. The pigment rhodoxanthin, which yields various colors depending on the acidity of the feather substrate, produces the beautiful blues, violets, and reds of the *Ptilinopus* fruit doves of New Guinea.

A comparison of the placement of carotenoid pigments in the feathers of male Red Crossbills and male White-winged Crossbills shows how different effects have been achieved (Figure 4 – 13). The feathers of the Red Crossbill are deep brick red, whereas those of the White-winged Crossbill are pink. Red pigment is present in both barbs and barbules of Red Crossbill feathers but only in the barbs of White-winged Crossbill feathers. The mixture of clear barbules and red barbs in the feathers of the White-winged Crossbill produces pink, but the bird's plumage reddens as the clear barbules abrade.

Melanins and carotenoids (xanthophyll, specifically) work together in one feather, most commonly producing olive-greens such as those of the European Greenfinch, the Great Tit, and the female Scarlet Tanager. Melanin, deposited at the tips of the barbules, overlies xanthophyll that is deposited at the bases of the barbules and in the barb rami. The reverse combination, yellow pigment in the barbule tips and melanin in the barbule bases, occurs in some green fruit pigeons.

Porphyrins, related chemically to hemoglobin and liver bile pigments, are the third major type of pigment. They show intense red fluorescence under ultraviolet illumination. Porphyrins are fairly common in the red or brown feathers of at least 13 orders of birds, notably owls and bustards. These pigments, chemically unstable and easily destroyed by sunlight, are found primarily in new feathers. Turacin, or uroporphyrin III, a unique

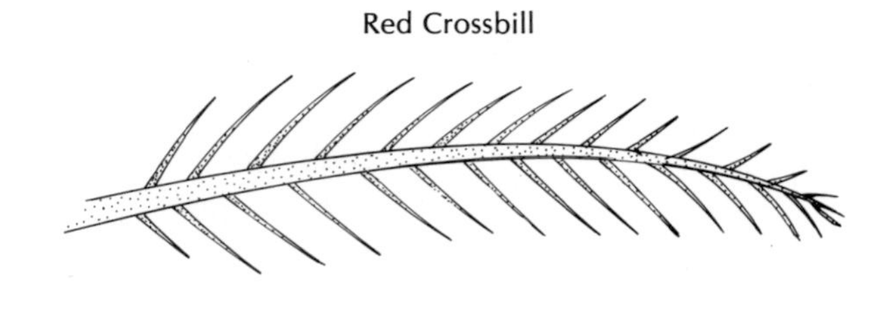

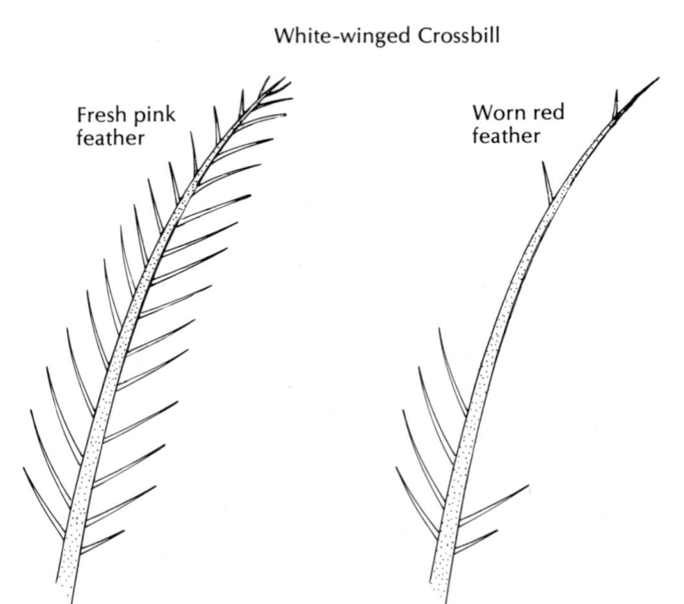

Figure 4–13 Red carotenoid pigment (stippling) is deposited in both the barbs and barbules of feathers of the Red Crossbill but only in the barbs of the similar White-winged Crossbill, leaving the barbules clear. New plumage of the White-winged Crossbill is therefore pink, not red; but as the unpigmented barbules wear off, the plumage reddens.

copper-containing pigment, produces the bright magenta in the wings of turacos, spectacular crow-sized birds of African forests (Figure 4–14).

We do not know the chemical structure of the pigments that create the green colors of the head feathers of pygmy geese and some eiders, the back and wing feathers of the Crested Wood-Partridge and the Blood Pheasant, or the wings of the Northern Jacana. A porphyrin pigment, turacoverdin, of uncertain chemical relationship to turacin, produces the brilliant greens of turacos.

Structural colors

Many of the brightest feather colors such as parrot greens, bluebird blues, and hummingbird iridescences are structural colors that result from the physical alteration of incident light at the feather surface. The classical view is that structural blues and greens result from the scattering of short (blue) wavelengths of incident light by tiny melanin particles in the surface cells on the feather barbs. The remaining longer (red and yellow) wavelengths pass through the surface layer to an absorbent melanin layer below. Blue is left as the apparent hue. Particles of various sizes scatter various wavelengths of light; those responsible for parrot greens are larger than

Figure 4–14 Hartlaub's Turaco, a bird with unique porphyrin pigments in its feathers. (Courtesy W. Conway)

those responsible for bluebird blues. This is not the only mechanism involved in structural colors. Analysis of the feather colors of the Peach-faced Lovebird, a small African parrot, suggests that structural blues and greens may also result from another process based on interference among different wavelengths of incident light (Dyck 1971; Brush 1972).

Carotenoid pigments can convert structural blues to green or violet. Wild Budgerigars, for example, are green because of an association of yellow pigment with structural blue. Mutant parakeets with a single recessive gene that blocks carotenoid pigment deposition are blue rather than green. If the carotenoid pigment is red rather than yellow, violet or purple results, as in the Pompadour Cotinga (Brush 1969). Structural blue from the barbs plus red pigment in the barbules is responsible for the purple head feathers of the Blossom-headed Parakeet.

Iridescence is the phenomenon of glistening colors that vary with the angle of incidence of illumination (Figure 4–15). Directional iridescence is seen from only one angle of view; from other directions the feather appears black. The iridescent colors of the "eyes" on a Peacock's tail and the quetzal's back feathers result from interference of light waves reflected from the outer and inner surfaces of a reflecting layer. The brilliant iridescences of hummingbird feathers come from 7 to 15 closely stacked layers of tiny melanin granules, located on barbules. Each is a flat, hollow platelet with two reflecting layers that create particular colors by light interference and reinforcement (Greenewalt 1960a; Greenewalt et al. 1960). The intensity of the iridescence increases with the number of granule layers.

Other types of reflecting layers have evolved in starlings and trogons. The iridescent colors of some African starlings are caused by reflections from the interfaces between melanin granules and keratin layers. Four different systems are known to produce the iridescent colors of five genera of trogons (Durrer and Villiger 1966). These controlling microstructures, which include air-filled melanin plates and hollow melanin tubes, are arranged in precise layers. As Alfred Lucas and Peter Stettenheim noted,

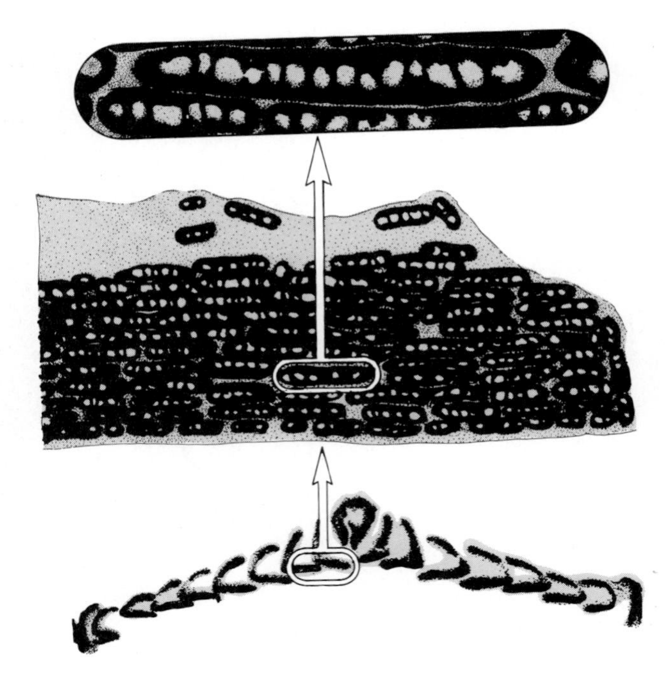

Figure 4–15 The bright, directional iridescent colors of hummingbirds result from interference and reinforcement of light components reflected by layers of hollow platelets on the surfaces of broadened barbules. Shown here are three electron photomicrographs of increasing magnification. A tiny portion of the cross section of a barbule (bottom) is magnified 16,000× to show the layers of platelets (middle); the cross section of one platelet is magnified 45,000× to show the internal air spaces (top). (From Greenewalt 1960a)

> These systems for creating iridescent colors are outstanding in their very high precision as well as their complexity. The dimensions of the tubular melanin granules in trogon feathers are uniform in some places to an accuracy of less than 0.01 microns . . . . Colors and patterns result from small variations in the refractive indices of keratin and melanin, the shape and measurements of the pigment granules, and the spacing within and between layers of granules . . . . These factors are subject to closely controlled variation in every part of every iridescent feather on a bird. As if that were not enough, they are created with the same precision year after year in subsequent plumages. The study of feathers poses no greater challenge than that of working out the genetic mechanisms by which these details are perpetuated. (Lucas and Stettenheim 1972 p. 409)

Plumage patterns

The intricacies of feather microstructure and pigments combine at times to produce stunning effects in plumage. Brilliant reds, greens, and blues are combined into bold plumage color patterns, as in the Painted Bunting. The tiny Many-colored Rush-Tyrant of South America is red, orange, blue, green, yellow, black, and white. At the other extreme, however, are drab gray-olive birds such as the Northern Beardless-Tyrannulet of Central America and the leaf-warblers of Europe and Asia. Bold or subdued, plumage color patterns evolve in concert with a bird's behavioral adaptations. The resulting concealing or signaling elements of plumage color usually have adaptive value (Cott 1940; Hamilton 1973; Hailman 1977; Burtt 1979).

Concealment is a main result of bird color patterns, not just of those that are obviously cryptic but also of many that are bright and bold. The predominant visual aspect of many plumage colors is their similarity to the

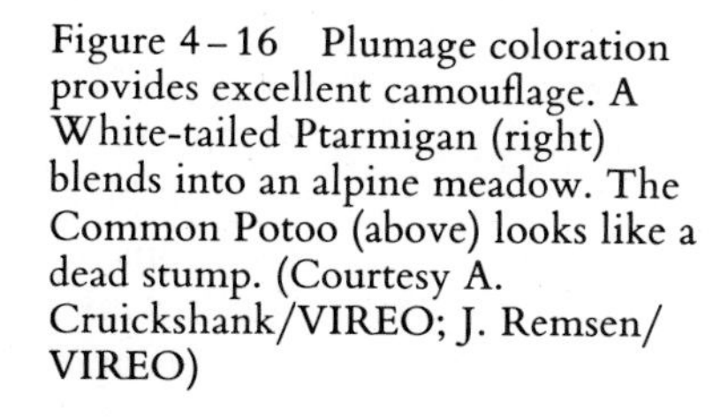

Figure 4–16 Plumage coloration provides excellent camouflage. A White-tailed Ptarmigan (right) blends into an alpine meadow. The Common Potoo (above) looks like a dead stump. (Courtesy A. Cruickshank/VIREO; J. Remsen/VIREO)

bird's usual environment. Ptarmigan are nearly pure white in winter, when they blend with the mountain snows. In spring, when patches of snow remain on the alpine meadows, the birds are white and brown. In summer, when herbs and lichen cover the rocks, ptarmigan are finely barred black and brown (Figure 4–16). Woodcock and Whip-poor-wills rest invisible to us on a forest floor of dead leaves. The American Bittern points its bill skyward, aligning its body contours and the stripes on its breast with the surrounding vertical marsh grasses. The Common Potoo of tropical America has the coloration and can assume the posture of a dead stump (Figure 4–16).

Some bold color patterns reduce the contrast between a bird's shape or outline and its background. The breast bands of the small plover, a classic example of a disruptive pattern, visually separate the outline of its head from that of its body. To be most effective, the contrast between disruptive patches on a bird's body should be as great as that between the bird and its background. Thus the color patches on a bird provide the most effective concealment when their sizes match those of light and dark elements in the background. The finely patterned summer plumage of a ptarmigan blends with the finely patterned alpine grasses and lichens, and the boldly patterned plumages of wood warblers blend with the small leaves, branches, and lighting particular to their arboreal niche.

Abbott Thayer (1909) was the first to identify the principle of countershading in concealment (Figure 4–17). Lower reflectivity of the dorsal surface of a bird interacts visually with contrasting light undersides to disguise its outline, helping it to match its background. The value of contrast increases with the intensity of illumination from above. Open-country birds, such as plovers, have strongly contrasting colors on their upper and lower surfaces. White underparts work particularly well in this regard as a neutral (achromatic) reflector that takes on the hue of the nearest surface. White breasts on small plovers function more effectively for countershading than do white breasts on large, long-legged shorebirds whose underparts are farther from the sand or muddy surface. Large shorebirds with long legs often have dark underparts rather than white.

(A)

(B)

Figure 4–17 (A) The plumage pattern of a Killdeer combines countershading, achromatic reflectance of substrate by white underparts, disruptive head and breast markings, and breast bands that help match horizontal breaks in the shoreline or horizon. (B) The breeding plumage of a Black-bellied Plover is an example of reverse countershading. (A, courtesy of A. Cruickshank/VIREO; B, courtesy of C.H. Greenewalt/VIREO)

Whereas countershading (dark upperparts and white underparts) abets concealment, reverse countershading (white upperparts and dark underparts) makes birds conspicuous. Breeding male Spectacled Eiders, Bobolinks, and Black-bellied Plovers have striking reverse countershading (Figure 4–17B). The triangular, black throat patches of Hooded Orioles and Golden-winged Warblers are more restricted forms of reverse countershading. The triangle points to the bill and thus may help focus attention on bill movements.

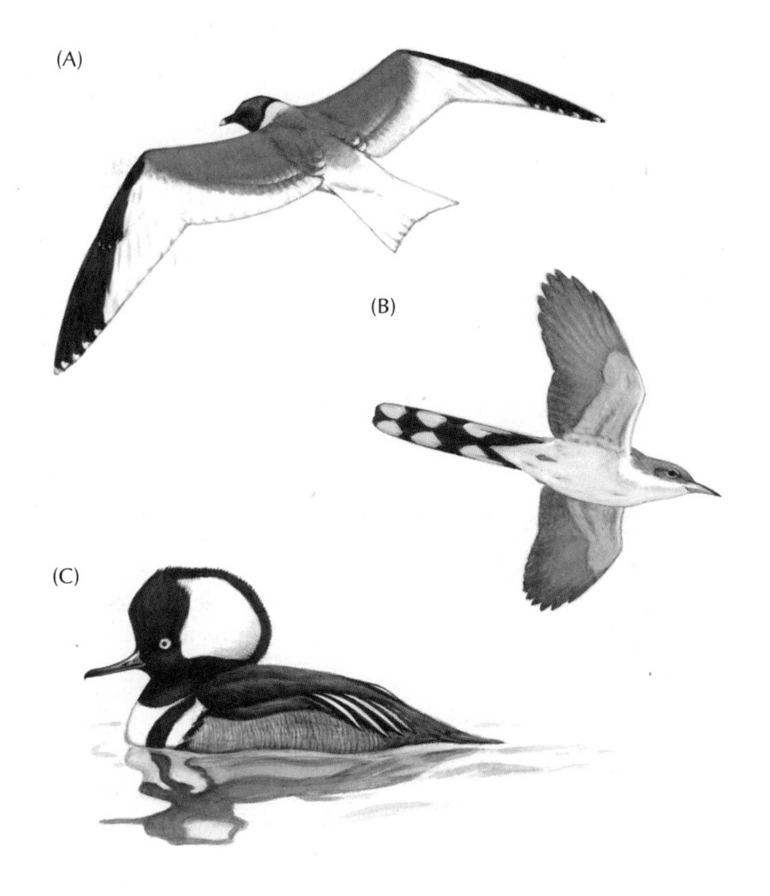

Figure 4–18 Conspicuous plumage signal patterns: (A) triangular wing pattern of adult Sabine's Gull; (B) repeated white tail spots of Yellow-billed Cuckoo; and (C) outlined crest of male Hooded Merganser.

The uniform coloration of the all-red Northern Cardinal enhances its outline and renders it more conspicuous than would a mixed color pattern, and the crest probably enhances this effect. Contrasting edgings enhance striking signal patches, such as the white crest of a Hooded Merganser (Figure 4–18A), the orange crown stripe of a Golden-crowned Kinglet, or a Mallard's blue wing speculum. Unusual shapes, especially those that are geometrically regular, such as the triangular white wing patches of an adult Sabine's Gull (Figure 4–18B) or the rectangular wing patches of ducks, are highly visible because they do not normally match the elements in a natural background. Regular repetition, such as in the tail spots of a Yellow-billed Cuckoo (Figure 4–18C) or the head stripes of a White-crowned Sparrow, achieve similar conspicuous results.

Summary

Feathers are a unique avian attribute. They apparently evolved from reptilian scales as an aid to temperature regulation and primitive flight. The tough, inert molecules that form the feather materials, called ϕ-keratin, evolved as a result of a change in the genes that control the synthesis of other kinds of keratin.

Feathers are extremely versatile in form and function. They provide insulation for maintenance of a high body temperature, are essential for flight, and serve in visual communication and camouflage. Modified feathers aid in swimming, sound production, protection, cleanliness, water repellence, water transport, tactile sensation, hearing, and support of the bird's body.

The basic structure of a contour feather consists of a stiff, central rachis with side branches called barbs and secondary side branches called barbules. The interlocking system of barbs and barbules forms a flexible but cohesive flat surface called the vane. Loose barbs and barbules at the base of the feather enhance insulation. Other major kinds of feathers include the flight feathers, down feathers, semiplumes, filoplumes, bristles, and powderdown.

The entire feather coat consists of thousands of individual feathers, which are arranged in groups called tracts. Linking the bases of adjacent feathers is a system of tiny muscles that control feather position. The feather coat of a bird typically weighs two to three times as much as its skeleton. The entire feather coat is replaced at least once a year in regular molts. Partial molts may supplement the main annual molt to produce composite plumages.

Feather coloration is controlled by carotenoid and melanin pigments, which are deposited in the barbs and barbules, and structural alteration of light at the feather surface, which is responsible for most of the blue and green colors of bird feathers. Iridescent colors result from interference patterns of light as it is reflected by special layers of pigment granules. Plumage color patterns, which evolve in concert with behavior and ecology, exhibit adaptations for concealment or signaling.

Further readings

Fox, D.L. 1976. Animal biochromes and structural colors, 2nd ed. Berkeley: University of California Press. *A survey of adaptive color patterns.*

Jacob, J., and V. Ziswiler. 1982. The uropygial gland. Avian Biology 6: 199–324. *A modern review of the structures of preen glands and the chemistry of their secretions.*

Lucas, A.M., and P.R. Stettenheim. 1972. Avian anatomy: Integument. Washington, D.C.: U.S. Gov. Print. Office. *The standard reference on feathers.*

Palmer, R.S. 1972. Patterns of molting. Avian Biology 2: 65–102. *A review of the details of molt patterns with special emphasis on ducks.*

Pettingill, O.S. 1984. Ornithology. New York: Academic Press. *A fine laboratory review of feather anatomy.*

Stettenheim, P. 1972. The integument of birds. Avian Biology 2:1–63. *A review of feather and skin anatomy.*

Stettenheim, P. 1974. The bristles of birds. The Living Bird 12: 201–234. *A well-illustrated review of these highly modified feathers.*

Stettenheim, P. 1976. Structural adaptations in feathers. Proc. 16th Intern. Ornithol. Congr.: 385–401. *An elegant summary of the adaptive modifications of feather structures.*

Stresemann, E., and V. Stresemann. 1966. Die Mauser der Vogel. Journ. für Ornithol. 107: 1–445. *The encyclopedia of molt patterns of birds of the world.*

chapter 5

Flight

Flight is the central avian adaptation. Yet birds do not merely fly. They are masters of the fluid that is air, just as fish are masters of the fluid that is water. Various kinds of birds can hover in one place, dive at breathtaking speeds, fly upside down and backward, and soar casually for days on end. In their evolution, birds have exploited an extraordinary range of specialized modes of flight (see Figure 5–1).

We begin this chapter by focusing on the anatomical bases of avian flight, particularly the skeleton and flight muscles. Then, because aerodynamic principles are fundamental to understanding avian flight, we review the nature of an airfoil, the phenomenon of lift, and the countering forces of drag. These affect the performance of a bird's wings and ultimately determine both mode of flight and the costs of flight power. We discuss hummingbirds as masters of flight control. At the opposite extreme are flightless birds, which have repeatedly evolved on islands without predators, and elsewhere. Some diving birds such as penguins have traded aerial flight for underwater flight using highly modified flipperlike wings.

Figure 5–1 Mastery of flight is a fundamental achievement of birds. (From Greenewalt 1960a)

The avian skeleton

The articulated skeleton of a bird is a vertebrate skeleton uniquely structured for flight. Fusions and reinforcements of lightweight bones make the avian skeleton both powerful and delicate. Unusual articulations of the joints not only make flight motions possible but also brace the body against the attendant stresses. The skeleton strategically supports the large muscles that provide the power for flight.

Cross sections of bird bones reveal light, air-filled structures unlike the relatively dense, solid bones of many terrestrial animals. The hollow, long bones of the wings are particularly strong (given their mass) and in many cases are strengthened further by internal struts. Instead of a heavy, bony jaw filled with dense teeth, birds have a lightweight, toothless bill designed to flex and to bite with great power (see Chapter 7). Because the bills are hollow, even the large, protuberant bills of toucans, for example, are not the burden they seem.

The avian skeleton is constructed to withstand the strain imposed by flight (Figure 5-2). The avian thorax is more rigid and better reinforced than that of a reptile. The fully ossified dorsal and ventral ribs provide a

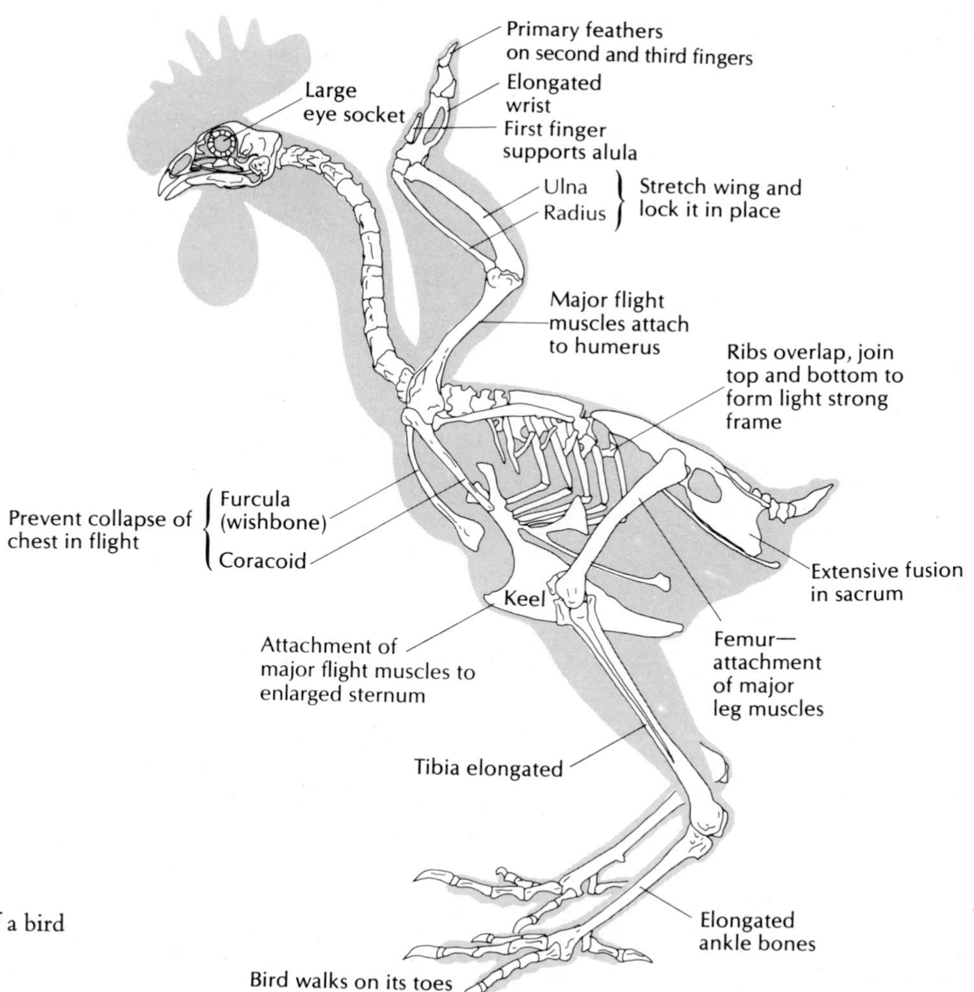

Figure 5-2 Major features of a bird skeleton. (After Lucas and Stettenheim 1972)

strong connection between the backbone and the breast bone. Horizontal bony flaps, called uncinate processes, extend posteriorly from the vertical upper ribs to overlap the adjacent rib and reinforce the rib cage. In addition, limited articular surfaces help to stabilize the partially fused thoracic vertebrae by restricting their movement.

The pectoral girdle includes the sternum, coracoids, and scapulae. The sternum, or breast bone, is a dominant feature of the avian skeleton and usually has a large keel, or carina, to which the major flight muscles attach. A bird's flying ability is correlated with the size of its keel; flightless birds lack the keel completely. The furcula (or wishbone, actually a fused pair of clavicles or "collar bones") and the coracoids (anterior elements of the pectoral girdle) are struts that resist the potentially chest-crushing pressures created by the wing strokes during flight. Dorsally situated on top of the rib cage are the long, saberlike scapulae, each of which joins anteriorly to the coracoid and furcula to complete a triangular system of struts. An acute angle between the attachment of the scapula to the coracoid reduces the stretch of the dorsal elevator muscles, which help pull the humerus (upper wing bone) upward, and thereby increases the force they can exert. This angle is oblique in flightless birds.

The avian wing is a modified forelimb. The humerus, ulna, and radius are homologous to the limb bones of tetrapods. Large, articulating surfaces at the joints between the limb bones allow the resting wing to fold neatly against the body. These elaborate joints also permit the wing to change positions and angles during takeoff, flight, and landing. When locked, these joints are strong enough to withstand the powerful torque created during wing strokes.

The fused hand and finger bones help provide strength and rigidity in the outer wing skeleton. Most of the wrist bones, the carpals and metacarpals, are fused into a single skeletal element called the carpometacarpus. There are only two free carpals in the avian wrist, compared with ten or more in most other vertebrates. The hand itself includes three digits, rather than the five found in most tetrapods. A similar condition is found in some dinosaur fossils. The alula, or bastard wing, originates from the first digit, the thumb, and moves independently of the rest of the wing tip. Within the wing itself are powerful tendons and compact packages of tiny muscles that control the subtle details of wing position.

Flight musculature

The two great flight muscles, pectoral and supracoracoideus, originate on the pectoral girdle and insert onto the expanded base of the humerus (Figure 5–3). The pectoral muscle complex is the largest and accounts for about 15 percent of the total mass of a flying bird. Contraction of this muscle pulls the wing down in the power stroke. The pectoral muscle attaches to the furcula and the strong membrane between the coracoids and the furcula. It attaches also to the peripheral portions of the sternum, including the outer part of the keel. In tree-trunk climbing birds with shallow keels, such as woodcreepers, the pectoral muscle spreads thinly over the rib cage for attachment.

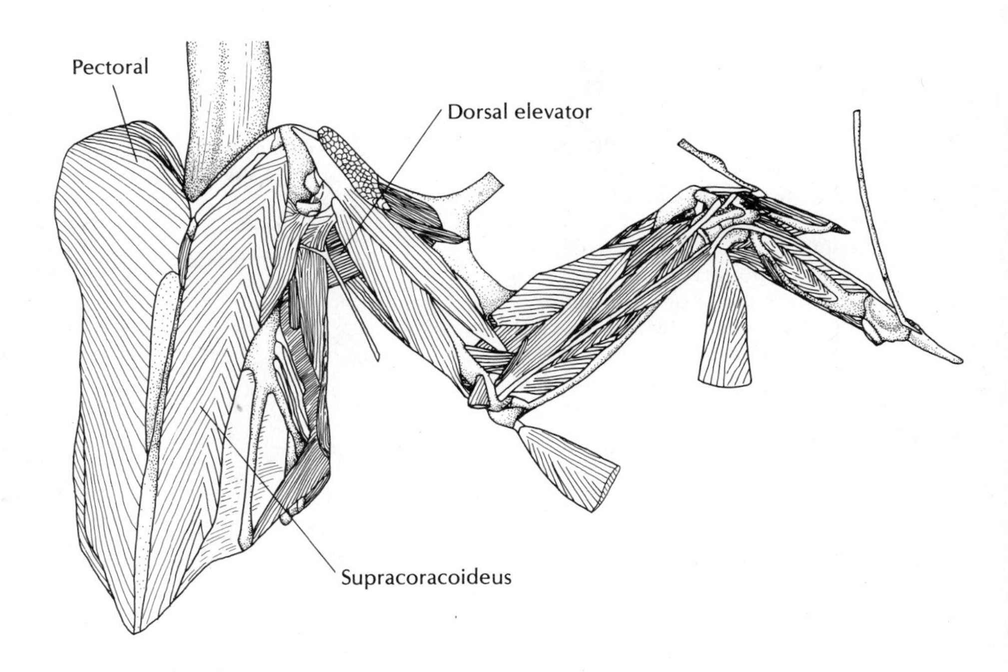

Figure 5 – 3 The flight musculature of birds consists of large pectoral muscles responsible for the wing's downstroke, the supracoracoideus muscles responsible for the wing's upstroke, and a host of small muscles on the wing itself. In this ventral view of the wing and breast muscles of a White Leghorn Chicken, the pectoral muscle has been removed from one side to show the underlying supracoracoideus muscle. (After Hudson and Lanzilloti 1964, as modified by Vanden Berge 1975)

The supracoracoideus muscles work in concert with smaller muscles, called the dorsal elevators, to lift the wings on the recovery stroke. The supracoracoideus muscle originates from the keeled sternum (Figure 5 – 4). The strong, supracoracoideus tendon, which enables a bird to raise its wing from the ventral position, passes upward and forward through the triosseal canal (formed by the junction of the coracoid, scapula, and furcula) and inserts on the dorsal side of the base of the humerus. The supracoracoideus is required for the rapid initial wingbeats that are essential for clearing the ground quickly and achieving a minimal air speed. A pigeon is unable to take off from the ground if its supracoracoideus tendon is cut experimentally (Sy 1936). Once launched and airborne, however, pigeons can fly quite well without a functional supracoracoideus, with the smaller dorsal elevators handling the less-demanding recovery strokes of sustained flight.

The supracoracoideus muscles are much smaller than the pectoral muscles. However, they are large relative to body size in certain birds, in which the upstroke of the wing is a propelling power stroke not just a recovery stroke. The supracoracoideus of hummingbirds is five times larger relative to body size than in most other birds. It is half the size of the pectoral muscle and comprises 11.5 percent of total body mass, more than in any other bird. The supracoracoideus muscle is unusually large also in penguins, whose flippers propel them forward with a powered upstroke.

Metabolism in flight muscles

The power for flight derives from metabolic activity in the cellular fibers of flight muscles, some of which have an extraordinary capacity for aerobic metabolism. Various types of muscle fibers employ particular metabolic processes suited to specific modes of flight. The extremes of the variation are red and white fibers, but intermediate fiber types exist.

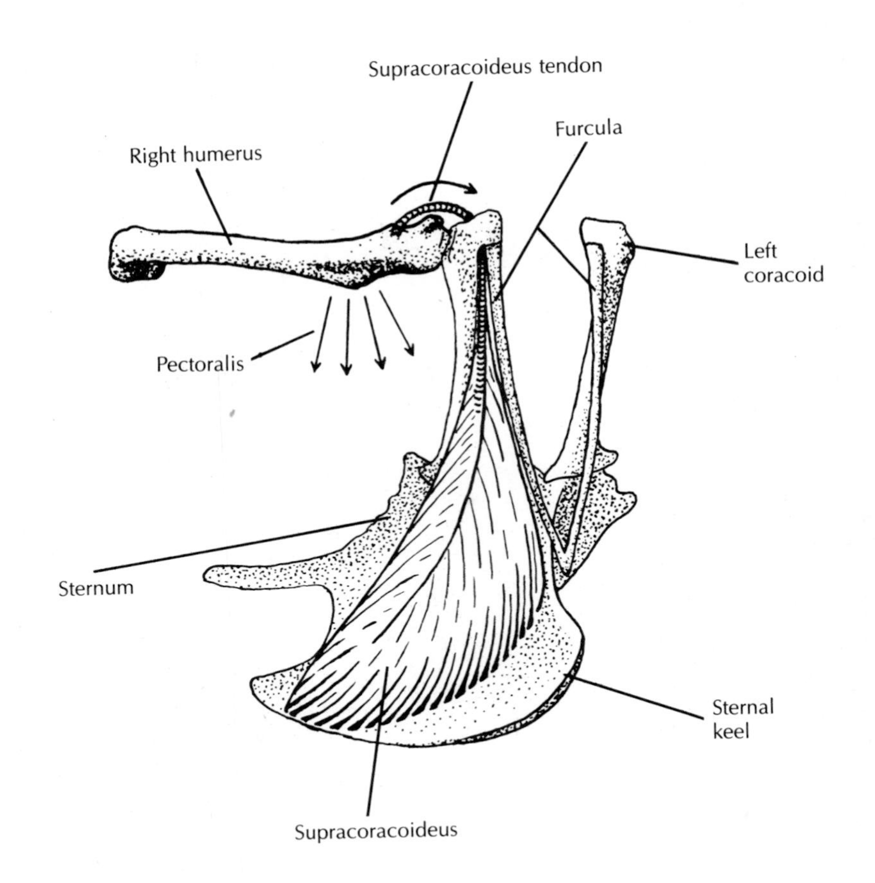

Figure 5–4 Right front view of the pectoral girdle of a pigeon. The ventrally located supracoracoideus muscle raises the wing by means of a pulleylike tendon that passes to the dorsal surface of the humerus through the triosseal canal at the junction of the scapula, coracoid, and furculum. The curved arrow indicates the action of this tendon. The pectoral muscle, which has been removed in this drawing, inserts onto the lower side of the humerus and pulls the humerus downward, as indicated. (From George and Berger 1966)

The sustained contractile power of red muscle fibers results from oxidative metabolism of fat and sugar (George and Berger 1966; Talesara and Goldspink 1978). These narrow fibers have high surface-to-volume ratios and short diffusion distances, which aid the uptake of oxygen. They also contain abundant myoglobin, mitochondria, fat, and the enzymes that catalyse the chain of metabolic reactions known as the Krebs' cycle. Experimental studies of extracts from pigeon breast muscle, which is rich in red fibers and the associated enzymes, laid the foundations of our present knowledge of aerobic metabolism.

Sustained flight power derives from a high concentration of red muscle fibers in the flight muscles. The aerobic, or free-oxygen, capacity of the pectoral muscles of small passerines and small bats is at the highest level known for vertebrates. Few birds have muscle that consists entirely of red fibers. Rather, blends of different fibers that combine short-term power with long-term endurance in flight are typical of most birds. The dark-red breast muscle of pigeons consists mostly, but not exclusively, of red fibers. Conversely, the white breast muscle of fowl has a low proportion of red fibers. Extreme cases include sparrows, which have only red fibers in the pectoral muscles, and hummingbirds which have only red fibers in both pectoral and supracoracoideus muscles (Chandra-Bose and George 1964).

White muscle fibers are powered by products of anaerobic (without free oxygen) metabolism. Unlike red fibers, they contain little myoglobin, few mitochondria, and a different set of enzymes. The white fibers are capable of a few rapid and powerful contractions, but they fatigue quickly as lactic acid, a product of anaerobic metabolism, accumulates. The light meat of the breast muscles of domestic fowl and grouse consists primarily of narrow, white muscle fibers, which enable these birds to take off with explosive power. The short-term power of white muscle fibers is useful as well for fast turns and evasive actions in flight, but the birds tire easily and cannot fly long or far.

Flight power

To stay aloft, birds must overcome the force of gravity with an equal and opposite set of forces. Obviously, wings are important in this effort, and the precise description of the aerodynamic functions of the avian wing is extremely complex. We can begin to understand how the wings of birds develop the countering forces required for flight by considering the nature of lift, an aerodynamic phenomenon that is caused by differential pressures created on the opposite surfaces of an airfoil, which is an asymmetrically curved structure that tapers posteriorly. The wings of birds are airfoils as are the wings of airplanes (Rüppell 1977).

Correct orientation of the airfoil with respect to passing air produces the net upward force called lift, which keeps a bird or an airplane airborne. Lift results in part from the different speeds at which air flows past the upper and lower surfaces of the airfoil. Owing to the faster air speeds on the curved upper surface of the wing, pressure on that surface is less than the pressure on the undersurface of the wing, generating a net upward force (Figure 5 – 5A).

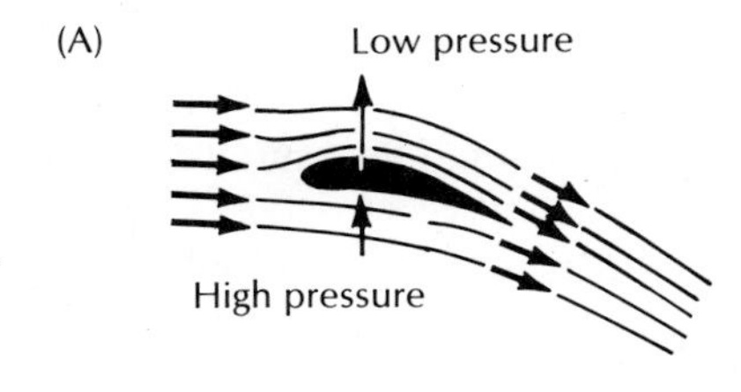

Figure 5 – 5 (A) The streamlined shape of an airfoil allows air to flow smoothly over both surfaces; it produces lift by reducing pressure on the upper curved surface relative to that on the lower surface. (B) Too much tilt destroys lift when the air stream separates from the upper surface of the airfoil. This causes a bird or plane to stall. (After Rüppell 1977)

The phenomenon of lift is an expression of the Bernoulli principle. Bernoulli's principle relates air pressure to air velocity. The movement of gaseous molecules in the atmosphere creates pressure on any surface they strike. The random Brownian motion of molecules in still air is equal in all directions, creating omnidirectional pressure called static pressure. Wind, on the other hand, imparts directional, or dynamic, pressure, and the force increases with air velocity. As the energy in dynamic pressure increases, that present in static pressure decreases. The sum of the two pressures remains constant. Thus, fast-moving air imparts less pressure against an adjacent surface than slower-moving air, causing a net force upward in the case of the paired surfaces of an airfoil. Proof of this can be seen if one blows gently over the upper surface of a piece of tissue paper. The tissue rises or straightens out because of the net upward pressure on the lower side. The unequal pressures that develop as air flows over the surfaces of a bird's wing have the same effect.

The orientation of a wing in the air current affects the balance and direction of the pressures on the surface and thus the amount of lift. Obviously, if the rear of the wing or airfoil is tilted upward so that air strikes the upper surface directly, surface pressures will oppose flight. As the rear edge of the wing is tilted downward, increasing the angle of attack, lift increases. If the wing is tilted too far downward, the airstream no longer follows the contours of the streamlined surface of the airfoil, and a partial vacuum develops next to the upper surface. Air then eddies into the low-pressure space from the rear edge of the wing, blocks the backward flow of air over the upper surface, and causes a total loss of lift, or a stall (Figure 5–5B). When an airplane lands, the pilot purposely stalls by increasing the airfoil's angle of attack just before the wheels touch the runway. Birds also control the angle of their wings and stall just before landing.

When a large bird, such as a gull, stands on the edge of a sea cliff facing into the wind, the flow of air across its outstretched wings generates lift, the amount of which increases by the square of the velocity of the airstream. If the wind is strong enough, the bird rises effortlessly into the air. Alternatively, in still air, the gull may have to jump off the cliff with wings outstretched; as it drops downward, its airspeed increases, and lift increases to the point of real flight. Birds that do not launch themselves from cliffs or trees may supplement the initial forward thrust by running as they take off. Diving ducks and coots skitter over the water until airborne.

In the outer half of the wing, each primary functions as a smaller, separate airfoil; together they produce forward thrust as does the propeller of an airplane. To produce forward thrust, the airfoils of propellers and of primaries move vertically rather than horizontally through the air. As the leading edge of the primary slices the air column during the downstroke, the net pressure on the back surface pushes the feather forward. The 90-degree rotation of a horizontal airfoil shifts the direction of net pressure from upward lift to forward "thrust." Control of the angle of attack of each primary by tendons and by the natural responses of the flexible vanes to air pressure results in a continuously integrated system of feather positions as the wing flaps through the air. The forward forces of thrust produced by the propellerlike primaries are transferred to the inner wing, the horizontal movement of which creates the upward forces of lift. The result is forward flight.

Slow-motion movies of birds in flight, especially as they take off or land, reveal the functional details of wing position. The wing is not just a flat, broad structure that flaps up and down. About 50 different muscles control the wing movements. Some muscles fold the wing, others unfold it. Some pull the wing upwards, others pull it down, and still others adjust its orientation. For example, when a goose takes off, its wrists rotate at the beginning of the downstroke so that the primaries drive forward and down at a sharp angle, creating forward thrust. The nearly horizontal angle of attack of the inner wing produces upward lift. These angular relationships are maintained through the downstroke. Next, at the beginning of the recovery or upstroke, its wing tips rotate backward so that the primaries provide some lift and some forward thrust. Lift decreases because air speed is reduced during this wing stroke. To compensate, the bird increases the inner wing's angle of attack. By maintaining a correct angle of attack, the wing supports the bird's weight through the entire wingbeat. Lift and thrust forces on various parts of the wing are continuously integrated during the normal wingbeat (Figure 5–6).

The inner and outer wing sections in smaller birds do not function so contrastingly with respect to lift and thrust. Typically, the wingbeats are faster, and recovery strokes, especially, are more rapid. The entire wing acts like a single large propeller with some accessory propellers in the wing tip.

Figure 5–6 Wing actions of a Canada Goose. (A) Downstroke with primary feathers positioned for forward thrust and inner section of wing positioned for maximum lift. (B) Start of forward stroke with both primaries and inner wing providing lift. (C) Start of backstroke with outer wing moving backward with primaries positioned to supply some lift, and inner wing moving forward with the bird supplying lift. (D) Middle of backstroke with inner section of wing still supplying lift as in C, but with outer section of wing sweeping backward against air, providing additional forward drive. (After Storer 1948)

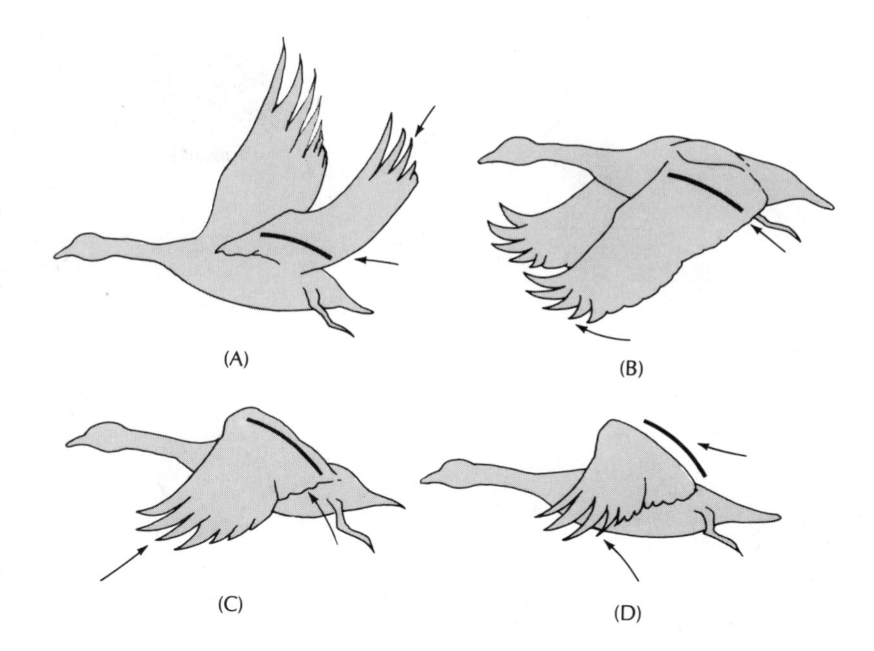

Little lift is achieved on the recovery stroke, during which the primaries are separated to minimize air resistance.

Slots between adjacent flight feathers aid fine control of the air moving over the wing surface and thereby aid in the extraction of energy. Slots are cracks or holes through which air squeezes, producing two kinds of results. First, some slots control the flow of air over the airfoil to maintain some lift, not only at slow speeds but also at high attack angles of the wing when a bird is stalling. The extended alula creates a slot that controls air flow, especially during landing and takeoff, when forward thrust is minimal and extra lift is essential. Second, air forced from beneath the wing through a slot expands on the upper side, reducing the pressure there and increasing lift. The slots in the wing tips of many soaring birds, such as the California Condor, function this way. So also does the crack between the rear edge of the wing and the fore edge of a spread tail (Figure 5–7).

Birds maneuver in flight by controlling the patterns of lift and thrust. No aircraft yet approaches the average bird's acrobatic maneuverability. The tail acts as a rudder and a brake. Even more important is the bird's control of each wing independently. Asymmetrical wing actions enable the bird to steer, turn, and twist. By flapping with one wing oriented for forward flight and the other wing oriented for backward flight, the bird can execute an abrupt turn. Setting the wings in a partially folded position reduces the amount of lift, enabling the bird to lose altitude gradually while gliding. By setting one wing back farther than the other, the bird adds curvature to its glide path. Slow-motion photographs of birds during aerial maneuvers, chases, and landings reveal the precise changes in wing position that control body orientation and air speed. Birds rarely crash.

Elaborate use of the wings in landing is a special feature of avian flight. Landing on elevated or arboreal perches is a feat that requires exceptional control of flight trajectory. Birds are unique among flying vertebrates

Figure 5–7 Slots aid fine control of air flow over the wing surface and prevent stalling at slow air speeds. (Right) Green Magpie about to land. The fully spread tail and wings and the extended alula prevent premature stallout. (Above) Slots constitute 40 percent of the wing area of a California Condor. Slight adjustment of the primaries and their associated slots control a condor's speed, lift, and aerial position as it searches the terrain for carcasses. (Courtesy W. Conway; Santa Barbara Museum of Natural History)

in the way they land (Caple et al. 1983, 1984). Aerial species such as bats, flying squirrels, and certain lizards make contact with their forelimbs and then rotate their bodies downward until the hind feet touch the landing surface. Variations exist, but only birds rotate their center of mass upward to stall directly over the landing site.

A study in control: Hummingbird flight

The control displayed by birds in flight greatly exceeds that achieved by piloted aircraft, and the control displayed by hummingbirds in flight is outstanding among birds. The hummingbird can move forward from stationary hovering just by changing the direction of the wingbeat because every angle produces a different combination of lift and thrust. Forward velocities increase as the wings beat in an increasingly vertical plane. This rotation of the wing is made possible by the unusual structure of the humerus and its articulation with the pectoral girdle. The secondaries of a hummingbird's inner wing are short, and the outer primaries are elongated to form a single, specialized propeller. The complete stroke of the wing tip describes a figure-eight pattern (Figure 5–8).

Crawford Greenewalt (1960a) took high-speed movies of hummingbird flight and then studied them at slow speeds to discover how hummingbirds achieve their remarkable control. He concluded that hummingbird flight resembles that of a helicopter or, more precisely, a novel combination of airplane and helicopter in which the propellers rotate about a horizontal axis to produce various combinations of lift and forward thrust. Greenewalt describes the action thus:

Figure 5-8 The time course of wing actions of a hovering hummingbird. Wing positions are shown at 4-millisecond (ms) intervals. The course of the wing tip during a complete wingbeat describes a flat figure eight. (From Greenewalt 1960a)

> In hovering flight the wings move backward and forward in a horizontal plane. On the down (or forward) stroke the wing moves with the long leading edge forward with the feathers trailing upwards to produce a small, positive angle of attack. On the back stroke, the leading edge rotates nearly a hundred and eighty degrees and moves backward, the underside of the feathers now uppermost and trailing the leading edge in such a way that the angle of attack varies from wing tip to shoulder, producing substantial twist in the profile of the wing. (Greenewalt 1960a)

Like the wings of insects, the wings of birds and their controlling musculature function as mechanical oscillators with intrinsic elasticity (Greenewalt 1960b). Two properties of mechanical oscillators, constant rhythms with symmetrical strokes and dependence of the rhythm on the size of the oscillator, are evident in the wingbeats of birds. As an example of the first property, the rate of wingbeat of a Ruby-throated Hummingbird is essentially constant at about 53 strokes per second, and the durations of the upstroke and downstroke are equal (Greenewalt 1960b). In support of the second property of mechanical oscillators, the wingbeat rates of various species of hummingbirds and most other birds decrease predictably with increasing wing length. These observations have important implications for the neuromuscular basis of avian flight. Once the wingbeat rate reaches its natural oscillating frequency, the nerves and muscle fibers responsible for sustaining the rhythm need to fire perhaps only once every four beats, like a child continues to swing back and forth with only an occasional push by the parent.

Friction, turbulence, and drag

Perfectly smooth, frictionless flow of air over an airfoil is only an ideal. Any slight air turbulence, or slowing of air molecules on contact with real surfaces, reduces the forces of lift and thrust. Negative forces that oppose a bird's movement through the air are called drag. Drag forces interact with lift and thrust to determine the power requirements for flight. The two primary categories of drag are parasitic drag and induced drag.

Parasitic drag includes all forms of friction between the air and a bird's body and wing surface. Profile drag, included by some in parasitic drag (Greenewalt 1975), is the reduction in thrust that results from friction between the wing surfaces and the airstream. If the leading edges of the wings were as thin as a sharp knife blade, profile drag would be minimized because a blunt surface presents a larger area for interaction with the air; for the same reason, it is easier to throw a frisbee than a soccer ball into the wind. Both parasitic and profile drag increase with air speed, making it harder to throw a frisbee or ball, or to fly, into a strong wind than into a light wind.

Induced drag is caused by disruptions of the airstream by the eddies and turbulence that inevitably arise from the differences in air pressure that produce lift. Air tends to move around the airfoil from regions of high pressure beneath the wing to regions of low pressure above the wing.

Figure 5–9 Snow Geese in V-formation flight. (Courtesy A. Cruickshank/VIREO)

Turbulence contributing to induced drag results from opposite flows of the airstream: toward the wing tip on the undersurface of the wing and toward the base of the wing on its upper surface. The resulting eddies on the rear edge of the wing and pronounced wing-tip vortices divert energy that would otherwise contribute to lift and thrust, thereby reducing flight efficiency. Pointed wings generate less induced drag than rounded ones because the small area of the pointed wing tip produces little pressure difference between upper and lower surfaces and therefore little compensatory air flow.

The eddies of air that whirl off the wing tips of a flying bird can be used to advantage by another bird. By flying in V formation, geese, pelicans, and cormorants, for example, take advantage of the rising eddies left by the outer wing of the preceding bird. All but the leader of the formation save energy in this way (Figure 5–9).

Induced drag decreases with increasing air speed because the negative influence of reverse eddies on lift decreases as the velocity of the main front-to-rear airstream increases. These relations stem from the different effects of laminar and turbulent air flow in an area called the boundary layer, which lies within one millimeter of the wing surface. Air flow changes from laminar to turbulent as velocity increases. Although laminar flow is smooth and streamlined, the airstream at the boundary layer tends to separate from the wing surface, causing large eddies to flow forward from the rear edge of the wing and disrupt the main lift-producing airstream. This can even cause stalling. Turbulent flow is disorderly, as its name implies, but its greater momentum keeps the boundary layer from separating from the airfoil surface. Therefore, fewer disruptive eddies form, and lift is maintained over more of the wing's surface.

Wing sizes and shapes

The speed at which a bird can fly, its agility in the air, and the rate at which it uses energy in flight all depend on the size and shape of its wings. Lift and drag forces vary with wing dimensions as well as with the patterns and intensities of air flow over the wing surfaces. Studies of variations in bird wings reveal the pervasive strength of evolutionary adaptation: Over the millenia birds have exploited the advantages of various wing designs.

The costs of flight are influenced by the relation between total wing area and body mass, that is, by how many grams of bird each unit area of wing surface must carry. The relationship between wing area and body mass, the wing loading, is given in grams per square centimeter of wing surface area.

Some birds have small wings relative to their body mass, and others have proportionately large wings. Small passerines have rather large wings and, consequently, low wing loadings of 1.7 to 1.9 grams per square centimeter. At the other extreme are large birds with small wings such as albatrosses (17 grams per square centimeter) and the Thick-billed Murre (26 grams per square centimeter), representing a 15-fold difference in wing-loading values. Loons, grebes, auks, diving ducks, and even flamingos have extremely high wing loadings. To take off, they must skitter over the water,

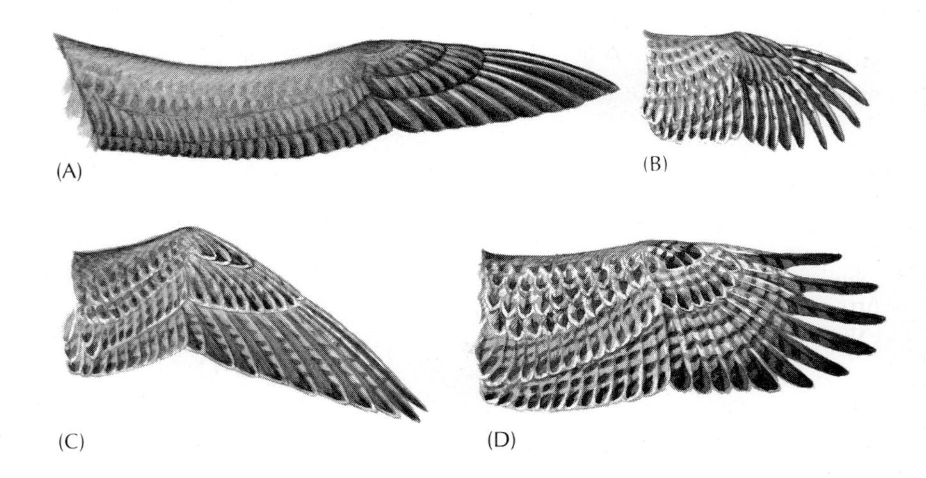

Figure 5–10 Flight abilities vary with the shapes of bird wings. (A) Long, narrow wings of high aspect ratio, such as those of an albatross, are best for high-speed gliding in high winds. (B) Short, rounded wings of low aspect ratio, such as those of a grouse, permit fast takeoffs and rapid maneuvers. (C) The slim, unslotted wings of falcons permit fast, efficient flight in open habitat. (D) Slots increase lift and gliding ability of the wings of buteos with intermediate dimensions.

flapping their wings to gain enough lift for flight. Other ducks, such as Mallards, have lower wing loadings and more lift per wingbeat, which enables them to take off vertically without a skittering start. Whereas long, high-lift wings enable Turkey Vultures to begin soaring early in the day, Black Vultures, which have short, rounded wings and high wing loading, usually wait for the assistance of rising warm-air currents.

Supplementing the wings are the tail's contributions to lift, which may be more important in young birds that are learning to fly than in skilled adults. Immature raptors, in particular, tend to have longer tails than do adults. The size difference (up to 15 percent) is most pronounced in short-tailed species such as the Bateleur and sea eagles. Immatures of these raptors have a more buoyant flight than do adults, which apparently reduces the chance of injury when they strike prey and also facilitates their mastery of flight and hunting skills (Amadon 1980).

Although wing lengths and areas increase with mass among birds, the increase is less than that required to maintain a constant wing loading. The wing area increases as the product (square) of two dimensions, whereas body mass increases with volume, or the product (cube) of three dimensions. Therefore, to maintain parity with mass, the wing area must increase 1.5 times for each unit increase in mass. Yet the slope of the relationship averages less than that for most birds, as little as 1.275 for birds such as passerines, herons, and raptors (Greenewalt 1975). Only in hummingbirds does the wing area increase in proportion to the mass increase, keeping the per-gram cost of hovering the same for hummingbirds of all sizes.

The shapes of wings also affect flight abilities. Long, narrow wings with more leading edge produce more lift than broad wings of equal total area because the wing's leading edge produces the most lift; the rear half of the wing produces the least. In addition, induced drag becomes less important as wing length increases because the disruptive air turbulences of opposite wing tips are farther apart. Long, narrow wings have small wing tips relative to total wing surface area, and this, too, reduces induced drag. Long, narrow, pointed wings, such as those of swallows, falcons, and albatrosses, have a high lift-to-drag ratio. Consequently, swallows can glide better than, say, sparrows, which have short, rounded wings of low lift-to-drag ratio (Figure 5–10).

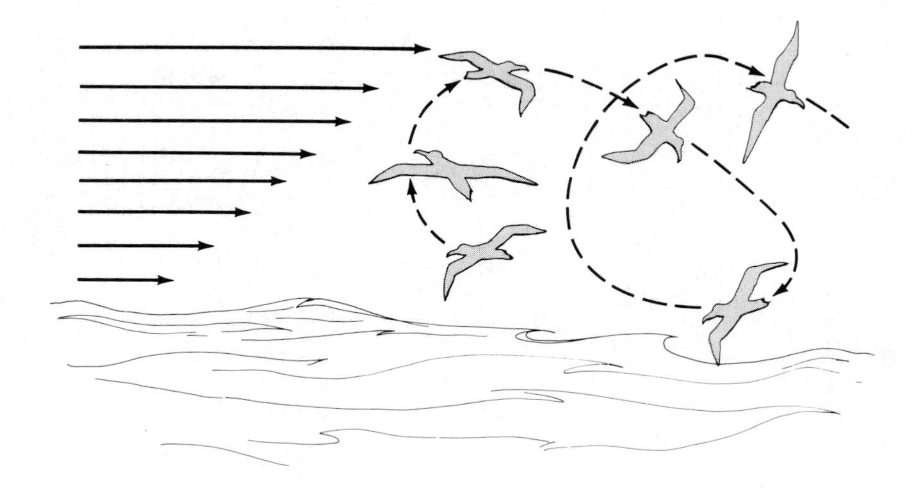

Figure 5 – 11 Dynamic soarers, such as albatrosses, take advantage of different wind-speed layers above the ocean. They accelerate downward in the fast-moving upper layers of air and then use momentum, gravity, and surface wind eddies to swing upwind in the slow, lower air layers. As they lose speed and lift, they bank high into the upper layers again. Arrows indicate relative air speeds. (After Rüppell 1977)

Long, narrow wings work best at high speeds. At low speeds, eddies from the rear edge interfere with the flow of air at the leading edge. Therefore, narrow-winged birds must fly fast to avoid stalling. Albatrosses and petrels achieve necessary speeds without apparent effort by exploiting air layers just above the ocean in a form of flight called dynamic soaring (Figure 5 – 11).

Aerial and open-country birds, such as shorebirds, swallows, and terns, typically have long, pointed wings, whereas species living in dense vegetation often have short, rounded wings. The wing shapes of migrant birds also tend to be longer and more pointed than those of related nonmigrant species. A falcon's pointed wings serve it well in high-speed chases in open country, whereas the short, rounded wings of a Sharp-shinned Hawk enable it to maneuver while chasing small birds in dense vegetation. Wrens and other passerines have short, rounded wings. Their rapid wingbeats enable them to maneuver without collision amid seemingly impenetrable networks of branches and vines. The short, rounded wings of gallinaceous birds, such as quail, pheasants, and grouse, permit short bursts of rapid acceleration, enhancing their chances of escaping predators.

Flight power

We rely on theoretical projections for flight power because the study of the aerodynamics of flapping flight has little experimental background (Greenewalt 1975). Equations for the power requirements of the flapping flight of birds are presented by Colin Pennycuick (1969, 1975) and Crawford Greenewalt (1975). An example of the equations devised by Greenewalt is provided in Box 5 – 1. Briefly, as a bird increases forward flight velocity from the hypothetical minimum of zero, induced power requirements decrease rapidly because lift on the horizontal wing surfaces increases with air speed and because induced drag at the wing tips decreases. A bird maintains velocity by generating sufficient thrust to overcome parasitic drag.

Box 5 – 1

Equations for flight power requirements of birds

C.H. Greenewalt's (1975) equation for the total power P_t, in watts, that is required by a bird flying at V_{mp} (minimum power speed) is

$$P_t = 2.071 \times 10^{-7}\, S^{0.7}\, b^{0.3}\, V^{2.7} + 7.879\, W^{1.91}\, b^{-2}\, V^{-1}$$

where S is the projected wing area in square centimeters, b is the wing span in centimeters, V is the velocity in km/h, and W is the body mass in grams. This equation can be used for any bird of known mass, wing span, and wing area.

The power required at V_{mp} is not the same as that required to fly as far as possible on a given amount of fuel at the optimum velocity V_{mr} (maximum range velocity). To get the equation for the energy required to fly a unit distance, that is, the cost of transport, divide the above power equation by V:

$$P_t/d = 2.071 \times 10^{-7}\, S^{0.7}\, b^{0.3}\, V^{1.7} + 7.879\, W^{1.91}\, b^{-2}\, V^{-2}$$

Therefore, equations for flight power requirements reflect the two main sources of the power required for flight, which are nearly opposite functions of velocity. Parasitic drag increases as the cube of velocity, whereas induced drag decreases as the inverse of velocity. This means that flight cost varies in a parabolic relationship to flight speed. The cost of flying is least at intermediate speeds and greatest at low and high speeds. Minimum power speed (V_{mp}), is the speed at which a bird uses fuel most slowly. Minimum power speed increases with a bird's size (Figure 5 – 12).

Estimates of flight (air) speeds indicate that most birds fly at 30 to 60 kilometers per hour (Rayner 1985). Common Eiders are among the fastest fliers (80 kilometers per hour) clocked in steady flight, whereas Common Swifts cruise slowly at only 23 kilometers per hour while feeding, close to predicted minimum power speed (V_{mp}). In general, birds fly at speeds with low power requirements consistent with aerodynamic predictions from the bird's mass and wing morphology (Rayner 1985). Yet it is unlikely that birds adhere strictly to flight speeds that minimize power costs.

To achieve the maximum flight range with a certain amount of fuel, for example, a bird should fly at maximum range velocity (V_{mr}), which is faster than V_{mp}. Adherence to V_{mr} is most characteristic of long-distance migrants that must maximize flight range on a particular amount of fuel. Common Swifts migrate at 40 kilometers per hour, close to V_{mr}. Speeds attained in short flights vary with purpose, which may not be fuel economy (Figure 5 – 12). Hummingbirds, for example, may hover in front of a flower to extract nectar, may fly slowly between adjacent flowers, or may fly faster than V_{mr} to beat competitors to nectar-filled flowers (Gill 1985). Peregrine Falcons are famous for the extraordinary speeds they attain when diving after prey. They apparently reach speeds of up to 180 kilometers per hour though accurate measurements have never been made.

Flightless birds

Not all birds fly. Besides the ratites (a group that includes ostriches and cassowaries), there are flightless grebes, pigeons, parrots, penguins, waterfowl, cormorants, auks, and rails. The original avifaunas of remote predator-free islands, such as the Hawaiian Islands in the Pacific Ocean and the Mascarene islands in the Indian Ocean, included a host of flightless birds: geese, ibis, rails, parrots, and the now-extinct Dodo. If flight and mobility are so clearly advantageous to the majority of birds, why are some birds flightless?

The answer lies largely in the costly development and maintenance of the anatomical apparatus required for flight. Large pectoral muscles, for example, are expensive to produce, and their maintenance requires much energy. In the absence of advantageous uses, such as the need to fly from predators, selection favors reduced investment in the material and energy for flight. Evolutionary reduction of the sternal keel and the mass of flight muscles is, in fact, a first sign of reduced flight ability. The angle between the scapula and coracoid also becomes more obtuse, and ultimately the wing bones become smaller. Kiwis, for example, have only vestigial wings.

One developmental route to flightlessness is evident in rails, a group that includes a great variety of flightless and near-flightless forms, mostly on isolated islands without predators. An enlarged, keeled, calcified sternum requires a major investment of energy and materials during early development. In rails, development of the sternum is not completed until the bird is nearly full grown (Olson 1973). Inasmuch as early development

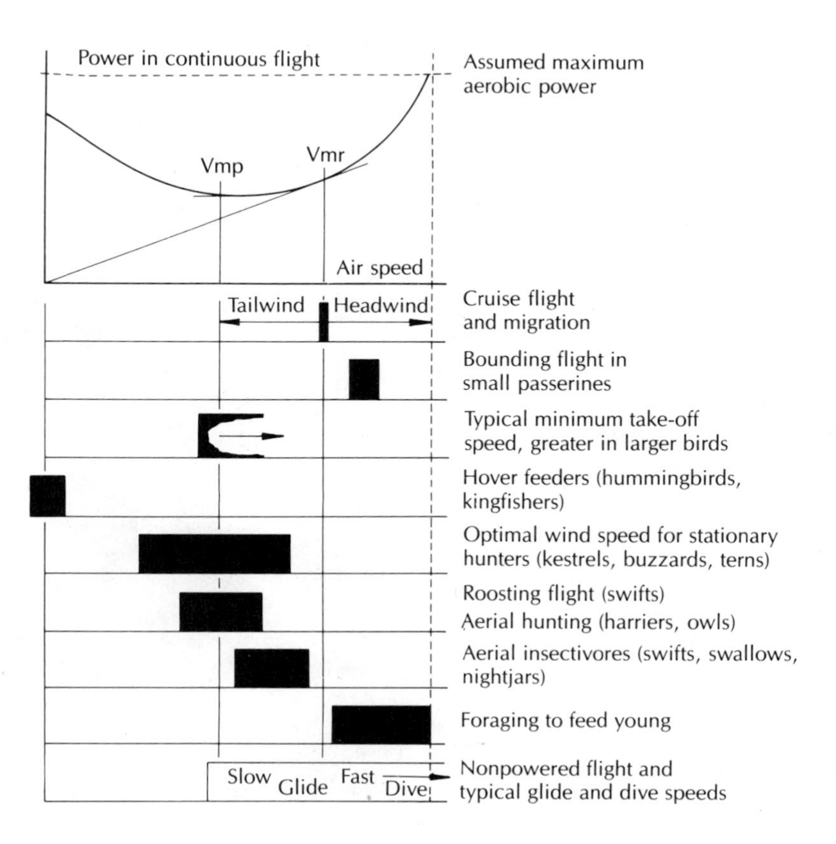

Figure 5-12 The power required for level flapping flight is least at intermediate speeds. Shown here are typical positions of V_{mp} and $V_{mr.}$ on the curve of power against air speed, plus some characteristic modes of flight as they relate to this curve. (From Rayner 1985)

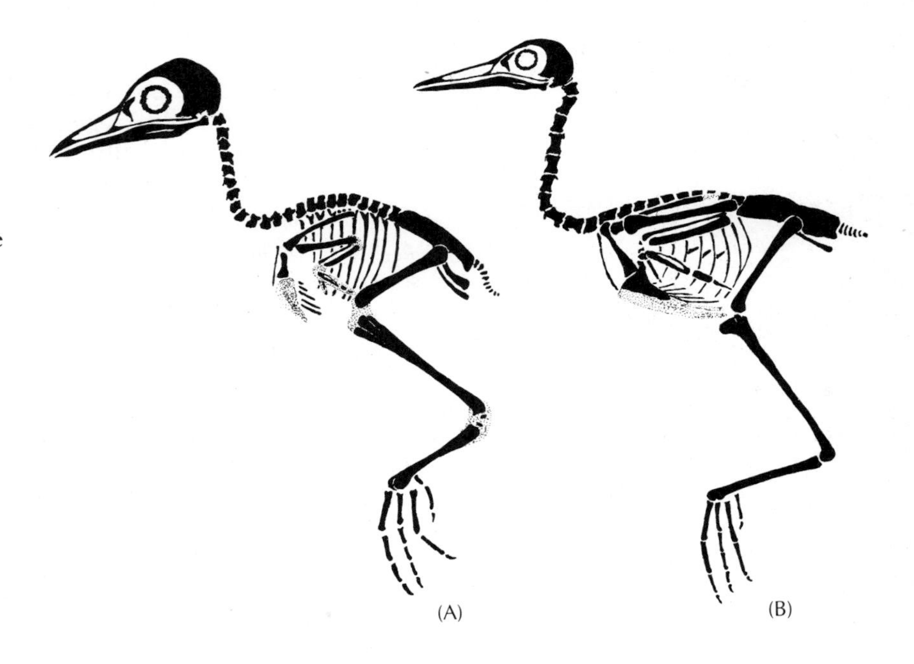

Figure 5 – 13 Skeletons of the King Rail, a flying rail, at 17 days (A) and (B) 47 days after hatching (size reduced so that femur lengths in the two drawings are equal). Stippled areas represent cartilage. Note the obtuse angle formed by the articulation of the scapula and coracoid in the younger form and the acute angle in the older form. (From Olson 1973)

of a large, keeled, calcified sternum requires a major investment of time and material, protracted sternal development of the rails apparently predisposes them to the evolution of flightless forms. The development of the flight mechanisms is permanently arrested at some intermediate stage (Figure 5 – 13). In contrast, chickens, pheasants, quail, and their allies have rarely evolved into flightless forms because, unlike rails, their chicks depend on a strong, calcified sternum and strong flight.

Other routes to the evolution of flightlessness are seen in specialized diving birds (Storer 1971a). Foot-propelled divers such as loons, grebes, and cormorants evolve powerful legs and feet that function as paddles. If

Figure 5 – 14 Flightless Cormorants preening and drying wings. (Courtesy E. and D. Hosking)

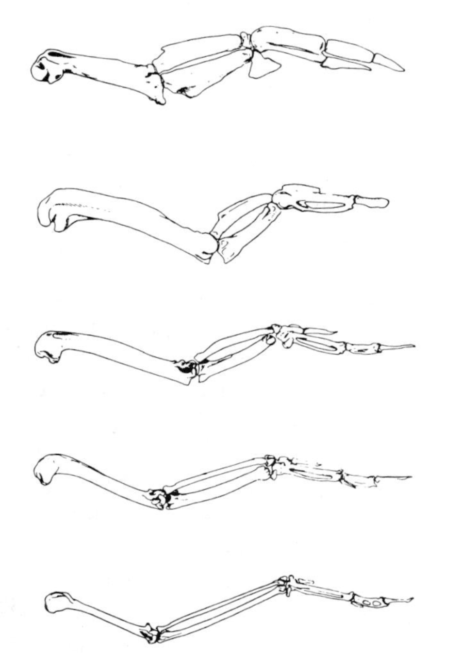

Southern hemisphere Petrel—Penguin Stock	Adaptive stage	Northern hemisphere Gull—Auk stock
Penguins	Wings used for submarine flight only Stage C	Great auk
Diving petrels	Wings used for both submarine and aerial flight Stage B	Razor-bill
Petrels	Wings used for aerial flight only Stage A	Gulls

Figure 5-15 Evolution of wing-propelled divers. (Right) Adaptive stages in the parallel evolution of two stocks of wing-propelled diving birds. (Above) Modifications of the wing skeleton in wing-propelled diving birds (bottom to top: an aerial gull, an auk, the flightless Great Auk, an extinct penguinlike auk, and a penguin). (From Storer 1960)

evolution favors hindlimbs for locomotion, wings and associated pectoral development may regress and render a diving bird nearly or completely flightless. Extreme cases are those of the flightless Short-winged Grebe of Lake Titicaca, Peru, the Flightless Cormorant of the Galapagos Islands (Figure 5-14), and the extinct flightless diving Cretaceous birds such as *Hesperornis.*

Penguins, which are wing-propelled divers, represent another route to flightlessness in specialized diving birds. Their wings propel them through water rather than through the air; their feet act as rudders rather than as paddles. The evolution of such forms has occurred not only in penguins but also among the auks in the northern hemisphere.

The evolution of wing-propelled divers from flying birds proceeds through an intermediate state in which wings are used for both submarine and aerial flight. Diving petrels represent the intermediate stage in the evolution from volant petrels to penguins. Smaller auks such as the Razor-billed Auk have dual-purpose wings that represent the intermediate stage in the evolution of specialized divers from volant gull-like ancestors to the flightless Great Auk of the North Atlantic. The progressive specialization of wing skeletal structure is evident in the changes from the slim wing bones of a gull through shorter and heavier bone structures to the broad, flat wing skeleton of a penguin's flipper (Figure 5-15).

Summary

The ability to fly is a central feature of the evolution of birds from reptiles. Structural adaptations for flight dominate avian anatomy. Fusions and reinforcements of lightweight bones are adaptations of the avian skeleton for flight. Of particular importance are the keeled sternum, which supports the powerful pectoral and supracoracoideus flight muscles, and the strutlike arrangement of the pectoral girdle. The tendons of the ventrally located supracoracoideus muscles pass through the triosseal canal to dorsal insertions on the humerus. The red fibers of avian flight muscles have an extraordinary capacity for aerobic metabolism and sustained work.

The form of the wing and of the individual flight feathers is that of an airfoil, which generates lift as air passes over the asymmetrical surfaces. Control of flight is achieved through changes in wing positions, through the use of slots between feathers, and through the use of the tail as a rudder. Birds are the only vertebrates that can land with precision on elevated or arboreal perches. Hummingbirds achieve extraordinary maneuverability in flight by beating their wings at different angles in a figure-eight pattern that includes a powered upstroke as well as a powered downstroke. Particular wing shapes adapt birds to specific modes of flight because they influence the penalties of induced and parasitic drag relative to the wing's ability to generate lift and thrust. Long, narrow wings sacrifice maneuverability for high-speed flight with low drag. The power costs of avian flight also depend on wing area relative to body mass, a ratio called wing loading.

Various birds have become flightless, particularly on remote islands that lack mammalian predators. Delayed ossification of the sternum in rails predisposes them to the evolution of flightlessness. Specialized diving birds rely either on hindlimb locomotion or wing-propelled underwater locomotion. Extremes of both kinds of diving birds have lost the power of flight. Penguins, for example, have flipperlike wings.

Further readings

Berger, M., and A.J. Hart. 1974. Physiology and energetics of flight. Avian Biology 4:415–477. *A detailed review of flight metabolism.*

Greenewalt, C.H. 1975. The flight of birds. Trans. Amer. Philosoph Soc. New Series, 65(4):1–67. *A classic review of the dimensions of birds and the aerodynamics of avian flight.*

Pennycuick, C.J. 1969. The mechanics of bird migration. Ibis 111:525–556. *An excellent summary of the power costs of long range flight.*

Pennycuick, C.J. 1975. Mechanics of flight. Avian Biology 5:1–75. *A technical summary of the aerodynamics of avian flight.*

Rüppell, G. 1977. Bird Flight. New York: Van Nostrand Reinhold. *The most readable, nontechnical introduction to avian flight.*

chapter 6

Physiology and the environment

The physiology of birds can be described in three words: power, endurance, and balance. Power and endurance derive from the maintenance of high constant body temperatures. Maintenance of core body temperatures within narrow limits is made possible by the excellent insulation of plumage and the balanced regulation of rates of heat loss. Temperature regulation demands energy and water, two resources that are often in short supply.

In this chapter we consider the environmental physiology of birds, specifically their metabolism as it relates to endothermy, the maintenance of a high body temperature by means of metabolic heat production, and to the power requirements of activity. Endothermy, although a vital adaptation, is energetically expensive and requires specialized physiology. Birds consume 20 to 30 times more energy than do similar sized reptiles. Supporting the demands of sustained aerobic metabolism are a highly efficient respiratory system coupled to a powerful heart and circulatory system. Water reserves, essential for evaporative cooling and the excretion of electrolytes, are easily depleted; therefore, birds maintain a delicate physiological balance of the

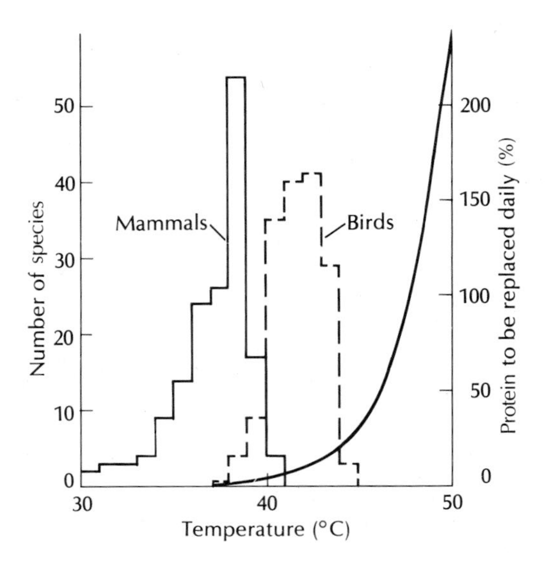

Figure 6–1 Birds and mammals regulate their body temperatures to be just below temperatures that destroy body proteins at lethal rates. Shown here are the body temperatures of many bird and mammal species and the rate of protein replacement as a function of body temperature (curved line). (After Morowitz 1968)

conflicting needs for temperature regulation, activity, and water economy. Other features of avian physiology such as digestion, the senses, reproduction, and hormonal control of the annual cycle, are discussed in later chapters.

Why do birds maintain high body temperatures? Birds are fully active in the early morning cold, in midwinter, and in the high mountains. Activity that is unconstrained by low ambient temperatures is one of the conspicuous advantages of endothermy. Nearly all birds, large and small, in the frigid arctic and in the hottest deserts, keep their core body temperature at about 40°C, which leaves only a narrow margin of safety. If the core body temperature rises above 46°C, most proteins in living cells are destroyed more rapidly than they are replaced (Figure 6–1), and changes in chemistry of the brain may cause death (Calder and King 1974). But below that lethal limit, the rates of physiological processes increase with temperature. This is called the Q_{10} effect. Transmission speed of nerve impulses, for example, increases 1.8 times with every 10°C increase in temperature. The speed and strength of muscle fiber contractions triple with each 10°C temperature jump. High body temperatures, therefore, enhance intrinsic reflexes and powers, enabling birds to be extraordinarily active, fast-moving creatures.

More important than speed or strength is endurance (Bennett 1980). Amphibians and reptiles can escape or strike with lightning speed, but they are quickly exhausted. Birds fly for hours, days, even weeks. Increased aerobic metabolism and insulation were among the major changes that accompanied the evolution of reptiles into birds. These changes apparently made possible regulated high body temperatures and the enormous advantages of constant and dependable rates of muscle function (see discussion of red versus white muscle fibers on p. 88). Higher activity levels coupled with endurance opened a new range of ecological opportunities for birds.

Insulation, heat loss, and the bird's climate space

A bird's thermal relations with its environment are central to its survival. Endothermy is part of a dynamic relationship between heat production inside the body and heat loss to the environment. Heat, a direct product of metabolism, is an inevitable result of the inefficiency of biochemical reactions that is predicted by the second law of thermodynamics. Rates of heat production or loss are expressed in watts, or joules, per hour. The average person sitting in a classroom, for example, produces heat at the same rate as a 100-watt lamp.

The rate of heat loss from the body is proportional in complex ways to (1) the difference between body temperature (or more exactly, surface temperature) T_b and the environmental temperature T_a (or $T_b - T_a$) and (2) the rate of heat transfer across the shell of surface layers. The rate of metabolic heat production H must balance the rate of this loss, as in the equation

$$H = \frac{(T_b - T_a)}{I}$$

where I is the insulation coefficient (resistance to heat flow). In special situations (for example, in a nest hole or a burrow free of wind in which wall temperature equals air temperature), mean environmental temperature can equal ambient air temperature; but in more complex environments, in which the sun shines and the wind blows, the true environmental temperature is a complex function of the intensity of radiation and convection. "Standard operative temperatures" make allowances for these variables (Weathers et al. 1984).

Insulation is one of the most important variables affecting the rate of heat loss, and bird feather coats are among the best natural lightweight insulations. The metabolism of frizzled chickens (Figure 6–2), whose abnormal feathers provide little insulation, is twice that of normal chickens at 17°C (Benedict et al. 1932). Contour feathers in the plumage contribute to a bird's insulation, but the true down feathers underneath the contour feathers are of primary importance. Thus, arctic finches have dense down, and tropical finches do not.

Insulation increases with the amount of plumage. Some birds enhance their insulation during cold seasons by molting into fresh, thick plumage (Calder and King 1974). House Sparrows increase plumage weight 70 percent, from 0.9 grams per bird in August to 1.5 grams in September after molting into fresh plumage. Seasonal adjustments in insulation are not to be expected in tropical birds or in migratory species that escape major seasonal changes in environmental temperatures. The Blackcap, which winters in chilly Europe, has 21 percent more plumage, and better insulation, than the similar-sized Garden Warbler, which winters in balmy north Africa.

Birds can adjust the positions of their feathers to enhance either heat loss or heat conservation. Fluffing the feathers in response to cold creates more air pockets and increases the insulation value of the plumage. A Ring-necked Dove, for example, continuously adjusts the position of its feathers in relation to air temperature (Figure 6–3). However, extreme

Figure 6–2 "Frizzled" chickens have high metabolism rates because their abnormal plumage does not provide as much insulation as that of normal chickens. The following description appeared in *Ornamental and Domestic Poultry* (Edmund Saul Dixon 1848): "It is difficult to say whether this be an aboriginal variety, or merely a peculiar instance of the morphology of feathers; the circumstance that there are also Frizzled Bantams would seem to indicate the latter case to be the fact. Schoolboys used to account for the up-curled feathers of the Frizzled Fowl, by supposing that they had *come the wrong way out of the shell.* They are to be met with of various colours, but are disliked and shunned, and crossly treated by other Poultry. Old-fashioned people sometimes call them French Hens. The reversion of the feathers rendering them of little use as clothing to the birds, makes this variety to be peculiarly susceptible of cold and wet. They have thus the demerit of being tender as well as ugly. In good specimens every feather looks as if it had been curled the wrong way with a pair of hot curling-irons. The stock is retained in existence in this country more by importation than by rearing. The small Frizzled Bantams at the Zoological Gardens, Regent's Park, are found to be excellent sitters and nurses."

elevation of the feathers exposes the bare apterial skin and enhances heat loss by convection. Tropical seabirds that nest in the open sun often elevate their plumage to avoid overheating.

Dark pigmentation aids temperature regulation by absorbing the energy-rich short wavelengths of the solar spectrum. When rewarming from mild overnight hypothermia, the Greater Roadrunner erects its scapular feathers and orients its body so that the morning sun shines on strips of black-pigmented skin on its dorsal apteria (Ohmart and Lasiewski 1971). Metabolism (and heat production) of gray-plumaged Zebra Finches decreases as a result of exposure to strong radiant energy, but that of white-plumaged finches does not (Hamilton and Heppner 1967; Heppner 1970). Light-colored plumage reflects, rather than absorbs, more of the impinging radiant energy than does dark plumage, but the net thermal effect is influenced by the wind.

Figure 6–3 The Ring-necked Dove adjusts the rate of heat loss from its body by subtly raising and lowering its feathers in reaction to ambient temperature. The higher the index, the greater the elevation of the feathers. (After McFarland and Baher 1968)

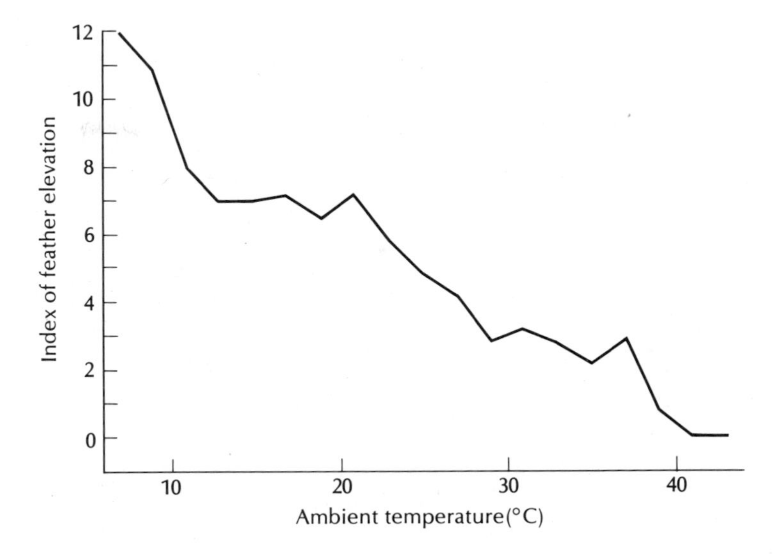

Wind increases the rate of heat loss (T_b is greater than T_a) and, therefore, a bird's metabolic requirements. A Snowy Owl's metabolic rate is directly proportional to the square root of wind velocity (Gessaman 1972). The thick plumage of this species provides excellent insulation, but the rate of heat loss triples in winds of only 27 kilometers per hour compared with the rate in still air. Small birds are particularly vulnerable to convective heat loss because they have more surface area relative to mass than do large birds (Dawson et al. 1983).

The cooling effects of wind are most pronounced on black feathers, which concentrate solar heat near the surface of the plumage and can increase the amount of heat a bird's body absorbs from the environment if there is no breeze. A light breeze, however, removes the accumulated surface heat and reduces further penetration of the radiant heat. The black plumage of desert ravens increases convective heat loss, providing the same effect as the black robes and tents of Saharan desert tribes.

Basal metabolism

Cellular metabolism never shuts down completely. Even resting birds use energy. Carefully controlled measurements of the minimal metabolic requirements of resting birds fasting at nonstressful (thermoneutral) temperatures give estimates of standard or basal metabolism. All birds have high basal metabolic rates, and passerine birds have the highest rates of any group of vertebrate animals. Basic metabolic rates of passerine birds average 50 to 60 percent higher than those of nonpasserines of the same body size. The reasons for the elevated metabolism of birds in general and of passerines in particular remain uncertain (Calder and King 1974).

Basal metabolism relates directly to mass, although not in a 1 : 1 relationship. Large birds require less oxygen per gram of mass than do small birds. Although a bustard may be 100 times larger than a small falcon, it

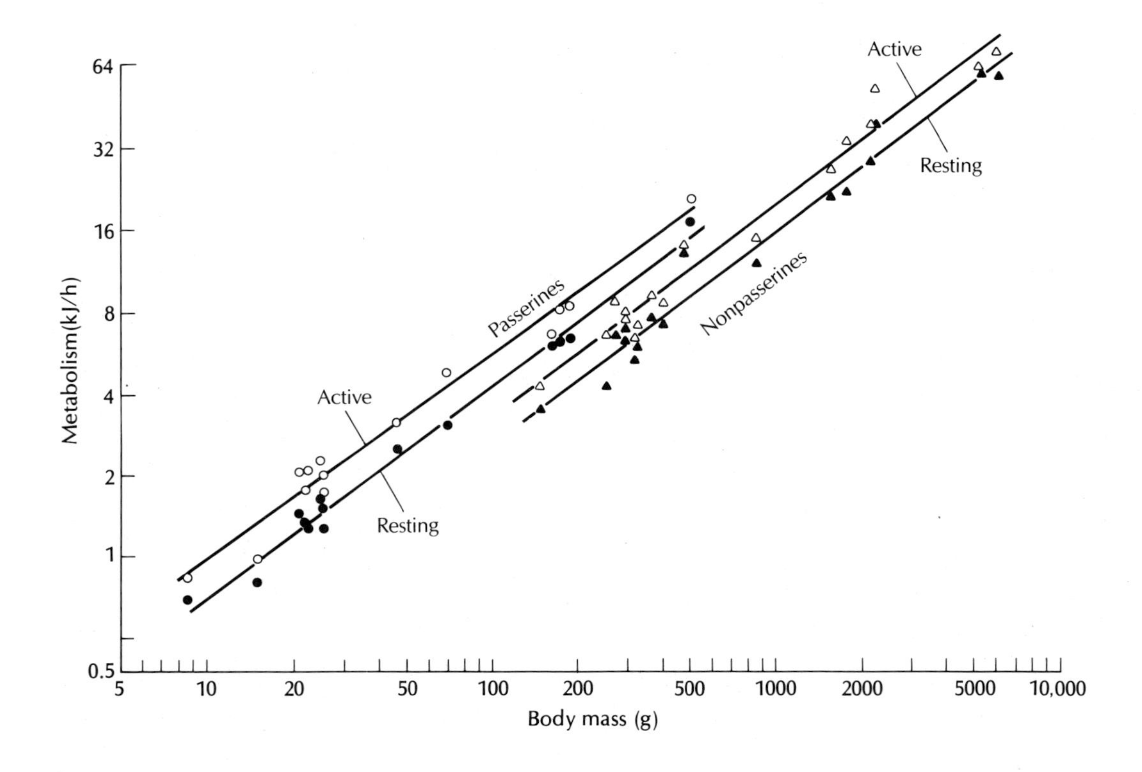

Figure 6–4 The relationship between body size and standard metabolism during active and resting portions of the daily cycle. Passerine birds have significantly higher metabolism than nonpasserines. (○ is active passerine, ● is resting passerine, △ is active nonpasserine, and ▲ is resting nonpasserine.) (After Calder and King 1974)

requires only 30 times as much oxygen per hour. The slope of this fundamental physiological relationship is predictably 0.72 to 0.73 for different-sized birds (Figure 6–4) as well as for different-sized mammals.

The surface area from which heat is lost partly explains the relationship between metabolism and mass. The surface area of a sphere increases as the 0.67 power of mass in accord with the basic two-thirds ratio of a two-dimensional area to the contained three-dimensional volume. A large bird cannot lose heat as fast as a small bird because it has less surface area per gram of heat-generating tissue. If an ostrich's tissues produced heat at the same rate as a sparrow's tissues, the ostrich would soon reach boiling temperatures internally because it could not dissipate heat from its body surfaces fast enough. Yet the simple 0.67 exponent of surface-to-volume relationships does not fully explain the 0.72 exponent of the relationship of metabolism to mass. Perhaps the difference relates to heat lost from other surface areas such as those of the respiratory tract. The precise relationship of metabolism to mass is not yet understood (Schmidt–Nielsen 1983).

Temperature regulation

Most birds do not have to change their rate of heat production to maintain a body temperature of 40°C over the range of air temperatures known as the thermoneutral zone. Birds can control the rates of heat loss by changing feather positions, by varying rates of return of venous blood flow from the skin, by manipulating blood circulation in their feet, and by changing exposure of the extremities, all of which require little direct energy expenditure. For example, fluffing feathers increases insulation. Constriction of veins and the flow of blood near the body's surface further reduces heat loss.

Outside the thermoneutral zone, however, temperature regulation requires an increase in metabolism. A cold bird first tenses its muscles and begins to shiver, which increases oxygen consumption. The pectoral muscles are the major source of heat production by shivering. Whether birds are capable of nonshivering thermogenesis is controversial at best (Calder and King 1974; Oliphant 1983). The temperature at which shivering begins is called the lower critical temperature (LCT).

The lower critical temperatures of large birds are lower than those of small birds. In the absence of special adaptations, therefore, small birds are more sensitive to cold than are large birds. The temperatures comprising the thermoneutral zone of bird species partly reflect adaptations to the average environmental temperatures in which they live (Weathers 1979). Birds living in colder northern climates do not shiver until air temperature drops below 9°C (Snow Bunting) or 7°C (Gray Jay). Similar-sized southern species, such as the Northern Cardinal and the Blue Jay, start shivering at 18°C (Calder and King 1974).

A simple corollary to thermoregulation is that daily energy requirements vary geographically and seasonally in relation to prevailing climates: With distance from the equator, ambient temperatures decrease and metabolic costs increase. For example, House Sparrows that live in the tropical climates of Panama at 9 degrees north latitude spend 58 to 67 kilojoules per day in summer and winter, respectively (Kendeigh 1976). House Sparrows in Churchill, Manitoba, at 50 degrees north latitude, spend 117 kilojoules per day in summer and 151 kilojoules per day in winter, twice as much as their Panamanian counterparts.

Natural physical adjustments to seasonal changes in temperature are called acclimatization. This process may go on over a period of weeks or months, but on a daily basis it reduces the costs of thermoregulation. Winter-acclimatized American Goldfinches, for example, can maintain normal body temperature for six to eight hours when subjected to extremely cold temperatures of −70°C. On the other hand, in summer the goldfinches cannot generate the metabolic heat required to maintain normal body temperature for more than one hour when exposed to frigid temperatures. The ability of winter-acclimatized goldfinches to withstand cold stress lies not only in increased stores of lipids and carbohydrates, but also in their ability to mobilize these stores for heat production more rapidly than can goldfinches in summer (Dawson and Carey 1976; Carey et al. 1978). The precise biochemical basis of acclimatization is not yet known, but it appears to involve emphasis on fatty acids rather than on carbohydrates as fuel for thermogenesis (Marsh and Dawson 1982; Dawson et al. 1983).

At air temperatures above the upper critical temperature (UCT), a bird must lose heat and to do so must increase oxygen consumption. Therefore, the amount of oxygen consumed by resting birds at various temperatures describes a curve that is flat in the middle region and that increases at both low and high temperatures. This simple model of endothermy, developed by Per Fredrik Scholander and his colleagues (1950), is one of the foundations of avian thermobiology (Figure 6–5).

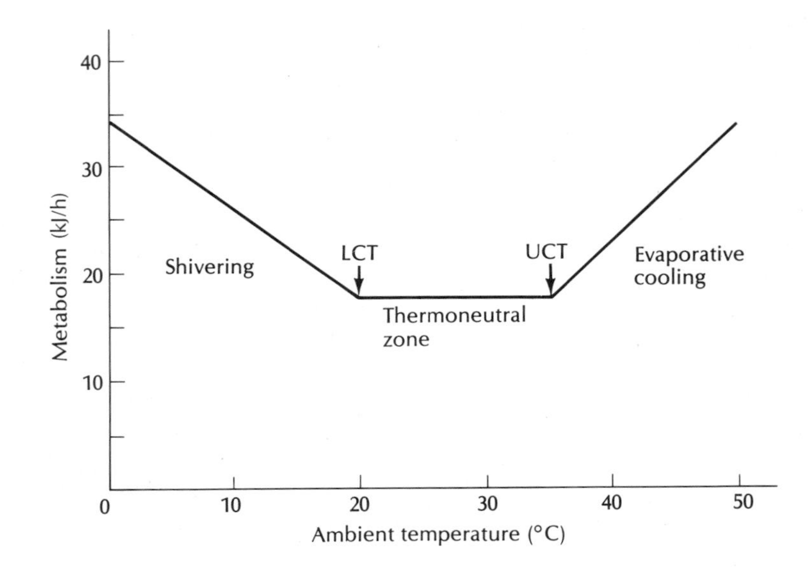

Figure 6–5 Scholander's model of endothermy. Metabolism increases below the lower critical temperature (LCT), primarily as a result of shivering heat production. Metabolism increases above the upper critical temperature (UCT) due to active loss of heat through panting and evaporative cooling, as well as the Q_{10} effect. Metabolism is relatively insensitive to changing ambient temperature in the zone of thermoneutrality. (After Calder and King 1974)

Metabolism increases above basal levels automatically as a result of rising body temperature (the Q_{10} effect). Metabolism increases also because of panting and other efforts that facilitate heat loss. Panting increases evaporative cooling from the nasal, buccal, and upper pharyngeal regions of the trachea. All birds studied so far ventilate faster during heat stress, when body temperatures rise above 41° to 44°C (Figure 6–6). Evaporative cooling is a highly effective method of heat loss, but it presents risk in the form of water loss.

Although birds do not have sweat glands, evaporative water loss may nonetheless take place directly through the skin. In most birds, the amount of water lost increases during heat stress and undoubtedly involves dilation of the blood vessels in the skin, as it does in humans. The precise mechanisms of control of this form of heat loss are still obscure. As our understanding of them improves, so may our appreciation of the role of apteria and the evolution of feather tracts.

Figure 6–6 The Sooty Tern, a bird that is subject to great heat stress at the nest. On a hot day, the bird uses a variety of heat-dissipating mechanisms: (1) exposing the bend of the wing, (2) panting, (3) ruffling crown feathers, (4) ruffling back feathers, (5) wetting abdomen periodically, and (6) exposing the legs. (From Drent 1972)

To supplement panting when they are hot, some birds rapidly vibrate the hyoid muscles and bones in their throats (Lasiewski 1972). This action, called gular fluttering, increases the rate of heat loss through evaporation of water from the mouth lining and upper throat. Common Poorwills, a kind of nightjar, achieve over half of their evaporative cooling this way. The young of many species, including herons, boobies, and roadrunners, which are vulnerable to the hot sun shining on exposed nests, also regulate body temperature by means of gular fluttering.

When necessary, herons and gulls can lose most of their metabolic heat through their feet (Fig. 6-7). They can also reduce this loss by more than 90 percent by controlling blood flow when heat conservation is important. Control of heat loss from the feet is made possible by special blood vessels in the avian leg, which act to conserve or dissipate heat, as needed. The arteries and veins serving the leg intertwine at the base of the legs in such a way that heat carried by arterial blood from the body core can be transferred directly to returning cool blood in the veins of an exposed extremity. This is called countercurrent exchange. It conserves body heat at low air temperatures. For cooling, the blood can completely bypass the network and go directly into the extremities. When overheated, Antarctic Giant-Petrels can increase the rate of blood flow through their feet from 5 to over 100 milliliters per 100 grams of tissue per minute.

There are several other methods of temperature regulation. The Wood Stork increases heat loss through evaporative cooling from its legs by excreting directly onto its legs (Kahl 1963). Helmeted Guineafowl have colorful, naked heads with large protrusions (helmets) and wattles that enhance convective heat loss, as do the wattles of chickens and other fowl (Whittow et al. 1964). Such heat loss may be so great that a guineafowl's head cools faster than its body, beyond the ability of increased blood flow from the body core to replace lost heat (Crowe and Withers 1979). Most birds regulate brain temperatures precisely and cannot survive much variation. Guineafowl brain temperatures, however, vary as much as 6.5°C without serious consequence.

Heat-stressed birds, especially if they are dehydrated, allow their body temperatures to rise 2° to 4°C above normal. Such controlled hyper-

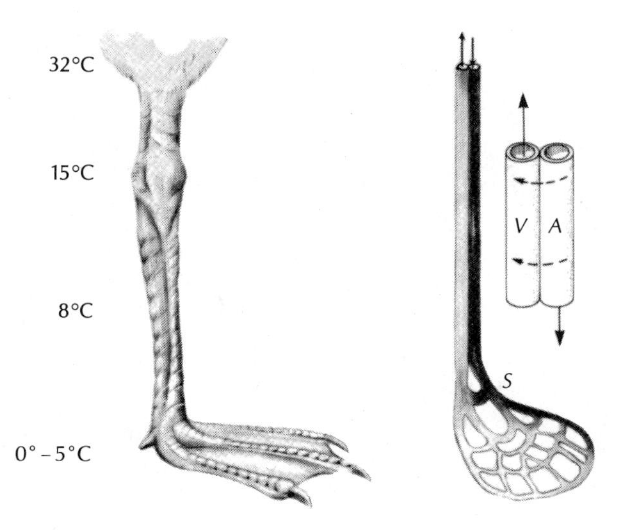

Figure 6-7 Gulls regulate the rate of heat loss from their feet by varying the amount of blood shunted directly to veins at the base of the foot, thus changing the circulation through the foot, where the rate of heat loss is high. Arrows indicate the direction of arterial *(A)* and venous *(V)* blood flow and dashed arrows the direction of heat transfer. A shunt *(S)* allows the constriction of blood vessels in the feet. (From Ricklefs 1979c)

thermia reduces the rate of heat gain from the environment by bringing body temperature closer to air temperature. A 3°C hyperthermic elevation of body temperature would enable a one-kilogram gull, for example, to store 75 percent of its basal heat. Changing the thermal gradient from body to air allows a new favorable equilibrium. In addition, if body temperatures exceed air temperatures, the hyperthermic bird can lose heat without evaporative cooling. Thus, significant water savings result from hyperthermic heat storage. Ostriches let their body temperatures increase 4.2°C during the daily cycle and thereby save liters of water per day that otherwise would be lost in evaporative cooling. Controlled hyperthermia during the warm daylight hours also allows for storage of extra heat needed to save fuel at cooler nighttime temperatures, especially in large birds.

In addition to internal physiological adjustments, birds use various microclimates to alter their rate of heat loss. Foraging by Gambel's Quail, for example, is controlled by sun and shade microclimates in the course of the desert day (Goldstein 1984). Roosting in holes or protected sites such as evergreen trees greatly reduces heat loss, which is especially important during cold winter nights for small passerine birds (Mayer et al. 1982; Dawson et al. 1983). Grouse and ptarmigan frequently burrow into the snow to insulate themselves from cold air temperatures; so do Willow Tits, Siberian Tits, and Common Redpolls (Cade 1953; Sulkava 1969; Korhonen 1981). Huddling together reduces heat loss, and sometimes, birds go to extremes: about 100 Pygmy Nuthatches roosted together in one pine tree cavity, so huddled that some suffocated (Knorr 1957).

Hypothermia and torpor

As an energy-saving measure, avian body temperatures fluctuate a few degrees during the normal, 24-hour activity cycle and may drop 2° to 3°C at night. Some birds, such as the Turkey Vulture, lower their body temperatures about 6°C at night until T_b equals 34°C, that is, they become mildly hypothermic. The Great Tit and the Black-capped Chickadee lower their body temperatures 8° to 12°C at night during extreme winter cold (Chaplin 1974). Some small birds, in particular hummingbirds, can enter a state of profound hypothermia, or torpor, in which they are unresponsive to most stimuli and incapable of normal activity. The oxygen consumption of a hummingbird drops by 75 percent when it allows its body temperature to drop 10°C (Calder and King 1974). Hummingbirds potentially save up to 27 percent of their total daily energy expenditures by allowing their nighttime body temperatures to drop 20° to 32°C below normal. However, a torpid hummingbird does not abandon control of its body temperature and let it come to equilibrium with air temperature (Hainsworth and Wolf 1970). Rather, it regulates a lower body temperature. Thus, a torpid hummingbird's oxygen consumption increases at air temperatures below its lowered body temperature (Figure 6–8).

Warming up is the main difficulty with hypothermic and especially deep torpid states. Birds waking from torpor begin to show good muscular coordination at 26° to 27°C, but normal activity requires body temperatures of at least 34° to 35°C. A small hummingbird requires about an hour

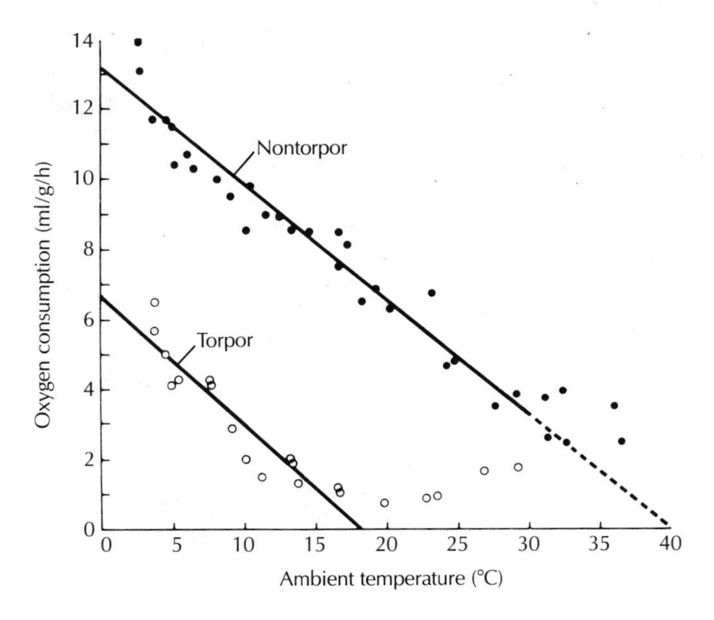

Figure 6–8 Metabolism of the Purple-throated Carib, a tropical hummingbird, during torpor and nontorpor. Nontorpid birds increase their metabolism (measured here in terms of oxygen consumption) as temperature decreases below the lower critical temperature of about 30°C. Torpid birds regulate their body temperature to about 17.5°C. (Adapted from Hainsworth and Wolf 1970)

to arouse from torpor at 20°C, but a medium-sized bird such as an American Kestrel would require 12 hours to warm up from a hypothermic body temperature of only 20°C. The extra costs of reheating a large, cool body tend also to be prohibitive. Thus, small birds such as hummingbirds can become torpid overnight, whereas kestrels cannot. Although full torpor is not practical or economical for short periods in larger birds, in the southwestern United States, Common Poorwills (55 grams) hibernate at a T_b of 6°C for two to three months during the winter. They are capable of spontaneous arousal at low ambient temperatures but require about seven hours to warm up fully (Ligon 1970).

Activity metabolism

Basal metabolism expends minimal energy in the thermoneutral zone. A bird usually spends only a fraction of its day at this low metabolic level, most of its time being spent in activities that require more energy and oxygen. These activities can be the simple digestion of a meal, the slight muscle actions associated with awareness and attention, a strenuous sprint, or a vertical takeoff. Birds are capable of extraordinary levels of aerobic metabolism. Small birds in flight can operate at 10 to 25 times their basal metabolic rate (BMR) for many hours, whereas small mammals can sustain an activity level of metabolism of only 5 to 6 times their BMR (Bartholomew 1972).

Just being awake increases metabolic rate by at least 25 percent above the basal rate. The increased metabolism characteristic of birds that are resting quietly in small cages without temperature stress apparently reflects only increased muscle tonus and mental activity. Some measurements of resting or fasting metabolism are 80 percent higher than basal metabolism (Wolf and Hainsworth 1971; King 1974).

Obviously, metabolic costs increase with exertion. The oxygen consumption of a Bronzy Sunbird increased directly with activity in a small cage (Wolf et al. 1975). A Greater Rhea's metabolism is 3.5 times BMR when it strolls along at one kilometer per hour but is 14 times its BMR when it trots at 10 kilometers per hour. The metabolism of swimming Mallards is 3.2 times their BMR at their most efficient swimming speed and 5.7 times their BMR when they swim as fast as they can; Mallards prefer to swim at the more economical speed (Prange and Schmidt-Nielsen 1970).

Flight metabolism

In general, flight is a more efficient form of locomotion than running. To fly one kilometer, for example, a 10-gram bird uses less than one percent of the energy that a 10-gram mouse uses to run the same distance. Estimates of flight metabolism range from 2 to 25 times the BMR, with variations that reflect flight mode, flight speeds, wing shape, and/or laboratory constraints (Farner 1970; Berger and Hart 1974). Low values of flight metabolism come from swallows and swifts in forward (partly soaring) flight, and high values come from finches and hovering hummingbirds. Hovering in one place is extremely expensive, costing hummingbirds, for example, an average of 0.286 watts per gram, or, depending on body size, 7 to 17 times their BMR (Figure 6-9).

The heat produced during flight could cause lethal increases in body temperature (Berger and Hart 1974). Rock Doves, for example, produce seven times more heat in flight than at rest, and their body temperatures quickly rise 1 to 2°C. The body temperatures of Budgerigars flying at 35 kilometers per hour (in 37°C air) rise to 44°C; they store 13 percent of the heat they produce. However, flight itself increases convective heat loss. The airstream compresses the plumage to the skin, and extension of the wings exposes the thinly feathered ventral base of the wing. As a result of such

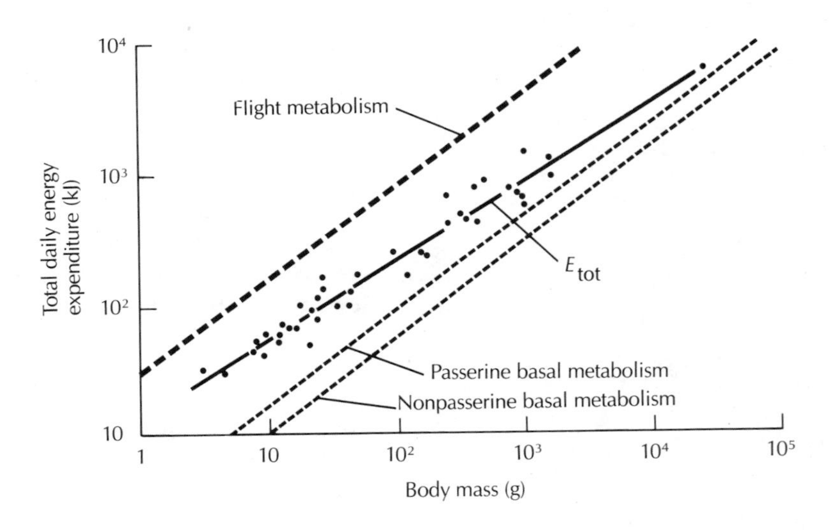

Figure 6-9 Oxygen consumption of birds during flight in relation to body mass and basal metabolism. See page 116 for explanation of total daily energy expenditure (DEE). (Adapted from Walsberg 1983)

changes, the rate of heat loss by flying parakeets increases to 3.1 times the resting value at 20°C and that of Laughing Gulls increases to 5.8 times the resting value. Nonevaporative heat loss directly from these bare skin areas accounts for 80 percent of total heat lost at low and medium temperatures (Berger and Hart 1974).

Daily time and energy budgets

Although time and energy resources can be separately defined, they are nearly inseparable considerations in the problem of daily energy balance and survival. The amount of energy a bird spends each day is roughly the sum of the hourly costs of sleeping, hopping, flying, and other activities, multiplied by the appropriate number of hours engaged in each activity. Such a conversion of time budget to energy budget provides a convenient interface between laboratory physiology and field ecology, but the figures are only approximations (see Weathers et al. 1984).

Time and energy budget studies of Anna's Hummingbirds provide an example of this method (Figure 6–10). Anna's Hummingbirds spend 15 percent of the daylight hours in September feeding on nectar, 2 percent of

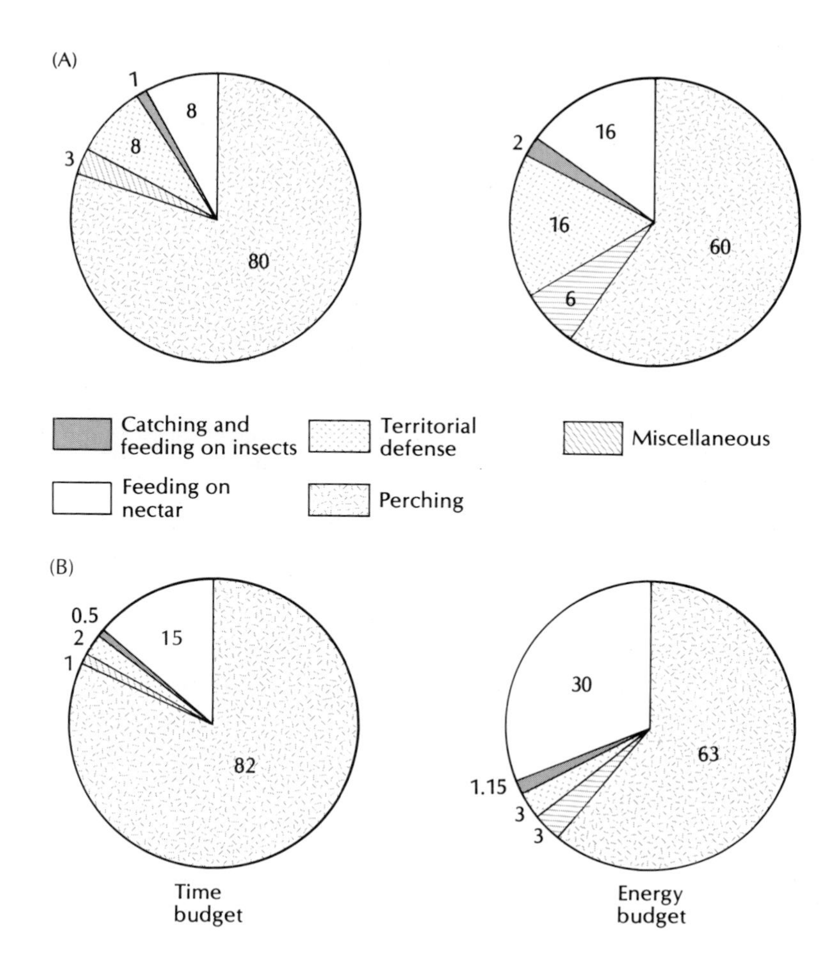

Figure 6–10 Time and energy budgets of the male Anna's Hummingbird. Fractions (%) of the daylight hours spent perching, feeding on nectar, defending territory, and catching and feeding on insects are shown in the circle diagrams for (A) the breeding season and (B) the nonbreeding season when they defend just a feeding territory. Time budgets are multiplied by metabolic rates appropriate to each activity to estimate total daytime energy expenditures. Note that though feeding takes little time, it is a costly activity that makes up a large part of the energy budget. Territorial defense requires a greater investment of time and energy in the breeding season than at other times of year. (After Stiles 1971)

the day chasing other hummingbirds from their feeding territory, 0.5 percent catching and eating insects, and most of the rest of the day (82 percent) perching. The estimated energy costs of each activity sum to a total of 25 kJ. Perching, an inexpensive activity, accounts for 63 percent of the energy budget, whereas feeding and aggression, which are less time-consuming but metabolically more expensive, account for 30 percent and 3 percent of the energy budget, respectively. During the breeding season in March, when male Anna's Hummingbirds defend breeding territories, which are larger than feeding territories, defense increases to 8 percent of the time budget and 16 percent of the energy cost. Total daily energy expenditures increase 13 percent to 28.5 kJ.

In general, daily energy expenditures (DEE) of free-living birds increase with body size (Walsberg 1983). The pertinent equation is

$$\ln \text{DEE} = \ln 13.1 + 0.61 \ln W$$

where ln DEE is the natural log of daily energy expenditures in kilojoules per day, and ln W is the natural log of body mass in kilograms. The slope of this equation (0.61) is lower than that of the equation relating basal metabolism to mass (0.73). It is not clear why this difference exists; perhaps small birds are more active than large birds, or perhaps the basic surface-to-volume relations prevail during activity but not at rest.

In the past few years, scientists have tried measuring daily energy expenditures of free-living birds with a more accurate and reliable field technique. They inject birds with radioactive water, release and recapture them, and then measure the rate of loss of the labeled water that normal metabolism converts into carbon dioxide (Ettinger and King 1980; Weathers and Nagy 1980; Williams and Nagy 1984). The results of studies using this effective technique are just starting to accumulate, but they will certainly lead to a refinement of the equation given above, as well as to accurate assessment of the true costs of daily activities at various stages of the annual cycle (see Chapter 10). This technique incorporates microclimate variables not reflected in the conversion of time budgets to energy budgets (Weathers et al. 1984).

The high metabolic demands of temperature regulation and of the daily activities of birds require extraordinary delivery rates of energy and oxygen to the body's cells as well as rapid removal of poisonous metabolic waste products. Powerful respiratory, circulatory, and excretory systems meet these demands and continuously keep a bird's body chemistry in balance. In the next three sections we discuss the major features of each of these three systems.

Respiratory system

The avian respiratory system is architecturally and functionally different from the mammalian respiratory system. Four anatomical features — the nostrils, tracheal system, lungs, and air sacs — shunt air between the atmosphere and the insides of a bird. With each breath, a bird replaces nearly all

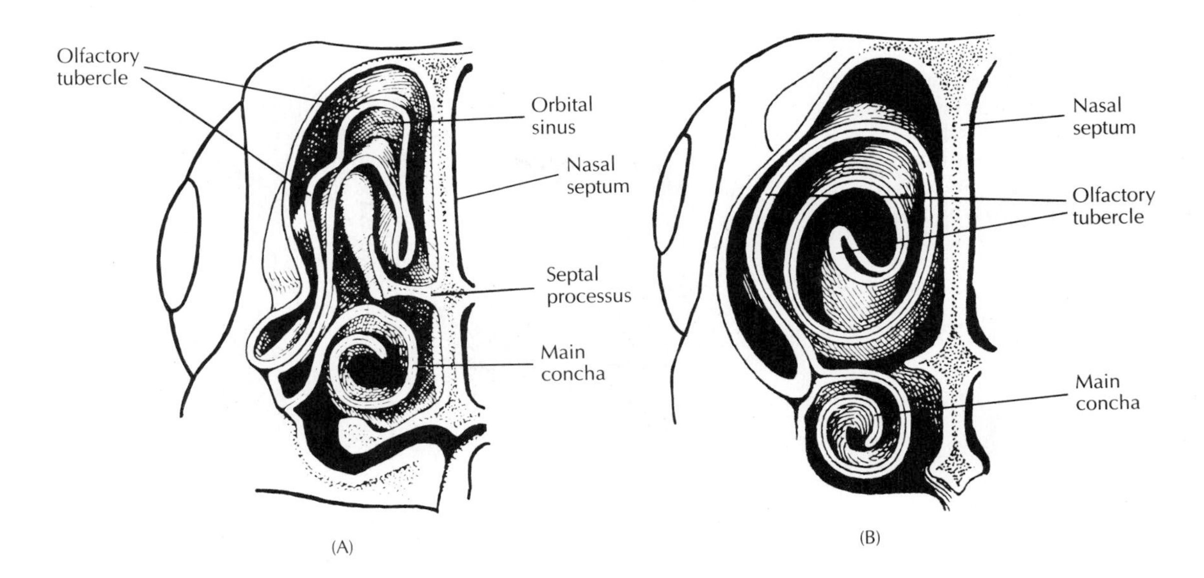

Figure 6–11 Cross section of the nasal cavities of (A) a Northern Fulmar and (B) a Turkey Vulture, showing the elaborate folds over which respiratory air passes. (Adapted from Portmann 1961)

the air in its lungs. Because there is no stagnation period during the avian ventilation cycle, as in that of mammals, birds transfer more oxygen during each breath and overall have a more efficient rate of gas exchange.

Most birds inhale through nostrils at the base of the bill. A flap, or operculum, covers and protects the nostrils in some birds, such as diving birds, which must keep water from entering their nostrils, and flower-feeding birds, whose nostrils might become clogged with pollen. Air passes through external nares into complex, paired nasal chambers (Figure 6–11). Each chamber has three sections containing tiny conchae, which are elaborate folds that greatly increase the epithelial surface area over which air flows. The conchae cleanse and heat the air before it enters the respiratory tract, and sample (or smell) its chemistry. The conchae are densely innervated and highly vascularized with a dense network of blood vessels *(rete mirabile)* that helps control the rate of water and heat loss from the body.

In contrast to human and other mammalian lungs, which are large, inflatable baglike structures that hang in the pleural cavities of the chest, avian lungs are small, compact, spongy structures molded among the ribs on the upper side of the chest cavity. Avian lungs weigh as much as the lungs of mammals of equal body weight, but because they have much greater tissue density, they occupy only about half the volume. Healthy bird lungs are well vascularized and light pink in color.

The internal structure of the mammalian lung resembles a bush with many subdividing bronchial stems and branches, whereas the bronchial branching patterns and connections in bird lungs more closely resemble the plumbing of a steam engine (Figure 6–12). Branching from each of the two primary bronchi that traverse the entire lung are about eleven secondary bronchi, four of which (the craniomedial bronchi) service the anterior and lower parts of the lung. Most of the lung tissue is made up of many (usually about 1800) smaller interconnecting tertiary bronchi; these lead into tiny air capillaries that intertwine with blood capillaries. Gases are exchanged here rather than in vascularized air sacs or alveoli as in mammals.

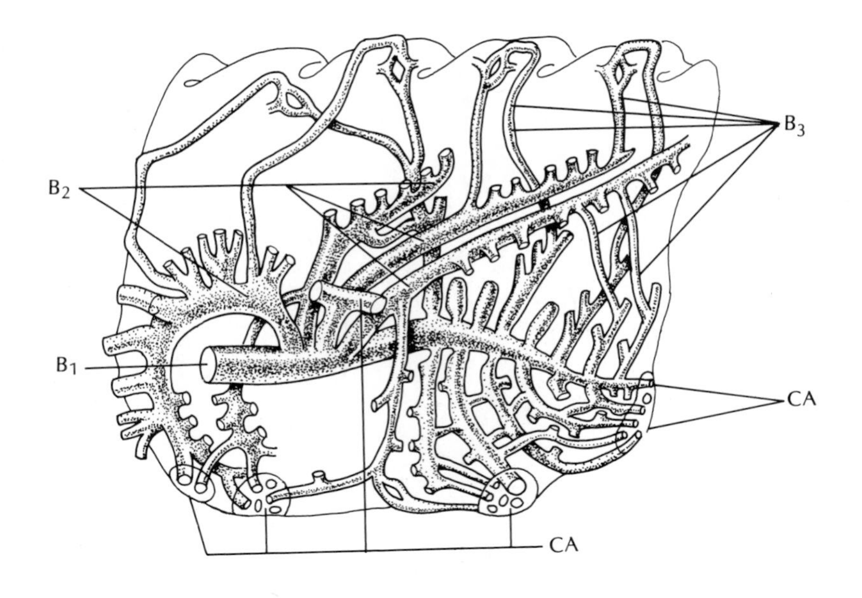

Figure 6–12 Interconnecting bronchial tubules form the internal structure of a bird's lung. Tertiary bronchi and fine air capillaries constitute most of the lung tissue. B_1, primary bronchus; B_2, secondary bronchi; B_3, tertiary bronchi; and CA, connections to air sacs. (After Lasiewski 1972)

The air sac system is an inconspicuous but integral part of the avian respiratory system (Figure 6–13). Air sacs are thin (only 1 to 2 cell layers thick) extensions of the lungs that extend throughout the body cavity and into the wing and leg bones. The air sacs make possible a continuous, unidirectional, highly efficient flow of air through the lungs. They not only help deliver the huge quantities of oxygen needed but also help remove the potentially lethal body heat produced during flight. Inflated air sacs also help protect the delicate internal organs during flight.

The air sacs connect directly to the primary and secondary bronchi and in some species connect indirectly to some tertiary bronchi. The number of air sacs varies from six in the House Sparrow and seven in the Common Loon and Wild Turkey to at least twelve in shorebirds and storks. Most birds, however, have nine air sacs:

1. the paired cervical sacs located in the neck, which are apparent in displaying male frigatebirds in the form of great, inflatable red sacs;

2. a pair of anterior thoracic sacs, which fill the forepart of the body cavity;

3. the large, paired postthoracic sacs, which fill the upper chest;

4. the large, paired abdominal sacs, which cushion the abdominal organs and carry air to pneumatic leg and pelvic bones; and

5. a single interclavicular sac, branches of which penetrate the wing bones, sternum, and syrinx. Pressure in the syrinx from the interclavicular sac is essential for sound production (see Chapter 10).

The avian lung, in contrast to the mammalian, is a model of efficiency. The pattern of respiratory air flow in mammals, though simple (down the trachea and bronchi into the bronchioles and alveoli during inhalation, and the reverse during exhalation), is relatively inefficient be-

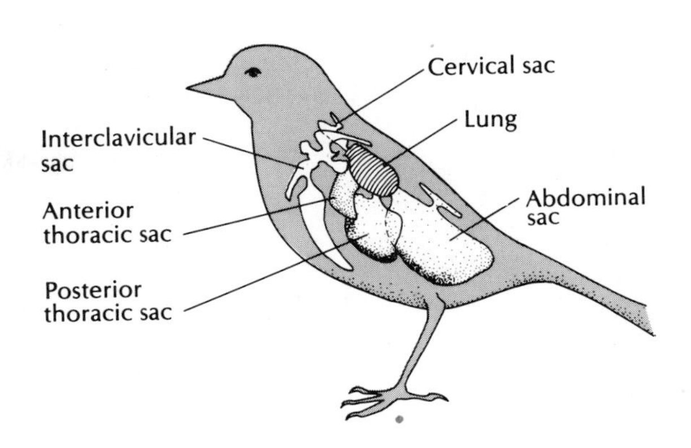

Figure 6 – 13 Positions of the air sacs and lung in a bird's body. (After Salt 1964)

cause a large fraction (20 percent) of the total volume of air in the respiratory pathways remains in contact with nonrespiratory surfaces. In the avian lung, air flows through one set of bronchi to enter the gas-exchange areas and exits through another set (Figure 6 – 14). Most of the air inspired during the first inhalation in the cycle passes via the primary bronchi to the posterior air sacs. During the exhalation phase of the first breath, the inhaled air moves from the posterior air sacs into the lungs via the caudodorsal system of secondary bronchi and flows through the air capillary system, the primary site of oxygen and carbon dioxide (CO_2) exchange. The next time the bird inhales, this oxygen-depleted air moves into the anterior air sacs via the cerebromedial secondary bronchi. During the second exhalation, the CO_2-rich air is expelled from the anterior air sacs, bronchi, and trachea to the atmosphere. The movement of successive volumes of air overlaps; that is, the second breath of one cycle is the first breath of the next one. The flow of air seems mostly passive in response to local pressure differentials; these, in turn, are modified by changes in the openings to the tertiary bronchi, which are controlled by smooth-muscle sphincters. We do not know of any other bronchial valves in birds.

Birds breathe more slowly while resting than do mammals. In both groups, however, the number of breaths per minute decreases with greater body weight. When not flying, a 2-gram hummingbird breathes about 143 times a minute, whereas a 10-kilogram turkey breathes only seven times a minute. In flight, birds meet the increased oxygen demand (10 to 20 times the resting demand) by increasing their ventilation rate (liters of air exchanged per minute of activity) to 12 to 20 times their normal resting rates. Early beliefs that wing and breathing movements were synchronous have proved wrong. Both respiration and wingbeats are controlled by the central nervous system, and wing action does not directly drive the breathing motions of the thorax.

Rapid breathing during exercise or at high, oxygen-poor altitudes expels large amounts of carbon dioxide. Loss of carbon dioxide increases the alkalinity of the blood (normally the pH lies between 7.3 and 7.4), which causes blood vessels to constrict, severely reducing the flow of oxygen-rich blood to the brain. In mammals, blood flow to the brain may drop 50 to 75 percent during such hyperventilation, which causes fainting, and sometimes death. Remarkably, for some unknown reason this does not happen in birds even at pH 8, which would kill a mammal (Grubb et al. 1978, 1979). Without this phenomenon, birds would be unable to fly at high altitudes.

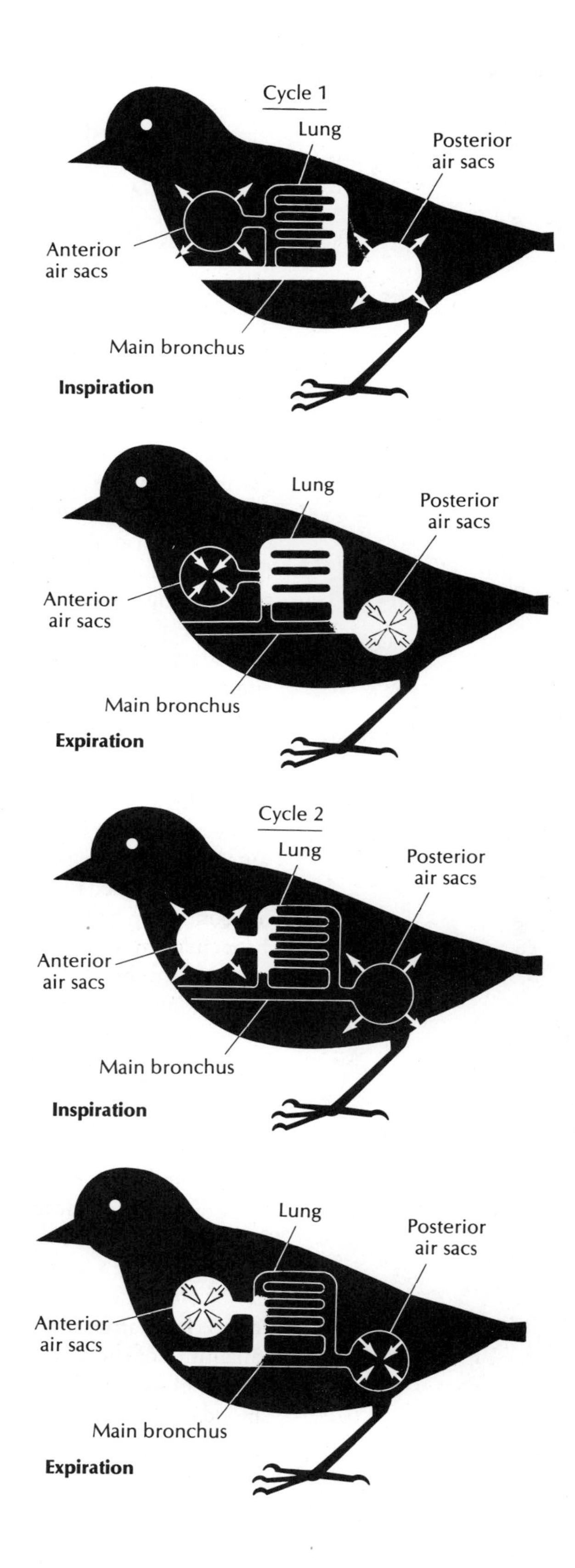

Figure 6–14 The unidirectional movement of a single, inhaled volume of air (shown in white) through the avian respiratory system. To move the gas through its complete path takes two full respiratory cycles: inspiration, expiration, inspiration, and expiration. (After Schmidt–Nielsen 1983).

Circulatory system

The avian circulatory system is matched to the demands of an extraordinary metabolism (Jones and Johansen 1972). The high metabolism of birds requires rapid circulation of high volumes of blood between sites of pickup and delivery of metabolic materials. The circulatory system must deliver oxygen to the body tissues at rates that match usage and must simultaneously remove carbon dioxide for exhalation. It must also meet the demands for fuel delivered in the form of glucose and elementary fatty acids and remove toxic waste products for excretion. Birds have high peak metabolic demands and sustain extraordinary levels of activity metabolism for long periods. The demands on the avian circulatory system are far greater than on those of reptiles, and even exceed those in most mammals.

Like mammals, birds have a double circulatory system and a four-chambered heart (Figure 6-15). The evolution of the four-chambered heart from the reptilian three-chambered heart (only one ventricle) relates to the increased oxygen requirements of endothermy. Circulation of blood to the lungs for gas exchange (pulmonary) is separated from general body (systemic) circulation. This separation allows the blood pressure to be lower in the pulmonary system than in the systemic system and insures that newly oxygenated blood is sent directly to the demanding tissues without mixing and dilution with blood returning from the tissues. The double circulatory systems of birds and mammals are different and, therefore, show evolutionary convergence: Blood passes to the lungs via the left systemic arch in mammals but via the right systemic arch in both birds and reptiles.

Birds have powerful hearts to meet the demands of their metabolism. A large structure in most birds, the heart accounts for 2 to 4 percent of the total weight of a hummingbird. Avian hearts are 50 to 100 percent larger and are more powerful than those of mammals of the same body size. The performance of the heart is measured in terms of cardiac output, or the rate

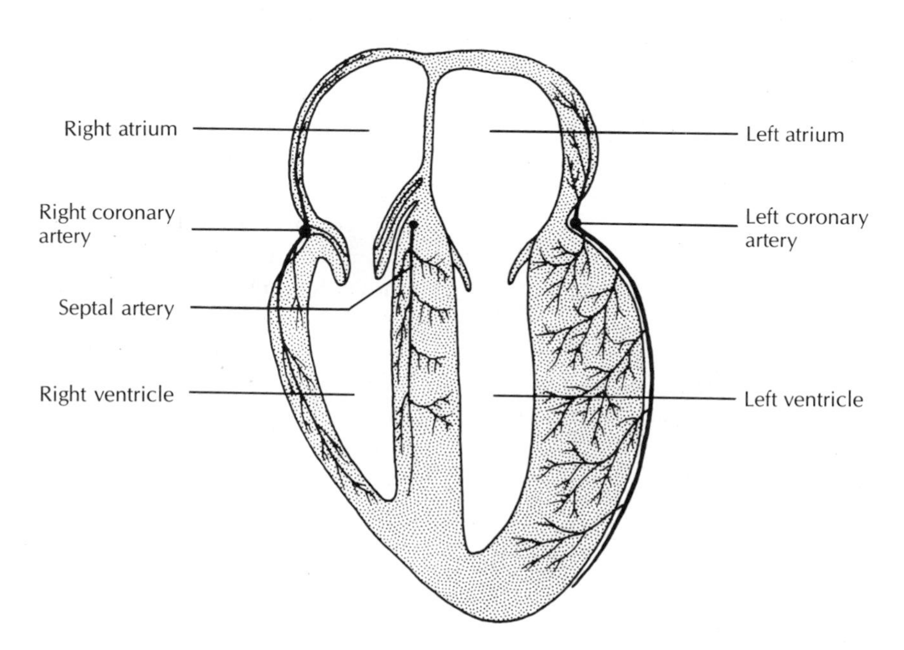

Figure 6–15 The bird heart is large, with four chambers that enable efficient oxygenation of blood. (From Jones and Johansen 1972)

at which the heart pumps blood into the arterial system. Defined as heart rate times stroke volume from one ventricle, it averages 100 to 200 milliliters of blood per kilogram per minute in birds. The resting cardiac output of ducks is apparently somewhat higher than that of other birds and averages 200 to 600 milliliters of blood per kilogram per minute.

Birds and mammals achieve similarly high cardiac outputs in different ways. The resting heart rates (HR) of birds, which are about half those of similar-sized mammals, relate to body mass *(M)* as $HR = 763\ M^{0.23}$. Normal resting heart rates in medium-sized birds range from 150 to 350 beats per minute and average about 220. Heart rates of small birds are much higher and exceed 1200 beats per minute in small hummingbirds (Lasiewski et al. 1967). Although bird hearts beat more slowly than mammal hearts at rest, their cardiac outputs are similar because of large avian stroke volumes. Not only is the avian heart large, but the ventricles empty more completely than those of mammals on each contraction. In addition, at high heartbeat rates the ventricle fills more completely between contractions.

The avian ventricle, or outgoing pump, is made up of more muscle fibers than in the mammalian heart. Reflecting a greater capacity for aerobic work and endurance at high activity levels, each fiber is thinner and its cells contain more mitochondria (energy-producing organelles that depend on the supply of oxygen) than mammalian heart muscle fibers. The thinness of avian fibers speeds the transfer of oxygen. In addition, the dorsal walls of the atria are strengthened by dense muscular cords, which enhances return of venous blood.

The Purkinje fiber system, which distributes and coordinates the impulses responsible for the contraction of heart muscle fibers, is more extensive in birds than in mammals. In some species, its elements lack a fibrous sheath, which permits faster propagation of impulses. A ring of Purkinje fibers around the right atrioventricular valve shuts that valve early during ventricular contractions to ensure that all blood exits through the aorta and is not forced back into the atrium.

A large proportion of the oxygenated cardiac output from the left ventricle goes directly to legs for the purpose of heat loss (see page 111). The legs get three times as much blood per heartbeat as the pectoral muscles and twice as much as the head, the next most imortant target. Together, the legs and brain receive 10 to 20 percent of the total cardiac output.

The high-performance design of the avian heart has its costs. The high tension of avian heart muscles and the strength of the ventricular contractions lead to high arterial blood pressures, up to 300 to 400 millimeters of mercury in some strains of domestic turkeys, the maximum known for any vertebrate. (A blood pressure of 150 millimeters of mercury is high for a human.) Not surprisingly, aortic rupture is a common cause of death in these turkeys, which are raised on high-fat diets for weight gain. Small, wild birds also die easily of heart failure, aortic rupture, or hemorrhage when frightened or otherwise stressed. A male Prairie Warbler, for example, is reported to have died of a heart attack during courtship (Ketterson 1977). Also contributing to such problems is the stiff structure of the avian arteries, which improves smooth, peripheral blood flow but increases susceptibility to atherosclerosis, especially in those fat domestic turkeys.

Specialized diving birds have an unusual tolerance to asphyxia, or loss of air. Mallard ducks, which dive only modestly, can reduce oxygen consumption by as much as 90 percent during prolonged dives. They ration the available oxygen sparingly to some sensitive tissues, especially the central nervous system, sensory organs, and endocrine glands. Blood flow to most other organs and skeletal muscles stops during a dive; they rely on anaerobic metabolism during this period. Selective vasoconstriction and a profound slowdown of heart rate starts within six seconds of immersion of the head, and the heart rate drops to half that of predive levels within eight seconds. The diving reflex is triggered by special receptors in the nares when water touches them. Small penguins show this diving reflex even while they are porpoising through the water (see Kanwisher et al. 1981, however).

Excretory system

Water economy

Satisfying daily energy requirements is only one side of the physiological coin. Equally important is the need to economize water. The potential for debilitating water loss is a corollary of the high body temperatures and activity levels of birds, especially during exposure to midday heat. Enhanced evaporative heat loss is essential to avoid heat stress during strenuous activity. Evaporative water loss in a desert finch, the Brown Towhee, for example, quadruples when ambient temperature increases from 30 to 40°C while oxygen use only doubles (Bartholomew and Cade 1963). A limited capacity for concentrating electrolytes in the urine and high evaporative water loss also enable birds to have high rates of water turnover.

Birds depend on several sources of water to replace that which is used. First, metabolic water is produced (this is water that is a byproduct of the oxidation of organic compounds containing hydrogen molecules). Because of their high metabolism, birds produce more metabolic water in relation to body size than do most vertebrates. Metabolism of one gram of fat yields not only 38.5 kilojoules of energy but also 1.07 grams of water. Metabolism of a gram of carbohydrate or protein produces about 0.56 and 0.40 grams of water, respectively. Metabolic water production is directly proportional to oxygen consumption, and the ratio of water evaporated to water produced is inversely related to body weight (Figure 6–16) (Bartholomew 1982). Metabolic water can thus supply most (up to 80 percent) of some birds' requirements.

It is doubtful, however, that any small birds, even those highly adapted to arid environments, can subsist on metabolic water alone, as do some desert reptiles and rodents; nor has any large bird been shown to do so, in part because most supplement metabolic water with the water contained in their food. All birds, large and small, evaporate more water than they produce metabolically at temperatures in the zone of thermoneutrality (Bartholomew 1982).

Much water is conserved by countercurrent cooling in the nasal and respiratory passages, a general function of the nares and pharynx of small birds that is important to both desert and nondesert species. The respiratory passages are cooled by the evaporation of water from the nasal passages during inhalation. During exhalation, the warm, humid air from the lungs

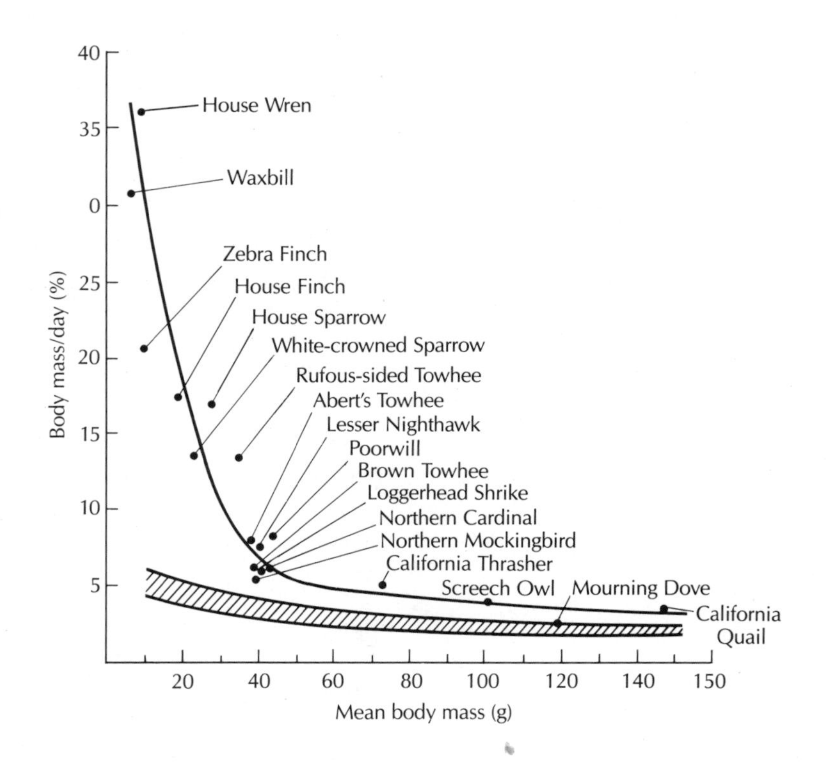

Figure 6–16 Evaporative water loss at nonstressful ambient temperatures (near 25°C) decreases sharply with increasing size of small birds. Evaporative loss is partially but not completely offset by metabolic water production, the projected range of which is indicated by the cross hatching. (After Bartholomew and Cade 1963)

comes in contact with the cool respiratory passages. Water vapor condenses on the nasal passages before the air is finally exhaled and is then reevaporated by the dry air that is inhaled when the bird takes its next breath.

The amount of water recoverable by these means increases with decreasing ambient temperature. Rates of evaporative water loss and metabolic water production (which, remember, increases with higher metabolism at lower temperatures) thus become more similar at colder temperatures, which relieves some of the need for water from external sources.

Water in food is the second source of water for many birds, particularly predatory birds such as owls and hawks. The body fluids of their prey satisfy most of their needs, which enables many predators to live in extreme desert heat. Peregrine Falcons and Sooty Falcons, which nest in arid parts of the Sahara where midday shade temperatures can reach 48.9°C (120°F), feed on migratory birds and bats. (However, they also need relief from the heat and often soar at high, cooler altitudes.) Likewise, insectivorous birds get most of the water they need from the body fluids of consumed insects; and unlike granivorous birds, they rarely visit water holes. Swallows are an exception, often drinking surface water, possibly because as aerial species that sustain flight metabolism for long periods, they lose water faster than most terrestrial species. Seed-eating birds experience the greatest need for free water. Some birds, such as California Quail and the Rock Wren, obtain adequate water by supplementing their diet of seeds with insects. Others, primarily small desert passerines such as Zebra Finches, can survive, drinking not a drop, on a diet of air-dried seeds containing less than 10 percent water.

Free water in streams, waterholes, dew, raindrops, and even snow is the third source of water used by birds. Drinking is a casual, incidental activity in most mesic habitats (those having a moderate amount of moisture); but in desert, or xeric, regions where water is restricted to isolated springs or water holes, daily commutes from the surrounding countryside may be necessary.

Dean Fisher and his colleagues (1972) conducted dawn-to-dusk watches at waterholes in the arid regions of western and central Australia. Over half of the 118 species of birds in the area appeared to be independent of surface water. Other species came in spectacular numbers to drink (Figure 6–17). In a two-hour period one hot day with a shade temperature of 32°C, 1500 to 2000 Spine-cheeked Wattlebirds came from all directions into the mulga trees that surrounded a waterhole. The drinking frequencies of parrots also correlated closely with maximum daily temperatures. One day, during an unusually dry period, Dean Fisher recorded 67,000 visits to one waterhole.

Figure 6–17 (Top) Huge flocks of birds regularly visit waterholes in arid Australia. (Bottom) Budgerigars at a waterhole. (Courtesy C.D. Fisher)

Excretion

The most conspicuous physiological adaptation for promoting water economy in birds is the excretion of nitrogenous wastes in the form of uric acid (rather than urea, as in mammals). The turnover of proteins in the maintenance of body structures produces nitrogenous products, which become toxic if allowed to accumulate. Excretion of nitrogen as urea in aqueous solution requires flushing by large quantities of water, but uric acid can be excreted as a semisolid suspension in which each molecule of uric acid contains twice as much nitrogen as a molecule of urea. Therefore, birds require only 0.5 to 1.0 milliliters of water to excrete 370 milligrams of nitrogen as uric acid, whereas mammals require 20 milliliters of water to excrete the same amount of nitrogen as urine. Birds can concentrate uric acid in the cloaca to amazing levels: up to 3000 times the acid level in their blood. Kangaroo Rats, one of the most efficient mammalian water conservationists, can concentrate urea to levels only 20 to 30 times those in the blood.

The avian kidney differs in structure and function from that of reptiles or mammals. It is unique in having separately controllable "mammalian" and "reptilian" type nephrons (single excretory units) and in its ability to adjust renal and peritubular capillary blood flow. In addition, the cloaca, the large intestine, and even intestinal ceca (cavities that open at only one end; see Figure 7 – 14) play major roles in the modification of urine from the kidneys.

Although avian kidneys can concentrate nitrogenous wastes very well, they usually cannot concentrate salt or electrolytes much above normal blood levels. Mammalian kidneys, especially those of the Kangaroo Rat, excel at this because of the Loop of Henle, which establishes the osmotic gradients necessary for withdrawing water molecules from solution. In contrast, the Loops of Henle in the avian kidney are short, presumably associated with the excretion of uric acid instead of urea. This presents a problem, particularly for oceanic birds, which must drink salty seawater. Seawater contains about three percent salt; the birds' body fluids contain one percent salt. The high salt content of their marine invertebrate foods further increases their need to excrete electrolytes. For this reason, seabirds, as well as other birds with water-conservation problems, rely on extrarenal structures called nasal salt glands (Figure 6 – 18).

Nasal salt glands enable seabirds to drink seawater and to unload the newly ingested salt rapidly (at a high energy cost) via concentrated salt solutions. For example, if a gull drank one-tenth of its body weight in seawater, it would excrete 90 percent of the new salt load within three hours (Schmidt – Nielsen 1983). These amazing glands produce and excrete salt solutions containing up to 5 percent salt, more concentrated than seawater.

Salt glands, large structures that sit in special depressions in the skull just above the eye, are special infoldings of the nasal epithelium, or lining. Inside the salt gland are many secretory tubules arranged in lobes. The tubules, each of which is composed of a modified epithelial cell, extract salt from blood in associated capillaries of the ophthalamic arteries (which also service the eyes) and empty directly into a central canal leading to the main duct. Salt concentration and removal by these tubules involve active transport of the ions, mediated by sodium (Na^+) activated and potassium (K^+)

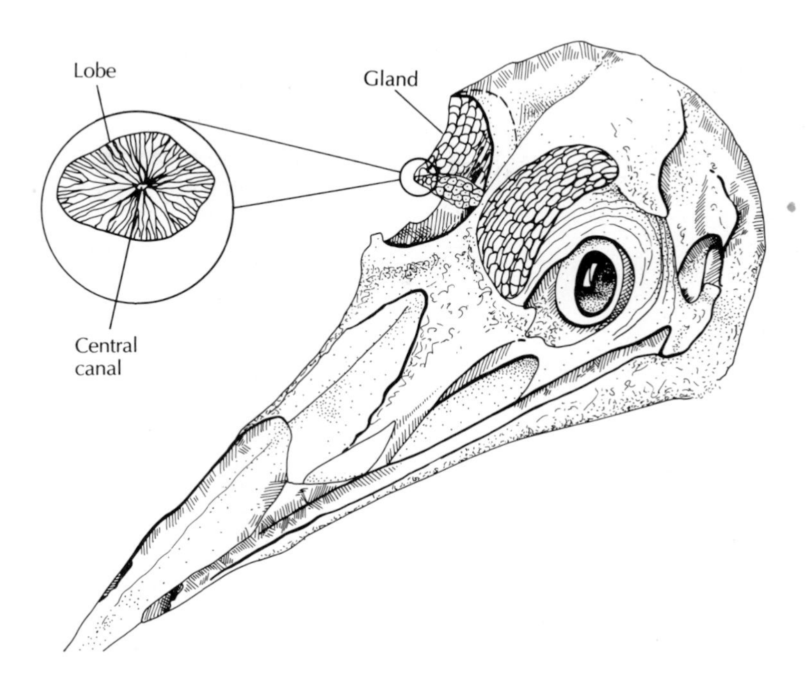

Figure 6–18 The salt glands of marine birds are located on top of the head in shallow depressions above each eye. (After Schmidt-Nielson 1983)

activated enzymes (ATPase). Each of the pairs of glands has a main duct that leads to the anterior nasal cavity. The salt concentrate runs out of the nostril and down grooves in the bill to the bill tip before dripping off. Some birds, such as storm-petrels, eject the fluid forcibly. Activity of the salt gland is stimulated directly by the intake of salt, or sometimes just by osmotic stress.

Salt glands are widespread among nonpasserine birds with potential electrolyte imbalance resulting from salty diets. Salt glands are largest and best developed in oceanic birds such as albatrosses, which must drink seawater. When individual birds, such as Mallards, drink saltwater instead of fresh water, their salt glands increase in size (Shoemaker 1972). The size of the gland reflects the number of lobes in it and varies in different birds from 0.1 to 1 gram per kilogram of body mass. The salt glands of marine species also have the greatest capacity for salt concentration. Auks and gulls have particularly large glands, with as many as 20 lobes. Surprisingly, and for no apparent reason other than evolutionary chance, no passerines have salt glands, not even those that live in salt marshes or feed on intertidal invertebrates on the seacoast.

Summary

Birds have high metabolism. Flight and the maintenance of high body temperatures use large amounts of energy. Both the circulatory and respiratory systems have evolved exceptional capacities for the delivery of fuel and the removal of metabolic products. Water loss linked to high metabolic rates also poses difficulties for birds.

Birds regulate their body temperatures at 40 to 42°C by adjusting plumage insulation, by increasing heat production through shivering when cold, and by evaporative water loss through panting and gular fluttering when hot. Regulation of blood flow through the feet aids heat loss or retention. To save energy some birds, notably hummingbirds, swifts, and nightjars, can lower body temperature and become torpid. Birds can also elevate body temperature a few degrees to reduce the need for evaporative water loss and to store body heat for cold nights. Birds, however, have little latitude for higher body temperatures: 46°C is lethal.

Birds depend on metabolic water as well as on that ingested in their food or drunk as free water. Desert waterholes attract huge aggregations of thirsty birds, though the excretion of nitrogenous wastes as uric acid rather than as urea promotes water economy in birds. Seabirds have well-developed salt glands imbedded in their skulls over their eyes. These glands excrete salt in extremely concentrated solutions and thereby enable the birds to drink seawater.

Further readings

Calder, W.A., and J.R. King. 1974. Thermal and caloric relations of birds. Avian Biology 4:259–413. *A comprehensive review of energetic aspects of avian physiology.*

Dawson, W.R., R.L. Marsh, and M.E. Yacoe. 1983. Metabolic adjustments of small passerine birds for migration and cold. Am. J. Physiol. 245 (Regulatory Integrative Comp. Physiol. 14): R755–R767. *An important summary of recent research.*

Jones, D.R., and K. Johansen. 1972. The blood vascular system. Avian Biology 2:158–256. *A detailed review of the avian circulatory system.*

Scheid, P. 1982. Respiration and control of breathing. Avian Biology 6:406–453. *A detailed review of the avian respiratory system.*

Shoemaker, V.H. 1972. Osmoregulation and excretion in birds. Avian Biology 2:527–573.

Schmidt-Nielsen, K. 1983. Animal Physiology. Cambridge: Cambridge University Press. *A basic text of comparative physiology.*

Sturkie, P.D. 1976. Avian Physiology. New York: Springer-Verlag. *A general review with an emphasis on poultry.*

Walsberg, G.E. 1983. Avian ecological energetics. Avian Biology 7:161–219. *A detailed review of recent research on the costs of reproduction.*

chapter 7

Feeding adaptations

Because birds expend energy at a tremendous rate, they must feed frequently to refuel themselves. Adaptations for feeding are a conspicuous feature of avian evolution. These adaptations include modes of locomotion birds use while feeding, structure of the bill, and the digestive system. Birds sit, walk, hop, fly, and dive in search of food. Sitting shrikes simply wait for prey, whereas crows walk methodically across fields, and warblers hop from twig to twig. Vultures soar, scanning for carrion below, falcons swoop down on fast-flying quarry. Ducks, grebes, loons, and auks dive in deep waters to catch fish or to pluck invertebrates from rocky moorings. A bird's bill is its key adaptation for feeding. The size, shape, and strength of the bill affect a bird's diet (Figure 7–1). The broad, flat bill of a duck is not suited to reach carpenter ants in a tree, whereas the chisellike bill of a woodpecker certainly is. Other bill types are designed to tear meat, spear fish, crack hard seeds, probe deeply into crevices or mudflats, or filter tiny creatures from the mud. The varied lengths and curvatures of shorebird bills determine which sorts of prey they can reach. Even slight variations in bill dimension greatly influence the rate at which food can be ingested.

Figure 7–1 The bills of birds reflect their feeding specialties. Red Crossbills extract seeds from pine cones. Northern Cardinals can crack large, hard seeds. Northern Shoveler ducks strain food from the mud. Golden Eagles tear apart the flesh of their prey. Reddish Egrets spear small fish.

Although seemingly specialized to one particular food or way of feeding, avian bills are multipurpose organs. Most birds feed on a variety of foods and may change diets with the season. The finch bill of a Song Sparrow, for example, is well adapted to its winter diet of hard seeds but also serves for catching and eating soft-bodied insects in the summer. Long-billed shorebirds can pick up prey from the surface as well as from deep in the mud (Figure 7–2).

Following a review of avian bills and digestive systems, we discuss feeding as a highly refined behavior. The study of the avian search for food and the choices inherent in feeding behaviors constitutes a new and particularly exciting field of exploration. We conclude with considerations of the balance between feeding effort and energy requirements, paying special attention to the provision of reserves to be used during difficult periods.

Bill structure

Three major features make up the general morphology of bird bills. The upper half of the bill, or maxilla, attaches to the brain case by a thin, flexible sheet of bone called the nasofrontal hinge. The lower jaw, or mandible, articulates with the quadrate, a large, complex bone at the mandible's posterior end. The large jaw muscles, which enable a bird to bite, attach to the posterior surfaces of the mandible. Covering both jaws is a horny sheath, or ramphotheca, which may have sharp cutting edges (as in boobies), numerous toothlike serrations (as in mergansers), or well-developed notches (as in toucans).

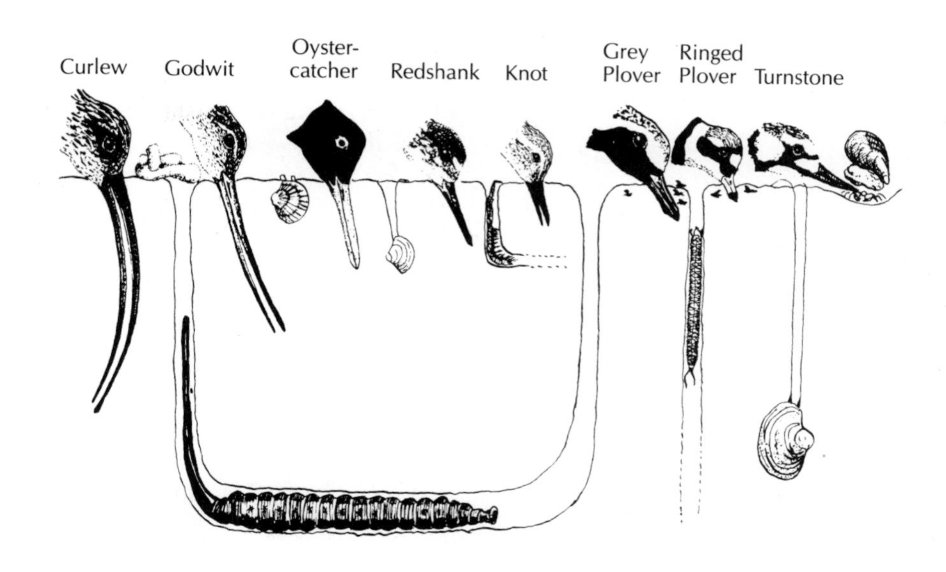

Figure 7–2 Varied bill lengths enable shorebirds to probe to various depths in the mud and sand for food. Plovers feed on small invertebrates, mainly by surface pecking with their short bills. Dunlin and other waders with moderate bill lengths probe the top four centimeters of the substratum, which contain many worms, bivalves, and crustacea. Only the long-billed birds such as curlews and godwits can reach deep-burrowing prey such as lugworms and ragworms. (Adapted from Goss-Custard 1975)

The avian bill is not rigid; birds can flex or bend the upper half of the bill, an ability called cranial kinesis (Zusi 1984). The upper jaws of most birds flex only at the nasofrontal hinge. In some birds the dorsal ridge of the bill itself bends (rhynchokinesis) at the base of the bill, near the tip of the bill, or at both sites. A woodcock, which is a large snipelike bird, can open just the tip of its bill to grasp an earthworm deep in the mud. The bone configurations that constitute the bill, jaws, and palatal region are an engineer's delight; their structure relates directly to the amount of force caused by using the bill (Bock 1966). The upper jaw is a flattened, hollow, bony cone reinforced internally by a complex system of bony struts, called trabeculae. The upper jawbone is reinforced where the greatest forces are manifest. The trabeculae are located near the nasofrontal hinge and help distribute the stress on the hinge that is caused by biting. The curvature of the continuous upper jawbone surface also adjusts for stress (Figure 7–3).

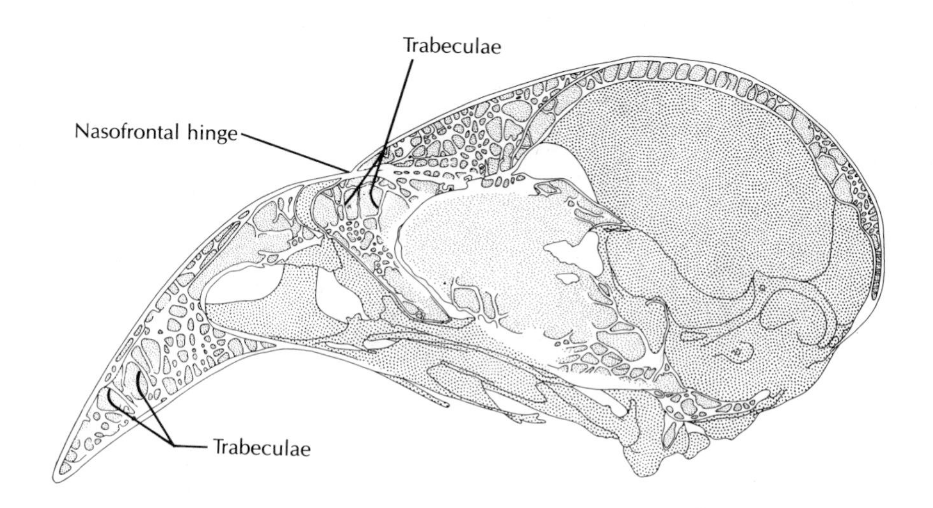

Figure 7–3 The form of their large bills enables finches such as the Northern Cardinal to bite hard seeds without straining the nasofrontal hinge (located between bill and skull) with excessive shear forces. Shown here is a cross section of a cardinal skull revealing the bony struts (trabeculae) in the upper jaw and forehead. The deeper, nontrabecular areas of the jaw are shown in fine stippling. other nontrabecular bone is shown in heavier stippling. (From Bock 1966)

Three specialized bill types

The functional refinements of bill structures can be seen in the details of the bills of flamingos, finches, and nectar-feeding birds. A flamingo's filter-feeding bill is one of the most distinctive and specialized of avian bills (Jenkin 1957). It consists of a large, troughlike lower mandible housing a powerful, fleshy tongue that creates suction as it pumps back and forth. Sets of lamellae, which vary in structure from hard ridges and large hooks to velvet-textured fringed platelets, sort appropriate-sized food from debris in the water. The flat, narrow upper mandible contains sets of lamellae, each opposing another set of lamellae on the inner surfaces of the lower jaw. With head and bill upside down, flamingos strain fine food particles from water or mud pumped by the tongue through the lamellae. The size of the filtering apparatus determines a flamingo's diet. The coarse filters on a Greater Flamingo's bill strain out small invertebrates. The Lesser Flamingo's fine filters strain out tiny blue-green algae. By virtue of this difference, the two species can feed side by side and eat different foods (Figure 7–4).

Seeds sustain a variety of birds. Doves and seedsnipe swallow seeds whole and grind them in the gizzard. Jays and titmice hammer them until they split open. Most passerines that specialize in seed eating, crack and shuck the seed husks with powerful bills. Evening Grosbeak and Hawfinches can crack the hardest seeds, such as olive and cherry pits. Many finches have elaborate, hard ridges and grooves in the inside upper mandible and the anterior palatal region, which enable them to hold the seed in place while it is cut by the sharp edges of the mandibles (Ziswiler and Farner 1972). Finches extract seed kernels by either crushing or cutting the seed hull. In the crushing method, one or both margins of the bill press the seed against a central ridge in the horny roof of the mouth to pop the kernel from the shell. In the cutting method, a finch uses its tongue to lodge the seed in special furrows of the hard palate and then cuts it with rapid forward and backward movements of the sharp edges of the mandible. In both cases, the cut husks fall out of the mouth, and the clean kernel is swallowed.

The husking speed of a finch depends on the relation of bill size and strength to seed size and hardness (Willson and Harmeson 1973). In general, large-billed finches can husk seeds of a wider range of size and hardness than can small-billed finches (Willson 1971; Abbott et al. 1975). The Northern Cardinal has a much larger bill than the White-throated Sparrow, which in turn has a bigger bill than the American Tree Sparrow. A Northern Cardinal can husk large, hard hemp seeds in 13.5 seconds, a White-throated Sparrow takes 13.9 seconds, and an American Tree Sparrow takes 19 seconds. Conversely, tree sparrows can husk small millet seeds in 1.6 seconds; the white-throats take 4.9 seconds.

A large finch bill can be so advantageous in times of food shortage that the average bill size in a population increases from one year to the next (Boag and Grant 1981). In 1976 and 1977, for example, a severe drought gripped Daphne Island in the Galapagos Archipelago. Plants failed to produce a new crop of seeds, and seed density dropped sharply, especially the density of small seeds. Many finches starved. Individual Medium Ground-Finches with particularly large, deep bills survived in greater numbers because they could crack the larger, harder seeds, which were less scarce. Thus, the large-billed individuals in the population prevailed. The result was a

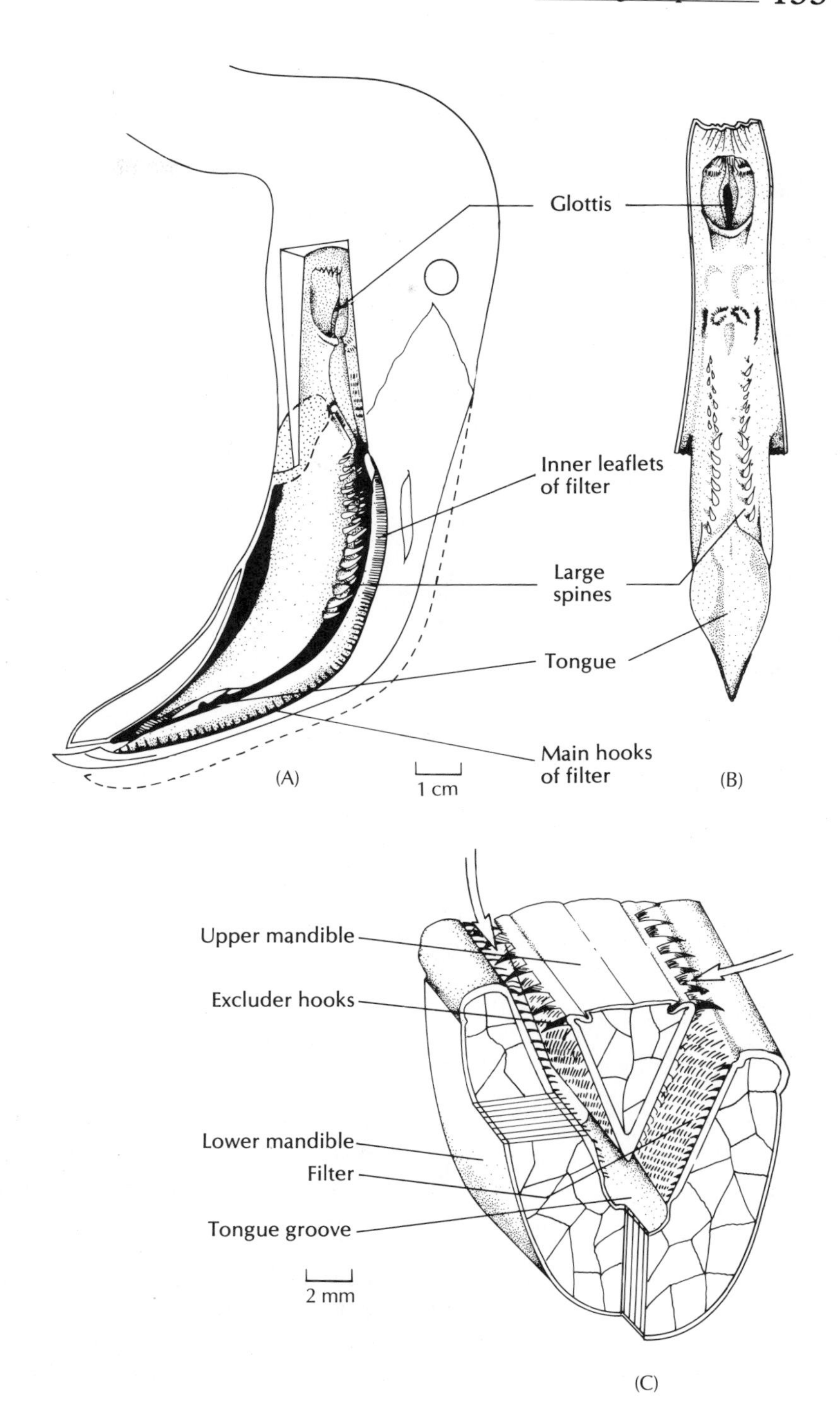

Figure 7–4 (A) Head of a flamingo in normal "upside down" feeding position. (B) The tongue. (C) Cross section of the bill of a Lesser Flamingo, showing flow of water currents caused by pumping the tongue. When the tongue (omitted from diagram) is depressed in the tongue groove, inflowing currents (arrows) are drawn through the large, hooklike lamellae of the upper mandible, which exclude large objects. When the tongue is elevated, outflowing currents pass through the succession of filters on both jaws formed by the smaller, velvet-textured lamellae, called fringed inner platelets. (After Jenkin 1957)

dramatic increase in average bill size over only one year's time, due to natural selection (Figure 7–5).

Nectar-feeding species, such as hummingbirds and sunbirds, probe their thin bills into floral nectar chambers and draw up nectar through capillarylike tongue tips. Bill forms tend to match the lengths and curvatures of preferred flowers, which, in turn, depend on the birds for pollina-

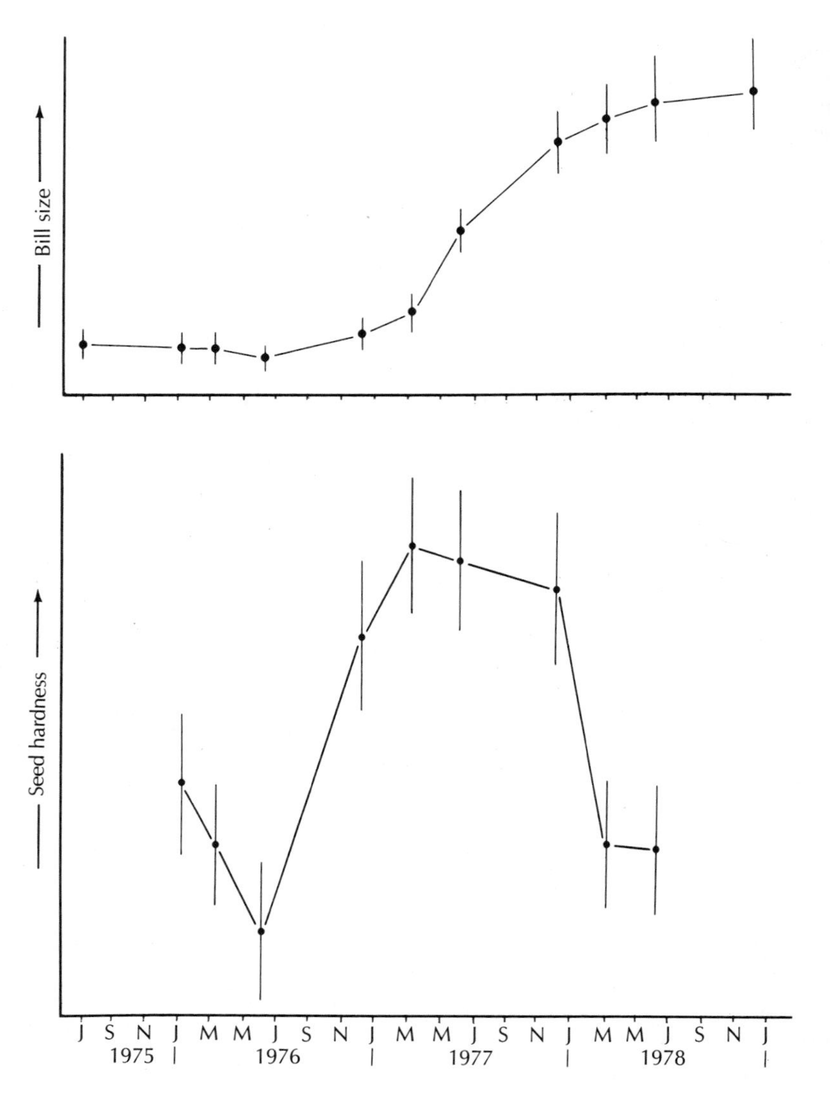

Figure 7–5 Increase in bill size (top) in the Medium Ground-Finch during a period of drought that caused intense natural selection. Failure of the usual seed crop on Daphne Island favored individuals with large bills able to crack the more abundant large, hard seeds (bottom). (Adapted from Boag and Grant 1981)

tion. Indeed, the bill morphologies of nectar-feeding birds appear to coevolve with the morphologies of particular flowers (Figure 7–6). Specialized nectar-feeding birds feed from their own well-matched flowers when these are available, but sometimes they must feed from flowers of other sizes and shapes. When this occurs, the effects of slight differences in bill dimensions on foraging are most apparent. Differences in the bills of three species of East African sunbirds, for example, affect not only the time they take to probe a mint flower but also how much of the available nectar they obtain (Table 7–1). The mint flower stores nectar in a chamber at the base of the corolla. The long, strongly curved bill of the Golden-winged Sunbird enables it to visit one mint flower and to extract 90 percent of the nectar in 1.3 seconds. This sunbird's bill inserts easily into the nectar chamber because its curvature matches that of the flower. The Malachite Sunbird, however, must force its straighter bill down the curved corolla with several jabbing motions and often misses the opening through the diaphragm that protects the nectar chamber. The Malachite Sunbird, therefore, takes more time per flower (1.8 seconds) and removes less nectar (82 percent) than the Golden-

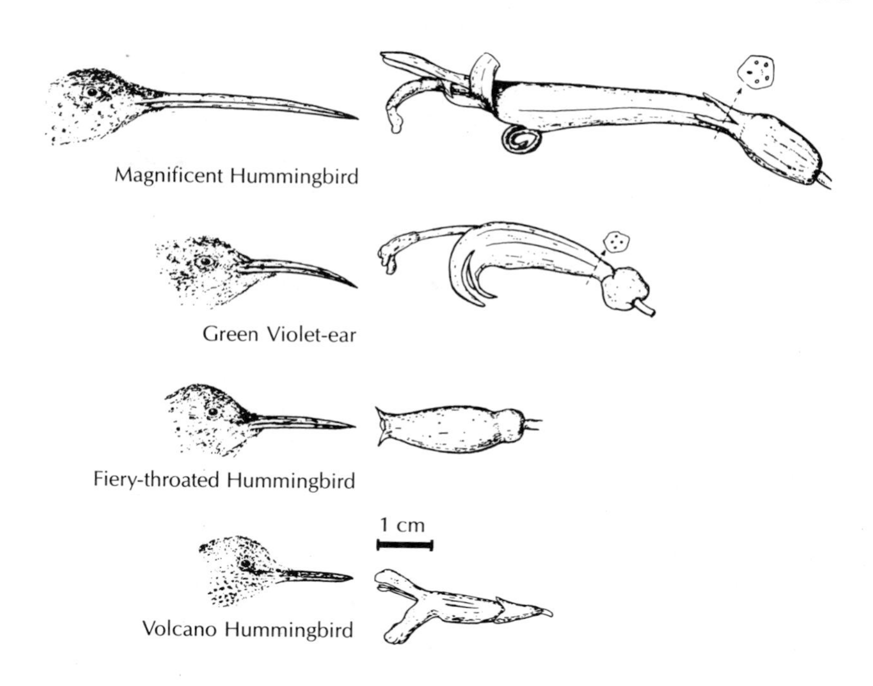

Figure 7-6 The lengths and curvatures of hummingbird bills match those of their preferred flowers. (Adapted from Wolf et al. 1976)

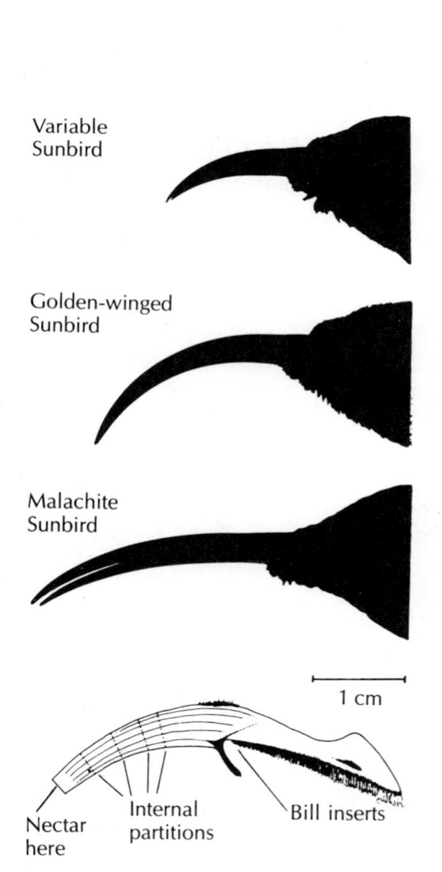

Figure 7-7 These three sunbirds all feed on flowers of a mint plant. The Golden-winged Sunbird can extract nectar faster than either of the other two species because its bill closely matches the length and curvature of the flower and thus enables it to reach nectar at the base of the flower more easily. (Adapted from Gill and Wolf 1978)

winged Sunbird. Malachite Sunbirds much prefer other flowers, such as those of aloes, which their bills fit more easily (Figure 7-7).

The Variable Sunbird also feeds on mint flowers, but because its bill and tongue are too short to reach down the full length of the flower, it must stab the base of the flower to gain access to the nectar chamber. This approach is slow (2.8 seconds per flower) and removes only 62 percent of the available nectar. Its adaptation enables the Golden-winged Sunbird to gather over three times as much nectar per second as can the Variable Sunbird, which prefers other flowers with short corollas.

There are, of course, many types of bill adaptations other than those we have described in detail. Carnivorous landbirds—eagles, hawks, falcons, and owls—have strong, hooked beaks with which they tear flesh and sinew. Among marine predators there is an array of bill adaptations. Penguin bills have slanted projections that direct fish towards the esophagus. Herons and bitterns, both waders, have tonglike bills. The bills of albatrosses, frigate-birds, and pelicans are hooked at the tips; and the anhingas, which are divers, spear prey with their swordlike beaks.

Table 7-1 Effect of Sunbird Bill Dimensions on Nectar Extraction from Mint Flowers

	Golden-winged	Malachite	Variable
Bill length (mm)	30	33	20
Time per flower (s)	1.3	1.8	2.8
Nectar removal (%)	90	82	62
Rate of extraction (μl/s)	3.6	2.4	1.1

(From Gill and Wolf 1978)

The digestive system

The avian digestive system is specialized to process unmasticated food (Ziswiler and Farner 1972; McLelland 1979). The major parts of this system — the oral cavity, esophagus, crop, two-chambered stomach, liver, pancreas, and intestine — are further specialized to accommodate particular types of diets and feeding practices (Figure 7–8).

The oral cavity houses taste buds, pressure receptors, and a tongue that is often specialized. Taste buds in the soft palate aid in food selection. Three major sets of salivary glands and a variety of smaller ones provide lubrication, essential for the passage of food toward the esophagus. The tongue aids in the gathering and swallowing of food. Most bird tongues have rear-directed papillae that aid in swallowing. Extremely sensitive structures, bird tongues are filled with tactile sensory corpuscles, especially at the tip. These corpuscles are best developed in the spoon-tipped tongues of seed-eating songbirds, which manipulate tiny seeds, and in the strong, club-shaped tongues of parrots. Bird tongues usually are not muscular structures but operate by means of the hyoid apparatus. Hummingbirds and woodpeckers, for instance, have long hyoidal extensions of the tongue that curl around and lie on top of the skull. These hyoids permit hummingbirds to extract nectar from long, curved flowers and woodpeckers to reach deep into tree trunks for insects. The woodpecker tongue is also fitted with barbs. Similarly, the tongues of penguins and other fish-eating birds often have

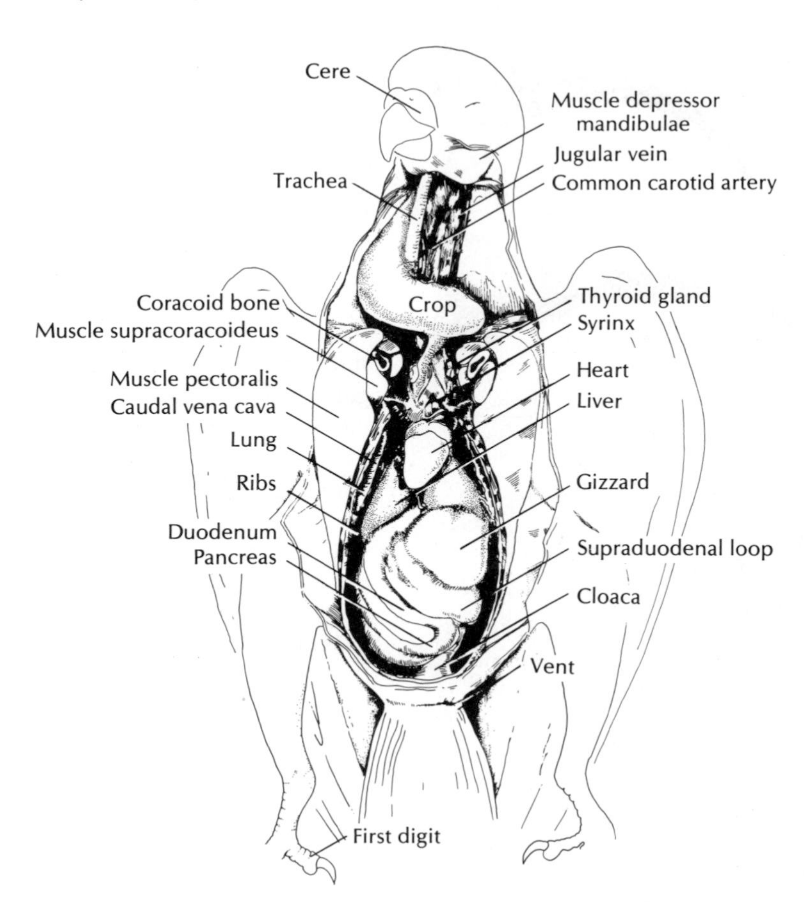

Figure 7–8 The internal anatomy and digestive tract of a Budgerigar. (After Evans 1969)

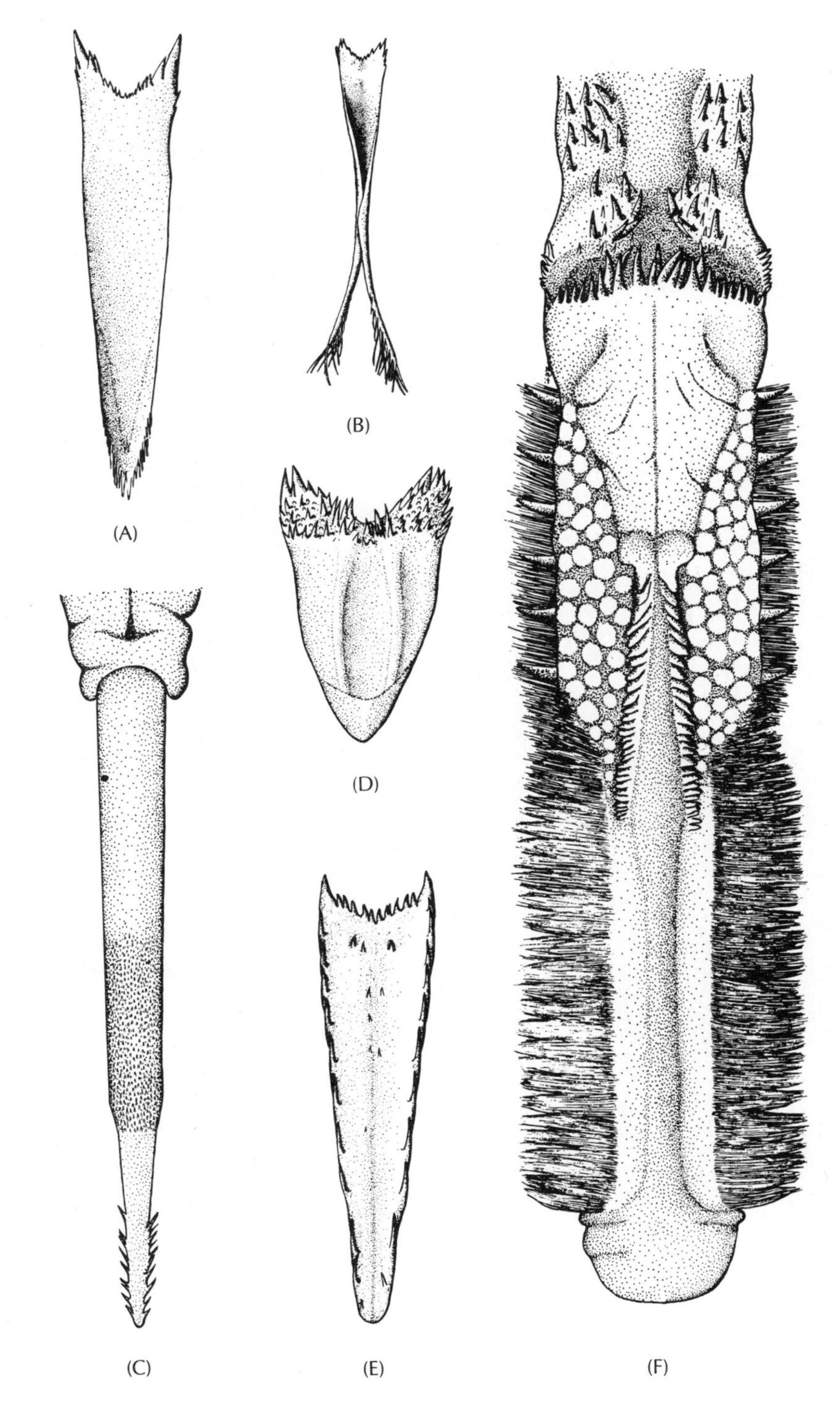

Figure 7–9 A variety of bird tongues: (A) generalized passerine tongue with terminal fringes (American Robin); (B) tubular, fringed nectar-feeding tongue (Bananaquit); (C) probing and spearing woodpecker tongue (White-headed Woodpecker); (D) short, broad tongue of a fruiteater (Diard's Trogon); (E) fish-eating tongue (Sooty Shearwater); and (F) food-straining tongue (Northern Shoveler). (Adapted from Gardner 1925)

rear-directed hooks that help keep slippery fish moving down the throat. Convergent with flamingos, the tongues of some filter-feeding waterfowl have fringes and peripheral grooves that help to strain tiny food particles from the mud. Ducks such as the Northern Shoveler draw mud and water into their mouth and then force it out through the filtering system. (Figure 7–9).

The lack of true teeth in the avian bill usually precludes the elaborate chewing and maceration of bulk that is characteristic of mammals. Instead, most birds swallow food whole without oral shredding or reduction. Exceptions include the plantcutters of South America, which chew vegetation with toothlike serrations on the edges of their bills, and some fruit-eating birds, such as the Eurasian Bullfinch, which crush berries before swallowing them. Most other fruit-eaters gulp fruits whole. The lack of teeth in birds appears to be a weight-reducing adaptation for flight because teeth require a heavy jawbone for support.

Although the primary purpose of the salivary gland is to provide lubrication, some birds produce salivary secretions for other uses as well. The salivary secretions of woodpeckers are sticky, which helps them extract insects from wood crevices and ants from nests. Gray Jays use their sticky salivary secretions to form lumps of food called boluses, which they store for future use (Bock 1961; Dow 1965). Swifts use salivary fluids to glue together and attach their nests to cave walls. Edible-nest Swiftlets of Southeast Asia make their nests almost entirely of hardened saliva, the primary ingredient of the gastronomic delicacy called Bird's Nest Soup (Figure 7–10).

In most birds, food passes from the mouth or pharynx to the stomach via the esophagus, a muscular structure lined with lubricating mucous glands. In birds that swallow large prey whole, fish-eating birds, for example, the esophagus is highly distendible. No mere passageway, the esophagus is a versatile organ. In pigeons, nutritious fluids for young, called pigeon milk, are produced in the esophagus. In pigeons, ostriches, bustards, the Sage Grouse, and the female Greater Painted-Snipe, the esophagus can be inflated for display and sound resonance.

Passage of food through the digestive tract — from the esophagus through the glandular stomach and gizzard into the intestine and finally out the cloaca as feces — varies from less than half an hour in the case of berries ingested by thrushes and the Phainopepla to half a day or more for less easily digested food. Each stage in this passage may include special processing. The

Figure 7–10 An Edible-Nest Swiftlet by its nest, which is made almost entirely of hardened saliva, the main ingredient of Bird's Nest Soup. (Courtesy the Director, Sarawak Museum)

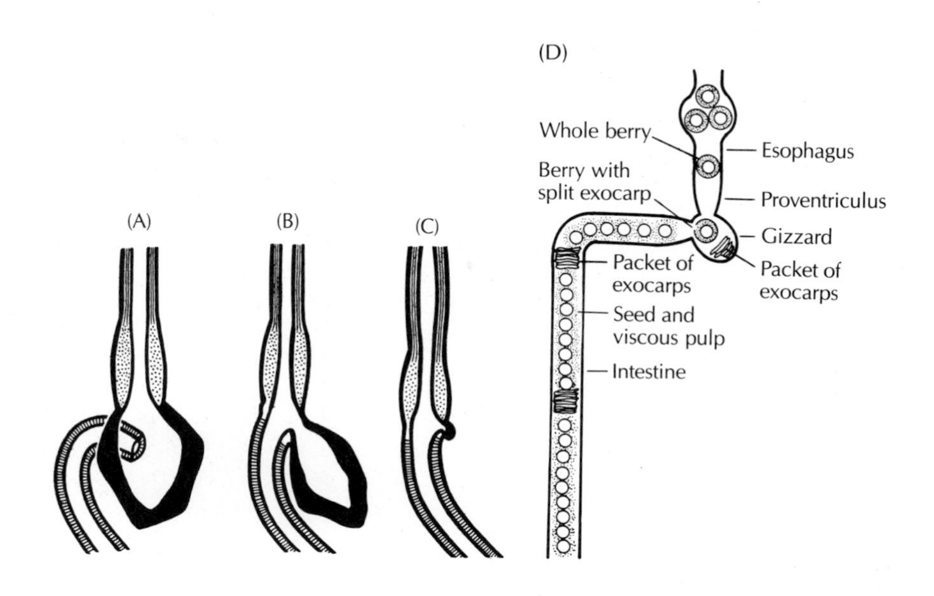

Figure 7–11 Specialized stomachs of fruiteaters: (A) unmodified gizzard of a primitive flowerpecker; (B) more specialized stomach of the Black-sided Flowerpecker that allows fruit to bypass the gizzard and insects to enter the gizzard for grinding; (C) rudimentary gizzard of the Violaceous Euphonia; (D) gizzard of the Phainopepla, which can shuck the outer layer skin (exocarp) from mistletoe berries and then defecate a pack of skins at intervals between the undigested parts of the berries. (Adapted from Desselberger 1931; Walsberg 1975)

sections of the digestive tract are adapted to the specific requirements of a bird's normal diet. Consider, for example, birds that feed principally on mistletoe berries, which include some flowerpeckers in the Old World tropics and euphonia tanagers and the Phainopepla in the New World (Docters van Leeuwen 1954; Walsberg 1975). The stomachs of these flowerpeckers and tanagers act as diverticulae for digesting insects only; the easily digested berries bypass the stomach after being shucked and go straight to the intestine. A Phainopepla's stomach shucks eight to sixteen mistletoe berries in succession, popping the seed and pulp directly into the intestine and retaining a stack of outer layers (exocarps), which then are passed as a group into the intestine and defecated between sets of undigested seed and pulp (Figure 7–11).

Feeding adaptations among nectar-feeding birds include a tubular tongue for nectar extraction, a distensible esophageal pouch (crop) for nectar storage, and juxtaposition of the entrance to the digestive area (proventriculus) and the opening into the intestine (pylorus). This anatomical arrangement allows nectar to bypass the stomach, quite like the arrangement in euphonia tanagers, while diverting insect food into the stomach for longer digestion. The gizzards of nectar-feeding birds are thin-walled structures. In contrast, finch gizzards are hard, muscular structures that pulverize ingested seeds before digestion can begin.

The crop, an expanded esophageal section in many birds, stores and softens food and regulates its flow through the digestive tract. In this respect, the crop serves the same function as the stomach of a monogastric (nonruminant) mammal. The crop varies greatly in shape from a simple expanded section of the esophagus, as in cormorants, ducks, and shorebirds, to a lobed, saclike diverticulus, as in fowl, pigeons, and redpolls (Figure 7–12). In two herbivorous birds, the Hoatzin of South America and the flightless Owl Parrot of New Zealand, the crop has evolved into a glandular muscular stomach, wherein the tough leaves that make up their spartan diets are digested. A concave section of sternal keel accommodates the Hoatzin's

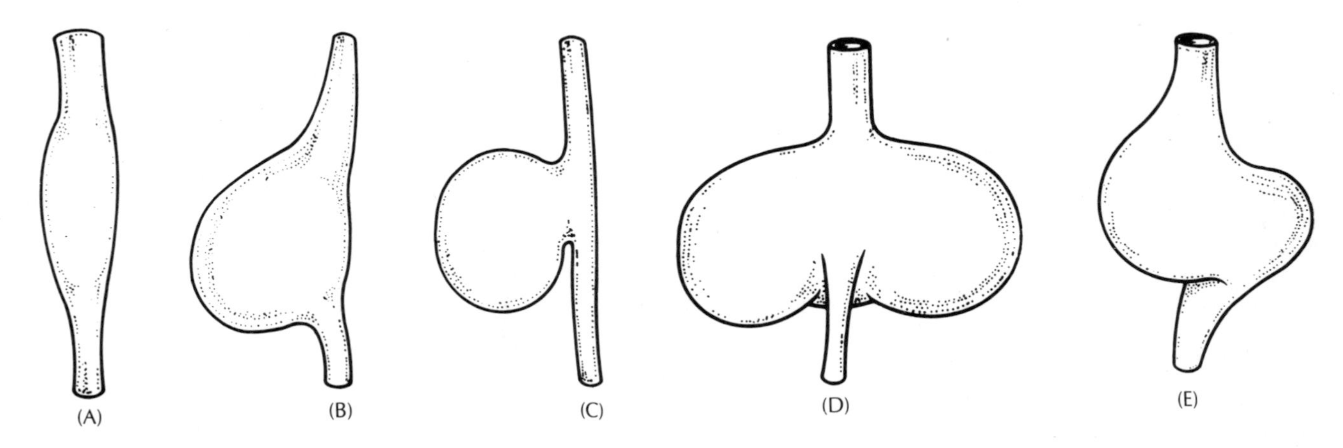

Figure 7–12 Some avian crops: (A) Great Cormorant; (B) Griffon Vulture; (C) Peafowl; (D) Rock Dove; and (E) Budgerigar. (From Pernkopf and Lehner 1937)

well-developed crop-stomach, which is 50 times the weight of its poorly developed true stomach (Böker 1929; Sick 1964).

Birds have a two-chambered stomach composed of an anterior glandular portion, or proventriculus, and a posterior muscular portion, or gizzard (Figure 7–13). The proventriculus, a structure not present in reptiles, is most developed in fish-eating birds and raptors. It secretes acidic gastric juices (pH 0.2 to 1.2) from its glandular walls, creating a favorable chemical environment for digestion. Peptic enzymes in the proventriculus dissolve bones rapidly. The Bearded Vulture can digest a cow vertebra in two days. A shrike can digest a mouse in three hours. Petrels use their well-developed proventriculus for a different purpose: to store oil byproducts of digestion, which they regurgitate as food for their young or spew at predators.

The avian gizzard is the functional analogue of mammalian molars. The gizzard is a large, strong, muscular structure shaped somewhat like a biconvex lens and used primarily for grinding and digesting tough food. The gizzards of grain eaters and seed eaters, such as turkeys, pigeons, and finches, are especially large and have powerful layers of striated muscles. Turkey gizzards can pulverize English walnuts, steel needles, and surgical lancets. Grebes swallow their own feathers, which accumulate in the pyloric region between the gizzard and intestine as a filter for sharp fish bones.

The internal grinding surfaces of the gizzard are covered with a strongly keratinized koilin layer, a rough pleated or folded surface with many grooves and ridges. In some pigeons it has strong, tooth-shaped projections. The gizzard can also contain large quantities of grit and stone that grind food (Meinertzhagen 1964). Quartz particles, the size of which correspond to the coarseness of diet, are a common form of grit. The gizzards of moas, extinct ostrichlike birds of New Zealand, have been found to contain as much as 2.3 kilograms of grit. The gizzard is not so muscular in birds that eat softer foods such as meat, insects, or fruit, and in raptors and herons they may take the form of a large thin-walled sac. In some birds, gizzard structure changes seasonally from large and hard to small and soft in relation to dietary changes. For example, in winter, when seeds are the main food of the Bearded Tit, its gizzard is a large, muscular, keratinized structure containing grit. In summer, when insects are the main food, its gizzard is smaller and less muscular (Spitzer 1972).

The length of the intestinal tract averages 8.6 times the body length but varies from three times the body length in the Common Swift to 20 times body length in the Ostrich. The intestine tends to be short in species that feed on fruit, meat, and insects and long in species that feed on seeds, plants, and fish. The detailed histology and patterns of relief of the absorption surfaces also vary in accord with diet (Ziswiler and Farner 1972; McLelland 1979).

Assimilation of digested food through the intestinal walls depends on the nature of the food ingested (Ricklefs 1974). Hummingbirds assimilate 97 to 99 percent of the energy in nectar, a high-quality food consisting primarily of simple sugars and water (Hainsworth 1974). Raptors assimilate 66 to 88 percent of the energy in meat and fish. Assimilation efficiencies of seed-eating Northern Cardinals and Song Sparrows vary from 49 to 89 percent depending on the types of seeds ingested. Herbivores assimilate as much as 60 to 70 percent of the energy in young plants but only 30 to 40 percent of the energy in mature foliage. Spruce Grouse assimilate only 30 percent of the energy in spruce leaves (Pendergast and Boag 1971).

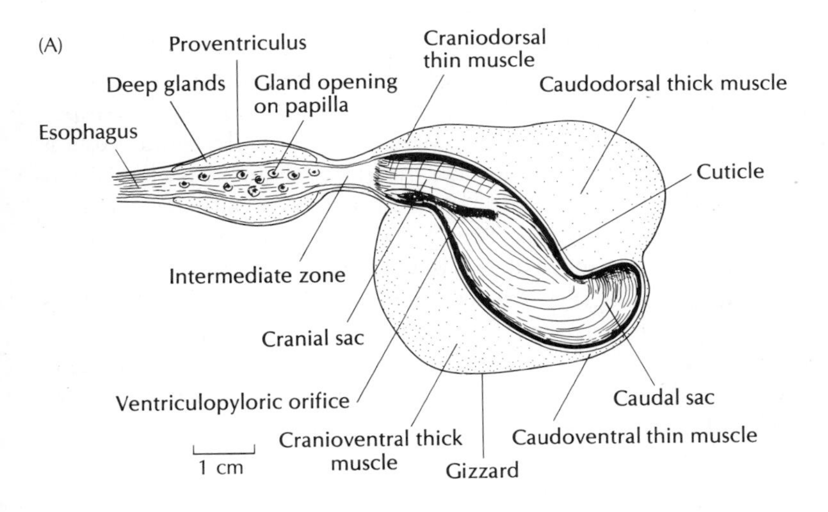

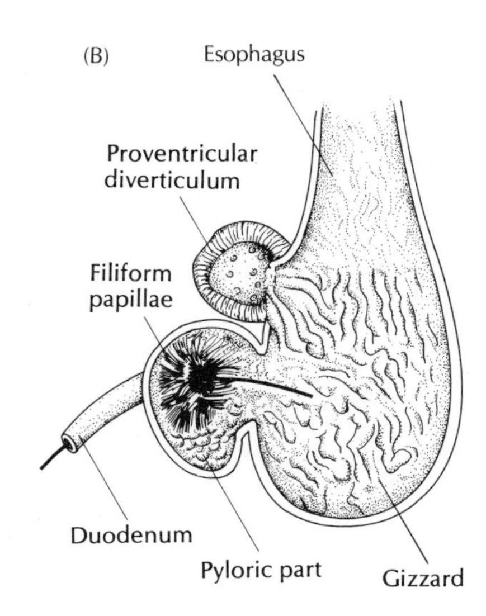

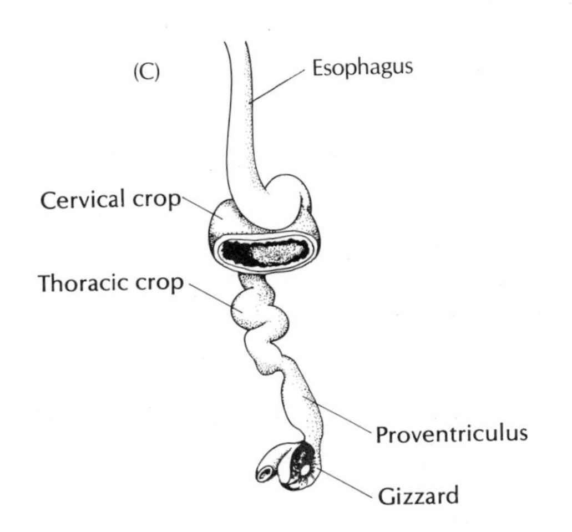

Figure 7–13 Three avian stomachs: (A) Domestic Chicken; (B) American Anhinga; and (C) Hoatzin, 66 percent of natural size. (After McLelland 1975; Garrod 1876; Pernkopf and Lehner 1937)

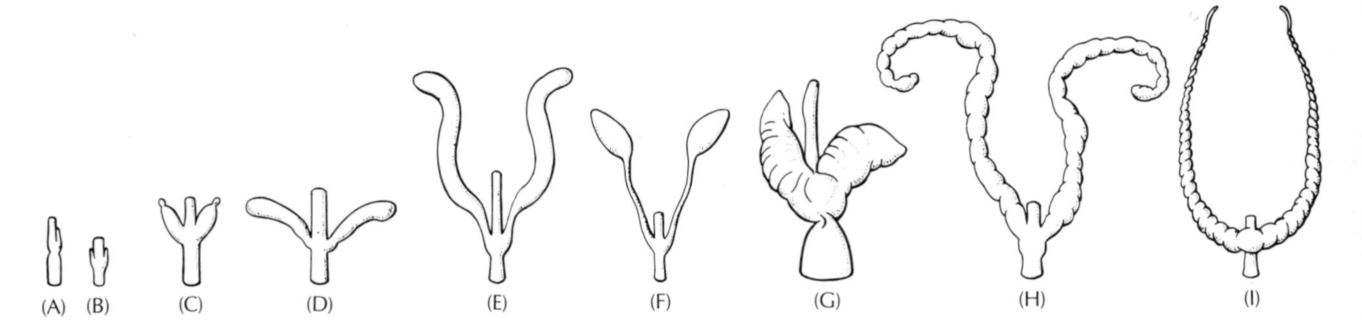

Figure 7 – 14 The variety of intestinal cecae: (A) Purple Heron; (B) Eurasian Sparrow Hawk; (C) Marabou Stork; (D) a rail; (E) Helmeted Guineafowl; (F) Barn Owl; (G) Northern Screamer; (H) Great Bustard; and (I) Ostrich. (From McLelland 1979)

Well-developed sacs, called cecae, aid digestion of plant foods. The cecae are attached to the posterior end of the large intestine of some birds. They are most prominent in fowl and ostriches, in which they functionally resemble the rumen of cattle. The precise role of cecae in digestion remains unclear, but it appears that bacteria in the cecae further digest and ferment partially digested foods into usable biochemical compounds that are absorbed through the cecal walls (see Gasaway 1976a, b). Cecae may also function to separate the nutrient-rich fluid in partially digested food from the fibrous portion, which is eventually eliminated (Fenna and Boag 1974). Cecae are poorly developed or nonexistent in most arboreal birds, perhaps because of the weight of watery digesta and the large structures required to handle them (King, personal communication). Indeed, well-developed cecal fermentation is restricted to cursorial and weakly flying birds and is much more common among mammals than birds (Figure 7 – 14).

Intestinal bacteria may also aid digestion of wax in the guts of honeyguides. These birds have long been famous for their ability to eat wax, usually from the honeycombs of bees but occasionally from candles on the altars of African missionaries (Friedmann and Kern 1956). Recent studies have shown that at least two seabirds also use the wax they ingest (Roby 1986). Least Auklets and Common Diving-Petrels metabolize wax found in the marine crustacea they usually eat. Wax may be a major source of energy for both adults and chicks of these seabirds.

Feeding behavior and energy balance

The feeding behavior of birds is influenced not only by their anatomical equipment but also by the variable availability of food. Prey may be scarce, cryptic, skilled at avoiding capture, or distasteful. Some birds use tools in feeding (Boswall 1977, 1983), and some, when faced with dramatic changes in resources, have developed innovative habits. The Woodpecker Finch of the Galapagos pries grubs from crevices with a stick or a cactus spine held in its bill (Millikan and Bowman 1967) (Figure 7 – 15). The nuthatchlike Orange-winged Sittella of Australia pries out grubs with sticks (Green 1972). Egyptian Vultures crack ostrich eggs with stones (Lawick-Goodall 1968). Green-backed Herons sometimes use pieces of bread as fishing bait (Lovell 1958). For example, a Green-backed Heron was seen dropping bait into pools of a stream and waiting for fish to gather. When currents began to carry off the bait, the heron retrieved it and used it again.

Figure 7–15 The Woodpecker Finch of the Galapagos uses twigs to pry insect larvae out of small holes. (Courtesy I. Eibl-Eibesfeldt)

Thirty years ago, a few Great Tits in the British Isles learned to rip open milk bottle caps to drink the cream (Fisher and Hinde 1949). Apparently this was a novel application of normal bark-tearing behavior (Morse 1980). The skill passed rapidly to other members of the Great Tit population, forcing milk companies to replace the cardboard caps with sturdier aluminum ones. The tits learned to open these too. Darwin's finches of the Galapagos Islands also are renowned for their novel feeding efforts (Bowman and Billeb 1965): The Sharp-beaked Ground-Finch on Tower Island pecks the pin feathers of nestling boobies and then drinks their nutritious blood. To uncover food, the Small Ground-Finch and Large Cactus-Finch have learned to push aside sizeable stones with their feet by first bracing their heads against a large rock for leverage; Paul DeBenedictis (1966) observed a 27-gram finch move a 378-gram stone in this manner.

In general, birds prefer familiar foods. Wood Pigeons, for example, search for familiar types of grain, and titmice choose familiar caterpillars (Murton 1971a; Boer 1971). The preference for familiar foods lessens the number of unpleasant surprises: distasteful, poisonous, or otherwise dangerous prey. Familiar food also can be found more readily than unfamiliar food if the foraging bird uses a specific "search image," as we do when we look for a friend in a crowd or for a jigsaw puzzle piece with a particular shape (Dawkins 1971). For example, when captive Blue Jays are shown color slides of tree trunks with and without cryptic moths, they learn to search quickly for a particular kind of moth and to peck a key ten times when they spot one (Pietrewicz and Kamil 1979). A Blue Jay's skill at spotting moths increases rapidly with experience, but only if it is shown a single kind of moth. If different kinds of moths are presented in successive slides, the Blue Jays cannot search effectively for the familiar kind.

Food supplies vary with site and time, and birds respond to this variation in several ways. Area-restricted search is one such response (Tinbergen et al. 1967). When food sites are concentrated, a bird improves its success by staying in or near sites of high food density and by moving rapidly

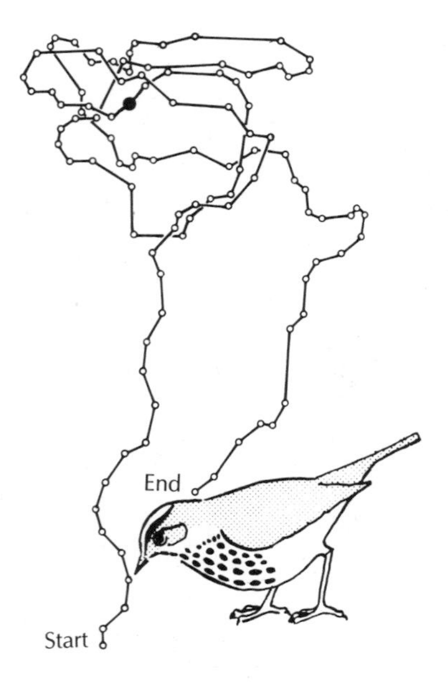

Figure 7–16 When prey distribution is locally dense rather than uniform or random, captive Ovenbirds search primarily near a site where they have been successful (solid circle). (From Zach and Falls 1977)

past sites of low food density. By turning sharply and moving short distances after food is found, the bird concentrates its search near concentrations of food. Eurasian Blackbirds, for example, adopt area-restricted searching when the low-density distribution of artificial prey (pastry caterpillars) is changed from uniform to random or clumped (Smith 1974). Caged Ovenbirds capture housefly pupae by moving shorter distances and turning more sharply, especially when the pupae are clustered (Zach and Falls 1977). Experienced Ovenbirds apply area-restricted searching to known productive sites even before they find prey (Figure 7–16).

Theoretically, birds might respond to varying food supplies by moving to a new feeding area as soon as the rate of food encounters declines to less than the average throughout a given habitat (Charnov 1976). If food in a habitat were scarce, birds would have to search longer. Consistent with theory were the results of experiments with captive Black-capped Chickadees that hunted for mealworms hidden in pine cones. How quickly they found the next mealworm determined when they stopped searching one group of cones and moved to the next group. The chickadees gave up after 11 seconds of unsuccessful searching in a rich habitat containing 1.9 mealworms per group of pine cones, but only after 16 seconds in a poor habitat containing 0.8 mealworms per group of cones (Krebs et al. 1977). Such behavior requires some knowledge of the quality of the habitat.

Another way that birds improve foraging success with varying food supplies is by avoiding places where they have recently harvested and concentrating on unvisited sites. Excellent spatial memory enables some birds to become refined in the harvesting of renewable food sources such as nectar. Territorial nectar-feeding birds, such as Golden-winged Sunbirds of Kenya and the unrelated Common Amakihi of Hawaii, concentrate their feeding efforts on unvisited flowers, which contain more than average nectar volumes (Gill and Wolf 1977; Kamil 1978). Nonrandom foraging increases the sunbird's intake per flower by 25 percent and the amakihi's by 15 percent.

Feeding success affects the exact location on the forest floor where a bird hunts for food or the choice of which flower it probes for nectar. Combined with morphological predisposition and experience, the choices of profitable feeding sites can lead to routine use of feeding stations that differ according to species. Thus, small species of titmice tend to feed at higher branches and on the outermost twiglets in British woods, whereas large species of titmice concentrate their efforts on the ground and on branches with more support (see Chapter 23). Using advanced statistical methods, Richard Holmes and his associates (1979) were able to define four groups of species in a New Hampshire forest:

1. ground feeders such as thrushes, Ovenbirds, and Dark-eyed Juncos;
2. tree-trunk feeders such as woodpeckers and nuthatches;
3. canopy feeders that tend to search for food in conifers and to glean insects and from twigs, such as Blackburnian Warblers and Black-capped Chickadees; and
4. general canopy feeders, such as warblers, Scarlet Tanagers, some vireos, and one species of flycatcher.

Increasing energy profit

Current research in the foraging ecology of birds focuses on energy profits rather than on essential nutrition. Little is known about the degree to which the diets and foraging behavior of wild birds are directed towards nutrition, though Willow Ptarmigan are known to prefer particular heather leaves, which are rich in the limiting nutrients nitrogen and phosphorus (Moss 1972; Moss et al. 1972). It is usually assumed that, in the course of their daily foraging to meet their energy needs, birds obtain adequate nutrition (Morse 1980).

Birds are highly sensitive to the net energetic profits of their feeding efforts. We can define the energetic profit of foraging as the rate of net energy gain, or

$$\text{profit} = \frac{\text{energy gain} - \text{energy cost}}{\text{foraging time}}$$

where energy gain is the assimilated caloric value of a bird's food; energy cost is the caloric value of energy expended in finding, capturing, and eating the food; and foraging time includes the time required to locate food and to consume it.

Birds tend to choose food of higher energetic profit. For example, White Wagtails prefer medium-sized flies, even though large flies with greater energy content are more common (Davies 1977). Medium-sized flies yield comparatively more energy per second of foraging time because large flies take too long to subdue and swallow relative to their higher energy content (Figure 7–17). On the other hand, Northwestern Crows living on Mandarte Island in British Columbia, Canada, selectively choose large whelks (Zach 1979). On the basis of size only, the crows select only whelks that supply about 8.5 kilojoules of energy, a yield that exceeds the

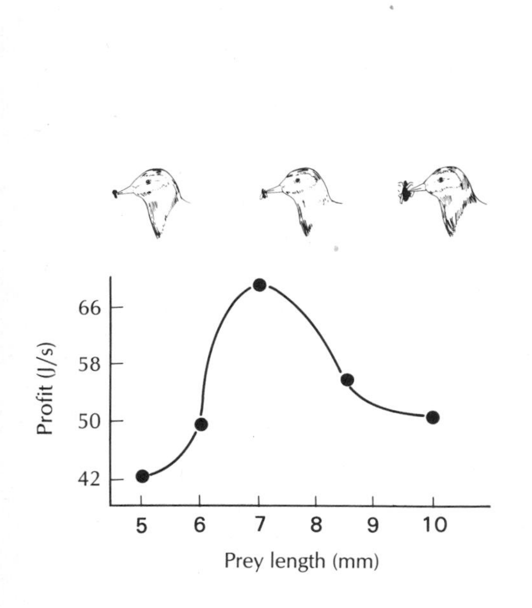

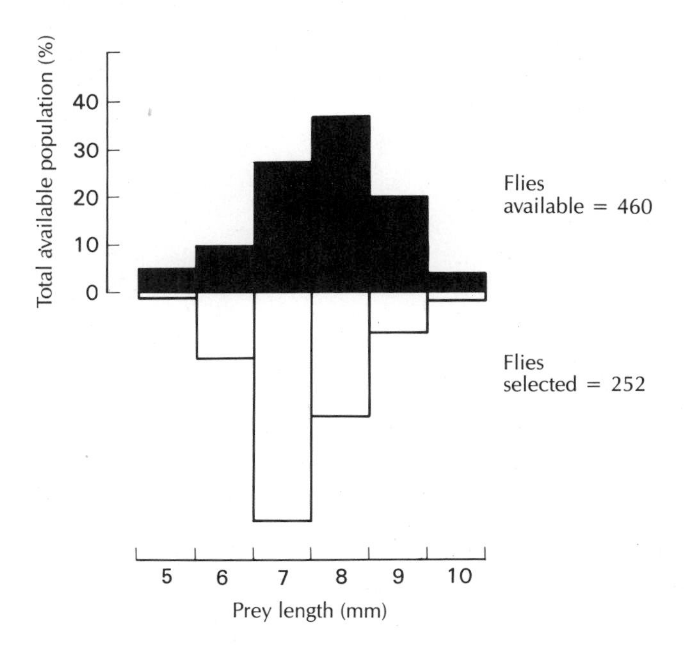

Figure 7–17 White Wagtails select medium-sized flies (seven millimeters) that yield the most energy per second of handling time. (Adapted from Krebs 1978)

foraging costs of 2.3 kilojoules per whelk. The high cost of extracting these animals from their shells is due to the effort required to drop them repeatedly from the air onto rocks until they break; 20 drops per whelk is not unusual.

When a hungry bird encounters an assortment of prey that differ slightly in energy yield, it must decide whether to pause long enough to eat low-yield prey or to continue seeking better prey. Captive Great Tits, allowed to select large and small mealworm pieces from a continuous supply on a little conveyer belt, ate large and small pieces when both were scarce but selected only large pieces when they were so common that the investment of time in the small pieces would lower their foraging efficiency (Krebs et al. 1977).

If a bird flies to distant feeding areas, commuting costs affect its foraging performance (Orians and Pearson 1979). The costs of flying far from a nest or a favorite perch to flowers in a field or to a nectar feeder are important energy considerations affecting the foraging habits of hummingbirds and sunbirds (Wolf 1975). Rather than fill their crop with nectar, hummingbirds stop feeding when the increasing flight costs imposed by the added weight of more nectar would reduce their net gain (DeBenedictis et al. 1978). Most birds face similar costs when they commute between nests and food sources to feed nestlings. Northern Wheatears, for example, compensate for increased travel to distant food cups containing maggots by spending more time at the cup to load up additional items for the return trip (Carlson and Moreno 1982) (Figure 7–18).

Birds do not always follow theoretical foraging patterns or fit them simply. A perplexing example of profit-sensitive foraging is that of Common Redshanks, which catch polychaete worms protruding from burrows in the mud (Goss-Custard 1977a, b). Large polychaetes yield more energy per unit of handling time than small polychaetes. When large polychaetes are abundant, the redshanks, as expected, eat them and ignore the small ones. On mudflats with fewer large polychaetes, the redshanks eat small ones. A computer model of redshank foraging behavior revealed that this behavior nets them the greatest possible rate of energy gain in both circumstances. But when prey other than the polychaete worms (namely, small

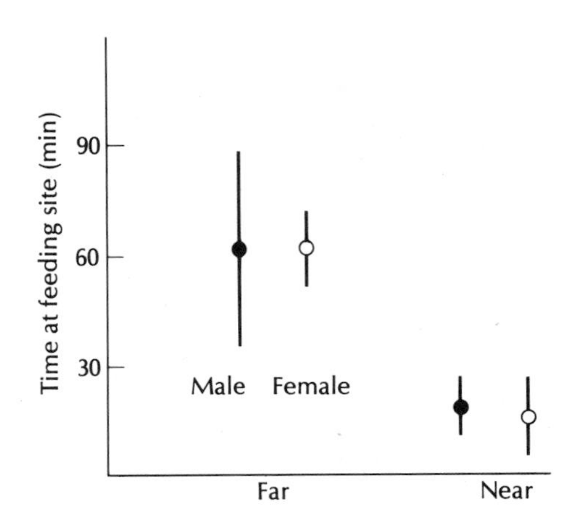

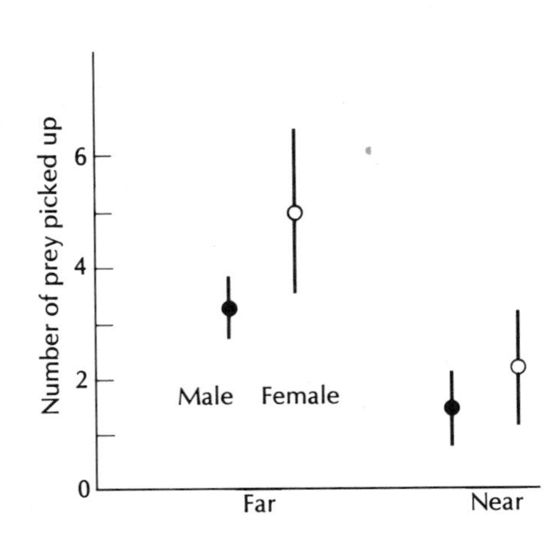

Figure 7–18 Northern Wheatears spend more time and pick up more prey (maggots) at feeding sites far from the nest than at feeding sites near to the nest. The bars are equal to standard deviations of the mean. (Drawn from data in Carlson and Moreno 1982)

Corophium amphipods) are included in the analysis, the redshanks prefer amphipods to worms, even though amphipods yield less energy per unit of time. Perhaps energy profit is the wrong currency in this case, and amphipods are more nutritious in other ways than are worms (Krebs 1978). A similar result is obtained in studies of seed preferences by finches with different-sized bills. Recall from our earlier discussion of these studies (p. 132), that seed husking times correlate with bill dimensions and seed size or hardness. The finches in those studies preferred seeds they could husk quickly, regardless of energy yield, rather than seeds that yield the highest rate of energy intake (Willson 1971). Most likely, prolonged exposure to predators is of paramount concern to such birds, which eat quickly lest they be eaten.

Another departure from the theoretically preferred, maximum-efficiency foraging effort is the tendency shown by birds to visit low-yield sites occasionally or to eat inferior food, despite obviously better options. Some degree of error may be inevitable in food selection. Possibly, birds select inferior food as a way of monitoring future foraging possibilities in case there is a change in food availability. Great Tits, for example, continue to explore low-yield patches while concentrating their search for mealworms in the site with the greatest prey density (Smith and Sweatman 1974). When the density of food of the highest-yielding patch drops, however, the tits switch to the best available alternative patch.

The foraging behavior of birds is not adapted only to increase short-term energy profits, which may be tempered by longer-term risks of inadequate intake. In experiments on "risk-sensitive" foraging, birds faced a choice of two food sources: one that always provided the same small number of seeds and another that sometimes produced abundant seeds and at other times produced nothing. Both options averaged the same number of seeds per visit, but the variable site potentially yielded more food faster. Yellow-eyed Juncos and White-crowned Sparrows preferred the reliable sources of food when they were close to meeting their daily energy requirements but risked starvation for greater rates of intake when they were hungrier (Caraco 1982; Caraco et al. 1980a).

Energy balance and reserves

Whether hungry or temporarily sated, all birds face the challenge of maintaining their energy balance. Energy balance is a dynamic relationship between energy intake and energy expenditure. Ideally, intake and expenditure are roughly equal so that the bird neither gains nor loses much weight. Preceding migration or winter, a bird may eat more than it metabolizes each day so that the excess can be stored as fat, which provides reserves needed to compensate for periods of inadequate intake.

The amount of time a bird must feed each day depends on its total energy requirements and achieved rate of energy intake. As requirements increase, so must foraging times or, alternatively, the rates of net energy gain. Roughly speaking, a bird's foraging time must double when its rate of net energy gain is reduced by half. If a short foraging time is sufficient for self-maintenance, individuals can build up energy reserves or undertake energy-expensive activities such as migration, molting, and breeding (see Chapter 12). Low foraging time also allows birds more time to remain

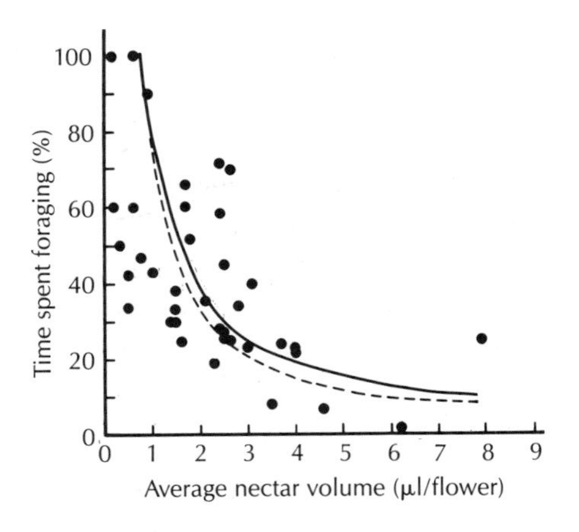

Figure 7–19 The amount of time a Golden-winged Sunbird feeds depends on the average amount of nectar it gets from a flower. The solid line is the predicted relationship, assuming that the sunbird visits only as many flowers as it needs to replace total daily expenditures. The dashed line is fit to the actual field measurements of foraging efforts. (Adapted from Gill and Wolf 1979)

hidden from predators, select favorable microclimates, establish dominance and property rights over other individuals, court mates, and rear young.

Birds vary their foraging time and effort in relation to their energy requirements and foraging success. Sunbird foraging efforts, yielding mainly short-term energy profits, for example, decline with an increase in floral nectar content (Wolf 1975; Gill and Wolf 1979) (Figure 7–19). The foraging times of Rufous-sided Towhees, breeding in oak–hickory forests where food is plentiful, are 28 to 40 percent less than those of towhees breeding in pine–oak forests where food is scarce (Greenlaw 1969). Foraging times also vary with seasonal changes in food availability. Anna's Hummingbirds forage 14 percent of the day in the nonbreeding season but only 8 percent of the day while breeding in the vicinity of nectar-rich flowers. Small titmice and goldcrests in England feed 90 percent of the day in winter when food is scarce, their metabolism is high, and days are short (Gibb 1960). They must find an insect every two seconds, on average, just to survive. Many fail to meet this requirement. Goldcrests in arctic Finland and Norway sometimes suffer high mortality rates in harsh winters because they are unable to balance their energy budgets (Österlöf 1966; Hogstad 1967).

Most birds maintain minimal fat reserves, probably because the survival benefits of large energy reserves do not offset the energy costs of carrying excess weight. Small temperate-zone passerines typically have fat reserves of no more than 10 percent of their body mass to cover their fasting needs during midwinter (King 1972a). Yellow-vented Bulbuls in Singapore maintain fat reserves throughout the year of only 5 percent of body weight, little more than is needed to survive overnight and to begin feeding the next morning (Ward 1969). In general, large birds can store more fat and can fast longer than smaller birds. At moderately low temperatures (−1° to −9°C), a 10-gram warbler, for example, may not survive a day without food, whereas a 200-gram American Kestrel can survive for five days (Kendeigh 1949; Calder 1974). Emperor Penguins fast for 60 frigid days during their incubation vigils of the Antarctic winter. Migrating raptors can fast for 30 days; their nearly effortless soaring requires little fuel (Smith 1980). Nestling Common Swifts can survive 10 days of fasting

Figure 7–20 Clark's Nutcracker, a bird with a remarkable memory. (Courtesy A. Cruickshank/VIREO)

during torpor but lose about half of their mass (Koskimies 1950). Adult swifts can survive for 4.5 days of fasting with a 38 percent weight loss.

Fasting birds draw first on their glycogen deposits. Only the lipid reserves of cardiac muscle are exempt from normal use. As a last resort, body tissues can be metabolized after fat deposits are exhausted. Pectoral muscles begin to atrophy during periods of food stress; even gonadal tissues may be sacrificed. Metabolism of these tissues causes free amino acid levels in the blood to increase rapidly. Blood levels of the amino acid lysine can increase 500 percent during a 36-hour fast (Fisher 1972).

Food caches

Hoarding food for future use is one way of preparing for food shortages. Acorn Woodpeckers, for example, build large granaries of acorns for the winter. Crows, jays, and nutcrackers are especially diligent hoarders (Balda 1980). A Clark's Nutcracker hides an average of two pine seeds in each of 1400 to 2000 caches in order to survive the winter and early spring (Figure 7–20). Titmice also depend on autumn caches for winter feeding; Crested Tits obtain up to 60 percent of their winter food from provisions built up earlier in the year. Titmice at high latitudes store more food than titmice in the south, where winter food stores last longer (Gibb 1954, 1960; Haftorn 1956; Morse 1980). Small titmice cache food more frequently than do larger species (Gibb 1954). The lower fat reserves and lesser fasting capabilities of smaller species apparently increase their need for cache reserves. Even meat-eaters such as raptors and shrikes routinely set aside a fraction of their prey for future use. Shrikes are notorious for impaling their prey on thorns for later consumption (Figure 7–21).

Figure 7–21 A Common Yellowthroat impaled on barbwire by a Loggerhead Shrike. (Courtesy S. Grimes)

Summary

Because they have such high energy demands, birds have a constant need for food. Specialized adaptations of their locomotory morphology, bill structures, digestive systems, and foraging behavior highlight the evolutionary responses of birds to this urgent need.

The avian bill consists of bony extensions of the jawbones, which are covered by a horny sheath, the rhampotheca. Modern birds lack teeth. The internal structure of the bill permits substantial bending and directs stress away from weak points. Subtle differences in bill structures affect the abilities of finches to husk seeds and hummingbirds to extract nectar from flowers.

The avian digestive system is specialized to process unmasticated food. Salivary glands lubricate food before it is swallowed. The gizzard pulverizes ingested food for digestion. The gizzard is particularly well developed in birds that eat hard or tough foods. Cecae, sacs attached to the posterior end of the intestine, primarily in terrestrial birds, also aid the digestion of foods.

Whereas the study of the functional feeding morphology of birds has a long history, the study of their foraging behavior and, particularly, their sensitivity to the costs and gains of alternative prey or feeding sites is a recent but popular topic. In subtle ways, birds bias foraging time to favor the most profitable feeding sites, select prey of the size that enhances their energetic profit, and innovate unusual foraging techniques, sometimes using tools. When they are very hungry, birds are more willing to risk starvation by spending time searching for an area with a high concentration of food. The amount of daily foraging time reflects a bird's energy needs. Birds usually maintain small to moderate fat reserves and so must balance their expenditures with new intakes on a daily basis. When food is not available, birds draw first on fat and glycogen reserves and then on other body tissues. Some birds, particularly nutcrackers, their relatives, and titmice, hoard food in caches for use during future periods of food stress.

Further readings

Krebs, J.R., and N.B. Davies. 1984. Behavioral Ecology. 2nd ed. Sunderland, Mass.: Sinauer. *Includes important reviews of research on the foraging behavior of birds.*

McLelland, J. 1979. "Digestive system." *In* Form and Function in Birds (A.S. King and J. McLelland, eds.), pp. 69–181. New York: Academic Press. *An excellent summary of literature on the anatomy and function of the digestive tract.*

Morse, D.H. 1980. Behavioral Mechanisms in Ecology. Cambridge: Harvard University Press. *An excellent summary of the literature on foraging behavior of birds.*

Storer, R.W. 1971. Adaptive radiation in birds. Avian Biology 1:150–188. *An elegant summary of the feeding adaptations of birds.*

Ziswiler, V., and D.S. Farner. 1972. Digestion and the digestive system. Avian Biology 2:343–431. *A detailed review of the digestive tract.*

part III

BEHAVIOR

chapter 8

The brain and sensory apparatus

Until recently, ornithologists assumed that birds perceive the world in much the same way as people do. The sensory experience of birds, however, extends far beyond human experience. Avian color vision is highly developed, reaching into the near ultraviolet range of the spectrum. The hearing range of birds encompasses infrasounds, and some owls can track prey in pitch darkness by their hearing alone. Birds navigate by means of patterns of the earth's magnetism and orient themselves in flight "automatically" owing to their extreme sensitivity to miniscule shifts in gravity and barometric pressure. In addition to new information about the sensory abilities of birds, recent research indicates that birds have well-developed brains and are quite intelligent.

Avian intelligence

When it comes to mastering complex problems in the laboratory, birds do very well, outperforming many mammals in advanced learning experiments (Bitterman 1965; Stettner and Matyniak 1968). Birds quickly learn to recognize the odd object, not only in a set of familiar objects but also in

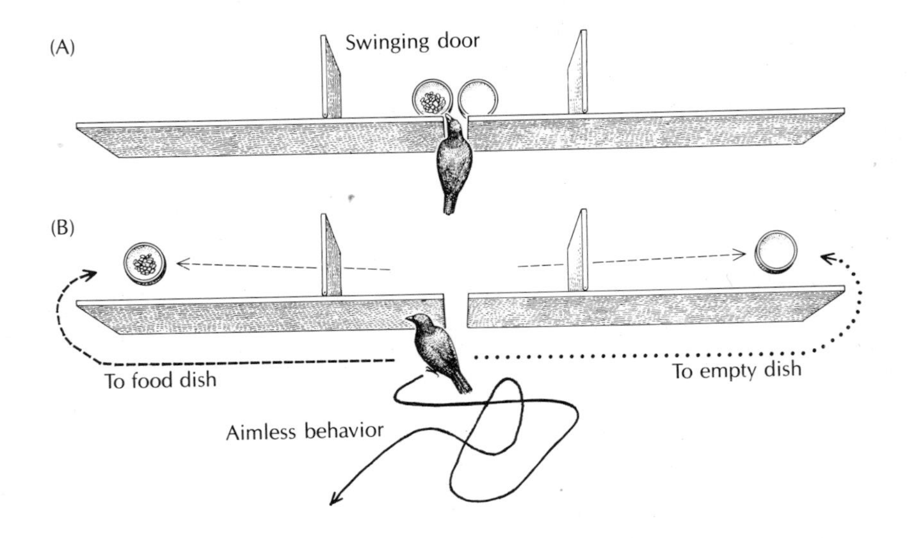

Figure 8-1 Crows and dogs perform best in the Krushinsky problem experiment, in which food dishes, (A) viewed by the subject through a slit in a wall, move out of sight behind swinging doors. (B) The subject must then choose to proceed left or right to find the food dish. (From Stettner and Matyniak 1968, with permission of Scientific American)

sets of unfamiliar objects; monkeys master this with difficulty. Corvids, such as crows and magpies, do especially well in laboratory experiments that test higher faculties. In one of these, the "Krushinsky problem," the bird looks through a slit in a wall at two food dishes, one empty and one full, that move out of sight in opposite directions. The bird must then decide which way to go around the intervening wall to get to the dish that contains food. Cats, rabbits, chickens, and pigeons do poorly in this test, but crows and dogs solve the problem immediately (Figure 8-1).

For most mammals, learning to count is a formidable problem. A monkey requires a training ordeal of 21,000 trials to learn to distinguish between two and three different tones. Rats cannot learn to discriminate between one and two tones. Birds, however, easily master complex counting problems. Ravens and parakeets can learn to count to seven; in one set of experiments they learned to choose the box containing food by counting the number of small objects in front of it (Koehler 1950).

Laboratory intelligence tests are of uncertain relevance to the problems that birds face in the wild. However, natural evidence of avian intelligence can be seen in the ability of nutcrackers to locate seed caches months after hiding them (see Chapter 7). In a series of experiments with Clark's Nutcrackers, Stephen Vander Wall moved the large objects that surrounded hidden caches, and the nutcrackers adjusted their search in the direction of the moved objects. He concluded that the nutcrackers memorize the locations of thousands of caches primarily by making spatial references to surrounding large objects and secondarily by noting cues such as soil disturbance (Vander Wall 1982). The ability to recall the precise locations of about 2000 caches for spans of as long as eight months demonstrates a phenomenal spatial memory.

The avian brain

"Birdbrain" is a familiar but inappropriate slur. Birds actually have large, well-developed brains, six to eleven times larger than those of like-sized reptiles and similar in size to those of small mammals. The brains of birds and mammals other than higher primates account for 2 to 9 percent of their total body mass.

The main divisions of the avian brain are the following:

1. the forebrain, which is responsible for complex behavioral instincts and instructions, sensory integration, and learned intelligence;

2. the midbrain, which regulates vision, muscular coordination and balance, physiological controls, and secretion of neurohormones that control seasonal reproduction; and

3. the hindbrain, or medulla, which links the spinal cord and peripheral nervous system to the major control centers of the brain.

Cranial nerves, except those controlling vision and smell, enter the brain through the medulla. The forebrain and midbrain in both birds and mammals are conspicuously more highly developed than those of reptiles. Otherwise, avian and mammalian brains are quite different (Figure 8–2).

A cross section of the avian forebrain reveals three major layers: the cerebral cortex on the dorsal surface, the hyperstriatum below it, and the neostriatum in the deep interior. Together, the hyperstriatum and the neostriatum comprise the corpus striatum, which accounts for the bulk of the forebrain. Small, deeper sections include the archistriatum and the paleostriatum. Parts of the corpus striatum control eating, eye movements, locomotion, and the complex behavioral instincts central to reproduction: cop-

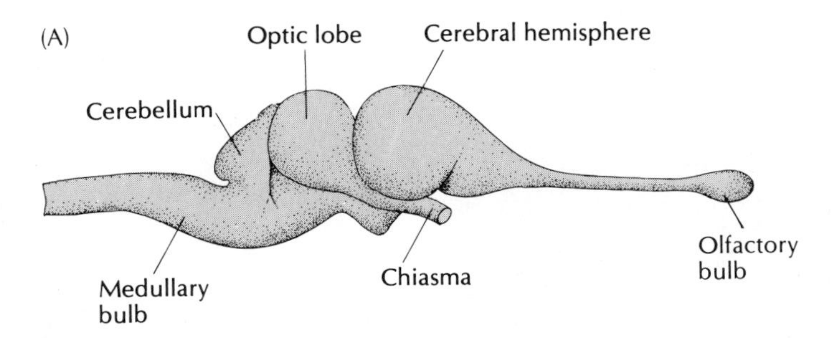

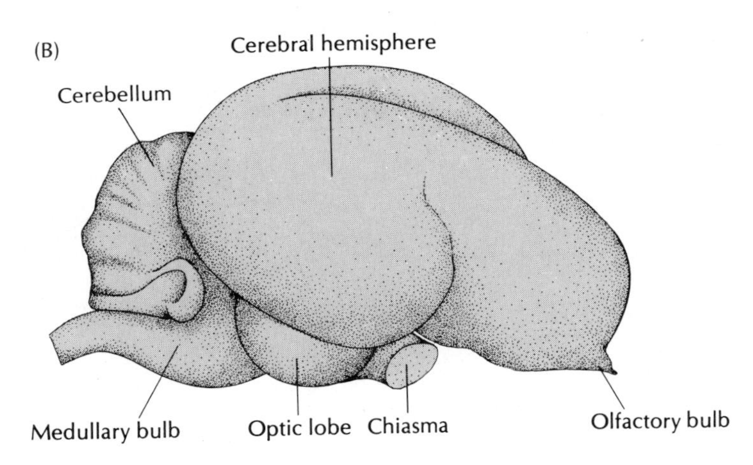

Figure 8–2 Brains of (A) a monitor lizard and (B) a macaw drawn to the same scale. Note the well-developed cerebral hemisphere and cerebellum in the avian brain. (After Portmann and Stingeln 1961)

ulation, nest building, incubation, and care of young. In general, the left cerebral hemisphere controls complex integration and learning processes and houses suppression mechanisms for sexual and attack behavior. The right cerebral hemisphere monitors the environment and selects novel stimuli for further processing, which may entail memorization by the left side.

Intelligence in birds has evolved quite differently than it has in mammals. In mammals the cerebral cortex is the principal feature of the forebrain; it overgrows the small corpus striatum and reaches its largest and most deeply fissured state in higher primates such as chimpanzees and humans. The thin cortex of the avian brain, on the other hand, has little to do with intelligence. Removal of the cortex has scarcely any effect on a bird's performance. A pigeon whose cortex has been removed still performs

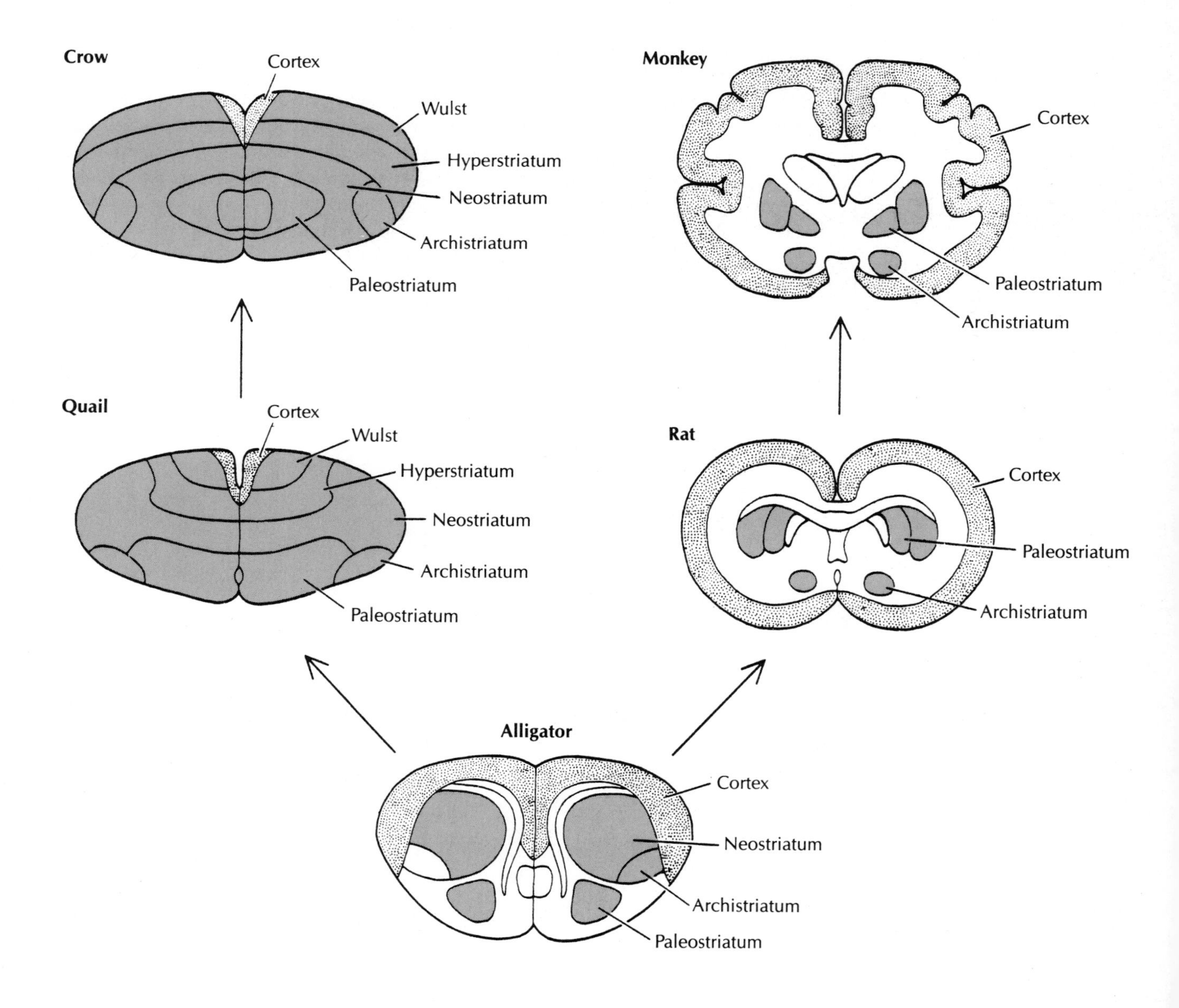

Figure 8–3 The brains of birds and mammals evolved in different directions after diverging from those of their reptilian ancestors (represented here by an alligator). The mammalian brain has been restructured by the enlargement of the cortex, which serves as the seat of intelligence. In birds, an alternative and unique feature, the hyperstriatum and associated Wulst, has become the seat of intelligence. The hyperstriatum is best developed in intelligent birds such as crows. (From Stettner and Matyniak 1968, with permission of Scientific American)

Figure 8–4 Electrical stimulation of the preoptic area of the brain causes a Ringed Turtle-Dove to adopt a series of display postures known as the bowing action. These postures include (from left to right): erecting the head and body, and standing at attention, ruffling the feathers, movements in the crop, and walking; walking in circles, bows, cooing calls, and lowering the tail; and standing upright with repeated turns of the head. (From Åkerman 1966a,b)

normally in visual discrimination tests and can mate and rear young successfully (Stettner and Matyniak 1968).

The hyperstriatum, unique in birds, is the center of avian learning and intelligence. It is best developed in intelligent birds such as crows, parrots, and passerines. Domestic Chickens, Japanese Quail, and Rock Doves, which do not perform as well in laboratory intelligence tests, have smaller hyperstriata. Damage to the hyperstriatum severely impairs a bird's behavior. The anterior hyperstriatum, actually a bump called the Wulst, may be the seat of higher learning processes in birds (Figure 8–3). Removal of the Wulst does not impair a bird's normal motor functions nor its ability to make simple choices, but it destroys its ability to learn complex tasks. For example, quail whose Wulst has been removed cannot master multiple reversal learning tests, in which they must learn to switch from one rewarding symbol to another that was previously nonrewarding.

Brain function

The brain analyzes incoming signals, integrates them with past experience, channels them through genetically programmed neural switches, and activates a series of motor instructions throughout the body. Classical studies of brain function stressed the results of direct electrical stimulation of the brain, which, for example, can cause doves to adopt particular display behaviors (Åkerman 1966a, b) (Figure 8–4). Recent studies of how the avian forebrain controls song (to be described later in this chapter) provide a deeper understanding of the functional relationship between brain cells and behavior (Gurney and Konishi 1980; Nottebohm 1980, 1981). Indeed, the results provide models of behavior control in the central nervous system.

Sensory input

Peripheral organs receive signals from all the senses — vision, hearing, touch, taste, and smell — and feed them to the brain for processing, integration, and response. Before reaching the main integration centers of the forebrain, sensory signals pass through their respective control centers. The various control centers in a bird's brain are distinctly compartmentalized, as they are in the primitive brains of fish and amphibians. Visual information goes to the optic lobes of the midbrain; information on body orientation and localized pressure goes to the cerebellum in the midbrain; acoustical information goes to its related processing centers in the hindbrain; olfactory information goes to the olfactory bulbs and then to the olfactory lobe in the forebrain (Figure 8–5).

The optic lobes and the cerebellum dominate the avian midbrain. The two optic lobes of birds are huge in relation to the rest of the brain.

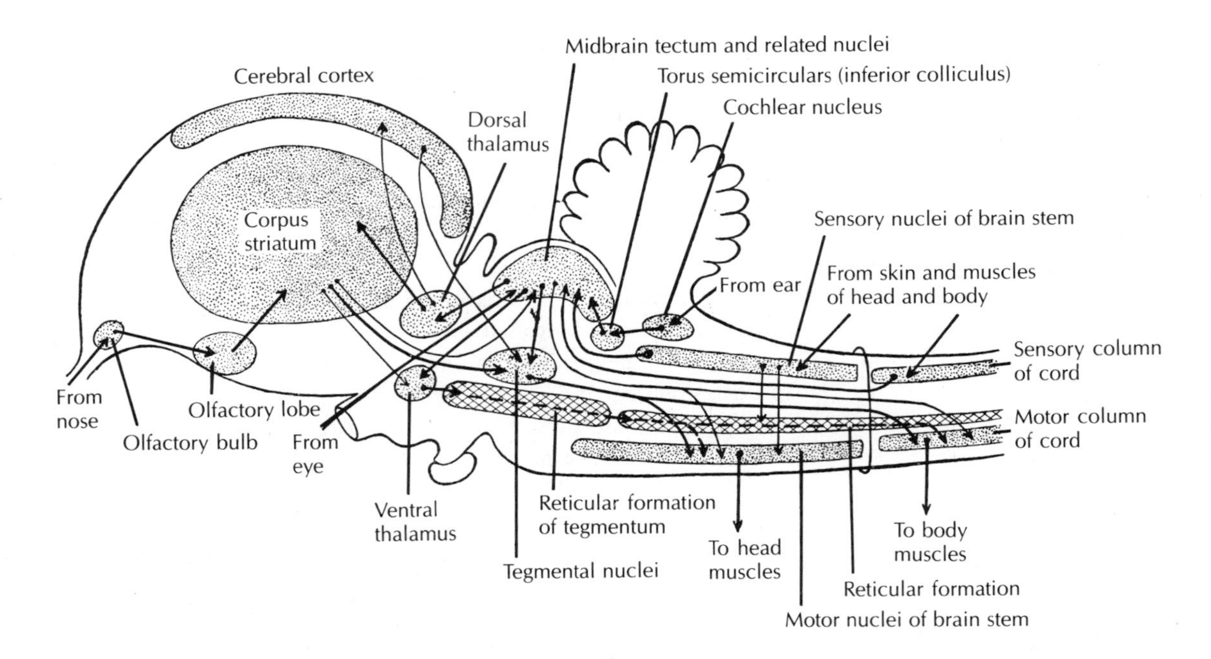

Figure 8–5 Schematic wiring diagram of the bird brain. The corpus striatum and the midbrain tectum are the primary control centers of bird behavior. (From Romer 1955)

Together with characteristically large eyes, this visual apparatus displaces the rest of the brain from the ventral and lateral portions of the skull. Balance and coordination during flight require extensive input from sensory receptors throughout the body and in the middle ear; the large size of the cerebellum reflects the importance of this input.

Acoustical information is processed primarily by auditory nuclei in the hindbrain. Specialized dark-hunting owls that rely on sound have an extraordinary number of ganglionic cells in the medulla for processing acousticospatial information. The Common Barn-Owl, for example, has about 47,600 ganglionic cells in one half of the medulla; the Carrion Crow has about 13,600; and the Little Owl, which hunts in the early morning light, has about 11,200 (Winter 1963). Oilbirds, which use sound to navigate in the dark, also have highly developed auditory centers.

Avian olfactory bulbs, located in front of the forebrain, are nearly identical in structure to those of mammals. They receive stimuli from receptor cells at the tips of the olfactory nerve fibers. The size of the olfactory bulbs in birds varies greatly in relation to olfactory ability (Bang 1971; Bang and Cobb 1968). Birds that have a well-developed sense of smell, such as honeyguides, petrels, vultures, and kiwis, have large olfactory bulbs. Birds that have a poorly developed sense of smell, such as pelicans, woodpeckers, and passerines, have small olfactory bulbs.

Sensory input passes from the primary control centers to intermediate centers in the midbrain and, in some cases, to the forebrain for further processing. One of these centers, the dorsal thalamus, also controls the body's water balance, temperature regulation, sleep, and most visceral functions. Two others, the tegmental nuclei and the ventral thalamus, transmit signals from the forebrain back to the body for action. In addition, the ventral thalamus, or hypothalamus, is the source of neurohormones that control seasonal activities, such as breeding and molt, through stimulation of the pituitary gland.

Control of song by the central nervous system

Bird song is controlled by a chain of distinct brain nuclei. Nerve impulses start in the forebrain (in the hyperstriatal nuclei of the corpus striatum), proceed to the cochlear nuclei of the midbrain, and pass through the hypoglossal nerves, which innervate the syringeal muscles, to control the sound-production apparatus, or syrinx (Arnold 1982). A variety of calls, some recognizable and some abnormal, can be evoked by electrical stimulation of the midbrain nuclei. Destruction of the hyperstriatal nuclei renders a songbird mute.

The level of development of the brain's song-control centers corresponds to the level of a bird's vocal activity. West coast Marsh Wrens, which learn three times as many songs as Marsh Wrens in New York, have a 50 percent larger volume of song-control nuclei (Canady et al. 1984). Among individual canaries, the level of development of brain tissue controlling song is directly related to the size of individual syllable repertoires (Nottebohm et al. 1981). Song-control nuclei are larger in male Zebra Finches, which sing, than in female Zebra Finches, which do not sing (Arnold 1980). Experimental exposure to the sex hormone estradiol at an early age affects the sizes of the nuclei as well as their sensitivity in the adult birds to sex hormones that stimulate singing during the breeding season (Arnold and Saltiel 1979; Gurney and Konishi 1980).

Bird song is normally controlled by the left hemisphere (Nottebohm 1980). The left hemisphere of a canary's forebrain is dominant (as is that of a human). The right cerebral hemisphere assumes control of the functions of the left hemisphere only if the left hemisphere is damaged. Destruction of a young canary's song control centers in the left hemisphere leads to formation of an alternative set in the right hemisphere and the acquisition of a new song repertoire. Such functional lateralization of the brain was once thought to be an exclusively human attribute, associated with extraordinary language abilities.

Vision

Birds are highly visual animals. They have large eyes, search visually for food, and spot predators at great distances. They also engage in complex, colorful courtship displays.

The true nature of avian vision, however, remains to be determined. Despite popular beliefs and historical conclusions about the excellence of avian vision, experiments have not yet confirmed extraordinary visual acuity (the ability to resolve fine detail at distance). Passerines and raptors, believed to have the keenest sight of all birds, can resolve details at 2.5 to 3 times the distance humans can, not 8 times as was once thought (Walls 1942; Pearson 1972). One of the distinctions of avian vision may be not in acuity but in the ability to capture at a glance a picture the human eye would need a laborious scan to piece together. Birds may also have an unparalleled system of color vision.

Avian eyes are large, prominent structures (see Figure 8–6). The European Starling's eyes account for 15 percent of its head mass, in contrast

Figure 8–6 The large, striking eyes of birds such as the hornbill suggest excellent vision, but the visual abilities of birds remain largely undetermined. (Courtesy S. Lipschutz/VIREO)

to human eyes, which account for less than 2 percent of head mass. The eyes of eagles and owls are as big as human eyes. Ostriches have the largest eyes of any terrestrial vertebrate (Walls 1942).

Unlike the round, rotating eyes of mammals, the eyes of birds are flat, globular, or tubular. They fill the orbits fully and are capable of only limited rotation, mostly toward the bill tip.

Because birds' eyes are generally set on the sides of their heads, birds see better laterally than forward. Penguins and passerines, for example, examine nearby objects with one eye at a time. The resulting image is relatively flat because monocular vision does not achieve depth perception with the same accuracy as binocular vision. To compensate, birds bob their heads quickly, viewing an object with one eye from two different angles in rapid succession. Some birds, such as swallows, goatsuckers, hawks, and owls, restrict lateral monocular vision to close objects and use forward binocular vision for distant viewing. Generally, binocular vision is atypical. Among ducks, only the Blue Duck of New Zealand can stare forward; other ducks use one eye at a time. Parrots have a binocular field of only 6 to 10 degrees (Walls 1942). Bitterns stare forward with binocular vision while pointing their bills skyward. Quite the opposite are woodcocks, whose huge eyes permit broad rearview binocular vision because they are set so far back on the head.

Eye anatomy

A cross section of the avian eye reveals a small anterior component that houses the cornea and lens and a larger posterior component that is the main body of the eye. The two sections are separated by a scleral ring composed of 12 to 15 small bones. Two striated muscles, Crampton's muscle and Brucke's muscle, attach to these ossicles and are responsible for focusing on objects. The lens is large and conspicuous. The pecten, a dis-

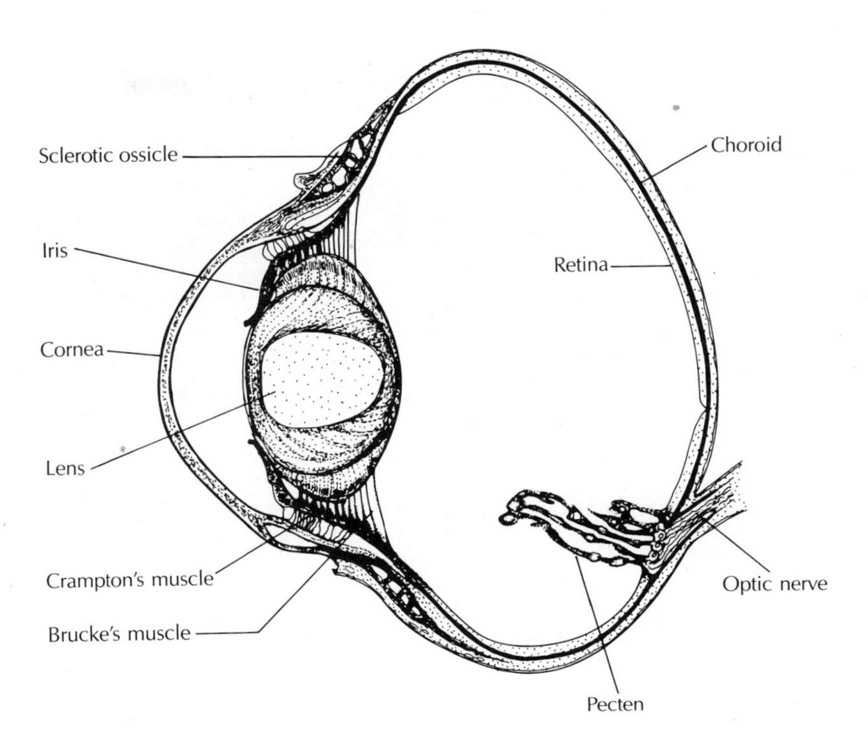

Figure 8–7 Cross section of the avian eye. (After Grassé 1950)

tinctive and intriguing feature of the avian eye, whose purpose remains unclear, projects from the rear surface of the eye near the optic nerve into the large cavity filled with vitreous humour, the clear substance that fills the eye behind the lens (Figure 8–7).

The cornea, lens, and iris In birds, both the cornea and the lens change their curvature in focusing; only the lens does this in mammals. Contraction of Crampton's muscle increases the cornea's curvature and thus its refractive power. A change in the cornea has little effect underwater, however, because the refractive index of the cornea is nearly the same as that of water. As we might expect, diving birds, such as cormorants, have weakly developed Crampton's muscles; instead, their strong Brucke's muscles change the shape of the soft, flexible lens.

Lens shape varies more among bird species than in other vertebrates. The lenses of parrots, storm-petrels, and the Hoopoe have a flat anterior surface and a strongly convex posterior surface. Ducks, owls, and nightjars have lenses that are strongly convex in both front and back, whereas in passerines and raptors the convex posterior curvature is noticeably greater than the anterior curvature. The reasons for these differences are not yet known.

Though the iris color in most birds is deep brown to black, there are striking exceptions. Adult Black-crowned Night-Herons have ruby red eyes; the eyes of Crowned Cranes are grey; cormorants' eyes are green; Northern Gannets' are pale blue; large owls' are yellow-orange; and many raptors' are bright yellow.

The pupil opening is round in all birds except the Black Skimmer whose pupil opening forms a catlike vertical slit in bright light, returning to the rounded shape in dim light (Zusi and Bridge 1981).

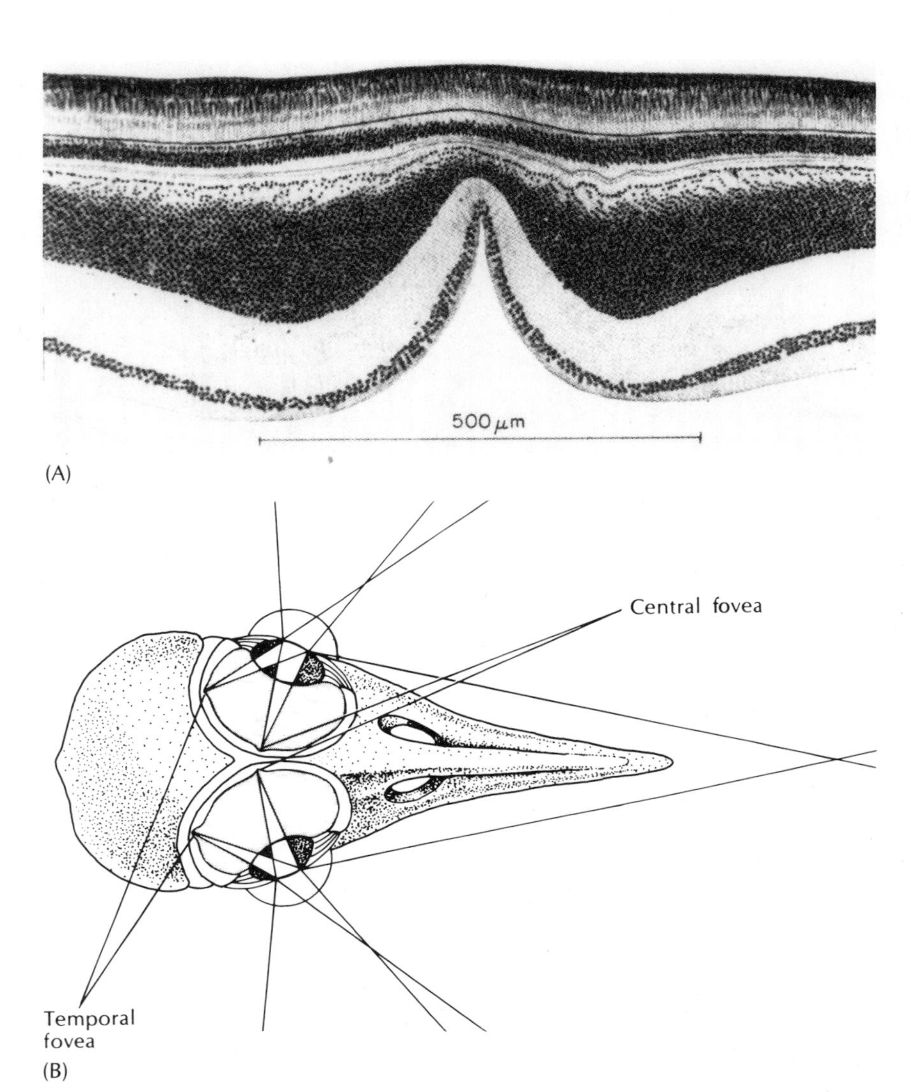

Figure 8–8 (A) Cross section of a Least Tern retina showing visual-cell layer with rods and cones, and the deep central fovea. (B) Some birds, such as raptors, have temporal fovea, which enhance forward binocular vision. (A, from Rochon-Duvigneaud 1950; B, after Wilson 1980)

At night, some birds' eyes shine bright red in the beam of a flashlight or automobile headlights. This is not the iris color, but rather the vascular membrane showing through the translucent pigment layer on the surface of the retina. Kiwis, thickknees, the Boat-billed Heron, the Kakapo, a flightless, nocturnal parrot, many nightjars, and owls share this distinctive trait.

Retina and fovea The highly developed anatomy of the avian retina and its light receptor cells suggest excellent vision. Apparently, a large number of cones enables birds to form sharp images, no matter where light strikes their retina. The number of cones, the daylight receptors of the retina, can be as high as 400,000 per square millimeter in House Sparrows and one million per square millimeter in the Common Buzzard. By way of comparison, the human eye has at most 200,000 cones per square millimeter (Walls 1942). Away from the densest concentrations in the fovea (to be defined), cone concentrations in the human retina drop sharply to only one-tenth of those of birds.

Foveae, concave depressions of high cone density, are known to be the sites of greatest visual sharpness in humans. Most birds have one fovea in each eye, located in the center of the retina near the optic nerve. This central fovea is deeper and histologically more complex in visually acute passerines, woodpeckers, and raptors than it is in pigeons and chickens. Whether deep foveae enhance avian visual acuity is not clear. They may, however, aid in detection of the movements of small images. Fast-flying birds that must judge distances and speeds accurately, such as hawks, eagles, terns, hummingbirds, kingfishers, and swallows, have temporal as well as central foveae. These birds also have forward-directed eyes and, therefore, good binocular vision. The images of their binocular vision are received by the temporal foveae and images of their peripheral or lateral monocular vision fall on the central foveae. The temporal foveae, therefore, enhance binocular vision (Figure 8–8).

Although cones are most abundant in the foveae, high cone densities also occur in horizontal, ribbonlike strips around the retina in albatrosses, grebes, plovers, and other birds. These ribbons apparently increase a bird's ability to perceive the horizon. Species with this ribbon tend to hold their heads so that both the retinal ribbon and the lateral element of the semicircular canal are horizontal. This suggests that the eye and the inner ear work together to achieve proper body orientation (Pearson 1972).

The pecten The pecten is a remarkable feature of the avian eye. It is a large, pleated, vascularized structure attached to the retina near the optic nerve. It protrudes conspicuously into the vitreous humour and, in some birds, almost touches the lens. The large, elaborate avian pecten is unique among vertebrates. In most birds the pecten has 20 or more accordian-pleated fins, giving it a superficial resemblance to an old-fashioned steam radiator (Walls 1942). The pectens of owls, nightjars, and the nocturnal Kakapo have only four to eight folds, whereas the kiwi's simple, reptilelike pecten, which has no folds at all, probably represents an evolutionarily degenerate condition (Sillman 1973) (Figure 8–9).

The avian pecten has fascinated scientists for centuries. At least 30 theories have been proposed to explain its existence. Some researchers believe the pecten is involved in the regulation of internal eye temperatures and hydrostatic pressures; some say it reduces glare; others hypothesize that it might be a sextant for navigation or a dark mirror for indirectly viewing objects near the sun. The majority opinion, however, holds that the avian pecten functions primarily as a source of nutrition and oxygen for the retina. Unlike its mammalian counterpart, the avian retina has no imbedded blood vessels. The assumption is that, instead, the vascular supply system is concentrated in a single structure, the pecten, which interferes less with visual functions than would a network of blood vessels.

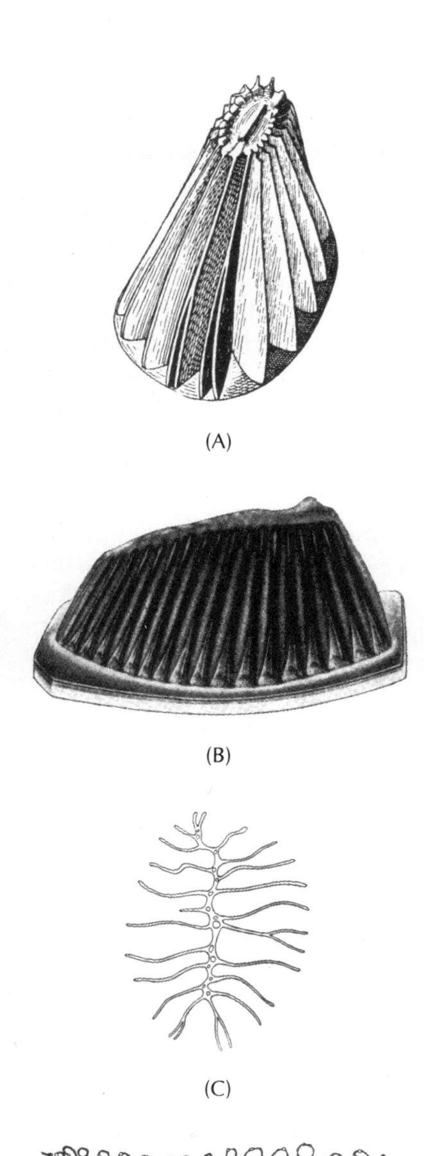

Figure 8–9 Structure of the pecten of (A) an Ostrich and (B) most modern birds. (C) Basal cross section of the structures of A, including central web and lateral vanes. (D) Dorsal view of the typical pleated structure of avian pectens. (Adapted from Walls 1942)

Color vision

The presence of large numbers of cone receptors, which contain the visual pigments, suggests that diurnal birds have well-developed color vision. The retinas of nocturnal owls contain mostly rods, which are simple light receptors. Color vision is based on visual pigments, which convert the electromagnetic energy of light into neural energy. In addition to visual pigments, the cones of diurnal birds often contain colored oil droplets. Carotenoid pigments (p. 75) in the oil act as red-yellow filters, but their contribution to color vision is not understood. Perhaps the yellow oil droplets enhance the contrast of objects seen against the sky by filtering out much of the blue background. Similarly, red oil droplets may enhance the contrast of objects against green backgrounds, such as fields and trees, by filtering out the prevailing green background. The yellow oil droplets are concentrated in the central and lower retina, where distant images such as those in the sky usually fall. Red oil droplets are concentrated in cones of the peripheral and upper retina, where nearby images such as those on land usually fall.

Unlike humans, birds are sensitive to light in the near-ultraviolet spectrum. In the human eye, the lenses absorb ultraviolet light; in birds, they transmit ultraviolet light to the retina, where some cones have peak sensitivity in the near-ultraviolet spectrum (Chen et al. 1984). Melvin Kreithen and Thomas Eisner (1978) demonstrated that homing pigeons, in addition to having the normal vertebrate sensitivity to blues and greens (at 500 to 600 nanometers), are sensitive to the near-ultraviolet spectrum (325 to 360 nanometers). Also tested recently were Black-chinned Hummingbirds, Belted Kingfishers, Mallards, and several passerines, all of which are sensitive to ultraviolet light (Goldsmith 1980; Parrish et al. 1984). Given the taxonomic diversity of those species tested, the majority of birds probably possess this trait.

The richness of avian color perception is probably beyond that of human experience (Goldsmith 1980). We speculate that primitive mammals, including the ancestors of primates, were nocturnal creatures that lost the retinal oil droplets associated with sensitive color vision. Once lost, these droplets did not evolve again in placental mammals. Instead, the color vision of humans and other primates was reevolved on a different basis without pigmented oil droplets. Very likely, the avian retina—with its high cone densities, deep foveae, near-ultraviolet receptors, and colored oil droplets that interact with several cone pigments—is the most capable diurnal retina of any animal.

Hearing

Sounds provide birds with essential information. From territorial defense to mate choice and recognition of individuals and from song learning and recognition to prey location, predator avoidance, and navigation, birds depend on their hearing for a wide range of activities. Although the anatomy of the avian ear has been fully described, we are just beginning to determine the full range of sounds that birds hear.

Ear structure

The three sections of the avian ear are the external ear, the middle ear, and the inner ear. The first two sections funnel sound waves from the environment into the cochlea, the fluid-filled, coiled section of the inner ear that is the base of the hearing organ. Hair cells in the cochlea monitor vibrations transmitted by the fluid and encode them into a temporal sequence of nerve impulses that register in the acoustical centers of the brain. The avian ear is structurally simpler than that of mammals. Its acoustical efficiency, however, is the same as that of the mammalian ear.

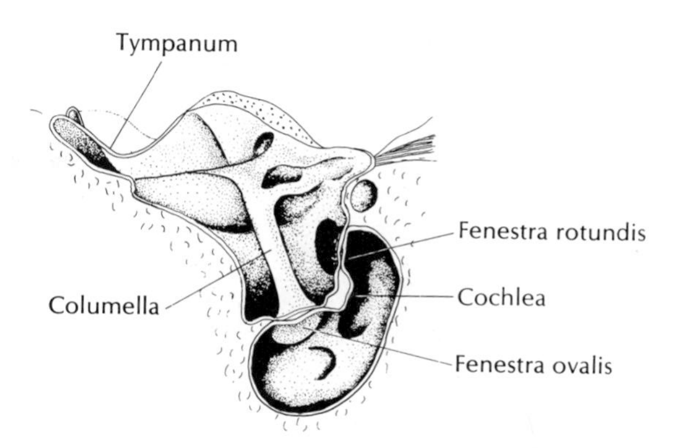

Figure 8–10 Middle-ear region of a chicken. A single bone, the columella, or stapes, transmits sound from the eardrum to the fluid-filled cochlea of the inner ear. (After Pohlman 1921)

The external ears of birds are inconspicuous structures lacking the elaborate pinnae, or projecting parts, of mammal ears. The middle ear has only one bone, the columella or stapes, which connects the eardrum, or tympanic membrane, to the pressure-sensitive fluid system of the inner ear. In contrast, the middle ear of a mammal has three bones. Located next to the attachment of the columella to the bony cochlea is the flexible round window, which protects the inner ear from pressure damage. The shape of the columella varies with taxon, but most birds have a simple columella similar to that of reptiles (Feduccia 1977). Compared with mammals, avian inner ears have a short basilar membrane, no division between inner and outer hair cells, and a simple system of cochlear nerves (Figure 8–10).

Specialized auricular feathers on the external ear protect the hearing organ from air turbulence during flight while permitting sound waves to pass inside (Il'ichev 1961). Diving birds, such as auks and penguins, have strong, protective feathers covering the external ear openings. These birds can protect the middle and inner ears from pressure damage in deep water by closing the enlarged caudal rim of the external ear. The entire muscular rim to which the auricular feathers are attached forms an enlarged, though inconspicuous, ear funnel in some birds, especially passerines, parrots, and raptors. Among raptors its shape varies from no funnel at all in the fish-eating Osprey to a well-developed funnel in harriers, which locate field mice acoustically (Dement'eve and Il'ichev 1963; Rice 1982). The superb hearing of nocturnal owls is related to their exceptional ear funnels (Norberg 1977). Large anterior and posterior ear flaps regulate the size of the ear opening and enhance acoustical acuity more than fivefold (Schwartzkopff 1973). In many owls, the external ears and in some cases the skull are bilaterally asymmetrical, a condition unique among freely moving vertebrates. As we shall see, this asymmetry aids precise location of prey.

Hearing ability

What do birds actually hear? How does their hearing compare with ours? Research on avian hearing abilities is still in its infancy, but already it is yielding some surprising results. Contrary to past impressions, birds may not have extraordinary acoustical acuity (Dooling 1982); humans can hear fainter sounds than most birds at most frequencies.

Furthermore, the frequency range of good hearing is narrower in birds than in mammals. Maximum sensitivity is confined to frequencies

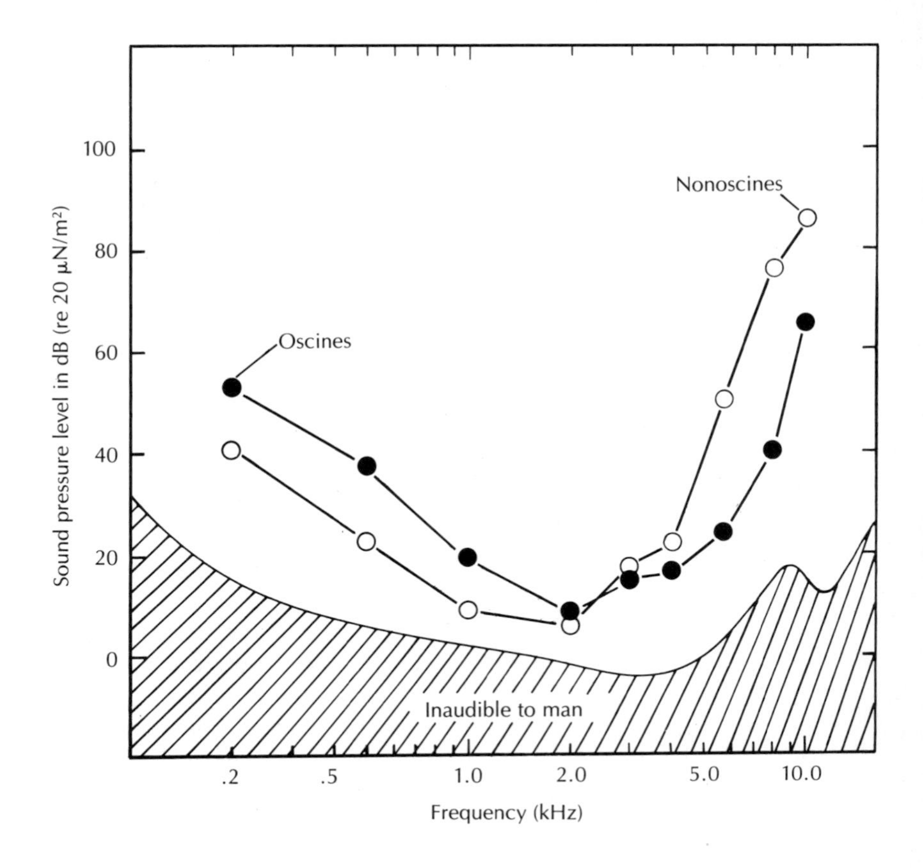

Figure 8–11 Hearing thresholds of nine species of oscine birds (solid circles) and seven nonoscine species (open circles). Compared with humans, birds hear well over a narrower range of frequencies. (From Dooling 1982)

between 1 and 5 kilohertz. Sensitivity decreases rapidly at both lower and higher frequencies. Owls are an exception: Great Horned Owls hear low-frequency sounds and Common Barn-Owls hear high-frequency sounds better than humans do. Passerines tend to hear high-frequency sounds better than nonpasserines, whereas nonpasserines hear low-frequency sounds better than passerines (Dooling 1982) (Figure 8–11).

Surprisingly, small songbirds are not particularly sensitive to high-frequency sounds. Unlike bats and some other mammals, birds do not hear high-frequency (ultrasonic) sounds. Some birds, however, hear very low frequencies (infrasound) extremely well; pigeons can hear sounds in the 1 to 10 hertz range that are 50 decibels lower than those humans can hear (Kreithen and Quine 1979). The significance of this ability is not yet known.

Birds are quite sensitive to small changes in the frequency and intensity of sound signals, but not unusually so, and are not as sensitive as humans. Birds can discriminate temporal variations in sound, such as duration of notes, gaps, and rate of amplitude modulation, as well as other vertebrates can, including humans. Laboratory tests do not support the idea that birds have exceptional powers of temporal resolution, a result that bears directly on the abilities of birds to recognize subtle call variations and on the evolution of vocalizations (see Chapter 10).

Budgerigars enhance reception of their own signals by filtering out environmental noises (Dooling 1982). Their greatest signal-to-noise sensitivity is at 2 to 4 kilohertz, the principal frequencies of their own vocalizations. Parakeet ears are also buffered against acoustical trauma; they are not temporarily deafened by loud noises as are most vertebrates.

Echolocation and prey detection

A few birds use echolocation, reflected vocalizations, for navigation. Some cave swiftlets of southeast Asia find their way through dark cave corridors by emitting short, probing clicks of one millisecond duration at normal frequencies (2 to 10 kilohertz), though they do not employ ultrasound as bats do (Medway and Pye 1977). Echolocation at low frequencies is at best only one-tenth as functional as the ultrasound sonar system of bats. For example, the cave-nesting Oilbird, or Guacharo, of South America echolocates with sharp clicks 15 to 20 milliseconds long over a broad frequency spectrum (1 to 15 kilohertz) (Konishi and Knudsen 1979) (Figure 8 – 12). They can avoid disks that are 20 millimeters or more in diameter but collide with smaller objects.

In complete darkness, owls have extraordinary abilities to locate prey by sound (Payne and Drury 1958; Knudsen 1981). The Common Barn-Owl can catch a running mouse in total darkness because it can pinpoint sounds to within one degree in both the vertical and the horizontal plane (Payne 1971). The owl can also determine the direction and speed of a mouse's movement. Humans can locate sounds in the horizontal plane (azimuth) about as well as a barn-owl but only one-third as well in the vertical plane (elevation).

Both owls and humans locate the sources of sounds by means of differences in the intensity and time of arrival of sounds at the two ears. Looking directly at the source equalizes these stimuli. The asymmetrical arrangement of the ears of some owls enhances reception differences and thus the ability to locate prey quickly and accurately. This ability is well-developed in the Common Barn-Owl, which locates sounds in the vertical plane by means of its asymmetrical ear openings and the troughs formed by the feathered facial ruff. The left ruff faces downward, thereby increasing

Figure 8 – 12 Oilbirds echolocate in the labyrinths of their nesting caves. The short series of sharp clicks they utter for this purpose are not nearly as sensitive as the ultrasound sonar used by bats. (Courtesy W.A. Conway)

sensitivity to sounds below the horizontal, and the right ruff faces upward, increasing sensitivity above the horizontal. The owl need only tilt its head up or down to equalize input to the two sides, and thus to pinpoint the location of a mouse.

The Common Barn-Owl's facial ruff focuses sounds in much the same way the human external ear does. It also efficiently gathers high-frequency sound waves (those above 3 kilohertz). High-frequency sound is subject to the greatest vertical time delay; low-frequency sound is subject to the greatest horizontal time delay. This explains why the experimental removal of the ruff reduces an owl's accuracy with respect to elevation (the vertical plane) but not to azimuth (the horizontal plane). A broad-frequency sound spectrum enables the barn-owls to target prey precisely; they do poorly with single-frequency tones (Figure 8–13).

The external system of structurally enhanced time delay is augmented by a powerful internal arrangement of specialized neurons in the auditory centers of the owl brain (Payne 1971; Knudsen 1980). These neurons are so sensitive to time delays of 40 to 100 microseconds that a difference of only 10 microseconds increases their sensitivity up to 75

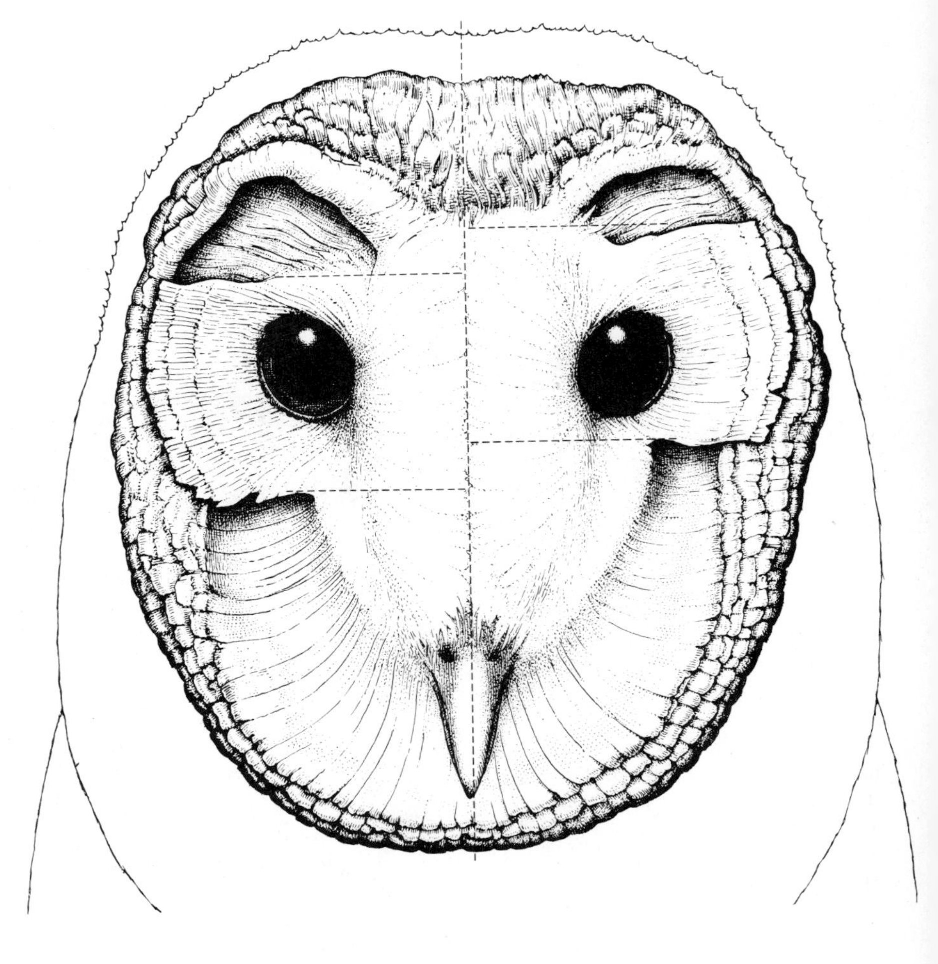

Figure 8–13 The heart-shaped face of the Common Barn-Owl is not perfectly symmetrical. The left ear, which is higher than the right ear, is most sensitive to sounds from below the horizontal. Conversely, the lower right ear is most sensitive to sounds from above the horizontal. This asymmetry causes a sound to arrive at each ear at slightly different times, enabling the owl to pinpoint the source of the sound. (From Knudsen 1981, with permission of Scientific American)

percent. The spatial arrangement of these neurons corresponds to the angle between the sound's source and the owl's focus: Outer neurons fire when the angle is 20 to 30 degrees, and inner ones fire when the angle is smaller. Thus, a turn of the head toward the source activates more inner neurons, which hold the owl's attention on its target.

The process of sound localization by songbirds is still poorly understood but may be based on mechanisms other than those used by either owls or humans. Of particular interest is the recent discovery of a connection between the two ears (Rosocoski and Saunders 1980). Sounds entering one ear reach the other ear's tympanic membranes and its inner ear. Such internal transfer of sound may add a new dimension to the hearing abilities of songbirds.

Mechanoreception

Birds are extremely sensitive to mechanical stimulation (Schwartzkopff 1973). Mechanoreception reaches its highest level in the hearing organ. Included among the senses of mechanoreception are tactile reception, equilibrium, or balance, and detection of barometric pressure.

Tactile corpuscles, the primary source of skin sensitivity, also monitor changes in muscle tension (proprioreception). Birds have three kinds of cells that are specialized for tactile response. These are all found at the ends of sheathed nerve fibers and include Merkel corpuscles, Grandry corpuscles, and Herbst corpuscles. Merkel corpuscles are common in the skin, tongue, and bill of a bird. Gandry corpuscles, which are similar but enclosed by a capsule of connective tissue, are found only in the bills of ducks.

Herbst corpuscles are the largest and most elaborate of the three kinds of tactile corpuscles. They are ellipsoid structures consisting of an outer, lamellated sheath and an inner core. The onionlike layers of the outer sheath are adapted for elastic reception and transfer of rapid pressure changes; the inner, cylindrical core is an elaborated sensory nerve fiber (Figure 8–14). Herbst corpuscles are abundant in the bill tips of birds, especially of sandpipers and snipes (Clara 1925; Bolze 1968), and in the tips of woodpecker tongues. They also are concentrated in feather follicles that have sensory functions, especially those of filoplumes and bristles (see

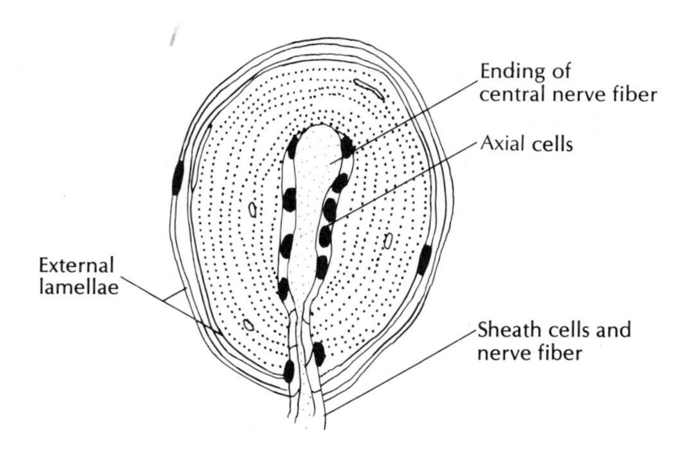

Figure 8–14 Herbst corpuscle from the bill of a duck. The most elaborate of avian tactile sensors, it consists of up to 12 onionlike layers of external lamellae that transfer slight pressure changes to the elaborated nerve ending of the receptor axion in the center. (After Portmann 1961)

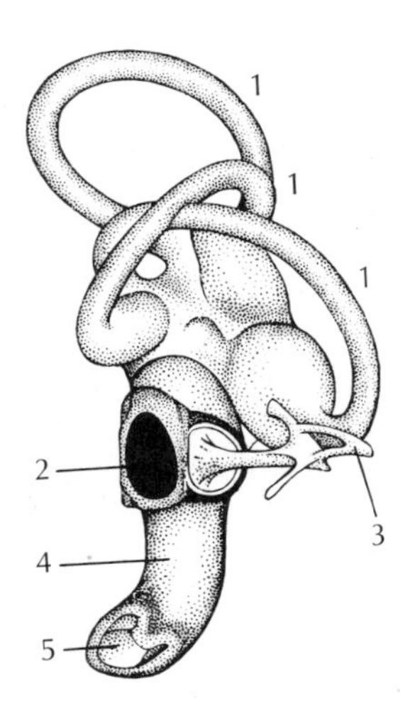

Figure 8–15 The semicircular canals (1), which are the organs of equilibrium, are located next to the apparatus of the middle ear [columella (2), round window (3)] and inner ear [cochlea (4), lagena (5)]. (Adapted from Pumphrey 1961)

Chapter 4) and are numerous in the wing joints of birds, where they help govern wing positions in flight.

Equilibrium

The organs of equilibrium are the semicircular canals located in the ears and the associated sets of specialized sensory cells. These are among a bird's most important sensory organs because they regulate the balance and spatial orientation so essential to skilled flight. They give birds an excellent sense of balance and body position, enabling them to reorient automatically with respect to gravity, even when blindfolded.

There are three semicircular canals, one oriented horizontally and two oriented vertically. They connect directly to the cerebrospinal fluid systems. When the position of the bird's head changes, fluid moves through the canals (Werner 1958). At the bases of the semicircular canals are delicate sets of membranes equipped with sensory hair cells, which detect the movements of small crystals of calcium carbonate (statoliths) floating in the fluid. Spatial variations in the pressure of the crystals on the hair cells cause different patterns of excitation, which enable the bird to sense the direction of gravity and linear and circular acceleration (Figure 8–15).

The sizes of the semicircular canals in various birds relate to flight performance: Pigeons, owls, thrushes, ravens, and falconiform birds have relatively larger canals than galliform birds and ducks. Among galliform birds, the size of the semicircular canals increases with the mobility of a species (Sagitov 1964); the size of the cerebellum of the avian midbrain, which is responsible for balanced muscular coordination, corresponds.

Barometric pressure and magnetism

Two of the most recently documented avian senses involve barometric pressure and magnetism. For years, ornithologists have appreciated that birds are aware of an approaching winter storm and feed actively to build their energy reserves. Birds also know how to choose altitudes for migration. These abilities suggest sensitivity to differences in barometric pressure. Homing pigeons, the only birds that have been studied in this regard, are, in fact, extremely sensitive to small changes in air pressure, comparable to differences of only 5 to 10 meters in altitude (Kreithen and Keeton 1974a). Future research will reveal the prevalence of this sense and whether the inner ear is the organ responsible.

Birds certainly use magnetic information for navigation (Chapter 14). Charles Walcott and his colleagues (1979) discovered tiny crystals of magnetite near the olfactory nerves between the eyes of pigeons. These may be the basis of a directional sensory system for the detection of magnetism. Similar crystals have been found in bacteria and honeybees that respond to magnetism.

Taste and smell

Birds can taste and smell, though how well birds taste is still unclear (Wenzel 1973). The few existing quantitative data on taste-acuity levels in birds suggest only that they may be equally or less sensitive than mammals with respect to some ingredients. A few taste buds are located on the rear of the avian tongue and on the floor of the pharynx: about 24 in the chicken, 37 in the pigeon, 46 in the Eurasian Bullfinch, and 62 in Japanese Quail. Avian taste buds are similar in structure to mammalian taste buds but negligible in number by comparison. Humans, for example, have roughly 10,000 taste buds.

Some birds have an excellent sense of smell. Northern Bobwhites, Common Canaries, Mallards, Domestic Chickens, Manx Shearwaters, Turkey Vultures, Brown Kiwis, and Humboldt Penguins all pass laboratory olfaction tests. Wild flying Turkey Vultures are attracted to the source of ethyl mercaptan fumes released into the air to simulate the smell of rotting meat (Stager 1964, 1967). Engineers were able to locate leaks in a pipeline 42 miles long by pumping the same chemical through it and then spotting where the Turkey Vultures gathered. In field experiments (Stager 1967), honeyguides have found concealed candles and burning beeswax, which suggests that they can smell wax. Various tube-nosed seabirds locate food by odors, which they follow upwind, but other seabirds in the same area do not exhibit the same behavior (Hutchinson and Wenzel 1980). Leach's Storm-Petrels use their well-developed sense of smell to locate their nesting burrows in the dark, conifer forests on islands in the Bay of Fundy (Grubb 1972, 1974; see also Chapter 14).

The avian sense of smell is based in the surface epithelium of the posterior concha of the olfactory cavities (see Chapter 6; also Bang 1971, for comparative descriptions of avian olfactory cavities). The flow of air through these cavities from the bird's mouth conceivably enables birds to smell their food from the inside as well as from the outside.

Senses and behavior: A projection

Reactions to external stimuli mediated by the primary senses underlie overt behavior and thus form a link between the bird's genetic heritage and its environment. The senses of a bird are an integral part of its feeding behaviors, affecting food detection, innovative feeding, and intelligent relocation of caches. These senses reach full expression in the process of communication via visual and vocal displays, the topics of the next two chapters. Finally, the outstanding navigational abilities of birds derive from high-level integration of their wide-ranging sensory capabilities, as we will discuss in Chapter 14.

Hormones set the physiological environment for behavior, including sensitivity to particular stimuli. Seasonal activities are controlled by photoreceptors in the midbrain that trigger the release of pituitary hormones; these hormones, in turn, govern details of reproductive behavior, molt, and migration (see Chapter 12). Also mediating the interaction between heritage and environment is a continuum of sensory experiences ranging from brief imprinting to prolonged intelligent learning (see Chapter 11). The sensory reactions of birds, therefore, only open the door to the full expression of the adaptive behaviors upon which we focus in the next chapters.

Summary

The evolution of the avian brain has taken a different course from that of mammals. The basis for avian intelligence lies in the hyperstriatum layer of the forebrain, primarily in tissue called the Wulst. Particularly exciting have been recent breakthroughs on the identification of the brain centers for control of song in passerine birds and the organization of neural control of this complex motor skill.

Birds have a full repertoire of well-developed senses. Large eyes and well-developed optic lobes of the brain enable excellent vision, including an ability to follow small moving objects. Birds may also have the most highly developed color vision of any vertebrate. The hearing of birds as a group is good but not extraordinary, except for the ability of Rock Doves to hear extremely low frequencies (infrasound) and the ability of Common Barn-Owls to pinpoint sounds made by potential prey. Birds are sensitive to slight differences in barometric pressure and to magnetism. Avian senses of smell, taste, and tactile sensation are also better developed than we once thought.

Further readings

Dooling, R.J. 1982. Auditory perception in birds. *In* Acoustic Communication in Birds, vol. 1, (D.E. Kroodsma and E.H. Miller, eds.), pp. 95–130. New York: Academic Press, *An excellent review of avian hearing abilities.*

Pearson, R. 1972. The Avian Brain. New York: Academic Press. *A fundamental reference.*

Schwartzkopff, J. 1973. Mechanoreception. Avian Biology 3:417–477. *A detailed review of the subject.*

Sillman, A.J. 1973. Avian vision. Avian Biology 3:349–388. *A detailed review of eye anatomy.*

Walls, G.L. 1942. The Vertebrate Eye and Its Adaptive Radiation. Bloomfield Hills, Mich.: Cranbrook Institute of Science. *The classic reference.*

Wenzel, B.M. 1973. Chemoreception. Avian Biology 3:389–415. *A detailed review of the subject.*

chapter 9

Visual communication

Communication is the transfer of information between a sender and a receiver; territorial defense, attraction and courtship of mates, and maintenance of flock structure all require birds to communicate. The resolution of conflicting individual purposes as well as cooperation between individuals depends on the exchange of information.

Birds communicate with each other by means of displays, which are acts that are specialized to make information available (Smith 1977). Plumage and postures evolved together, each accentuating the other and thereby enhancing the impact of the display itself and enriching its informational content regarding identity, status, intentions, and potential as a mate. Consider the elaborate courtship display of the King Bird-of-Paradise (Figure 9–1).

> He always commences his display by giving forth several notes and squeaks, sometimes resembling the call of a quail, sometimes the whine of a pet dog. Next he spreads out his wings, occasionally quite hiding his head; at times stretched upright, he flaps them, as if he intended to take flight, and then, with a sudden movement, gives himself a half turn, so that he faces the spectators, puffing out his silky-white lower feathers; now he bursts into his beautiful melodi-

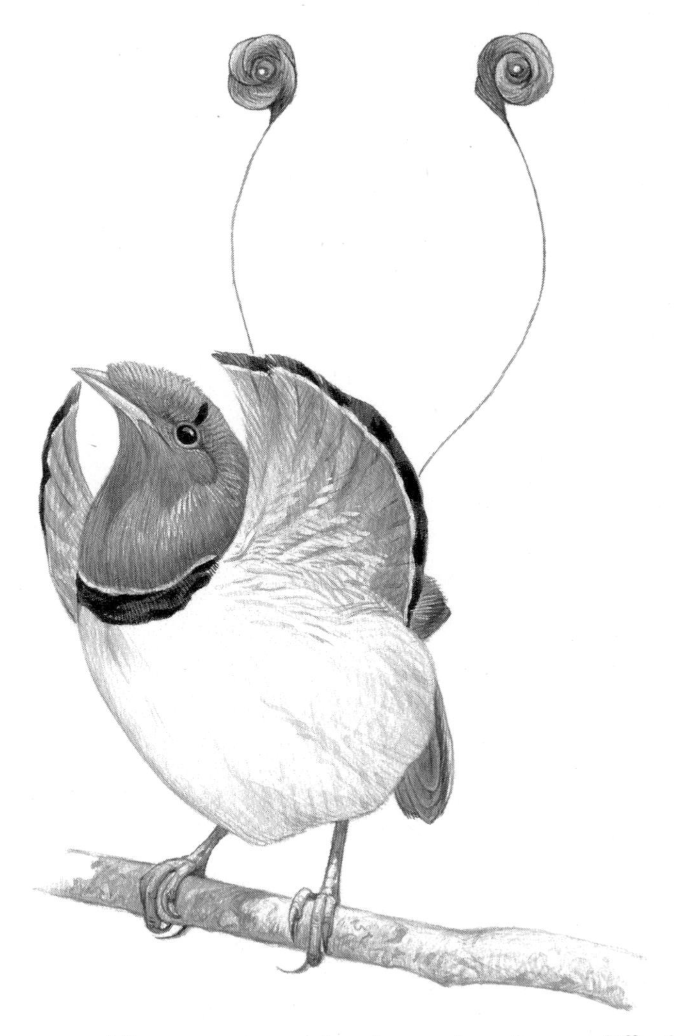

Figure 9–1 King Bird-of-Paradise in full display.

> ous warbling song, so enchanting to hear but so difficult to describe . . . . Then comes the finale, which lasts only a few seconds. He suddenly turns right around and shows his back, the white fluffy feathers under the tail bristling in his excitement; he bends down on the perch in the attitude of a fighting cock, his widely opened bill showing distinctly the extraordinary light apple-green colour of the inside of the mouth, and sings the same gurgling notes without once closing his bill, and with a slow, dying-away movement of the tail and body. A single drawn-out note is then uttered, the tail and wings are lowered, and the dance and song are over. (Ingram 1907)

The comparative study of bird displays has contributed much to our understanding of animal communication; Niko Tinbergen and Konrad Lorenz received Nobel prizes for their pioneering contributions to this field. It is well established that avian displays serve primarily to:

1. identify an individual;
2. signal attack, escape, and/or neutral locomotive intentions and
3. communicate location or the desire to play, mate, or take over a territory.

The traditional view holds that bird displays convey information unambiguously and that birds are forthright in their communications. Challenging this view is the recently developed position that avian communication may include intentional bluff, ambiguity and perhaps even deceit.

In this chapter we first review recognition by means of visual cues, a very basic form of communication. Then we discuss the evolutionary origin and ritualization of elaborate visual displays. Using the well-studied pair formation displays of the Great Blue Heron, we consider the kinds of information transmitted and the ways in which herons enhance the process of communication. Finally, we consider the attack and escape elements in the agonistic behavior of birds and the current controversy over whether these displays are "honest" or "devious."

Visual identity

Many birds, such as male North American wood warblers and African plovers, have distinctive plumage color patterns, often illustrated in identification guides (Peterson 1980). Although we usually assume that birds use distinctive color pattern differences to recognize members of their own species, experimental demonstrations of this performance are few. The imprinting by young birds on foster parents of another species provides one line of proof (see Chapter 11); another comes from experiments using direct manipulation of the visual characters of breeding adults.

One of the most intriguing experimental studies of visual species-recognition signals is that of Neal Smith (1966) on large gulls that breed in the Canadian Arctic. These gulls, which include the Herring Gull and the Glaucous Gull, are approximately the same size and differ only slightly in the amount of gray on the back and black in the wing tips. Even though they appear quite similar to us, recall that birds do not necessarily see themselves as we see them. These species use the color of their fleshy eye rings to identify one another. Smith caught large numbers of these gulls near their nesting grounds and changed the eye-ring colors that serve as a means of identification. He gave some birds the eye rings of another species and painted other birds with their natural color to be sure that paint per se had no effect (Figure 9–2).

Alterations of the eye rings at the beginning of pair formation interfered with normal pair formation and even promoted interspecific pair formation. Females would not pair with a male with the wrong eye-ring color but, in some cases, actually paired with males of another species with eye rings painted to match theirs. Alterations after pair formation caused the pairs to separate. Although the female's eye-ring color was not important to the male during early pair formation, it became so just before copulation: A male would not mate with a female that had the wrong eye-ring color. Thus, the importance of eye-ring color to both males and females ensured a successful pair formation process.

Birds also can distinguish among individuals by means of subtle variations in plumage patterns, size, voice, and behavior. Ruffs have perhaps the most striking individual variations in head-color patterns of any species. More subtle are the variable, harlequin head-color patterns of Ruddy Turn-

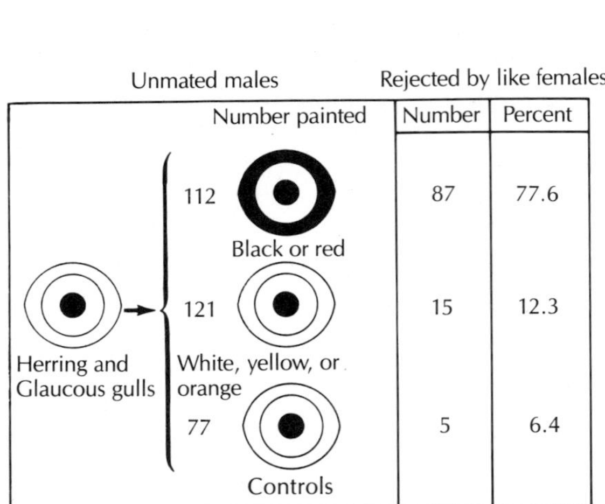

Mated females (Herring and Glaucous gulls)

Number painted		Pairs broken: Number	Percent
173	Black or red	132	76.3
163	White, yellow, or orange	10	6.1
71	Controls	5	7.0
93	Not caught	2	2.1

Mated males (Herring and Glaucous gulls)

Number painted		Pairs broken: Number	Percent
164	Dark	4	.02

Figure 9–2 Experimental alterations of the fleshy eye rings of Glaucous Gulls and Herring Gulls caused females to reject altered males in the early stages of pair formation and males to reject their altered mates just prior to copulation. Experimental controls were drugged for capture but not painted before release. (Adapted from Smith 1966)

stones, which would provide a simple basis for individual recognition (Figure 9–3). These are extreme examples. Field ornithologists learn quickly to recognize individuals by more subtle differences: plumage wear, a missing feather, or odd habits, in combination with eye colors or plumage colors typical of certain age and sex classes. Doubtless, birds are even more sensitive than we are in the use of such information.

Evolution of displays

Virtually any nonsignal behavioral pattern can evolve into a ritualized display with particular functions. Feather positions initially used to control heat loss from the body are a common source of display features (Morris 1956). Most displays, however, may have evolved from incomplete locomotor movements, called intention movements, such as the initial postures associated with leaping into the air to fly or flexing the head and bill to peck. The head-throw display of courting male Common Goldeneye ducks seems to have evolved from the locomotory movement of leaping out of the water (Figure 9–4A).

Figure 9–3 The harlequin face and neck patterns of Ruddy Turnstones vary distinctly among individuals. (From Ferns 1978)

Seemingly inappropriate behaviors, such as beak wiping, feather preening, or drinking, often appear in aggressive situations (Tinbergen 1952; Ziegler 1964). In the middle of a fight, a Blue Jay may suddenly wipe its bill several times as if it had a compelling itch. Alternatively, a bird may redirect its actions: Instead of attacking a mate, it may attack another individual. Herring Gulls that are facing combat redirect their pecking at

(A)

(B)

Figure 9–4 Two ritualized displays. (A) Head-throw display of a Common Goldeneye. (B) Tidbitting display of a Peacock. (Courtesy San Diego Zoological Society)

the ground. Such redirected behaviors may be incorporated into the formal display repertoire of a species. Preening movements, for example, have become a ritualized element of the courtship displays of terns and ducks (Lorenz 1941; Van Iersel and Bol 1958).

Ritualization is the process of the evolution of signals and displays from nonsignal movements. Ritualized feeding movements of various forms are incorporated into the courtship displays of quails, pheasants, and peacocks (Schenkel 1956; Williams et al. 1968). In these so-called tidbitting displays, the male bows before the female, spreading wings and tail to varying degrees, and, in some species, gives a food call (Figure 9–4B). Male Northern Bobwhites feed their mates as part of the tidbitting display just before copulation. Ring-necked Pheasants and Domestic Chickens display by manipulation of food and mock pecking but do not feed their mate. The Himalayan Monal Pheasant bows low before the female and pecks vigorously at the ground amidst full sexual display of its wings, tail, and head feathers. In the most simplified and ritualized form of the tidbitting display, the Common Peafowl merely points its bill at the ground. The origin of the display movement would not be apparent without comparison among related species.

Behavior, therefore, reflects evolutionary heritage. In fact, we can infer the phylogeny of behavioral traits in much the same way as we infer the phylogeny of morphological traits. The two are often correlated: Closely related birds tend to have similar behaviors. Behavioral traits, for example, distinguish the main groups of pelecaniform seabirds (Van Tets 1965). Members of the suborder Pelecani—pelicans, boobies, anhingas, and cormorants—have a similar bowing courtship display that is absent from the courtship behavior of tropicbirds and frigatebirds. The closely related boobies and gannets have characteristic head-wagging displays, whereas the closely related anhingas and cormorants have kink-throating and pointing displays. Pelicans, on the other hand, lack these displays as well as sky-pointing, wing-waving, and the hop used by boobies, anhingas, and cormorants (Figure 9–5).

Figure 9–5 Some displays of pelecaniform birds. (A) Bowing displays: 1, quiver-bowing of Brown Booby; 2 and 3, front-bowing of Common Cormorant; 4 and 5, wing-bowing of Northern Gannet; 6 and 7, front-bowing of Red-footed Booby. (B) Pointing (left) and kink-throating (right) displays of male American Anhinga. (C) Sky-pointing displays (left to right): Brown Booby, Masked Booby, Red-footed Booby, and Blue-footed Booby. (D) Wing-waving displays (left to right): Great Cormorant, Olivaceous Cormorant and Pelagic Cormorant. (Adapted from Van Tets 1965)

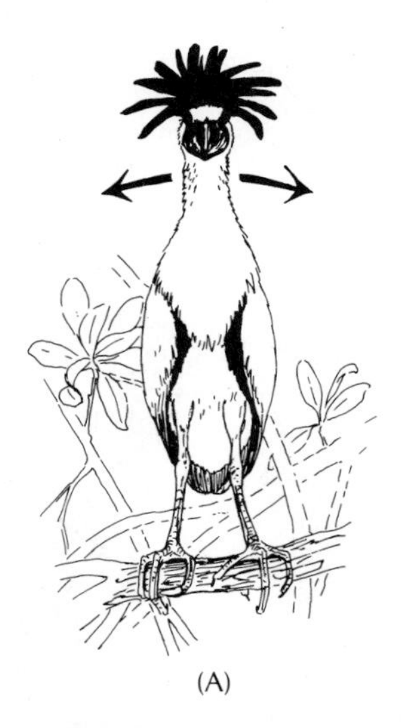
(A)

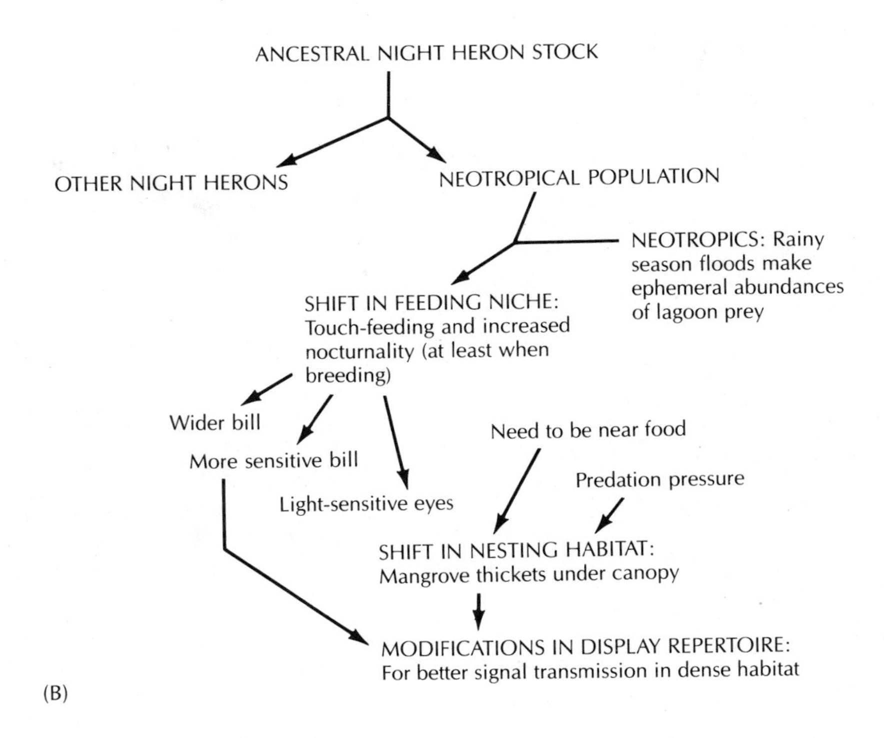

(B)

Figure 9–6 Tall-rocking display of the Boat-billed Heron (A), a species in which shifts in ecology (B) have led to major display modifications. (Adapted from Mock 1975)

In some birds, such as the Boat-billed Heron, a shift in feeding ecology has led to transformation of displays. The Boat-billed Heron is a remarkable neotropical night heron with a wide bill evolved for touch-feeding in seasonally flooded lagoons (Mock 1975). Its repertoire of social displays differs substantially from that of other herons. For example, the displays emphasize acoustical signals made with the bill. The heron also uses dramatic visual displays such as the tall-rocking display (Figure 9–6). It appears that a shift in the Boat-billed Heron's feeding niche fostered the evolution of such morphological developments as a wider, more sensitive bill and more sensitive eyes for nocturnal feeding. The ecological and accompanying morphological transformations then led to radical modifications of displays for better communication in the dense nesting habitat where visibility was poor.

The extrinsic displays of bowerbirds

In the bowerbirds of Australia and New Guinea, the evolution of courtship displays has had an unusual result, namely a separation of the relationship between the bird's morphology and its displays. Courtship display has somehow been transferred from an association with plumage to an association with external nestlike structures, called bowers, which are essential for courtship and copulation. Male bowerbirds build bowers of sticks and grass in elaborate forms of two general kinds: maypole bowers and avenue bowers. Maypole bowers consist of sticks built around a central sapling, or maypole; avenue bowers are walled structures placed on a south side of a

(A)

display court. Bowers of both kinds are decorated with brightly colored objects (Figure 9–7).

The decorations are as extraordinary as the bower structures themselves. Some species paint the walls of their bowers with fruit pulp, charcoal, or shredded dry grass mixed with saliva. The Satin Bowerbird applies the paint with a twig brush. Other species decorate their bowers with mosses, living orchids, fresh leaves turned upside down, or colorful fruits. They restore experimental changes in arrangements, discard unsuitable items placed near their bowers by experimenters, and replace wilted flowers or leaves with fresh ones daily. Males often steal each other's decorations.

The Satin Bowerbird complements its brilliant blue eyes with anything blue it can find: One bower was decorated with parrot feathers, flowers, glass fragments, patterned crockery, rags, rubber, paper, bus tickets, candy wrappers, fragments of a blue piano castor, a child's blue mug, a toothbrush, hair ribbons, a blue-bordered handkerchief, and blue bags from domestic laundries (Marshall 1954). The Spotted Bowerbird of Australia is notorious for household and camp pilferage of scissors, knives, silverware, coins, jewelry, ignition keys, and even a glass eye snatched from a man's bedside.

(B)

Figure 9–7 (A) Bowers of bowerbirds include simple forest clearings with ornaments on the ground (1); a mat of lichens decorated with snail shells (2); a maypole built of sticks about a central sapling or fern and surrounded by a raised, ornamented court (3–5); and a decorated avenue built with varying complexity of stick walls opening onto a platform (6–9). (B) Breeding behavior of the Satin Bowerbird. The male builds an avenue bower of sticks (top left), at which it courts visiting females (top center). Females judge bower quality and then may solicit copulation by crouching in the bower (top right). Other males destroy the bower in the absence of the owner (bottom left) and may try to interrupt copulation (bottom center). Mated females lay their eggs and rear their young without male help at nest sites away from the bower (bottom right). (From Borgia 1986, with permission of Scientific American)

Particularly intriguing is the apparent trade-off or transfer of function between plumage elaborations and bower displays. Modestly colored bowerbirds tend to have more elaborate bowers than do brightly colored species (Gilliard 1956, 1969). The Striped Bowerbird, which has only a short orange crest, builds an elaborate, well-decorated hut around the maypole, whereas the related MacGregor's Bowerbird, which has a long and conspicuous orange crest, builds only a simple column of sticks without much decoration.

We see, therefore, that avian displays involve various signals and coordinated behaviors. The elaboration and emphasis of visual signals usually emphasizes the bird's plumage coloration. The process of evolutionary ritualization transforms explicit actions into symbolic ones.

The value of ritualization

Ritualized displays are packages of signals evolved for purposes of communication. The traditional view is that ritualization increases the efficiency and clarity of information transfer (Cullen 1966; Zahavi 1977). To this end, ritualization also increases the stereotypy of a display, sometimes to astonishing precision. The 1.3-second head-throw display of the Common Goldeneye, for example, varies with a standard deviation of only 0.08 second (Dane et al. 1959). Clarity rather than ambiguity would seem to be the overriding value of stereotyped displays.

Clarity may be enhanced further by the increased discreteness or contrast of the display. The elaborate cartwheel display of several male manakins is a well-delineated, all-or-nothing courtship display with clear intention. In the case of the displays used by the Common Black-headed Gull, the contrast between the upright-threat display and normal posture defines unambiguously whether the recipient is welcome.

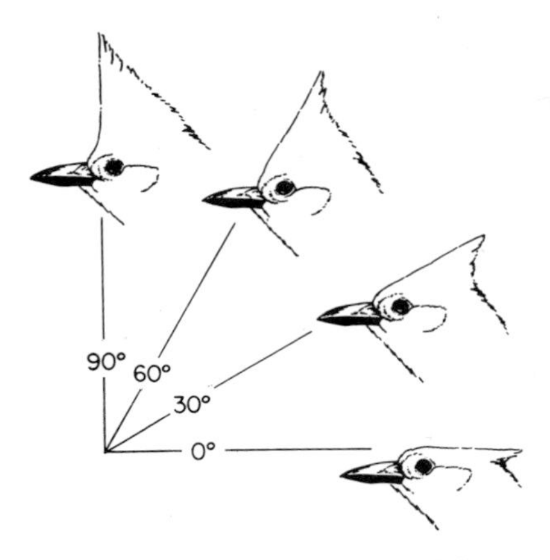

Figure 9–8 The positions of the crest of a Steller's Jay signal the likelihood of attack (high crest) or escape (low crest). (From Brown 1964a)

Graded or variable displays convey information about intensity of motivation and the probability of a sender's subsequent actions. The high-crest positions assumed by a defensive Steller's Jay indicate that it will probably attack rather than flee its opponent (Brown 1964a) (Figure 9–8). Variations in stereotyped displays do not necessarily make for confusion; they may enhance communication once a reliable standard has been established (Zahavi 1980).

Deciphering displays

Deciphering the information transmitted by a display remains one of the greatest challenges in the study of bird behavior. Ornithologists can only guess at the message of a display from correlations between preceding and succeeding actions of sender and receiver. We now recognize not only that a single display may contain several messages but also that the messages themselves may vary with the context of the display (Smith 1969, 1977).

Douglas Mock's studies of the displays of Great Blue Herons and Great Egrets illustrate the ways ritualized displays are used to communicate with mates or potential mates (Mock 1976, 1978, 1980). The two species of herons nest side by side in open habitats and communicate primarily by means of visual displays. Each of the fifteen displays in the repertoire of the Great Blue Heron has its particular set of contexts (Table 9–1 and Figure 9–9). The highly stereotyped stretch display is a long, conspicuous display that exhibits a heron's bright bill colors and chestnut wing linings to the fullest. The erected neck plumes enhance the visual impact of the neck motions during this display. The stretch display occurs in several contexts: in male advertisement; as part of the nest relief ceremony, when mates relieve each other of incubation duties; and when the female sends her mate to collect another stick for the nest, a form of pair-bonding cooperation that Mock refers to as intrapair appeasement.

The forward display of the Great Blue Heron (see Figure 9–9) is a variable display used in contexts of aggression, particularly in nest defense and female-to-female encounters. Douglas Mock describes a typical performance:

> The heron moves its wrists out from the sides of the body, retracts its neck part way onto the shoulders, and erects all plumes of the head, neck, and back. In this position it either stabs at another bird or walks toward it before stabbing. The stab is performed with a rocking motion: the legs straighten, the neck extends, and the head passes through a short arc as the heron emits a sharp "squawk" and clacks its bill at the point closest to its opponent. (Mock 1976, p. 206)

Males refrain from using the forward display in the early stages of pair formation, when attraction of a potential mate is the goal, but use it more often as courtship proceeds through phases of critical assessment and possible rejection. The use of this display subsides once the pair bond is firmly established.

Table 9–1 Contexts and Messages of Great Blue Heron Displays

	Displays														
	Stretch	Snap	Wing Preen	Circle Flight	Landing Call	Twig Shake	Crest Raising	Fluffed Neck	Arched Neck	Forward	Supplanting	Bill Duel	Bill Clappering	Tall Alert	Static-optic
I. Uses/contexts:															
External disturbances						X	X	X						X	
Nest defense						X	X	X	X	X				X	X
Male advertisement	X	X	X	X		X	X								X
Female-female encounters							X		X	X	X			X	
Greetings at nest*	X				X		X	X	X				X		
Intra-pair appeasement	X				X								X		
Intra-pair aggression							X					X			
II. Messages (Smith 1969)															
Identification	X	X	X	X	X	X	X	X	X	X	X	X	X	X	X
Probability	X	X	X	X	X	X	X	X	X	X	X	X	X	X	X
General set							X								
Locomotion					X	X	X				X				
Attack							X	X	X	X	X	X			
Escape							X								
Nonagonistic	X												X		
Association	X	X	X	X									X		
Bond-limited	X		X		X								X		
Play															
Copulation															
Frustration															

*Includes greeting ceremony, nest relief ceremony, and stick transfer ceremony
(From Mock 1976)

Each display in the repertoire of the Great Blue Heron conveys a particular message. The various forward displays project the probability of an attack. The stretch display, frequently used in courtship, projects a message of withdrawal or submission. Great Blue Herons use display signals in a variety of contexts to achieve a variety of ends. They may increase the

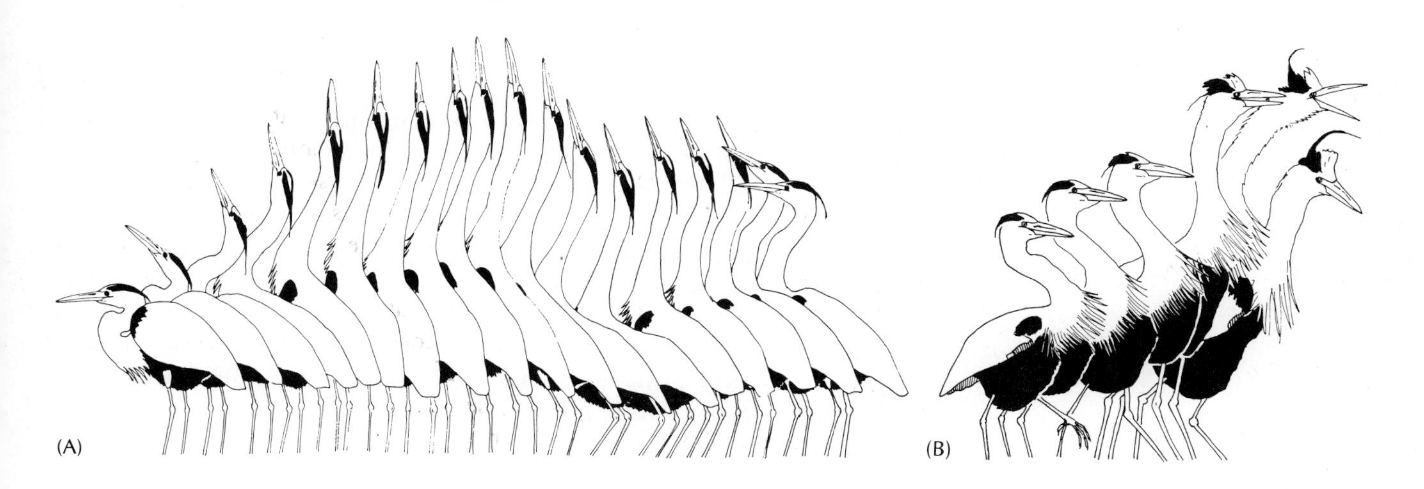

Figure 9–9 (A) Stretch display and (B) forward display of the Great Blue Heron. These sequences are drawn directly from movie frames of a filmed display. (From Mock 1976)

complexity of a signal by adding vocalizations, by erecting particular plumes, and by coding information into the variability of the display itself, in the same way that the Steller's Jay reveals the probability of attack with the position of its crest. Plume positions, body positions, duration of the display, and repetition of movements are some of the display variables that provide information.

The Great Egret achieves similar ends by different means. The displays in its repertoire of 16 are each less variable in form than those of the Great Blue Heron and are less likely to be accompanied by acoustical signals. Whereas the Great Blue Heron uses the stretch display in a variety of contexts, the Great Egret uses it in only one, spontaneous advertisement by unpaired males. Instead of varying the presentation of one display, the Great Egret increases its power of communication by varying the sequence in which different displays are presented.

Agonistic behavior

When two birds interact, they both have selfish purposes that can foster either hostility or cooperation. Birds can manipulate one another to individual or sometimes mutual benefit. Inherent in all social interactions governed by rules is the threat of cheating by those that would take advantage of the existing system. For many years, students of bird behavior have tended to assume the morality of truthfulness in their interpretations. Now it appears that avian social communication may not be as straightforward and honest as we once supposed. Individual birds serve their own interests in many ways. The nature of communication between rivals as well as between partners, therefore, invites our attention. We begin this discussion with a traditional view and finish with the provocative idea of avian deceit.

The competitive encounters between rivals, complex mixtures of aggression (attack, threat) and escape (submit, flee), are called agonistic behavior. When birds fight over something — mates, food, or territory — they usually avoid direct contact and risk of injury by using threat and appeasement displays.

Threat displays, which emphasize the bird's best weapons, its bill and wings, herald a real attack if the issue is not resolved quickly (Figure 9–10). Appeasement or submission displays signal the opposite intent, a willingness to yield on the point at issue, which defuses the conflict and thereby protects the yielding individual from direct attack. Often the submissive bird turns its head and bill away from a threatening rival, which reduces the level of provocation and avoids a physical attack. An appeasing avocet, for example, hides its long bill beneath its back feathers and adopts a sleeping posture. Other species fluff their feathers, in contrast to the sleeked postures associated with threat displays.

Communication of aggressive intent or submission is a central function of the social displays of birds. Even courtship usually starts with aggression by the male toward the female. If the female stays, the actions of the male shift from hostility to appeasement, subordination, solicitation, and ultimately establishment of a pair bond with regular contact and copulation.

Figure 9–10 Threat and submissive postures. (A) Threat display of Great Tit. (B) Great Tit (right) about to attack a Blue Tit (left) in submissive posture. (Courtesy E. Hosking)

The courtship of Common Black-headed Gulls illustrates this process (Moynihan 1955). Prior to the arrival of the females, male Common Black-headed Gulls gather in large areas near the nesting colony called "clubs," where each bird establishes a small, temporary pairing territory. Rival males avoid physical battle and risk of injury by ritualized aggression. As they await the arrival of a potential mate, they threaten each other with the upright-threat display, with the combination long-call and oblique display, and with the forward display attack posture. If stylized threats do not succeed, physical attacks may ensue (Figure 9–11).

Females visit pairing territories in the club to find a mate. The male greets an approaching female as a potential rival with the aggressive oblique display and long-calls. Instead of fleeing or challenging him as would a rival male, the female Common Black-headed Gull stretches her neck upward and faces away (facing-away display), revealing her sex and her potential as a

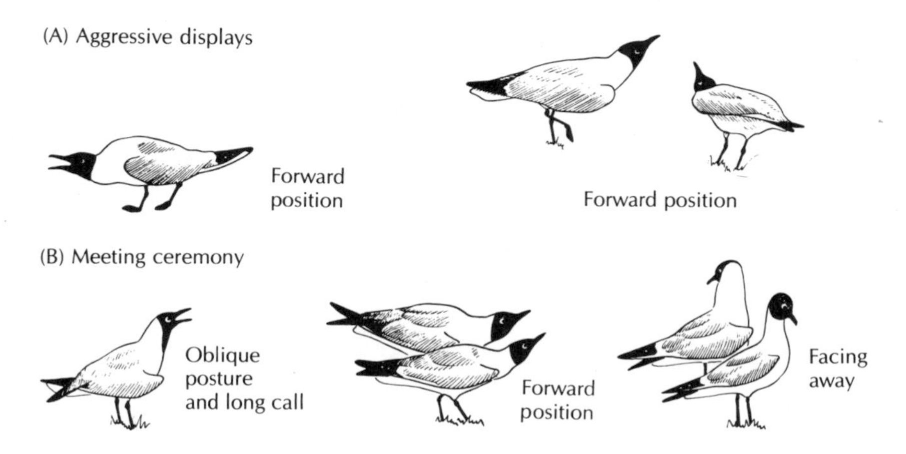

Figure 9–11 Some displays of the Common Black-headed Gull. (A) Aggressive displays using forward postures. (B) Greeting ceremony display, including oblique display with long-call (top), forward display (center), and facing-away display (bottom). (From Tinbergen 1959)

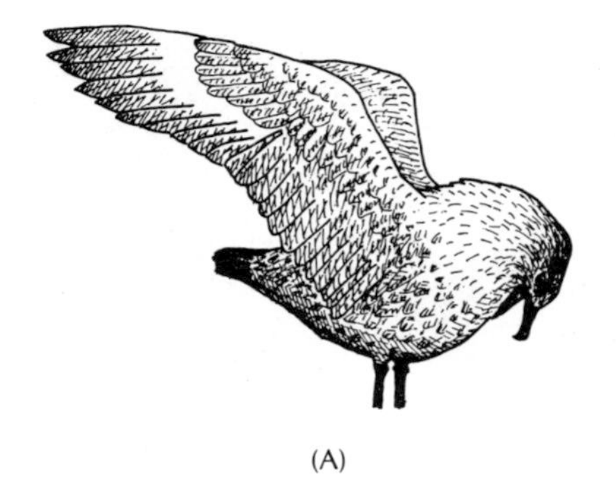

(A)

(B)

(C)

Figure 9–12 Threat displays of the Great Skua. (A) Bend posture with long-call and wing-raising, which indicates that the skua will stay and will probably attack. (B) Bend posture, which indicates that the skua will stay but is less likely to attack than in A. (C) Relaxed posture with neck withdrawn. (Adapted from Andersson 1976)

mate. In response, the aggressive male reduces the severity of his threat by redirecting the oblique display to the side. Transient at first, the female keeps returning to a selected male and stays longer each time. Ritualized appeasement replaces ritualized aggression. The potential mates engage in mutual displays, such as facing-away. Gradually, the female moves closer and begs for food, which the male regurgitates onto the ground in front of her. Copulation follows. The male then deserts its pairing territory and assists the female in nesting and rearing of young.

What information do Common Black-headed Gulls actually communicate by such displays? The traditional view is that threat and appeasement postures signal the probability that an individual will attack or escape. Reduced ambiguity through ritualization of agonistic displays should lessen the possibility that a bird would misread its opponent's intentions (Cullen 1966). Agonistic displays, therefore, should accurately communicate the probability of attack or escape and thus reliably inform a receiver of what will follow.

In fact, varied threat displays, such as the upright and the forward threat displays of the Common Black-headed Gull, do convey different probabilities of subsequent attack, but these probabilities may not be very high. Critical studies reveal that attacks do not reliably follow even the highest-ranked threat displays (Caryl 1979). For example, the most intense threat displays of the Blue Tit and the Great Skua are followed by an attack only about half of the time (Stokes 1960; Andersson 1976) (Figure 9–12). Furthermore, the degree of the reactions that threatened individuals show is not directly related to the degree of threat in the displays, as we would expect them to be if "probability of attack" was the message being sent.

Information about the probability of escape, however, is a more trustworthy aspect of agonistic displays. Submissive or appeasement displays predict escape more reliably than threat displays predict attack. Escapes almost always follow certain postures, such as the crest-erect-facing-rival display of the Blue Tit. Such instances suggest that we might view complex sequences of agonistic displays more productively as contests between individuals, as games of bluff. It appears that escape signaling is straightforward and that threat displays tend to be bluffs. It follows that the submissive bird determines the outcome in many ritualized contests.

The game of communication

Is avian communication fundamentally different from human communication? To what degree do birds use their repertoires of visual and vocal signals to manipulate other birds in conscious or unconscious ways? In later chapters, we shall mention cases in which birds apparently have evolved convergent plumage color patterns, mimicking other birds to manipulate them. South Pacific orioles thereby gain access to fruit trees that are controlled by aggressive friarbirds; tanagers in the Andes flock together with reduced interspecific strife (see Chapter 15). Cases of social mimicry illustrate the evolutionary games that guide the development of visual signals. Still more striking examples of trickery in avian communication are found in brood parasites that mimic their hosts (see Chapter 20).

The view that avian communication may have elements of deceit or ambiguity challenges traditional constructs that the evolution of displays is toward a reliable transfer of truthful information (Krebs and Dawkins 1984). In fact, aggressive displays do not carry a reliable message, as we have just seen. Rather, they signal some possibility of continuing the encounter, and may be only a bluff. In the game of communication, senders may try to manipulate the attention and muscle power of their observers to some selfish advantage; other individuals are considered tools to be used if possible. For example, the unmated male not only announces his availability with displays and song but also does his best to get prospective females interested enough to move into his territory. Listeners try to decipher the true intentions of a sender and respond accordingly in their own best interests. If the female Common Black-headed Gull approaching a male on a club territory concludes that, despite her submissive response, she will be attacked rather than tolerated, she will leave and perhaps try again, or she will move to an alternative prospect.

Summary

Birds communicate with each other via displays, which are acts specialized to make information available. Avian displays serve primarily as a means of identification, to communicate locomotory intentions (attack, escape, move, or stay still) or other intentions such as a desire to mate, play, or claim ownership. The comparative study of displays of related species of birds has played a major role in our understanding of animal behavior.

Visual displays function in concert with plumage color and plumage elaborations. The bowerbirds of Australia and New Guinea use external displays of elaborate accumulations of colorful objects. Displays evolve from nonsignal behavior patterns through a process called ritualization, which leads to increased uniformity of performance as well as modification of behavior patterns. Avian display repertoires are used in a variety of contexts or only in specific contexts.

Agonistic displays inform opponents about probabilities of attack or escape, but birds may also bluff. Birds may not always be truthful with each other.

Further readings

Alcock, J. 1984. Animal Behavior, 3rd ed. Sunderland, Mass.: Sinauer. *An excellent text with many bird examples.*

Hailman, J.P. 1977. Optical Signals: Animal Communication and Light. Bloomington, Ind.: Indiana University. *An intriguing review of avian color patterns in the context of the science of semiotics.*

Krebs, J.R., and R. Dawkins. 1984. Animal signals: Mind-reading and manipulation. *In* Behavioural Ecology (J.R. Krebs and N.B. Davies, eds), pp. 380–402. Sunderland, Mass.: Sinauer. *A provocative new view of avian communication.*

Lorenz, K. 1965. Evolution and Modification of Behavior. Chicago: University of Chicago Press. *A classic.*

Marler, P., and W.J. Hamilton. 1966. Mechanisms of Animal Behavior. New York: John Wiley & Sons. *An excellent older text that emphasizes communication.*

Smith, W.J. 1977. The Behavior of Communicating. Cambridge: Harvard University Press. *A pioneering work on the role of context and nature of messages in avian communication.*

chapter 10

Vocalizations

Animals communicate by means of visual, acoustical, tactile, chemical, and even electrical signals. Two of these, visual and acoustical communication, dominate in bird behavior. Complementing the use of visual displays to mediate social interactions are rich vocabularies of sounds. Vocalizations serve birds especially well for communication over long distances, at night, and in dense cover.

The scientific literature on avian vocalizations began almost 400 years ago with the observation by Ulyssis Aldrovandus that ducks and chickens called even after their heads were chopped off; the source of the vocalizations was apparently sited in the body and not the head (Greenewalt 1968). The source of avian vocal abilities is, in fact, a unique organ, the syrinx, which operates with nearly 100 percent physical efficiency to create loud, complex sounds and which can even produce two independent songs simultaneously. In other respects, we now know that bird song has much in common with human music and speech, sharing similar sounds, tones, and tempos. Furthermore, bird song is produced by a series of rapid and complex

Figure 10–1 Northern Mockingbird singing. (Courtesy A. Cruickshank/VIREO)

motor actions, such as those controlling the tongue during speech or the fingers of a skilled violinist playing an intricate passage (Marler 1981). In recent decades, the study of bird song has greatly benefitted from advances in recording technology and our ability to analyze sounds electronically (Figure 10–2).

In this chapter we discuss the physical characteristics of bird vocalizations and examine how the syrinx produces these sounds. We then turn to the functional aspects of acoustical communication by birds: What kinds of information do particular songs and calls convey, and what information do birds use to recognize their own species or to discriminate among individuals? The sizes of song repertoires vary greatly among species and even among males of the same species. The advantages of song variety and the role of vocal mimicry are considered at the end of the chapter. How young birds learn their songs is one of the most profound areas of modern ornithological research; this topic and related topics such as geographical dialects will be treated in Chapter 11.

Discussion of bird vocalizations requires a small, specialized working vocabulary of terms from music and from the scientific study of sounds (acoustical physics). Included in Box 10–1 are some of these terms.

Physical attributes

Bird vocalizations are not easily classified. They range from the short clicks of swifts, to the quavering whistles of the tropical, partridgelike tinamous, to the long, tinkling melodies of wrens. The tremendous variety of sounds emitted by birds reflects adjustments of the potential attributes of sound to enhance information content and to enhance the physical transmission of information to listeners.

A traditional distinction exists between "songs" and "calls." The term *song* connotes long vocal displays with specific, repeated patterns often pleasing to the human ear (Pettingill 1984). Song refers primarily to the vocal displays of territorial male birds. The term *call* connotes a short, simple vocalization. Various calls include distress calls, flight calls, warning calls,

Box 10–1

Vocabulary for vocalizations

Amplitude: Loudness or maximum energy content of a sound.

Frequency: Number of complete cycles per unit time completed by an oscillating sound wave form; usually expressed in hertz.

Harmonic: Tone in the series of overtones produced by a fundamental tone; the frequencies of the tones in a harmonic series are consecutive integral multiples of the frequency of the fundamental.

Hertz (Hz): Unit of frequency equal to one cycle per second.

Modulation: Defining the form of a carrier wave by variation of either frequency or amplitude.

Oscillograph: Device that records oscillations as a continuous graph (called an oscillogram) of corresponding variations in an electric current, as would be generated by a tape recording of a sound (see Figure 10–2).

Pitch: Relative position of a tone in a scale, as determined by its frequency.

Sinusoidal wave form: Simple, pulsed cycles of energy that describe a regularly rising and falling sine curve, defined by the equation $y = \sin x$.

Sonagram: Visual display of the frequency content of a sound distributed in relation to time (see Figure 10–2).

feeding calls, nest calls, and flock calls (Thorpe 1961). In actuality, there is no real dichotomy between songs and calls in either their acoustical structure or their function. Yet the term *song* is so entrenched and alternatives so lacking that continued use seems certain (Greenewalt 1968).

A fundamental dichotomy, unlinked to the perception of songs versus calls, does exist in the acoustical structure of bird vocalizations: whistled songs versus harmonic songs (Greenewalt 1968). Whistled songs lack harmonic content. The repertoires of nonpasserine birds include both whistled and harmonic songs. Although most passerine bird songs are whistled, there are many exceptions including the harmonic-rich calls of crows and jays and the scolds of Black-capped Chickadees.

Whistled songs consist of nearly pure (lacking harmonics) sinusoidal wave forms, which result from the varying compression of escaping air. The higher the pitch, the more frequently the sound waves oscillate. Both the basso profundo (80 to 90 hertz) of a Spruce Grouse and the high, thin notes (9000 hertz) of Blackpoll Warblers are, technically speaking, whistled songs (Figure 10–2).

Harmonic songs employ harmonics, or overtones, that have multiple frequencies of the fundamental frequency. One harmonic is dominant

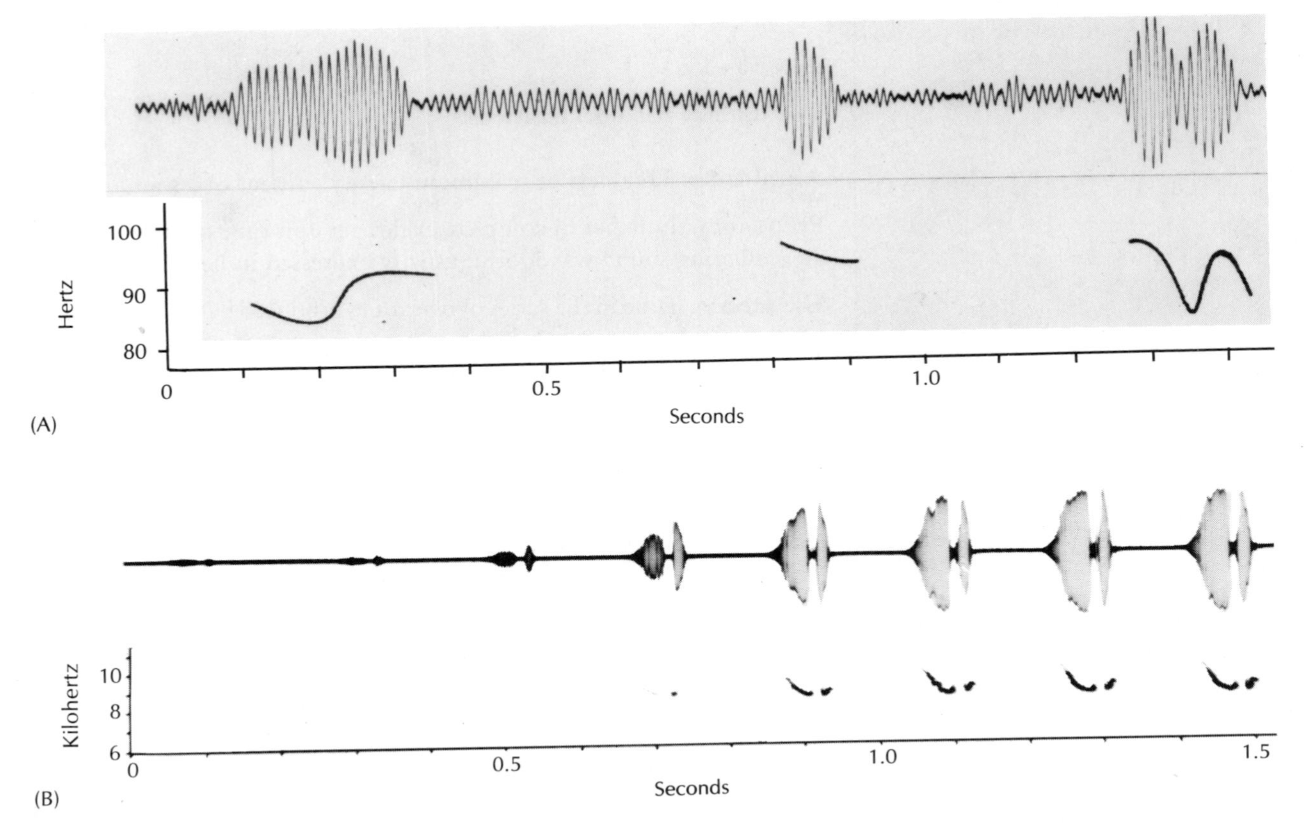

Figure 10–2 Paired oscillograms and sonagrams of (A) the bass notes (90 cycles per second = 90 hertz) of the whistled song of a Spruce Grouse and (B) the high, thin notes (9 kilocycles per second = 9 kilohertz) of the whistled song of a Blackpoll Warbler. The oscillogram (upper) displays patterns of amplitude modulation as the vertical deflection (above and below the midpoint) of the sinusoidal wave form; frequency is calculated from the number of complete cycles per second. The sonagram (lower) displays the distribution of energy (kilocycles per second = kilohertz) in a song with respect to time. (From Greenewalt 1968)

(has more energy than others) in the spectrum. The number of harmonics and their relative amplitudes determine the timbre, or general tonal quality, of the notes of bird songs (and musical instruments). Qualities such as clarity, brilliance, and shrillness, as well as nasal and hornlike tones, reflect combinations and emphases of harmonics (Figure 10–3). The distinctive sounds of a clarinet and a Hermit Thrush result from an emphasis on the odd-numbered (3-5-7 etc.) harmonics (Marler 1967, 1969).

The physical structure of a sound affects the ease with which a listener—predator or neighbor—can locate its source (Marler 1955a). The calls that birds use to locate or attract each other, for example, are made up of short notes with a broad frequency range. The assortment of frequencies in such notes enriches the information about direction and distance (p. 167). In contrast, alarm calls are faint, thin (narrow frequency range), high-pitched calls of long duration, designed to conceal the sender's whereabouts (Figure 10–4). The physical structure of a particular sound also determines the distance it will travel and how much physical distortion it will sustain

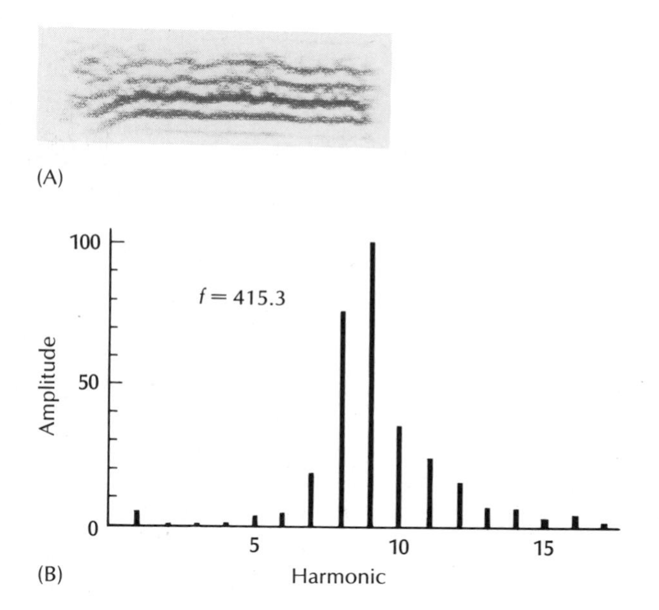

Figure 10–3 The *dee-dee-dee-dee* scold call of a Black-capped Chickadee consists of a series of harmonic phrases. (A) Sonagram of one *dee* phrase. (B) Amplitude of the tones of the harmonic series; the fundamental frequency *f* is 415.3 hertz. (From Greenewalt 1968)

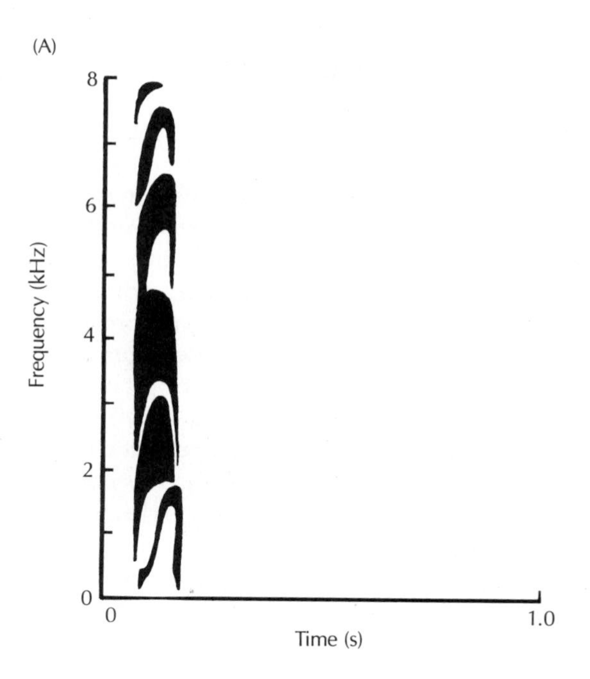

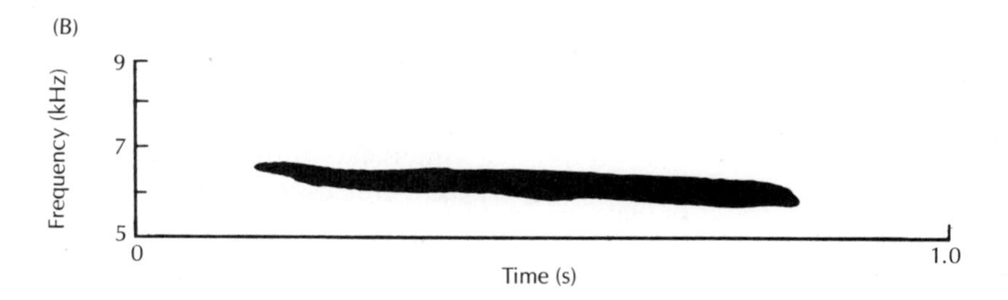

Figure 10–4 Sonagrams of two vocalizations of a Eurasian Blackbird. (A) The call that is used when mobbing an owl is of short duration and broad frequency range; it is easy to locate and attracts other birds to the site. (B) The alarm call, used when a hawk flies over, is of long duration and narrow frequency range; it is difficult to locate and thus does not reveal the blackbird's location to the hawk. (Adapted from Marler 1969)

before reaching the listener. Interference, absorption, and scattering of the sound waves by vegetation, the ground, and air progressively distort a sound (Morton 1975; Wiley and Richards 1982).

Low-frequency sounds, such as the calls of grouse, cuckoos, doves, and large owls, are the most effective for long-distance communication; they are less subject to attenuation and interference than are high-frequency sounds. The calls of tropical forest birds, which depend on long-distance communication through dense vegetation, are usually lower in frequency than those of tropical species living in open habitats (Chappuis 1971; Morton 1975). Birds of the forest floor, such as antbirds and curassows, have low-pitched calls that suffer minimal distortion from ground reflections. The complex buzzlike songs of open field birds such as Clay-colored Sparrows contrast with the simpler whistles of forest birds. Reverberations in forests mask the fine temporal structure of bird songs. Forest-dwelling birds, therefore, tend to produce simple sounds (Wiley and Richards 1982). Conversely, broadband songs, rich in temporal structure (with complex frequency modulations), are advantageous in open habitats because simple, sustained tones tend to be distorted by strong temperature gradients and air turbulence.

Sound production by the syrinx

Birds range from virtually silent to garrulous. At one extreme, Mute Swans, Turkey Vultures, and Common Rheas only hiss and grunt occasionally. At the other extreme, mynas, parrots, mockingbirds, and skylarks possess seemingly unlimited vocabularies.

The vocal virtuosity of birds stems from the structure of their unusual and powerful vocal apparatus. All songs and calls come from the syrinx, a unique avian organ located in the body cavity at the junction of the trachea and the two primary bronchi. The syrinx may form from tracheal tissues, as in Neotropical woodcreepers, antbirds, and their relatives, from bronchial tissues, as in most cuckoos, nightjars, and some owls, or from tissues of both structures, as in most birds. The avian larynx, located at the top of the trachea, does not include vocal cords but serves only to open and close the glottis and thereby keep food and water out of the respiratory tract (Figure 10–5).

Crawford Greenewalt (1968, 1969; see also Gaunt and Wells 1973) outlined the main principles by which the syrinx produces sound. Contraction of thoracic and abdominal muscles forces air from the main air sacs through the bronchi to the syrinx. On each side of the syrinx is a thin, glass-clear membrane, the internal tympaniform membrane *(membrana tympaniformis interna)*. Sound is caused by the vibration of the air column as air passes through the narrow (syringeal) passageway between the internal tympaniform membrane on each side of the syrinx and a corresponding projection on the opposite wall, called the external labium. Vibrations of the internal tympaniform membrane — regulated by its mass, internal tension, and protrusion into the adjacent air column — determine the sound characteristics. The efficiency of sound production is extraordinary: Nearly 100 percent of the air passing through the syrinx is used to make sound, com-

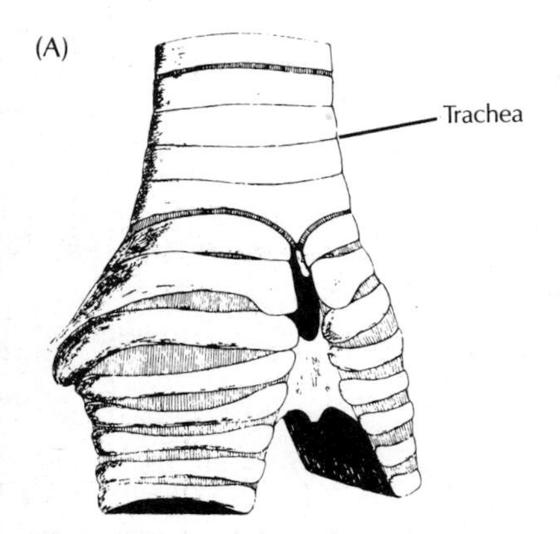

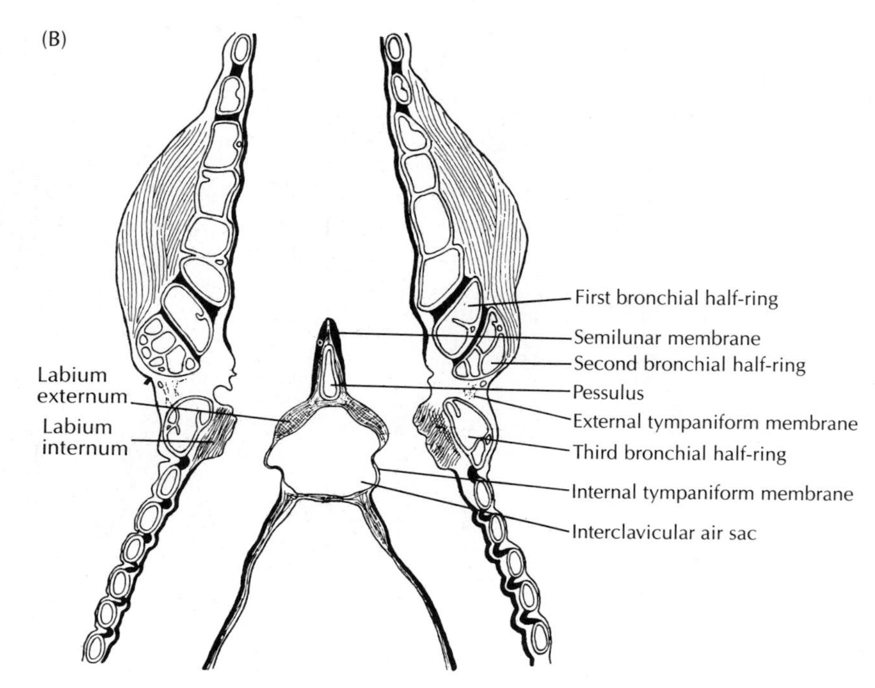

Figure 10–5 (A) Bird vocalizations originate from the syrinx, an organ located at the junction of the base of the trachea and the two bronchi. (B) The main elements of the syrinx are its vibrating tympaniform membranes, the muscles that control tension in these membranes, and the supporting cartilage. (Adapted from Häcker 1900)

pared with only two percent in human sound production (Figure 10–6).

Surrounding the syrinx is a single interclavicular air sac. Pressure in the interclavicular air sac pushes the thin membranes into the bronchial air space, in position for vibration and creation of sounds. A needle puncture of the interclavicular air sac prevents buildup of the necessary pressures, which affect the position of the tympaniform membranes; it thereby renders a bird voiceless. Sound tone depends on the precise tension of the membrane. Syringeal muscles change the tension of the tympaniform membrane as a bird sings. When the membrane vibrates without constraint, it produces a whistled song without harmonics. Rippling distortions of the membrane's free movement result in the production of harmonics, but the exact origin of these distortions is still under investigation.

The syringeal muscles control the details of syrinx action during song production. Species that lack functional syringeal muscles, such as ratites, storks, and New World vultures, can only grunt, hiss, or make similar noises. Most nonpasserine birds have only two pairs of narrow muscles on the sides of the trachea above the syrinx; these are called extrinsic muscles because they originate outside of the syrinx. More elaborate musculature is characteristic of oscine songbirds, which have up to six pairs of intrinsic syringeal muscles in addition to the extrinsic muscles. These in-

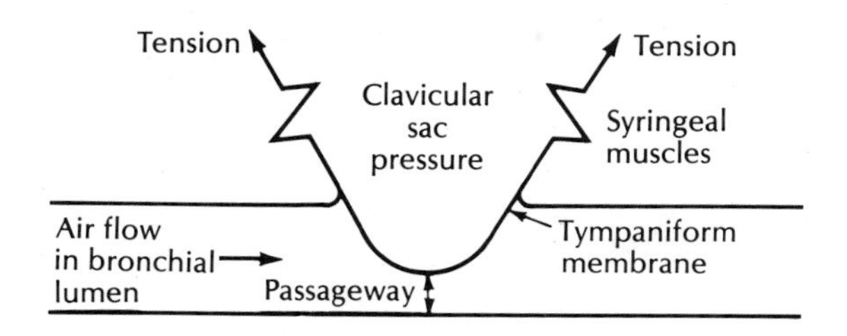

Figure 10–6 Sound production in the syrinx depends on tension of the internal tympaniform membrane, which is controlled by pressure in the interclavicular sac, contraction of the syringeal muscles, and the diameter of the air passageway. (After Greenewalt 1968)

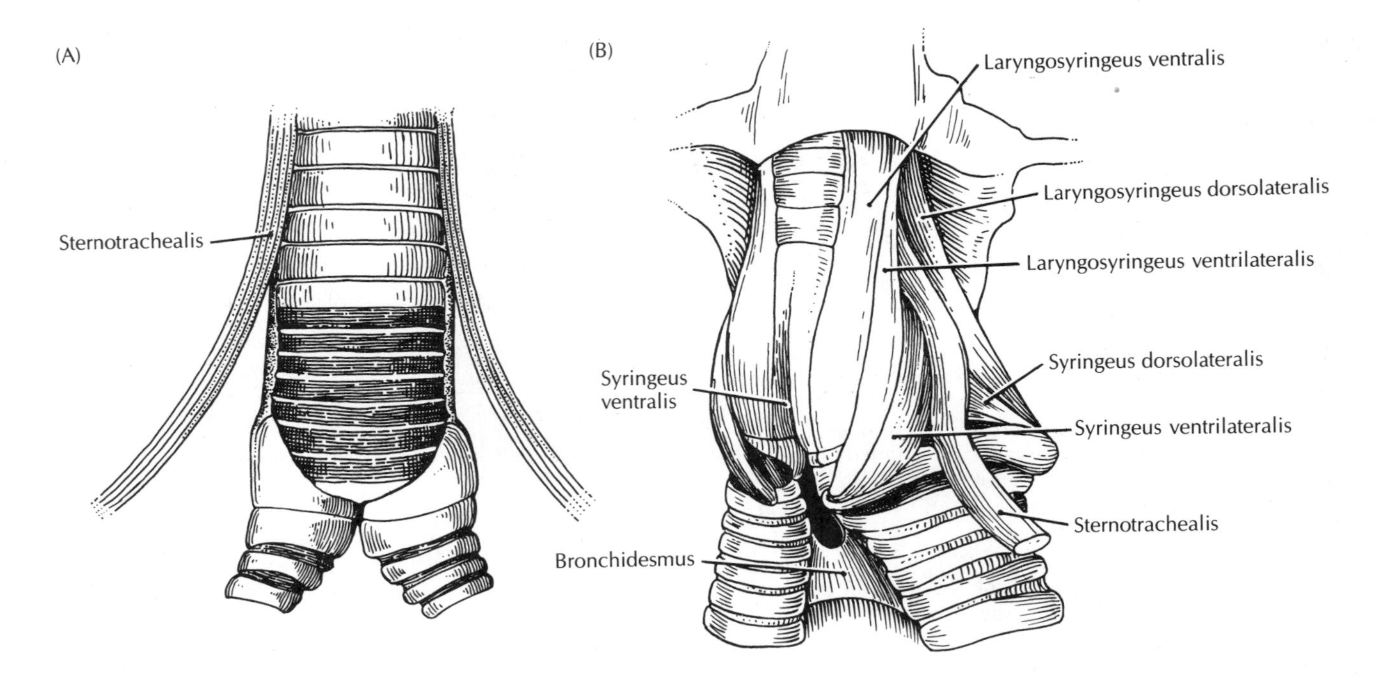

Figure 10–7 Simple and complex syringeal musculature. (A) The simple tracheal syrinx of the Chestnut-belted Gnateater with its pair of extrinsic syringeal muscles; (B) the elaborate tracheobronchial syrinx of the Little Spiderhunter, with six pairs of intrinsic syringeal muscles. The bronchidesmus is a wide band of tissue that ties the two bronchi together. (Adapted from Van Tyne and Berger 1976)

trinsic syringeal muscles originate within the syrinx and insert onto the bronchial rings, the internal and external tympaniform membranes, and onto the syringeal cartilage. Despite such well-developed syringeal muscles, the songs of oscines are not acoustically much more complex than those of species with simpler syringeal muscle arrangements. Possibly, the elaborate syringeal musculature of oscines enhances the frequency range of their songs (Miskimen 1951) (Figure 10–7).

The syrinx consists of two independent halves that can produce different, complex songs simultaneously. R.K. Potter and his colleagues (1947) discovered the phenomenon of two independent voices in an analysis of the song of a Brown Thrasher, and a few years later Donald Borror and Carl Reese (1956) described two voices in the extraordinary song of the Wood Thrush (Figure 10–8). Similar ability has since been discovered in grebes, bitterns, ducks, sandpipers, bellbirds, and many songbirds (Greenewalt 1968; Miller 1977). In addition to having different frequency con-

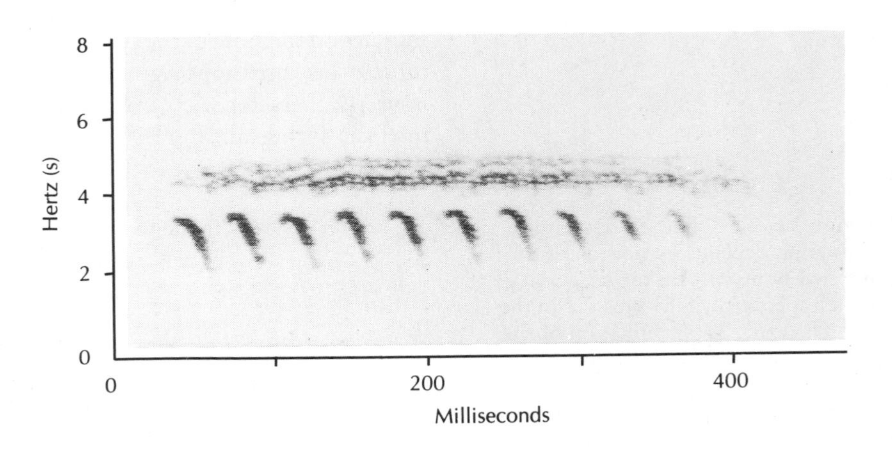

Figure 10–8 The Wood Thrush can sing a duet by itself, using two separate voices. Shown here is a sonagram of the final double phrase of the song. One voice sings a continuous series of complex, modulated phrase elements while the other voice sings a steady trill at a lower frequency. (From Greenewalt 1968)

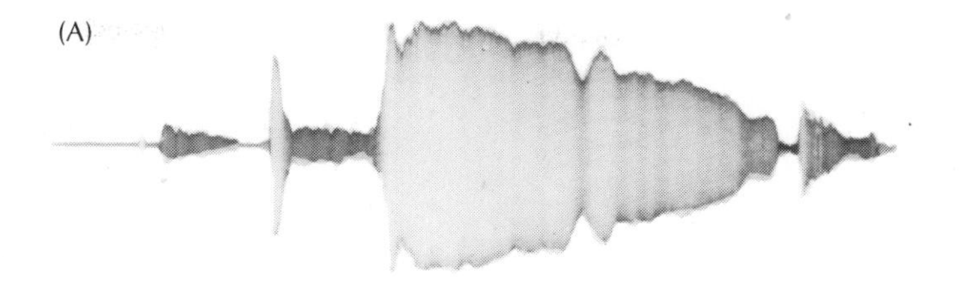

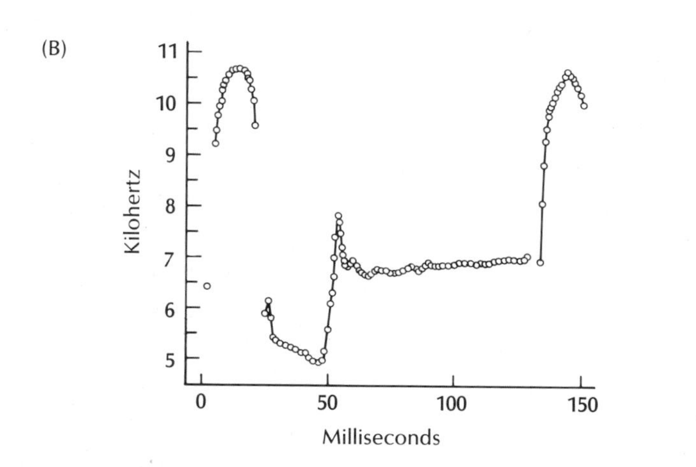

Figure 10–9 Complex modulations of frequency and amplitude characterize the song of the Brown-headed Cowbird. (A) Oscillogram of the *glee* phrase, in which the rapid cycles of the sinusoidal wave form cannot be individually distinguished; (B) a summary of the succession of frequencies composing the phrase. Note the rapid frequency modulations at 50 and 130 milliseconds. (From Greenewalt 1968)

tent, the notes produced by the dual voices can be modulated independently of one another. The existence of two separate sound sources negates both musical instrument and human voice analogies of bird song because it is physically impossible to modulate two sounds separately in a single trachea, oral cavity, or instrumental sound chamber.

Song complexity is due to modulation of the frequency and/or amplitude of a sound signal over time and is achieved through changes in the diameter of the passageway and the tension of the tympaniform membrane. Modulation in bird song is defined as a modification of the constant amplitude or frequency of a phrase. The rapidity of the modulations, their continuous orchestration, and the variable coupling of frequency and amplitude changes enable the rather simple syringeal anatomical system to produce a variety of modulated sounds.

The extraordinary frequency and amplitude modulations of bird songs are unique among animal sounds. Simple tones, such as the notes of a White-throated Sparrow, contain little modulation, whereas the variable songs of a Song Sparrow and the brief notes of a Tree Swallow contain complex, rapid modulations. Even short phrases within songs may include rapid modulations. The brief *glug glug glee* song of the Brown-headed Cowbird, for example, encompasses a four-octave interval from 700 to 11,000 hertz, the greatest frequency range in a single bird song. In one 4-millisecond fraction of the *glee,* the signal rises continuously from five to eight kilohertz, an amazingly rapid glissando (Greenewalt 1968) (Figure 10–9).

Rapid, pulsed bursts of air, when expired, cause amplitude modulations in the songs of some species. The trilled whistles of young chicks, for example, are produced by rapid vibrations of the abdominal muscles (up to 50 cycles per second) and the resulting pressure pulses of air forced into the

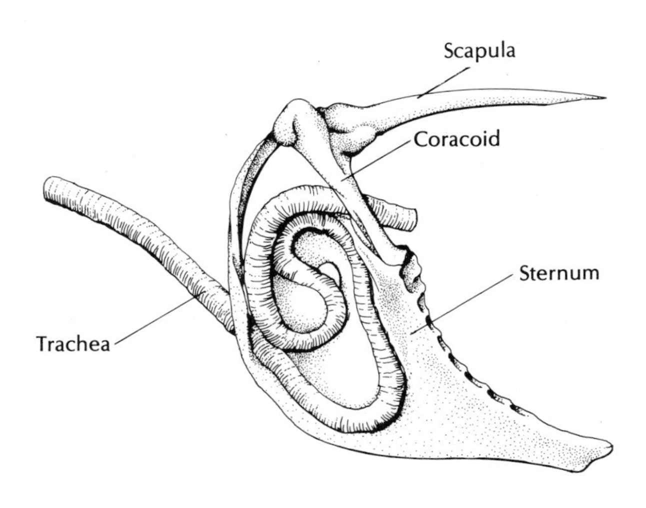

Figure 10 – 10 A crane's elongated trachea is coiled inside the sternum. (After Grassé 1950)

syrinx (Phillips and Youngren 1981). The detailed relationships between song and respiration are extremely complex. Birds such as the Pale Grasshopper-Warbler, which have sustained songs, apparently breathe and sing simultaneously by using shallow "minibreaths" (Brackenbury 1982).

To what degree the trachea modifies sound produced in the syrinx is still debated (Greenewalt 1968; Brackenbury 1982; Gaunt and Gaunt 1985). Loud, trumpetlike calls that carry great distances are characteristic of swans, cranes, some curassows, and guineafowl, all of which have an extraordinarily long trachea that is coiled in the body cavity or in the bony sternum itself. Two genera of birds-of-paradise have elongated tracheae coiled between the skin and breast musculature (Clench 1978) (Figure 10 – 10). Passage through this coil apparently enhances the resonance of the call, but not its fundamental characteristics.

Vocal communication

Just as a bird can have a repertoire of visual displays, it also can have a repertoire of calls and songs. The Blue Jay gives the familiar *jay jay* alarm call but also has a large repertoire of distinct calls that include a gurgling "pump handle call" and excellent imitations of some hawks. Birds generally have five to fourteen distinct calls with a variety of overlapping functions (Thorpe 1961; Armstrong 1963). These functions include proclamation of territorial ownership, attraction of mates, broadcast of personal characteristics (species, age, sex, competence), warning of potential dangers, and maintenance of social contact.

Most birds have some calls that are used only for occasional special purposes. Contact or association calls, for instance, help birds to keep track of one another while flocking or when in dense vegetation. Alarm calls signal danger and advise escape flight. Tyrant flycatchers emit complex "dawn songs" during the breeding season. The precopulatory trills and postcopulatory grunts integral to mating ceremonies are heard at no other time. The Common Chaffinch has twelve adult sounds, seven of which are

Table 10–1 Repertoire of the Common Chaffinch

Vocalization	Transcription	Context
Flight call	*tupe* or *tsup*	Flight or flight preparation
Social call	*chink* or *spink*	Seeking companion of unknown whereabouts
Injury call	*seeee*	Injured in fight
Aggressive call	*zzzzzz* or *zh-zh-zh*	Fighting (captive males only)
Alarm calls	*tew*	Danger, used especially by young birds
	seee	Escaping a real threat, just after copulation (breeding males only)
	huit	Moderate danger or after real danger (breeding males only)
Courtship calls	*kseep*	Active courtship (breeding males only)
	tchirp	Ambivalence toward approach and copulation with female (breeding males only)
	seep	Ready for copulation (females only)
Subsong		Practice of real song (see Chapter 11)
Song		Territoriality, identification, and courtship; average is 2 to 3 per male, up to 6.

(From Marler 1956)

used only in the breeding season, six by the male and one by the female (Table 10–1).

The loud, complex territorial songs of birds are among their most conspicuous and familiar vocal displays. Usually these are long-distance communications (carrying 50 to 200 meters or more) and convey information about the identity, location, and motivation of the singer. Territorial songs serve as signals to potential rivals that the territory is occupied by a resident male prepared to protect his exclusive use of that space and any associated females. When a territorial male Great Tit, for example, is removed from its territory, another male moves in within ten daylight hours, unless territorial song is broadcast from loudspeakers on the territory (Krebs 1977). When a song is broadcast, rival males take three times as long (30 daylight hours) to exploit the vacancy.

Inseparably coupled to the warning message is advertisement to unmated females. Female attraction to territorial male song is the first step toward courtship and pair formation. Strong but indirect evidence lies in observations that males sing less frequently after they acquire a female than before. Experimental removal of females from paired, male White-throated Sparrows results in a resurgence of song, which then subsides after a second mate is obtained (Wasserman 1977).

Territorial song also includes information about a singing male's location in the territory. The spectrum of frequencies in a call note reveals or conceals the singer's location, as will be explained. So do variant song forms

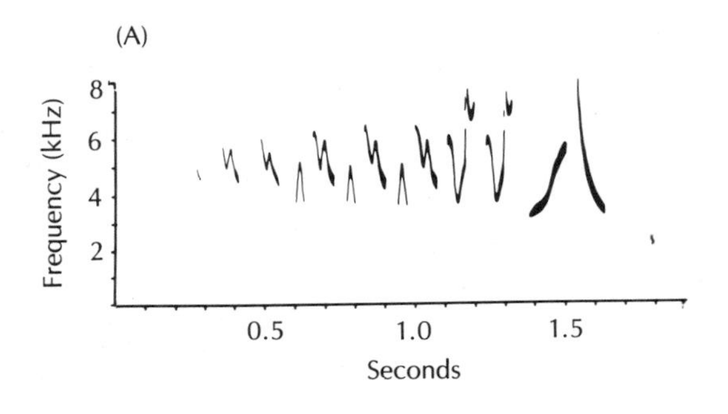

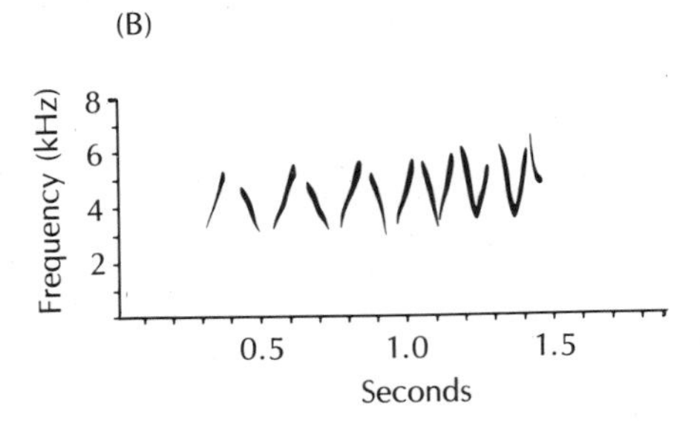

Figure 10–11 Chestnut-sided Warblers use (A) mostly accented forms of their territorial song (phoneticized as *I wish to see Miss BEECHER*) in the centers of their territories and (B) unaccented forms near the edges of their territories. (From Lein 1978)

in a male's repertoire. Male Chestnut-sided Warblers, for example, reveal their location by using accented song forms in the centers of their territories, an unaccented form exclusively at the territorial boundaries, and two other song forms at intermediate sites (Lein 1978). Territorial Yellow-throated Vireos also indicate their location, and whether they are about to change locations, by the selection of elements from their repertoire (Smith et al. 1978) (Figure 10–11).

Vocal identity

The identity of the sender is a message that pervades both visual and vocal displays of birds. After species identity is established, individual identity becomes important for reasons of status, pair bonds, and family relationships.

Birds respond readily to playback of tape-recorded songs and discriminate in field experiments between songs of their own and those of other species. The two similar kinglets of Europe, Goldcrest and Firecrest, have high-pitched, warbling songs. Despite this similarity, males always approach the tape-recorded broadcast of the song of their own species and only occasionally show interest in the song of the other species (Becker 1976, 1982). Song playback not only enables birders to draw out and view secretive species but also has clarified the taxonomic status of difficult species, such as *Myiarchus* flycatchers and *Empidonax* flycatchers. The Willow Flycatcher and Alder Flycatcher, for example, can be distinguished by their songs but not by size or coloration (Figure 10–12).

Features that birds use for species recognition are embedded in the acoustical structure of the song. In the case of the White-throated Sparrow, regularity of the pattern—pure, sustained tones without harmonics—and

Figure 10–12 The Alder Flycatcher, shown here at its nest, can be distinguished reliably from the virtually identical Willow Flycatcher only by its song. The Alder Flycatcher calls *fee-beeo,* whereas the Willow Flycatcher calls *fitz-bew.* (Courtesy A. Cruickshank/VIREO)

pitch itself determine species recognition (Falls 1969). The song of this common North American sparrow consists of a series of spaced, clear whistles differing in pitch, which fit the words *old Sam Peabody.* Male sparrows do not respond strongly to songs with wavering notes, notes with multiple tones or harmonics, frequent changes in pitch between notes, or long intervals between notes. The Goldcrests previously mentioned also use acoustical structure to aid recognition, relying on frequency range, frequency changes between notes, and alternation of long and short elements to discriminate their songs from those of Firecrests (Becker 1982).

Syntax, the sequence of particular notes, is sometimes an important recognition feature in songs. Brown Thrashers, for example, distinguish their songs from those of Gray Catbirds by the number of repetitions of each syllable (Boughey and Thompson 1976). Syntax, however, is not an essential recognition feature of the songs of either White-throated Sparrows or Indigo Buntings. Rearrangement of the notes and intervals of the sparrow song does not diminish a male's interest in the playback. Male Indigo Buntings use the rhythmic timing and patterns of frequency change in individual notes, but not the distinctive couplet arrangement of notes, to recognize other buntings (Emlen 1972).

Most experimental studies of species recognition in songbirds test the aggressive responses of territorial males to male song. Because females do not usually respond aggressively to song playback, it is difficult to study the vocal cues they use to identify others of their own species. Yet female responses to male vocalizations must be especially important if song plays a role in pair formation. Females sometimes respond directly to male song with either precopulatory trills or copulatory postures (King and West 1977; Searcy and Marler 1981). Female Song Sparrows and Swamp Sparrows whose sex drive has been experimentally enhanced with the hormone estradiol will respond more strongly to conspecific than to alien songs. They discriminate between the two by recognizing distinctive syllable structures and patterns of syllable delivery (Searcy et al. 1981).

Individual recognition

Details of song pitch, phrase structure, syntax, and composition serve as signatures of individuals enabling birds to identify young, parents, mates, and neighbors (Beer 1970; Falls 1982). White-throated Sparrows, for example, use variations in pitch to this end. Ovenbirds use variations in the structure of the phrase that can be verbalized as *teacher,* and Indigo Buntings use groupings of repeated syllables as individual signatures.

Recognition of parents by the young or of young by the parents provides an essential bond for feeding and protection, especially in dense nesting colonies in which confusion is likely. Mutual voice recognition between parents and young has been documented in penguins, alcids, swallows, and the Pinyon Jay. Although young gulls recognize their parents by voice, parents recognize their young by behavioral responses to parental calls rather than by the youngster's calls (Beer 1979).

Discrimination of individual vocalizations enables mates to recognize each other. Colonial seabirds, in particular, need to distinguish their partners from the hordes of potentially antagonistic neighbors. Black-legged Kittiwakes, among others, use frequent individually distinctive calls to maintain productive perennial pair bonds and coordinate sharing of a narrow nest ledge (Wooller 1978). Emperor Penguins, Northern Gannets, Least Terns, and various species of gulls respond quickly to playback of their mates' calls but not to those of other individuals (Jouventin et al. 1979; White 1971) (Figure 10–13).

Individual vocal differences enable birds to distinguish neighbors from strangers and to respond accordingly. Careful playback experiments have shown that birds recognize the specific calls of their neighbors (Falls 1982). Territorial males use this information to concentrate their defense efforts against strangers who are attempting to locate a mate or a territorial vacancy. Territorial males do not react aggressively to a neighbor as long as the neighbor is where it belongs, in its own territory, but they will attack a singing neighbor if it happens to trespass; they will also respond aggressively if the neighbor's songs are experimentally broadcast from the wrong side of the territory, which may indicate a greater threat of trespass.

Some birds use distinctive vocal duets for individual recognition. Vocal duets are coordinated, overlapping bouts of sounds by members of a mated pair or extended family group (Farabaugh 1982). At least 222 species

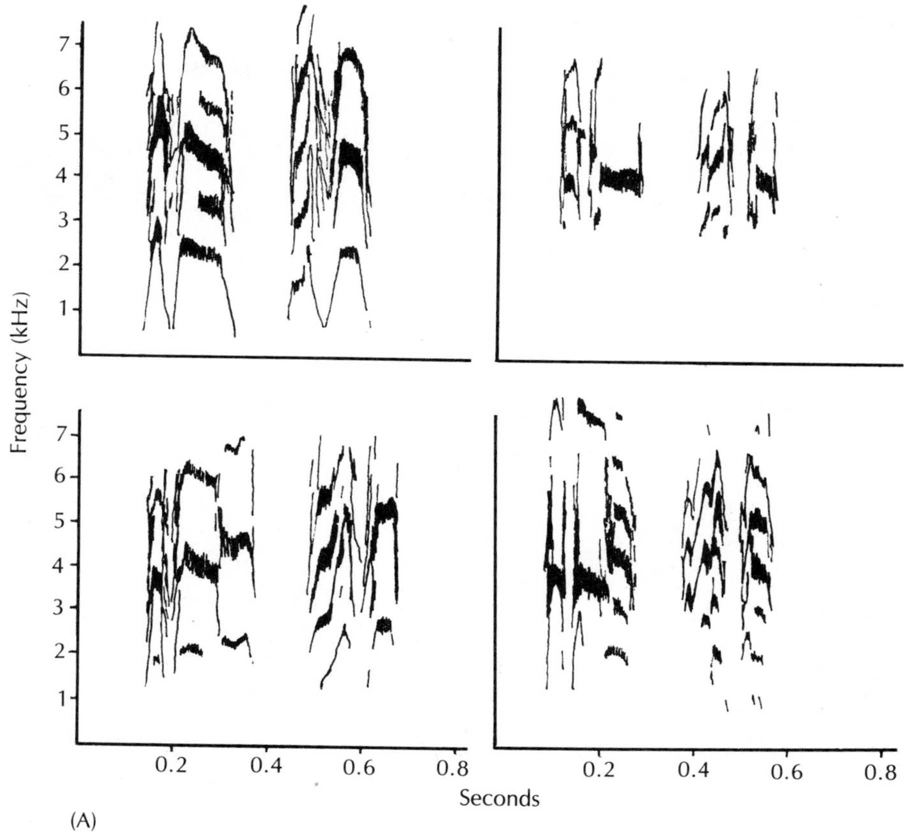

Figure 10–13 Individual differences in the "purrit-tit-tit" call enable Least Terns to recognize each other. Shown here are (top) sonagrams of the calls of four individuals and (bottom) two Least Terns at a nest (Top, from Moseley 1979; bottom, courtesy A. Cruickshank/VIREO).

in 44 families are known to sing duets. Most of these are monogamous, tropical birds that defend year-round territories. The duets function both in joint defense of territorial space against encroaching neighbors and in maintenance of the pair bond.

For example, each pair of Tropical Boubous, a kind of African bush shrike, develops a unique set of dueting patterns, which they use to keep

Figure 10–14 Tropical Boubou, an African bird famous for its duets.

track of each other in dense vegetation, to synchronize their reproductive cycles, and to maintain their territorial integrity (Thorpe and North 1966). Either member of the pair can initiate the duet. The respective note contributions are so well synchronized that few people realize that two birds, not one, are singing. A pair of Tropical Boubous increases the complexity of its duet patterns as the density of shrikes and, perhaps, the need for distinction increase (Figure 10–14).

Duetting bush shrikes respond to cues (preceding notes) in only a fraction of a second and with astonishing precision (Thorpe 1963). These reaction times can be measured quite accurately in the duets of the Black-headed Gonolek, a bush shrike that has a simpler duet than that of the Tropical Boubou. The female gonolek responds to the male's lead *youck* with a sneezelike hiss. The average response time of one female was only 144 milliseconds, with a standard deviation of 12.6 milliseconds (standard deviation is a measure of how much variation there is above and below an average). Another female responded in 425 milliseconds, with a standard deviation of 4.9 milliseconds. These values (12.6 and 4.9 milliseconds) are exceedingly low. Human auditory reaction times, not nearly as precise, have a standard deviation of 20 milliseconds.

Song repertoires

The vocal repertoires of birds are among the richest and most varied in the animal kingdom and are comparable to those of nonhuman primates (Marler and Hamilton 1966). Yet there exists a tremendous range in the sizes of avian vocal repertoires. The repertoires of territorial songs vary in size from the single song type of a White-throated Sparrow and the two distinct territorial songs of many species of wood warblers to hundreds of songs used by some wrens and mockingbirds (Kroodsma 1982a). Huge repertoires constructed from assorted syllables are typical of some wrens, several thrushes, and accomplished mimics such as the Northern Mockingbird and the Superb Lyrebird of Australia. Among wrens, the Canyon Wren has but three simple songs per individual, whereas Sedge Wrens and Marsh Wrens have over 100 songs (Kroodsma 1977). Even though Winter Wrens in Oregon have a relatively small repertoire of roughly 30 songs per individual, the songs are extraordinary: Lasting a full eight seconds, they are composed of organized sets of syllables, each consisting of 50 notes selected from a pool of 100 (Kroodsma 1980).

Various rules govern the formal sequences of bird song delivery (Lemon and Chatfield 1971; Lemon 1977). Swainson's Thrushes sing one song type after another (Dobson and Lemon 1977). Song Sparrows sing one of their 10 to 20 patterns a dozen times before switching to another pattern (Mulligan 1966). Northern Cardinals and Marsh Wrens string different songs in rigid sequences, in which each song (except the first) is determined by the preceding one.

Determining the functions of different-sized repertoires is a relatively new field of study. Large repertoires may enhance a male's attractiveness to females and his ability to compete with neighbors, discourage would-be territorial males, or stimulate continued interest by listeners. A male's reproductive success may increase with repertoire size. Female Swamp Sparrows and Song Sparrows that are treated with the sex hormone estradiol adopt copulation postures more readily in response to a male repertoire that varies (Searcy and Marler 1981). In experiments with taped song sequences, female Common Canaries respond to large repertoires by building nests faster, laying the first egg sooner, and laying larger clutches than females exposed to taped song sequences with only five different syllables. Females may also use large repertoires to identify older, more experienced mates (Nottebohm and Nottebohm 1978; Yasukawa 1981).

Male Great Tits with large repertoires achieve greater reproductive success (in terms of heavier young) than those with small repertoires. This may indicate that they were able to stimulate females more, as just mentioned, or to acquire and protect better territories (Krebs and Kroodsma 1980). The broadcast of large tape-recorded repertoires on territories that were experimentally emptied of Great Tits in one study and of Red-winged Blackbirds in another study resulted in a slower reoccupancy rate than repeated broadcasts of simple song types (Krebs 1977; Yasukawa 1981). It is not clear whether large repertoires discourage potential settlers because they mimic the presence of several males rather than one or because they signal the high social status of the singer.

Eugene Morton (1982) suggests that varied repertoires make it more difficult for neighbors to accurately assess one another's location. The patterns of signal degradation due to interference and distortion are so predictable that males of at least one species, the Carolina Wren, can use that information to judge the distance to a singing rival (Richards 1981). This ability requires detailed familiarity with a neighbor's distinctive song phrases. Variable repertoires might make critical evaluation of each phrase much more difficult, forcing neighbors to be more attentive to their territorial boundaries. The cost to rivals of time and energy, and therefore survival and reproductive output, may eventually work to the versatile singer's advantage.

Large repertoires make up the base for vocal duels between neighboring territorial males. When a Northern Mockingbird imitates a Red-winged Blackbird, for example, its neighbor will try to match that imitation (Derrickson, unpublished). Marsh Wrens also engage in vocal duels (Kroodsma 1979). Each male wren has a repertoire of over 100 slightly different songs, which are sung in complex, varied sequences. Neighboring males try to match each other's sequences or take the initiative in a duel. Leadership in such ritualized vocal duels, which draw on skill and repertoire size, can promote social dominance and its benefits.

It has not been definitely established that large repertoires prevent declining interest, or habituation, by the listener (Hartshorne 1973; Kroodsma 1978), but Great Tits, Red-winged Blackbirds, and Song Sparrows show renewed interest in playbacks of taped songs when they hear new song types (Dobson and Lemon 1975). If habituation by listeners helps to promote diversity of song types, continuous singers might project a greater variety of songs than those that sing periodically. In fact, continuous singers tend to be most versatile in their use of distinctive song forms. North American wrens, in particular, are likely to present new songs after short intervals of silence (Kroodsma 1982a).

Vocal mimicry

Imitating the calls of other species is one way that some birds enlarge their vocal repertoires (Baylis 1982). The most renowned vocal mimics include the Northern Mockingbird, European Starling, Marsh Warbler, Australian lyrebirds, bowerbirds, scrub birds, and the African robin chats. Many other birds also practice vocal mimicry, 15 to 20 percent of the passerine birds in most regions of the world do so (Vernon 1973).

One Northern Mockingbird may imitate dozens of different species, broadcasting in sequence, for example, the songs of an American Robin, a Blue Jay, a Northern Cardinal, and a variety of other common species of the eastern United States. In Texas, mockingbirds may broadcast the calls of Bell's Vireos, Great-tailed Grackles, and Dickcissels, among others. Some mockingbirds sing the songs of birds found only hundreds of miles away. Jim Tucker of Austin, Texas, for example, was surprised one morning to hear a mockingbird imitate a Green Jay, a species that is found only in the Rio Grande valley 500 kilometers to the south. Whether this song was learned directly from a Green Jay in the Rio Grande valley or was passed northward through a series of mockingbird generations remains a mystery.

Migratory species may have international repertoires. Unlike Northern Mockingbirds, which are not migratory, Marsh Warblers, among Europe's most versatile vocal mimics, spend much of the year in Africa (Dowsett-Lemaire 1979). Although they imitate some European species, most of the songs a Marsh Warbler broadcasts are those of African birds, which it hears during migration and on the wintering grounds. Territorial male Marsh Warblers thus potentially inform potential mates where they spend the winter. It may be to a female's advantage to pair with males adapted for wintering in the same part of Africa as she does, and thus to produce young with similar tendencies.

Although ornithologists have long described and appreciated the phenomenon of vocal mimicry in birds, little is known about its precise functions. For example, aside from simple expansion of repertoire size and vocal versatility, do Northern Mockingbirds perhaps communicate directly with the species they imitate? Whether mockingbirds use vocal mimicry to help exclude other species from a breeding territory remains uncertain. It is unlikely that a mockingbird could actually deceive listeners of other species and deter them vocally as it can other mockingbirds (Dobkin 1979; Howard 1974). Northern Mockingbirds, however, do chase a variety of species and also countersing alternate phrases with some of the species they imitate (Baylis 1982). Possibly, therefore, vocal mimicry mediates subtle forms of interspecific communication and reduces hostile encounters.

Some birds use vocal mimicry to attract help in the mobbing of predators. Thick-billed Euphonias, a tropical tanager, imitate the mobbing calls of other tropical birds when a nest is threatened (Morton 1976). Neighbors of the species being imitated then gather to scold and help discourage the predator.

The ability to imitate new sounds is important to the development of a young bird's vocal repertoire. Copying the vocalizations of neighbors also leads naturally to the formation of regional dialects, quite like the local accents of humans. These two important topics of current research, song development and avian dialects, are considered in detail in the next chapter.

Summary

Birds use vocalizations to mediate social interactions, particularly over long distances, at night, and in dense cover. The physical characteristics of vocalizations affect their information content and their transmission through the environment.

The vocal virtuosity of birds stems from the structure of the syrinx, a sound-producing organ located at the junction of the two bronchi at the base of the trachea. Sound results from the vibration of a thin membrane, the tension and position of which is controlled by syringeal muscles and air pressure in the interclavicular air sac. Many birds can control the two sides of the syrinx independently and thus sing two songs simultaneously. The role of the trachea in sound production, especially the long, coiled tracheae of cranes, swans, and some birds-of-paradise, remains uncertain.

The vocal repertoires of birds are among the richest in the animal kingdom. The loud broadcasts of territorial birds, which are among the most familiar vocal displays, convey information about the identity, location, and motivation of the singer, including ownership of territorial space. More varied song repertoires help to attract females and foster superiority in vocal duels between competing males. Included in the acoustical structure of songs are features that birds use for both species and individual recognition. Precise duets used by mated pairs also serve as distinctive vocal signatures. Vocal mimicry is one way some species increase the size of their vocal repertoire.

Further readings

Gaunt, A.S., and S.L.L. Gaunt. 1985. Syringeal structure and avian phonation. Current Ornithology 2:213–245. *A review of recent work on the mechanics of syrinx function.*

Greenewalt, C.H. 1968. Bird Song: Acoustics and Physiology. New York: Random House (Smithsonian Inst. Press). *A pioneering acoustical analysis of bird song.*

Kroodsma, D.E., and E.H. Miller, eds. 1982. Acoustic Communication in Birds. New York: Academic Press. *A rich collection of papers on all aspects of avian vocalizations.*

Nottebohm, F. 1975a. Vocal behavior in birds. Avian Biology 5:289–331. *A solid review of the subject.*

Thorpe, W.H. 1961. Bird Song. London: Cambridge University Press. *A classic.*

chapter 11

Learning and the development of behavior

Heritage and experience affect all behavior. Brief imprinting exposures and prolonged learning link the individual's genetic heritage of nerves, hormones, muscles, and bones to social and ambient environments. We do not concern ourselves in this chapter with the intense debates of past decades as to whether a particular behavior is innate or learned. The dichotomy was false, irrelevant to the study of bird behavior (Lehrman 1953, 1961; Lorenz 1961, 1965; Hinde 1970; Smith 1983). Instead, we shall examine evidence that behavioral patterns range continuously from those modified slightly by experience to those derived entirely from experience. Evidence comes from investigation of such diverse aspects of behavior as a bird's recognition of its own species, its mating and nesting, its choice of habitats and search for food, its social relationships, its escape from predators, and, perhaps most strikingly, its development of song.

At one extreme are the seemingly inherited behaviors young birds exhibit before they have had time to learn or modify them. Hand-raised Great Kiskadees and Turquoise-browed Motmots, for example, are frightened by sticks painted with black, red, and yellow bands to look like coral snakes (Smith 1975, 1977). Such a reaction is clearly adaptive: Coral snakes are dangerous. Rather than having to learn to associate this color pattern with danger by direct experience, birds are programmed from the outset to avoid the risk.

Birds also use experience to recognize predators by observing the mobbing behavior of other birds (Curio et al. 1978). Owls and snakes are scolded vocally and attacked when discovered. Inexperienced birds quickly associate potential danger with this commotion. Eurasian Blackbirds that watch others mob an owl in an aviary will later copy the behavior. Blackbirds will mob a stuffed honeyeater (harmless even when alive) or even a Clorox bottle if, in experiments, they have seen other birds appear to mob them.

The natural responses of young birds include positive responses of clear adaptive value, such as the reaction to coral snake colors. As soon as they are able, hatchling Herring Gulls peck at the red spot on the bill tip of a parent to receive food (Tinbergen and Perdeck 1950). The apparently simple stimulus of red near the end of the bill is in reality quite complicated, involving several ingredients such as shape and color contrast. Experiments with color-patterned billlike sticks in this species and in the Laughing Gull (Hailman 1967) revealed that the most effective stimulus for eliciting pecking was a red or blue nine-millimeter-wide oblong rod shape, held vertically at the chick's eye level and moved horizontally 80 times a minute. Hatchling Herring Gulls react faster to a red knitting needle with three white bands near the tip than they do to a parent's bill. The contrasting red and white borders of this stimulus enhance the most important stimulus features of a real bill.

All behavior shows some refinement with age. Accuracy in pecking increases with age as a Laughing Gull chick's depth perception, motor coordination, and ability to anticipate the parent's position improve (Hailman 1967). Older, more experienced chicks restrict their pecking to stimuli most similar to the head and bill of a real adult (Figure 11–1).

(A)

(B)

Figure 11–1 The accuracy of a baby Laughing Gull in pecking an adult's bill improves with age, as shown in this experiment using a painted card. Dots indicate pecks. (A) The record of a newly hatched chick and (B) two days later. (From Hailman 1969, with permission of Scientific American)

Experience improves the construction of the typical nest of a species. In the nest, many young birds manipulate the materials with interest but little dexterity, performing some of the nest-shaping movements used by their parents (Skutch 1976). The role of experience, however, is manifest much later when they actually try to build a nest. Improvement in nest construction is particularly evident in species that build elaborate nests, such as the Village Weaver (Collias and Collias 1964). Immature males build crude structures; as they mature they become more skilled in the arts of knot tying and weaving. Older males can build refined products.

When nesting for the first time, the Common Jackdaw, a small European crow, rapidly improves its nest-building skills from clumsy movements with a variety of sometimes inappropriate nest materials to efficient constructions with a range of suitable nest materials. At first, the inexperienced young jackdaw tries to shove almost anything into the nest platform. Sticks of the right size and texture insert easily and firmly into the matrix, but objects such as lightbulbs do not. By the time the nest is complete, the range of materials gathered has narrowed to the types of twigs that are most suitable for nest construction (Lorenz 1969).

Most passerine birds build nests with architectural features so typical that we can identify the builder. How does a young bird know how to build a complex nest similar to that built by its parents? Quite possibly, birds of some species inherit a mental image of their own type of nest, which guides them in construction. A male Village Weaver, hand-raised in isolation without ever seeing a nest, can build a nest that is typical of its species. Early experiences can also play a role. Zebra Finches build their first nests in the same habitats and of the same colors of materials as those of the nest in which they were raised (Sargent 1965).

Raptors imprint on their natal nest sites so that they choose a similar situation several years later when they reach maturity (Temple 1977). This ability has proved important to the conservation of endangered species. The Mauritius Kestrel, for example, ancestrally nested in tree cavities that became vulnerable to predation by introduced monkeys, causing the population to decline to only a few pairs in the 1960s. One pair then nested on a cliff ledge, out of reach of the monkeys, as did its successful young in later years. Similarly, the Peregrine Falcon reestablishment program of the Cornell Laboratory of Ornithology relies on the tendency of young Peregrines to imprint on special nesting towers for future breeding. The falcons now prefer these nest sites to traditional cliff-site eyries, which are vulnerable to Great Horned Owl predation. Imprinting, it turns out, is an extremely important interconnection between heritage and environment.

Imprinting

> It is a fact most surprising to the layman as well as to the zoologist that most birds do not recognize their own species "instinctively," but that by far the greater part of their reactions, whose normal object is represented by a fellow-member of the species, must be conditioned to this object during the individual life of every bird. (Lorenz 1937)

The process of imprinting is fundamental to the development of behavior in many birds (Gottlieb 1968, 1971; Bateson 1976; Hess 1973; Immelmann 1975; Smith 1983). Imprinting is a special kind of learning that occurs only during a restricted time period called the critical learning period; it is irreversible, that is, once learned it is persistent and cannot be forgotten (Smith 1983). The moving objects that ducklings follow in the first 24 hours after hatching define their future acceptance of comrades and mates. Imprinting determines adult habitat preferences (Klopfer 1963), the prey-impaling behavior of the Loggerhead Shrikes (Smith 1972), and the selection of nest materials and sites by adult Zebra Finches (Sargent 1965).

A well-defined period of critical learning is a distinguishing aspect of imprinting (Bateson 1976). An early sensitive period enables young precocial birds (see Chapter 19) to establish the critical concept of "parent," on which their survival depends. Ducklings, for example, imprint most strongly on a moving and calling object when they are 13 to 16 hours old (Hess 1959a, b). Young of species that leave the nest shortly after hatching, such as ducklings, must learn to distinguish their parents from inanimate or inappropriate objects. Two particular stimuli help define a parent to ducklings: movement and short, repetitious call notes. Imprinting is enhanced if both stimuli are present, but movement alone is sufficient. Chicks, ducklings, and goslings will follow and imprint on a human, a moving box containing a ticking alarm clock, or even a moving shadow on a wall. The strength of imprinting increases with the conspicuousness and variety of stimuli presented by the parent (Smith 1983).

The next step in the behavioral development of a young bird is to learn to distinguish its parents from other adults. The parents' visual appearance alone may be an important distinguishing factor (Collias 1952). In experiments with various breeds of hens, baby chicks followed the hen that looked most like their mother, on whom they had imprinted initially. Aggressive rebuffs by other adults may reinforce this process; Eurasian Coots learn to avoid menacing adults when 8 to 11 days old. Baby birds may also imprint quickly on a parent's voice, which is one of the first sounds they hear, perhaps in the egg. Baby Common Murres recognize their parents' voices upon hatching (Tschanz 1968). Accurate parent-chick recognition is most important in birds that gather in large colonies and yet whose chicks require parental attention. Parent King Penguins and their young achieve mutual voice recognition in one week (Stonehouse 1960).

A chick's early imprinting on its parents has an impact on its eventual mate choice. Cross-fostering experiments using a variety of birds including ducks, doves, finches, and gulls cause the sexual interests of young birds, particularly males, to shift to the foster species (Schutz 1965, 1971; Immelmann 1969, 1970; Harris 1970; Brosset 1971; Brown 1975). Captive birds tend to imprint on their human keepers rather than on their species, causing normal adult reproductive interests to fail. Disguises and model intermediaries help establish the development of proper species-recognition behavior (Figure 11–2).

Improper recognition behavior has some scientific advantages. Captive birds that have imprinted on their human keepers will ejaculate onto the keeper's hand, providing sperm for artificial insemination. This technique has been used for captive propagation of endangered species, as in Tom Cade's program to restock eastern populations of the Peregrine Falcon.

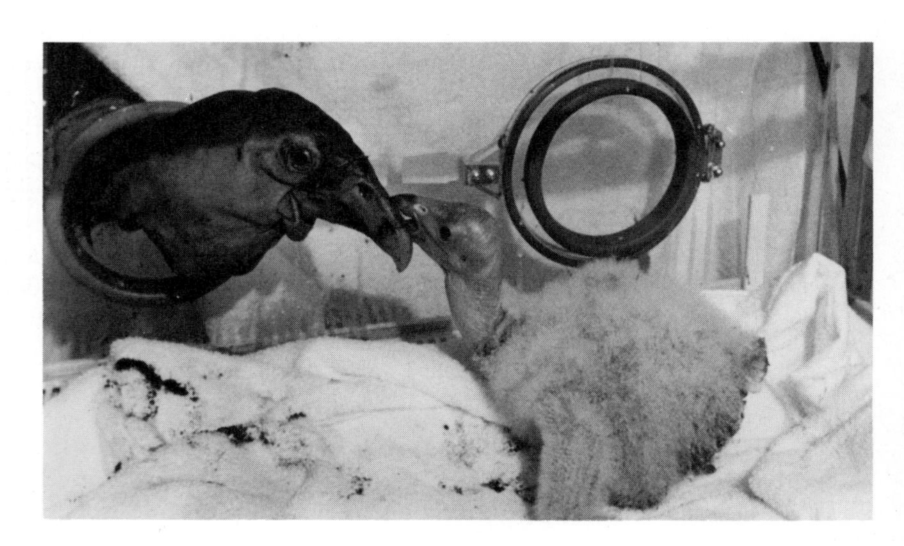

Figure 11–2 To prevent hand-raised California Condor chicks from imprinting on their human keepers, a model condor head was created as the surrogate parent. (© Zoological Society of San Diego)

Learning to feed

Efficient feeding, as described in Chapter 7, requires practice. First, the chick must develop adequate motor abilities for pecking, refined flight, and, in some species, the ability to stalk prey. Young Royal Terns, Brown Pelicans, and Little Blue Herons, for example, miss the fish they dive at more often than adults (Orians 1969; Recher and Recher 1969; Buckley and Buckley 1974; Schnell et al. 1983). The young bird must also learn what is food and what is not, and which prey are potentially harmful. For example, young European Reed-Warblers and Spotted Flycatchers progressively improve their foraging efficiency with age; they peck less often at inedible objects, increase the complexity of capture techniques, and fly longer distances to catch prey (Davies and Green 1976; Davies 1976a).

The detailed patterns of adult foraging behavior may be set by early experiences. Hand-raised nestlings of the Chestnut-sided Warbler are reluctant to take food in novel situations, but not in the situations they were exposed to while very young (Greenberg 1983, 1984). Adult Chestnut-sided Warblers are also reluctant to feed in novel situations. They are neophobic, that is, they exhibit little opportunism in the way they look for food on their breeding grounds, on their tropical wintering grounds, or in the laboratory. Early foraging experiences can even cause adults to prefer particular configurations of leaves and branches.

The period of parental care after young have left the nest can be an important training period that relates to the success of a young bird's foraging techniques, among other survival skills. Young raptors need parental training and much practice to become skilled in prey capture and killing. The training period relates directly to diet in Eurasian Oystercatchers, which by family tradition are either polychaete worm specialists or crab and bivalve mollusc specialists. Feeding on marine worms is relatively easy; they simply have to be caught peeking out of their burrows. Successful spearing of bivalve molluscs, however, is difficult. Young oystercatchers that are learning to feed on worms require only six to seven weeks to become independent of parental feeding. Learning to feed on molluscs, however, takes months, and adult oystercatchers that specialize on them help feed their young for up to a year (Norton-Griffiths 1969). Similarly, the long parental care periods of tropical insect-eating birds may be necessary to train young to find cryptic insect prey (Fogden 1972). To develop the skills of feeding on the wing, swifts and swallows practice taking meals from parents in flight. White-fronted Nunbird fledglings take food by hovering in front of a sitting parent (Skutch 1976). Juvenile Inca Terns practice picking up pieces of seaweed or trash from the ocean surface and may repeatedly release and retrieve their trial objects, or "toys," as if they are playing (Ashmole and Tovar 1968).

Play

The phenomenon of play is not nearly as well documented in birds as it is in mammals (Ficken 1977; Fagen 1981; Smith 1983). Play is a form of practice of essential locomotory and social skills. Some avian antics fit our concept of play, which includes the repeated performance of incomplete, reordered, or exaggerated behavior sequences, often out of normal contexts. For example, Common Eiders sometimes float down a series of rapids over and over again, and Adelie Penguins repeatedly ride a group of ice floes on a tide run (Roberts 1934; Thompson 1964). When young Garden Warblers discovered that dropping pebbles into a glass made a ringing sound, it became a group activity. Corvids (i.e., crows, ravens, jackdaws, and their relatives) frequently play and even create elaborate social games similar to "king of the mountain" or "follow the leader." Stick balancing and manipulation or exchange of sticks, sometimes while upside down, and taking turns sliding down a smooth piece of wood in a cage are among the many games that these intelligent birds play (Gwinner 1966).

One of the most advanced forms of learning, insight learning or learning by observation and imitation of others, may be routine among birds. Blue Jays, for example, learn the difference between edible and inedible butterflies by watching the feeding behavior of jays in another cage (Brower et al. 1970). A European Greenfinch that had been trained to discriminate so well between palatable and unpalatable foods that it always chose the first and ignored the second, was so influenced by watching an untrained companion that it began to take unpalatable food again (Klopfer 1959). The spread of the milk bottle feeding habit among English tits (page 143) is attributed to learning by imitation. Doubtless, too, novel behavior involving tool use, such as throwing stones at ostrich eggs by Egyptian Vultures and stick-probing by the Galapagos Woodpecker Finch, spread by imitation of innovative individuals. Cultural transmission of novel behavioral traits can thus be important elements in the evolution of behavior in birds.

Learning to sing

Study of the ways in which young birds learn their songs is one of the most fascinating forefronts of ornithological research. Learning guides the development of the songs or calls of parrots, some hummingbirds, and nearly 300 songbirds — all that have been studied so far (Kroodsma 1982b). Though we do not yet fully understand the complex physiological dynamics and neural mechanisms involved, the stages of song development are well documented. In fact, our current knowledge of the ontogeny of bird song may provide the best picture to date of how a complex, learned motor skill develops (Marler 1981).

Avian vocalizations can be inherited, learned, or innovated. The calls of chickens and doves, as well as the songs of certain sparrows, are inherited. When these birds are raised in acoustical isolation or are deafened before they hear their fellows sing or call, they nonetheless sing normal songs as adults (Konishi 1963; Nottebohm and Nottebohm 1971; Kroodsma 1984). The call notes of the Eastern Meadowlark are also inherited. Their songs, however, are learned from other meadowlarks (Lanyon 1957). Even in species that learn songs after much practice, inherited mechanisms play an important role in song recognition. The hearts of young Song Sparrows actually beat faster the first time they hear the song of their species. There is no change in their heartbeat when they hear the songs of another kind of sparrow. Inherited mechanisms screen out irrelevant sounds, such as those made by insects, frogs, waterfalls, and trains, and target appropriate song models.

Ornithologists have marked four stages in avian song learning:

1. An early critical learning period;
2. a comparatively long silent period;
3. subsong, a practice period that is analogous to infant babbling; and
4. crystallization, a practice period during which preliminary patterns develop into songs.

Although some birds such as Gray Parrots and Northern Mockingbirds add new vocalizations to their repertoires throughout their lives, vocal learning is most intense during, and is often restricted to, an early age. The critical learning period can be defined as the period during which information is stored for use in later stages of learning. In most species, the critical learning stage lasts less than a year. Most Marsh Wrens, for example, learn their songs before they are two months old (Kroodsma and Pickert 1980). An experimentally isolated White-crowned Sparrow, exposed to its species song by a loudspeaker, has a critical learning period of from 10 to 50 days of age. In its natural habitat, however, this sparrow has a critical learning period that may last several more months.

Some other songbirds, represented by the well-studied Common Chaffinch (Thorpe 1958), are typically receptive to song models for 10 to 12 months. The critical learning period in such birds lasts about a year, or into the first breeding season, at which time first-year males have a chance to learn songs from more experienced males. Termination of the critical learning period of the Common Chaffinch corresponds to the rise of testosterone levels in the spring. Castrated chaffinches can learn new songs for up to two years (Nottebohm 1967) (Figure 11–3).

Isolation experiments demonstrate the importance of experience during the early critical learning periods (Nottebohm 1975). Isolation

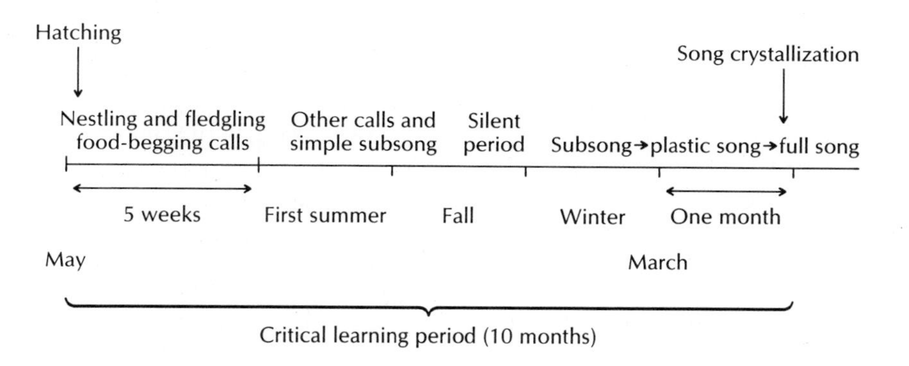

Figure 11–3 Stages of song development in the Common Chaffinch. (Adapted from Nottebohm 1971)

from the model songs of adults at this stage permanently handicaps a bird's singing ability; it will never develop a normal song. Although individuals that are isolated at an early age still sing, their songs are less complex, with fewer notes per syllable and with less frequency modulation than the normal songs. Nevertheless, the innate songs of isolated birds resemble those of their species. The syllables are fairly similar in form and are repeated in approximately the same timing, and the tonal quality also resembles that of normal song.

The second stage of song development, the silent period, can be characterized as a time during which syllables that have been memorized during the critical learning period are stored in the brain. Swamp Sparrows store memorized song syllables for 240 days (Marler and Peters 1981). When this period has elapsed, young sparrows start listening to themselves and matching some of their vocalizations to previously memorized syllables. The initial, sensitive perceptual phase of song learning is well separated from the later sensorimotor phase by this silent period.

The practice stages begin with subsong, a long, soft, unstructured series of syllables and ill-formed sounds. Subsong is known only in birds that learn their songs. It apparently bridges the gap between the perceptual and sensorimotor stages of vocal learning. Subsong is a period of practice without communication, perhaps a form of vocal play. Distinctly formed sounds begin to emerge, some of them recognizable as syllables heard during the sensitive period. Within a month or so, depending on the species, subsong develops into the first attempts at producing mature song. This is called plastic song and has only rudiments of the final structure. In a matter of weeks, during the song crystallization stage, the young bird transforms plastic song into real song. It selects a few syllables from its unstructured repertoire, perfects them, and organizes them into a correct pattern. Even those that have been isolated as nestlings develop the timing of their species. In their final songs, young male Swamp Sparrows use only one-fourth of the syllables they learned and practiced in the earlier phases of song development (Marler and Peters 1982b).

Auditory feedback is essential for song development. No passerine produces a normal song if it was deafened before song crystallization begins. In the deaf bird, recognizable structural entities do not often appear or, if they do, they deteriorate quickly. Frequency modulation of syllables is extremely poor in deaf birds, and they do not repeat sounds accurately (Konishi and Nottebohm 1969). Experimental deafening of male White-crowned Sparrows during their silent period (70 to 100 days of age) erases their original song memory or interferes with a necessary matching process. Songs of such males do not differ from those of males that have been deafened so early that they never heard model songs. Deafening after song is crystallized, however, has little effect. Apparently, auditory feedback becomes less important after the correct motor patterns are developed (Marler and Mundinger 1971).

Learning and imitation are not the only elements of song acquisition. Individuality is important, too. Young birds transform and improvise in developing individual signatures in their songs. They subject memorized themes to systematic transformations or mix syllables from several models into unique themes. A single song of the Swamp Sparrow, for example, may contain invented, improvised, and imitated elements. However, in this

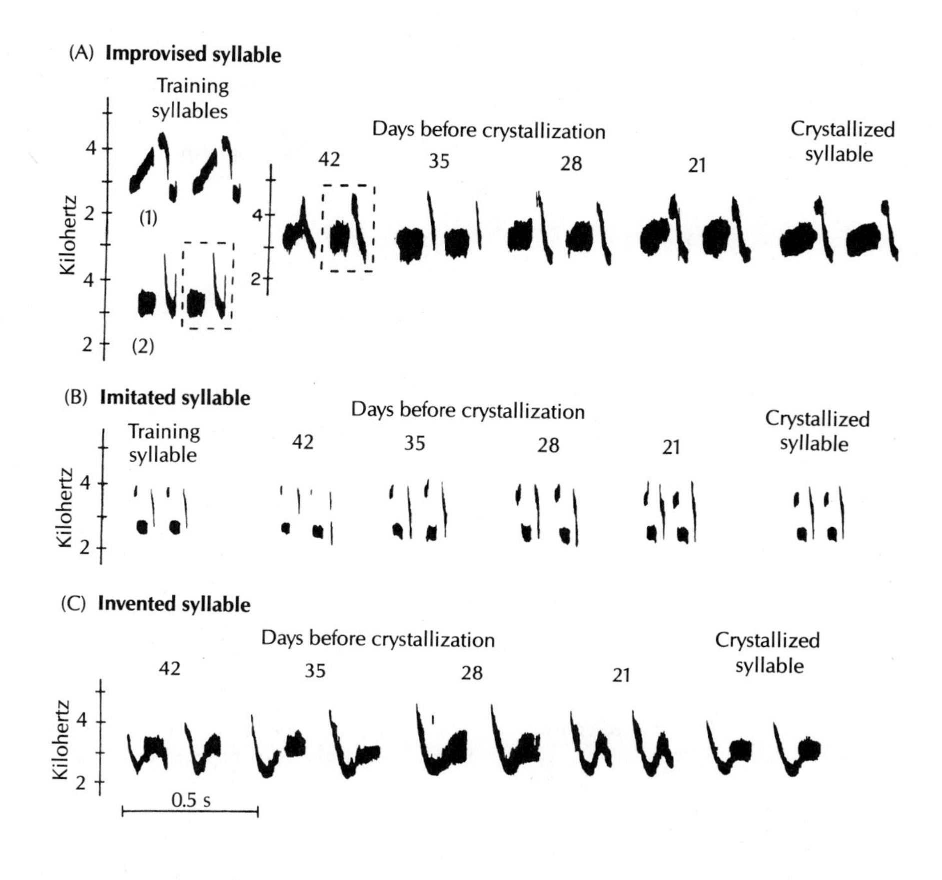

Figure 11–4 Development of three Swamp Sparrow song syllables by means of (A) improvisation, (B) imitation, and (C) invention. Training syllable A2 was improvised by substituting the second note of training syllable A1. The invented syllable was unlike any training syllable, but proceeded through the same developmental stages as the improvised and imitated syllables to a final crystallized form. (From Marler and Peters 1982a)

effort the sparrow rarely breaks up series of notes that constitute a syllable. In fact, the syllable may be a natural perceptual unit, designed to map readily onto patterns of sound production (Marler 1981) (Figure 11–4).

Song learning is mediated and constrained by inherited sensory templates, neural filters that pass only particular sounds. A young bird must select appropriate song models with precision from a rich sound environment. Recent studies of Swamp Sparrows and Song Sparrows illustrate this aspect of song learning (Marler and Peters 1977, 1981, 1982a). A Swamp Sparrow's song is a repetitious trill of a single syllable, whereas a Song Sparrow's song uses a pattern of several complex syllables. To discover how the young of these species learn their own songs despite the fact that they grow up hearing both songs, Susan Peters and Peter Marler isolated nestling sparrows and then exposed them to taped songs during the critical period.

Syllable structure is the key to song learning for young Swamp Sparrows, whereas temporal pattern is the key for young Song Sparrows. Young Swamp Sparrows reject Song Sparrow syllables, but they accept Swamp Sparrow syllables, even when these are presented in the temporal pattern of the Song Sparrow song. Young Song Sparrows can learn Swamp Sparrow syllables, but only if they are presented in the Song Sparrow temporal pattern. Therefore, in the natural setting, Swamp Sparrows do not learn the Song Sparrow song because they cannot learn its syllables; and Song Sparrows do not learn the Swamp Sparrow song because, though they can learn the syllables, they cannot learn them unless they are presented in the Song Sparrow temporal pattern.

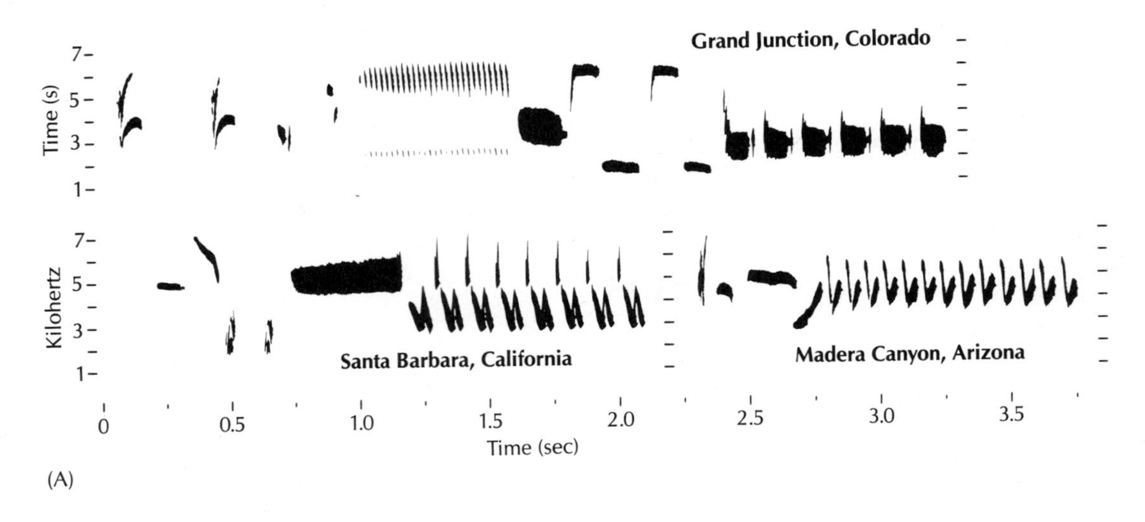

Figure 11–5 Geographical variations among song dialects and within a dialect. (Above) Bewick's Wrens sing strikingly different songs in Colorado, California, and Arizona. (Opposite) Local song dialects of White-crowned Sparrows in central California. (Above, from Kroodsma 1982a; opposite, courtesy of Peter Marler)

Dialects

Bird songs can vary within a species from coast to coast, or from one hilltop to the next. The importance of imitative learning in the song acquisition of young birds leads naturally to local dialects similar to those typical of human societies. The song traditions of birds may, in fact, provide excellent models for the analysis of rates of culturally related change in human language and in bird behavior (Payne 1981a, b; Payne et al. 1981).

Regional variation in song characteristics, such as syllable structure or delivery pattern, is typical of oscine birds (Mundinger 1982). Carolina Wrens in Ohio, for example, sing faster than those in Florida. Bewick's Wrens in California, Arizona, and Colorado have regionally distinctive song patterns (Kroodsma 1982a) (Figure 11–5). The song dialects of White-crowned Sparrows on the central California coast are restricted to areas of only a few square kilometers (Marler and Tamura 1962; Baker 1975).

How persistent are the local song dialects of birds? The *Berkeley* dialect of the White-crowned Sparrow has persisted for at least several decades. The distinctive *Egge* dialect of Common Chaffinches in northwestern Germany persisted for at least 20 years (Conrads 1966). The average lifespan of local song neighborhoods occupied by Indigo Buntings in southern Michigan was three times that of the individual buntings (or roughly 15 years) (Payne 1983). Random drift and copying errors were the primary forces of change in these instances.

To the degree that vocal dialects are epiphenomena of the process of learned song acquisition, patterns of geographical song variation may simply reflect recent history. New song traditions arise when young birds colonize new areas. The distinct songs of Common Chaffinches in each valley of northern Scotland presumably arose this way, as did the mosaic distribution of local song types of Common Creepers (Thielcke 1961, 1969). This theory states that the historical model of song dialect formation parallels that of geographical speciation, in which small, isolated populations are founded and then evolve differences (see Chapter 25).

An alternative theory is that local song dialects arise because a young male's reproductive success increases if it sings like a neighbor. The young

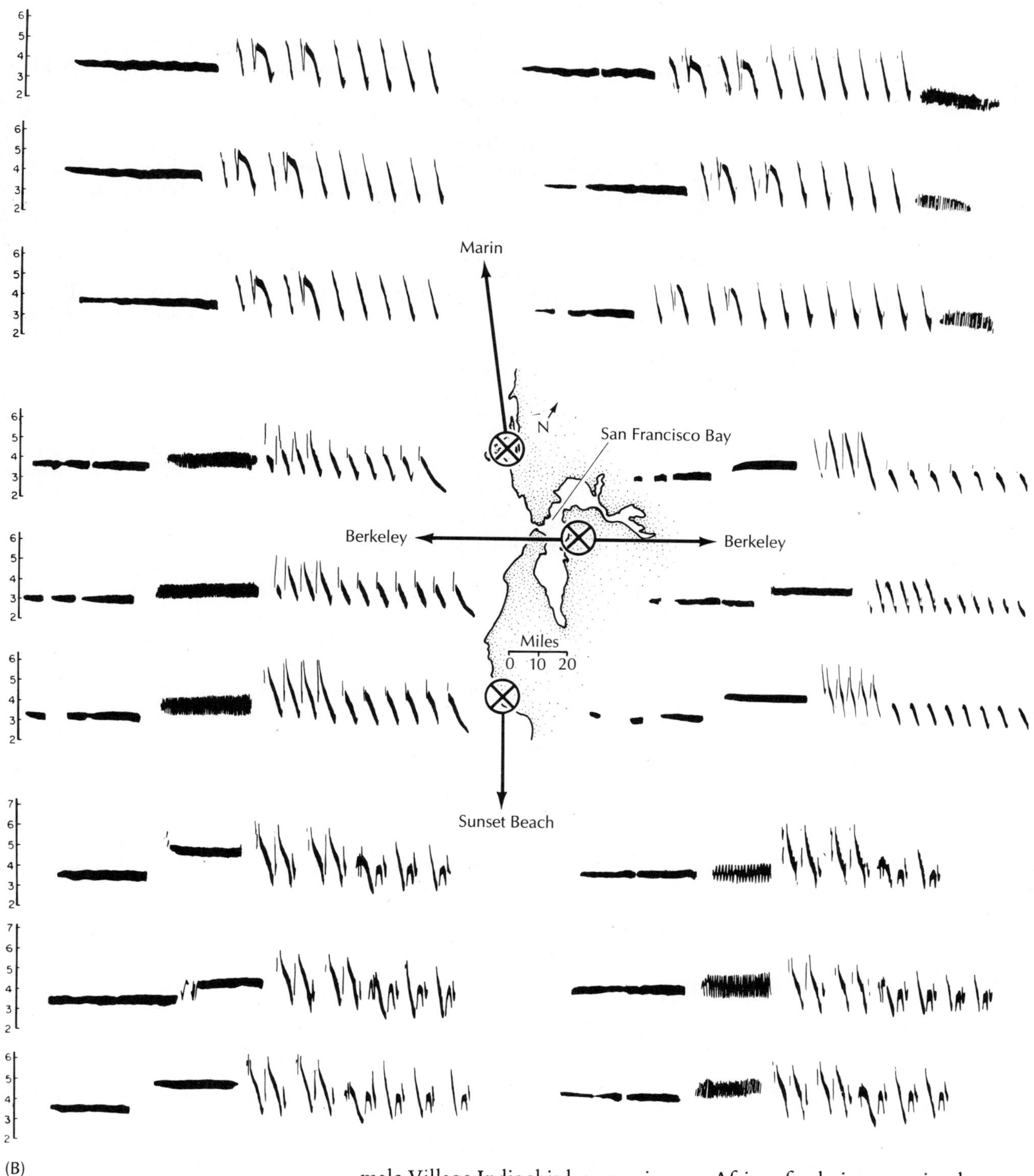

(B)

male Village Indigobird, a promiscuous African finch, increases its chances of attracting females by mimicking the song details of the dominant males that do most of the mating in a local area (Payne and Payne 1977). In North America, young Indigo Buntings, a species unrelated to the African indigobirds, also increase their reproductive success through social mimicry. First-year males copy the song of an established neighbor and thereby increase their chances of holding a territory, pairing with a female, and fledging young of their own (Payne 1982b). Social forces may thus mold the patterns of local song variation.

A third theory, the ecological hypothesis, holds that song dialects influence the genetics of local populations (Baker 1982). To the degree that local song dialects mark an environment, such as the one in which a young bird was raised successfully, they could potentially guide an individual's choice of territory and mate. The only evidence possibly linking dialects to the genetic structures of populations is the observation that genetic differences do exist among California White-crowned Sparrows of differing dialects (Baker and Fox 1978), but this conclusion is contested (Petrinovich et al. 1981; Zink and Barrowclough 1984). Nevertheless, the functional relationships among song dialects, the genetic structures of local populations, and speciation will be important topics for future research.

Summary

Both heritage and experience affect the behavior of birds. The nature of formative experiences varies from brief imprinting exposures during critical sensitive periods early in life to prolonged learning and cultural exchanges of information. Imprinting affects many aspects of avian behavior from recognition of species to choice of nest sites and habitats. However, young birds must use experience to learn to find appropriate food efficiently and to avoid danger. Avian play is an important way for birds to practice essential locomotory and social skills.

Avian vocalizations may be inherited, learned, or innovated. Learning guides vocalization development in songbirds, parrots, and hummingbirds. Four stages of song learning are evident: an early critical learning period, a long silent period, and two practice periods, called subsong and song crystallization. Guiding this process are inherited templates of song characteristics that screen out irrelevant sounds.

The formation of song dialects in a local culture is one possible consequence of the process of song learning. Dialects may reflect accidents of history and cultural change, may be used to enhance the reproductive success of young males, or may foster the evolution of local genetic differences among bird populations.

Further readings

Hess, E.H. 1973. Imprinting. New York: Van Nostrand Reinhold. *An important treatise.*

Kroodsma, D.E., and E.H. Miller. 1982. Acoustic Communication in Birds. New York: Academic Press. *Includes excellent review papers on song learning.*

Skutch, A.F. 1976. Parent Birds and Their Young. Austin: University of Texas Press. *A highly readable account of the natural history of avian reproduction, with good chapters on young birds growing up.*

Smith, S.M. 1983. The ontogeny of avian behavior. Avian Biology 7:85–159. *An excellent current review of the subject.*

Staddon, J.M. 1983. Adaptive Behavior and Learning. Cambridge: Cambridge University Press. *A psychologist's viewpoint of what laboratory and field studies have in common.*

part IV

BEHAVIOR AND THE ENVIRONMENT

chapter 12

Seasonal efforts

The singing behavior of birds reaches full expression during the breeding season. In both northern and southern latitudes this season coincides with long days and favorable food conditions. The relationship between day length and singing behavior is not coincidental. Variations in day length guide the internal physiology of birds through the annual cycle, preparing them for seasonal efforts such as breeding (of which singing is an integral part), molting, and migration.

Birds face seasons of stress and opportunity. Seasonal changes in day length, climate, and resources are inherent in the earth's daily spin and annual solar orbit. Although temperate ecosystems have seasons that are tied to temperature, the seasons in tropical ecosystems are tied to rainfall. The behavior that is necessary to survive periods of stress and to exploit the opportunities during periods of plenty has led to the evolution of physiological rhythms that correspond to environmental cycles. A bird is prepared for each season by internal hormonal controls, guided by cellular clocks that operate on daily and annual cycles in accordance with external cues such as day length, food availability, and rainfall.

In this chapter we discuss the seasonal activities of birds with particular emphasis on the timing of reproduction, molting, and migration, the three main efforts of a bird's life which are scheduled during the most favorable time of the year. We begin by describing the typical annual cycles of birds, pointing out that breeding, molting, and migration place conflicting demands on a bird's ability to obtain the resources required for survival. Then we consider the physiological clocks of birds, their circadian and circannual rhythms, which help synchronize their internal states with their environments. Photoperiod, the length of daylight, is an essential environmental cue that adjusts three clocks. Finally, we examine some factors of the timing and costs of breeding and molting. Energy for these activities must be above and beyond that required for self-maintenance.

Annual cycles

The conflicting demands of molt, migration, and reproduction, combined with seasonal differences in resources and opportunities, define a bird's annual cycle. The typical year of some permanent residents has only three main phases: breed, molt, and survive until the following breeding season. In equatorial Borneo (Sarawak), for example, where the climate is unusually stable and day length is unchanging, small birds normally start to nest when the heavy rains begin in December (Fogden 1972). Adults begin to molt shortly after the young have left the nest in May and continue molting until the beginning of the two-month "dry" season when food starts to become scarce. When heavy rains resume and food supplies increase, their gonads increase in size and the cycle repeats itself (Figure 12–1).

Similar sequences of reproduction and molt are typical of permanent residents of northern temperate localities, including Song Sparrows in Ohio, Black-capped Chickadees in Wisconsin, and Common Chaffinches in Britain. After the quiescent winter months, sex hormones flow, gonads increase in size, and males proclaim their territories with conspicuous songs and, if necessary, brutal fights. Pair bonds are established or reaffirmed and mating occurs. Young are hatched in May and June and generally reach independence by late July. Molt follows in August and September. At this time, young birds leave their natal territories, and some residents, among them chickadees and chaffinches, aggregate into well-organized flocks for the winter.

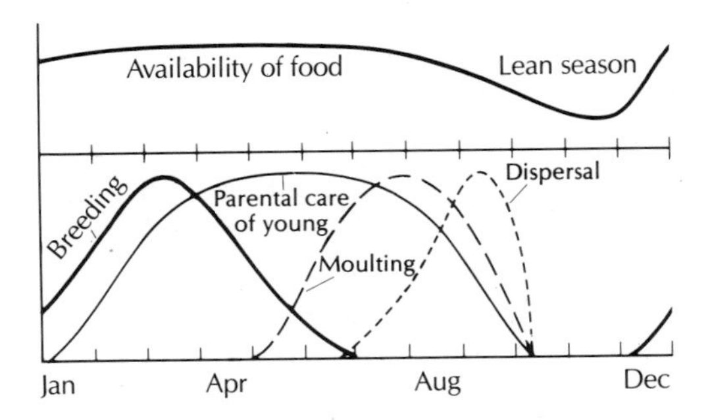

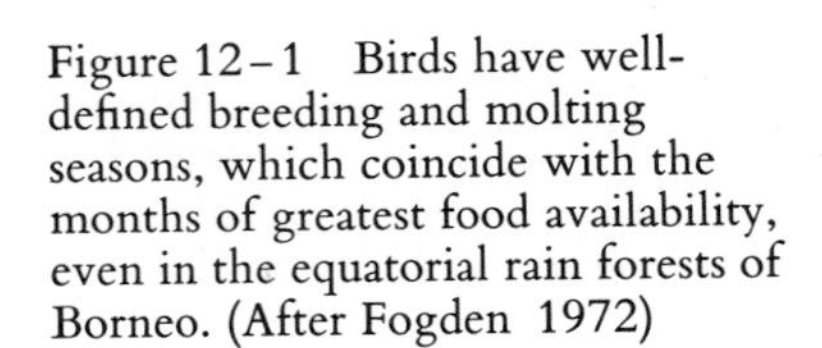
Figure 12–1 Birds have well-defined breeding and molting seasons, which coincide with the months of greatest food availability, even in the equatorial rain forests of Borneo. (After Fogden 1972)

Migration complicates the annual cycle. After molt, migratory birds generally gather in flocks and eat tremendous amounts of food, fueling themselves for their trip. As the date to depart for the south approaches, they become restless after dark. After migration, when they have reached their wintering grounds, their physiology returns to "normal" (see Chapter 13). Migratory preparations are repeated the following spring for the return to the north, where the cycle of reproduction, molt, and preparation for migration repeats. Many temperate-zone birds, especially those that migrate, molt twice a year, once after breeding and again in late winter or early spring. The spring molt does not usually include the flight feathers (see Chapter 4).

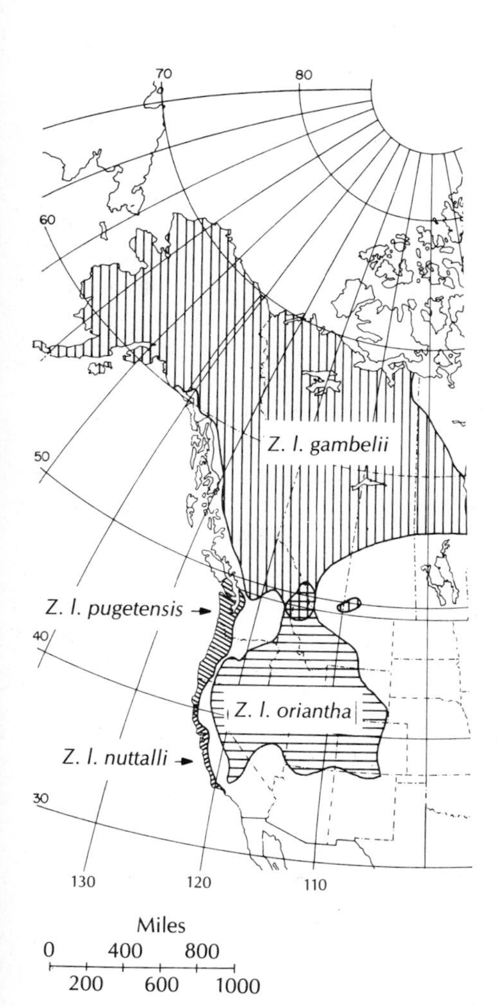

Figure 12–2 Breeding ranges of four western subspecies of White-crowned Sparrows *(Zonotrichia leucophrys)*. The most northern races, *Z. l. gambelii* and *Z. l. pugetensis,* migrate to central California, where they winter with resident *Z. l. nuttalli.* The Rocky Mountain race, *Z. l. oriantha,* migrates south to Arizona and Mexico. (From Cortopassi and Mewaldt 1965)

Variations in the annual cycle occur not just among species, but also among populations of the same species. White-crowned Sparrows *(Zonotrichia leucophrys)* on the Pacific coast of North America illustrate such variation (Cortopassi and Mewaldt 1965; Mewaldt and King 1978). The White-crowned Sparrow, a large, handsome sparrow with bold black and white head stripes and pearly gray underparts, breeds throughout northern Canada and from southern Alaska to central California (Figure 12–2). Populations on the Pacific coast differ in the extent of their annual migrations and in other aspects of their annual cycles. Those that breed in Alaska and in northwestern Canada *(Z.l. gambelii)* are long-distance migrants that winter primarily in California. In central California, they mix with winter flocks of the local nonmigratory sparrows *(Z.l. nuttalli)*. Members of another population *(Z.l. pugetensis),* which breed on the coasts of Washington, Oregon, and British Colombia, also mix with *nuttalli* flocks during the winter.

White-crowned Sparrows from northern localities nest later in the spring than those from southern localities. The resident *nuttalli* come into breeding condition first, then *pugetensis,* and finally the *gambelii* of the far north. Differences in the timing of gonadal enlargement and breeding activities characterize not only the three major subspecies but also the geographical gradient of populations within each subspecies.

Some, but not all, of these White-crowned Sparrows molt in the spring before breeding. This extra molt is known as the prealternate molt (page 72). Most of the migratory northern coastal sparrows *(pugetensis)* undergo a complete heavy prealternate molt in the spring before migration and nesting, whereas southern resident *nuttalli* merely molt some head feathers in the spring, if any. All *gambelii* molt before migrating northward, and northern populations of this race molt faster than southern populations. Their molt is not faster because the growth rate of individual feathers is faster but because the amount of plumage replaced at any one time is proportionally greater. Northern populations also start this spring molt later than do southern populations, about 3.4 days for each latitudinal degree north (Mewaldt and King 1978).

Circadian and circannual rhythms

A network of physiological controls regulates the schedules of reproduction, molt, sleep, feeding, and migration. Biological clocks in the cells of all plants and animals control the release of hormones and other chemicals that determine metabolism, reproduction, and behavior. Birds possess an elaborate system of biological clocks (Gwinner 1975). In addition to regulating the basic daily functions of general activity and cycles of body temperature, these internal clocks are essential to the proper functioning of the sun compass by which birds navigate (see Chapter 14) and are used for the measurement of day length itself. Some biological clocks are synchronized to the daily 24-hour cycle of the earth's rotation on its axis. Others are synchronized to the annual cycle of the earth's revolution around the sun.

Circadian cycles

Daily cycles of daylight and darkness have a profound impact on the biology of all animals. Twilight triggers a switch in animal physiology from diurnal to nocturnal systems. Every individual has an intrinsic rhythm approximately 24 hours in length in which the rate of metabolism, body temperature, and level of alertness cycle in predictable ways. Because they are not exactly 24 hours in length, these internal cycles tend to gradually depart from real time, starting slightly earlier or later each day, unless they are somehow synchronized or entrained by external cues, or *Zeitgebers.* For example, when Common Chaffinches are kept in constant dim light, their endogenous rhythms (e.g., activity and metabolic rate) function in a period of about 23 hours and therefore drift about one hour per day (Figure 12–3). White-crowned Sparrows have a regular cycle of activity and sleep of just under 24 hours when they are kept in a dimly lit experimental cage. Natural, external light-dark cycles, however, entrain the endogenous rhythm a little each day, keeping it synchronized with the 24-hour cycle.

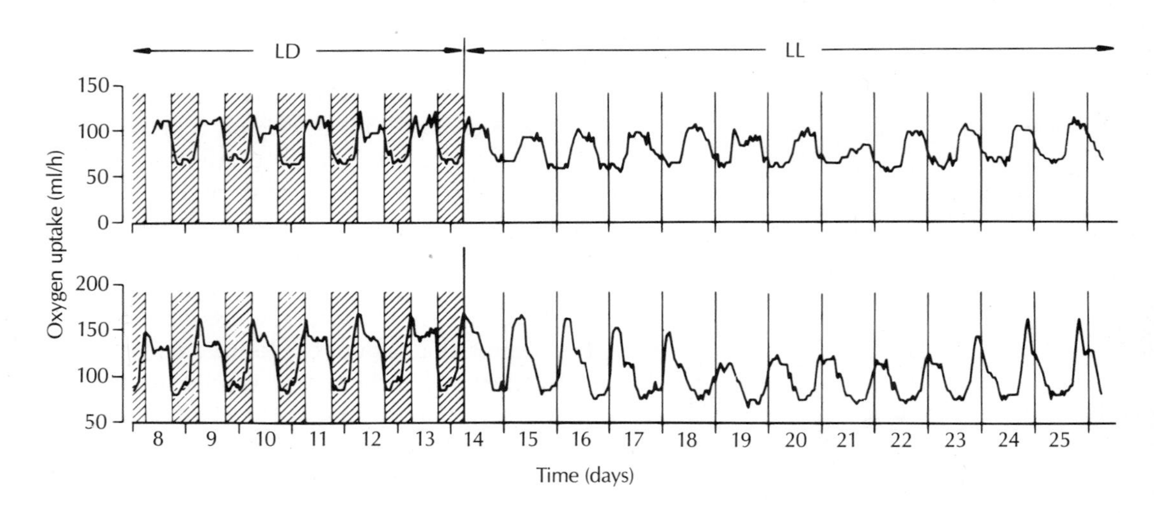

Figure 12–3 Common Chaffinches kept in a dimly lit environment have a daily activity cycle (measured here in terms of oxygen uptake) of just under 24 hours. This experiment demonstrates that under constant dim illumination (LL) the cycle drifts one hour of clock time unless it is synchronized by an external stimulus such as regular 24-hour light-dark cycles (LD). (Adapted from Aschoff 1980)

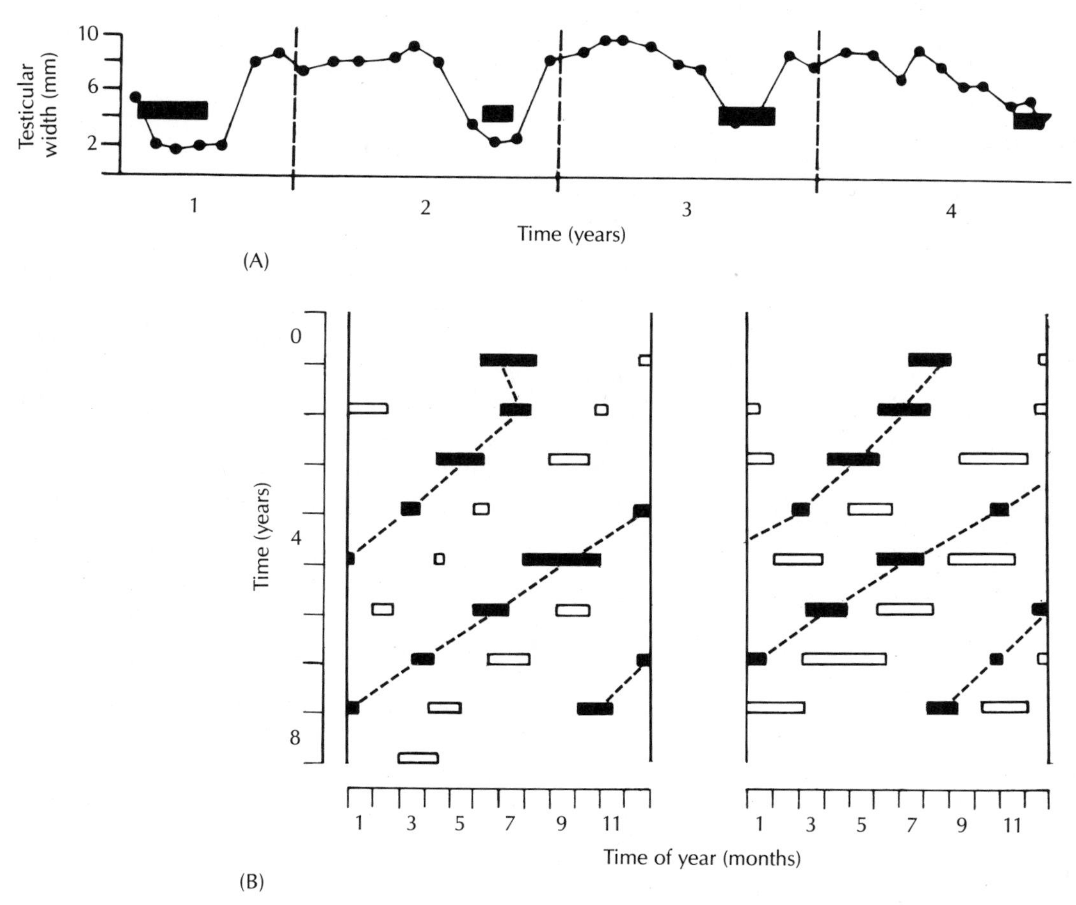

Figure 12–4 Circannual rhythms under constant photoperiodic conditions: (A) Rhythms of testicular width (curves) and molt (bars) in a European Starling. The undamped oscillations in testes size and the intervals between successive molts deviate irregularly from 12 months. (B) Rhythms of summer molt (black bars) and winter molt (white bars) in a Garden Warbler (left) and in a Blackcap (right), both kept for 8 years. Both molts occur progressively earlier each year because the birds have an internal rhythm with a mean period of about 10 months. (Adapted from Gwinner 1977; Berthold 1978)

Circannual cycles

Endogenous rhythms control the annual cycles of some birds. These self-sustaining circannual rhythms have a period of approximately one year. For example, when captive European Starlings, Garden Warblers, and Blackcaps are kept in a constant daily environment of 12 hours of light and 12 hours of dark, they come into breeding condition and molt in a predictable annual cycle (Figure 12–4). Drift characterizes circannual cycles in constant environments as it does circadian cycles. We suspect that seasonal changes in day length entrain the endogenous circannual rhythms just as daily light-dark cycles entrain the circadian rhythms (Aschoff 1955; Immelmann 1967; Gwinner 1977), but this hypothesis requires more study (Farner 1980b). In certain equatorial birds, such as the Sooty Tern, which functions on a 9.6-month internal cycle rather than a 12-month cycle, natural selection has favored the uncoupling of the internal annual cycle from seasonal change (King 1974).

Role of photoperiod

Avian seasonal reproduction has favored the evolution of a control system that synchronizes the physiologies of individuals with the environment. Day length, or photoperiod, plays a key role in this control system. Specifically, avian photoperiodic control systems use two kinds of information (Farner 1980c): environmental light, which stimulates neural receptors, and clock information from an internal circadian cycle, which enables the bird to measure day length. This meets the four basic requirements for control of the annual reproductive cycle:

1. prediction of mean optimal time for reproduction,
2. accommodation of annual variations in weather,
3. synchronization of reproductive function in mating pairs, and
4. termination of reproductive function.

In recent years the pineal gland has been at the center of discussion as the possible location of the biological clock and the mechanisms of photosensitivity in birds, partly because the pineal glands of many reptiles are photosensitive organs. The pineal glands are essential to normal circadian rhythms in at least some species of birds. Experimental removal of the pineal gland causes normal 24-hour cycles to disappear. Unlike the glands in reptiles, however, avian pineal glands do not house the primary light receptors.

The pioneering work on photoperiod control of avian gonad cycles was done by William Rowan (1929). He showed experimentally that increases in photoperiod of only 5 to 10 minutes per day caused the testes of Dark-eyed Juncos to increase in size, an effect that was reversible and repeatable up to three times before spring (Figure 12–5). His results stimulated photoperiod research throughout Europe and the United States. The phenomenon of photoperiodic control of gonad cycles has since been recognized in over 60 north temperate bird species in various orders and families.

The control of the White-crowned Sparrow's annual cycle is one of the best examples of this physiological phenomenon (Farner 1980c). Increasing photoperiods during late winter and early spring stimulate events in the annual cycle. The longer days of early spring stimulate gonad development, the prealternate molt, and spring migration. Supplemental information such as warmer temperatures, rainfall, and the springtime display behavior of other sparrows provide fine tuning of physiological events later in the year; they stimulate the final stages of gonad development on the breeding ground and, as a result, the increased secretion of sexual hormones. After the birds breed, the shortened days of late summer stimulate the main

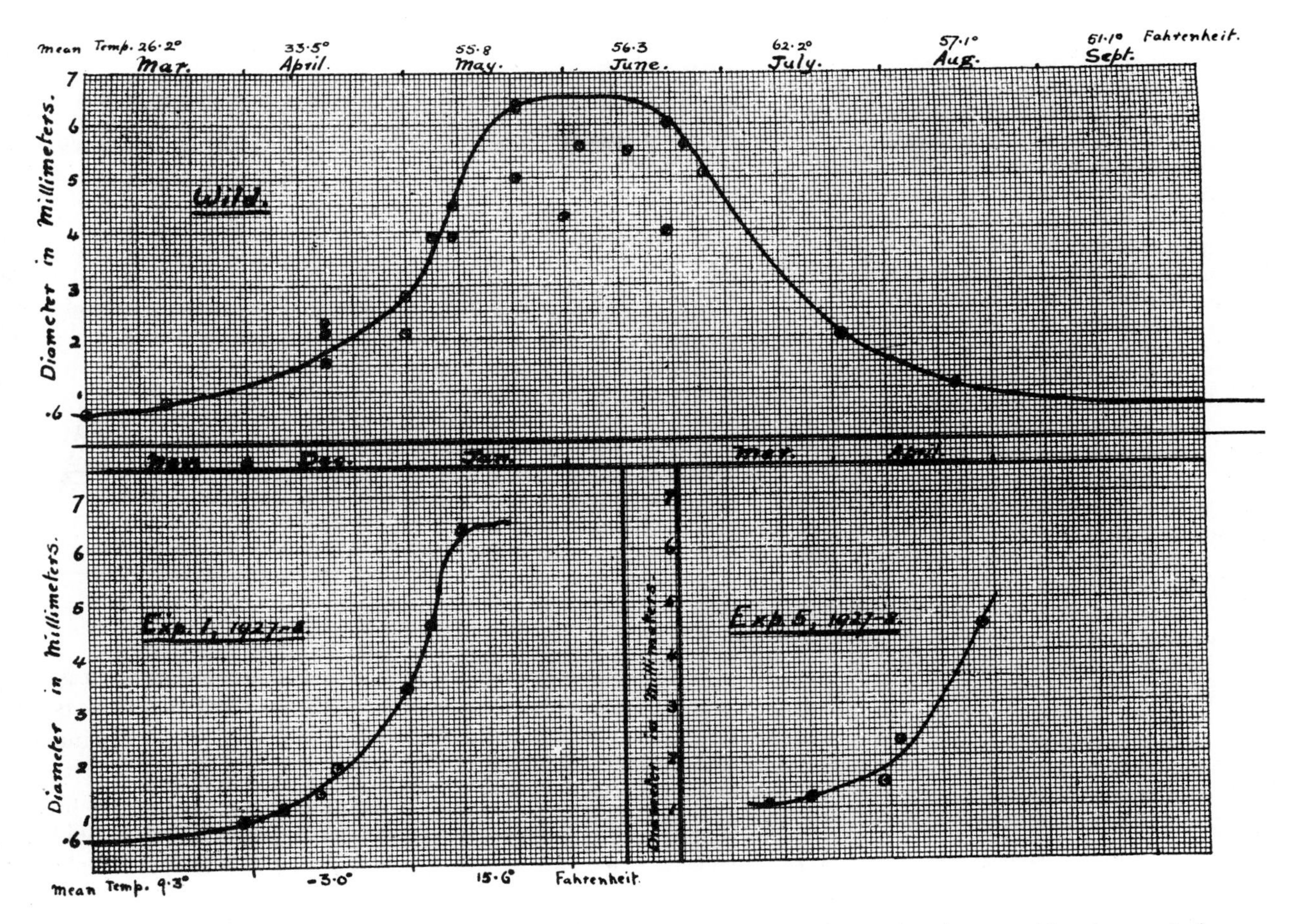

Figure 12–5 In the pioneer study of annual cycle control by photoperiod, William K. Rowan demonstrated that longer day lengths caused the testes of captive Dark-eyed Juncos to increase prematurely to full size in January (lower left) and again in April (lower right), instead of in May and June as in wild juncos (upper). Mean temperature is the average air temperature in that month. (From Rowan 1929)

(prebasic) molt. The timing of this molt, as well as of a light-insensitive or photorefractory period of the testis and changes in body fat in preparation for fall migration, is primarily determined during the preceding spring by a series of physiological events triggered by the longer photoperiods. Finally, the very short days of early winter reset sensitivity to the stimulus of long photoperiods, and the cycle begins anew. Short winter days are essential to the control of the annual cycle: The testes will not grow in response to the long days of spring unless the bird has experienced a period of short day length. Thus, White-crowned Sparrows can be maintained in nonbreeding condition for several years by exposing them constantly to long photoperiods.

Circadian rhythms of photosensitivity are probably a basic adaptation of cellular organisms to the 24-hour light-dark cycle of the planet (Farner 1980a). Circadian rhythms include a limited photosensitivity period, during which external light stimulates receptors in the brain, which in

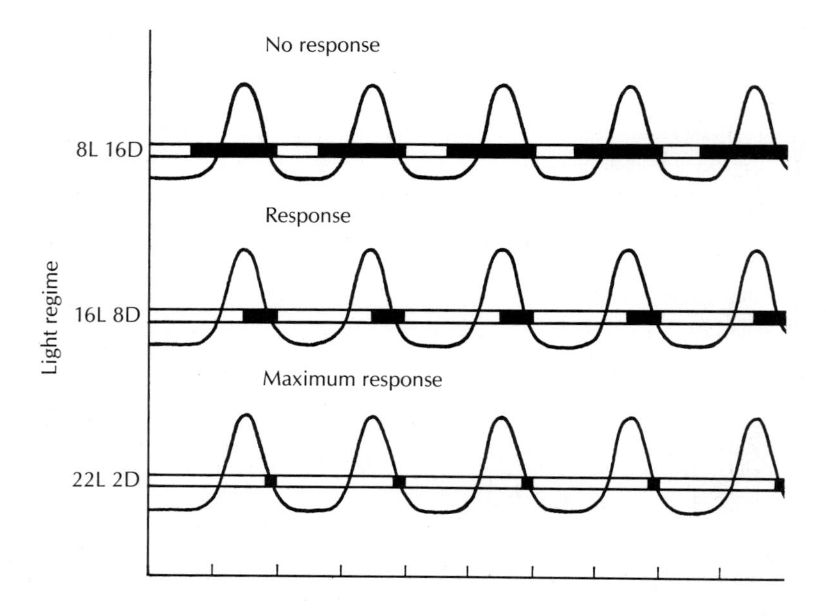

Figure 12–6 The external coincidence model suggests that day length is measured via the increased amount of time that daylight periods (white bar lengths) coincide with the photosensitive phase of the circadian rhythm (oscillation peaks). L = number of hours of light; D = number of hours of dark. Response was measured in terms of gonadal enlargement. (From Farner 1980a)

turn stimulate a series of physiological reactions. As day length increases, so does the chance that there will be daylight during the photosensitive period (Figure 12–6). Not only does the chance of coincidence or overlap increase with increasing day length but also the duration of the period of overlap also increases. It is the amount of overlap that enables birds to measure day length. We now have evidence of this "external coincidence" model for at least ten species of birds. The phenomenon is also known to take place in insects and plants. The external coincidence model is the simplest working hypothesis about the mechanisms of circadian rhythm. Other, more complicated models exist, and their refinement is a high priority in the study of the control of biological rhythms.

The receptor-control system

Birds do not monitor day length visually, as do mammals, but by means of special receptors in the hypothalamus of the brain. Longer day lengths induce gonad development and migratory behavior even in eyeless birds. The light receptors of the White-crowned Sparrow, for example, lie in the ventromedial hypothalamus of the lower midbrain. They are structurally unspecialized elements that are sensitive to extremely low light intensities such as those that directly penetrate brain tissues. Pinpoint illumination of the hypothalamic receptors by means of a single, thin, light-conducting optical fiber induces both testicular growth and migratory behavior (Yokoyama and Farner 1978). This technique facilitates precise determination of the locations of the receptor cells.

After stimulation of the photoreceptors, neurosecretory cells in the hypothalamus induce the release of neurohormones from the ends of axons in the median eminence, the neural portion of the pituitary that links this organ to the midbrain. The released neurohormones are then carried in the

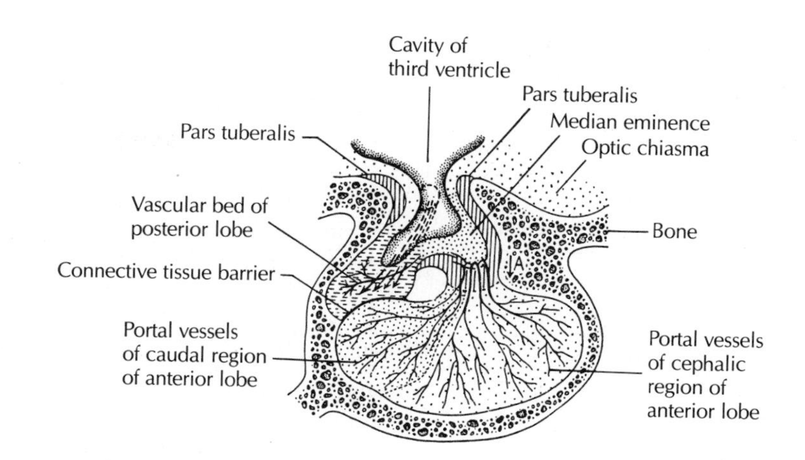

Figure 12–7 Avian pituitary gland. Daylight stimulates special photoreceptors in the tuberal region (pars tuberalis) of the lower hypothalamus of the midbrain. Neurohormones are released in the median eminence and carried to the anterior pituitary gland via the hypophysial portal blood vessels. They stimulate gonadal hormone production and, as a result, gonadal activity. (From Höhn 1961)

blood to the anterior pituitary gland where they induce the synthesis and release of hormones that directly affect the activity of the gonads themselves. Thus, a series of neural and physiological events translates increasing day length into sexual activity (Figure 12–7).

The avian control system is different from those of tetrapod vertebrates, not only in the use of hypothalamic photoreceptors and a circadian system for measuring day length but also by morphological separation of function within the median eminence of the hypothalamus, within the system of portal blood vessels, and within the pars distalis, or main part of the pituitary. The division of these organs may be ultimately related to the evolution of separate controls of follicle-stimulating hormone (FSH) and luteinizing hormone (LH), the two hormones that are most important in controlling gonadal development and function (see Chapter 17). Recent research on male birds has shown that photostimulation during the photosensitive phase of the circadian rhythm causes an increase in the pulsed release of LH. After release, pulses of plasma LH travel throughout the bird's body and stimulate gonadal activity. Release of FSH also increases with photostimulation though the speed of its release lags behind that of LH. In combination with LH, FSH stimulates the ovaries and testes to make gametes.

It appears that after photoperiodic regulation of the annual cycle evolved, some additional safeguards and corrections were essential. Photorefractory physiology is one of these. The gonadal cycle normally concludes with a rapid collapse and reabsorption of gonadal tissue. The photorefractory period follows, during which long days do not induce gonadal regrowth. The mechanisms of photorefractory physiology are still a mystery. The photorefractory period is best developed in migratory, temperate-zone species such as the White-crowned Sparrow. Some nonmigratory, temperate-zone birds also have a photorefractory period, but in most photosensitive tropical species examined to date it is weak or absent. The photorefractory physiology of adults, however it works, seems to be an adaptation for the purpose of scheduling molt and migratory preparations during the favorable conditions of late summer by discontinuing reproductive activity while days are still long (Miller 1959; Farner 1980c). Reproductive activity by yearling birds in response to late summer photoperiods would likewise be disadvantageous.

Breeding seasons

We have seen how physiological cycles and hormone controls prepare individuals for reproduction, migration, and molt. Cues of day length or rainfall link these controls to the external resources needed by a bird for each effort. This elaborate system has evolved because, over generations, its various parts have improved the chances for reproductive success with minimal risk to survival.

Guiding the evolution of seasonal physiological controls have been such factors as the timing of adequate food supplies for both parents and their young, the availability of nest sites, the locations of favorable climates, and areas or times of low predation risk, all of which ornithologists call ultimate factors (Perrins 1970; Skutch 1976). Ultimate factors favor hormonal control systems that optimally guide a bird through the annual cycle of its environment. Over many generations, the control systems are tuned to the best time for reproduction. However, they provide no guarantee against the vagaries of particular years. Drought or parasites may cause widespread nesting failure in some years. The birds cannot predict such disasters before starting to nest, but they can, of course, make last minute adjustments. Changes in the levels of particular hormones prepare an individual bird's physiological machinery for the particular tasks.

Proximate factors are the external conditions that actually induce reproduction. The correct habitat, green vegetation, the ritualized displays and aggression among neighbors, and social stimulation in general are all proximate factors that help to bring on the final stages of gonad enlargement and ovarian development. For desert species, the trigger may be good rains. Temperature is probably the most important modifier of annual gonad cycles (Farner and Mewaldt 1952). A 10-degree increase in average spring air temperatures stimulates Eurasian Skylarks to lay their eggs (Delius 1965).

Pinyon Jays begin their breeding activities when they see young green pinecones (Ligon 1974), their annual cycle being closely tied to the availability of the seeds of the Pinyon Pine, one of their primary foods. In southwestern New Mexico, Pinyon Jays breed sporadically, sometimes in the autumn if there is a bumper crop of pine seeds. To determine whether green pine cones, the first visual evidence of future food abundance, actually triggered breeding by these jays, David Ligon isolated two groups of ten male jays for a year. In the summer, when their testes decreased in size after the normal spring enlargement stimulated by photoperiod, he gave one group fresh green pine cones daily. This offering caused a dramatic reversal of their reproductive state. Whereas the testes of the control group of jays (no pine cones) continued to shrink, the testes of the jays that were given green pine cones enlarged again. Thus, food abundance is an important modifier of annual gonad cycles.

The joint action of proximate and ultimate factors delimits the characteristic breeding seasons for the majority of species in a particular region (Skutch 1976). April, May, and June, the spring months when temperatures rise and insects emerge, comprise the primary breeding season in the north temperate region, and September to December are the primary breeding months in the south temperate region (Moreau 1950). A few species nest before the weather seems to be suitable. Red Crossbills in the Rocky

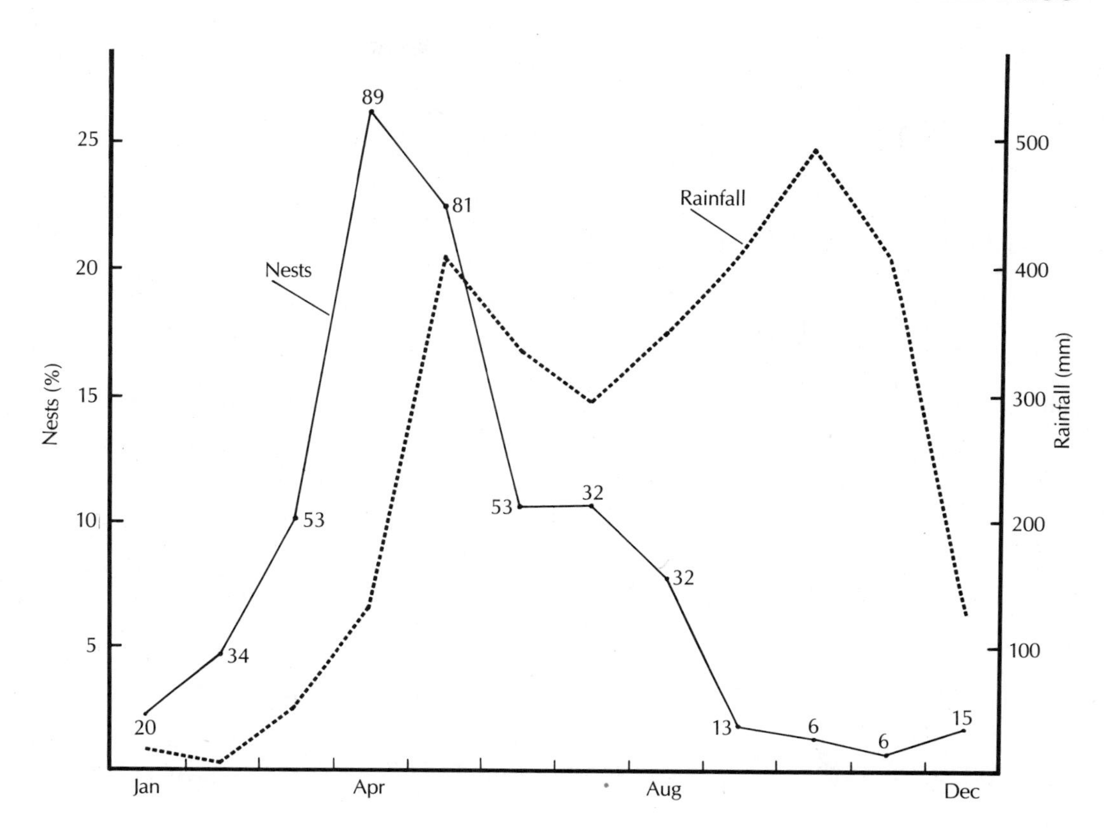

Figure 12–8 Nesting of Costa Rican birds in relation to rainfall in the valley of El General (600 to 900 meters in altitude), based on records of 1357 nests of 140 species. The numerals along the solid line (nests) indicate the number of species in each monthly total. The broken line indicates the average monthly rainfall. (From Skutch 1976)

Mountains, for example, will nest in January and February, surrounded by snow, if their primary food, conifer seeds, is abundant. Omnivorous Common Ravens and Eurasian Rooks, as well as Gray Jays in Canada and their counterparts in Siberia, also nest early in the spring if they are able to find adequate food. Many large raptors can find food more easily in the open winter woods than after new leaves emerge; Great Horned Owls for example, incubate their eggs in January and February.

It is rainfall that usually defines the seasons in the tropical lowlands, especially in arid environments (Bourne 1955; Marchant 1960). Although some individuals can be found breeding in most months in the tropics, nesting activity in lowland Costa Rica peaks for most birds at the end of the dry season and early in the rainy season (Figure 12–8). Kingfishers are an exception, preferring to breed during the dry season when streams run shallow and clear, making fish easier to capture. Hummingbirds, too, nest at the beginning of the dry season when flowers begin to bloom (Skutch 1976).

Tropical nesting seasons last longer than those in the temperate zone (Ricklefs 1969a). In the high arctic, where only a month or so is suitable for breeding, birds must start nesting immediately after migration, sometimes using old nests. Nesting seasons at temperate latitudes usually last three to four months. Less stringent tropical climates permit nesting for six to ten months, or even, in some cases, throughout the year. On Trinidad, for example, nests of the Ruddy Ground-Dove, the Barred Antshrike, and the Palm Tanager are among those found throughout the year (Snow and Snow 1964). It is unlikely, however, that a single pair breeds continuously at such locations though they may breed several times in succession.

Figure 12–9 (A) Projected breeding seasons of seabirds on Christmas Island, Pacific Ocean. Black indicates the presence of eggs, slanted lines the presence of chicks but not eggs, and dashed lines the presence of adults without eggs or chicks. (B) The time and length of the breeding season of the eastern race of Brown Pelican varies geographically as shown by the date that eggs are laid. The thicker portion of the lines indicates the probable presence of eggs. (A, adapted from Schreiber and Ashmole 1970; B, adapted from Schreiber 1980a)

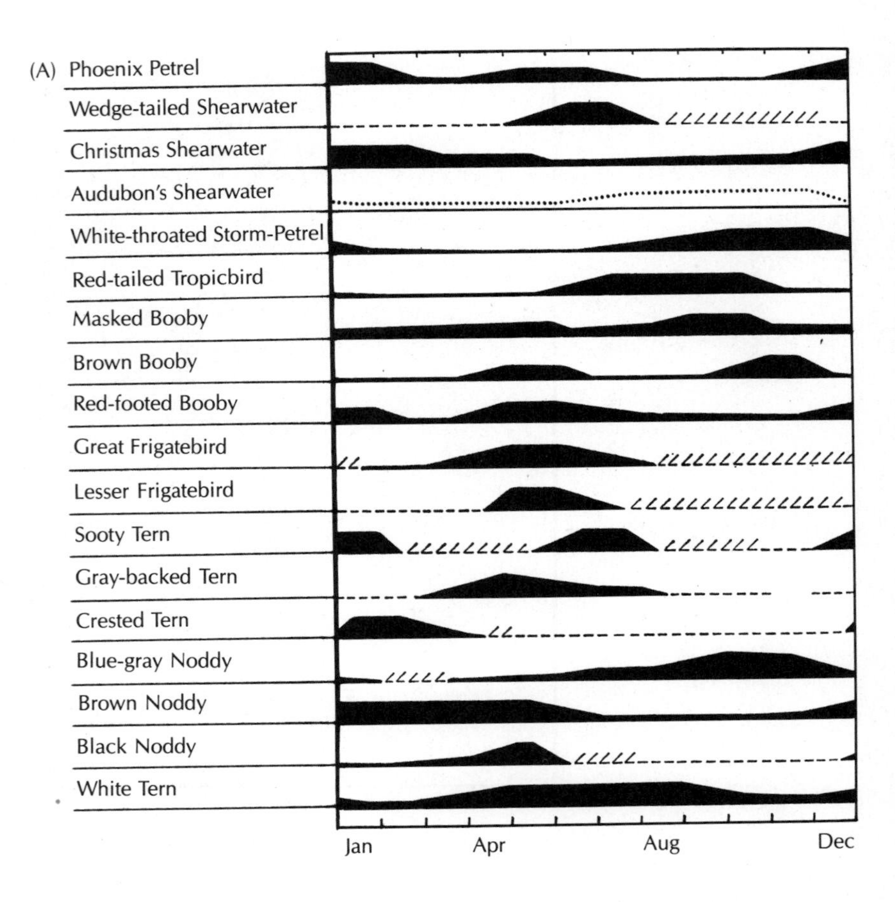

(B)

Location	Latitude (°N)
North Carolina	35
South Carolina	31–33
Louisiana	29
Texas	28
Florida, East Coast	27–28
Pelican Island	27
Tarpon Key	27
Alafia Banks	27
Charlotte Harbor	26
Keys	25
Marco Island	25
Dry Tortugas	24
Cuba	22
U.S. Virgin Islands	19
St. Martin	18
Puerto Rico	18
Lesser Antilles	17
Aruba	12
Venezuela	11
Trinidad and Tobago	11
Costa Rica	10

Jan Apr Aug Dec

Each of the eighteen species of seabirds that nests on Christmas Island in the tropical Pacific Ocean does so when the specific foods it requires to raise young are easiest to find (Figure 12–9A). The response of each species to seasonal changes in feeding opportunities and to the specific demands of its young is tempered by internal physiological rhythms and is independent of the responses of other species (Schreiber and Ashmole 1970). Gray-backed Terns and Wedge-tailed Shearwaters have well-defined spring and late summer nesting seasons, respectively, whereas Christmas Shearwaters and White Terns nest throughout the year with ill-defined peaks at opposite seasons. Sooty Terns and Brown Boobies have two distinct nesting seasons each year, the terns in winter and summer, the boobies in spring and fall. In the case of the terns, the second peak reflects another breeding attempt by individuals that were unsuccessful in the first season.

Local populations of a species respond to local conditions. Nesting by Brown Pelicans, for example, is strongly seasonal at northern sites but prolonged at tropical sites (Schreiber 1980a). Cold water temperatures, which depress food supplies, appear to delay the onset of nesting at all sites. After food availability, it seems that the hurricane season is the second most important factor controlling the onset of nesting in these pelicans (this holds true for seabirds in general). Pelicans nest irregularly throughout the year in the Caribbean and northern South America, more predictably during the winter and spring in Florida, and from March to June in Louisiana and the Carolinas (Figure 12–9B).

The general correspondence between breeding season and food availability brings us back to the central issue of the annual energy budgets of birds and the energetic costs of reproduction and molt that favor segregation of these stages during the annual cycle. Birds can assume the costs of migration, reproduction, or molt only after they have first met the costs of self-maintenance, their highest priority, and such basic social interactions as may be necessary to obtain food or a roost site, their second-highest priorities (King 1974). Some seasons, such as a northern winter, permit only self-maintenance, whereas others accommodate additional activities. Reproduction and molt must be scheduled during the months when a bird's requirements for self-maintenance are lowest. Usually, the costs of only one extra activity can be accommodated (Kendeigh 1949; Farner 1964).

Energetic costs of reproduction

Peak reproductive activities increase total daily energy expenditures by as much as 50 percent (Ricklefs 1974; Walsberg 1983). Daytime activity costs may actually double or even triple, but overnight costs remain constant. At the beginning of the breeding season, courtship, territoriality, and nest building demand significant effort. Only minor amounts of productive energy are channeled into growth of the gonadal tissues, but subsequent egg formation and egg laying by females impose new demands on energy and nutrition (see Chapter 15). The production of large eggs by waterfowl is especially expensive and may temporarily double a female's total daily energy requirement. Incubation can also create an energy shortage because it limits the amount of time a bird can forage for its own maintenance.

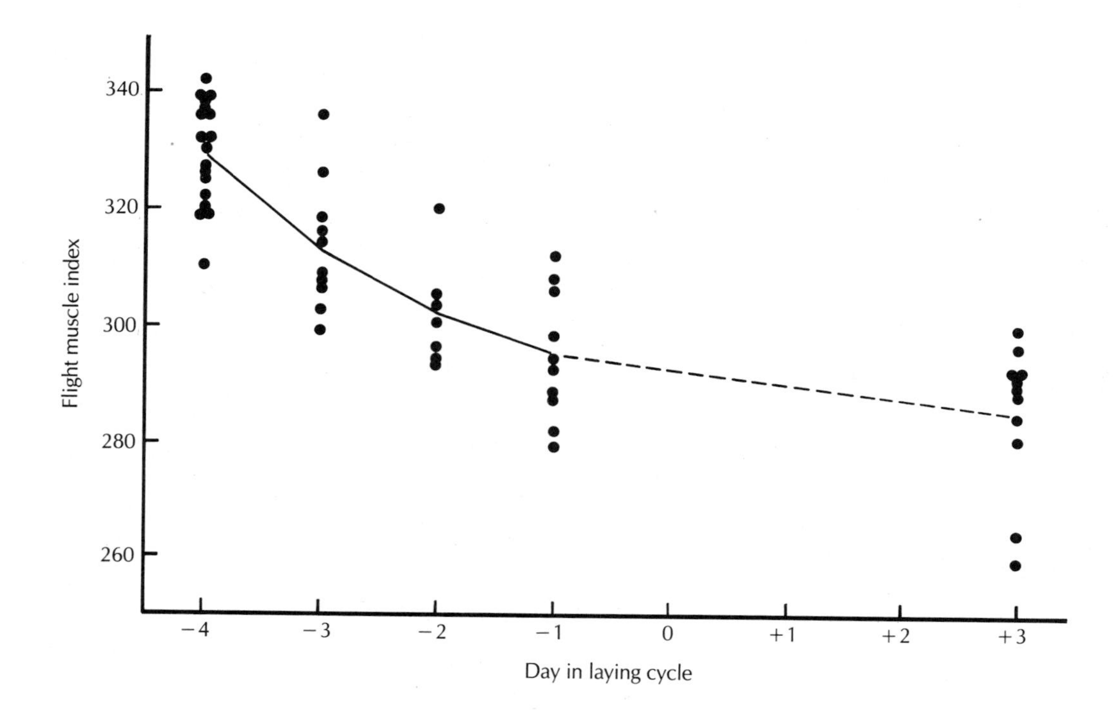

Figure 12–10 Changes in the protein reserve of adult female Gray-backed Camaropteras, an African warbler, in relation to each day of its egg-laying cycle. Each black circle indicates a measurement of one female warbler. The flight muscle index, which is based on the lean dry mass of the pectoral muscles relative to body size, estimates a female's protein reserve. The dashed section of the curve is an extrapolation of the trend for days when no information was obtained. (From Fogden and Fogden 1979)

Finally, the parents face yet another surge of demands on their time and energy when the chicks hatch and require food and brooding.

Birds draw on their reserves to get through periods of peak energy demand. In some species, a female's protein reserve may control when she breeds (Jones and Ward 1976; Fogden and Fogden 1979). The protein reserve of female Gray-backed Camaropteras and Red-billed Queleas falls substantially with egg laying, apparently because the females metabolize their protein reserve to extract essential amino acids not available in adequate amounts in their diet (Figure 12–10). Protein reserves can be estimated from the lean dry mass of the pectoral muscles. If food is not plentiful enough to permit birds to build a reserve, they will not breed.

Timing of molt

Molt is a period of intense physiological change. The bird sheds and then regenerates thousands of feathers, roughly 25 to 40 percent of its lean dry mass (i.e., excluding fat and water content). Accompanying this expenditure are the synthesis of keratin by the skin, increased amino acid metabolism, increased cardiovascular activity to supply blood to the growing feathers, the shunting of water to the developing feathers, and many other subtle exertions. Muscle tissue may be broken down to provide the amino acids needed for the chemical bonds in keratin (Dolnik and Gavrilov 1979; Payne 1972).

The complete annual molt is a major undertaking. The required increase in general metabolism potentially conflicts with the energetic demands of breeding or migration. Adult Common Chaffinches invest a total of 905 kilojoules of energy into their annual molt and up to 21 kilojoules per day at the beginning of the molt (Dolnik and Gavrilov 1979). Thomas Bancroft and Glen Woolfenden (1982) estimated that adult Scrub Jays and Blue Jays must increase daily metabolism 15 to 16 percent during peak periods of feather production. The metabolic rate of a Domestic Chicken increases 45 percent during molt (Perek and Sulman 1945). The metabolism of an Ortolan Bunting, a migratory bird, increases 26 percent, and the metabolism of the nonmigratory, slow-molting Yellowhammer increases 14 percent.

Feather production is the primary reason for energy expenditure during molt, unless a bird molts at cool temperatures (Lustick 1970). At moderate air temperatures, molting Brown-headed Cowbirds consume 13 percent more oxygen than do nonmolting cowbirds. At lower air temperatures, however, when greater heat production is needed to maintain a constant body temperature, molting cowbirds consume 24 percent more oxygen than do nonmolting birds. The difference of 11 percent reflects the cost of poorer insulation. In Common Chaffinches, the daily metabolic costs of molt at cooler temperatures include 9 percent of total energy expenditure for feather production and 11 percent for temperature regulation (Dolnik and Gavrilov 1979). Thus, molting during the summer is obviously advantageous for birds of northern climates.

Although molt is classified as an activity that is secondary in priority, it is essential to the annual cycle. To ensure that the molt is finished before the end of the summer, birds may not nest a second time (Pitelka 1958). Molt is so important for the Sooty Terns of Ascension Island that it may prevent continuous year-round nesting (Ashmole 1965).

Molt typically follows breeding and precedes migration. Few species breed and molt at the same time, and the exceptions to the rule are instructive. Some female hornbills molt while imprisoned in sealed nest cavities, incubating eggs and brooding young. The energy requirements for self-maintenance are minimal for these domestics, and as a result, the added costs of molt can be accommodated. Loss of feathers and reduced insulation may, in fact, be advantageous at this time because of the high temperatures that build up inside the nest cavity. Male hornbills, which feed the incubating females, wait until their families fledge before molting.

Some female raptors, including Peregrine Falcons, Snowy Owls, Northern Goshawks, and Ospreys, molt while incubating their eggs. The daily energy budget of these females is minimized by inactivity. Their mates (which do not molt until after the breeding period) feed the females regularly on the nest.

Gulls and sandpipers that breed in the high Arctic, where the reproductive season is short, must start molting before they finish breeding to be ready for migration. The Dunlin, for example, begins to molt its primaries just before incubation and then finishes four to five weeks later. It leaves the breeding grounds later than sandpipers, which winter farther south. Resident petrels of Antarctica molt and breed at the same time during the brief summer (Maher 1962). Thus, molt and breeding may occur at the same time in the absence of the usual stresses or when necessitated by the season, but the general pattern is to separate molt from breeding and migration.

Cases of synchronous molt and breeding occasionally occur in birds that live in productive, tropical environments with minimal seasonal variation. Three to four percent of the African birds examined by Robert Payne (1969a) were molting while breeding. Eight to ten percent of the Costa Rican birds examined by Mercedes Foster (1975) bore signs of both molt and reproductive activity. The prolonged molts of tropical birds apparently minimize daily costs in the absence of strong seasonal constraints. Tropical birds molt more predictably than they breed because reproduction may be tied to irregular periods of rain or may involve several renesting attempts. Because occasional overlap between molting and nesting is likely, it is sometimes advantageous to have a flexible system of physiological controls (Foster 1975).

Desert birds such as Darwin's finches of the Galapagos and the Zebra Finch of Australia molt on a regular schedule but nest whenever unpredictable rains begin, stopping the molt temporarily to do so. No additional feathers are replaced until after nesting is completed. Molt interruption also occurs in the Gray-backed Camaroptera in Uganda (Fogden and Fogden 1979), which breeds in the rainy season, and in Pinyon Jays of New Mexico, which breed when food is abundant. Tropical terns such as the White Terns on Christmas Island turn the molt on and off to accommodate breeding whenever possible (Ashmole 1968). This bird has no pigment in its flight feathers, which consequently wear easily and must be replaced more often than in most other terns. Wave after wave of molt is initiated in the flight feathers. In fact, the innermost primaries often begin to molt again before the outermost primaries are replaced in the preceding molt. As many as three successive molts may be in progress simultaneously. When a White Tern starts to nest (actually it simply lays an egg precariously on a bare branch), the molt suddenly stops, no matter which feathers may be missing, the molting equivalent of musical chairs. After the tern has finished nesting, molt resumes in the correct place in the complicated pattern, as if there had been no interruption (Figure 12–11).

Figure 12–11 White Terns molt almost continuously to replace their worn, unpigmented feathers, but they interrupt the molt upon laying an egg. (Photograph courtesy Ralph W. Shreiber)

A molt may also be interrupted when the bird migrates. Arctic Peregrine Falcons and Lesser Golden-Plovers begin their molt on the breeding ground but are unable to complete it in time to leave for the south. Molt of the flight feathers stops just before migration and then resumes for several more months on the wintering grounds.

The pace of molt and its costs vary in relation to the time available before migration becomes imperative. In Denmark, Common Ringed Plovers molt their entire plumages in the late summer, in about 1.5 months, after breeding and before migrating a short distance to southwestern Europe. Their relatives that breed in the Arctic leave immediately after breeding, fly directly to southern Africa, and then molt at a leisurely pace for nearly four months (Stresemann 1967). The northernmost populations of the White-crowned Sparrow *(pugetensis)* complete their molt in 47 days, compared with the 83 days typical of the slow-molting southern populations *(nuttalli)*. Thrush Nightingales, Lapland Longspurs, and renesting White-crowned Sparrows molt so fast at high latitudes that, temporarily, they become virtually flightless.

Physiological control of molt

Molt and preparations for migration are triggered by changes in day length and can be experimentally manipulated. Stephen Emlen (1969), for example, accelerated the annual cycle of Indigo Buntings, inducing an extra molt into the year by suddenly increasing the photoperiods to which captive birds were exposed. The endocrine pathways that tie molt directly to photoperiod, or those that tie it indirectly to photoperiod via the gonad cycles, are not as well defined as the links between the gonads and the hypothalamus through the pituitary gland. Yet the endocrine hormones clearly affect the timing and course of molt. Historically, thyroid hormone was thought to regulate molt in birds, but it is now clear that the relationship is not a simple or direct one; rather, indirect interactions between thyroid activity and the gonad cycle are probably involved.

The gonadal hormones, androgens and estrogens, appear to inhibit molt because molt begins as the influence of the hormones on the bird's breeding physiology wanes. Nonbreeding and reproductively unsuccessful individuals begin to molt earlier than successful breeders. Experimental injections of gonadal hormones into molting birds slow or even stop molt. Molt cycles, however, are not directly coupled to gonad cycles because castrated birds continue to molt on schedule. Thus, the inhibition of molt by gonadal hormones may control the timing of molt to some degree, but they cannot be responsible for the initiation of the process itself. Furthermore, Rock Doves, Anna's Hummingbirds, and Great-tailed Grackles, among others, start to molt in the spring while their gonads are enlarging and their sex hormone levels are high. Although some ornithologists and endocrine physiologists have made major advances in our understanding of the physiological control of annual cycles, they are just beginning. Researchers have begun monitoring and manipulating hormone levels in wild birds, and there is exciting potential for future breakthroughs (Wingfield and Farner 1978; Wingfield 1984).

Nonannual cycles

Not all birds follow a 12-month cycle. A relative of the White-crowned Sparrow of North America, the Rufous-collared Sparrow, which ranges from Mexico to Chile and has been studied in Colombia near the equator, has been found to breed and undergo a complete molt twice a year (Miller 1962). These cycles correspond to the two dry seasons each year. In Costa Rica, south of Mexico, Rufous-collared Sparrows also breed twice a year, but like the White-crowned Sparrows in Washington, they only have one complete molt and one partial molt each year (Wolf 1969).

Double breeding seasons are more common among tropical bird species than previously suspected. Year-round availability of adequate food fosters such cycles. The Sooty Terns of Christmas Island breed every six months, though the individuals that breed twice in the same year are those that failed during the first breeding season. Successful individuals wait eight to nine months before breeding the following year (Ashmole 1963a). In another case of double breeding seasons, two populations of Band-rumped Storm-Petrels of the Galapagos Islands alternate use of nesting burrows (Harris 1969). Good nest sites may be limited for this species.

In only a few cases is the breeding cycle independent of calendar year. Unlike Sooty Terns on Christmas Island, the Sooty Terns on Ascension Island in the tropical Atlantic nest every 9.6 months, in different months in successive years. Ample food is available every month, so successful nesting is possible at any time of the year (Ashmole 1965) (Figure 12–12). Brown Boobies on Ascension Island nest at 8-month intervals, and White-tailed Tropicbirds nest at 10-month intervals if they are successful and renest in 5 months if they are not. Successful Audubon's Shearwaters and Swallow-tailed Gulls on the Galapagos Islands nest at 9-month intervals.

Figure 12–12 Sooty Terns on Ascension Island do not have a regular 12-month breeding cycle, but instead breed approximately every 9.6 months and, consequently, in different months in successive years. (Photography courtesy A. Cruickshank/VIREO)

A few very large birds cannot fit their extended reproductive efforts into a single year and hence may skip a year between nestings. Frigatebirds, Crowned Eagles, and Wandering Albatrosses nest once every two years. King Penguins take 2 months to incubate their eggs, 10 to 13 months to raise their nestlings, and then molt. As a result, they breed only twice every 3 years (Stonehouse 1960).

Many tropical and desert birds cannot breed on a regular schedule but do so opportunistically whenever unpredictable rains permit. Zebra Finches, which begin nesting as soon as the rainy season starts, maintain partially developed gonads in the nonbreeding season, thus minimizing the time required to develop functional organs. Increased water intake and, possibly, the consequent effects of changes in the salt concentration of body fluids on the hypothalamic receptors, rather than photoperiod, induce the final stages of gonad development. Termination of reproductive gonad activity may be induced by the unavailability of drinking water after the rains end. Donald Farner (1967) proposed that in such species the hypothalamus stimulates the release of gonadotropins from the pituitary on a steady, predictable basis unless inhibited by unfavorable external information such as cold weather or drought.

Birds adapt to local opportunities whether they exist as predictable seasons or irregular occasions, probably by means of a variety of physiological mechanisms. The physiological adaptations of only a few species of birds have been documented, perhaps just a small sample of the full spectrum of avian adaptation to changes in the environment.

Summary

Birds face seasonal cycles of stress and opportunity. Physiological cycles, guided by internal cellular clocks, prepare the bird for each season. In general, seasonal change in day length, or photoperiod (which controls gonad activity and therefore reproductive efforts), directly stimulates receptors in the midbrain and, in turn, the secretion of gonadal hormones by the pituitary gland.

Ultimate factors such as food supplies, nest sites, climate, and predator risk determine the evolution of breeding seasons in birds. Proximate factors such as temperature, rainfall, and green vegetation adjust the actual onset of reproduction to local conditions. Warm spring and summer months constitute the main breeding season in the temperate zone. Rainfall usually defines tropical breeding seasons.

Birds generally do not breed and molt at the same time, but undertake these efforts, which require substantial energy, in different months. In some exceptional cases, molt and breeding take place simultaneously; for example, female hornbills, confined to the nest and fed by the males, can afford to molt, and some sandpipers must molt and nest to accommodate the short arctic summer. Opportunistic breeders such as the White Tern interrupt molt while they nest.

The simplest annual cycles proceed from breeding to molting to surviving seasons of reduced food availability to breeding again. Seasonal migrations and extra molts complicate the annual cycles of many birds. A few, mostly tropical, birds have 6-month cycles, breeding twice a year. Others have 9- or 10-month cycles, thus breeding in a different month each year.

Further readings

Farner, D.S. 1980. The regulation of the annual cycle of the White-crowned Sparrow, *Zonotrichia leucophrys gambelii.* Acta XVII Congressus Internationalis Ornithologicus 1:71–82. *An excellent review of the physiological controls of the annual cycle.*

Gwinner, E. 1975. Circadian and circannual rhythms in birds. Avian Biology 5:221–287. *A detailed review of biological clocks in birds.*

Gwinner, E. 1977. Circannual rhythms in bird migration. Annual Review of Ecology and Systematics 8:381–405. *A summary by one of the field's pioneers.*

Immelmann, K. 1971. Ecological aspects of periodic reproduction. Avian Biology 1:342–389. *A fine summary of ecological correlations of breeding by birds.*

King, J.R. 1974. Seasonal allocation of time and energy resources in birds. Avian Energetics Publication of the Nuttall Ornithological Club 15:4–70. *A basic reference for the topics discussed in this chapter.*

Meier, A.H., and A.C. Russo. 1985. Circadian organization of the avian annual cycle. Current Ornithology 2:303–343. *A technical review of current research.*

Payne, R.B. 1972. Mechanisms and control of molt. Avian Biology 2:104–156. *A detailed review of the subject.*

Skutch, A.F. 1976. Parent Birds and Their Young. Austin: University of Texas Press. *A readable review of the natural history of avian reproduction, with a chapter on breeding seasonality.*

Walsberg, G.E. 1983. Avian ecological energetics. Avian Biology 7:161–220. *A comprehensive review of avian energy budgets, with an emphasis on reproductive activities.*

chapter 13

Migration

There are ancient records of the seasonal appearances and disappearances of birds. Early naturalists were not certain whether birds migrated or hibernated. Aristotle understood that cranes moved seasonally from the steppes of Asia Minor (then Scythia) to the marshes of the Nile, but he believed that small birds, such as swallows, larks, and turtle doves, hibernated. Later anecdotes of swallows that, found frozen in the marshes, flew off after being thawed, fueled this misconception and postponed a general appreciation of the phenomenon of bird migration (Dorst 1962).

We now know that every fall an estimated five billion landbirds of 187 species leave Europe and Asia for Africa (Moreau 1972), and that a similar number of over two hundred species migrate from North America to the New World tropics. Migration permits year-round activity, unlike dormancy, hibernation, and diapause, the means by which many animals live through severe seasons. The advantage of migration is that birds can exploit seasonal feeding opportunities while living in favorable climates throughout the year.

The costs of migration are great, and it takes radical physiological adjustments and sustained fine tuning to survive such extended travel.

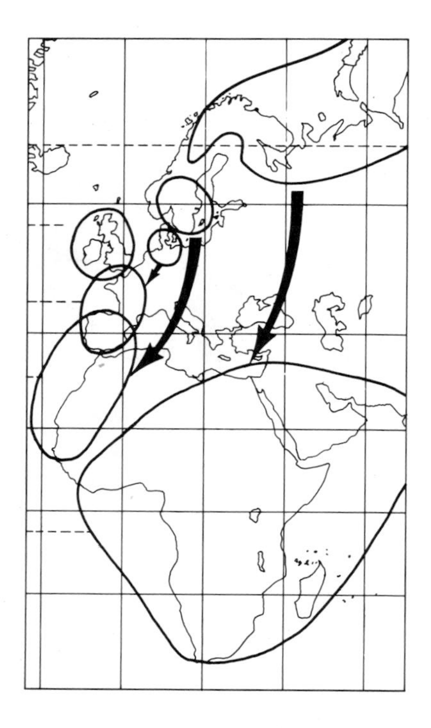

Figure 13–1 Various populations of the Common Ringed Plover maintain distinct wintering as well as breeding ranges. The populations that breed farthest north, winter farthest south. The Common Ringed Plovers of the British Isles do not migrate at all. (After Dorst 1962)

Because we do not yet fully understand the evolutionary origins of migration and have not been able to document the behavior of long-distance migrants en route, the delicate balance between the costs and advantages of migration awaits full disclosure. In this chapter we first present the leading theories on the origin of migration. Then we establish the dimensions of migration, which sometimes requires extraordinary physiological endurance and large fuel supplies. Direct extensions of the physiological and ecological controls that manage other aspects of the annual cycles of birds also control the timing of migration.

Why birds migrate is still one of the most challenging questions in ornithology, despite a century of effort to frame a satisfactory answer (Wallace 1874; Dorst 1962; Gauthreaux 1982). Although the goal is to formulate a convincing theory of the evolutionary steps a sedentary species might have taken in becoming a migratory species, seemingly simple questions, such as why some populations of a species migrate and other populations of the same species remain sedentary, also beg for answers. Populations of Common Ringed Plovers, for example, differ in their migratory habits. Those that breed farthest north, on the arctic coasts of Norway and the Soviet Union, winter farthest south, in central and southern Africa and the Middle East, whereas populations that breed in Sweden winter on the northwest coast of Africa and in Spain; and plovers that breed in Denmark winter primarily in France but also appear in Spain. The Common Ringed Plovers of the British Isles are resident rather than migratory (Figure 13–1).

Why do some birds migrate farther than others? Sanderlings, the common sandpipers that scurry back and forth with the waves on sandy beaches, fly from their breeding sites in the high Arctic to wintering sites as near as the state of Washington and as far as southern Chile (Myers et al. 1985) (Figure 13–2). The energetic price of the 230-hour, 7500-kilometer flight to Chile matches the cost of living one midwinter month in California. Why should a Sanderling invest so much to go so far when apparently suitable beaches line the Pacific coast from Washington to Chile? Ornithologists cannot answer this question, but they have started to identify the main forces involved in such behavior. Early efforts to understand the migration of birds focused on the biogeographical origins of the migrants. The current view that migrants evolve in strongly seasonal environments has caused ornithologists to shift their focus to the evolution

Figure 13–2 A flock of migrating Sanderlings. The lead birds are Black Turnstones. (Courtesy J.P. Myers/VIREO)

of the migratory habit itself and to the balance of cost and benefit that affects the survival and breeding success of migrants. Identifying the sources of selection that have molded the life histories of migrants is the essential first step. Much more difficult is the accurate measurement of selection forces manifest at the far-flung residencies of the migrants.

Migration, which is tied to predictable seasonal opportunities, is different from nomadic wandering, which is tied to unpredictable opportunities. Sporadic, scattered pine seed crops or insect infestations attract opportunistic feeding from nomadic species such as Red Crossbills, which wander great distances in search of pine seeds and may breed wherever food is abundant. In the tropics, fruit-eating and nectar-feeding birds wander locally in search of their unpredictable sources of food. In contrast, predictable seasonal cycles of climate or insect abundance attract corresponding cycles of breeding, flocking, and migratory relocation. To take advantage of predictably favorable conditions, birds undertake both local and long-distance movements between seasonal residencies. On a local scale, tropical hummingbirds simply migrate up and down mountain slopes. On a global scale, Arctic Terns leave their nesting colonies in the far north Atlantic and Arctic oceans for the waters of the opposite pole. Buff-breasted Sandpipers fly 13,000 kilometers from breeding grounds in the vast high Arctic to winter on the pampas of Argentina. More common are migrations to less distant wintering grounds. Many species of wood warblers that breed in the northern United States and southern Canada spend the winter in Central America and in the West Indies.

The benefits of migration must be substantial because migration is clearly costly. Long-distance travel is a hazardous undertaking because of its energy costs and risk of exposure, exhaustion, and other physical calamities. More than half the birds that leave each year on their southbound migration never return. Ocean and desert crossings, aside from posing extreme physical demands, carry the threat of devastating hurricanes and sandstorms. Weakened migrants are also vulnerable to diurnal predators such as Eleanora's Falcon, which breeds in the fall so that it can raise its nestlings on migrants trying to cross the Mediterranean (Figure 13–3). Losses on the wintering grounds and on the return trip also raise the mortality figures.

Figure 13–3 Eleanora's Falcon breeds in the fall, when small birds that migrate past its Mediterranean breeding areas provide an abundant source of food for its nestlings. (Courtesy Harmut Walter)

The potential benefits of migration are species- or population-specific. Discussions in the past have emphasized the need to escape from inhospitable climates, probable starvation, intense nest predation, social dominance, shortage of nest or roost sites, or competition for food. A more positive view of the same ecological forces is that migrants aggressively exploit temporarily favorable opportunities. Following this view, we should think of many northern temperate-zone migrants as tropical birds that temporarily exploit the long days and abundant insects of high-latitude summers, rather than as temperate birds that tolerate the tropics to escape the northern winter (Stiles 1980). Hummingbirds, flycatchers, tanagers, and many wood warblers are tropical American birds that have expanded their seasonal activities northward. Such tropical elements are more pronounced among the migrant birds of North America than among those of Europe and western Asia, apparently because in the New World there are few barriers such as the Mediterranean Sea and Sahara Desert to discourage northward travel (Dorst 1962). In eastern Asia, the absence of such barriers has permitted a variety of tropical birds—kingfishers, orioles, pittas, drongos, white-eyes, and minivets—to migrate north. Attractive nesting opportunities also invite migration to temperate latitudes. The large expanses of northern temperate-zone habitats facilitate dispersed, low-density breeding. Reduced predation of nests may be one result of low densities, breeding opportunities for yearlings another. Several years' wait for a breeding space is often the case in the tropics. The high densities of shorebirds on their Argentine wintering grounds apparently prohibit breeding; defensible areas are too small to provide an adequate food supply for more than a single bird (Myers 1980). Such factors are often incentives to migrate.

The list of ecological incentives to migration is a long one (Welty 1982; Gauthreaux 1982; Cox 1968, 1985). Nevertheless, they do not explain the evolution of the migratory habit. For this we turn to comparisons of resident and nonresident populations of the same species. Species in which some populations migrate and some do not, such as White-crowned Sparrows (page 225), are called partial migrants. Migratory behavior has a genetic basis in at least one partial migrant, the European Robin (Biebach 1983). Resident individuals, which make up about one fifth of the robin population in southwestern Germany, remain within five kilometers of their breeding territories, do not put on large reserves of premigratory fat, and do not exhibit sustained *zugunruhe* behavior (migratory restlessness) in the laboratory. In contrast, migrant individuals fatten in the fall, exhibit intense *zugunruhe* behavior, and travel an average of 1032 kilometers to their winter habitat. The migratory behavior of offspring has a significant correlation with that of the parents (heritability = 0.52). Because the climatic conditions swing from mild winters that favor residents to severe winters that favor migrants, migratory behavior in the population is maintained as a behavioral polymorphism. As long as winter conditions are unpredictable, both migrants and nonmigrants are favored (Figure 13–4).

Populations can acquire or lose the migratory habit. The Common Serin, for example, has spread throughout Europe from the Mediterranean in the last 100 years, and whereas the ancestral Mediterranean populations have been resident, the new northern populations are now migratory (Mayr 1926). Conversely, the resident Fieldfares of Greenland are recent

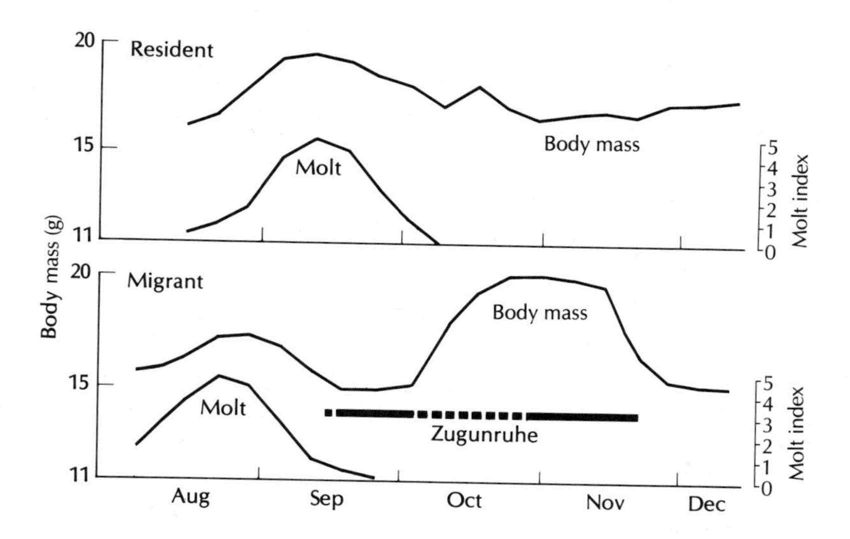

Figure 13–4 Body mass, molt, and *zugunruhe* behavior (migratory restlessness) of (top) a young resident and (bottom) a young migratory European Robin in the laboratory. Breeding experiments revealed a genetically based polymorphism for migratory behavior, including early molt, premigratory fattening, and migratory restlessness in the two forms of this species. A molt index of 1 indicates the beginning or end of the molt. A molt index of 5 indicates a heavy molt that includes most of the feather coat. (After Biebach 1983)

colonists from the migratory populations of Europe. Several Palearctic species that winter in South Africa, including the Common House-Martin and White Stork, have now established resident breeding populations in South Africa (Moreau 1966). Similarly, Barn Swallows wintering in Argentina during the austral spring and summer have lately started to nest there (Martinez 1983). The migratory habit thus changes in relation to new geographic and ecological circumstances.

Social and ecological forces seem to sponsor trade-offs between winter survival and reproductive success (Fretwell 1968, 1980). Field Sparrows winter in seed-rich pine and broomsedge fields in North Carolina. In the spring, some stay there to breed, and others migrate north to nest. How do these alternatives compare? Stephen Fretwell's studies suggest that resident individuals survive three times longer than migratory individuals but have lower nesting success due to crowded conditions that attract high nest predation. During the difficult winter months the residents control the best feeding areas. As a result, the winter visitors suffer greater mortality.

Generally speaking, birds that migrate to the tropics survive the winter better than do those that stay in the temperate zone (Table 13–1), but temperate-zone residents achieve higher per capita reproductive success than do returning migrants. Tropical residents trade low productivity for high survivorship. Few nests succeed, clutch sizes are small, and each pair attempts several costly nests a year, but adults are long-lived. The expected annual survival rate in the tropics is 80 to 90 percent compared to 50 percent for migrants and 20 to 58 percent for northern temperate residents.

Table 13–1 Life History Traits of Residents and Migrants

Trait	Temperate Resident	Migrant	Tropical Resident
Productivity	High	Moderate	Low
Adult survival	Low	Moderate	High
Juvenile survival	Low	Moderate	Moderate to high

The evolution of a migratory species presumably has two stages: partial migration, the division of a species into migratory and resident populations, and then fully disjunct migration due to the gradual elimination of resident populations (Cox 1968, 1985).

Partial migration evolves when local conditions promote opportunistic movement of birds to better conditions in nearby regions. In the case of the Field Sparrow, competitive pressures in North Carolina favor the northward movement of some segments of the population. In the case of the European Robin, harsh winters sometimes favor southward migrations. Periodic droughts can also foster partial migration by birds, including some species that were once residents of the arid Mexican Plateau and southwestern United States but now migrate annually to the north. Apparently, George Cox suggests, increased severity of the dry, summer season during the Pleistocene Epoch eliminated the resident populations of these species, which were replaced by other species that are better adapted to the new desert conditions, leaving only the migratory populations. At first such populations migrate short distances of a few hundred miles, but the distance between summer and winter ranges gradually increases to thousands of miles as a species expands its latitudinal range or as the habitats it requires become more separated by changes in the earth's climates.

Individuals of fully migratory species vary in the distances they migrate. From their breeding range throughout the northern United States and southern Canada, juncos migrate south to wintering grounds throughout the eastern United States to the Gulf Coast, but the migration distances of individual Dark-eyed Juncos translate into different average wintering distributions of males and females, adults and young (Ketterson and Nolan 1982, 1983). Adult females migrate farthest south to the southernmost states, young males stay farthest north in Indiana and Ohio, and adult males and young females settle at intermediate latitudes. Careful study of the patterns of annual survival suggests that each class of juncos chooses its own balance of three factors, which vary with age and sex:

1. mortality during migration,

2. overwinter survival, and

3. reproductive success as a function of time of arrival on the breeding grounds.

Greater mortality among the young selects for shorter migrations by the young of both sexes. Territory establishment by the earliest arrivals on the breeding grounds favors shorter migrations by males than by females, and especially by young males, which are at a disadvantage in the competition for breeding territories. Adult females migrate farther south to regions of lower population density, where overwinter survival is greatest (Figure 13–5).

So far, ornithologists have studied the benefits of migration only within a species such as the Dark-eyed Junco that migrates short distances. Comparable studies of long-distance migrants will be the next step in the effort to understand the evolution of migration in birds.

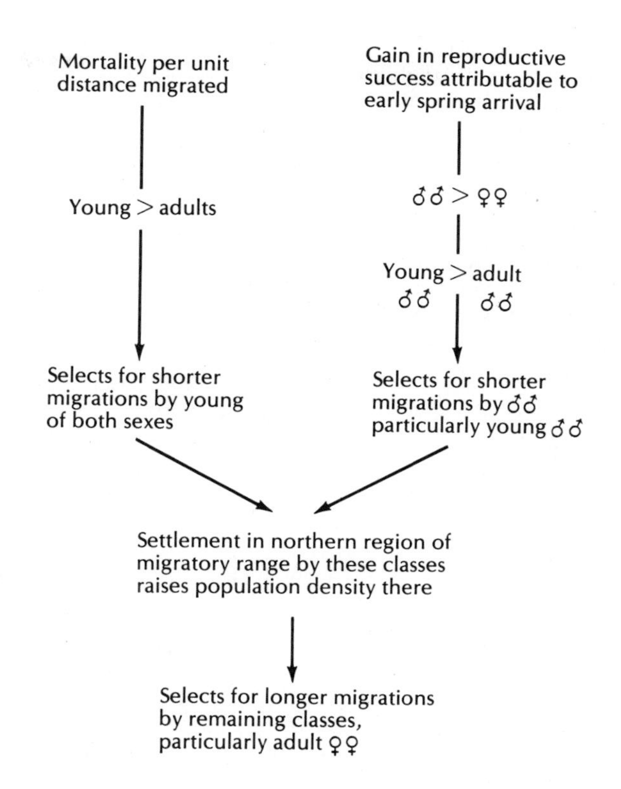

Figure 13–5 Evolution of differential migration by age and sex in Dark-eyed Juncos. (After Ketterson and Nolan 1983)

Migratory feats

Long-distance migratory flights are extraordinary feats of physiological endurance. The migrations of arctic shorebirds regularly exceed 13,000 kilometers one way from the high Arctic to distant South America. Red Knots, for example, fly from Baffin Island to Tierra del Fuego. A banded Lesser Yellowlegs flew 3220 kilometers from Massachusetts to Martinique, West Indies, in 5 or 6 days. These and other migrants cross thousands of kilometers of open ocean or inhospitable terrain without stopping, stretching their fuel reserves and physical abilities to the limit. Dangerous as nonstop crossings may be, they are often the only way to reach a destination or they may be preferable to longer, safer routes because of shorter flight time.

Every fall, vast numbers of migrants leave the coasts of New England and the maritime provinces of Canada, heading southeast over the ocean. The capacity and predeliction of larger, faster shorebirds such as the Lesser Golden-Plover for such flights has been known for many years, but radar studies now reveal parallel efforts by millions of small landbirds (Williams and Williams 1978). Up to 12 million birds have passed over Cape Cod in one night, embarking on a tremendous nonstop journey of 80 to 90 hours. Wave after wave of the migrants, such as the Blackpoll Warbler, depart at intervals of several days, heading past Bermuda and from there continuing on to the Lesser Antilles. Radar stations on Bermuda and Antigua pick up the approaching and passing waves of migrants. As these migrants reach the latitudes of Florida, they encounter strong trade winds from the northeast. The migrants then fly with the wind to the southwest toward the north

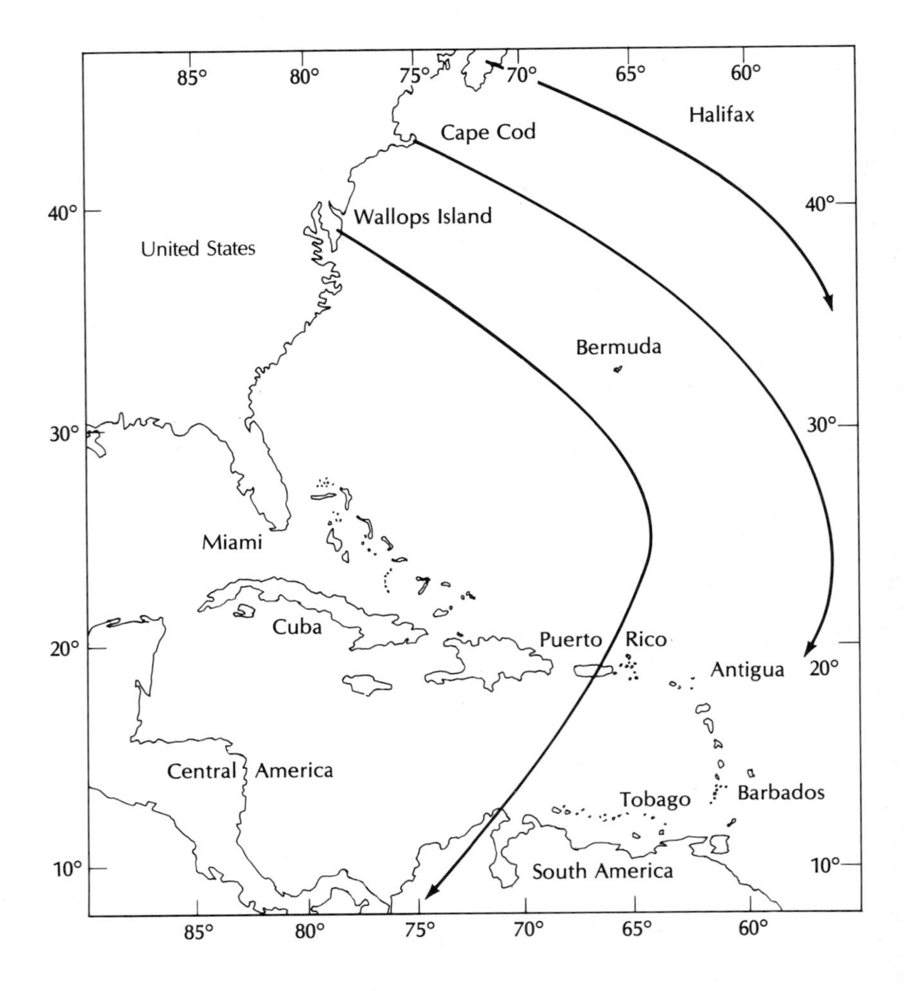

Figure 13–6 Millions of fall migrants such as Blackpoll Warblers fly directly from northeastern North America to northeastern South America. This 86-hour marathon flight takes them southeast past Bermuda to the trade winds that assist them on a southwesterly course to the Lesser Antilles and the coast of South America. (After Williams and Williams 1978, with permission of Scientific American)

coast of South America. The strong tail winds enable the tired travelers to make the last half of the journey somewhat more easily (Figure 13–6).

Evidence of the strenuous nature of the trip and of the way that the migrants stretch their physical capabilities to the limit can be seen in the exhausted condition of birds that stop at Curaçao, short of their destination, when flight conditions have been poor. Little more than feathered skeletons, they have depleted their fat reserves, metabolized much of their protein, and drained the remnants of their precious body water (Voous 1957). Tim and Janet Williams (1978) point out that "the trip requires a degree of exertion not matched by any other vertebrate; in man the metabolic equivalent would be to run a 4 minute mile for 80 hours. If a Blackpoll Warbler were burning gasoline instead of reserves of body fat, it could boast of getting 720,000 miles to the gallon."

Eurasian migrants also face herculean challenges (Moreau 1972). Northern Wheatears from Greenland start their journey to the British Isles by crossing 2000 to 3000 kilometers of open ocean with no assurance of favorable winds. Many European migrants fly 1100 kilometers directly across the Mediterranean and then, almost immediately thereafter, 1600 formidable kilometers nonstop across the Sahara desert (Moreau 1961). In the spring, they return across the Sahara, proceed 400 kilometers across the eastern Mediterranean, fly 600 kilometers over bleak, foodless Anatolia (in

Table 13-2 Fuels for Migration

Fuel	Energy Yield (kJ)	Metabolic Water (g)
Fat	38.9	1.07
Carbohydrate	17.6	0.55
Protein	17.2	0.41

Turkey), and finally travel another 650 to 1100 kilometers across the open water of the Black Sea. Still another route between Asia and Africa, used by birds that breed in northern Russia, includes traversing 1600 kilometers of Caspian desert plus 1700 kilometers of Saudi Arabian deserts and the intervening water passages. Some migrants such as falcons and bee-eaters cross from India directly to East Africa over 4000 kilometers of the Indian Ocean.

Fat as the fuel for migration

Migrants develop stores of fat especially for migration. Fat yields twice as much energy and water per gram metabolized than does either carbohydrate or protein (Table 13-2). Fat is stored in adipose tissues at various sites under the skin, in the muscles, and in the peritoneal cavity. For example, in White-crowned Sparrows subcutaneous fat is deposited initially at fifteen separate sites, and with continued deposition, the fat stores spread laterally and coalesce into a continuous layer between the skin and muscles (King and Farner 1965). Some fat is also stored in most muscles and in internal organs. Unlike the human heart, the avian heart does not accumulate much fat, even when the migrant reaches peak obesity (Odum and Connell 1956; King and Farner 1965) (Figure 13-7).

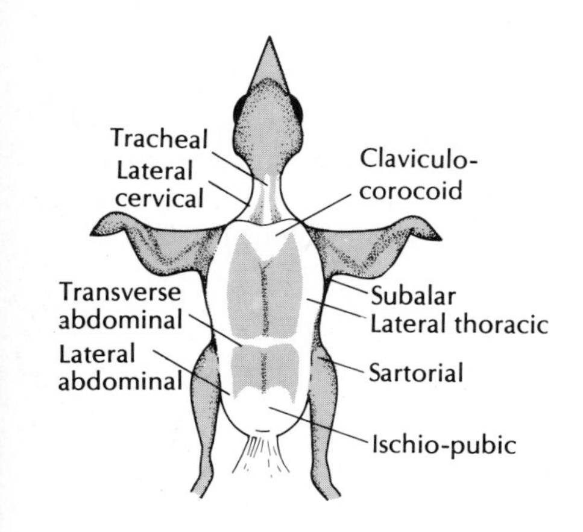

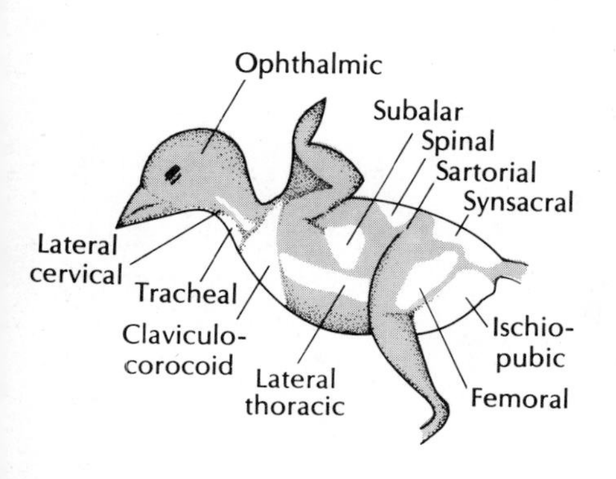

Figure 13-7 Principal sites of subcutaneous fat deposition in the White-crowned Sparrow. (After King and Farner 1965)

Adipose tissue does not consist simply of large, inert globs of fat, but supports a dynamic system of synthesis, storage, and release of lipids (George and Berger 1966). The enzyme lipase breaks down fat into free fatty acids and glycerol for transport to sites of use. Fatty acids are transported by the blood to mitochondria in the muscle cells for oxidation. In the mitochondria, fatty acids are progressively chopped apart into 2-carbon (acetyl) fragments, which are oxidized further in the Krebs cycle. The key result of this is the production of adenosine triphosphate (ATP) molecules, which power the contractions of muscle fibers. In the muscles, lipase activity, which is a good index of the capacity of muscles for fat metabolism, increases in relation to migratory activity. Today, specialists in the physiology of endurance exercise are very interested in the metabolic abilities of the pectoral muscles of migrant birds (Dawson et al. 1983).

Migrants fatten rapidly just before migration, sometimes to the point of obesity, by consuming enormous quantities of energy-rich food. Adult Lesser Golden-Plovers fatten themselves in Newfoundland each fall before making their flight to South America. Blackpoll Warblers nearly double their weight, from an average of 11 grams to an average of 21 grams. Ruby-throated Hummingbirds, which cross 500 to 600 miles of open water in the Gulf of Mexico, also nearly double their normal weight of 3 grams to make this trip.

How much fat migrants store reflects their requirements. Fat comprises 3 to 5 percent of the normal mass of small nonmigrating birds. Short-range migrants, which can refuel regularly, carry low to moderate fat reserves of 13 to 25 percent (Berthold 1975). In one study, the average reserves of White-throated Sparrows increased to 17 percent of total body weight just before migration, dropped to 6 percent when they reached their destination, and then increased to 12 percent as the winter progressed (Odum and Perkinson 1951). In contrast, Dunlins wintering at Teesmouth, England, develop moderate midwinter fat reserves of 15 percent and then fatten rapidly prior to spring migration to Arctic nesting grounds (Davidson 1983). Such long-range and intercontinental migrants as the Dunlin may build up fat deposits that comprise 30 to 47 percent of their total weight, mainly in preparation for long, nonstop flights (Berthold 1975).

Regular refueling usually accompanies long-distance migrations. Songbirds typically fly only several hundred kilometers and then pause for one to three days of rest and refueling. Some songbirds, however, press on several nights in succession until their reserves are nearly exhausted. Three to four refueling stopovers are a strategic aspect of the extraordinary migrations of arctic shorebirds, which congregate by the millions at key staging areas such as the Copper River Delta in Alaska, the Vendee in France, and the Bay of Fundy in eastern Canada. Five to twenty million shorebirds pass through the Copper River Delta every spring (Senner 1979; Isleib 1979). Their migratory movements are timed to coincide with the appearance of abundant food at these sites, where they build up fat reserves required for the next leg of their journey. Today, conservation programs for shorebirds are directed toward the protection of these staging areas (Figure 13–8).

Flight ranges of migrants

How far migrants can fly nonstop depends both on their fat reserves and on how quickly they use their fuel. David Hussell and his associates at Long Point Observatory on the north shore of Lake Ontario captured and weighed nocturnal migrants arriving at various times of the night after

Figure 13–8 Millions of shorebirds gather at key staging areas such as the Copper River Delta in Alaska to refuel for the next (in this case, final) leg of their migration to northern breeding grounds. (Courtesy D. Norton)

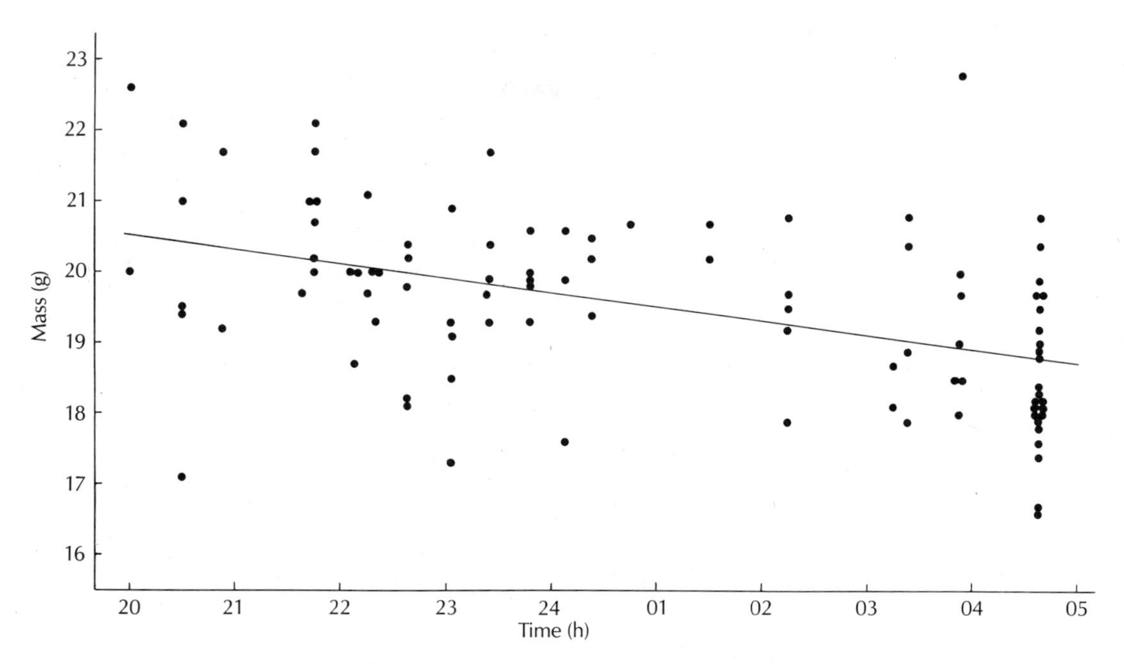

Figure 13–9 Ovenbirds, weighed on arrival at Long Point Observatory on the north shore of Lake Ontario, decreased in mass by an average of 0.2 grams per hour as the night proceeded. Assuming that those that arrived later had flown longer than those that arrived earlier, one can make an estimate of the energy costs of migratory flights. (Adapted from Hussell 1969)

flying north across Lake Erie. If one assumes that all the birds that were weighed had taken off at the same time and that those arriving later at Long Point had flown longer, then weight loss is seen to relate directly to time in the air. These data suggest an average weight loss of 0.9 percent of body weight per hour of flight (Hussell and Lambert 1980). This loss can be extrapolated to rate of fuel use if the loss is attributed to fat metabolism only, and if other components of body weight, especially water, are assumed to be in balance (Berger and Hart 1974). Thus, weight losses of about 1 percent, typical of the small migrant passerines weighed, project to expenditure of about 418 joules of energy per gram of body weight per hour of flight (Figure 13–9). The Blackpoll Warbler appears to be more fuel efficient than most other migrants: Hussell's studies indicate weight losses of only 0.6 percent per hour of flight for them, or energy expenditures of 318 joules per gram per hour of migration.

The potential flight range of a migrant is a function of its fat load (lipid index). The various equations available for projecting flight ranges of migrants are based on various assumptions about physiology and aerodynamics and therefore produce variable results (Berger and Hart 1974; Pennycuick 1969, 1975; Greenewalt 1975; Summers and Waltner 1979; Davidson 1983). Peter Berthold (1975) concluded that small birds that expend 418 joules per gram per hour during migratory flight and that have fat reserves of 40 percent of total live weight can fly about 100 hours and cover about 2500 kilometers. At that rate, they should be able to cross the most extensive barriers with energy to spare, unless they encounter strong headwinds. Migrant shorebirds such as the Dunlin have estimated flight range potentials of 3000 to 4000 kilometers (Davidson 1983) (Figure 13–10).

Efforts to understand the migration flights of Ruby-throated Hummingbirds illustrate how knowledge of flight physiology has yielded increasingly realistic projections of flight range. Ornithologists had long wondered how such a tiny bird could carry enough fuel to cross the Gulf of

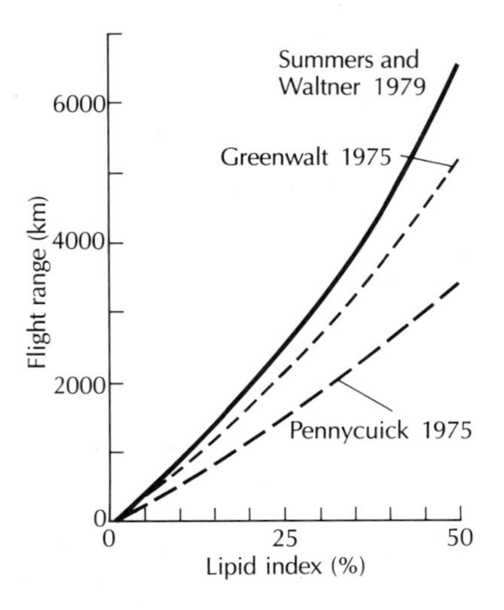

Figure 13–10 The potential flight range of birds, such as the Dunlin shown here, increases with fat load (lipid index); but estimates of flight range vary greatly, depending on authors' assumptions about physiology and aerodynamics. (Adapted from Davidson 1983)

Mexico, and some doubted that hummingbirds crossed at all, suggesting that they took a less direct route overland to Central America; others even suggested that hummingbirds hitched rides on the backs of larger migrants. The problem intensified when Oliver Pearson (1950) and Eugene Odum and C. E. Connell (1956) projected an inadequate maximum flight range of only 640 kilometers for this hummingbird. However, subsequent studies of hummingbird flight metabolism suggested much lower values (roughly half) for the caloric costs of flight, so that if a hummingbird consumed fat at the rate of 9.18 watts, carried 2 grams of fat, and flew at a velocity of 40 kilometers per hour, it should be capable of flying over 1000 kilometers nonstop in about 26 hours, more than enough to cross the Gulf of Mexico (Lasiewski 1962). By a slightly different means, Odum and his colleagues (1961) came to a similar conclusion.

As mentioned earlier, it is known that some migrants save energy by utilizing tailwinds. The landbirds that fly to South America orient their trip so that they pick up the southwesterly trade winds as they enter the tropical Caribbean region. They backtrack to land during their first night out at sea if wind conditions seem unfavorable for intercontinental flight (Richardson 1978). The observation that the ground speed of migrants does not increase in proportion to the known strength of tailwinds suggests that some, perhaps many, birds throttle back and coast with tailwinds (Bellrose 1967), thereby saving energy and potentially increasing flight range by 30 to 40 percent (King 1972).

Timing of migration

Precise arrival and departure dates are an impressive feature of migration in some species. Every year, after their transequatorial migration, Short-tailed Shearwaters arrive at their breeding colonies off southern Australia within a week of the same date. The traditional return of Cliff Swallows on March 19 to the San Juan Capistrano mission in California has become a symbol of this precision, as well as of spring itself.

Internal rhythms that are linked to other aspects of the annual cycle guide the timing of migration. Caged migratory passerines predictably become restless just before the time they would migrate in the wild. This phenomenon, called *zugunruhe* or migratory restlessness, has been familiar to bird fanciers for at least 200 years. Typically, a captive bird wakes shortly after dark and then jumps or flutters in the cage until at least midnight. Because the amount of activity is easily measured, it lends itself to experimental study of both the physiology of migration and orientation behavior (see Chapter 12). Nonmigratory birds do not exhibit *zugunruhe* behavior (see Figure 13–4). Adrenocortical hormones are known to act in concert with prolactin in stimulating this behavior in White-crowned Sparrows, one of the few species in which the links between specific hormones and migratory behavior have been demonstrated.

We now know that increasing day length in winter stimulates early spring restlessness, hyperphagia (eating to excess), fat deposition, and weight increases in many migratory birds. Extending Rowan's findings about the photoperiodic control of the annual cycle (see Chapter 12), Albert

Wolfson showed that Dark-eyed Juncos from migratory populations respond to increasing day length by adding fat stores, whereas sedentary juncos do not (Wolfson 1942). Spring fat deposition and migratory activity of Gambel's White-crowned Sparrows are under the direct control of increasing day length, mediated precisely by an internal clock. The average date of onset of springtime premigratory fat deposits in captive Gambel's White-crowned Sparrows has been shown to remain constant over the course of eight years, with a standard error of only one day (King 1972).

The timing of preparations for fall migration is indirectly set by the spring activities. The normal fall sequence of photorefractory testes, prebasic molt, and preparations for migration in White-crowned Sparrows, for example, depends on prior exposure to long photoperiods, but the pace is proximately influenced by shortening days (Farner and Lewis 1971). Rowan suggested some causal relations between gonadal cycles and migration, but the available evidence now indicates that sex hormones do not directly regulate migration. In one set of experiments, for example, castration did not prevent male Golden-crowned Sparrows from becoming restless and putting on their premigratory fat deposits at the appropriate time of the year (Morton and Mewaldt 1962).

The timing of migration relates first to internal physiological rhythms (see Chapter 10), but extrinsic weather factors also play a role, primarily fine tuning (Saunders 1959). Northward movements of migrants in the spring correlate with the warming of the higher latitudes. Both American Robins and Canada Geese move north in the eastern United States, just behind the main spring thaw, along regions that have a mean temperature of 2°C (Figure 13–11). A line connecting these points is called the 2°C isotherm. Willow Warblers in Europe move north with the 9°C isotherm.

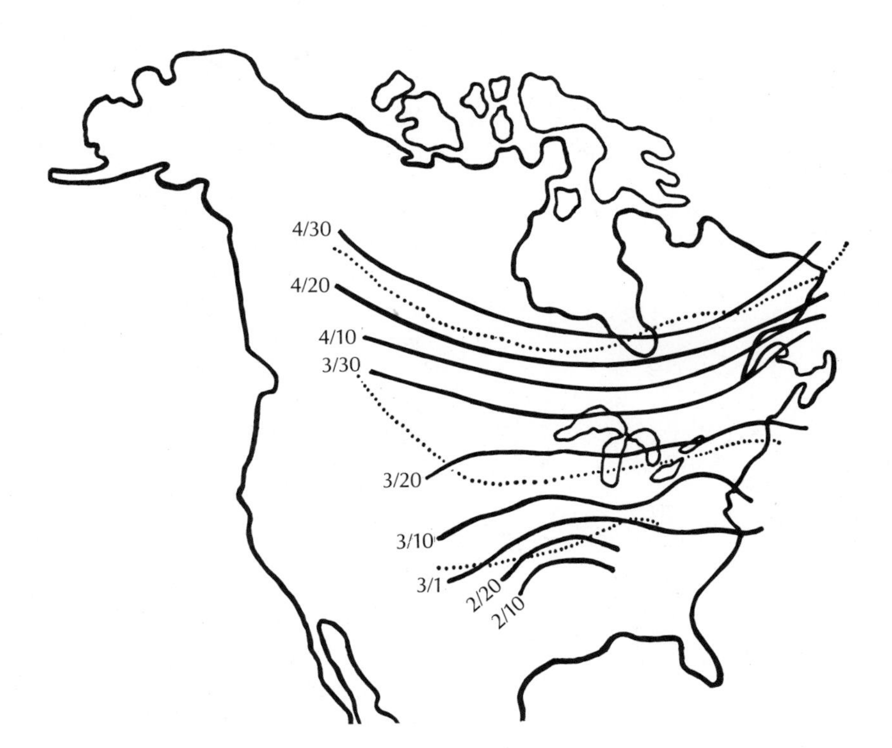

Figure 13–11 The rate of spread of Canada Geese through North America in spring (dotted lines) corresponds to the northward movement of the 2°C isotherm (solid line). (Adapted from Dorst 1962)

Daily weather conditions, in particular, favorable winds, also influence departure times (Raynor 1956; Richardson 1978). In spring, major northward movements in the United States coincide with a depression (lowering of barometric pressure) toward the southwest, followed by a strong flow of warm southern winds from the Gulf of Mexico toward the northeast. The sizes of migration waves relate directly to the intensity of the depression and strength of the favorable winds (Bagg et al. 1950). The value of favorable winds is clearly seen in records of arrivals of northbound migrants at Baton Rouge, Louisiana (Gauthreaux 1971). Migrants from Central America usually reach Louisiana in midafternoon after crossing the Gulf of Mexico; but when they have strong southern tailwinds, they arrive several hours earlier, in the late morning. On rainy days with adverse winds, they arrive later in the evening and do not arrive at all on days when there are cold fronts or east winds.

Fall migration departures are also stimulated by favorable weather conditions. Good flights of raptors at Hawk Mountain, Pennsylvania, and of landbirds generally at Cape May, New Jersey, are the result of strong northwest winds due to a barometric depression moving east from the Great Lakes region. Departures from the New England coast are related to favorable tailwinds (Richardson 1978), and peak flights south across the Gulf of Mexico in early October coincide with improved flight conditions to the north (Buskirk 1980).

Exactly how migrants forecast weather conditions is a mystery, but birds are sensitive to changes in barometric pressure (see Chapter 7) and feed more intensely as storms approach and barometers fall. Wind directions aloft, however, are not easily judged from the ground. Meteorologists track weather fronts by monitoring infrasound (low-frequency sound, see page 166) with a special system of microphones. Pigeons, too, seem to be sensitive to infrasound and may use this source of information in some way (Kreithen and Quine 1979).

Some variations in the timing of migration relate to age and sex. Males generally migrate north before females to compete for breeding territories. Male Red-winged Blackbirds, for example, arrive on the breeding grounds one to five weeks before their potential mates. In Europe, male Eurasian Skylarks spend a month alone on their territories. In the fall, young Least Flycatchers, Common Swifts, and Common Chaffinches migrate south before their parents. The females of these species tend to move south ahead of the males. Adult shorebirds generally leave before their young, a whole month earlier in the case of Hudsonian Godwits.

Some birds migrate by day and others by night, and still others, such as waterfowl and shorebirds, at both times. Diurnal and nocturnal flights offer different advantages. Hawks migrate during daylight hours when they can take advantage of warm rising air currents. Swifts and swallows, which feed on the wing, also migrate by day. However, many small land birds, including most flycatchers, thrushes, warblers, and orioles, as well as rails and woodcocks, depart shortly after dark and migrate through the night. These nocturnal migrants can sometimes be seen through a telescope as silhouettes against a full moon. Predation by hawks and gulls is less likely at night, and the migrants have the next twelve daylight hours to feed. Also, stable night air creates favorable flight conditions, and cool nighttime temperatures favor heat loss and water retention. Shorebirds fly higher than landbirds, reaching altitudes of several kilometers. Most migrating ducks travel below 5000 meters, as do nocturnal landbirds. Single migrants from the Yucatan approach Louisiana flying low over the Gulf of Mexico at 244 to 488 meters, but when they reach land, they climb to 1220 to 1524 meters and coalesce into small flocks of about 20 birds (Gauthreaux 1972). The reasons for these variations remain unknown.

Patterns of migration

Migration routes and patterns are almost as varied as the migrants themselves, reflecting the histories of populations, their abilities to cross large barriers, the positions of topographical barriers, and the relative locations of summering and wintering grounds. Details are now available for hundreds of species in both the New World and the Old World as a result of the extensive marking and recovery programs of the past 50 years (Lincoln 1939; Dorst 1962; Moreau 1972). The main migration routes in North America are oriented north-south, partly because wintering ranges of most species lie south of breeding ranges and partly because the coasts, major mountain ranges (Appalachian Mountains, Rocky Mountains, and the Sierra Nevada), and major river valleys (Mississippi) trend north-south. In the Old World, birds migrate east-west in accordance with the longitudinal displacement of seasonal ranges and the east-west orientation of the Alps, the Mediterranean Sea, the North Sea coasts, and the great deserts of North Africa and the Middle East. In general, birds of the southern hemisphere do not migrate as far north as north-temperate birds migrate south. In South America, the Kelp Goose, and the Buff-necked Ibis, all of which nest at the southern reaches of the continent, migrate north only as far as central Chile and Argentina. Some swallows and a few flycatchers, such as the Crowned Slaty-Flycatcher, move north into tropical South America, but they are exceptions to the rule. Similarly, only about 20 South African species winter as far north as equatorial Africa, in contrast to 183 Palearctic species that move to subsaharan Africa.

Migration routes sometimes reflect the recent distributional histories of birds. Those individuals that inhabit newly colonized areas tend to retrace the population's historical expansion routes. Northern Wheatears that colonized Greenland from the British Isles return there and then head south on their way to Africa for the winter. Pectoral Sandpipers from Alaska recently established a new breeding population in Siberia. Instead of migrating south through the Orient as do most Siberian shorebirds, these "Siberian" Pectoral Sandpipers fly back to Alaska and then south with the rest of their species to South America. Conversely, Arctic Warblers, Yellow Wagtails, and Northern Wheatears, species that have spread recently into Alaska from Siberia, return to Siberia before migrating south.

In addition to the physical feats involved, migration via particular routes between precise breeding territories and wintering stations represents a navigational feat. What cues do birds use to maintain a steady southward course to fly steadily south for thousands of miles? What kinds of information do young birds use on their first migratory flight? Such questions are the topic of the next chapter.

Summary

Billions of birds migrate every fall and spring to exploit seasonal feeding and nesting opportunities. Why some populations migrate and others of the same species are sedentary is an age-old and unresolved question. The migratory habit may appear in newly established populations of nonmigratory species or, in contrast, may be lost by colonizing populations of migratory species. In general, migrants achieve moderate levels of reproductive success and adult survivorship, whereas residents sacrifice productivity for high survivorship (tropical residents) or survivorship for high productivity (temperate residents). Trade-offs between the costs and benefits of migration determine how far individuals migrate.

The flights of many long-distance migrants require extraordinary physical endurance. Nonstop three-to-four-day journeys across the open ocean or desert regions are fueled by reserves of fat. Small landbirds have a maximum flight range of about 2500 kilometers, and shorebirds can fly 3000 to 4000 kilometers. Regular refueling stops, however, are typical of most migrants. Shorebirds, for example, gather in vast numbers at critical enroute staging areas such as the Copper River Delta in Alaska. In addition to the major migrations between North and South America, and between Eurasia and Africa, many species migrate short distances up and down mountain slopes or between southern Canada and the central United States.

Precise arrival and departure dates are an impressive feature of migration. Internal rhythms, linked to other aspects of the annual cycle, guide the timing of migration. Changes in day length stimulate preparations for migration such as accumulation of fat deposits and migratory restlessness. Weather conditions such as favorable winds and changing temperatures control the day-to-day efforts of migrants. Many birds migrate at night, when flight conditions are more favorable and predators are few.

Recapturing birds wearing numbered aluminum rings has enabled ornithologists to map the migration routes of many birds. Corresponding in part to the topography of the continents, major migration routes orient north-south in North America and east-west in Europe. The migratory routes of some birds retrace the history of the expansion of the range of their species.

Further readings

Berthold, P. 1975. Migration: Control and metabolic physiology. Avian Biology 5:77–127. *An excellent review of the physiology of migration.*

Gauthreaux, S.A. 1982. Ecology and evolution of migration. Avian Biology 6:93–167. *A comprehensive review of the historical literature on origins of migration.*

Dorst, J. 1962. The Migrations of Birds. Boston: Houghton Mifflin. *A classic review.*

Keast, A., and E. Morton. 1980. Migrant Birds in the Neotropics: Ecology, Behavior, Distribution, and Conservation. Washington. D.C.: Smithsonian Institution Press. *A rich set of varied papers.*

Moreau, R.E. 1972. The Palearctic-African Bird Migration Systems. London: Academic Press. *The final work of the dean of Old World migration systems.*

chapter 14

Orientation and navigation

How a bird finds its way across vast, unfamiliar terrain has intrigued us for centuries. Recently ornithologists have started to analyze the navigational systems of birds and have found that birds use several sources of information, often preferring one source if it is available and using the others only when necessary (Emlen 1975a). In addition to using visual landmarks such as landscapes and buildings, migrants use the sun by day and the stars by night. Birds also use the earth's magnetic field, olfactory cues, and perhaps very low sound frequencies (infrasound) as well. The evolution of extraordinary navigational abilities has been a corollary of avian flight and migration.

Individual birds have an astonishing ability to return to a particular tree in Europe after wintering in Africa or to migrate annually between particular sites in North America and South America (Moreau 1972; Rappole and Warner 1980). In one of the earliest experiments, an Eastern Phoebe, wearing a silk thread placed on its leg by John James Audubon in 1803, returned the next spring to Audubon's house in Mill Grove, Pennsylvania, presumably after wintering somewhere in the southern United

Figure 14–1 The black-chequered homing pigeon. (From Lumley 1895)

States. In a more recent study of migrants on their winter territories in South America, banded Northern Waterthrushes returned predictably every year to the same sites in Venezuela (Schwartz 1964).

The homing feats of displaced birds also testify to their navigational abilities. Homing pigeons can return to their lofts by flying as much as 800 kilometers per day from unfamiliar places. Ancient Egyptians and Romans developed these messengers by breeding for the natural orientation abilities of feral pigeons (Figure 14–1). Shearwaters and sparrows, as well as a variety of other birds, can return to a home site after being transported thousands of miles away (Table 14–1). A Manx Shearwater, for example, returned to its nest burrow in Wales only twelve and a half days after being released in Boston (Mazzeo 1953). Marked White-crowned Sparrows that were shipped to Baton Rouge, Louisiana, returned the following winter to their wintering grounds in San Jose, California, where they were recaptured. They returned again after a second displacement to Laurel, Maryland (Mewaldt 1964) (Figure 14–2).

In this chapter we review the kinds of information that birds use to navigate while migrating, while commuting between nest sites and feeding grounds, and while flying home after being displaced by a curious ornithologist. We also consider how young birds learn to navigate.

Use of visual landmarks

It appears that birds rely heavily upon visual landmarks for both local travel and long-distance migration. Diurnal migrants, for example, follow water courses and coastlines, gathering en route in great numbers where restricted corridors function as funnels (Bellrose 1967). The Strait of Gibraltar and the Bosphorus at Istanbul are major funneling points for Eurasian migrants that detour around the Mediterranean Sea. The coasts of Central America funnel thousands of migrating raptors — Broad-winged Hawks, Swainson's Hawks, and Turkey Vultures — over Panama City. Crowds of birdwatchers gather to view the spectacle of migrants funneled to the tips of peninsulas such as Point Pelee, Ontario, and Cape May, New Jersey.

Table 14–1 Abilities of Birds to Return to the Site of Capture after Transport to a Distant, Unfamiliar Release Site

Species	Number of Birds	Distance (km)	Return (%)	Speed (km/day)
Leach's Storm-Petrel	61	250–870	67	56
Manx Shearwater	42	491–768	90	370
Laysan Albatross	11	3083–7630	82	370
Northern Gannet	18	394	63	185
Herring Gull	109	396–1615	90	112
Common Tern	44	422–748	43	231
Barn Swallow	21	444–574	52	278
European Starling	68	370–815	46	46

(From Griffin 1974)

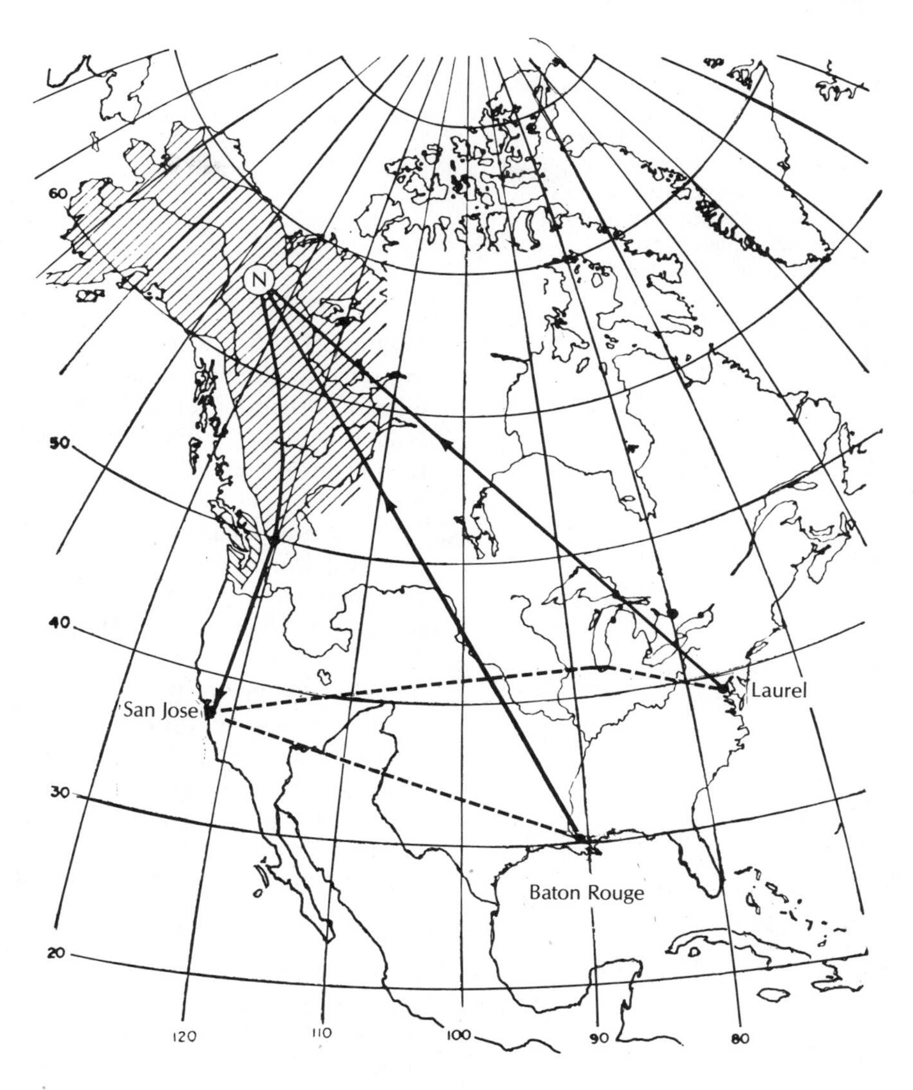

Figure 14–2 White-crowned Sparrows returned to their wintering grounds in San Jose, California, after being displaced by aircraft (broken lines) to Baton Rouge, Louisiana, and to Laurel, Maryland. These marked sparrows apparently spent the intervening summers on their nesting grounds in Alaska. The solid lines show their probable flight paths. (Adapted from Mewaldt 1964)

The sun compass

Several lines of evidence have revealed that birds use navigational cues other than landmarks and, perhaps, senses other than sight. In one study, homing pigeons ignored obvious landmarks until they came within sight of a tall building near Boston, which they used to make a final correction toward home (Michener and Walcott 1967). In another experiment well-trained homing pigeons were fitted with frosted contact lenses that eliminated image formation beyond three meters (Schlichte and Schmidt-

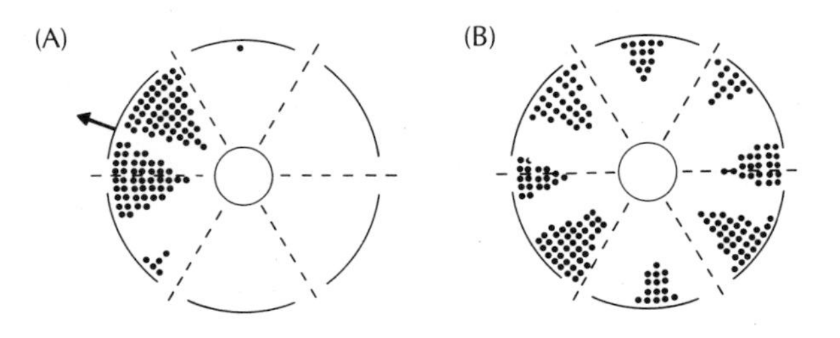

Figure 14–3 European Starlings use the sun to orient in a circular cage. (A) As long as they could see the position of the sun in the sky, they oriented their restless migratory behavior to the northwest. (B) On overcast days when they could not see the sun, they showed no directional orientation. Each dot represents 10 seconds of fluttering activity. (From Kramer 1951; Emlen 1975a)

Koenig 1971; Schlichte 1973). These severely myopic birds flew "blind" for over 170 kilometers directly back to their lofts. When they reached the vicinity of their lofts, they hovered and then landed much like helicopters. Not all such pigeons performed perfectly, some crashed and some missed the loft altogether, but many oriented well without being able to see landmarks.

Scientists had long suspected that birds navigate by the sun, but proof of this ability awaited experiments conducted with starlings and homing pigeons in the 1950s. In Germany, Gustav Kramer (1950, 1951) studied the orientation of migratory restlessness *(zugunruhe)* in European Starlings. The birds were housed in circular cages and placed in a large pavilion with windows through which they could see the sun, including its change of position as the day progressed. As long as they could see the sun, they focused their attention to the northwest, the correct direction for spring migration. On overcast days, however, the starlings showed no such directional tendency (Figure 14–3).

In Great Britain, Geoffrey Matthews (1951, 1953, 1968) established that homing pigeons use the sun to guide them back to their lofts. He released them from unfamiliar sites away from the loft under a variety of weather conditions. The pigeons flew directly home when they could see the sun, but they fared poorly under overcast skies. As a result of releasing the pigeons at various times of day, Matthews also discovered a key feature of this orientation behavior: Not only could the pigeons use the sun for directional information, but also they were able to compensate for its changing position as the day progressed as if they could "tell time."

The position of a point on the earth relative to the sun changes continuously by 15 degrees per hour. To orient consistently in one direction, a bird must somehow understand the changing position of the sun relative to direction throughout the day; that is, the sun compass must be time-compensated by the birds, as Matthews proposed. In an important experiment that extended his other pioneering experiments with the starlings, Kramer and his colleagues showed that birds compensated for the apparent motion of the sun by training starlings (and some other birds) to feed from the northwest cup of a series of cups placed around the perimeter of a circular cage. The birds had no trouble choosing the correct cup if they could see the sun. When they were trained to accept a stationary light bulb as a substitute for the sun, they fed from cups farther and farther to their left as they continued to compensate for the progress of the "sun" over the course of the day (Kramer 1952).

Adelie Penguins, which swim and walk to their destinations rather than fly, also use a time-compensated sun compass. John Emlen and Richard Penney (1964) took penguins from their coastal breeding rookeries to

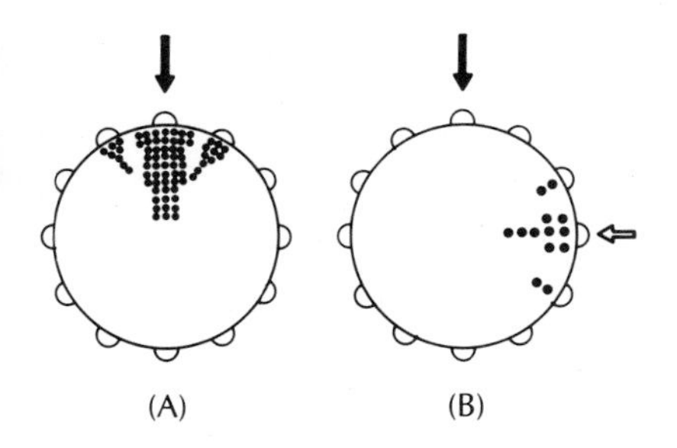

Figure 14–4 When the internal clock of a European Starling is set 6 hours behind natural time (by changing the schedule of light and dark), it misreads the sun's position and looks for food 90 degrees (white arrow) from the correct location (black arrows). (A) Behavior during training, showing correct orientation; (B) behavior after the 6-hour clock-shift in internal schedule. Each dot shows an attempt to find food. (Adapted from Hoffmann 1954; Emlen 1975a)

the interior of Antarctica and released them. On cloudy days, the penguins wandered about randomly without significant orientation. When the sun was shining, however, they headed north-northeast, compensating for the sun's apparent counterclockwise movement in the southern hemisphere by correcting their orientation 15 degrees per hour clockwise relative to the sun's position.

The next step in the study of a time-compensated solar compass was to change a bird's internal clock and thereby trick it into misreading the sun's position. Konrad Hoffmann (1954) did this by studying orientation behavior in "clock-shifted" European Starlings trained to find food in a particular compass direction. He kept these birds on a 12-hour-dark and 12-hour-light cycle that was 6 hours out of phase with natural daylight (the lights went on at 1200 instead of 0600). Accustomed to this schedule, the starlings predictably misread the sun's position in the sky. When they awakened, for example, it was actually noon, not 0600, and the sun was in its southern midday position, not its eastern dawn position. The clock-shifted starlings, however, interpreted this to be the dawn position of the sun; therefore, their "east" was really south. As a result they looked for food at a position 90 degrees clockwise from the correct bearing (Figure 14–4). This is a standard result: A 6-hour clock shift causes a 90-degree disorientation. Experiments with many other clock-shifted birds, including homing pigeons, have since confirmed the widespread use by birds of time-compensated solar cues.

Navigation at night

A great many landbirds and waterfowl migrate at night by using the stars as a source of directional information. An example of this ability was provided by William Cochran and his coworkers (1967). They captured a migrating Gray-cheeked Thrush in central Illinois one afternoon and attached a tiny radio transmitter to it. At dusk the thrush took off on the next leg of its journey, followed by the ornithologists in a small plane. A severe thunderstorm and lack of fuel forced their plane down during the night, but the thrush flew on. After refueling, the Cochran group took off again and were fortunate enough to find the thrush by dead reckoning. The thrush landed at dawn in Wisconsin after flying 650 kilometers on a firm compass bearing all night, without refueling.

Franz and Eleanore Sauer (Sauer 1957, 1958) first demonstrated the ability of migrating passerine birds to use the stars for navigation in experiments with hand-reared Garden Warblers. The birds were kept in circular, experimental cages in a planetarium. When these birds are ready to migrate, they become hyperactive and restless; even when caged, they tried to fly or hop in their migratory direction. The Sauers watched the birds' orientation through the glass bottom of their cage. The warblers oriented north in the "spring" and south in the "fall" of the simulated night sky of the planetarium. When the "stars" were turned off, the warblers became disoriented. When the Sauers rotated the north-south axis of the planetarium sky 180 degrees, the warblers also reversed their compass headings.

Stephen Emlen (1967a, 1975b) duplicated the Sauers' results with a North American migrant, the Indigo Bunting. The cage in which Emlen

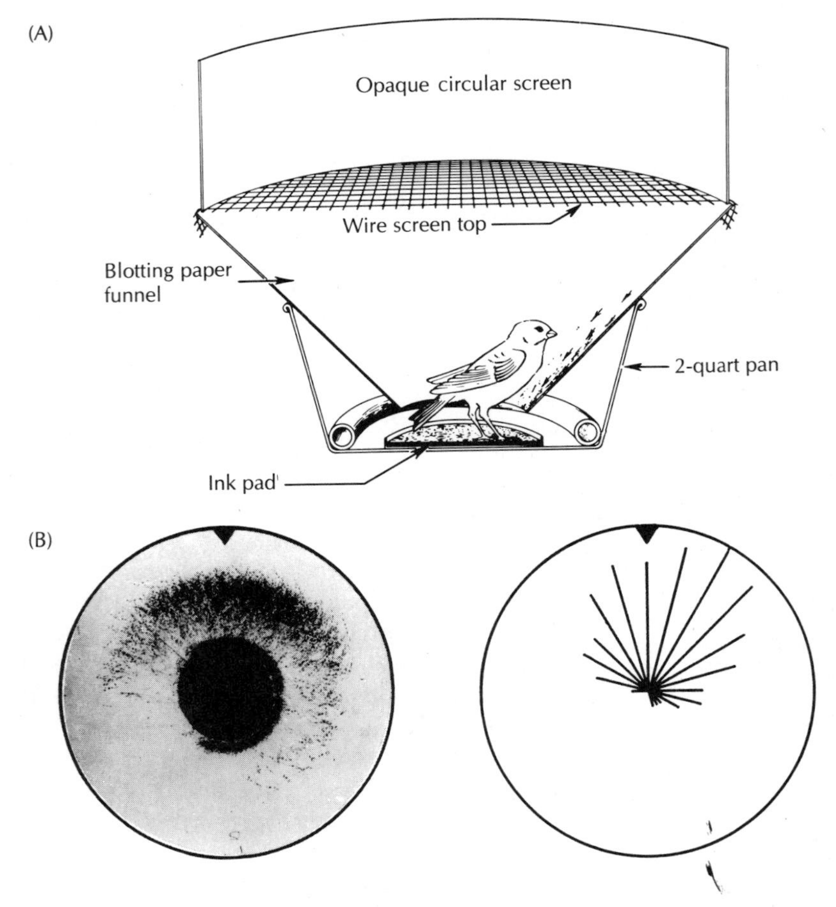

Figure 14–5 (A) The Indigo Bunting migrates at night between its summer range in the eastern United States and its winter range in Central America. Buntings in a state of migratory restlessness orient by the stars at night, even when confined to a funnel-like cage placed under a planetarium sky. (B) Inky footprints record the orientation direction; the lengths of line vectors measure the intensity of ink left in each 15-degree sector. (Adapted from Emlen and Emlen 1966)

placed this bird was a paper cone with steep sides sloping downward to an ink pad at the apex. A restless bird would hop up the side of the cone, leaving inky foot marks, and slide back down onto the pad. The intensity of ink on various sections of the cone indicated the relative degree of activity in each direction (Figure 14–5). Even in these experimental cages, Indigo Buntings oriented north when a spring night sky was simulated in a planetarium, and south when a winter night sky was simulated. Like the warblers, the buntings became disoriented when the planetarium sky was turned off and reversed their orientation when the axis of the sky was reversed (Figure 14–6).

Whether nocturnal migrants compensate for the movement of the stars as other birds do for the movement of the sun is not yet clear. The warblers studied by the Sauers changed their orientation when the planetarium sky was set ahead or behind the correct time, which suggests that they navigate by a time-compensated stellar compass. Indigo Buntings, on the other hand, maintain their northward orientation under phase-shifted planetarium skies, which means that the buntings do not use a specific star but refer instead to star patterns or constellations. Studies of White-throated Sparrows and Mallard Ducks also suggest that use of stellar cues does not include, and perhaps does not require, time compensation. Short-distance migrants such as the buntings may use the stars differently, perhaps with less sensitivity and less time compensation, than do long-distance migrants such as the warblers. Transequatorial migrants not only may need greater preci-

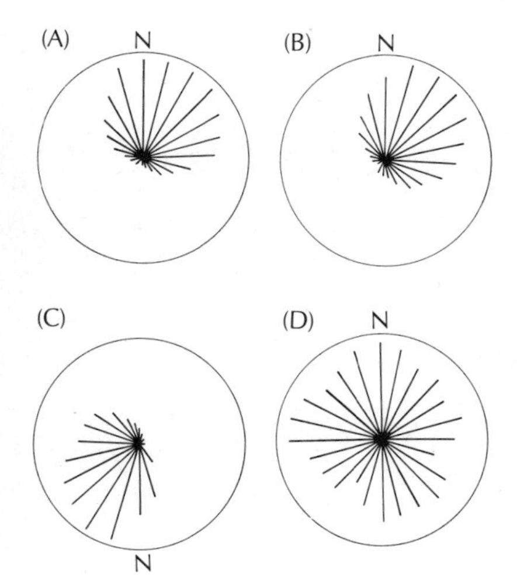

Figure 14–6 Line vectors, such as the ones described in Figure 14–5, show how Indigo Buntings use the stars to orient north in the spring. They do so under (A) natural night skies and (B) simulated night skies in a planetarium. (C) When the planetarium stars are shifted so that the North Star, N, is at true South, the birds reverse their orientation. (D) When the stars are turned off and the planetarium is diffusely illuminated, the buntings do not orient. (Adapted from Emlen 1975b)

sion in their initial orientation because of the long distance they fly, but they also must orient under the different skies of the southern and northern hemispheres.

Stephen Emlen (1967b) attempted to identify the stars that buntings use for orientation by systematically blocking out various constellations. It was logical to assume that the buntings orient by the North Star, the one obvious point in the night sky that is unmoving. But they did not. Instead, they used the constellations that were within 35 degrees of the North Star. Moreover, the buntings were familiar with most of the major constellations in the northern hemisphere, including the Big Dipper, the Little Dipper, Draco, Cepheus, and Cassiopeia; if one of these was blocked from view, they used the others. Such redundancy is useful when portions of the sky are overcast; it also allows the birds to be flexible in their choice of guideposts in the complex ever-changing night sky. In addition, different birds appear to use different parts of the sky, their use apparently based on experience.

As we discussed in Chapter 12, it is easy to change a bird's hormonal physiology by changing day length or photoperiod. Simulating the various seasons by increasing or decreasing day lengths can bring caged birds into breeding condition, cause them to molt more often than is natural, and cause them to accumulate premigratory fat at the wrong time of the year. Using unnatural photoperiod regimes, Emlen (1969) manipulated the seasonal physiology of two groups of Indigo Buntings. He induced readiness for northward spring migration in one group and readiness for southward fall migration in the other group. Exposed to the same planetarium sky, buntings in the two groups oriented north and south, respectively. These results show that migratory orientation is under physiological control.

Geomagnetism

Some migrating birds can maintain their flight directions under totally overcast skies or while flying through clouds that obscure both celestial information from above and landmark information from below. Radar studies show that under these conditions orientation is not as accurate as under clear skies and that some birds shift from direct flight paths to looping, twisting paths. Yet on the average they maintain their correct orientation over many hours.

Ornithologists were slow to accept the hypothesis that birds use the earth's magnetic field for orientation. Despite their recognition of potential geomagnetic navigational cues, biologists regarded the idea of the existence of a magnetic compass in birds with about the same skepticism that geologists once had for the idea that the continents floated apart over the course of geological time. An early report that magnets disrupted a pigeon's homing ability (Yeagley 1947) was discredited, largely because the results could not be repeated.

Then, Frederick Merkel, Wolfgang Wiltschko, and Roswitha Wiltschko (1965) showed that captive European Robins can orient in experimental solid steel cages without celestial cues. In continued experiments that aroused general interest, these researchers showed that a robin reverses its orientation when the magnetic field imposed on the steel cage is

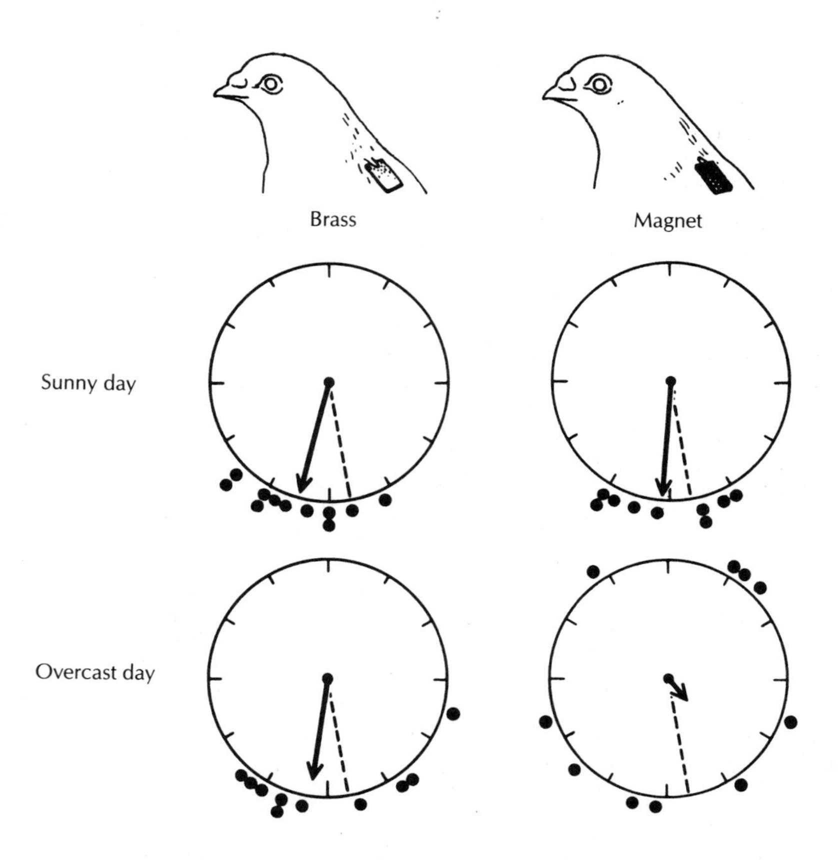

Figure 14–7 A bar magnet interferes with a homing pigeon's ability to return to its loft on overcast days. On sunny days, pigeons wearing magnets and control pigeons wearing brass bars both adopt accurate home bearings at unfamiliar release sites. On overcast days when they cannot orient by the sun (their preferred cue), the pigeons wearing magnets become disoriented. The control group, however, orients by means of the earth's magnetic information. Vectors (arrows) show mean direction and consistency of orientation among individuals: Long vectors show consistent orientation, and short vectors show variable orientation. Dots represent bearings recorded for each pigeon tested. The dashed line represents the correct orientation. (From Keeton 1974)

reversed. Several years later William Keeton (1971, 1972) showed that free-flying homing pigeons wearing bar magnets often could not orient properly on cloudy days, whereas control pigeons wearing brass bars usually could (Figure 14–7). These results resurrected the possible validity of Yeagley's earlier results. An important revelation of Keeton's work was the lack of difference between the two experimental groups on sunny days, when they both used the sun compass in preference to the magnetic compass. This result accounts for some of the failures to repeat Yeagley's experiments. Finally, in experiments that swayed even the skeptical, Charles Walcott and Robert Green (1974) fitted homing pigeons with electric caps (containing Helmholtz coils) that produced a magnetic field through the bird's head. Under overcast skies, reversing the field's direction by reversing the current caused free-flying pigeons to fly in the direction opposite to their original course (Figure 14–8).

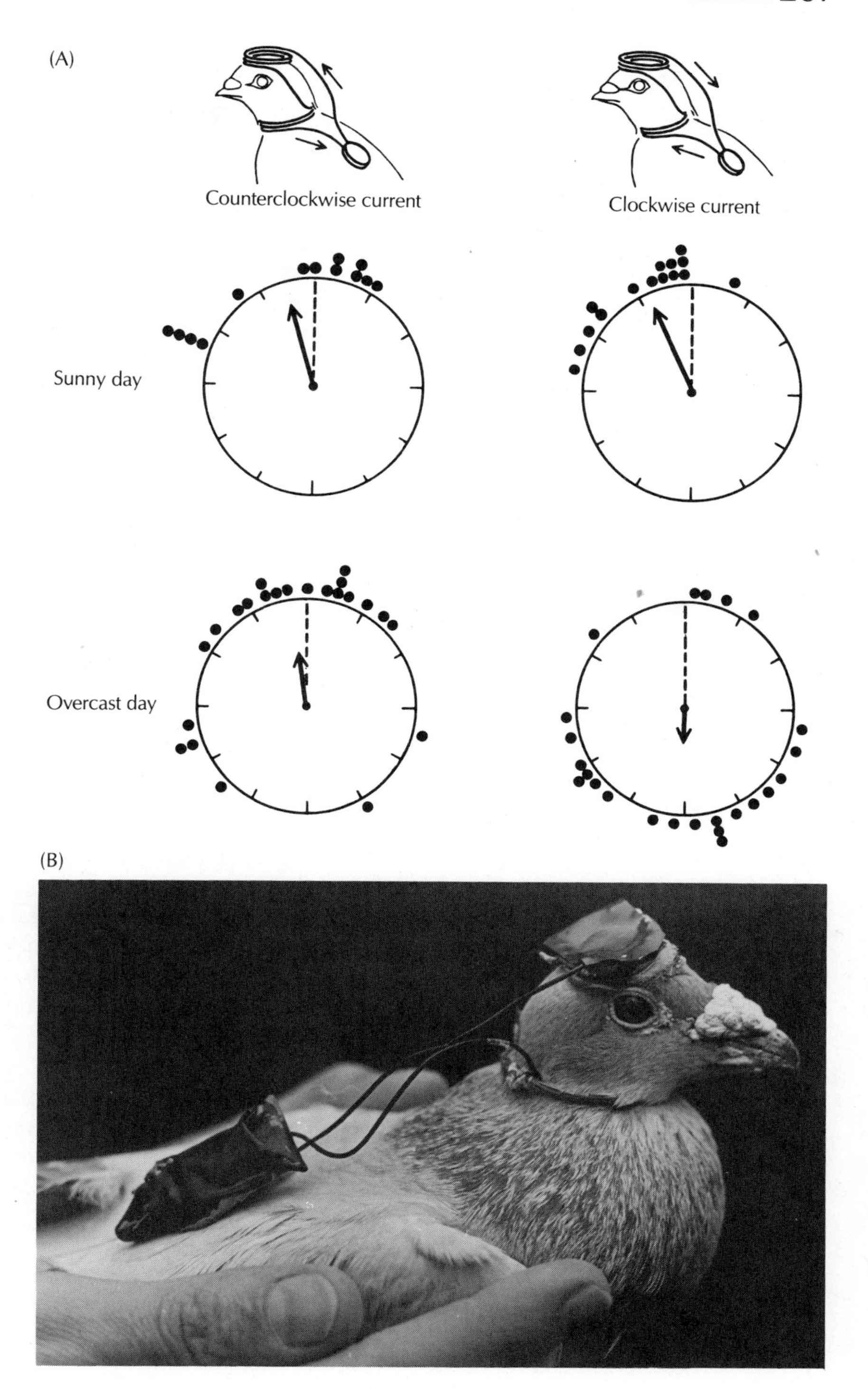

Figure 14–8 (A) Attaching Helmholtz coils to the heads of homing pigeons, Charles Walcott and Robert Green reversed magnetic fields generated by an electric current flowing through the coils. The reversal of electric current caused the pigeons to reverse their orientation direction on overcast days. Vectors are portrayed as in Figure 14–7. (B) A homing pigeon equipped with Helmholtz coils. (A, adapted from Walcott and Green 1974, Keeton 1974; B, courtesy of C. Walcott)

Further experiments have revealed that many migrating bird species can navigate by using the earth's magnetic fields. Experiments with European Robins and several species of European warblers have also revealed much about the magnetic fields that these birds can sense. Birds are sensitive to extremely weak fields, such as the earth's, including levels as low as 10^{-7} to 10^{-9} gauss, which correspond to the natural fluctuations in the earth's magnetic field caused by sunspots and hills of iron ore. Natural fluctuations in the earth's magnetic field seem to disrupt the orientation of passerine birds migrating at night (Moore 1977), and orientation in Ring-billed Gull chicks under cloudy skies is disrupted by solar storms that slightly change the earth's magnetic field (Southern 1971, 1972). In the laboratory, a slight change (± 10 percent) in a field's intensity disrupts orientation for three days, the time it takes for the bird to adjust to the new field. Many early laboratory experiments failed because researchers used magnetic fields that varied in intensity, not allowing the birds enough time to adjust, or that were too strong, exceeding the birds' range of sensitivity.

Olfaction

Preliminary studies of pigeons and petrels indicate that they may be able to smell their way home, using olfactory cues for navigation. In one experiment, pigeons with their nostrils plugged with cotton or with their olfactory nerves cut did not find their lofts as well as normal controls (Papi et al. 1971, 1972). However, other experiments with homing pigeons have failed to show that impaired olfaction affects their navigation abilities (Keeton et al. 1977). Some tube-nosed seabirds (Procellariiformes) use olfactory cues to find food and to locate their nesting burrows. Leach's Storm-Petrels, for example, nest on forested islands in the Bay of Fundy. On return from the sea, they fly into the forest a short distance downwind from their nest burrows, which they then locate precisely by the smell of the nests (Grubb 1972, 1974). These petrels can find their dismantled nest material even when it is placed in one area of an experimental maze. Storm-petrels with an experimentally impaired sense of smell do not return to their burrows after one week, whereas controls do.

Learning to navigate

It appears that navigational abilities are partly innate and partly dependent on early experience, with the result that inexperienced young migrant birds become lost more often than experienced adults. The rare visitors that excite birders, for example, are often lost immature birds (DeSante 1983). Young European Starlings are capable of simple directional orientation but not of orientation to a particular geographical location, or goal orientation. A.C. Perdeck (1967), a Dutch ornithologist, captured thousands of starlings that were migrating to their wintering grounds and took them southeast from the capture site in Holland to Switzerland where he released them. Normally these birds move west-southwest through The Hague from breeding grounds as far away as Denmark and Poland to wintering grounds in southern England, Belgium, and northern France. Recoveries of the displaced adult starlings revealed that they had reoriented and flown northwest toward their wintering grounds. The young starlings, however, failed to compensate for their geographical displacement and continued west-southwest, ending up in Spain and southern France. Apparently, adults use some form of goal orientation rather than simple directional orientation (Figure 14–9).

More recent studies of the ontogeny of orientation abilities in birds emphasize the importance of brief experiences early in life (see Chapter 11). Homing pigeons reared without seeing the sun and exercised only on overcast days do not, and probably cannot, use the sun compass (Keeton and Gobert 1970). Six-hour clock-shift experiments with such pigeons do not produce the 90-degree orientation error characteristic of young birds that have seen the sun while growing up. Exposure to the sun for less than one hour, however, is enough to activate the sun compass, which becomes increasingly refined with experience. Pigeons do not inherit a knowledge of sun compass positions but calibrate the sun compass from experience, possibly by using information from the magnetic compass (Wiltschko 1982).

A pigeon's ability to use magnetic compass information develops before sun compass ability is manifest. On their first flight, juvenile homing pigeons cannot orient under overcast skies and so fly off in random direc-

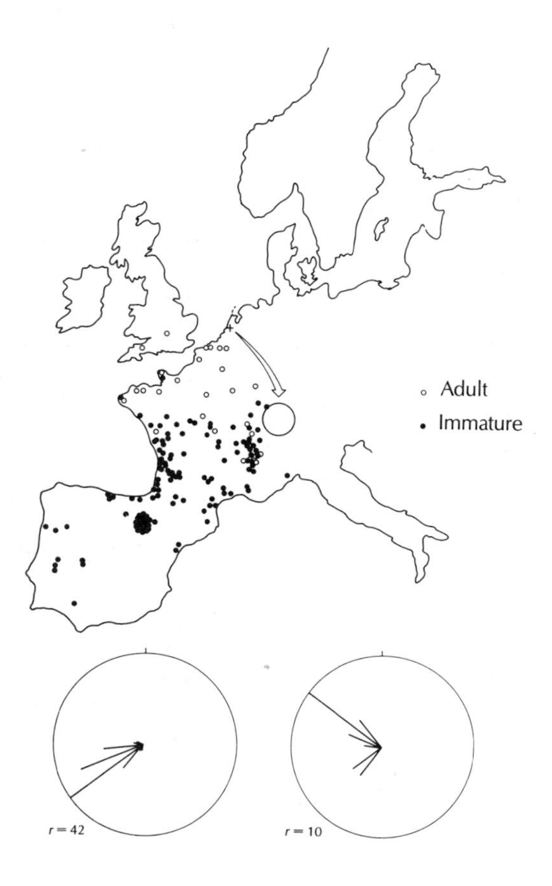

Figure 14–9 Recapture locations of adult and immature European Starlings displaced from Holland to Switzerland during autumn migration. The immatures (left circle) continued their migration to the southwest, whereas the adults (right circle) oriented to their correct wintering grounds in southern England, Belgium, and northern France. Vector diagrams summarize recoveries in 15-degree orientation sectors. (From Emlen 1975a, after Perdeck 1958)

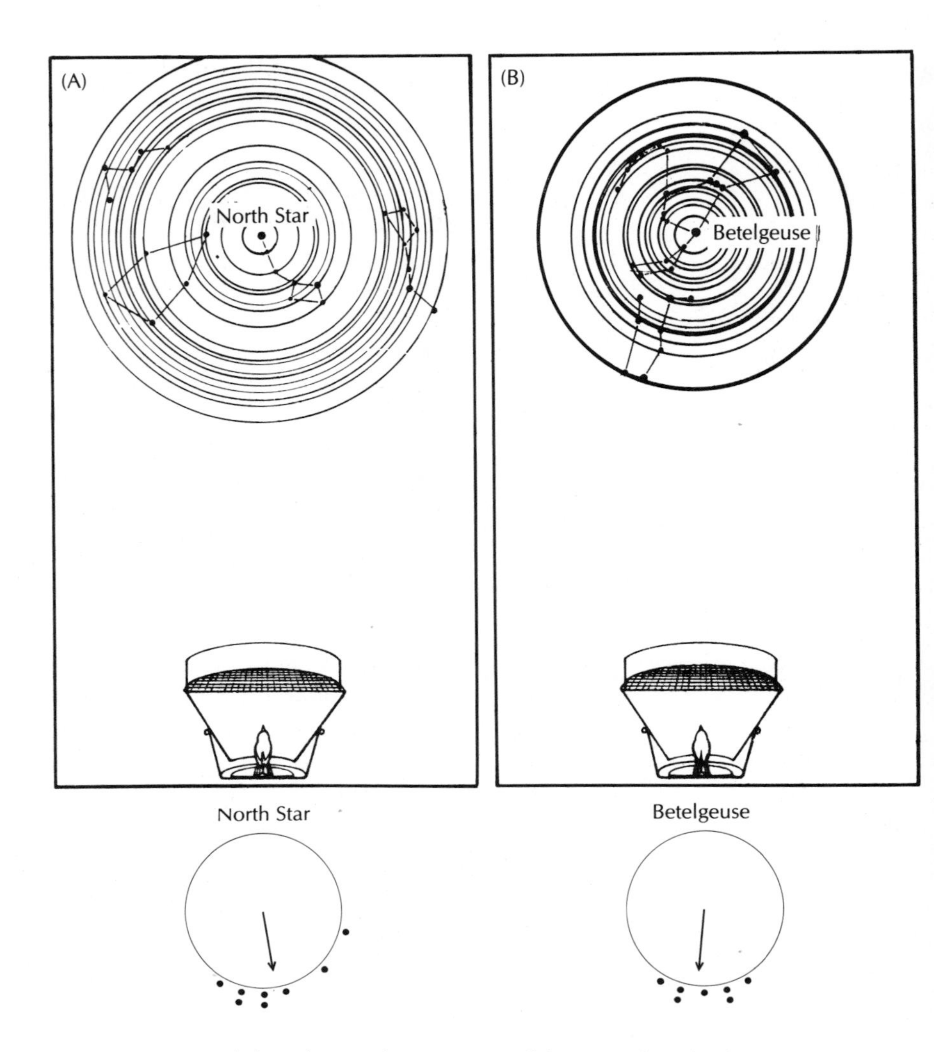

Figure 14–10 (A) Early visual experience of the natural night sky entrains an Indigo Bunting's use of the stars for orientation. (B) Buntings raised under a modified night sky that rotated around Betelgeuse, instead of the North Star, adopted Betelgeuse as the pole star and consistently oriented from it. Each dot represents the direction selected by one young bunting. The vectors (arrows) show the general direction of orientation. (Adapted from Emlen 1975b)

tions. Limited training under overcast skies or in the late afternoon activates the magnetic compass whether or not a sun compass has been established (Wiltschko et al. 1981). Young homing pigeons, however, simply do not orient as well as adults; time and experience are needed to develop full sensitivity to the polarity and declination lines of the earth's magnetic field.

Celestial navigators must learn the sky. Baby Indigo Buntings, hand-reared without seeing the stars, cannot orient when exposed to the night sky. In fact, they must see the sky regularly during the first month of life to be able to choose their migratory direction. The axis of rotation of the night sky, which centers on the North Star, establishes their north-south frame of reference (Emlen 1970). They then learn the constellations associated with this axis. If the axis of rotation of the planetarium sky is switched from the North Star to a star such as Betelgeuse, the brightest star in the constellation Orion in the southern sky, the baby buntings orient south in line with the new axis of rotation. Emlen points out that the genetic program for imprinting on the polar axis will serve future generations of buntings better than would a genetic program for recognition of particular patterns of polar stars. The stars associated with the polar axis change with time as a result of the earth's "wobble"; those we now see in the autumn will appear in the spring some thousands of years from now, and Vega, not Polaris, will be the north star. Thus, by using the polar axis rather than specific polar stars, Indigo Buntings have a reliable, long-term compass that will not require future changes (Figure 14–10).

Maps and bicoordinate navigation

Choice and maintenance of a compass direction are only part of the challenge of navigation. If a bird is to reach a goal, such as a loft in the case of the homing pigeons, it must also know where it is relative to its goal. It must have a sense of location, or a map: A bird displaced to the north must fly south to the loft, a bird displaced to the west must fly east. To get our bearings in an unfamiliar place humans would look at a map or ask someone which way to go. It is not known how birds solve this dilemma. Sun position potentially provides information on longitude as well as latitude. For example, at northern temperate latitudes, the sun is higher at noon in the south than in the north. The sun rises progressively later as one travels west and, therefore, will be at a different position in its arc relative to any given absolute clock time (people adjust for this by having official times that compensate roughly for later sunrises in the west). A simple rule for westward navigation in North America would be: If the sun at your present location is higher than it would be at your goal at this time of day, fly away from it; if it is lower, fly toward it (Griffin 1974). The "sun-arc" hypotheses of Geoffrey Matthews (1968) and Colin Pennycuick (1960) embrace these possibilities, which require a bird to have both an accurate memory of sun positions and an acute sense of time.

Clock-shift experiments have failed to show that homing pigeons can use sun position for anything more than simple compass direction (Emlen 1975a). Clock-shift should drastically alter a bird's interpretation of the sun's height above the horizon at a particular time of day. A bird whose internal clock had been advanced experimentally would view the sun as too low and therefore should fly east to compensate. But homing pigeons do not compensate in this way. It could be that birds use geomagnetism as a basis for bicoordinate navigation. The earth's magnetic field varies predictably with latitude and longitude in ways that potentially form a navigational grid. But whether birds can use this information has not yet been demonstrated.

Internal programming

In the previous chapter we mentioned that migratory restlessness and premigratory fattening appear to have a genetic basis in such partial migrants as the European Robin. This is just one of a series of recent revelations about the internal programming of migratory behavior. Caged migrants not only exhibit well-defined orientation behavior, they also change their compass direction in ways that correspond to their natural migration routes. Garden Warblers change direction during their fall migration from southwest to south-southeast on their route from Spain to southern Africa. Devoid of cues other than magnetism, the orientation of migratory restlessness in the laboratory shows a corresponding shift (Gwinner 1977). Restless, caged Garden Warblers orient southwest in August and September and then shift their heading to south-southeast from October to December.

Temporal patterns of migratory activity also have some endogenous basis. Not only are migratory preparations and migration itself linked directly to endogenous circannual rhythms (see Chapter 12), but the duration and pace of migration may be linked to these rhythms as well. One such indication is that quantitative laboratory measurements of migratory restlessness in eight species of European warblers correlate extremely well with their typical migratory distances. Long-distance migrants such as Garden Warblers show more total nocturnal activity than do short-distance migrants such as the related Marmora's Warbler. Similar patterns are apparent in the duration of migratory restlessness. The Willow Warbler normally takes 3 to 4 months to migrate from Europe to southern Africa; intense migratory restlessness of this warbler in the laboratory lasts over 4 months. The Chiffchaff takes only 1 to 2 months to migrate from southern Europe to northern Africa; intense migratory restlessness in the laboratory lasts 60 days. Relating the length of nocturnal activity in the laboratory to the distances these warblers migrate accurately projects the distance traveled to their respective winter ranges (Gwinner 1977) (Figure 14–11). These re-

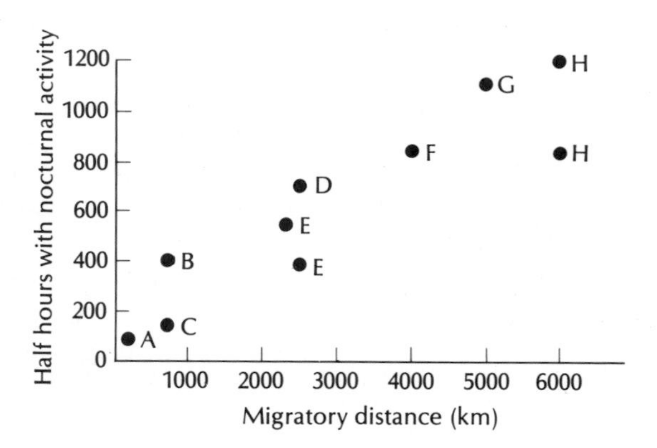

Figure 14–11 The lengths of time of nocturnal restlessness in the laboratory are well correlated with the migration distances covered by eight species of European warblers: (A) Marmora's Warbler, (B) Dartford Warbler, (C) Sardinian Warbler, (D) Blackcap, (E) Chiffchaff, (F) Subalpine Warbler, (G) Garden Warbler, and (H) Willow Warbler. Results for Willow Warblers and Chiffchaffs tested under different conditions are shown separately. (After Gwinner 1977)

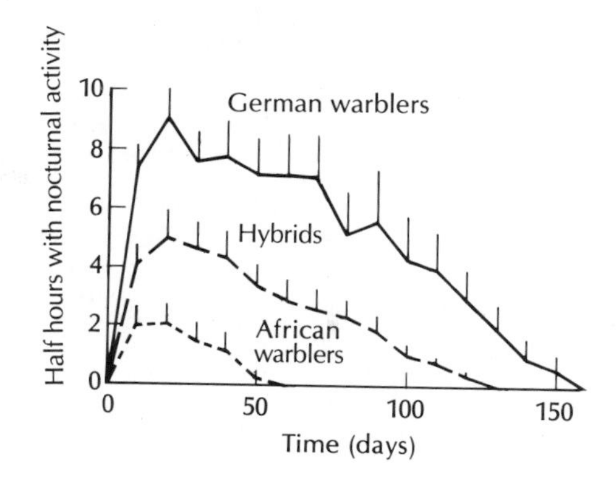

Figure 14–12 Blackcaps from migratory populations in Germany show intense and prolonged migratory restlessness, whereas individuals from a nonmigratory population in Africa show very little. Hand-raised hybrids of these forms have intermediate migratory behavior. (After Berthold and Querner 1981)

sults suggest a basis for the simple directional navigation shown by the young starlings mentioned earlier. Programmed for general orientation and the approximate amount of time they should migrate in that direction, a young bird should reach some part of its species' winter range. It can then develop the more refined abilities required for goal orientation on subsequent migrations.

Like the warblers, populations of the Blackcap differ from each other in the seasonal course and magnitude of *zugunruhe,* and the differences also correspond directly to the distance each population normally migrates. Moreover, the differences in *zugunruhe* suggest a genetic basis (Berthold and Querner 1981). Hybrids of the migratory German population and the nonmigratory African population exhibit intermediate activity, suggesting direct genetic control of the programming for directional migration (Figure 14–12).

Exciting as these revelations are, they do not mean that internal programs actually guide migrants to their winter residencies. Certainly, external forces including food availability, climate, and competitive interactions come into play at various stages of the journey and may in fact be dominant factors, especially in short-distance migration.

Summary

In the last two decades ornithologists have made great progress documenting the cues by which birds navigate great distances across unfamiliar terrain. It is well established that birds rely on acute visual memories for short-distance travel and local orientation. Birds also use the positions of the sun by day and the stars by night. Recently we have learned that birds use the earth's magnetic fields. Young birds refine their use of celestial and magnetic cues through experience. To some degree the distances and directions they migrate seem to be genetically determined.

Further readings

Emlen, S.T. 1975. Migration: Orientation and navigation. Avian Biology 5:129–219. *An excellent, comprehensive, and critical review of the literature.*

Gwinner, E. 1977. Circannual rhythms in bird migration. Annual Review of Ecology and Systematics 8:381–405. *Details of the role of the clocks in avian migration.*

Keeton, W.T. 1980. Avian orientation and navigation: New developments in an old mystery. Acta XVII Congr. Int. Ornithol. 1:137–158. *An extraordinary summary of the topic.*

chapter 15

Social behavior

The needs for food and for protection are the most pressing requirements of any living creature, determining where and how to live and the degrees to which behavior is social or asocial, and cooperative or competitive. It is sometimes advantageous for an individual to go it alone; at other times there is safety in numbers. Among birds we find many variations in the spacing of individuals. At one extreme, Solitary Eagles live alone on exclusive expanses of tropical mountain forest. At the other extreme, Social Weavers cluster together in gigantic communal nests. Whether a bird lives alone or with others, the fact remains that space, the intelligent use of which is crucial for survival, is limited. Ultimately birds must share space, and they have evolved various ways of doing so. Whether breeding or not, birds may space themselves at regular intervals over large territories, congregate in large numbers, or cluster in small groups. In this chapter, we examine the spacing behaviors of birds, and outline the specific costs and benefits of territoriality, coloniality, and flocking.

Individual spacing behavior

Most birds maintain a small individual space around them wherever they go. Swallows, for example, space themselves at regular intervals on a telephone wire (Figure 15–1). Sparrows and sandpipers feeding in large flocks also maintain small distances from one to another, as if each were surrounded by an invisible force field. This space increases their individual foraging efficiencies and reduces the frequency of hostile interactions (Morse 1980).

The tendency of individuals to separate promotes uniform patterns of spacing. If birds landed on a field at random, some sites in the field would remain empty, and others would receive several birds in succession, resulting in random patterns of association. In all probability, the birds would not sit quietly after landing. Individuals close to one another would move apart and fill the unoccupied spaces. Such regular, or uniform, dispersion patterns are typical of birds that occupy relatively uniform habitats. Killdeers residing in large fields, American Robins nesting in suburbia, and American Kestrels wintering along roadsides space themselves in a regular manner.

Individuals may space themselves uniformly in small areas, but in larger areas they may tend to separate by greater distances or to clump together. When birds fly from a field in a flock, the distances between individuals within the flock are small and uniform. The distances between different flocks are substantial. Flocking Snow Geese in winter fields clump together, but on a larger scale, the distributions of the flocks themselves may be random, uniform, or clumped.

Figure 15–1 Cliff Swallows space themselves at regular intervals on a telephone wire. (Courtesy A. Cruickshank/VIREO)

Territoriality

Birds aggressively establish, maintain, and protect their spatial relationships; aggressive individual assertions of status or rights to resources are normal parts of avian social life. Assertion of spatial rights is very apparent in territorial birds, which must win and continually maintain exclusive rights to particular areas, food supplies, or mates. Territorial behavior is a primary form of aggressive spacing behavior that has intrigued naturalists since Aristotle. H.E. Howard's *Territory in Bird Life* (1920) formally introduced scientific inquiry into the subject. Research on avian territoriality has now established three major aspects of territorial behavior (Brown and Orians 1970):

1. Acts of display or defense discourage rival birds that would otherwise enter or approach the territorial space.

2. Primary if not exclusive use of a territory is thereby limited to the defending individual and, perhaps, its mate and progeny.

3. A territory is a fixed area defended continuously for some period of time, even if only hours, in either or both the breeding and nonbreeding seasons.

Ornithologists once thought that the territorial behavior of birds was genetically programmed and static. In fact, territorial behavior is flexible and dynamic. Great Tits, for example, forego defense of their winter territories on the coldest days to save essential energy (Hinde 1956). The territorial behavior of Sanderlings is manifest only at low tide; at high tide this sandpiper feeds or roosts in flocks (Figure 15–2). However, in years when Merlins take up residence in their area, Sanderlings are often not territorial because isolated individuals would be too vulnerable to the predatory falcons (Myers et al. 1985).

The simplest territories are those with only one type of resource, such as the feeding territories of hummingbirds in fields of flowers or those of sandpipers on a beach at low tide. At the other extreme are the one- to two-acre all-purpose nesting territories of landbirds, which are used for male display, courtship, nest seclusion, and feeding. These territories enable

Figure 15–2 Sanderlings may defend exclusive feeding territories or feed in large flocks. (Courtesy A. Amos/VIREO)

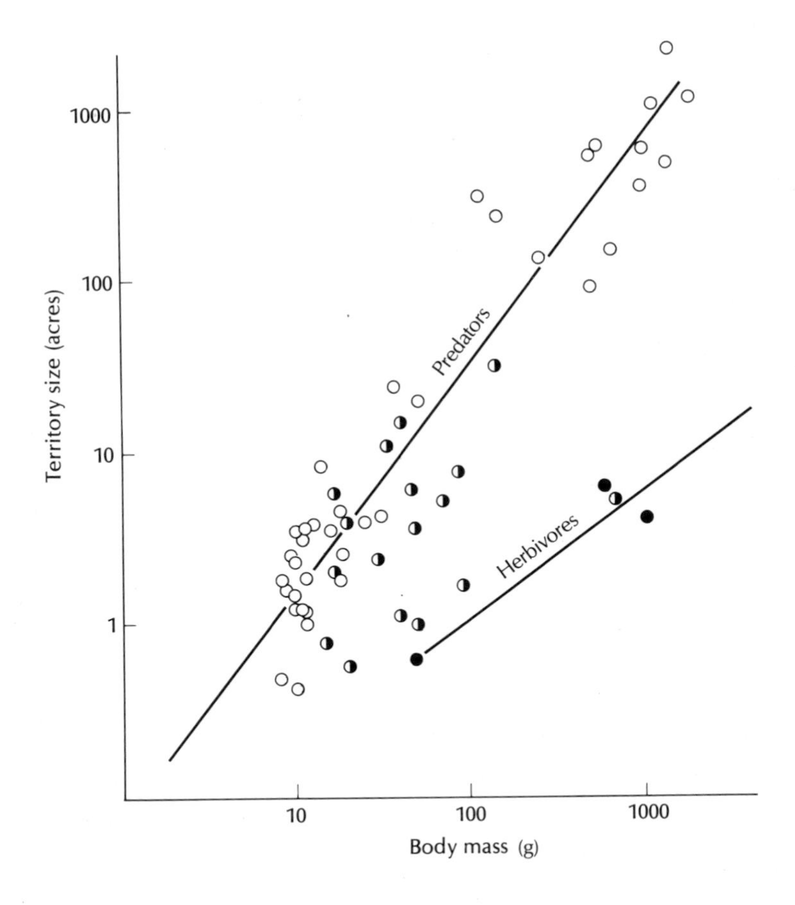

Figure 15–3 Territories or home ranges of birds increase directly in relation to body size, energy requirements, and selection of food types. The correlation suggests that territory size is geared to the food and energy requirements of the bird. Predators have higher daily energy requirements than do herbivores, which have correspondingly smaller territories. Half-shaded circles indicate species with mixed diets. (After Schoener 1968)

individuals to space themselves rather uniformly to reserve essential resources, reduce predation, and control sexual interference by neighbors and vagrants (Lack 1968). In suitable habitats, territories are usually contiguous areas separated by boundaries that, though invisible to us, are well defined.

The average sizes of territories increase directly in relation to body size, energy requirements, and food habits of the various species of birds (Figure 15–3). This suggests a general importance of food resources to the territorial individual (Schoener 1968). Variations within species are even more revealing. Pomarine Jaegers, for example, defend small breeding territories of 19 hectares when lemmings, their principal food, are abundant, and territories of 45 hectares when lemmings are scarce (Pitelka et al. 1955; Maher 1970). The feeding territories of Rufous Hummingbirds and Golden-winged Sunbirds decrease in size as flower density, and thus the quantity of nectar, increases (Gill and Wolf 1975; Kodric-Brown and Brown 1978).

Simple relationships between food abundance and territory size, however, do not necessarily demonstrate that food and energy requirements alone control territory size. Territory size also depends on the density of competitors for the available space (Myers et al. 1979; Ewald 1980). When population density is low, territorial American Tree Sparrows regularly use only 15 to 18 percent of their large territories (Weeden 1965).

They concentrate their activities in the core section but also defend a less frequently used buffer zone. In years of high population density and increased competition for breeding space, denser packing of smaller territories eliminates the buffer zones. The nest territories of Royal Terns actually pack into a hexagonal configuration resembling the cells in a bee's honeycomb (Grant 1968; Buckley and Buckley 1977).

Territorial defense incurs costs as well as benefits. Conspicuous display can attract predators. The time and energy required to display, patrol territorial boundaries, and chase intruders can be a major investment. Territoriality is favored when the resulting benefits outweigh the incurred costs. The central requirement is that adequate resources be economically defensible (Brown 1964b). Two features of resource distribution, temporal variability and spatial variability, determine whether territories are economically defensible. Resources that change rapidly in time invite opportunistic use, not site-specific investment or long-term commitment. Aerial insects whose locations and densities shift frequently, for example, are usually not defensible food resources. Territorial sunbirds, which do not tolerate each other near chosen flowers, will sit side by side in a bush while they catch passing insects.

Sites that are extraordinarily rich in resources attract hordes of competitors and may be indefensible as a result. No gull would attempt to maintain a feeding territory on a garbage dump where thousands of other gulls vie for the same scraps. Similarly, Sanderlings do not always defend their feeding territories on California beaches (Myers et al. 1979). Beach space with few prey is not worth defending, and beach space with dense concentrations of prey (isopods) is not defensible because no single Sanderling can keep the hordes of other Sanderlings away. Thus, Sanderlings defend only territories on beach sections with intermediate densities of prey. The size of the territories they defend on the controllable beach sections also reflects the necessary defense effort: Where there is more competition, smaller territories are formed (Figure 15–4).

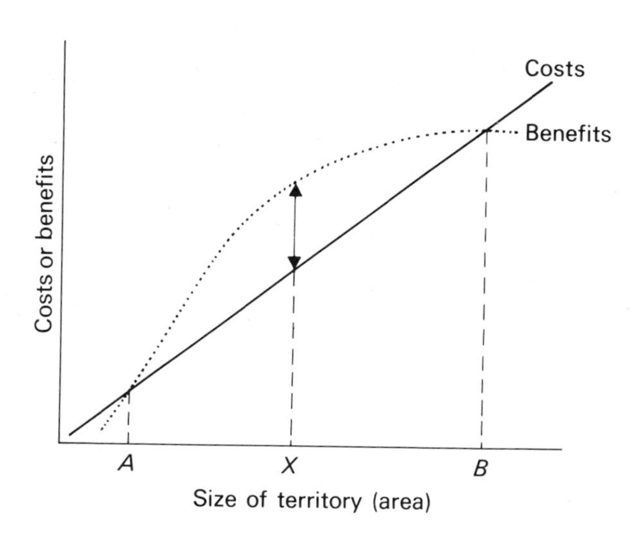

Figure 15–4 Territories of intermediate sizes (*A* to *B*) are economically defendable because the benefits exceed the costs. The costs of defense increase as territory size increases. The benefits relative to need (dotted line) increase rapidly at first but then reach a maximum value when needs are filled, as would be the case when food is in excess. Optimum territory size is at *X*, where the net benefit is greatest. (From Davies 1978a)

Figure 15–5 Golden-winged Sunbird, a species that often defends territories of nectar-rich flowers. (Courtesy C.H. Greenewalt/VIREO)

The costs and benefits of the feeding territories of nectar-feeding birds are unusually straightforward and easily defined. Hummingbirds and sunbirds defend particular clumps of flowers for several days to several weeks or longer. Golden-winged Sunbirds in Kenya, for example, defend about 1600 flowers of a mint, which produce enough nectar each day to satisfy an individual's energy requirements. (Figure 15–5). Golden-winged Sunbirds defend these territories when the benefits exceed the costs (Gill and Wolf 1975, 1979). The primary cost is the energy required to chase intruders, approximately 12.5 kilojoules per hour. The territorial sunbird benefits by having an assured, adequate food supply. The sunbird also saves energy by feeding at nectar-rich flowers on its territory rather than at nectar-poor, undefended flowers visited frequently by other sunbirds. The territorial sunbird can satisfy its feeding requirements in less time each day than a nonterritorial sunbird and thus can spend more time sitting, which costs less energy (1.7 kilojoules per hour versus 4.0 kilojoules per hour). When a defense investment of 3 kilojoules per day causes the average nectar volume to increase from 1 to 2 microliters per flower, a 6-kilojoule net savings of energy is realized (Table 15–1). When the projected savings are less than the investment, the territory is not defended.

Although birds usually defend territories against others of the same species, interspecific territorial defense is not uncommon. Golden-winged Sunbirds defend their territories against a variety of nectar-feeding birds, as do territorial hummingbirds. In the winter, Northern Mockingbirds defend berry-rich feeding territories against other species, especially those that would eat some of the berries. The intensity of a mockingbird's defense increases with the potential threat to its food supplies (Moore 1978). Some other species defend nesting territories against other closely related species (Orians and Willson 1964; Murray 1971).

Territories may be occupied and defended by a single bird, a mated or cooperating pair of birds, an extended family, or even a group of unrelated individuals. Small groups of wintering tits and chickadees, for example, defend woodlot territories containing both food and roosting holes. Groups of four unrelated Willow Tits establish common winter territories by late summer (Ekman 1979). Group membership, which includes male and female pairs of both resident adults and newly settled first-year birds, is stable throughout the winter. In addition to protection of food stores for the winter, spring territorial breeding opportunities emerge from the winter communal effort.

Table 15–1 Energy Costs of Feeding on Undefended and Defended Flowers, for the Golden-winged Sunbird

	Undefended Flowers (1 μl nectar/flower)			Defended Flowers (2 μl nectar/flower)		
Activity	Time Spent (h)	Energy Rate (kJ/h)	Energy Spent (kJ)	Time Spent (h)	Energy Rate (kJ/h)	Energy Spent (kJ)
Foraging	8	4.0	32.0	4	4.0	16.0
Sitting	—	—	—	3.7	1.7	6.3
Defense	—	—	—	0.3	12.5	3.7
Total energy spent			32.0			26.0

Energy saved by feeding on defended flowers: 6.0 kJ

Dominance

Birds assert themselves more effectively on familiar ground or home territories than when they are strangers in a new place. Territorial owners usually win encounters with intruders. For one thing, during high-speed attacks and chases, the owner can use familiar details of the territory to its own advantage. Because territorial owners have an investment to protect, they do not usually give up a fight as easily as a newcomer. Acorn Woodpeckers, for example, vigorously defend their tree granaries against squirrels, jays, and other Acorn Woodpeckers (MacRoberts and MacRoberts 1976; Koenig 1981a). These granaries hold valuable stores of winter food; in addition, each of the many holes (up to 11,000) represents an investment of 30 to 60 minutes of drilling time. These woodpeckers defend trees that are riddled with empty holes as well as those with holes that contain acorns (Figure 15–6).

Territoriality is related to the more general phenomenon of dominance behavior. Dominance and aggressive reinforcement of status are a normal part of the social lives of birds. Individuals that win aggressive encounters achieve dominance, and consistent losers become subordinate. As social ranks are established in new groups of birds, losers cease challenging dominant individuals. Dominants use threat displays to assert their status and reserve their access to mates, space, and food. They move without hesitation to a feeder or desirable perch, supplanting subordinates and pecking those that do not yield at their approach. Subordinates are tentative in their actions and frequently adopt submissive display postures (see page 185).

Rank has its advantages. High-ranking Dark-eyed Juncos and Field Sparrows survive longer than low-ranking ones (Baker and Fox 1978; Fretwell 1968). Subordinate Wood Pigeons obtain less food per hour than dominants, which increases their probability of starving (Murton 1967; Murton et al. 1971). Low-ranking individuals have less access to good

Figure 15–6 The granaries of Acorn Woodpeckers are valuable defendable resources that contain essential supplies of acorns for the winter. (Courtesy M.H. MacRoberts and W. Koenig)

feeding sites and are usually the first to emigrate. Weakened physical condition plus the extra costs and dangers of travel through unfamiliar situations all increase the risk of death.

Dominance status is directly related to age and sex. Generally, large birds dominate small ones, males dominate females, and older birds dominate younger ones. Within an age group or gender, physiology and genetics greatly affect dominance. Aggressive tendencies and dominance status are correlated with slight differences in adrenal gland activity and brain chemistry (Brown 1975). Aggressive, dominant strains of domestic chickens can be developed by artificial selection (Craig et al. 1965).

The dominance status of individuals changes with location. The ability of territorial male Steller's Jays to win fights, for example, decreases with distance from their nesting areas rather than ceasing abruptly at a territorial boundary (Brown 1975). Similarly, the point of parity among Bicolored Antbirds (the place at which each pair wins 50 percent of the encounters) has been observed to be the approximate boundaries of their overlapping territories (Willis 1967). Although expression of dominance and territoriality both relate to specific resources such as food and may be initiated over rather large distances, the two behaviors differ with regard to the site defended, which is fixed in the case of territoriality and movable in the case of dominance. Dominance and territoriality, however, become indistinguishable in the site-dependent dominance systems of Steller's Jays and Bicolored Antbirds. Cases of temporary residency also show more vague lines of definition between the two behaviors. Roving male Bronzy Sunbirds, for example, shift from dominance behavior to territoriality through intermediate states of aggressive behavior. They often displace subordinate sunbirds to feed on certain flowers and then leave (Wolf 1978), but also they may defend flowers for an hour or so of exclusive access and then leave, only to return later for another period of temporary residence. When conditions are poor and flowers scarce, they defend the territory constantly for several days to several weeks.

Sometimes territorial birds defend a nonstationary resource. Constant defense of a female and her immediate area, for example, borders on territorial defense of a well-defined resource. Such behavior is typical of the Cassin's Finch and other carduelline finches, particularly when an excess of males competes for mates (Newton 1972; Samson 1976). Glaucous Gulls and Glaucous-winged Gulls defend feeding eiders, a kind of sea duck that brings food to the surface, against other gulls (Ingolffson 1969; Prys-Jones 1973). Sanderlings will defend Willets from other Sanderlings when the Willet has a large sand crab, bits of which fall to the defending Sanderling (Myers et al. 1979).

Coloniality

Whereas territoriality and dominance behavior reflect an emphasis on competition for resources, coloniality reflects an emphasis on tolerance and, sometimes, cooperation. The two main disadvantages to colonial living are that large groups require large amounts of food and that they may attract predators, parasites, and diseases. The advantages, however, far outweigh

Figure 15–7 Brewer's Blackbirds exploit food supplies in large, undefended areas and nest in loose colonies. Such behavior may exemplify the evolutionary roots of true coloniality. (Courtesy A. Cruickshank/VIREO)

the disadvantages. Individuals can improve their foraging by watching others. Colonies also provide protection, which is of paramount importance when birds are breeding, brooding, and nurturing young. The alternative to high-density breeding colonies and well-spaced territories are loose colonies such as those of Brewer's Blackbirds. Unlike pairs of other species that feed in exclusive territories where food is uniformly distributed, pairs of Brewer's Blackbirds congregate at good locations central to large, undefended areas in which the exact location of food varies irregularly (Horn 1968). In Washington state, Brewer's Blackbirds nest in defended clumps of greasewood or sagebrush near ponds and marshes. In the morning, they feed on aquatic insects emerging from the ponds, and during the rest of the day they commute to adjacent, undefended fields to feed. Such behavior seems to be the evolutionary basis for true coloniality (Figure 15–7).

Avian breeding colonies range in size from a few to millions of pairs. On the Peruvian coast, black and white Guanay Cormorants pack together at densities of 12,000 nests per acre and may attain a total colony size of 4 to 5 million birds. In Africa, 2 to 3 million pairs of the sparrowlike Red-billed Quelea nest in less than 100 hectares of Acacia savanna. Colonial birds choose isolated islands, beaches, rookeries, or cliff faces, safe from predators, in which restricted distribution of inaccessible sites favors a high concentration of individuals. Hence, the burrows of nocturnal auklets and petrels riddle the hillsides of oceanic islets; the nest holes of swallows, swifts, and bee-eaters riddle dirt embankments; and caciques and weaverbirds crowd their nests into tall trees over water or into spiny Acacias (see Chapter 18).

To support large congregations of birds, suitable nest sites must be near rich, clumped food supplies. The huge colonies of Guanay Cormorants and other seabirds that nest on the coast of Peru, for example, depend on the productive cold waters of the Humboldt current. The combination of the abundance of food and the vastness of oceanic habitat can support enormous populations of seabirds, which concentrate at the few available nesting locations. Inland, colonies of Pinyon Jays and crossbills settle near conifer forests, and weaver colonies settle near rich grain fields. In spite of food abundance, large colonies sometimes exhaust their local food supplies and abandon their nests (Payne 1969b; Brown and Urban 1969; Johnson and Sloan 1978; Jones and Ward 1979).

When the precise location of good feeding sites varies from hour to hour, colonial individuals use each other as clues for finding food (Zahavi 1971a, Ward and Zahavi 1973; Krebs 1974). Seabirds track the locations of small schools of fish by following the line of individuals returning to the colony with food. Bank Swallows, which feed on aerial insects that concentrate in the eddies of shifting breezes, may derive a similar advantage (Emlen and Demong 1975). Observations that seem to support this "information center hypothesis" have been reported for birds as diverse as Tricolored Blackbirds, Bank Swallows, Phainopeplas, and Great Blue Herons. Such advantages are probably side benefits, rather than the principal reasons for coloniality (Wittenberger 1981).

Individuals are safer in colonies. Large numbers of colonial birds detect predators more quickly than small groups or pairs and can drive them from the vicinity of the nesting area. The effectiveness with which Common Black-headed Gulls mob predators increases with the number of participants (Kruuk 1964). Nests at the edges of breeding colonies are more vulnerable to predators than those in the centers, and the preference for advantageous central sites promotes dense centralized packing of nests even in ample areas. Synchronized nesting further decreases risk to a particular nest because the sudden abundance of eggs and chicks exceeds the daily needs of predators.

Studies of the Bank Swallow document the advantages and the disadvantages of coloniality (Hoogland and Sherman 1976). Bank Swallows nest in colonies ranging from a few to several hundred nests, which are built in dirt embankments throughout North America. The disadvantages include increased competition for nest sites, stealing of nest materials, increased physical interference, and increased competition for mates. Burrows in large colonies are more likely to be infested by fleas than those in small colonies. Young swallows in large colonies are apt to wander into the wrong burrow and perish because they are not fed. Adults of this species of swallow learn to recognize their own young by means of individually distinctive calls and thereby do not accept young other than their own (Beecher 1982) (Figure 15–8). In contrast, Northern Rough-winged Swallows, a related

(A)

(B)

Figure 15–8 (A) Colonies of Bank Swallows riddle dirt embankments with their nesting tunnels. (B) A brood of three young swallows, almost ready to fledge, waits for food at the entrance. (Courtesy A. Cruickshank/VIREO)

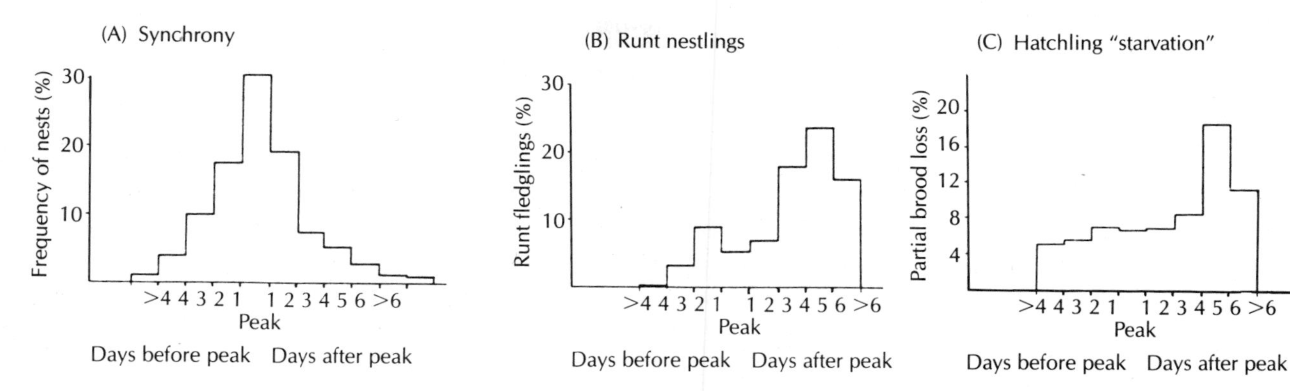

Figure 15–9 Young Bank Swallows in nests started later in the season relative to most nests in the colony, the peak in (A), are more subject to (B) retarded development ("runts") or (C) starvation. (Adapted from Emlen and Demong 1975)

but solitary nesting species, do not discriminate between their own offspring and those of others placed in their nests. There are two primary advantages of coloniality for the Bank Swallow. First, predators are more quickly detected and mobbed. John Hoogland and Paul Sherman (1976) demonstrated this by placing a stuffed weasel near colonies of various sizes and recording the consequences. Second, colonial nesting seems to enable the swallows to keep track of their aerial insect food supplies (Emlen and Demong 1975). Synchronized breeding is apparently important in this regard because those pairs that nest several days later than the majority have trouble feeding their young, many of which die of starvation or are runts. The apparent reason for this is that late breeders are left to find food on their own after most pairs have departed with their fledged young (Figure 15–9).

Flocking

Whereas stable food resources and defensible spaces promote territoriality, unstable food resources and indefensible areas promote flocking. In certain ways, though, flocks resemble colonies, and flocking behavior has features in common with territoriality. For example, even though a flock member benefits from the group effort, it is subject to a dominance system. White Wagtails in Britain join flocks when food on their territories becomes inadequate (Davies 1976b). They also stop defending territories at experimental piles of food and join feeding flocks when the same amount of food is evenly dispersed over a large area (Zahavi 1971b). Similarly, crows, jays, and magpies abandon territories and form flocks when feeding conditions deteriorate, mainly when food supplies become less stable and more patchy in distribution (Verbeek 1973).

Flocks range in composition from loose temporary aggregations to organized foraging associations of diverse species. At one extreme are the millions of blackbirds in the United States or the Bramblings in Europe that converge each evening at traditional roost sites. Temporary feeding aggregations of herons and seabirds are also open gatherings of individuals responding opportunistically to special situations. Multispecies flocks of tropical birds, which feed together daily as a group throughout the year and actively exclude new individuals from membership, are closed social systems, similar in many ways to much smaller family units.

The stable composition of flock membership facilitates recognition of individuals and the development of a dominance hierarchy. Most dominance hierarchies in stable flocks are linear, or "peck right" hierarchies, in which each individual clearly ranks above or below a set of others. If an individual must win merely the majority of encounters rather than virtually all of them to be dominant, the hierarchy is called a "peck dominance" hierarchy (Allee 1938). In closed, stable social units, for example, those of Common Jackdaws, social rank increases gradually in relation to time, individual tenure, and occasional changes in group composition. The advantage of stable dominance relationships is that they lower the frequency and intensity of overt hostility. Aggressive peck rates in stable groups of caged hens, for example, averaged 14 per 15 minutes when a bowl of food was placed in the cage. When group composition changed weekly, peck rates increased to 72 per 15 minutes (Guhl 1968). Social dominance relationships in flocks are based on mutual recognition among individuals. Hens easily recognize up to 10 other individuals, and one hen could reliably recognize 27 individuals in four different flocks (Welty 1982). Physical details serve for individual recognition and, if altered, disrupt the social organization. Hens whose combs are dyed or covered with bonnets are regarded as strangers when returned to their flock (Schjeldrup-Ebbe 1935).

The marked variations in plumage color of Harris's Sparrows serve as badges of their social status (Rohwer 1977, 1982) (Figure 15 – 10). Dominant individuals have conspicuous, contrasting black markings on the plumage of the head and neck; subordinate individuals have no such markings. Many individuals are intermediate in appearance. Such variations facil-

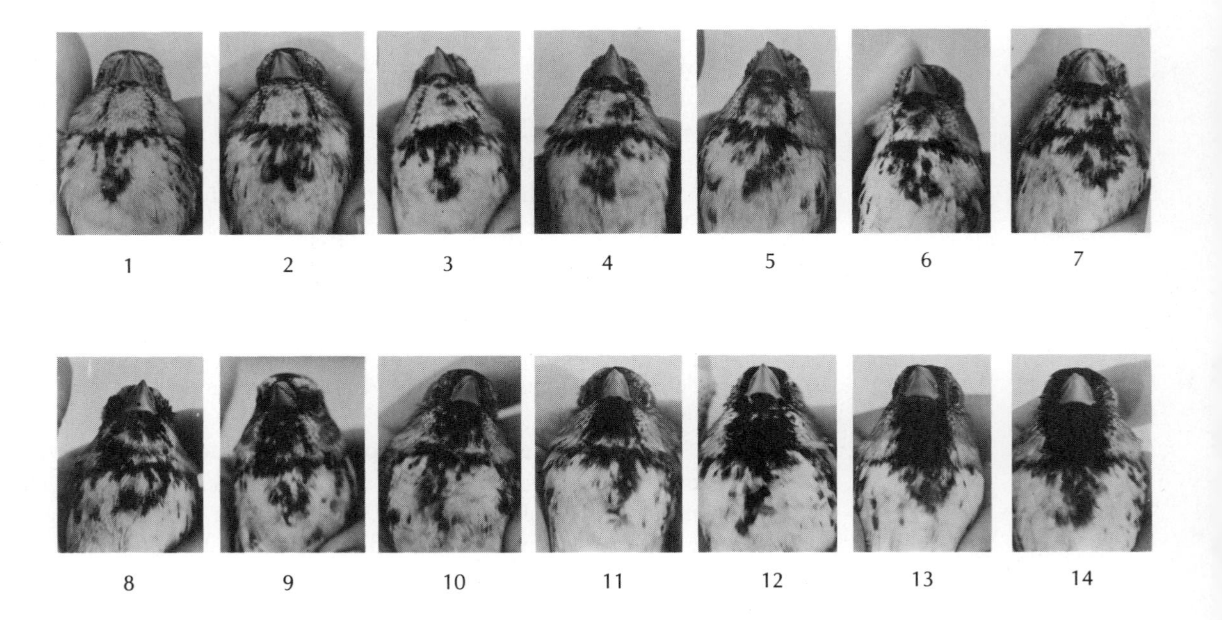

Figure 15 – 10 Plumage variations in male Harris's Sparrows reflect their social status. Dominance is correlated with an increasing extent of black markings on the head and neck. (Courtesy S. Rohwer and *Evolution*)

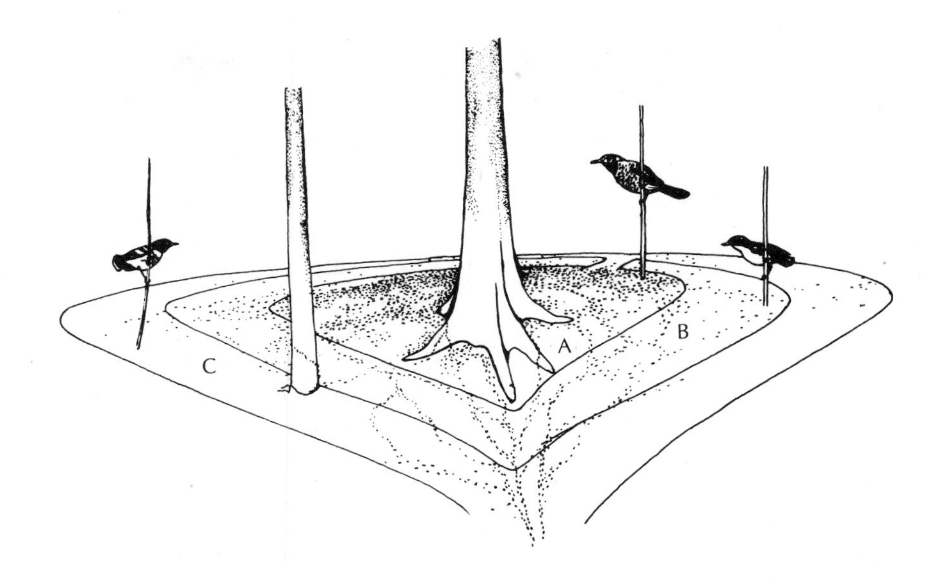

Figure 15–11 The hierarchy of interspecific dominance among birds that follow army ants. Large, dominant species such as the Ocellated Antbird control central sites (zone A) where foraging for flushed insects is best and displace smaller species to outer zones, for example, Bicolored Antbirds to zone B. In turn, the Bicolored Antbirds displace Spotted Antbirds to zone C. Sometimes a subordinate species can infiltrate the central zone, but only such zone-C antbirds as the White-plumed Antbird do this regularly. (From Willis and Oniki 1978)

itate individual recognition among the members of the large flocks that this sparrow typically forms during winter. The evolution of the variability seems directly tied to the advantages of being dominant versus the advantages of being subordinate. Dominant individuals assert the prerogatives of their rank, including access to food. Subordinates of plain appearance, on the other hand, benefit from flock membership, which they can maintain because they do not threaten the dominant individuals with visual badges of high status. In addition, subordinates are tolerated because they help find food, which dominants can usurp. When dyed with black to look like a dominant individual, subordinates suffer more frequent attacks but do not rise in status because they are not inherently aggressive.

Dominance hierarchies are a conspicuous feature of the associations of birds that follow raiding parties of tropical army ants, which flush large numbers of insects and small reptiles that are usually camouflaged and hard to find. Tropical antbirds and woodcreepers habitually associate with ant swarms (Willis and Oniki 1978) and over 50 species of neotropical birds are "professional" ant followers, that is, they obtain over 50 percent of their food from the vicinity of ant swarms. Large dominant species, such as the large Ocellated Antbird, control the central zone of the ant swarm where prey are most likely to be flushed by the dense, leading columns of ants. Smaller, subordinate species, which are chased from this zone, take up stations in peripheral, less productive foraging zones but move towards the center when opportunity arises. In the presence of Ocellated Antbirds, Bicolored Antbirds occupy the intermediate zone, and the small Spotted Antbirds are shunted to the edge (Figure 15–11).

Feeding in flocks

Casual aggregations of individuals at rich feeding grounds are obviously fortuitous, but why do unrelated individuals form stable foraging partnerships? Social tensions and the frequency of fights increase with group size. Subordinate individuals could avoid dominant "bullies" by feeding alone. Competition for conspicuous or rich food items also increases in groups. What then are the advantages of feeding together in organized flocks? The answers lie in the inescapable daily concerns of foraging efficiency and predation risk. Some of the advantages are straightforward, practical ones. Flocks of pelicans encircle and trap schools of fish in shallow water; groups of cormorants and mergansers drive fish toward the shore where they are more vulnerable (Bartholomew 1942; Emlen and Ambrose 1970). Common Ravens steal Black-legged Kittiwake eggs more easily when hunting in groups than when alone, and subordinate birds profit by moving together onto defended food sources where they can overwhelm the territorial individual. Autumn migrants crossing the Mediterranean may escape the attack of one Eleonora's Falcon, but they have less chance of evading several falcons hunting cooperatively near their colonial breeding areas (Walter 1979).

Flock members also benefit from the "beater effect": Prey that is flushed (and missed) by one bird can be grabbed by another. Ground Hornbills in Africa, for example, walk in a line across fields to catch insects flushed by each other (Rand 1954). Drongos and flycatchers participate in mixed foraging flocks and specialize in prey flushed by other birds. Flock membership also improves foraging in more subtle ways. Group foraging by pigeons and titmice helps them locate food because members can join successful individuals at rich clumps or concentrate their search efforts nearby (Murton 1971b; Krebs et al. 1972; Krebs 1973). Groups of four titmice in captivity, for example, found more hidden food together than alone. They watched each other's successes and modified the intensity and direction of their searches accordingly. Dominant individuals tend to benefit most because they can usurp the sites discovered by subordinate members of the flock.

Instead of increasing or maximizing individual foraging efficiency, foraging in flocks may help insure that an individual finds at least some food on a regular basis, before its reserves are exhausted. This would be most important during the stressful winter period. The observation that small birds with limited fasting abilities tend to flock more than large birds supports this idea (Morse 1970, 1978; Thompson et al. 1974). The security of a large group also enables an individual to relax its personal vigil for predators and hence to feed more actively and to spend more time feeding.

Safety in flocks

Joining a flock theoretically decreases the risk of being caught and eaten because there is safety in numbers. Predator confusion, the difficulty a predator has in focusing on one bird when many flush, is one advantage. A falcon, for example, risks injury from incorrect contact during a high-speed strike and is reluctant to swoop down on a fast-moving, swirling flock of birds. It cannot crash full speed into the center of a flock and hope to strike safely (Tinbergen 1951). When European Starlings sight a hawk, they flock more tightly together. An individual's chances of being a victim decrease as the number of potential victims in the flock increases and, as in nesting colonies, decrease even further for individuals near the center of the flock (Hamilton 1971) (Figure 15–12). The success of a Merlin varied according to the size of sandpiper flock it attacked. It fared poorly with medium-sized sandpiper flocks, but did well with isolated individuals and

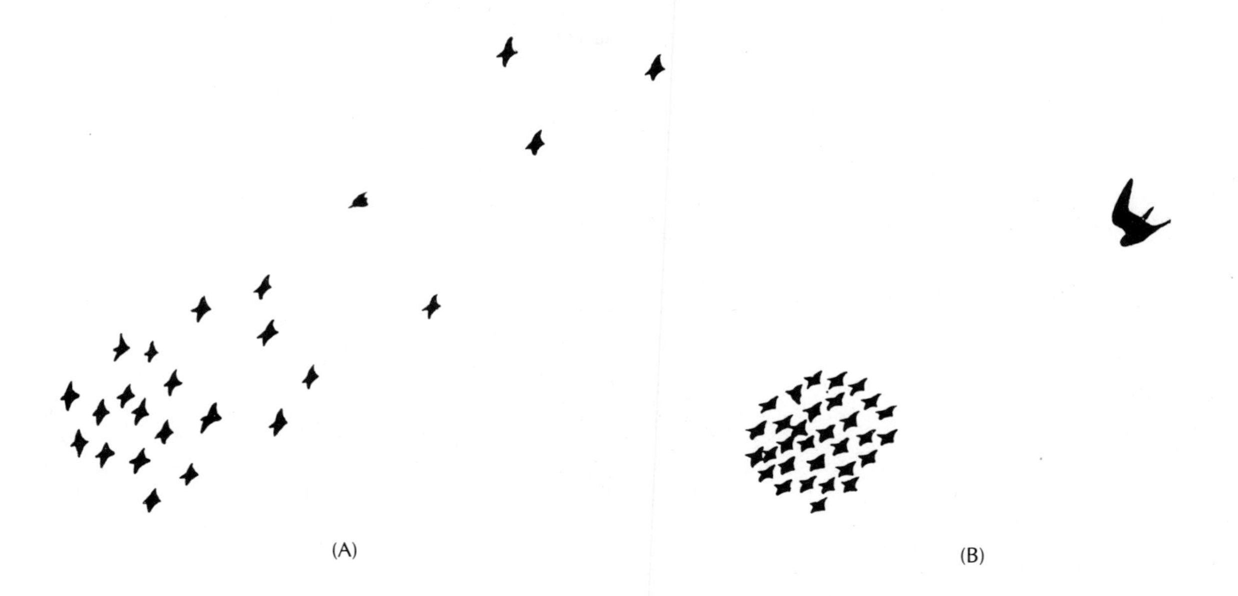

Figure 15–12 (A) European Starlings, which normally fly in loose flock formations, (B) form tight formations when threatened by a hawk. (From Tinbergen 1951)

with large flocks which were less able to maintain a tight formation (Page and Whitacre 1975).

Predator detection also improves in flocks; greater individual security is the result. Spotted Antbirds are distinctly less nervous and wary when in flocks than when they are alone (Willis 1972). Northern Goshawks could not close in on large flocks of Wood Pigeons without being detected (Kenward 1978). Ostriches stick their heads up randomly to look for approaching lions; at any given time, at least one in the flock functions as a lookout (Bertram 1980). Flock members warn each other of danger and communicate so that they can take off at the same time. Ducks flush together at the approach of a predator because the individuals synchronize their takeoffs with a series of flight-intention movements that prime every duck's readiness for flight. Flight calls enable longspurs to flush as a group rather than singly. Contact calls enable birds to associate and to maintain a cohesive flock structure even in dense vegetation. Alarm calls serve to alert other members of the social group to possible danger. When one member of a flock spots a predator, it gives an alarm call, and the rest of the flock either freezes or dives for cover. Giving an alarm call would seem advantageous to all but the one that thus revealed its position. Warning calls may seem to be heroic or altruistic acts, but they carry benefits for the caller as well if others in the flock are genetic relatives, such as siblings, parents, or offspring. Each flock member also can count on a certain degree of reciprocity. Most importantly, by calling loudly the potential victim robs a predator of the element of surprise and thereby reduces the likelihood of attack. The intended victim reduces its own danger as it alerts kin and neighbors.

By relying in part on such mutual protection, each individual in a flock can spend less time looking for predators and more time feeding than when alone. However, the time an individual in a flock saves because of decreased surveillance is offset to some degree by the time it loses to involvement in aggressive interactions, which increase in frequency with group size (Caraco 1979). The amount of time available for feeding should,

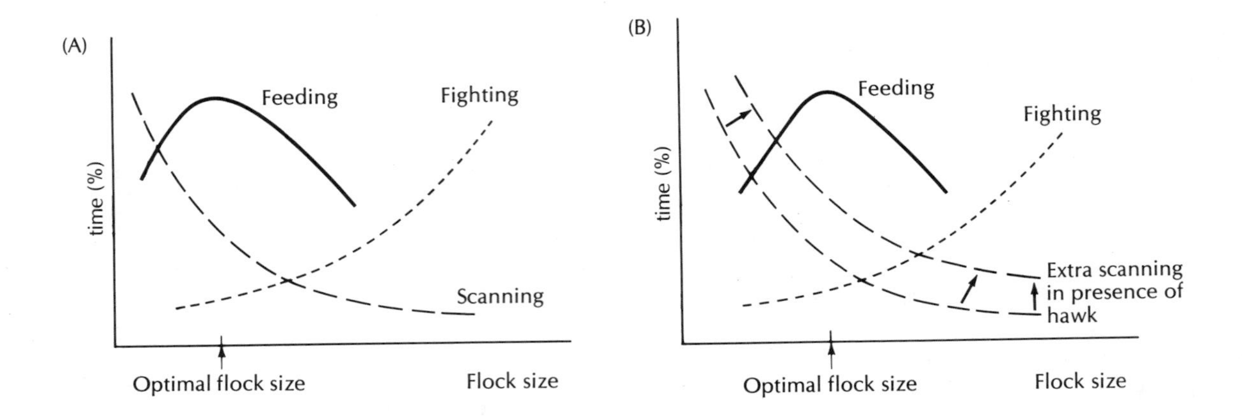

Figure 15–13 (A) The optimal flock size theoretically results from a balance between time spent fighting other members of the flock, time spent scanning for predators, and time devoted to feeding. An intermediate flock size permits the most feeding time. (B) When a predator hawk is present, more time must be spent scanning and the optimal flock size increases. (After Caraco et al. 1980)

therefore, be greatest in flocks of intermediate size. Moreover, optimum group size should increase when predators are near and when each bird must spend more time in surveillance. Thomas Caraco and his colleagues (1980b) confirmed this in studies of Yellow-eyed Juncos in Arizona. In one experiment, average flock size increased from 3.9 to 7.3 juncos when a tame Harris's Hawk flew over the feeding grounds (Figure 15–13).

Mixed-species flocks and social signals

Flocks are not limited to members of the same species. Rich assemblages of species forage together. Flocks of chickadees, titmice, nuthatches, woodpeckers, creepers, and other associates are familiar in both the United States and in Europe, and several species of warblers may join them in the warmer months. Noisy gatherings of antbirds, antwrens, woodpeckers, greenlets, flycatchers, and honeycreepers surge through the rainforests of South America. Tropical flocks may include sixty birds of thirty different species whereas temperate-zone flocks average ten to fifteen birds of six or seven species. Curiously, flock size increases primarily as a result of the addition of new species, not more individuals of a few species (Powell 1979). Furthermore, flock composition changes regularly as the flock moves along, a result of new individuals joining and others leaving (Powell 1979; Munn and Terborgh 1979; Greenberg and Gradwohl 1983). Individuals join the flock as it moves through their territory, only to be replaced by neighbors as the flock moves out of it.

Nuclear species provide the main element of a flock structure. In temperate-zone woodlands of North America, for example, titmice and chickadees are nuclear species. Large antbirds and greenlets take this role in lowland tropical forests. In eastern Peru, the Bluish-slate Antshrike and the Dusky-throated Antshrike assemble 30 other species with their loud rallying calls early every morning (Munn and Terborgh 1979). Plain-colored Tanagers and Blue-gray Tanagers are the usual nuclear species in canopy flocks in lowland Panama but are replaced by bush-tanagers in highland habitats (Moynihan 1962). Other species, the "followers," join the flocks opportunistically for varying periods and are subordinate to the nuclear species.

Why do birds of various species assemble to feed together and, in particular, why do subordinate species join the nuclear species? In multispecies assemblages, the protection inherent to flocks can be achieved without the costs of increased conspecific competition for food (Moynihan 1962; Morse 1980). Furthermore, territorial or rare species that are unable to put together a flock of their own kind can benefit by flocking with other species even though they are subordinate (Buskirk 1976). Foraging success increases in such flocks also (Krebs 1973). Mixed flocks of several species, each with its own searching skills, increase total scanning efforts for clumped, unpredictable prey. Social learning thus enables each participating individual to profit from the successes and failures of its associates. Individuals of different species can monitor each other's foraging success and modify their search efforts accordingly.

The advantages of interspecific feeding associations are so marked that unrelated bird species have evolved similar plumage color patterns that promote flock cohesion. Subordinate species may gain acceptance by resembling dominant flock members. In the South Pacific, for example, certain orioles and friarbirds are confusingly similar in plumage coloration. Orioles are usually bright yellow, but where they inhabit the same region as the dull brown friarbirds, they have the same coloration. Moreover, the appearance of both orioles and friarbirds varies in concert from one island to another (Diamond 1982). It is possible that by adopting a dull plumage resembling that of the friarbirds, orioles are able to gain access to fruiting trees controlled by friarbirds.

The color patterns of birds that flock together in the mountains of Central and South America offer even more striking examples of social adaptations (Moynihan 1968) (Table 15–2). Whereas the neutral, nonthreatening plumage colors of nuclear species such as the Plain-colored Tanager may promote flock cohesion, species that habitually flock together tend to have similar brightly colored plumage patterns. Those species that participate regularly in the montane flocks of western Panama are typically black and yellowish, sometimes variegated with brown and white, whereas bright blue or combinations of blue and yellow prevail in the humid temperate zone of the northern Andes. Further south in Bolivia, the flock colors switch to blue or blue-gray above and chestnut below. Conceivably, such distinctive color patterns serve as flock "badges," which enhance the social integrity of multispecies flocks. The evolution of plumage colors of flocking birds, however, remains a controversial topic (Hamilton 1973; Powell 1985). Countering the potential value of cohesive "social mimicry" is

Table 15–2 Birds that Flock Together in the Mountains of Tropical America

Western Panama (black and yellow)	Northern Andes (blue, blue and yellow)	South Central Andes (blue, chestnut)
Yellow-thighed Brush-Finch	Blue-and-Black Tanager	White-browed Conebill
Yellow-throated Brush-Finch	Masked Honeycreeper	Blue-backed Conebill
Sooty-capped Bush-Tanager	Blue-capped Tanager	Chestnut-bellied Mountain-Tanager
Silver-throated Tanager	Black-cheeked Mountain-Tanager	Black-eared Hemispingus
Slate-throated Redstart	Hooded Mountain Tanager	Golden-collared Tanager
Collared Redstart	Masked Mountain Tanager	Plush-capped Finch
Black-cheeked Warbler	Blue-winged Mountain Tanager	
	Buff-breasted Mountain Tanager	

(From Moynihan 1968)

the need for species distinctiveness. The bold color patterns of many tropical flocking birds may well promote recognition of conspecifics, a phenomenon known to occur in flocks of tropical reef fish.

Summary

The defensibility of a given space, the variability of food resources, and the probability of attack are crucial factors in determining avian spatial relationships and social behavior. Territoriality, the most aggressive avian social behavior, is characterized by acts of display intended to discourage the presence of rivals, and by the exclusive continued use of a defined area for an individual and perhaps its mate and progeny. A central feature of territorial behavior is dominance, which also comes into play in the less aggressive social behaviors of coloniality and flocking. The main advantages of coloniality are feeding efficiency and safety from predators. Territoriality and dominance reflect an emphasis on competition, and coloniality reflects an emphasis on tolerance and perhaps cooperation. Flocking shares features with territoriality and coloniality. Individual flock members observe a hierarchical dominance system but benefit from group feeding and defense efforts.

Further readings

Krebs, J.R., and N.B. Davies. 1984. Behavioral Ecology. 2nd ed. Sunderland, Mass.: Sinauer Associates. *The current forefront of a fast-moving field.*

Morse, D.H. 1980. Behavioral Mechanisms in Ecology. Cambridge: Harvard University Press. *A scholarly review of the literature.*

Wilson, E.O. 1975. Sociobiology. Cambridge, Mass.: Belknap Press. *A classic book that helped establish a new field of study.*

Wittenberger, J.F. 1981. Animal Social Behavior. Belmont, Calif.: Wadsworth. *A comprehensive review of the ornithological literature and the evolution of social behavior.*

part V

REPRODUCTION AND DEVELOPMENT

chapter **16**

Mating systems

Competition and territoriality reach full expression during the breeding season when birds compete with each other for the right to breed, for mates, and for the resources needed to raise young. The vast majority of birds attempt to maximize the number of offspring they leave to succeeding generations through union with a single mate and cooperative parental care. A minority of birds, however, exercises alternative options with multiple mates, in which one sex assumes most of the responsibilities of parental care.

In this chapter we define the various avian mating systems and review the reproductive roles of males and females in the context of monogamy and in the context of the diverse alternative mating systems of weavers, birds-of-paradise, and sandpipers. Competition for mates not only pits rivals against one another but also fosters the process of courtship itself, which enables birds to assess each other as prospective partners. The resulting evolutionary process of sexual selection leads to differences between the sexes in size and ornamentation. Polygamous mating systems and, particularly, behavior on communal display grounds, or leks, are a most elaborate result of this evolutionary process.

Kinds of mating systems

Sexual bonds vary from sustained mutual efforts to brief unions. Both the duration of the association and the number of sexual partners help define mating systems (Oring 1982). The reproductive success of males relative to females also varies from system to system . The principal kinds of mating systems, as defined in Box 16–1, are *monogamy, polygamy, polygyny, polyandry,* and *promiscuity.*

Box 16–1

Mating systems

Monogamy (Greek: *monus,* single; *gamos,* marriage): A prolonged and essentially exclusive reproductive association with a single member of the opposite sex. Monogamous males and females are parents of the same number of offspring. Over 90 percent of all birds are considered monogamous though strict monogamy may be the exception rather than the rule among birds. Philandering, or casual extra matings outside the primary pair bond, is not unusual (Haartman 1969; Oring 1982; Ford 1983). In these cases, the most productive course for males may be to help one female raise young but without passing up opportunities to inseminate other females. Monogamous female birds also engage in occasional outside sexual liaisons.

Polygamy (Greek: *poly,* many; *gamos,* marriage): Any mating system involving pair bonds with multiple mates of the opposite sex.

Polygyny (Greek: *poly,* many; *gyna,* woman): That kind of polygamy in which a male pairs with several females. It is called bigamy if he pairs with only two females. For polygynous birds, male breeding success is more variable than that of females. Sustained associations distinguish this mating system from promiscuous behavior. About two percent of all birds are polygynous.

Polyandry (Greek: *poly,* many; *andros,* man): That kind of polygamy in which a female pairs with several males (i.e., the opposite of polygyny). Each male may tend a clutch of eggs. For polyandrous birds, female breeding success is more variable than that of males. If a female lays full clutches of eggs for successive mates, the system is called sequential polyandry. Fewer than one percent of all birds are polyandrous.

Promiscuity (Latin: *pro,* for; *miscere,* mix): Indiscriminate, casual sexual relationships, usually of brief duration. Male hummingbirds, manakins, and grouse, which mate with any receptive visiting female, are technically promiscuous. Variance in male reproductive success reaches its maximum value. About six percent of all birds are promiscuous.

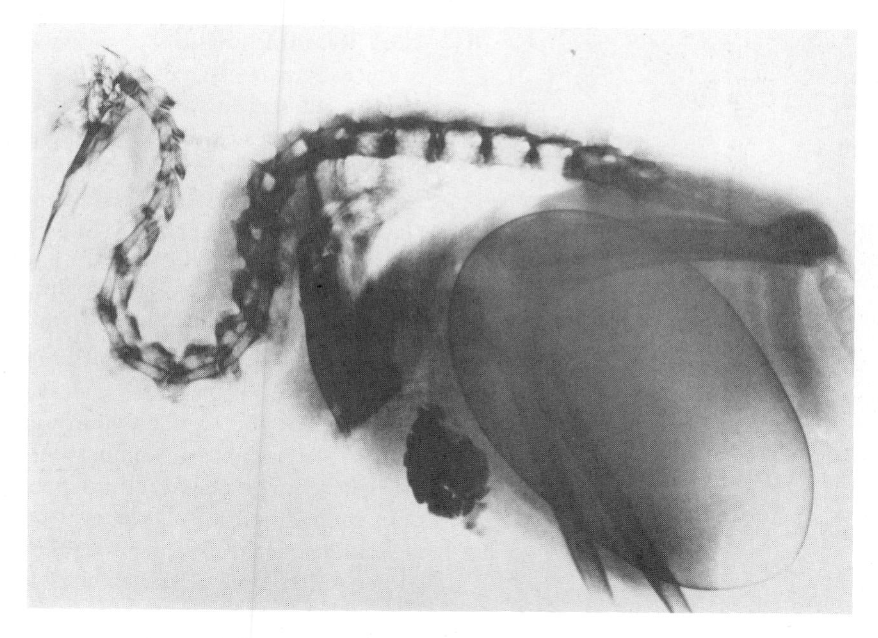

Figure 16–1 The Brown Kiwi, which produces an enormous egg relative to its body size, is an extreme example of the great investment of reproductive energy that female birds put into egg production. (Copyright 1978 by the Otorohanga Zoological Society, used with permission)

Males and females differ fundamentally in their reproductive options, particularly in the degree to which individuals vary in reproductive success. Female birds can produce only a limited number of eggs, which must be hatched and the offspring nurtured to independence. The Brown Kiwi is an extreme example of the demands that reproduction puts on female birds. Its eggs are huge in relation to its body size (Calder 1979) (Figure 16–1). Males produce vast numbers of sperm, which can fertilize the eggs of many females. Thus, potential variance in reproductive success among males is greater than the potential variance in reproductive success among females. This basic difference in gamete production controls patterns of competition for mates and energy investment in parental care. Inseminated females usually concentrate on care of their limited zygotes to maximize their reproductive contribution to the next generation. Males, on the other hand, may benefit more from additional matings than from caring for offspring.

Monogamy

Monogamy refers to a prolonged and essentially exclusive pair bond with a single member of the opposite sex for purposes of raising young. The rituals of pair formation and the practices of biparental care also serve to reduce energetic strain on adults. Pair bonds may last for a breeding season or for life. Pairs of parrots, albatrosses, eagles, geese, and pigeons all reputedly sustain life-long associations though separation is not uncommon. Among Mute Swans, for example, roughly 5 percent of breeding pairs and 10 percent of nonbreeding pairs separate each year (Minton 1968). A 16-percent separation rate is typical of old pairs of Adelie Penguins, and nearly half (44 percent) of young pairs do not stay together more than one breeding season (Ainley et al. 1983).

Still, birds are the most monogamous of organisms. Less than 10 percent of all birds engage in other kinds of mating relationships. Traditionally, ornithologists have viewed monogamy as the mating system of choice: A pair can raise more young than can a female without a mate (Lack 1968). Most birds spend weeks or months tending their eggs and young; in contrast, most reptiles simply lay their eggs and leave them. Not only do avian eggs and chicks require more parental care than the offspring of most vertebrates, but the participation of both sexes frequently appears to be essential. Biparental care of eggs and young is the rule among monogamous birds. Most monogamous males help their mates build nests, incubate, and feed young (Verner and Willson 1969). A male's contribution to raising young can also be indirect. Defense of territorial space, which provides adequate food supplies for the female and young, generally falls to the male.

Cold temperatures and risk of predation may prevent even brief exposure of the eggs or young and mandate a shared vigil. Mandatory continuous incubation and prolonged close care of their young by penguins in Antarctica practically eliminate any possibility of solo parenthood. Predators such as South Polar Skuas would quickly consume unguarded eggs or chicks. For seabirds in general, both parents are needed to offset the dangerous combination of the extended absences required for feeding, a harsh climate, and the presence of predators.

In many cases, biparental care enhances fledging success of larger broods. This is particularly apparent in temperate-zone species, in which clutch size seems tied to effort by both parents. Such observations suggest that the male's principal role is to feed young. Male help (bringing extra food to the female) may also be essential during egg formation, incubation, or early brooding. The conflict between the necessary time for parental care and the time required for self-maintenance emerges as one of the key constraints on solo parenthood in birds (Walters 1984). Peak breeding activity increases total daily energy expenditures by up to 50 percent, the fueling of which requires some combination of increased foraging time or food supplies, use of accumulated reserves, or help by mates or fully grown offspring (Ricklefs 1974; Walsberg 1983). Studies of nest helpers (see Chapter 20) have shown that helpers' efforts go primarily toward increasing the parents' chances of survival and their ability to produce additional clutches of eggs. Assistance of mates by monogamous males undoubtedly has similar positive effects.

Nest building itself can be strenuous. Female Great Tits and Blue Tits devote one to three hours a day to building their nests, time that could be devoted to feeding for self-maintenance and egg production. Common House-Martins that build new nests lay one less egg than neighbors that recondition old nests, apparently because they devote metabolic reserves to building instead of egg production (Lind 1964). Most monogamous male North American passerines contribute to the nest-building effort (Verner and Willson 1969). A male's presence at the nest site in the earliest stages of nesting, however, may be primarily to protect his female from insemination by other males (to guard his paternity).

Incubation is potentially a difficult period, not because of increased energy expenditures, but because of the constraints on foraging time. An individual must find adequate food during brief absences from the nest. Eggs must not be allowed to cool nor must they be exposed to predators. Raptors cannot be certain of feeding quickly because they cannot be sure of finding and capturing suitable prey; thus male raptors, such as the Northern Goshawk, regularly feed their incubating mate and may be responsible for almost all of her food intake. Males of most North American passerine birds (211 of 250 species for which information is available) help incubate or feed their mates during incubation (Verner and Willson 1969). A few monogamous males apparently make little direct contribution to the nesting effort; perhaps they are "on call."

We now recognize that "courtship feeding" is not merely a ritual of pair formation; it is of considerable energetic benefit to the female. Courtship feeding by male Common Terns of their mates during egg laying directly affects the timing of laying and the size of the eggs laid (Nisbet 1973). Male Dot-winged Antwrens provide 40 percent of their mate's daily intake during egg laying and incubation (Greenberg and Gradwohl 1983). Male Pied Flycatchers contribute nearly half of the incubating female's food (Curio 1959; Haartman 1958). When the male is experimentally removed, the female spends less time incubating (58 percent versus 79 percent with a male present) and more time foraging, but she still loses weight. Male Great Tits and Blue Tits feed their mates more frequently during egg laying and incubation than during courtship, up to 160 feedings per day for the Great Tit and over 1000 for the Blue Tit, which eats smaller prey (Royama 1966a). Such supplementary feeding more than doubles a female's rate of energy intake and helps her maintain body weight while producing eggs in a timely fashion (Krebs 1970). Male European Robins feed their mates nearly every five minutes during incubation (East 1981).

The parental roles of monogamous males are substantial. Hence, mutual assessment of prospective partners is a subtle but vital aspect of the early stages of courtship and pair formation. A female must assess her prospective mate's commitment and ability to sustain efforts in raising young. Female pigeons, for example, prefer experienced but not-too-elderly males (Burley and Moran 1979). The ritualized display of food or nest materials during courtship may reveal an individual's skill in gathering these essential items. Courtship feeding by the male not only helps a female build the nutritional reserves needed for egg production but also may serve to show a male's food-gathering abilities. In Common Terns, for example, a male's ability to feed young correlates well with the intensity of his courtship feeding efforts (Nisbet 1973) (Figure 16–2).

Figure 16–2 A male Common Tern feeds its newly hatched young while the female continues to incubate the second egg. (Courtesy E. and D. Hosking)

Conversely, a male must ascertain during courtship how receptive the female is, and in particular, he must be certain that he alone will be the father of the chicks he will care for. Unreceptive females do not tolerate the male's initial aggression, or at least do not return his attention readily. In some birds, courtship stimulates the last phases of gonadal activity and helps pairs to synchronize their readiness to mate. Prolonged courtship, even of receptive females, may also help insure paternity. Experiments with Ringed Turtle-Doves, for example, have shown that aggression by the male delays ovulation in its prospective mate. Since most sperm die within six days of their release, delayed ovulation helps ensure the demise of any sperm remaining from earlier inseminations and increases the likelihood that the current suitor will be the father (Erickson and Zenone 1978; Zenone et al. 1979).

Uniparental care

When male participation is essential for raising young or when males cannot commandeer the resources necessary for supporting extra mates, monogamy may be the only option; but as a female's ability to take care of young by herself increases, polygyny becomes a more viable option for those males that are able to control the best territories or to attract the most females. Also, if males become superfluous or even liabilities during parental care, they no longer are limited by their pair bonds and can compete with other males for extra matings. What conditions release a parent from parental care responsibilities? One of the most obvious is mode of development of the young. Young that leave the nest soon after hatching and that can feed themselves require less care than those that remain dependent in the nest. Seventeen percent of birds with young that leave the nest soon after hatching (nidifugous young) are polygamous or promiscuous, compared with seven percent of birds with young that remain in the nest (nidicolous young). Abundant or easy-to-find food may also relieve parental pressures. Savannah Sparrows are sometimes bigamous where food is abundant and are always monogamous elsewhere (Welsh 1975). Female marsh-nesting blackbirds, wrens, warblers, and sparrows care for young alone by exploiting emerging aquatic insects on prime territories. Temporary food abundance on the Arctic tundra and on the African savannah also eases the need for dual parental effort.

Males of many tropical, fruit-eating birds do not help care for their young. Fruit and floral nectar are conspicuous food sources that require little searching; once a bird locates them, regular revisitation minimizes foraging effort. Incubating females can easily slip off the nest to feed quickly. As long as the energetic requirements of nestlings can be partially satisfied with fruit or nectar, one parent can raise them successfully. Thus, males of these species devote themselves to display to attract additional mates.

Like variations in social behavior (see Chapter 15), pair bonds and avian mating systems reflect the availability of key ecological resources—space, food, and protection—as well as the availability of mates and the feasibility of uniparental care. The relations of mating systems to ecology are evident in weavers, birds-of-paradise, and sandpipers.

African weavers include both monogamous and polygynous species (Crook 1964). Those, such as the Forest Weaver, that are adapted to stable forest environments with uniform food distributions, tend to be territorial ("solitary") and monogamous. Savannah weavers, such as the Golden-backed Weaver and Red Bishop, which exploit ephemeral, unpredictable, occasionally superabundant foods, tend to be colonial or polygynous species in which males control limited safe nest sites near good food supplies. The most abundant of the colonial savannah weavers, the Red-billed Quelea, is an instructive exception. It is monogamous even though it nests in huge colonies near abundant food. Quelea colonies are so large that their members deplete nearby food stores during nesting and must commute farther and farther to gather food for their young. Male assistance becomes essential to ensure that older nestlings are fed (Ward 1965) (Figure 16–3).

Figure 16–3 African weavers have differing mating systems: (A) Forest Weaver, a territorial, monogamous species; (B) Red Bishop, a territorial, polygynous species; (C) Golden-backed Weaver, a colonial, polygynous species; (D) Red-billed Quelea, a colonial, monogamous species.

As noted earlier, fruit diets favor the evolution of polygynous mating systems in birds. Details of fruit dispersion patterns, abundances, and nutritional value are basic considerations (Ricklefs 1980a). Males of 34 of the 43 birds-of-paradise are known or presumed to be promiscuous (Beehler and Pruett-Jones 1983). The two best studied species, Magnificent Bird-of-Paradise and Count Raggi's Bird-of-Paradise, feed on a variety of predictable, nutritious fruits. Monogamous species such as the Trumpetbird, however, specialize on locally abundant figs that have little nutrition. Males of such species apparently must help deliver the quantities of the low-quality food that the young require (Beehler 1985). In birds-of-paradise, therefore, food quality overrides food distribution in determining mating systems.

Shorebirds have also evolved diverse mating systems (Pitelka et al. 1974; Myers 1981a, b). The three phalaropes are polyandrous whereas woodcock and some snipe are promiscuous. Fourteen species of arctic-breeding sandpipers are monogamous whereas three species (the White-rumped Sandpiper, the Curlew Sandpiper, and the Sharp-tailed Sandpiper) are polygynous, and three others (the Pectoral Sandpiper, the Buff-breasted Sandpiper, and the Ruff) are promiscuous. Buff-breasted Sandpipers and Ruffs show lek behavior. Males and females of three species (the Little Stint, the Temminck's Stint, and the Sanderling) incubate separate clutches (protopolyandry), and all three are sometimes polyandrous. As in weavers, most of the monogamous sandpipers defend large territories whereas polygynous species tend to cluster opportunistically in favorable localities (Pitelka et al. 1974), but more study is required to explain the evolution of mating systems in shorebirds.

Sexual selection

Striking sexual differences in plumage or size are typical of nonmonogamous birds. Compare, for example, the ornate plumages of male birds-of-paradise with those of their sombre females or the grand size of a strutting tom Wild Turkey with its smaller dame (Figure 16–4). Darwin concluded that elaborate sexual differences usually evolve as a result of what he called sexual selection, namely, contests among males for mates and female preferences for particular males. Sexual selection usually causes males to become the fancier sex. Males rather than females tend to compete for mates, and their reproductive success varies more than that of females (Payne 1979). The reproductive success of monogamous males varies from zero to one female. At the other extreme are promiscuous males and masters of harems with a dozen or more females. As potential male reproductive success increases, so does the value of the characteristics — such as large size, fancy plumage, intricate songs, and striking displays — that are responsible for the success.

Before proceeding with the details of sexual selection as the process that is mainly responsible for differences between males and females, we must note that both ecological and physiological, as well as reproductive, forces may promote sexual differences in color and size. For example, cryp-

Figure 16–4 Strutting male Wild Turkeys compete for mates. (Courtesy A. Cruickshank/VIREO)

tic coloration helps to protect females during incubation. The different sizes of males and females may reduce ecological competition between the sexes. Sexual differences in size could result simply from the evolution of small females, which may be favored because of their ability to sequester energetic reserves required for egg formation and early breeding (Downhower 1976). Fertility advantages of small females are partly responsible for the five percent size difference of male and female Medium Ground-Finch, a monogamous species of Darwin's finch (Price 1984). Small females were observed to have a three percent advantage over large females in the production of young, whereas large males had a two percent advantage over small males in mating success.

Darwin's insights into the evolutionary role of sexual selection are now largely confirmed, but the precise nature of the process invites more study. The effects of competition among males, female choice, and resources other than mates intertwine in ways that only well-designed field experiments can tease apart. The North American Red-winged Blackbird, a species that is marked by large variation in male sexual success and by striking sexual dimorphism, has become a focus of these studies (Searcy and Yasukawa 1983). Male Red-winged Blackbirds are jet black with bright red and yellow shoulder patches, or "epaulettes"; females are smaller and plainer, streaked brown. Males establish and defend large territories in marshes throughout North America. Those with the best territories attract harems of up to 15 females. The resulting spread in male reproductive success is greater than that of females and increases directly in relation to the

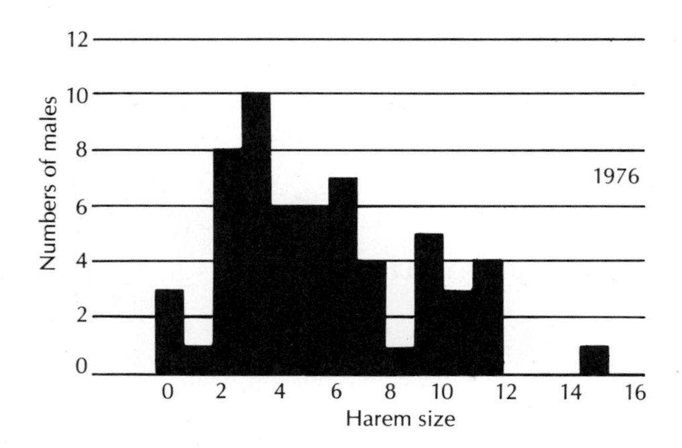

Figure 16–5 Variation in harem sizes of male Red-winged Blackbirds in Washington state. (Adapted from Searcy and Yasukawa 1983)

size of a male's harem (Figure 16–5). What are the foundations of male reproductive success in this species, and is male competition or female choice responsible?

Female Red-winged Blackbirds consistently choose high-quality territories rather than particular males. Water level, nest cover, and food abundance determine territory quality and, therefore, a female's nesting success. Male age is a secondary criterion for females in some parts of the country. Females prefer older, more experienced males in Indiana where males help feed the young, but not in Washington where males do not help feed the young. Thus, sexual selection among male Red-winged Blackbirds operates through competition for the best territories (Figure 16–6). Experiments have demonstrated that the male's red epaulettes are essential in this regard. Males on which the red is dyed black suffer more frequent challenges and usually lose their territories, though those that are not challenged still attract mates. We can conclude that the epaulettes have evolved in relation to male competition, not female choice. Song, which functions as the first line of territorial defense, has also evolved because of competition among males for territories. Intruders less often invade territories in which song is broadcast (whether natural or artificial). Large size may also have some advantage in territory control.

An extraordinary counterpart of the Red-winged Blackbird lives in East Africa. The male Long-tailed Widowbird, another polygynous species

Figure 16–6 A territorial male Red-winged Blackbird in aggressive display posture. (From Orians and Christman 1968)

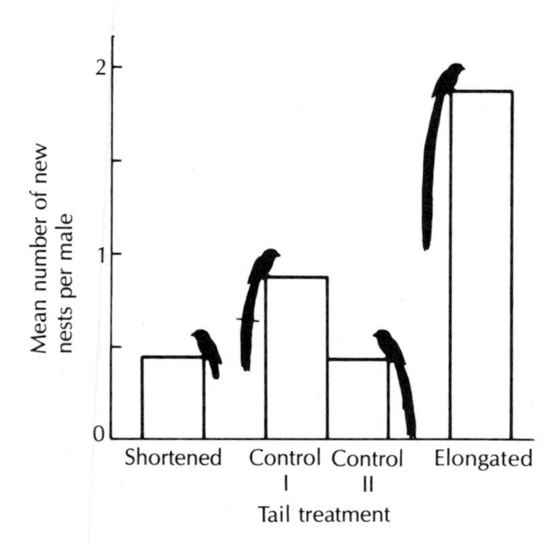

Figure 16 – 7 Female Long-tailed Widowbirds prefer males with long tails. In this experiment, the tails of some males were shortened and the tails of others were extended. Control I males had their tails cut off and then restored; control II males had unaltered tails. The ability of males to attract females to their territories directly reflected their tail length. (Adapted from Andersson 1982)

that is jet black with bright red epaulettes, defends marshland territories in the highlands of Kenya. Sexual selection seems to have gone a step or two farther in the enhancement of the male display that is characteristic of this species: The Long-tailed Widowbird has an enormous tail, up to half a meter long. Like female Red-winged Blackbirds, female whydahs are small, brown-streaked birds. Sexual selection favors the long tail of this bird because it enables females to spot males from afar (Andersson 1982). Humans can spot a displaying Long-tailed Widowbird from over a kilometer away. In a particularly elegant experiment, Malte Andersson increased the tail lengths of some males by 25 centimeters and decreased the tail lengths of others by that same amount. Males with "super" tails attracted more females to nest on their territories then males with shorter tails or tails of normal length. These experimental manipulations did not, however, affect a male's ability to hold his territory. Female preference, rather than male competition, was the source of sexual selection (Figure 16 – 7).

Courtship displays reveal plumage signals in ways that females cannot possibly ignore. Female preferences must somehow influence the evolution of ornate display by solitary birds-of-paradise, hummingbirds, and cotingas, but the role of female preferences for large or elaborate males remains poorly documented. Controlled experiments showing that female birds prefer brighter, fancier, or larger males are still few, though three recent studies support the sexual selection theory. In one case, female Cactus Finches tended to choose large males for mates (Price 1984); and in another, captive female Zebra Finches apparently preferred males with red color bands on their legs (Burley 1981).

The preferences of female Satin Bowerbirds are the most intriguing case study currently available, partly because of the clear role of female choice in a natural population that is subject to strong sexual selection, but particularly because the preference of females is for well-made and well-decorated bowers rather than for male size or plumage (Borgia 1985; Borgia et al. 1985). Recall our discussion of bowers as an extrinsic form of courtship display and of the lengths to which a male Satin Bowerbird will go to decorate its bower (page 180).

Although early research emphasized the evolution of bower displays as ritualized courtship nests, it now appears that the evolution of bowers probably had little to do with nests (Borgia et al. 1985). Rather, bowers have been added to male display courts as markers of male social status and ability. The construction of an elaborate bower and its provision with fresh decorations require experience and considerable effort. Male Satin Bowerbirds tear apart each other's bowers, if they can, and steal each other's prized decorations. Decorations that are rare in the environment, such as blue parrot feathers in northern Queensland, are particularly prized and subject to theft (Borgia and Gore 1984). Dominant males, better able to protect their bowers, have more time to visit and degrade the bowers of nearby competing males, which must constantly rebuild and struggle to keep up a minimally acceptable bower. A male's ability and status is directly reflected in the quality of its bower.

A female Satin Bowerbird visits an average of 3.6 bowers in a local area before copulating with a particular male. If the relative success of males is any indication, females clearly prefer well-made bowers with special decorations. Five of 22 males accounted for 56 percent of the 212 copulations recorded by Gerry Borgia and his assistants in 1981. These males had the most blue parrot feathers, snail shells, and yellow leaves as decorations, as well as the best bower structures, judged in terms of symmetry, stick size, stick density, and quality of construction. Males whose yellow leaf decorations were experimentally removed from their bowers obtained fewer matings than control males.

Exactly why females prefer larger males or fancier displays is not obvious from a theoretical point of view, unless these qualities somehow signal genetic or physiological superiority or indirectly ensure better resources for nesting. Some males outcompete others, and it should be to the females' advantage to have "preferred sons" that will achieve similar success in later generations. Once the process of favoring slightly more elaborate displays or plumages begins, it may go to extremes, as in the case of the bizarre plumage displays of the birds-of-paradise (Figure 16–8).

The large size and conspicuous plumage favored in reproductive display may be a liability in other regards. The same bright colors that announce a male's presence to potential rivals or mates may attract predators. Fancy display plumage such as that of male birds-of-paradise may also hinder escape. Large size carries the liability of greater energy expenditures. There is some evidence in Red-winged Blackbirds that large males are at a disadvantage because they must sacrifice display time for feeding. Among species of North American blackbirds, males that are much larger than females tend to suffer greater mortality (Searcy and Yasukawa 1983). Male Common Capercaillies, a huge European grouse, grow twice as fast as females to reach full adult size by the end of their first summer. Their higher energy requirements render them more vulnerable to starvation when food is scarce (Wegge 1980).

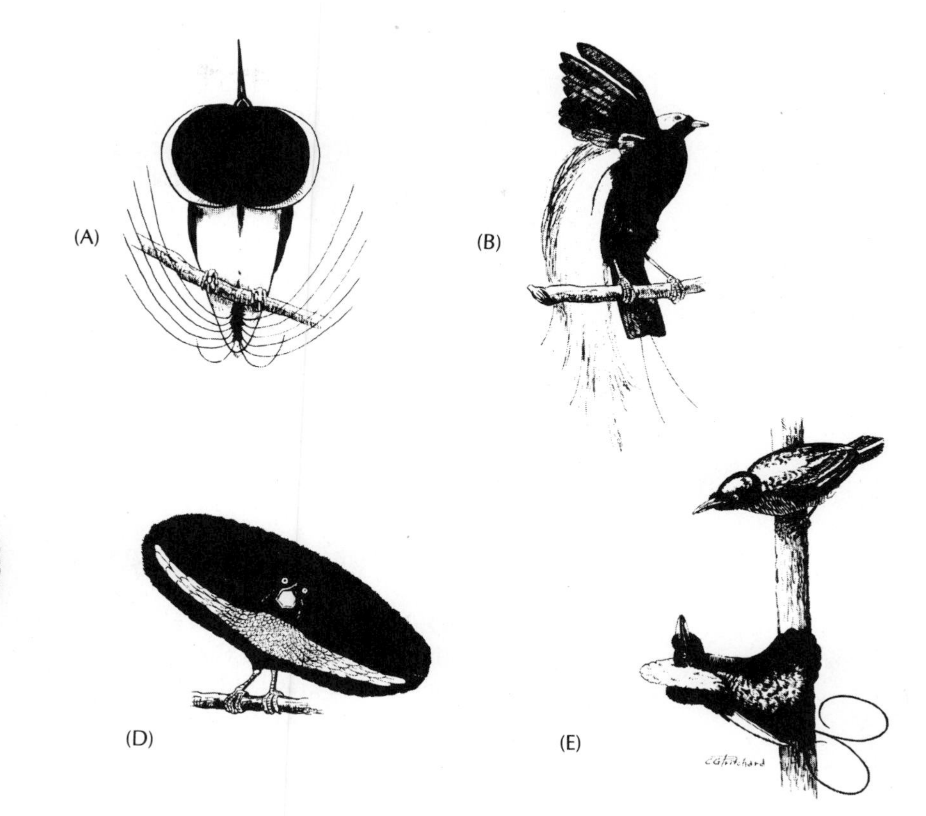

Figure 16–8 Some elaborate plumages and displays of male birds-of-paradise: (A) Twelve-wired Bird-of-Paradise; (B) Lesser Bird-of-Paradise; (C) Magnificent Rifle-bird; (D) Superb Bird-of-Paradise; (E) Magnificent Bird-of-Paradise. (Adapted from Johnsgard 1967)

The effect of male traits on females may be subject to social suppression. The potency of male song in Brown-headed Cowbirds is easily defined by its effect on females, which solicit copulation (West and King 1980; West et al. 1981). Male cowbirds, even those hand-raised in isolation, are capable of singing potent songs, but only the top-ranked dominants in a group actually do so. If a subordinate dares to use potent vocalizations while displaying, it invites attack by the dominant male. As a result, subordinate males downgrade their vocalizations and apparently must wait for an opportunity to sing their best songs without risk.

A recent hypothesis suggests that elaborate plumage and sustained display reveal a male's state of health and particularly its resistance to pathogens and parasites (Hamilton and Zuk 1982). Females might be able to spot the resistant males because increased levels of infection due to lower genetic-based resistance would lower the quality of display plumage or sap a male's ability to display and compete for long hours against other males. Rigorous tests of this hypothesis are not yet available, but the range of superficial patterns of support are intriguing. If resistance to pathogens and parasites is the underlying basis of sexual selection in birds, we might expect the results to be most pronounced in the tropics, where disease abounds, and least pronounced on islands, where diseases are relatively few. Future studies of sexual selection in this context will be most interesting.

The intensity and direction of sexual selection depend on the degree to which one sex is a limited resource. When males greatly outnumber females, they must compete more intensely than when the sex ratio is equal, which often leads to polygyny. Alternatively, females may compete for males when the latter are scarce, which may lead to polyandry. The effective sex ratio depends less on the absolute numbers of males and females than on the actual ratio of fertilizable females to sexually active males, called the operational sex ratio (Emlen and Oring 1977). If the numerical sex ratio is 1 : 1 and all individuals of both sexes come into breeding condition at the same time, the operational sex ratio is also 1 : 1. If the breeding season is prolonged, however, and female receptivities are brief and asynchronous, fewer females will be available to each male, the operational sex ratio may be, perhaps, 2 : 1 and competition among males for mates increases. Skewed operational sex ratios increase sexual selection and favor polygynous or polyandrous mating systems.

Evolution of polygyny

Only two percent of all birds are regularly polygynous. In North America, these include 14 of the 278 breeding passerine species, 11 of which nest in marshes or grasslands (Verner and Willson 1969; Ford 1983). Throughout the tropics, birds that nest colonially in "safe" trees or in marshes tend to be polygynous. These figures for the frequency of polygyny, however, are probably conservative. Careful study of color-marked individuals often reveals a few bigamous males in otherwise monogamous species (Ford 1983).

Why should a female prefer to share a male with another female rather than to pair monogamously? Females that join a harem presumably do so because they can do as well as or better than when paired alone. Some male Lark Buntings, a grassland bird of the western United States, for example, mate with both a primary and a secondary female (Pleszcynska 1978; Pleszcynska and Hansell 1980). The primary female gets all or most of the male's help in raising the young whereas the secondary female gets some protection and use of the territory but no direct assistance. The advantage is directly linked to territory quality, specifically the availability of shaded nest cover. Nestling Lark Buntings easily overheat and die in poorly shaded nests on poor territories. Therefore, some females choose to raise young by themselves in shaded nests on good territories rather than with male help in unshaded nests on poor territories. The cost of sharing male help is clearly evident in Great Reed-Warblers, a few of which are polygynous each year (Dyrcz 1977). Nestlings of polygynous males die more often from starvation than those of monogamous males. Starvation is most frequent during cold, wet spells when food is scarce and the young depend on food delivered by the male as well as by the female parent. Losses to predators, on the other hand, were fewer on polygamous territories than on monogamous territories, which gave females on polygamous territories an overall advantage in reproductive success.

Variation in territory quality, to which marshlands and grasslands seem particularly prone, leads to polygyny (Verner and Willson 1966; Orians 1969b). Polygynous male Marsh Wrens, Red-winged Blackbirds, and Indigo Buntings, among others, all control better quality territories than do unmated or bigamous males in the same area (Verner 1964; Searcy and Yasukawa 1983; Carey and Nolan 1975). The mating success of Marsh Wrens, for example, reflects the proportion of the territorial area with emergent marsh vegetation that provides good nest sites: 54 percent on territories of unmated males, 80 percent on territories of monogamous males, and 95 percent on territories of polygynous males.

The relationships between variations in territory quality and patterns of female settlement that lead to polygyny can be seen by relating female breeding success to territory quality (Verner and Willson 1966; Orians 1969b). First-arriving female Marsh Wrens or Red-winged Blackbirds, for example, pair with those males that control the best territories. Later-arriving females may choose unmated males on inferior territories or become secondary females of males on good territories. Whatever the choice, the potential reproductive output of these females is less than that of the first-arriving females. The point at which a female chooses to join a harem rather than to nest alone on a poor territory is called the "polygyny threshold." Female Red-winged Blackbirds that are members of a harem seem to do as well as or better than lone females of this species; female reproductive success even increases slightly with harem size. The females in a harem do well because the territories they occupy are superior despite multiple use (Figure 16–9).

Control or monopolization of quality resources is a key to the evolution of polygyny (Emlen and Oring 1977). Clumped resources are easier to monopolize than uniformly distributed resources; thus, extending Brown's concept of economic defensibility (see Chapter 15), the environmental potential for polygyny increases with clumped resource distributions. Male Red-winged Blackbirds compete with each other for defensible, high-qual-

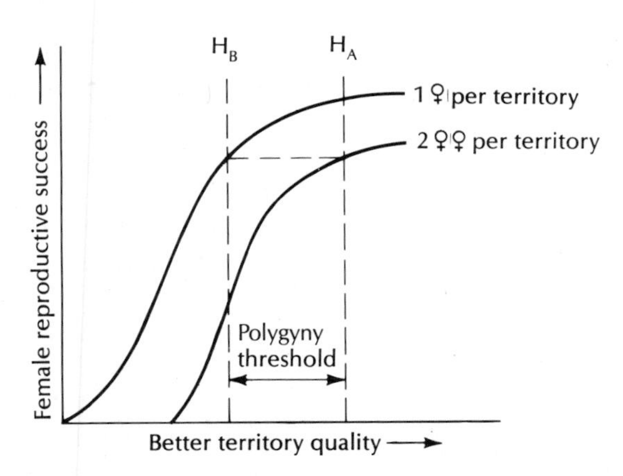

Figure 16–9 Polygyny threshold model. Reproductive success of females increases with territory quality. The secondary female on a good territory may have reproductive success equal to or better than a single female on a poor territory (H_A = best available breeding habitat; H_B = marginal breeding habitat). A female should choose to become a secondary female of a male in H_A when all males holding territories in habitats better than H_B are mated. The difference in territory quality that favors joining a harem rather than nesting alone is called the polygyny threshold. (After Wittenberger 1981)

ity territories that attract females. Defense of specific feeding sites by males takes resource-defense polygyny one step further. Male Yellow-rumped Honeyguides, for example, commandeer the best bee nests in an area and await females that come to feed on the wax (Cronin and Sherman 1976). Dominant males that control the largest bee nests in an area attract the most mates. Male honeyguides, which are specialized brood parasites (see Chapter 20), have no parental duties and therefore occupy themselves fully with control of prime sites and promiscuous mating behavior.

Evolution of polyandry

In only a few birds do females pair with several males, which then incubate the eggs and take care of the young. Such females defend territories, compete for males, and take the lead in courtship. They are also the larger or more brightly colored sex. Polyandrous mating systems have evolved primarily in two orders of birds. In the order Gruiformes, the buttonquails, roatelos, and some rails are polyandrous; in the order Charadriiformes, the jacanas, a few sandpipers, and painted snipes are polyandrous. Male ratites incubate mixed clutches of eggs from several females, which deposit eggs successively with different males. This mixed mating system, which is common among fishes, is called polygynandry. Sex-role reversal has led to the evolution of large and brightly colored females. Sexual selection is apparently intense among polyandrous female phalaropes, which compete for males in congregations at productive feeding sites and initiate courtship with males. Males incubate the resulting clutch of eggs by themselves and do not tolerate the female near the nest after the clutch is complete. Females then lay clutches for other males.

The familiar Spotted Sandpiper of North America is also polyandrous (Hays 1972; Oring and Knudsen 1972). Females, which are 25 percent larger than males, defend large nesting territories and fight each other for males (Figure 16–10). During each breeding season, they lay a clutch of four eggs for each mate, up to four clutches in succession. Each male incubates his clutch of eggs; the female may incubate the final clutch of the season herself. When a male loses his clutch of eggs to a predator, the female quickly replaces the clutch with a new set of eggs. Northern Jacanas, the colorful, long-toed, raillike birds that trot lightly over floating lily pads in open tropical marshes, offer another example of polyandry. Females dominate the much smaller males, each of which defends a small nesting territory in limited, prime habitats. The females compete with each other to mate with as many males as possible (Jenni and Collier 1972). The males incubate eggs and care for the young with little help from the females. As with Spotted Sandpipers, the female quickly replaces lost clutches.

Why should such systems ever evolve? Why do they not evolve more often? The development of a theory of the evolution of polyandry has lagged far behind that for the evolution of polygyny. Current thinking stresses responsibility for incubation and risk of predation (Jenni 1974; Oring 1982). Some male commitment to incubation is an essential prerequisite. If this prerequisite is met, and if the probability of predation is high

Figure 16–10 Female Spotted Sandpipers, which are polyandrous, defend territories and lay clutches for each of several males, which tend them. (Photography by S.J. Maxson, courtesy L. Oring)

and food is abundant, multiple clutches are fertilized and incubated by other males because this behavior enhances a female's reproductive success.

Simultaneous incubation of separate clutches by each member of a monogamous pair represents the possible first step to full parental care by the male and satisfies the first prerequisite. At least five species of ground-nesting birds with precocial young do this: two partridges (the Red-legged Patridge and the California Quail), two sandpipers (the Sanderling and the Temminck's Stint), and a plover (the Mountain Plover). In short breeding seasons or unpredictable environments in which renesting is impossible, two simultaneous clutches potentially increase the number of chicks and reduce the risk of total loss of young to a nest predator. Fertilization and incubation of the second clutch by a second male is the final evolutionary step to polyandry. The second male liberates the female completely from incubation responsibilities, enabling her to recruit additional males.

Lek displays of promiscuous birds

The behaviors of promiscuous males on their display grounds represent the extreme consequences of sexual selection in birds, and the evolution of communal display grounds, or leks, is a matter of vigorous discussion. Promiscuous males display in courtship arenas that contain no resources. Females visit for one purpose only, fertilization, and then build their nests and raise their young elsewhere by themselves. The display grounds of promiscuous birds vary from isolated courts to large aggregations of males on traditional communal display grounds (Figure 16–11). At the one extreme, male Great Argus Pheasants in Malaysia, Superb Lyrebirds in Australia, and Common Capercaillies in Europe hold forth on isolated deep forest courts; less well known are the drab promiscuous birds such as the Ochre-bellied Flycatchers that display solitarily or in loose aggregations in tropical forests (Willis et al. 1978; Snow and Snow 1979). At the other extreme, male Guianan Cock-of-the-Rocks gather like glowing orange ornaments in the forest understory, and golden-plumed Lesser Birds-of-Paradise shake their silky pectoral feathers while dancing in the open forest canopy of New Guinea.

Lek displays have been incorporated into the ceremonial dances of human societies (Armstrong 1942). The Jivaro Indians of South America copy the Andean Cock-of-the-Rock in a sensual dance ceremony. Siberian Chukchees mimic the notes of the Ruff. Blackfoot Indians of the western United States mimic the foot stomping, bowing, and strutting of the Sage Grouse, and their costumes are interpretations of the grouse's spread tail.

Some lek displays are quite spectacular (Figure 16–12). David Snow describes the "grunt-jump" and other dances of the White-bearded Manakin:

> Landing transversely on one of the uprights within a few inches of the ground, it becomes momentarily tense, with beard extended — a slowed down film shows the bird quivering as if bracing itself for the effort — then . . . it projects itself at lightning speed headfirst down to the ground, turns in the air to land on its feet for a split

Figure 16–11 The mating grounds of promiscuous birds include (A) communal leks of Black Grouse and (B) isolated display courts of Great Argus Pheasants. (Adapted from Lack 1968)

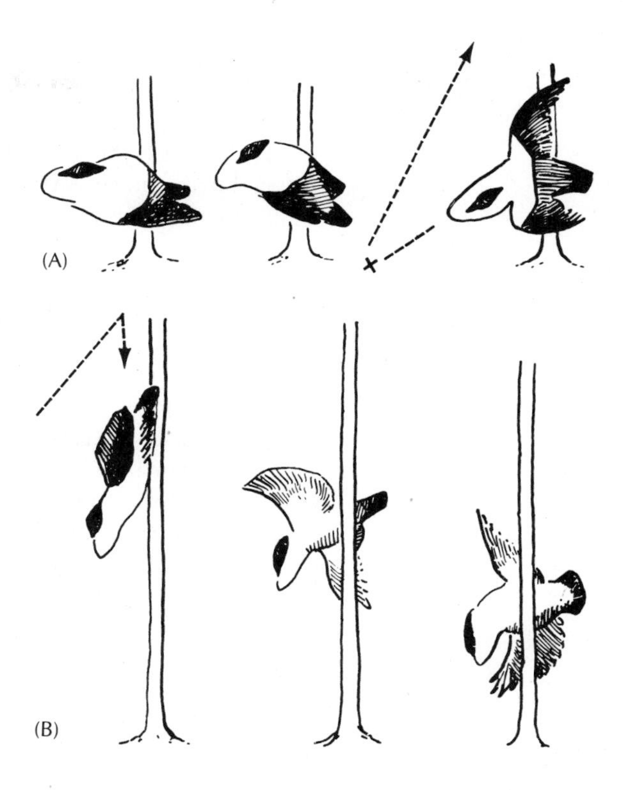

Figure 16–12 (A) The "grunt jump" display of the White-bearded Manakin; (B) the "slide-down-the-pole" display. (From Snow 1976)

second, and with a peculiar grunting noise rockets up to land in a higher position than the one it has just left. The . . . "grunt jump" lasts about a third of a second. It may then do . . . its "slide-down-the-pole": with fanning wings and taking such short rapid steps that it seems to slide, it moves down the perch for a foot or so and remains near the bottom of the upright for a moment, usually to resume its to-and-fro leaping and snapping. (Snow 1976)

Attraction of females for insemination is the single-minded purpose of promiscuous males on leks. Intense competition ranks males in a dominance hierarchy that determines who sires most of the next generation. In the well-studied Black Grouse of Europe and the Sage Grouse of the western United States, less than 10 percent of the males on large leks achieve 70 to 80 percent of all matings (Kruijt et al. 1972; Wiley 1974). This is also the case for the White-bearded Manakin, Red-capped Manakin, and Lesser Bird-of-Paradise (Lill 1974a, b; Beehler 1983). In the case of the Lesser Bird-of-Paradise, one male on a lek of seven made 24 of the 25 observed copulations. Dominance is a matter of age, experience, and ability. By mating with a dominant male, females obtain for their offspring the genes responsible for the male's superior traits. The dominance hierarchy, in effect, selects among males and thus simplifies the selection of a good male.

Clustering of males on leks is partly a natural consequence of the tendency of young, inexperienced males to gather near older or successful males. In this way, the young males get occasional matings and gradually achieve a controlling position in the system. Extreme cases of such associations are seen in *Chiroxiphia* manakins and in Ruffs. Several species of *Chiroxiphia*, gorgeous, light-blue-backed manakins with red caps, engage in circuslike, cooperative routines (Foster 1977, 1981). When an interested female visits the lek, two or three males line up together on a single branch and perform the cartwheel dance (Figure 16–13), described for the Swallow-tailed Manakin by Helmut Sick:

> The males perch closely side-by-side, in a row, on a slightly sloping (or horizontal) twig, face the same direction, all crouched, tripping [moving back and forth with tiny steps], forming a vibrating mass. They call in the recurrent rhythm of a perfectly synchronized "frog chorus." Suddenly the lowest male on the twig rises straight into the air one to two feet and hangs momentarily suspended facing the female. He delivers a sharp "dik dik dik," then lands at the upper end of the row of males at the side of the motionless female. He pivots immediately in the direction of the other males and joins the other males in tripping. Now the lowest bird performs in a similar manner and so on. The entire performance occurs rapidly, giving the impression of a turning wheel. (Sick 1967)

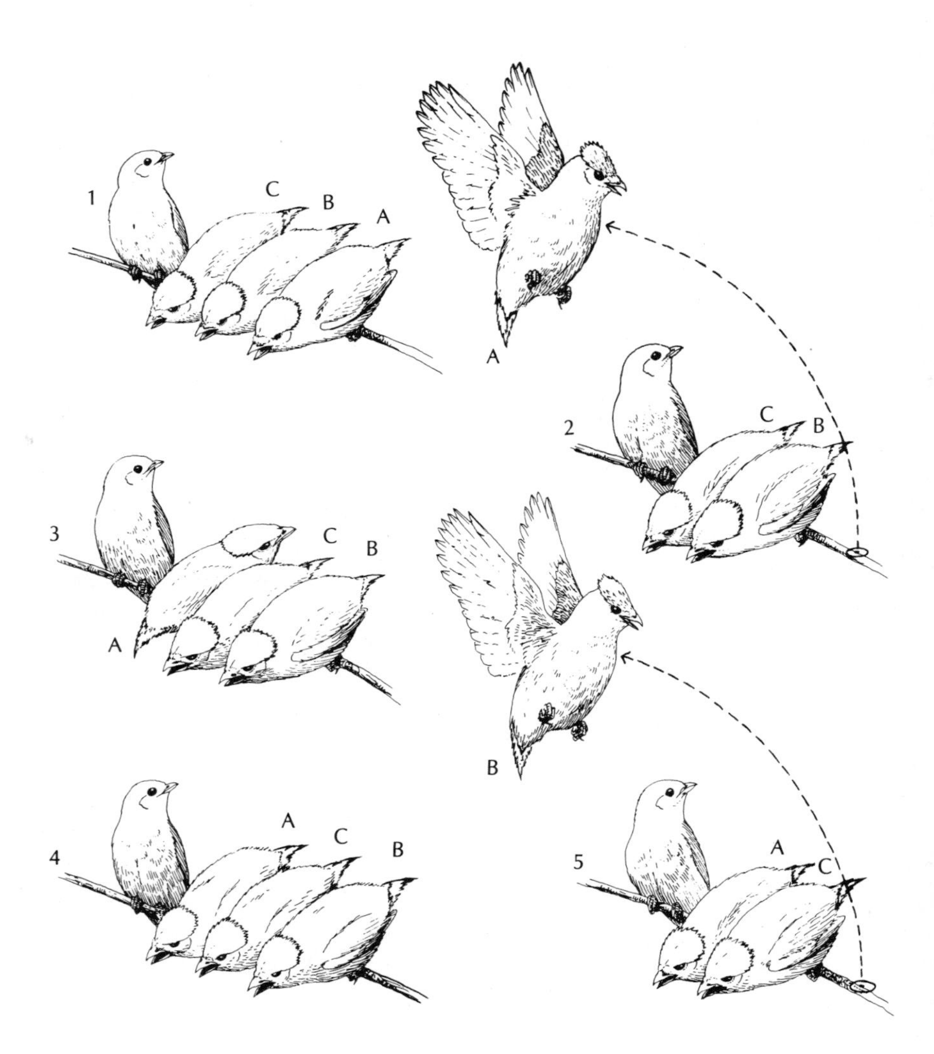

Figure 16–13 Cooperative courtship display of the Swallow-tailed Manakin. Males leap over each other in rapid succession before a waiting female, which may then copulate with the oldest, dominant male. (Adapted from Sick 1967)

Figure 16–14 The Ruff is an unusual species, having variable male plumages. Two social classes of males act as partners on a lek display territory. The white-ruffed satellite males are subordinate to the variably colored dark-ruffed resident males. (Courtesy E. and D. Hosking)

This team performance becomes more and more frenzied, then suddenly stops. The oldest, dominant male does a brief, precopulatory solo display and then mounts the female (Foster 1981). Only cooperative group displays attract and excite females, but subordinate males are not being altruistic. They occasionally copulate when the dominant male is absent. They also develop their expertise, and some of them eventually achieve a more dominant status.

The leks of Ruffs in Europe also contain an intriguing element of cooperation between territorial and satellite males (Hogan-Warburg 1966; Rhijn 1973; Shepard 1975). Ruffs are large sandpipers with prominent neck feathers, called ruffs, and ear tufts, the colors of which vary greatly among males; no other bird is individually so variable in color pattern. Before mating, the females (called Reeves) nibble at the male's feathered ruff, which is flared in a courtship dance. Territorial lek males have dark ruffs (black, brown, or variously patterned) whereas satellite males have white ruffs. The satellite males, which are not so aggressive, associate themselves meekly with an aggressive territorial male and then steal copulations with visiting females when their "partner" is busy defending his territory. Territorial males tolerate the satellite males because their conspicuous white ruffs attract the attention of females (Figure 16–14).

Evolution of leks

Why should promiscuous males gather in leks, in which a few dominant individuals mate most frequently? What are the conditions that favor lek clusters over dispersed nonmonogamous males trying to attract the same females? No one knows for sure, but several possibilities stand out. Reduced risk of predation is one possibility. As in the case of flocking (see Chapter 15), predators may have less chance of surprising a group of males that display together than an isolated male preoccupied with sexual display.

Haven Wiley (1974) suggests that this is the primary reason that open-country grouse display in leks whereas forest grouse tend to display solitarily. Birds-of-paradise that display in leks are species that inhabit forest borders and second-growth forest, where predation risk tends to be higher than in primary forest (Beehler and Pruett-Jones 1983). Conspicuous display on traditional sites, however, may attract predators and thereby counter any possible advantages. The Tiny Hawk of Central America, for example, seems to specialize on lek hummingbirds (Stiles 1978).

The two favored hypotheses for the evolution of leks are that males gather at sites where they are most likely to encounter roaming females, and that males gather because females prefer to choose mates from large aggregations of males because the groups facilitate comparisons (Bradbury and Gibson 1980; Bradbury 1981; Vehrencamp and Bradbury 1984). In the male-initiated model, good regional positioning more than offsets competition within the lek, especially if a male is dominant or has a chance of attaining dominant status. In the female-initiated model, groups of males permit more rapid, efficient comparisons than scattered individuals. Young females also could avoid naive mistakes by learning from the choices of older females. The observation that leks of Sage Grouse tend to be closer together than the average diameter of female home ranges supports the male-initiated model rather than the female-choice model, but not yet with much authority.

Equally demanding of an explanation are the intermediate dispersion patterns of promiscuous male display courts, or exploded leks, in which small numbers of males display 50 to 150 meters apart out of sight of each other but usually within earshot. More numerous than classical lek aggregations, exploded leks are used by some promiscuous sandpipers, manakins, flycatchers, parasitic finches, bowerbirds, birds-of-paradise, and hummingbirds. The dispersion of the display courts of birds-of-paradise, for example, is clearly correlated with diet. Frugivorous species aggregate, insectivorous species disperse, and species with mixed diets are intermediate (Beehler and Pruett-Jones 1983). The reasons for such dispersion patterns remain obscure.

Failure to consummate copulations because of disruption is one of the major liabilities of joining an aggregation of eager males (Foster 1983; Trail 1985). Destruction of nearby bowers clearly discourages clustering of display courts of the Satin Bowerbird. Thus, the advantages of dispersed display with little disruption counter the potential advantages of displaying together. The final dispersion of male display grounds probably represents an equilibrium between such competing alternatives. Intermediate dispersion patterns, or exploded leks, are to be expected when either female preferences for aggregations or the advantages for males are not too strong.

Future of mating systems research

In this chapter we have expressed the traditional thinking, which has stressed the correlations between mating systems and ecology, mediated by parental care requirements. A new wave of study of avian mating systems is now upon us (Mock 1983; Vehrencamp and Bradbury 1984). It is distinguished by its focus on the components of survival and reproductive success that are improved by alternative mating systems, by the interaction of male and female options, and by a reexamination of monogamy itself. Instead of a spirited, cooperative partnership, monogamy may, in fact, be a grudging, temporary truce between selfish, competitive individuals (Mock 1983).

Monogamy, as we usually see it, is a social relationship between members of the opposite sex that is built on the assumption that the offspring are truly their genetic offspring (Gowaty 1981). Mixed genetic paternity of broods due to cuckoldry and forced copulations is proving to be more common than once suspected. Female Red-winged Blackbirds whose mates are vasectomized manage to achieve a high rate of fertility in their clutches through liaisons with other males. Forced copulations are a way of life among breeding waterfowl (Barash 1977; McKinney et al. 1978). Even male Eastern Bluebirds, which symbolize devoted monogamy, cannot be absolutely certain that they are the fathers of the offspring in their nest box. Adding to such uncertainties is regular intraspecific brood parasitism, which is difficult to detect (see Chapter 20) and casts doubt on genetic maternity as well as paternity. The disparity between social and genetic monogamy has direct implications for the study of avian mating systems. The degree to which a male's genetic responsibility for a brood of young can be proven is directly related to the chance of male desertion, and perhaps to acquisition of additional females. Both are more advantageous than investing parental care in a competitor's offspring. Close consortship of mates to ensure paternity is a major feature of early stages of nesting and must continue for renesting. The basis of decisions to guard or desert, manipulated perhaps by the female to her own best interest, will be the focus of future work on avian mating systems.

Summary

Over 90 percent of birds are essentially monogamous though liaisons outside the pair bond are by no means rare. The main advantage of monogamy is biparental care of young: It serves the needs of chicks and helps parents (especially females) to reduce their own energy costs. Polygyny becomes a viable system when females can take care of young without the assistance of males. Species with precocial young and those that feed on easily accessible resources tend to be polygynous. Polyandry, found primarily in the orders Gruiformes and Charadriiformes, is a system with competitive, territorial females, which are generally larger than their male counterparts. Promiscuity, characterized by brief matings and immediate separation, often leads to spectacular display behavior, especially among males that establish themselves on leks, or communal breeding grounds. We are not entirely sure why leks evolve. The advantage from the male viewpoint is that leks may increase their chances of encountering females; from the female viewpoint, leks may facilitate comparison of prospective mates.

The study of avian mating systems is a particularly active area of research. Monogamy, once thought to be an example of sustained cooperative behavior, is being reexamined. Ornithologists now suspect that monogamy may be a grudging arrangement entered into only because of biological necessity. Different reproductive courses are available to males and females, and uncertainties about genetic parentage put stress on the pair bonds and commitment to care of the young.

Further readings

Ford, N.L. 1983. Variation in mate fidelity in monogamous birds. Current Ornithology 1:329–356. *An important review of avian monogamy and infidelity.*

Lack, D. 1968. Ecological Adaptations for Breeding in Birds. London: Metheun. *A classic review of the diversity of avian mating systems.*

Mock, D.W. 1983. On the study of avian mating systems. *In* Perspectives in Ornithology (A.H. Brush and G.A. Clark, eds.), pp. 55–91. Oxford: Oxford University Press. *An important, provocative essay.*

Oring, L.W. 1982. Avian mating systems. Avian Biology 6:1–91. *A comprehensive review of the literature.*

Skutch, A.F. 1976. Parent Birds and Their Young. Austin: University of Texas Press. *A readable review of the natural history of reproduction in birds.*

Snow, D.W. 1976. The Web of Adaptation: Bird Studies in the American Tropics. New York: New York Times Book Co. *A lively introduction to the mating systems of cotingas and manakins.*

Vehrencamp, S.L., and J.W. Bradbury. 1983. Mating systems and ecology. *In* Behavioral Ecology (J.R. Krebs and N.B. Davies, eds.), 2nd ed., pp. 251–278. Sunderland, Mass.: Sinauer Associates., *A critical review of classical thinking and a call for a new approach.*

chapter 17

Reproduction

Courtship and competition for mates are only the prelude to reproduction. We now address the nature of sex in birds: the physiology and anatomy of the gonads, copulation and fertilization of the ovum, and the production of a fully formed egg in the oviduct. Hormones from the pituitary gland and from the gonads themselves regulate reproduction. In addition to the primary function of gamete production, the gonads produce sex hormones that control the development of the secondary sexual characteristics, the external distinctions of males and females. After fertilization, the mature ovum proceeds down the oviduct through an assembly line of provisions for the embryo's use after the egg is laid. The high cost of egg production influences both the timing of breeding and the number of eggs a female lays.

The completed avian egg is an elaborate reproductive cell. It is a self-contained environment in which the embryo develops into a chick that is capable of breaking its way out. Inside the egg are food and water provisions. The shell protects the delicate egg contents and allows the exchange of water vapor and respiratory gases with the atmosphere.

Sexual physiology

The gonads of birds consist of paired testes in the male and a single ovary in the female. These sex organs are responsible for both the production of gametes and the secretion of sex hormones. The testes of male birds are bean-shaped organs that are attached to the dorsal body wall at the anterior ends of the kidneys. Avian testes are usually cream-colored but may be dark gray or even blackish. Initially only a few millimeters long in small birds, they swell rapidly at the beginning of the breeding season, often reaching 400 to 500 times their inactive mass. The testes of a mature Japanese Quail, for example, increase from 8 to 3000 milligrams in just three weeks. Fertility in domestic geese is directly related to the weight of the mature testes, which is an inherited trait (Szumowski and Theret 1965) (Figure 17–1A,B).

The avian ovary resembles a small bunch of grapes. Most birds have only a single ovary (the left) with an associated oviduct. Functional right

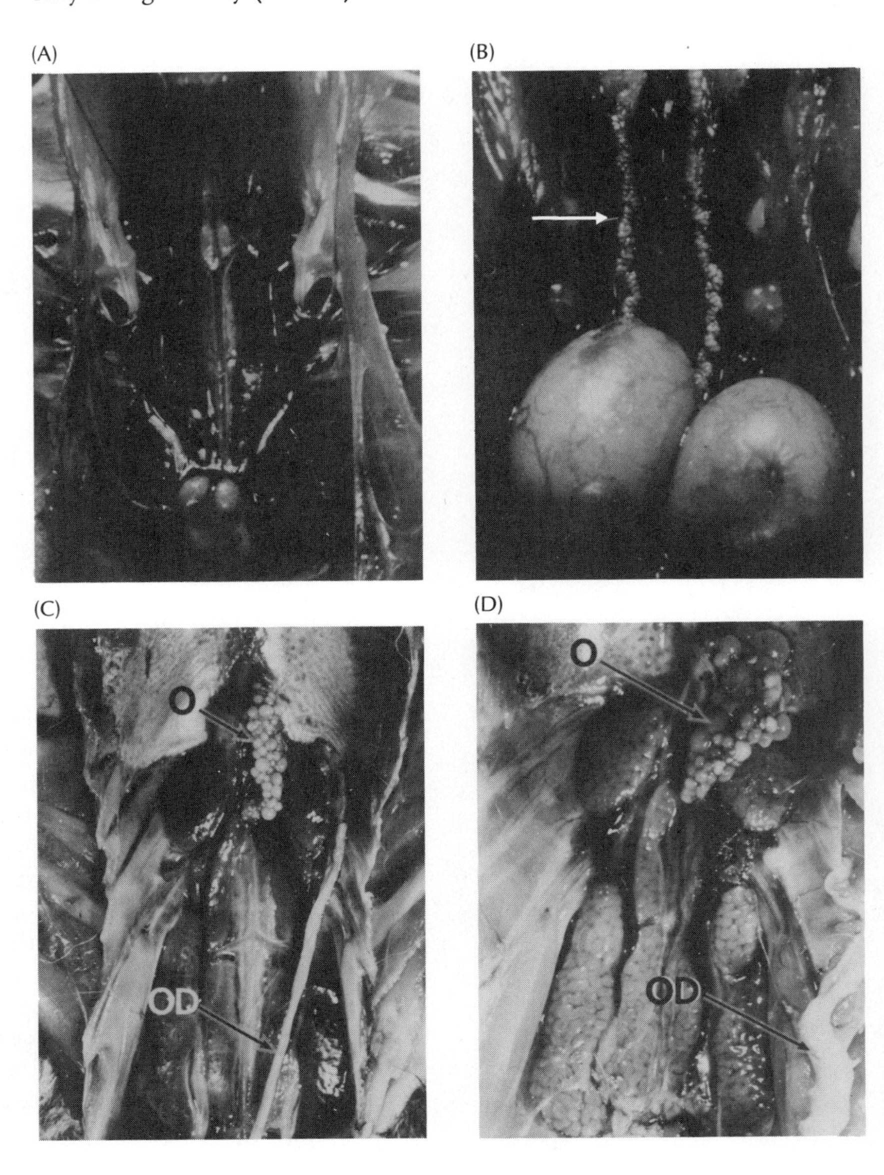

Figure 17–1 Avian reproductive systems. (A) Testes of the male Eurasian Tree Sparrow in winter and (B) at full size during the breeding season. Note also enlarged vas deferens indicated by the arrow (magnification 5×). (C) Ovary (O) and oviduct (OD) of the female Eurasian Tree Sparrow in winter and (D) during the breeding season (magnification 4×). (Courtesy B. Lofts)

ovaries, however, are typical of raptors in the families Accipitridae, Falconidae, and Cathartidae, and of Brown Kiwis (Kinsky 1971); they also occur occasionally in pigeons, gulls, and some passerines. At maturity, the microscopic ovarian granules of the immature bird have increased in size 10 to 15 times (Figure 17–1C,D). The total number of ova in a wild bird is at least 500; more often there are several thousand, certainly many more than are ever used to produce functional eggs.

During early embryogeny, primordial germ cells migrate to the site where the gonads will develop. More of these germ cells settle on the left side than on the right, which establishes the asymmetry that persists throughout subsequent gonadal development, and leads to the development of an unpaired left ovary in female birds. The primordial germ cells first generate the medullary tissue, which (though found in both testes and ovaries) is the principal active tissue of the testes. A second phase of proliferation creates the cortex, the principal active tissue of the ovary. Removal of the left ovary, particularly from young female birds, prompts the right gonadal tissue to develop into an intermediate structure, the ovotestis. Ovarian medullary tissue normally becomes more active with age in females, which in extreme cases causes overt masculinization in older females. Sombre female Golden Pheasants, for example, acquire the spectacular plumages of males as a result of this phenomenon.

Follicle-stimulating hormone (FSH) regulates gamete formation in both the testes and the ovary, and luteinizing hormone (LH) regulates hormone secretion in the testes and maturation of ova in the ovary. Gonadal secretion of the two principal steroid hormones, testosterone and estrogen, directly activates reproductive behavior and controls the development of secondary sexual characteristics. Although testosterone is well known as the male hormone and estrogen as the female hormone, both hormones are in fact present in both males and females, but the proportion of the two hormones and the ways in which body tissues react to each of them are the causes of an individual's sexual attributes.

Secondary sex characteristics

Sexual distinctions in plumage, body size, and voice are influenced by testosterone and estrogen. Acquisition of male breeding plumage in many species results from increasing amounts of testosterone in the blood. Removing the source of this hormone by castration prevents Ruffs, for example, from acquiring their fancy neck feathers. Testosterone causes the bills of European Starlings to turn bright yellow in the breeding season, whereas estrogen causes the red bills of female Red-billed Queleas to turn yellow in the breeding season. The growth of wattles and combs on roosters and the development of the bill ornamentations of breeding auklets also depend on testosterone. Testosterone is responsible for a variety of sex-related peculiarities. Phalaropes (Phalaropidae), for example, are unusual sandpipers in that the bright-plumaged females defend breeding territories and the less-colorful males assume the duties of incubation and parental care. Growth of colorful feathers by either sex is triggered by injection of testosterone. Female phalaropes normally have higher concentrations of testosterone than do males, whose maximum levels of testosterone remain below the threshold required to produce colorful feathers. In a similar case,

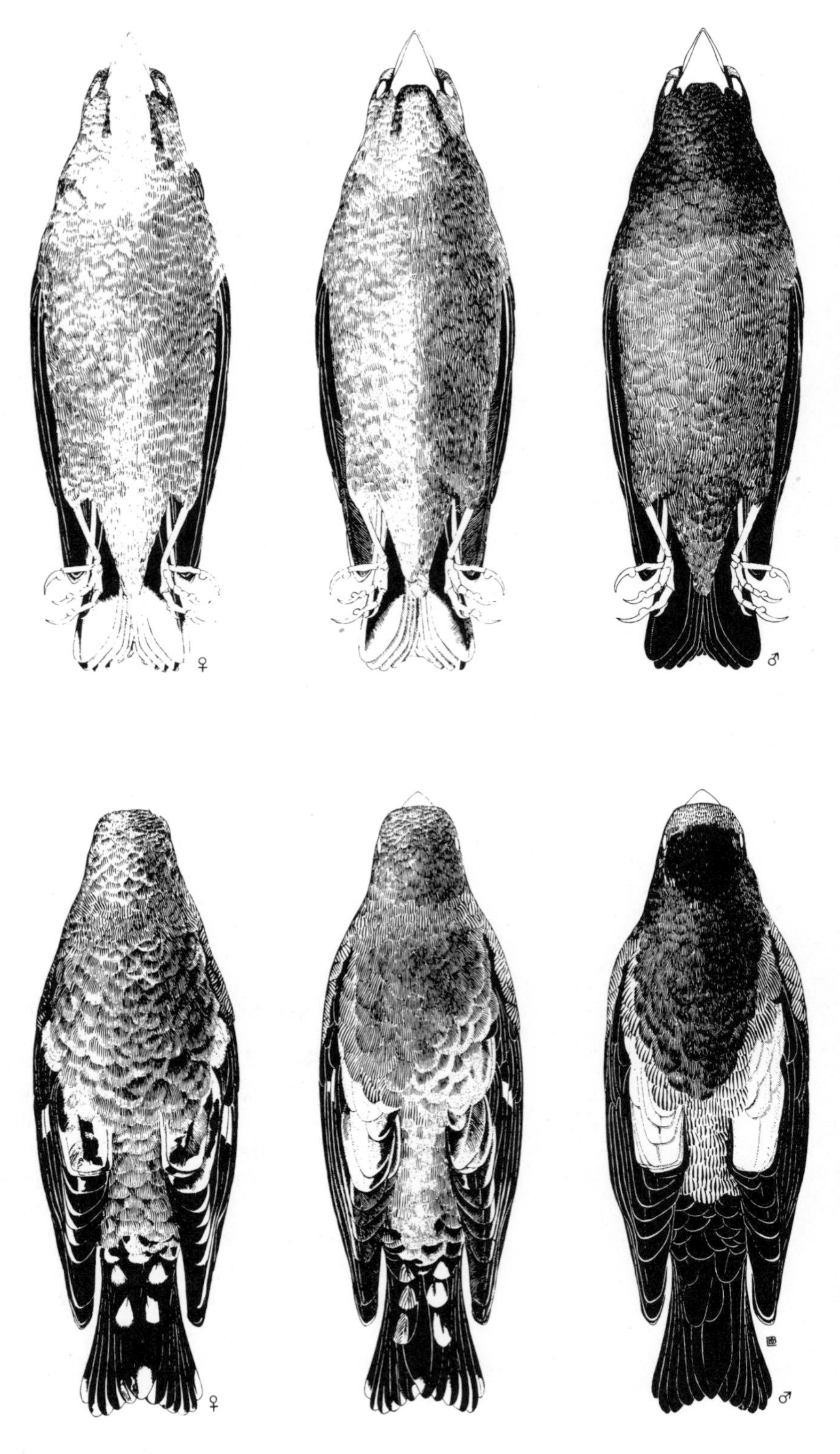

Figure 17–2 Rare individual Evening Grosbeaks are male on one side and female on the other as a result of an aberration in the first cell division. These are called bilateral gynandromorphs (center, top and bottom). (From Laybourne 1967)

males of some breeds of chickens have femalelike feathers because the cellular chemistry in the skin actively converts testosterone into estrogen. When castrated, they grow male feathers. Injection of testosterone into these castrated males causes them to revert to the female type of feathering (George et al. 1981).

Testosterone and estrogen are not responsible for all sex differences. In weavers, the colorful breeding plumages of males result from responses of feather follicles to luteinizing hormone (LH) secreted by the pituitary gland. LH controls breeding physiology in both sexes, but its potential influence on female plumage is inhibited by the presence of estrogen.

As in reptiles and butterflies, but not mammals, female birds are the heterogametic sex; they have the inactive Y sex chromosome and only one X sex chromosome (XY). Male birds, on the other hand, have two X chromosomes (XX). Occasionally, due to an aberration in the first mitotic division of the fertilized ovum, one half of the embryo becomes XY and the other half becomes XX. Such individuals are bilateral gynandromorphs, or half male and half female. Internally they have a testis on one side and an ovary on the other; externally they have male and female plumages on the corresponding right and left sides of the body, with a sharp division down the center. Bilateral gynandromorphs are reported occasionally among Evening Grosbeaks (Laybourne 1967). Isolated cases also are known in Orchard Orioles, American Kestrels, Eurasian Bullfinches, and Common Chaffinches. Nothing is known about the breeding activities of such birds (Figure 17–2).

Sperm production

The thick, outer fibrous sheath, or *tunica albuginea,* of the testes encases a dense mass of convoluted tubules, called seminiferous tubules, which are lined by active germinal epithelia that produce sperm and also secrete sex hormones. Local, synchronous transformation of the cells of the germinal epithelia into mature sperm proceeds in waves down the tubule. The entire length of a seminiferous tubule produces spermatazoa, or sperm, in strongly seasonal breeders such as arctic shorebirds. Secretion of steroid sex hormones, notably testosterone, occurs both in the Sertoli cells lining the tubules and, particularly, in the Leydig cells which are packed between the tubules. These cells undergo well-defined seasonal cycles in the accumulation of lipid and cholesterol used in spermatogenesis. Mature sperm quickly leave the testis through a series of thin tubules: rete tubules, vasa efferentia, epididyma, and vasa deferentia (Figure 17–3).

The testes of mammals and reptiles are found in cooler external scrota, whereas the testes of birds are housed at body temperature inside the abdominal cavity. To compensate for the extra body heat, spermatogenesis occurs primarily at night when body temperature is slightly lower. New sperm are then stored in swollen seminal vesicles, which are responsible for the conspicuous cloacal protuberance of breeding males. Temperatures in this structure, a functional analogue of the mammalian scrotum, are 4°C cooler than internal body temperatures (Wolfson 1954). Typical vertebrate spermatazoa, including those of birds, consist of three sections. The head (acrosome and nucleus) contains the male genetic materials. The midpiece provides metabolic power. The tail (axial filament and tail membrane)

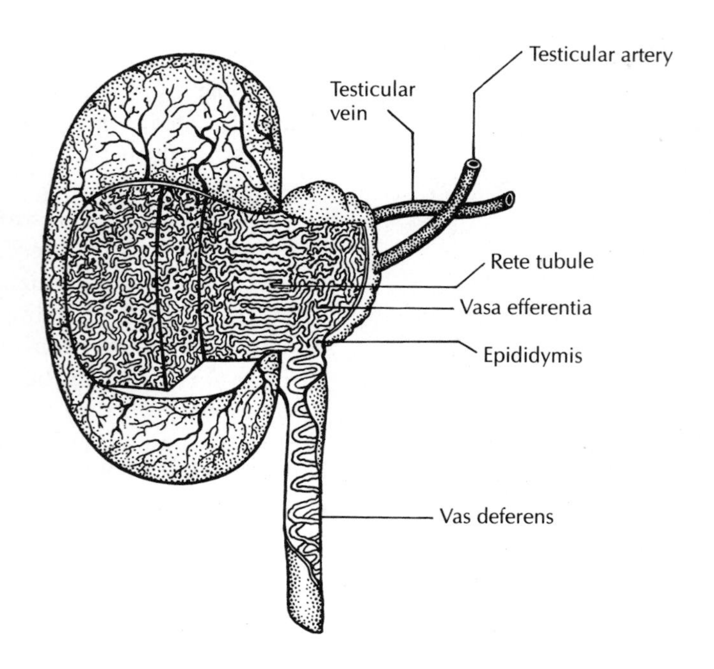

Figure 17–3 Internal anatomy of the avian testis. (From Marshall 1961)

propels the sperm forward. Distinctive sperm structures characterize various kinds of birds, such as the oscine passerine birds (McFarlane 1963, Henley et al. 1978). Unlike nonpasserine sperm, which are generally long and straight like those of mammals, passerine sperm have a spiral head and a long, helical tail membrane. Instead of swimming by beating the flagellum-like tail, they spin (Figure 17–4).

The seminal vesicles, which are the expanded bases of the two vasa deferentia, swell with accumulated semen awaiting discharge. In mammals, the seminal vesicles and the prostate gland add nutritious ingredients to the semen. In birds, the seminal vesicles do not supply many of these nutrients, and the prostate gland is nonexistent. Both fructose and citrate are absent from bird semen, and chloride concentrations are low (Sturkie 1976).

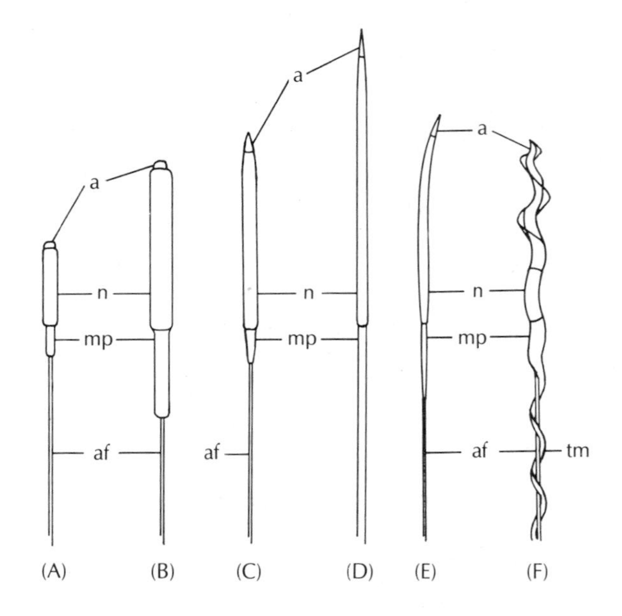

Figure 17–4 Sperm structures characterize the orders of birds (a, acrosome; af, axial filament; mp, midpiece; n, nucleus; tm, tail membrane): (A) Collared Trogon, (B) Great Black-backed Gull, (C) Common Eider, (D) Blue Ground-Dove, (E) Domestic Chicken, and (F) Yellow-rumped Warbler. (From McFarlane 1963)

Maturation of ova

A bird's egg is one of the most complex and highly differentiated reproductive cells achieved in the evolution of animal sexuality. The freshly laid egg consists of:

1. the ovum, a minute cell of life if fertilized,
2. a full supply of food to nourish the embryo, and
3. protective layers to ensure the security of the internal environment.

Initially only microscopic in size, an ovum swells over 1000 times in volume by the time it is laid. The infusion of yolk, the deposition of egg white, or albumen, and the shell layers all contribute to the enlargement. The yolk is added to the ovum prior to ovulation, and the rest of the components of the egg are added during the egg's passage through the oviduct. Development of a mature ovum includes two different yet interdependent processes: the formation and deposition of yolk layers, and the differentiation, growth, and maturation of the germ cell itself. Primary oocytes, the cells that give rise to ova, are already present in a hatchling bird, but distinct ova do not appear until the bird is two months of age, when the first small amounts of true yolk are added to the oocyte. Most of the yolk is added to an ovum much later, during the week prior to ovulation, when the ovum swells to its functional size.

Each ovum is contained by a follicle, which consists of layers of vascularized cells. In the early stages of ovum maturation, the granulosa cells of the follicle actually extrude intracellular organs called Golgi bodies into the oocyte. Fragments of these bodies become centers of yolk formation (Figure 17 – 5). Later, these cells pass materials necessary for yolk formation

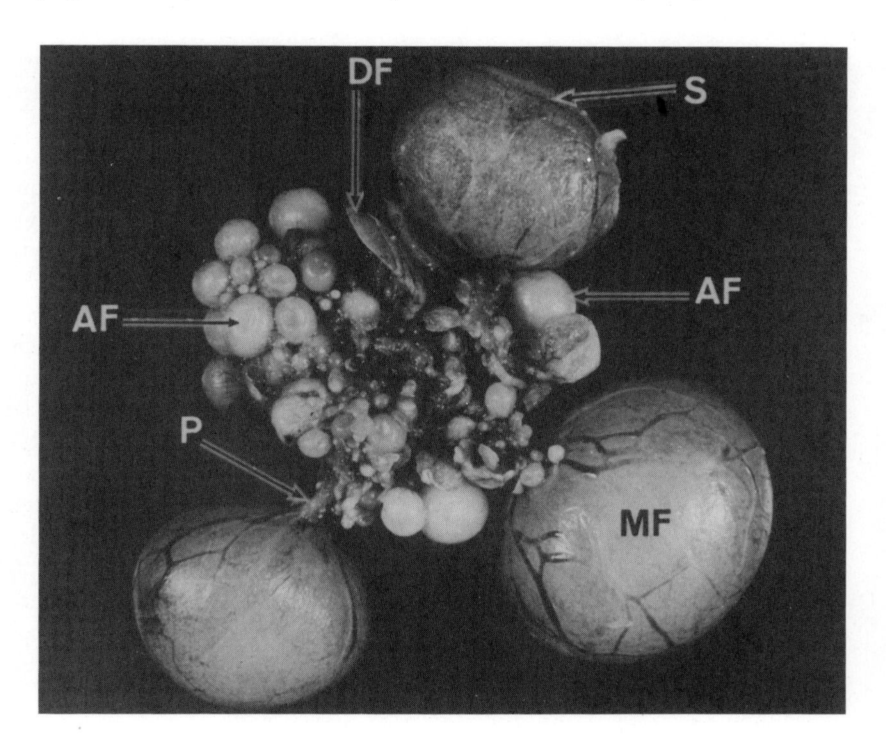

Figure 17 – 5 The ovary of a sexually mature chicken showing mature follicles (MF) with basal stalk, or pedicel (P), atretic follicles (AF), and a recently discharged follicle (DF). S indicates a stigma, the scarlike area where the follicle will rupture during ovulation. (Courtesy Lofts and Murton 1973)

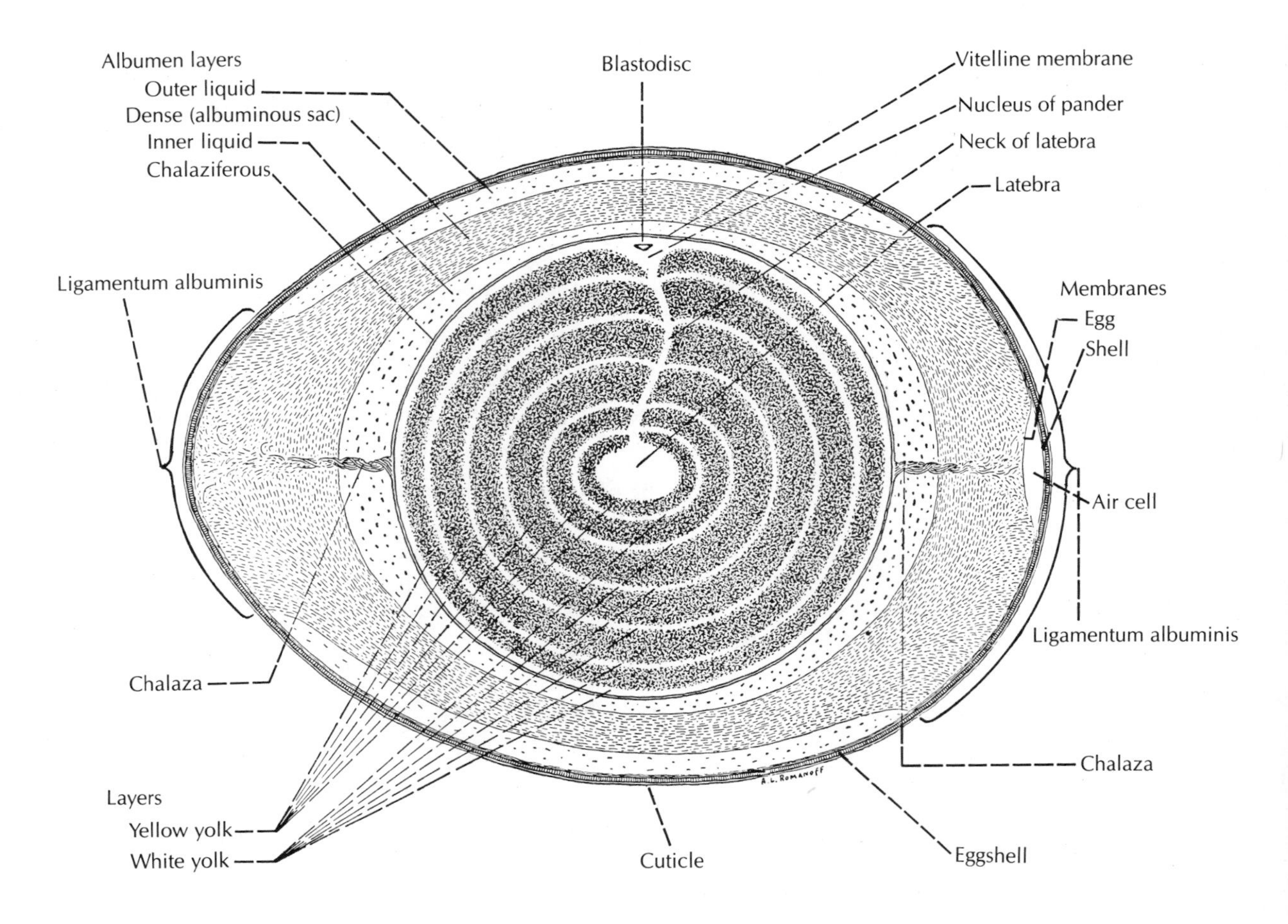

Figure 17–6 Structure of a freshly laid hen's egg. Note alternating layers of white yolk and yellow yolk. The components of egg structure are discussed throughout this chapter. (Adapted from Romanoff and Romanoff 1949)

from the blood to the oocyte across intercellular bridges. The passage of substances into the oocyte is strictly controlled by a process of selective filtration.

The yolk is an inhomogeneous structure, consisting of alternating layers of yellow yolk in large globules (0.025 to 0.15 millimeters in diameter) and white yolk in smaller globules (0.004 to 0.075 millimeters in diameter). The layers reflect daytime (yellow) versus nighttime (white) yolk deposition. When stained with dichromate, these layers can be counted like a tree's growth rings to determine the time required for yolk formation (Grau 1976; Roudybush et al. 1979). The center of the yolk, or central latebra, is composed of a fluid, white substance called vitellin, which extends to the periphery through a distinct, narrow passage. A thin vitelline membrane encases the yolk and separates it from the albumen (Figure 17–6).

The period of yolk formation, or follicular maturation of the ovum, lasts 4 to 5 days in passerine birds such as the Great Tit, White-crowned Sparrow, and Common Jackdaw, 6 to 8 days in larger birds such as ducks and pigeons, and 16 days in some penguins (King 1973; Grau 1982). A study of yolk formation periods among charadriiform shorebirds and seabirds, using yolk ring patterns, revealed that small shorebirds take 4 to 7 days, small gulls 5 to 8 days, small alcids 8 to 10 days, and large gulls 10 to 13 days to form a yolk (Roudybush et al. 1979). Most of the nutrients that supply the energy present in the completed egg are added during this formation period.

When the full-sized ovum is swollen with yolk, it is ready for its passage down the oviduct. Only a few ova actually make it to this stage. Many abort development during the early stages of maturation and are resorbed or become atretic. During ovulation, when the egg is released from the ovary, the follicle enclosing the mature ovum ruptures at the stigma, a layer of smooth muscle fibers, and the enlarged ovum pops out and falls into the ovarian pocket, an irregular cavity formed around the ovary by the surrounding organs. Entry into the oviduct is not simply a matter of chance. The open upper end of the oviduct, called the infundibulum, pulses back and forth toward the new ovum, partially engulfing it and then releasing it for up to half an hour, before finally taking it in (Romanoff and Romanoff 1949). When it is finally inside the infundibulum, the ovum is ready for fertilization.

Copulation

Most birds lack external genitalia, so mating normally involves only brief cloacal contact, usually described as a "cloacal kiss." Standing or treading precariously on the female's back, the male twists his tail under hers, and she in turn twists into a receptive position. The fluttering male may slip off while trying to maintain contact for the few seconds required (Figure 17–7). Swifts and swallows copulate in midair. Sperm transfer occurs when each partner's cloaca is turned inside out and tiny papillae protruding from the posterior walls of the male's sperm sacs into his cloaca are brought into

Figure 17–7 Copulation in Common Black-headed Gulls (left) and Little Blue Penguins (right). (Courtesy E. and D. Hosking and J. Warham)

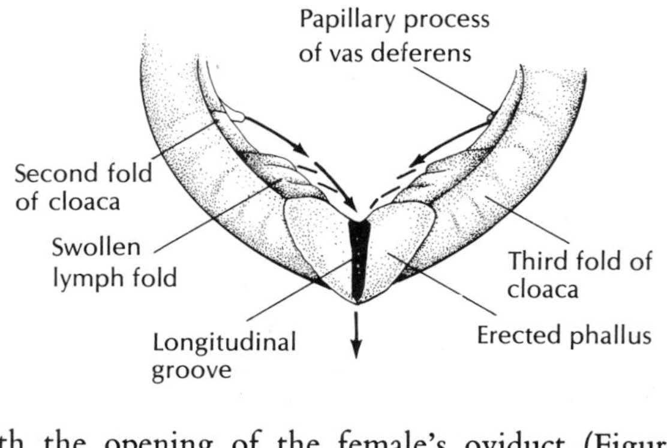

Figure 17–8 Ejaculation of semen by a rooster. Semen, ejected from the papillary process of the vas deferens, combines with transparent fluid from the swollen lymph fold and passes externally via the longitudinal groove of the erect phallus. (After Nishiyama 1955 and Sturkie 1976)

contact with the opening of the female's oviduct (Figure 17–8). In chickens, average concentrations of spermatozoa are 3.5 million per cubic millimeter of semen. A single ejaculation passes 1.7 to 3.5 billion spermatozoa (with records of 7 to 8.2 billion by Brown Leghorn cocks); however, the concentration of spermatozoa drops rapidly after three or four ejaculations. A minimum of about 100 million spermatozoa are required for the proper fertilization of hens (Sturkie 1976). Little is known about sperm counts in wild birds.

A few birds do have an erectile, penislike intromittent organ, which is a special modification of the ventral wall of the cloaca. The list of species so endowed includes tinamous, most waterfowl, curassows, storks, and ostriches. The fully extended penis of an Ostrich may be 20 centimeters long and is bright red in color. Chickens and turkeys have a small penis, which enlarges with lymph fluid that is added to semen in the vas deferens; ejaculation of this fluid occurs through a longitudinal phallic groove. Only one passerine bird, the Buffalo Weaver, has this structure. The reasons why such different birds have evolved intromittent organs is not clear, though the nonmonogamous habits of some of them may favor an apparatus that increases the probability of insemination during the brief copulation period. The waterfowl penis probably facilitates sperm transfer underwater. Some species, including ducks and grebes, have elaborate postcopulatory displays. Immediately after copulation, a male Mallard, for example, suddenly flings his head upward and backward, in what is called the bridling display, and gives a whistled call (Figure 17–9). Then he swims around the female, holding his head low to the water, in what is called nod-swimming (Lorenz 1951). These displays apparently announce successful intromission (Hailman 1977).

Figure 17–9 Bridling, a postcopulatory display of the male Mallard. (Courtesy F. McKinney)

Avian sperm swim directly to the upper end of the oviduct, where they may encounter the ripe ovum. They can reach the infundibulum in less than 30 minutes. Normally, eggs are fertilized within a few days of copulation, but a few sperm occasionally remain viable for weeks. Domestic chickens and turkeys, in particular, can produce fertile eggs 30 to 72 days after copulation. However, in most birds, the probability of laying fertile eggs decreases rapidly one to two weeks after copulation. Some female birds have special sperm storage areas. Parthenogenetic development of unfertilized eggs occurs regularly in domestic turkeys; 32 to 49 per cent of infertile eggs may initiate development, but the embryos usually die at an early stage. Surviving parthenogenetic turkey chicks are always males, have a full diploid set of chromosomes, and may even be sexually competent (Olsen 1960; Sturkie 1976).

Egg production

Passage through the oviduct

After fertilization, the ovum begins its passage through the oviduct to complete the process of egg formation. The oviduct is a long, convoluted tube with elastic walls able to accommodate the egg as it enlarges and approaches completion. Peristaltic contractions of smooth muscle layers propel the egg from the infundibulum to the vagina through distinct sections in which a glandular epithelial lining adds the albumen, shell membranes, and pigmentation in succession. The passage of the egg through the oviduct takes about 24 hours. After only a brief stay in the infundibulum (20 minutes), the egg enters the main length of the oviduct for three to four hours, progressing at a rate of 2.3 millimeters per minute. The albumen is added during this period. The membranes of egg and shell are added next, in a one-hour passage through the isthmus section of the oviduct at a rate of about 1.4 millimeters per minute. Shell formation in the uterus then takes 19 to 20 hours (Figure 17–10).

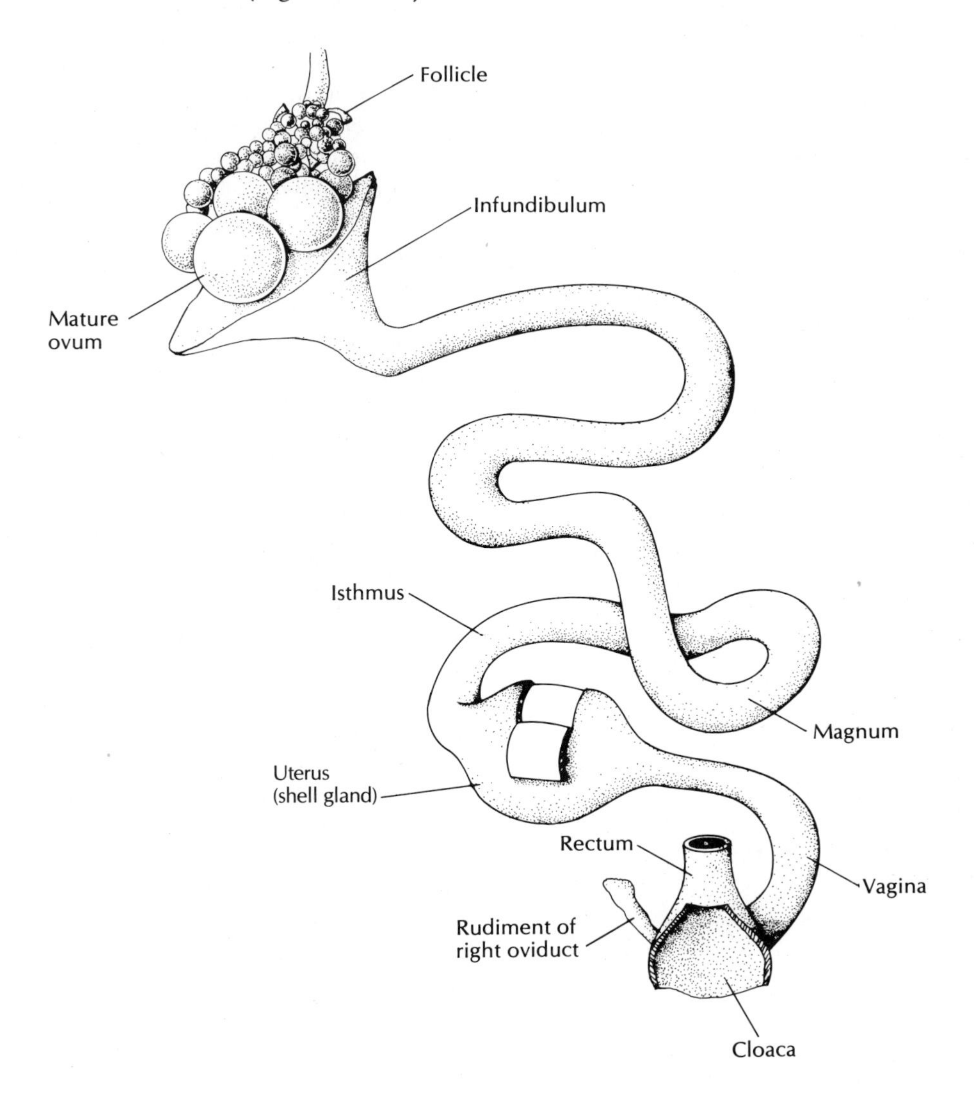

Figure 17–10 Sections of a chicken's oviduct. (After Taylor 1970, with permission of Scientific American)

Albumen is secreted in the anterior section of the oviduct, called the magnum, where four layers of egg white are added that differ in viscosity and material composition. The innermost liquid layer is literally squeezed from the other albumin by the tightening of a meshwork of microscopic fibers. The yolk rotates gently in response to the slight spiral arrangement of the cellular ridges that line the oviduct's interior. The twisted strands of albumen, called chalazae, that form as the yolk rotates act as small built-in springs that help stabilize the yolk position, keeping the dividing cells that form the embryo oriented upward in the finished egg.

Covered with albumen, the egg then enters the isthmus, where the shell membranes are added. The inner membrane, or *membrana putaminis,* surrounds the albumen. The outer, or shell membrane, is usually firmly attached to the shell itself. This pliable and tough membrane is generally made of felted protein fibers strengthened by albuminous cement, which is riddled with tiny pores that permit the passage of gases and liquids by osmosis and diffusion. The thin inner-egg membrane is formed when the cells lining the isthmus apply sticky keratin fibers to the egg surface. Small amounts of pigments which are also added to the shell membrane, may impart a pinkish hue. Shell colors are added first as pigments deposited during shell formation (the ground color) and then later as superficial markings in the cuticle, the thin transparent coating of protein molecules that covers the entire shell. The shell pigments are porphyrins (page 75), which derive from the hematin of old blood cells that have been broken down in the liver and transformed into bile pigments.

The final stage of egg production, the addition of a hard shell consisting mostly of calcium carbonate ($CaCO_3$) in the form of calcite crystals, occurs in the uterus. Magnesium and phosphate are minor components of the shell structure, but even slight variations in their concentrations affect the strength and hardness of the shell in dramatic ways. Slight excesses of phosphate may hinder calcite formation by preventing $CaCO_3$ precipitation; slight excesses of magnesium hinder calcite crystal growth (Cooke 1975). Thinning and increased fragility of the eggshell results, and the delicate balance of gas and water required by the embryo may be altered. Magnesium is usually concentrated in a very thin layer of the inner shell, where it plays a role in the reclamation of eggshell salts by the embryo. Fowls (Galliformes) are distinguished from other orders of birds in that they have two layers of magnesium deposition (Board and Love 1980).

DDT and DDE pesticides affect normal eggshell formation by increasing magnesium and phosphate levels, with fatal consequences. The normal level of magnesium in Common Tern eggshells is 1.54 percent; the normal phosphate level is 0.25 percent. Exposure to DDT and DDE increases these concentrations to 2.1 percent and over 0.6 percent, respectively, causing denting and developmental failure (Fox 1976). An even higher phosphate level (0.86 percent) has been associated with dead embryos. These pesticides were responsible for widespread eggshell thinning and reproductive failure in the 1960s of Brown Pelicans and a variety of raptors (see Chapter 21).

When the process of egg provisioning is complete, the egg is ready to be laid. In a few cases, an egg may first rotate 180° so that the blunt end exits first (Figure 17–11). Birds can eject an egg voluntarily with their powerful vaginal musculature. Most birds lay their eggs early in the morn-

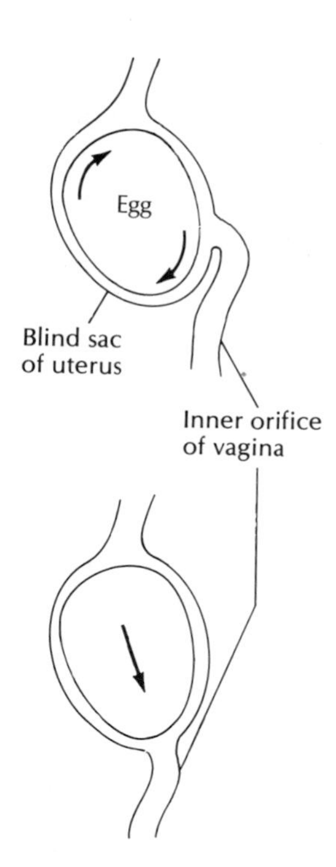

Figure 17–11 Rotation of the egg in the uterus before laying: Some eggs turn 180° so that they are laid blunt end first (top); most, however, proceed directly through the uterus (bottom). (After Romanoff and Romanoff 1949)

ing, probably to avoid the risks that daytime activity could pose to a bird carrying a heavy, fragile egg in its oviduct. Brood parasites such as cuckoos and cowbirds quickly deposit their eggs into the unguarded nests of their hosts. Most passerines, ducks and some geese, hens, woodpeckers, rollers, small shorebirds, and small grebes can lay an egg a day. At the other extreme, megapodes require 4 to 8 days to produce their huge eggs. As a rule, egg production takes 1 to 2 days. Ratites, penguins, and large raptors take 3 to 5 days, and boobies and hornbills take up to 7 days.

Bird eggs vary in size from the tiny eggs of hummingbirds (0.2 grams) to the enormous, half-gallon eggs of elephantbirds (9 kilograms) (Heinroth 1922; Schonwetter 1960–1980). Although egg size increases with body mass, small birds lay much larger eggs relative to their body mass than do large birds. Most birds lay small eggs relative to their own body size; the eggs vary from 11 percent to as little as 2 percent of body mass. There are some dramatic exceptions, however. Kiwis and the Crab Plover, for example, lay exceptionally large eggs. The Brown Kiwi lays a 500-gram egg, which is 25 percent of its mass (Calder 1979). On the other hand, mousebirds, swifts, and parasitic cuckoos lay small eggs for their body masses. Occasionally, birds lay dwarf eggs that are less than half the size of their normal eggs (Ricklefs 1975a); most of these lack a yolk and result from aberrant stimulation of the oviduct by an object such as a blood clot.

Clutch size

The number of eggs a bird lays is subject to proximate constraints, such as energy available for egg formation, and to long-term considerations of lifetime reproductive success. Shearwaters, petrels, penguins, and albatrosses lay only a single egg each breeding season, and arctic sandpipers lay four eggs; but not all birds have such fixed clutch sizes. Average clutch sizes range from 3 to 12 among species of waterfowl and from 2 to 23 among species of gallinaceous birds (Lack 1968). Clutch sizes can also range within a single species, for example, from 4 to 14 in the Northern Flicker and from 8 to 19 in the Blue Tit. Some of these variations reflect heritable differences between individuals, but age, food availability, and season also affect how many eggs a female lays. The inheritance of egg-laying ability is well known to poultry farmers, who increase egg production by artificial selection (Kinney 1969; King and Henderson 1954). Great Tits inherit a tendency to lay larger-than-average clutches, average clutches, or smaller-than-average clutches, but not a particular number of eggs (Perrins and Jones 1974). The evolution of such variations in clutch sizes is an exceedingly complex topic (see Chapter 21). Here we touch on some of the physiological constraints on egg production.

The high costs of egg production can strain a female's daily energy balance and slow egg formation (Ricklefs 1974; Walsberg 1983). The daily energy requirements for egg production average 40 to 50 percent of the basal metabolic rate (BMR) for altricial land birds, and 125 to 180 percent of the BMR for precocial birds. Expressed simply in terms of the extra daily cost of egg production, altricial landbirds increase their daily energy budget by 13 to 16 percent, galliformes by 12 to 30 percent, and waterfowl by 51 to 70 percent (King 1973). In addition, the lack of protein and minerals, such as ash and calcium, may limit egg production, especially in birds that eat fruit and seeds, which are poorly supplied with these elements.

The greatest costs of egg formation are incurred during the period of yolk production. The peak daily energy expenditure for total egg production depends on the amount of overlap in growth cycles of separate ova and the number of follicles growing simultaneously (King 1973). In the Fiordland Crested Penguin, for example, the peak occurs on day 20 as the albumen is added to the first egg at the same time as the last of the yolk is added to the second egg. Generally speaking, the amount of energy transferred to the egg varies from 4.2 kilojoules per gram in passerine birds such as the Eurasian Tree Sparrow to as much as 8.4 kilojoules per gram in the fat-rich eggs of waterfowl. The efficiency of energy transfer is only about 20 percent; a laying female passerine bird, for example, must eat 5 kilojoules of food for every kilojoule that is transferred to her eggs.

The resources required for egg production come from increased daily intake, which is sometimes coupled with reduced activity and with use of a female's stored energy reserves (Walsberg 1983). Increased foraging by laying female waterfowl and shorebirds is particularly conspicuous. Reduced activity in female Willow Flycatchers and Black-billed Magpies enables them to shunt the energy conserved into egg production (Ettinger and King 1980; Mugaas and King 1981). Use of stored reserves for egg production by passerines is not yet well studied except in the Red-billed Quelea and the Brown-headed Cowbird (Jones and Ward 1976; Ankney and Scott 1980). Red-billed Queleas draw heavily on breast muscle for some of the protein they need for their eggs. They also use their fat reserves as fuel for the activity required to find additional protein, as well as for the energy required to produce their eggs.

Wood Ducks lay large clutches of about 12 richly provisioned (and, therefore, energy-expensive) eggs at a total metabolic cost of 6000 kilojoules (Drobney 1980). A hen's fat reserves provide most of the energetic requirements of egg production (88 percent). The protein content of these eggs comes from invertebrates that the hen eats during the laying period. The cost of feeding on invertebrates, which is not profitable in terms of energy, is also supported by the hen's fat reserves.

Female Snow Geese, among many other birds, depend on their reserves for egg production and also for incubation (see Chapter 18). The number of very large follicles in a female's ovary, and hence her projected clutch size, is directly related to the quantity of protein and fat reserves on her arrival on the Arctic breeding grounds (Ankney and MacInnes 1978). After laying their clutches of varying sizes, all females have about the same reserves, which indicates the limits set on clutch size by the reserves (Figure 17–12). Females that arrive with low reserves fail to lay at all.

Food shortages can reduce or stall egg production and thus affect clutch size (King 1973; Ricklefs 1974). Year-to-year variations in average clutch size in the Great Tit, for example, relate directly to food abundance (Perrins 1965). Among species that do not feed their young, clutch size also seems limited by the ability of females to procure sufficient food to produce eggs (Lack 1966, 1968). Furthermore, the resources available for production must be apportioned between egg size and clutch size; eggs in large clutches tend to be smaller than those in small clutches in both ducks and

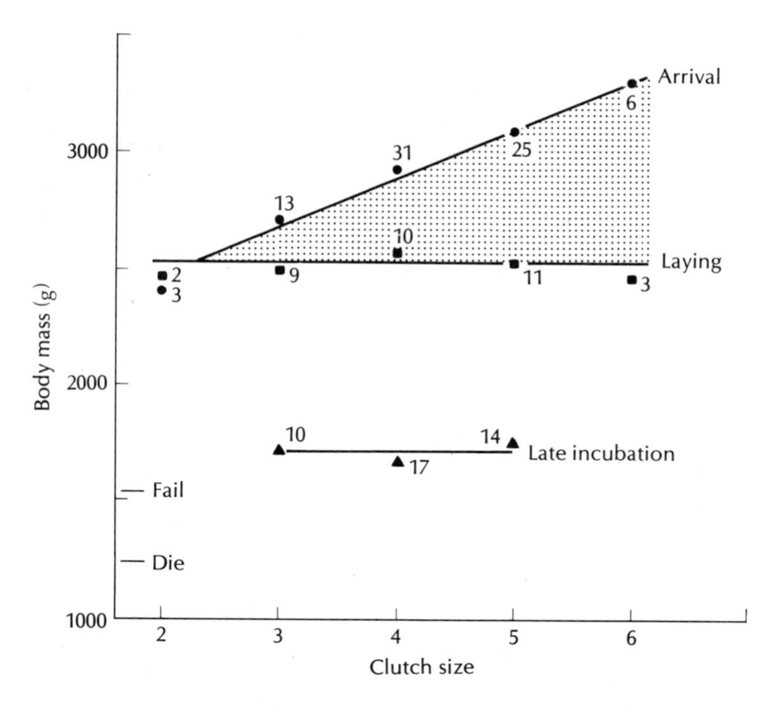

Figure 17–12 Relation of reserves of female Snow Geese arriving on Arctic breeding grounds to their projected clutch size. Females use some reserves (measured by loss in body mass) to produce eggs and then use more reserves during incubation. The number of eggs a female lays is directly related to its reserves. Most females finish laying and start incubating with approximately the same body mass and hence similar reserves. Females that start incubation with inadequate reserves may abandon their eggs to avoid starving but sometimes do not do so in time. (After Drent and Daan 1980, from data in Ankney and MacInnes 1978)

gallinaceous birds. Both clutch sizes and egg sizes of California Gulls reflect variations in food supplies, which are mediated by the frequency of courtship feeding (Tasker and Mills 1981; Winkler and Walters 1983).

Historically, a distinction has been made between determinate layers, species that lay a fixed number of eggs, and indeterminate layers, species that lay extra eggs if some are removed from the nest early in incubation. The classic example of an indeterminate layer is a prodigious female Northern Flicker that laid a total of 71 eggs in 73 days to replace those removed as soon as they were laid (Bent 1939). Domestic hens and Japanese Quail can produce an egg a day all year long. In contrast are the determinate layers, which do not replace eggs removed from their nests. Shorebirds and gulls are usually classified as determinate layers, but they do replace eggs that are removed as soon as they are laid. Galliformes and ducks lay a full complement of replacement eggs if all but one egg of the original clutch is removed as soon as laying is complete. The importance of genetic, nutritional, and psychological factors controlling replacement egg production needs study, and the basic concept of determinate versus indeterminate laying ability needs reevaluation (Winkler and Walters 1983). No clear classification of species can be made until all the species to be compared are subjected to the same experimental regimen.

The avian egg

> I think, that, if required on pain of death to name instantly the most perfect thing in the universe, I should risk my fate on a bird's egg. (T.W. Higginson 1863, p. 297)

Bird eggs fascinated the earliest ornithologists. Their variable sizes, shapes, tints, and textures inspired naturalists to collect them, and, like shells, rare and beautiful eggs commanded great prices. Interest in the avian egg influenced the development of ornithology as a comparative science. Nineteenth century ornithologists published enormous monographs illustrating the eggs of British and African birds (e.g., Seebohm 1885), and serious students of oology (the study of eggs) undertook detailed studies of

Figure 17–13 Egg shell patterns (1/5 more than normal size): **1,** Patagonian Tinamou; **2,** Violet-tipped Courser; **3,** Chilean Tinamou; **4,** Lesser Bird-of-Paradise; **5,** Cape Rook; **6,** Red-headed Rockfowl; **7,** Three-banded Plover; **8,** African Jaçana; **9,** Stripe-backed Bittern ; **10,** White-throated Laughing Thrush; **11, 12,** Brown Babbler; **13, 14, 15, 16,** Tawny-flanked Prinia; **17,** Grayish Saltator; **18, 19, 20,** Winding Cisticola; **21,** Yellow-green Vireo. (From Winterbottom 1971; painting by A. Hughes)

the microscopic structure of eggshells and embryos (Romanoff and Romanoff 1949) (Figure 17–13).

We have reviewed the physiology of sex and of egg production, so now we turn to the egg itself as a self-contained chamber that nourishes and protects the growth and development of the embryo. The avian egg is closed, or cleidoic, the type of egg that freed the reptiles from the aquatic mode of life. It contains all the nutrients and water the embryo requires for its early development. Cleidoic eggs evolved from the naked, amniotic eggs of ancestral reptiles in response to predation by soil invertebrates and microbes (Packard and Packard 1980). Increased calcification of the eggshell provided better protection for eggs laid in the soil. What was sacrificed was the ability of the egg to absorb the water needed by an embryo. The flexible shell membranes of primitive reptilian eggs were water-permeable, but the harder, calcified, avian eggshells are less so. To compensate, water was added to the egg contents in the form of albumen (the egg white). Among reptiles, rigid-shelled eggs supplied with albumen and functionally cleidoic are typical of some turtles and crocodilians, but the evolution of the cleidoic egg proceeded even further in birds.

Egg contents

Three extraembryonic membranes support the life and growth of the avian embryo (Figure 17–14). The *amnion* surrounds the embryo, which floats in a contained environment of water and salts. The *chorion* is a protective membrane that surrounds all the embryonic structures. The *allantoic sac* functions in both respiration and excretion and increases in size as development proceeds, while a growing network of fine capillaries keeps it well supplied with blood.

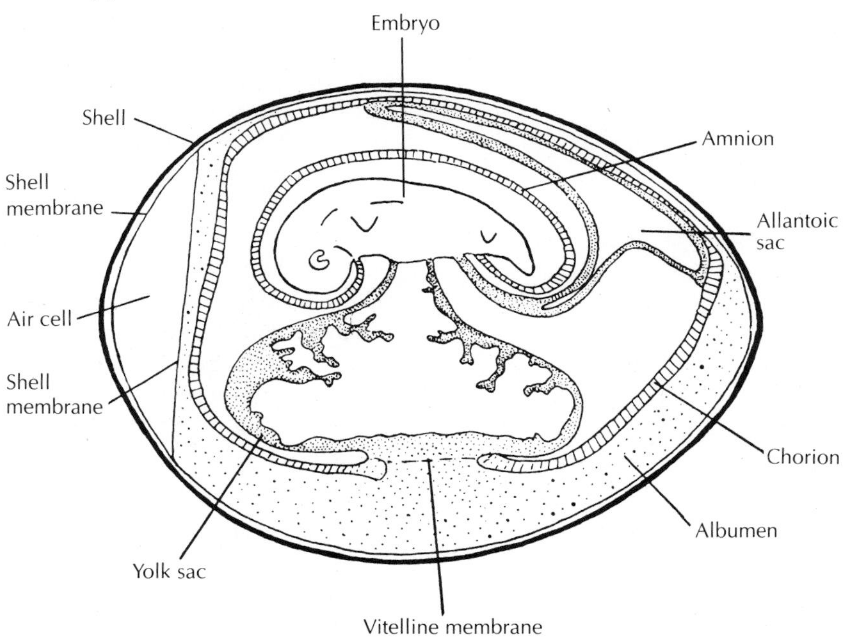

Figure 17–14 The developing embryo and the extraembryonic membranes. (Adapted from Bellairs 1960)

Pressed tightly against the chorion and the shell membranes, the resulting "chorioallantois" exchanges carbon dioxide produced by the embryo for oxygen from the outside world. The allantois also acts as a reservoir for storage of nitrogenous wastes that would pose a serious problem if they were allowed to accumulate inside the amnion. Although the avian egg provides a secure, self-contained environment for embryonic development, it also imposes restrictions on the kind of nitrogenous waste the embryo can produce. Ammonia is not a suitable waste product because the embryo, confined in its shell, cannot excrete it, and unexcreted ammonia would rapidly reach toxic concentrations. Nor is the water-soluble compound urea acceptable because the egg does not have the space required to store large volumes of this dilute waste. Birds (embryonic and adult) have found an excellent solution to their waste-disposal problem: uric acid. Their nonsoluble form of nitrogenous waste can be deposited safely as tiny crystals inside the allantois; it is not toxic nor does it require large volumes of water to flush it from the system.

In addition to the blastoderm, the center of growth of the embryo, and its associated extraembryonic membranes, the freshly laid avian egg consists of three principal components: the yolk, the albumen, and the shell. The yolk is an energy-rich food supply for the embryo. Lipids comprise 21 to 36 percent of the yolk materials, and proteins make up another 16 to 22 percent. The rest is primarily water. The yolk sac, or vitelline membrane, which functions as the early analogue of a stomach and intestines, contains yolk, the rich embryonic food supply, and is ultimately absorbed into the body cavity. In addition to providing nutrition, the yolk initially cradles the tiny embryo in a small pocket. Yolks vary in color from pale yellow or light cream to dark orange-red or even brilliant orange. Within a species, such variations partially reflect diet. Hens that eat red peppers, for example, lay eggs with red yolks instead of the normal yellow yolks (Fox 1976). The albumen, or egg white, consists primarily of water (90 percent) and protein (10 percent). Besides being the embryo's water supply, the albumen is an elastic, shock-absorbing sac that protects the embryo when the egg is moved or jolted. It also insulates and buffers the embryo from sudden changes in air temperature and slows the cooling rate when the parent is not incubating. Albumen constitutes 50 to 71 percent of the total egg weight.

The external layers of the egg shield the embryo, conserve food and water, and facilitate the respiratory exchange of gases. Above all, the hard shell provides structural support and protection of the egg contents from soil invertebrates and microbes. Eggshells vary in thickness from paper-thin in small landbirds to as much as 2.7 millimeters thick in ostriches. They are strong enough to withstand the weight of an incubating adult yet delicate enough to allow the chick to break out. The shell constitutes 11 to 15 percent of an egg's total weight.

Egg shapes

The term *egg-shaped* brings to mind a rounded structure, longer than it is wide and slightly more pointed at one end than at the other, as is the familiar hen's egg. Physiological factors influence the variations in egg shapes of domestic hens, but males do not come from pointed eggs nor females from more rounded ones, as Aristotle once suggested. Egg shapes

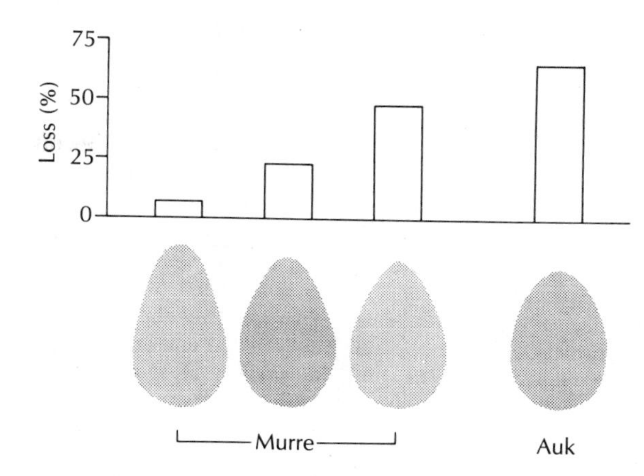

Figure 17–15 Pointed eggs such as those of the Common Murre are less likely to roll off a cliff ledge than are the more rounded eggs of Razor-billed Auks. Data presented here are from 400 trial experiments in which eggs of each type were pushed gently on a nesting ledge. (Adapted from Drent 1975 and Tschanz et al. 1969)

vary from the nearly spherical eggs of petrels, turacos, owls, and kingfishers to the pointed (pyriform) eggs of plovers and murres. Between these shapes are the ellipsoidal or biconical eggs of grebes, pelicans, and bitterns.

Egg shapes are a compromise between structural advantages, clutch volume, and egg content (Andersson 1978). Spherical eggs maximize shell strength, conservation of heat, and conservation of shell materials by maximizing volume relative to shell surface. Pointed eggs, in the clutches of three to four large eggs typically laid by shorebirds, further enhance the volume or content of large eggs within the limits set by the area an incubating parent can cover with its body. The pointed eggs of murres and other cliff-nesting birds offer an additional advantage; they roll only in a tight arc, which lessens their chance of falling from their precarious positions on nest ledges (Drent 1975) (Figure 17–15).

Eggshell color and texture

Many bird eggs, especially those of hole- or burrow-nesting species, are dull white. The need for camouflage is minimal in such nest sites, and enhanced visibility of the eggs in the dark interior of the nest cavity may reduce accidental breakage by the parents. Most eggs laid in open nest sites, on the other hand, are exquisitely colored and patterned. Shaded ground colors, superficial blotches, and fine specklings or scrawls help blend the smooth contours of an egg into its background. In exceptional cases in which the eggs of some ground-nesting species, such as nightjars, are conspicuously white, the well-camouflaged incubating parent shields them from the eyes of potential predators. The whitish eggs of grebes are quickly camouflaged by brownish stains from mud and rotting nest vegetation. A variety of birds such as American Robins lays bright blue eggs. The brightest blue eggs of all are those of the Great Tinamou of Central and South America. The function of blue coloration is still not known though there is a general correlation with nest site. British thrushes that lay blue eggs usually nest in open-forked branches, whereas related species that nest on the ground or in dense thickets lay brown-speckled eggs (Lack 1958).

Various shell textures characterize the various families of birds. Accentuating the bright blues, greens, and violets of tinamou eggs is their polished, enamellike texture. The eggs of ibises and megapodes, in contrast, have dull, chalky textures, whereas the eggs of ducks are oily and waterproofed, and the eggs of cassowaries are heavily pitted. The functions of such shell structures are not well understood.

Eggshell structure

Eggshells are made of inorganic calcium and magnesium salts (carbonates and phosphates) imbedded in a network of delicate, collagenlike fibers (Board and Scott 1980; Carey 1983). There are two distinct layers of shell microstructure: an inner cone layer with basal protuberances that adhere to the shell membrane, and a palisade layer that comprises most of the shell material. Crystalline calcite ($CaCO_3$) is the principal construction material. This inorganic salt is gradually mobilized from the shell and used as calcium for bone growth by the embryo. Covering the outer surface of the eggshell is the cuticle, a thin, proteinaceous froth of air bubbles a mere 0.5 to 12.8 nanometers thick (Figure 17–16).

Eggshell textures reflect a porous microstructure that regulates the passage of water vapor, respiratory gases, and microorganisms between the embryonic environment inside the egg and the external world. The eggshell is permeated by thousands of microscopic pores (Becking 1975; Board and Scott 1980; Tullet and Board 1977) (Figure 17–17). The ordinary hen's egg has over 7500 pores, most of which are at the blunt end of the egg. Most birds have simple, straight canals that widen slightly toward the openings on the exterior surface. The eggshell pores of swans and ratites, on the other hand, branch from their origins near the shell membrane into a more complex network (Tyler and Simkiss 1959). Covering the exterior openings of the pore canals of all birds except pigeons and doves are tiny plugs or caps, which may act as pressure-sensitive valves.

Diffusion through the shell membranes allows the exchange of water vapor and gases, which are vital to embryonic life (Ar and Rahn 1980; Carey 1983). Eggs breathe passively. No active, regulated

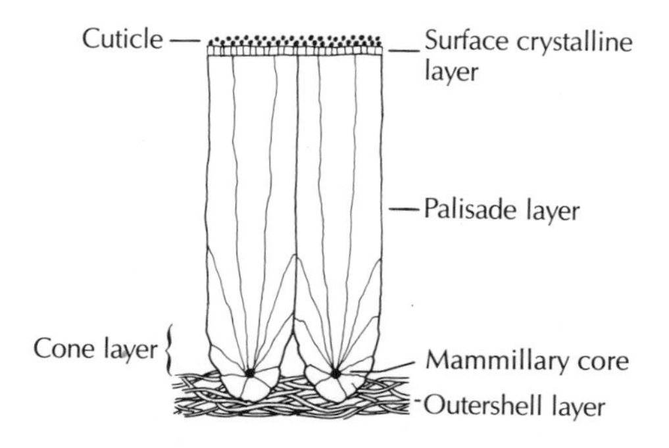

Figure 17–16 Structure of an eggshell. (Adapted from Board and Scott 1980)

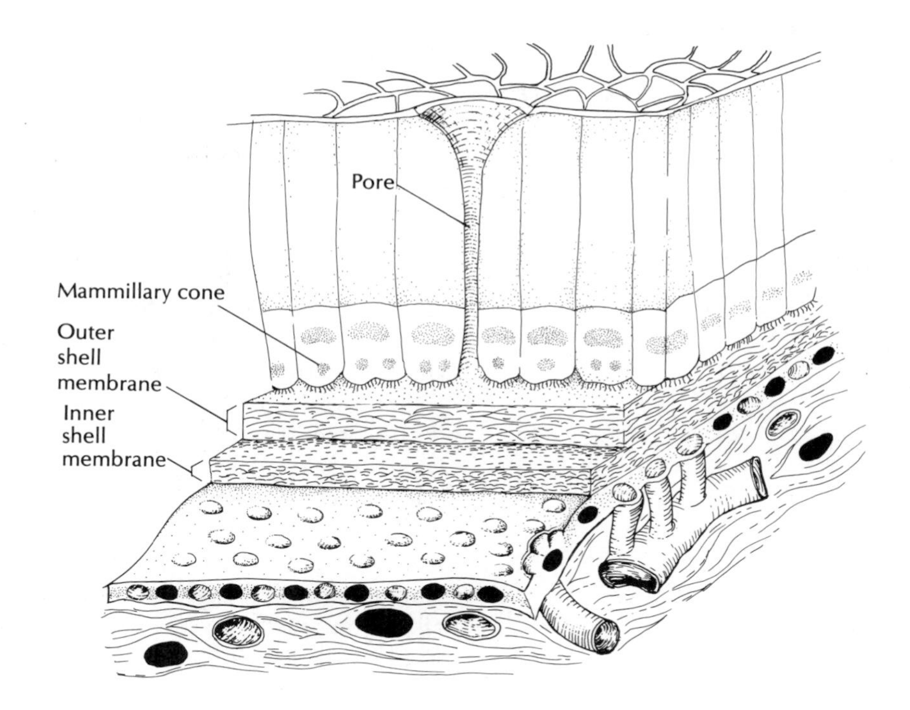

Figure 17–17 Pore canals allow gas exchange through the eggshell. Oxygen enters the eggs via pores that traverse the cuticle and passes through columns of crystals to the permeable shell membranes. Carbon dioxide and water vapor escape to the outside environment through these same pores. Blood vessels in the capillary bed of the chorioallantois link the developing embryo to the gas exchange pathway. (After Wangensteen et al. 1970; Drent 1975; and Rahn et al. 1979, with permission of Scientific American)

exchange is known, nor is it required to account for the known rates of gas and water vapor exchange. The density of pores is a compromise between the optimal high densities that would facilitate rapid gas exchange and the low densities that would minimize water loss. The surface cuticle prevents the entry of microorganisms, without impairing gas exchange. The dynamics of gas exchange varies as incubation proceeds. Removal of calcium from the shell itself for incorporation into the embryonic skeleton promotes progressive thinning of the shell, which increases the rate of gas exchange. The permeability of the shell membranes to oxygen increases as they dry out, and the rate of inward movement of oxygen increases as the growing embryo draws increasing amounts of oxygen from the chorioallantois.

The characteristics of eggshells could limit the altitudinal or geographical distributions of birds. The high rates of potential water loss in xeric (dry) habitats, where relative humidity is low, or at high altitudes, where barometric pressures are low, might limit the hatchability of eggs without some adjustment in pore density or length. Domestic chickens apparently change their eggshell microstructure in response to altitude (Rahn et al. 1982). Some compensation for altitude also occurs in the eggs of swallows (Carey 1980), but how much eggshell microstructures vary as a form of environmental adaptation is not yet known.

In conclusion, the embryo inside the avian egg is not isolated from the external environment. Its survival requires an active exchange of oxygen, carbon dioxide, and water vapor through the shell membranes. Its growth and well-being depend on the egg's provisions and also on its temperature. Its chances of hatching greatly depend on the ability of the parents to regulate the egg's immediate environment within narrow limits. In addition to caring for the egg's physiological requirements, parents must also protect the egg from predators. Birds respond to these challenges by incubating their eggs in nests, the topic of the next chapter.

Summary

Birds reproduce sexually. The gonads consist of paired testes in males and a single ovary in females. Avian testes are located internally, attached to the dorsal body wall at the anterior ends of the kidneys. The avian ovary, which resembles a bunch of grapes, is comprised of hundreds, sometimes thousands, of ova. The gonads are controlled by two hormones secreted by the anterior pituitary: follicle stimulating hormone (FSH), which regulates gamete formation, and luteinizing hormone (LH), which regulates hormone secretion by the testes and maturation of ova in the ovary. Estrogen and testosterone control sexual distinctions in plumage, body size, voice, and behavior.

Because most birds lack external genitalia, copulation normally involves only brief cloacal contact. Sperm swim directly to the upper end of the oviduct, where eggs are fertilized, usually within a few days of copulation. After fertilization, the egg passes through the oviduct, a process which generally takes about 24 hours. Albumen, egg and shell membranes, and a hard shell made of calcium carbonate surround the yolk. Egg formation takes from one to seven days, depending on the species. Some birds have fixed clutch sizes, others do not. Energy requirements, food supplies, egg size, and parental care requirements all influence clutch size.

The avian egg is one of the most complex and highly differentiated reproductive cells achieved in the evolution of animal sexuality. Not only does it provide nourishment for the developing embryo, but ventilation, insulation, warmth, and protection as well. Pores that permit gas exchange and water loss permeate the microstructure of the avian eggshell, which evolved to protect the embryo from soil invertebrates and microbes.

Further readings

Carey, C. 1983. Structure and function of avian eggs. Current Ornithology 1:69–103. *An excellent summary of recent work on the physiology of eggs.*

Lofts, B., and R.K. Murton. 1973. Reproduction in birds. Avian Biology 3:1–108. *A detailed review of the anatomy and physiology of reproduction.*

Ricklefs, R.E. 1974. Energetics of reproduction in birds. *In* Avian Energetics; Publications Nuttall Ornithological Society no. 15, pp. 152–292. *A detailed review of the subject.*

Romanoff, A.L., and A.J. Romanoff. 1949. The Avian Egg. New York: John Wiley & Sons. *A classic reference*

Sturkie, P.D. 1976. Avian Physiology. New York: Springer-Verlag. *A general review with an emphasis on poultry.*

chapter 18

Nests and incubation

Birds must nurture their eggs and young outside the body. Eggs are formed in the oviduct, but the embryos inside them do not normally begin to develop until after the egg is laid; no birds give birth to live young. Birds must prepare a place to lay their eggs and must care for them and for the dependent young that hatch from them.

The nesting behaviors of birds provide excellent studies of complex adaptations and the variety of solutions birds have evolved to meet the challenges of reproduction. Eggs, nests, and the incubation behavior of adult birds constitute an integrated system of reproductive adaptations. Eggs and attending parents tempt a host of predators. Nest structure and location, as well as adult behavior, determine the risks of predation; nest structure and adult behavior also determine the microclimate in which the embryo develops and in which the adult must maintain itself.

Nest building

Site selection

A nest may be built by either member of a pair of birds, or it may be built jointly during courtship and pair formation. Nest site selection, accompanied by displays, may be an integral component of pair formation. Male Pied Flycatchers, for example, sing long, soft songs as they lead females to potential nest holes. A male Blue Tit flies with slow, mothlike wingbeats to the hole of his choice and then, as he clings to the edge of the hole, flashes his white cheek patches by wagging his head or by repeatedly inserting his head into the hole and withdrawing it (Stokes 1960). A male Common Redstart, a kind of thrush, also displays at the nest hole. He either inserts his head into the hole and spreads his bright rufous tail outside, or he goes inside, turns around, and extends his head outside to display the contrast of his black throat against his white forehead (Buxton 1950) (Figure 18–1).

Wrens and weavers construct nests for evaluation by prospective mates. If a prospective mate rejects the nest, Village Weavers tear it down and build a new one. Male Marsh Wrens may build more than 20 nests for comparison by prospective mates; bigamous males build an average of 24.9 nests, monogamous males 22.1, and unmated males 17.4 (Verner and Engelson 1970). Those nests not used serve as dummy nests that help confuse nest predators.

Figure 18–1 Male Common Redstart at a nest site. (Courtesy C. Greenewalt/VIREO)

Gathering nest materials

Nests are made of various kinds of plant matter, including twigs, grass, lichens, and leaves. Certain kinds of plants apparently help combat disease and ectoparasite infection, which can be a serious problem in fouled, unsanitary nests and in reused nest sites such as cavities and artificial nest boxes (Wimberger 1984; Collias and Collias 1984). Green vegetation seems to be particularly useful in this regard. In general, hole nesters incorporate fresh, green vegetation more regularly into their nests than do open nesters (Clark and Mason 1985). European Starlings, in particular, select certain plants, such as Red-dead Nettle and Yarrow, which contain high concentrations of chemical compounds that discourage herbivorous insects from feeding on the plant's leaves. When placed in a nest, these chemicals inhibit bacteria growth and the hatching of arthropod nest parasite eggs.

Inorganic materials, including mud pellets, rocks, tinfoil, and ribbons are also used in nest construction, and animal products are an important supplement; some birds use spider webs for mooring or cohesion, feathers and hairs for the final lining, and snakeskins for external embellishment. In Europe, Long-tailed Tit and Goldcrest nests may contain 2000 or more feathers, in addition to other materials. Birds go to extremes to get these prime materials, which may be in short supply. Many birds pluck hair, a prized nest-lining material, from livestock. Galapagos mockingbirds occasionally take hair from the heads of tourists. Waterfowl pluck down from their own breasts, and the Superb Lyrebird plucks down from its flanks. Thievery is common, especially in large seabird, heron, and penguin colonies. Tanagers and oropendolas living in tropical forests, where there may be shortages of prime materials, often steal nest materials for nest construction.

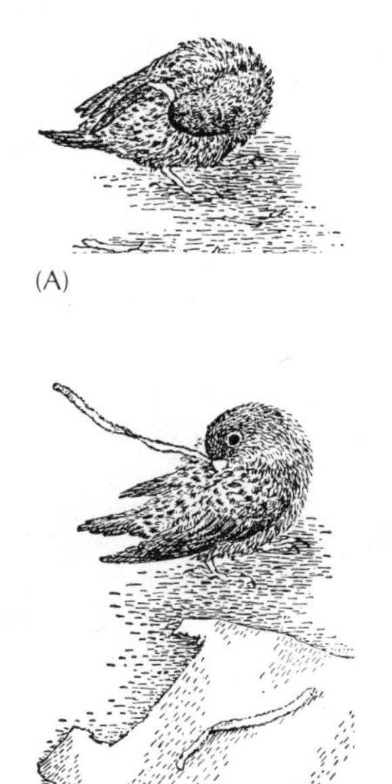

Figure 18–2 Lovebirds carrying nest strips. (A) The Peach-faced Lovebird tucks them into its rump feathers, whereas the Masked Lovebird (not shown) carries them in its bill. (B) Hybrids of these two species try to tuck strips but usually fail. (From Dilger 1962, with permission of Scientific American)

The behavior associated with nest building varies from the simple accumulation of materials to elaborate construction. The nonincubating parent may simply toss materials in the direction of the nest site, creating a mound of debris or a conspicuous rim near the eggs, leaving the incubating parent to delineate the nest site by drawing the materials toward itself. Deliberate transport of suitable materials to the nest site was a major step in the evolution of nest building behavior among birds (Collias and Collias 1984); it led to the modification and design of the nest site and to more complex nest architecture.

Birds usually carry nest materials in their bills or feet. Some lovebirds, which are small African parrots, transport nest materials in an unusual way that apparently is genetically determined. The Masked Lovebird carries one strip of nesting material at a time in its bill, but the related Peach-faced Lovebird tucks the ends of several strips beneath its rump feathers and flies to the nest with the strips in tow (Dilger 1962). Experimentally bred hybrids of these two species try to tuck strips into their rump feathers but cannot do so correctly. Sometimes the hybrids fail to complete the tuck; more often they hold the strip by the middle instead of the end, fail to let go of the strip after tucking it, or tuck it into the wrong place. Many strips do not reach the nest box. The hybrid's genetic program for carrying nesting material apparently contains conflicting, species-specific instructions (Figure 18–2).

Nest construction

Bills and feet are the essential nest-building tools. Bills serve as wood chisels and drills, as picks for digging into the ground, as shuttles for weaving, as needles for sewing, as trowels for plastering, and as forceps (Skutch 1976). Bill structures have evolved toward efficiency in food gathering (Chapter 7) rather than nest construction, though the Grosbeak Starling may be an exception: Its heavy bill seems unnecessary for consuming the soft fruit it eats but is essential for digging its nest into the trunks of old trees (Gilliard 1958). Nests are also built by stamping, scraping, kneading, and scratching as reptilian ancestors did. Burrow nesters dig by kicking loose soil backward; they then mold the internal nest dimensions by using their breasts and feet.

The cup nests of small arboreal landbirds are usually built from the bottom up. However, open cup nests suspended by the rim, such as those of vireos, are built by wrapping nest materials around the supporting twigs first and then looping strands of material from side to side to form the framework of the cup. The long, hanging nests of tropical flycatchers such as the Sulphur-rumped Flycatcher begin as an accumulation of materials stuffed into a tangled mass. The flycatcher forces its way into the center and gradually expands the nest cavity from the inside out; then it reinforces and lines the hollowed-out cavity (Skutch 1976) (Figure 18–3).

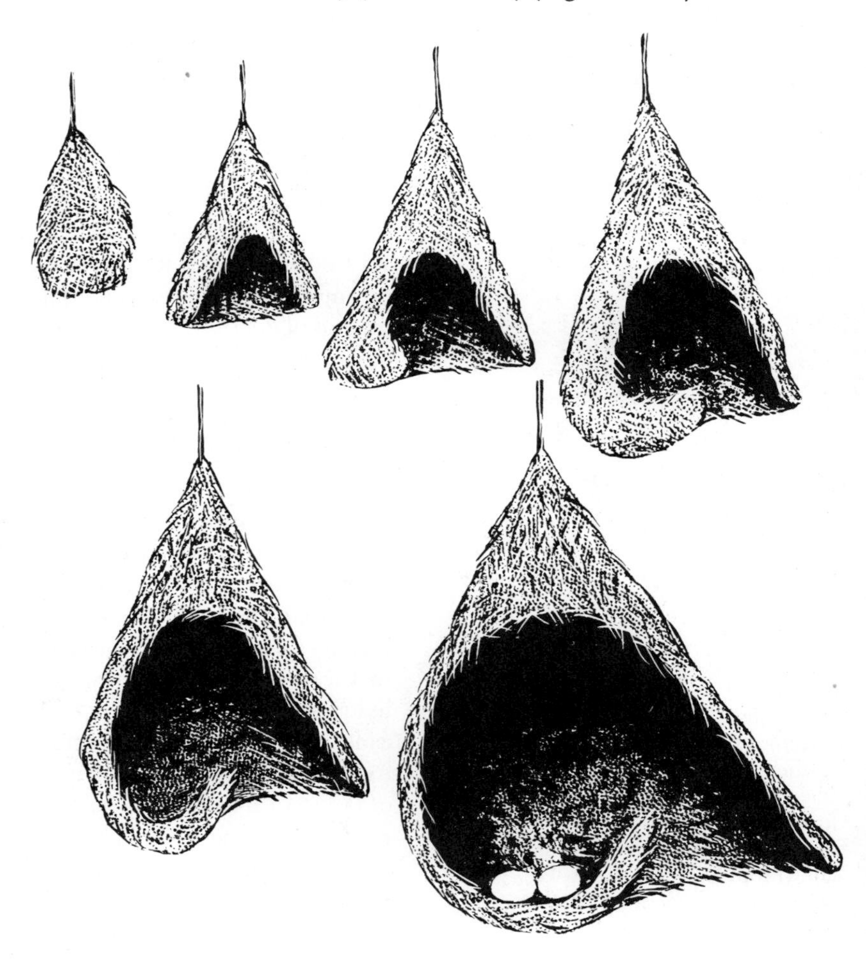

Figure 18–3 Stages of nest building in Sulphur-rumped Flycatchers. (From Skutch 1960, 1976)

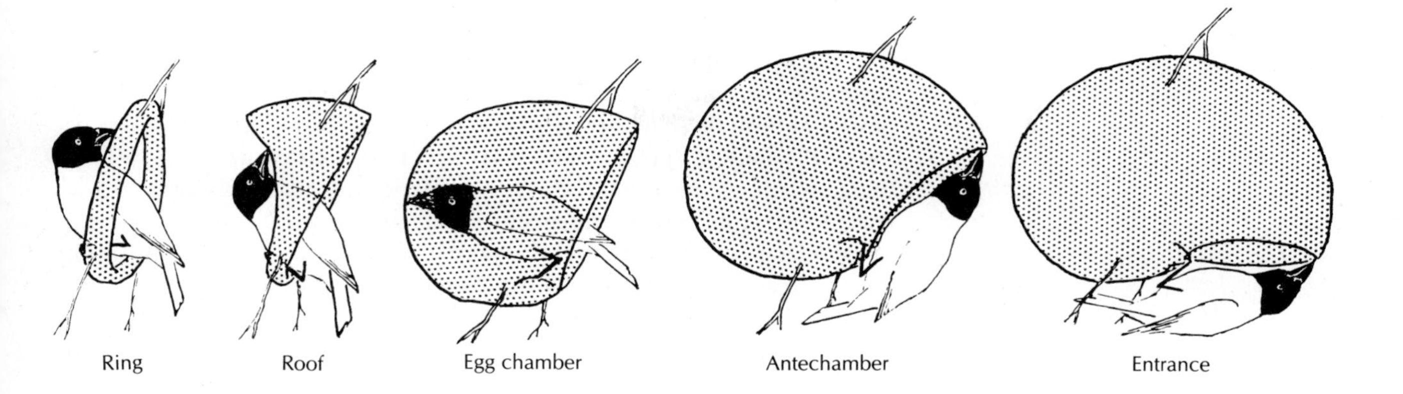

Figure 18–4 Stages of nest construction by the male Village Weaver. (Adapted from Collias and Collias 1964)

The hanging nests of orioles and weavers are more elaborately woven. The male Village Weaver, for example, begins with a vertical ring, to which it adds in succession a roof, the walls of the main nest chamber, an antechamber, and finally the finished entrance (Figure 18–4). The structural features of these nests are woven into their final positions using special knots. The types of knots used are species-specific. Some weavers tie simple knots; others tie knots such as half hitches and slip knots (Figure 18–5).

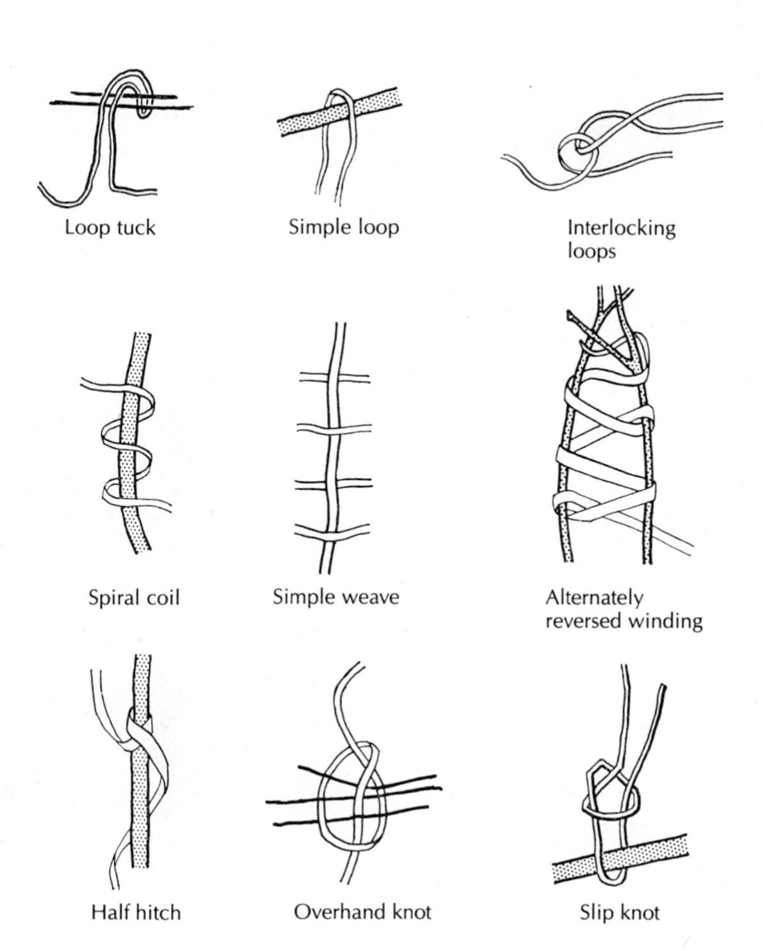

Figure 18–5 Some knots and stitches used by weavers in the construction of their nests. (From Collias and Collias 1964)

Nest structure

Birds build nests to protect themselves, their eggs, and their young from predators and from adverse weather. Other animal species also build nests, but birds do so in a greater variety of forms, from a greater variety of materials, and on a greater variety of sites. Bird nests range from the bare branches on which the White Tern precariously places its eggs to the enormous communal apartments of Social Weavers, and from the simple ground scrapes of auks and nighthawks to the elaborate stick castles of South American ovenbirds (Figure 18–6). Nests may be constructed from ready-for-use pebbles and sticks or laboriously woven from natural fibers. The most elaborate nests are constructed by passerine birds. Bird nests range in size from the few sticks assembled by some doves to the gargantuan eyries of eagles. One of the latter, that of a Bald Eagle, weighed over two tons when it finally fell in a storm after 30 years of annual use, repairs, and additions (Herrick 1932).

(A)

(B)

(C)

(D)

(E)

Figure 18–6 The nests of birds vary from simple to elaborate, large to small. Ground nests: (A) floating platform nest of Western Grebe; (B) sand scrape nest of Wilson's Plover; (C and D) down-lined, camouflaged nest of Cinnamon Teal; (E) mud nest of Rufous Hornero; (F) mud nests of Cliff Swallows; (G) hole nest (in cactus) of Gila Woodpecker; (H) straw nest of Cactus Wren; (I) stick nests of Great Blue Herons; (J) stick nest of Rufous-fronted Thornbird; (K) cup nest of Broad-tailed Hummingbird; (L) suspended cup nest of Warbling Vireo; (M) suspended nests of Crested Oropendola; (N) intricately woven nest of Cassin's Malimbe. (Courtesy A. Cruickshank/VIREO, A–C, F–H, K, L; O. Pettingill/VIREO, E, M; T. Fitzharris/VIREO, I; P. Alden/VIREO, J; E. and N. Collias, N)

(F) (G) (H)

(I) (J) (K)

(L) (M) (N)

Pensile nests represent the apex of nest architecture. These baglike nests hang by silky cobwebs or, in some cases, by wiry, black fungus fibers. Some are suspended far below a branch. Others such as those of orioles are hung from the thin, outermost branches of large trees. The integrity of pensile nests derives from their tightly woven construction, tough knots, and strong binding materials. The intricately woven, meter-long nest of the Cassin's Malimbe, a West African weaver, may well be the epitome of avian nest construction (Collias and Collias 1984).

Although many birds nest colonially, only a few actually build compound, communal nests divided into individual compartments. The Palmchat of Hispaniola and the Monk Parakeet of Argentina are in this group. Instead of nesting in excavated cavities or burrows like most parrots, the Monk Parakeet builds huge stick nests that are occupied by up to 15 pairs and sometimes also by Speckled Teal and Spot-winged Falconets (Martella and Bucher 1984). The nest of the Social Weaver of southwest Africa is certainly one of the largest and most spectacular of all avian nests; it resembles a large haystack in a thorny tree. A common roof covers 100 or more separate nest chambers. The pairs that will occupy the structure share in building the roof. The geographical distribution of this species is limited to the extremely arid sections of southwest Africa, probably because any rain would waterlog the nest and create an unsupportable weight.

Nest function

Nests have four primary functions:

1. The foremost function is safety from predators.

2. Nests provide a microclimate suitable for incubation; this microclimate must meet the needs both of the eggs and of the adult that must sit on them for long periods of time.

3. Nests may serve as a cradle for dependent young until they fledge.

4. Nests serve as roosting chambers for adults while they attend their eggs or young.

Structure and function are inseparable features of nest architecture. Nest structures include conspicuous features for protection from predators as well as for insulation from adverse climates, and more subtle features for regulation of energy-efficient microclimates.

Safety from predators

Nests and their contents, including the incubating parents, are vulnerable to predation (see Ricklefs 1969b for detailed analysis). Data gathered on the fates of nests reveal that predation is the primary cause of nest failure in birds generally. Long-term studies show that predation accounted for 88 percent of nest losses in deciduous scrub vegetation in Indiana (Nolan 1963), 75 percent of nest losses in Britain (Lack 1954), and 86 percent of nest losses in the tropical White-bearded Manakin (Snow 1962). In a two-year study, Field Sparrows sustained a 90 percent loss of nests with eggs, mostly to snakes (Best 1978).

Safety depends on the selection of nest sites as well as the design of the nest itself. Invisibility, inaccessibility, and impregnability all contribute to safety (Skutch 1976). The camouflage of incubating nightjars and of shorebird eggs renders nest sites nearly invisible, as do the lichen decorations on the sides of a hummingbird's nest. Cryptic sites in dense clumps of grass, vine tangles, or hidden crevices also minimize the chance of discovery. Seabirds that nest on sheer cliffs and swifts that nest in deep caves or behind waterfalls achieve safety through inaccessibility. Horned Coots pile up stones in the middle of high Andean lakes to build their own nesting islands out of reach of terrestrial predators (Figure 18–7).

Nests on the ground are more vulnerable to mammalian predators than nests in trees or bushes. In Oklahoma, for example, 71 percent of 130 Mourning Dove nests placed on the ground were destroyed, compared with only 51 percent of 167 dove nests in trees in the same area (Downing 1959). Although Ospreys and American Robins usually nest in trees, on Gardiner's Island, New York, where there are no predators, these species often nest on the ground. Conversely, Tooth-billed Pigeons once nested on the ground on Samoa but shifted to tree nesting after cats were introduced to this South Pacific island by whalers (Austin and Singer 1985). Many birds build globular, enclosed, domed, or pensile nests to discourage predators. Nonaerial predators such as snakes cannot easily reach pensile nests. Protruding entrance tubes thwart attempts by nonaerial predators, especially snakes, to

(A)

(B)

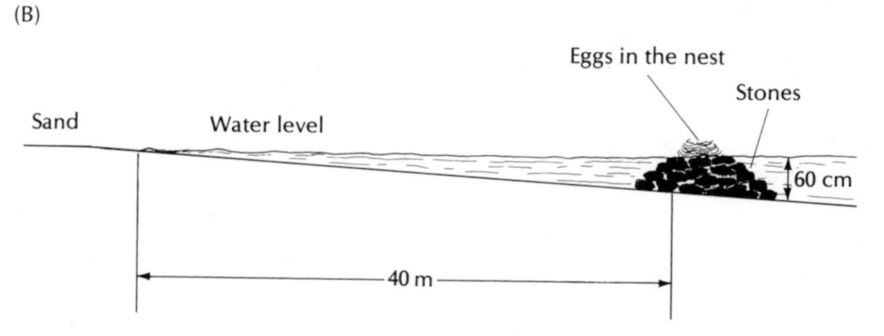

Figure 18–7 Horned Coots build their nests on foundations of stones, which they assemble in high-altitude Andean lakes. (A) Horned Coot on nest, (B) diagram of structure of nest structure. (A, courtesy P. Canevari/VIREO; B, after Ripley 1957)

crawl inside. Tropical passerines build domed nests, often with an entrance tube on the sides. Eggs in a covered nest are less visible to potential predators when the parent is absent than are eggs in a nest without a roof.

Cavity nesting is safer than open nesting, yielding a 66 percent chance of fledging young compared with the 50 percent chance yielded by open nests (Nice 1957). Half of the avian orders, among them all parrots, trogons, and kingfishers and their relatives, nest in cavities or holes. Owls, parrots, and Australian frogmouths nest in natural cavities; trogons, titmice, and piculets excavate cavities in the soft or rotten wood of old trees. Kingfishers, ovenbird miners, and penguins excavate burrows in the ground. The most developed evolutionary stage of cavity nesting is found in woodpeckers, which can chisel cavities into living hardwoods. Abandoned woodpecker holes are in great demand among other birds unable to make their own holes. The intrinsic values of a good nesting cavity promote intense competition. The availability of nest holes limits the population sizes of some species (see Chapter 22). European Starlings are notorious for their aggressive displacement of Northern Flickers and bluebirds in the New World and of other species in Europe. The Eurasian Nuthatch responds to the threat of takeover by European Starlings by making the entrance hole to its nesting cavity too small for the starlings to enter.

Some birds increase their safety by nesting near protective insects or animals. Reddish Hermit hummingbirds in South America and Barn Swallows in Mississippi often place their nests next to those of wasps, which seem to be undisturbed by the birds' presence. Violaceous Trogons excavate a nest cavity inside the large paper nest of a tropical wasp. A variety of tropical birds nest in acacias with stinging commensal ants or wasp nests. Fifty-eight of 61 nests of Dull-colored Seedeaters in Argentina were sited next to wasps (Contino 1968), and in Australia the Black-throated Gerygone, a warblerlike bird, is called the "hornet-nest bird" because of its association with bees, wasps, and other Hymenoptera (Collias and Collias 1984).

European Starlings and House Sparrows nest on the fringes of Imperial Eagle eyries. The Water Dikkop, a ploverlike bird of Africa, nests on sandy shores beside brooding crocodiles. For protection from the Arctic Fox, Snow Geese, Brant, and Common Eiders often nest near Snowy Owls, which can easily discourage the approach of this mammalian predator. Trees near human habitations, where snakes will not be tolerated, are prime nesting sites for tropical birds. At least 49 species, including 25 percent of all kingfishers, excavate nest cavities inside the mounds of social termites, which then seal off the nest from the rest of the mound (Hindwood 1959). The Orange-fronted Parakeet nests exclusively in the termitary of one species, *Eutermis nigriceps.* The geographical distribution of this parakeet is restricted to that of its termite host.

Birds rely on cryptic coloration and distracting and surreptitious behavior to thwart predators. Many incubating parents avoid discovery by not flushing until the last possible moment, relying on their own camouflage and that of the nest for protection. Female Common Eiders, for example, are camouflaged from Arctic Foxes and weasels by their finely patterned plumage. If discovered, however, they defecate a noxious repellent fluid over their eggs as they flush (Swennen 1968). Many birds guard their nests more directly by boldly attacking trespassers. Eastern Kingbirds

(A)

(B)

Figure 18–8 Distraction displays: (A) by feigning injury, a Common Nighthawk distracts predators and thus protects its young. (B) the rodent-run display in the Tasmanian Native Hen in contrast to its normal upright walking posture. (A, courtesy S.A. Grimes; B, from Ridpath 1972)

chase anything that violates nearby airspace. Northern Mockingbirds, Blue Jays, and Arctic Terns can draw blood and bits of fur from cats that come too close to their nests or young. They may attack people as well. Great Horned Owls and various large eagles with powerful feet and sharp talons can seriously wound approaching climbers.

A parent flushed from the nest may try to distract a predator's attention away from the nest site with elaborate distraction displays. The two most common are "injury flight" and "rodent run" (Simmons 1952, 1955; Brown 1962). Using the injury-flight display, feigning a broken wing and calling in great alarm, an adult sandpiper can easily draw a fox away from its nest. The sandpiper may then switch to the rodent-run display to keep the fox's attention, running in a low crouch, which appeals to the mouse-catching instincts of the fox. Distraction displays are risky, but more often than not the parent escapes and the predator loses track of the original nest location. Variations on these displays are virtually universal (Skutch 1976) (Figure 18–8).

To avoid attracting attention when returning to the nest after a recess, parents adopt surreptitious behavior. Meadowlarks land some distance from the nest and sneak back to it through the grass, using one of several indirect routes. Bearded Tits pretend to look for food as they get near their nests and then enter rapidly if they perceive that the coast is clear. The female Long-tailed Hermit, a tropical hummingbird, behaves similarly. Upon returning from foraging, she searches intensively for spiders on the buttresses of large trees before suddenly flipping onto her nest and sitting very still (Gill personal observation).

The selection of safe nest sites broadly molds the behavior and morphology of a species (Cullen 1957; Haartman 1957). Most gulls, for example, nest on the ground, where they are vulnerable to predation by birds, such as other gulls and crows, and sometimes by mammals. Black-legged Kittiwakes, however, cling to narrow, predator-free nesting ledges on windswept sea cliffs, using their strong claws and toe muscles (Figure 18–9). The absence of nest predation in this species has fostered the loss of

Figure 18–9 Unlike most other gulls, Black-legged Kittiwakes nest on cliffs. (Courtesy E. Hosking)

antipredator behaviors, including alarm calls, predator mobbing, and removal of the eggshells from the nest site. Young kittiwakes, a conspicuous silvery white in color, stand still and hide their beaks when frightened, rather than run and hide as do the cryptically colored chicks of ground-nesting gulls.

Likewise, the physical restrictions of a kittiwake's narrow nesting ledge have favored aggressive and courtship displays that differ from those of other gulls. Instead of using the "long call," for instance, kittiwakes announce territorial ownership with a modest choking display (see Chapter 9). Aggression between males is expressed by bill jabs and grabs from a fixed position; it does not extend to the flamboyant charges of other species. Females commonly hide their beaks to minimize attack and physical displacement. Courting males do not regurgitate food onto the ground in front of the female (there is no place to do so) but give it to her directly.

Nest microclimate

The microclimate of a nest is crucial to the successful incubation of the eggs and the health of baby birds. Nest microclimate also influences the adult's daily energy requirements, as well as the amount of time parents must spend on the nest incubating eggs and brooding young. Nest warmth, like that of a house, is determined by the thickness of its insulation (Skowron and Kern 1980). In addition to keeping the eggs warm, nest insulation reduces the energy requirements of the incubating parent (Walsberg and King 1978). The outstanding insulating properties of breast down used by eiders and other waterfowl greatly reduce the cooling rate of eggs not covered by the parent. Mallard eggs, for example, cool twice as fast uncovered as when they are buried in down. Experimental reduction of the lining thickness of a Village Weaver nest results in a corresponding increase in the amount of time the female stays on the nest to keep the eggs warm and, thereby, a reduction in the time she can feed (White and Kinney 1974). The weavers adjust the amount of nest insulation in relation to need, increasing it with the altitude of their nests (Collias and Collias 1971). Seasonal divergences in the thermal conductance of songbird nests result from variations in the tightness of the weave (Skowron and Kern 1980).

Nest placement in or out of the sun, shade, or wind has a major effect on the nest microclimate and, therefore, on a pair's breeding success. Early in the season in Arizona, Verdins and Cactus Wrens build nests where they are protected from cool winds and bathed in the warm morning sun. Later in the season, when it is hot, they build well-shaded nests, exposed to cooling breezes (Ricklefs and Hainsworth 1969; Austin 1976). The enormous communal nests of the Social Weaver have great thermal inertia, which keeps them cool in the daytime and warm at night. Temperatures inside the nest at night remain 18° to 23°C above external temperatures, owing in part to heat absorbed during the day and in part to heat generated from the bodies of large numbers of roosting birds. F.N. White and his colleagues (1975) estimated that communal roosting of Social Weavers reduces adult metabolism by 43 percent.

Thick nest insulation and the careful selection of sites reduce heat loss by incubating hummingbirds. Their energy requirements can drop 13 percent as a result of a minor 0.05-centimeter increase in nest thickness (Smith et al. 1974). Heat loss from the exposed upper surface of the hummingbird's body can be substantial, especially at night. Consequently, hummingbirds choose nest sites beneath branches, which reduces heat loss to open air by half (Calder 1974; Southwick and Gates 1975). Careful selection of nest sites to reduce heat loss is especially important in species such as the tiny Calliope Hummingbird, which nests at high elevations in the Rocky Mountains, where night temperatures are cool (Calder 1971).

Cavity nests and burrow nests also help conserve energy. Like the haystack nests of Social Weavers, cavity nests and burrow nests buffer eggs, parents, and young against fluctuations in external temperatures. The temperatures inside the burrows of Cape Penguins, for example, stayed between 17° and 20°C despite an outside temperature range of 12° to 36°C; the burrow nests of European Bee-eaters remained at a mean of 25°C despite an outside range of 13° to 51°C (White et al. 1978). The adobe walls of Rufous Hornero nests function like the thick walls of human habitations that are constructed of the same materials. Deep, cool burrow nests have their drawbacks, however. Poor ventilation limits the amount of time parents can spend inside with growing young. In the case of the tunnel-nesting European Bee-eater, diffusion of gases through the soil and the nest tunnel usually keeps the air in the nest chamber adequately ventilated. Convection at the tunnel entrance aids air exchange, as do the movements of adults in and out of the nest, which affect the inside air much as a moving piston would. On windless days, however, the ammonia and carbon dioxide tend to build up as a result of decay of excreta amidst unsanitary nest conditions, and oxygen levels occasionally decline until the occupants have difficulty breathing.

Incubation

The narrow temperature range that the embryo can tolerate commits parents to rigorous incubation patterns because incubation transfers the heat necessary for embryonic development to eggs in the nest. Highly developed incubation behavior is a distinctive avian characteristic; among reptiles, only crocodiles and Indian Rock Pythons incubate their eggs, and they do so in primitive ways. No birds incubate their eggs internally as do ovoviviparous animals that bear live young. Most birds delay incubation until their clutch is complete. This insures that the embryos begin development and hatch at the same time even though some eggs are laid earlier than others. Pigeons and doves sit on their first egg before the second is laid but do not bring it up to the required temperatures for incubation. Owls and raptors, on the other hand, begin incubation before the clutch is complete, with the result that young hatch at different times and the older offspring are conspicuously larger than their siblings.

Brood patches

Birds transfer body heat to their eggs through the brood or incubation patch, which is a bare, flaccid section of skin on the abdomen or breast. This area may be a single median patch, as in most passerines and pigeons, or

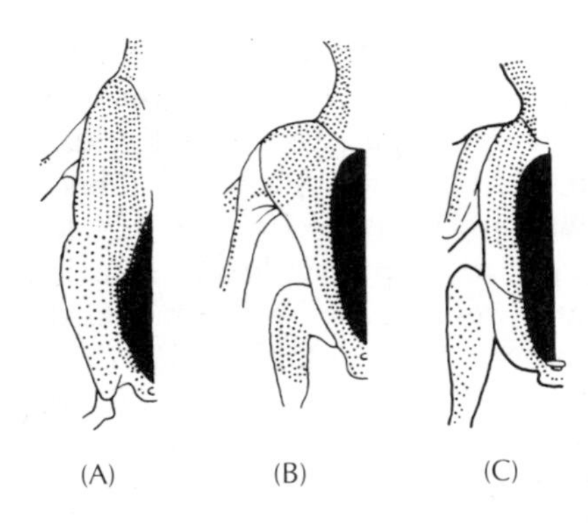

Figure 18–10 Incubation patches (in black) of (A) a grebe, (B) a hawk, and (C) a passerine bird. Stippling indicates feather tracts. Clear areas indicate areas without feathers, called apteria. (Adapted from Drent 1975)

a two-sided lateral patch, as in most shorebirds, gulls, and quail. Incubation patches develop just prior to the incubation period and then regress. If both parents incubate, the patches occur in both sexes; otherwise, individuals of the nonincubating sex usually have the potential for brood-patch development in the event that they should have to incubate for some unusual reason such as a mate's death.

The four anatomical features of incubation patches are as follows (Figure 18–10):

1. The skin area is devoid of feathers. Pigeons and doves use a normally bare apterium, or featherless region (page 66), but most birds lose the feathers from the incubation patch for the purpose of brooding.

2. Addition of fluids (edema) and the infiltration of white blood cells makes the skin flaccid and wrinkled, which allows better contact between the surfaces of the incubation patch and the egg.

3. The epidermis thickens into a callused surface that is not damaged by sustained contact or friction with the eggs.

4. Blood vessels, which deliver body heat to the eggs, proliferate throughout the patch. The arterioles of the network of blood vessels have a well-developed musculature that directs the flow of warm blood to the skin surface during incubation and stops it when the parent is not actively incubating.

Seasonal development of incubation patches is under direct hormonal control (Bailey 1952). Prolactin or estrogen, or both, depending on the species, stimulate defeathering and vascularization of the incubation patch. Progesterone stimulates thickening and increased sensitivity of the epidermis. Most birds develop brood patches in response to experimental hormone treatment. However, Brown-headed Cowbirds, brood parasites that never incubate (see Chapter 20), are insensitive to hormonal stimulation of brood patches (Selander and Kuich 1963). Although incubation patches are the typical mode of heat transfer among birds, some birds (waterfowl, penguins, and pelecaniform birds) lack them. Waterfowl line their nests with down that they pluck from their breasts, placing their eggs between the layer of down insulation and their bare skin. Gannets and boobies, pelecaniform birds that also lack a brood patch, incubate with their feet. They may grasp a single egg in their well-vascularized, webbed feet or even hold two eggs, one in each foot. Murres and penguins incubate their eggs on the top surface of their feet. Some penguins have a muscular pouch of belly skin that holds the egg in this position.

Keeping eggs warm

The first priority of incubation is to keep the eggs close to the optimum temperature for development: 37° to 38°C. Serious problems result if the embryo is exposed to temperatures outside the range of 35° to 40.5°C. Exposure to higher temperatures is lethal, and even a short exposure to cool temperatures between 26° and 35°C can disrupt normal development. Below 26°C the development of young embryos simply stops, potentially for a long time. Thus, frequent or continuous warming is necessary unless air temperatures are hot. Internal egg temperatures are low at first, but they increase steadily during incubation as a result of increased parental incubation, further incubation patch development, and heat generated internally by the metabolism of the embryo. The temperature increase has been demonstrated by regularly measuring the temperature of naturally incubated eggs of the Herring Gull (Drent 1975) (Figure 18–11).

Incubating parents are able to keep the internal temperatures of their eggs remarkably stable, despite the fact that incubation behavior itself is

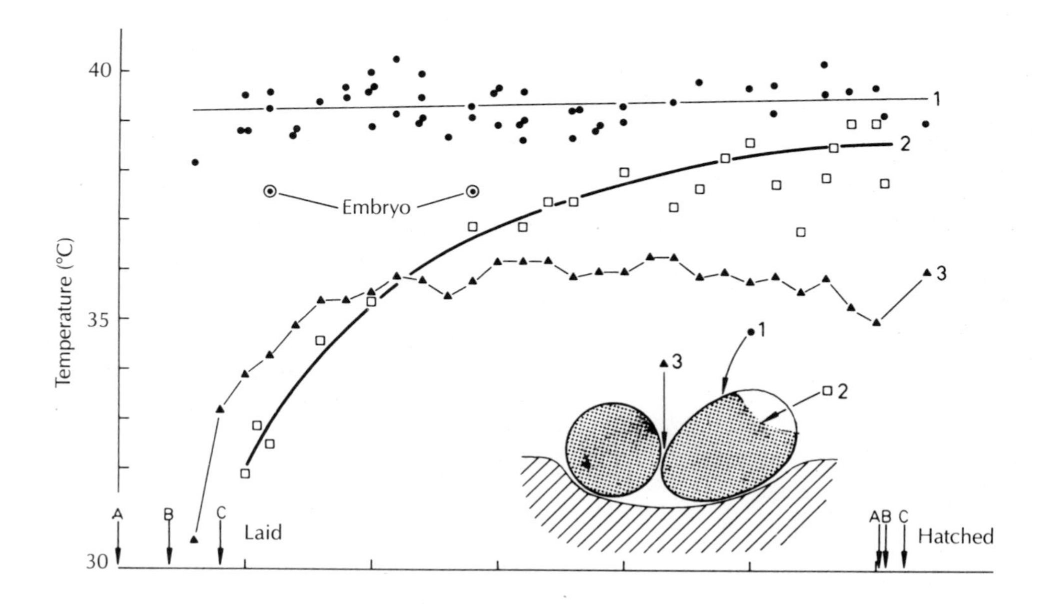

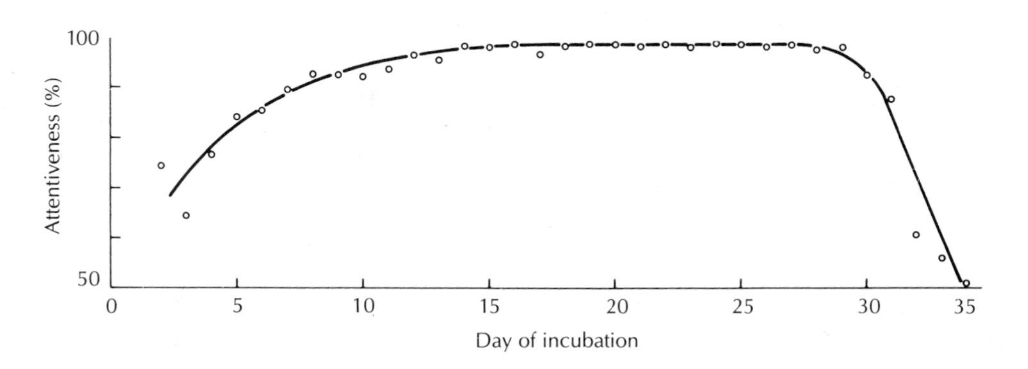

Figure 18–11 Nest and egg temperatures (top) during natural incubation by a Herring Gull. Sites of measurement included (1) the egg surface, (2) inside the egg, and (3) between eggs A, B, and C. Points labeled "embryo" indicate measurements taken beside embryo on days 6 and 14. The constancy of incubation (attentiveness) of adults increased steadily during the first two weeks of incubation (bottom). (Adapted from Drent 1975)

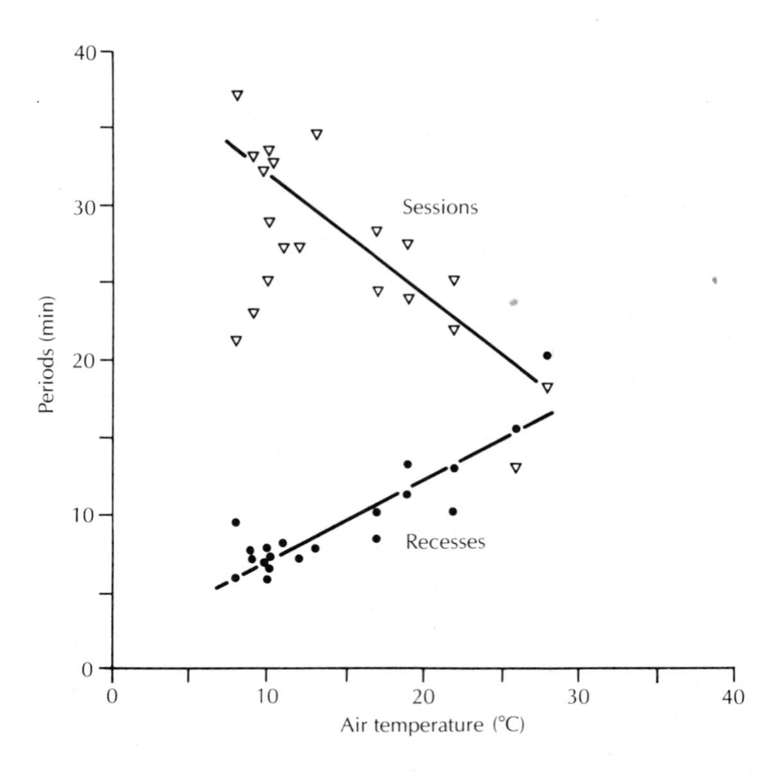

Figure 18–12 Incubation rhythms of the Great Tit are directly related to the air temperature in the nest box. Time on the eggs (sessions) increases when the air is cooler, and time off the eggs (recesses) decreases. (Adapted from Drent 1972 and Kluijver 1950)

comprised of many conflicting options. Central to the question of egg temperature regulation is the pattern of attentiveness, or incubation sessions versus incubation recesses (Kendeigh 1952; Haartman 1956; Drent 1975; Haftorn 1978a,b). The natural incubation rhythm of a species (Figure 18–12) can be measured by activity recorders placed at the entrance to the nest box (Kluijver 1950; Drent 1975). The incubation rhythms of the parents are geared directly to the maintenance of critical egg temperatures. At cooler air temperatures, the lengths of sessions on the eggs increase and the lengths of recesses decrease. Experimental increases in the air temperature inside the nest boxes of Pied Flycatchers cause the adults to shorten their sessions on the eggs (Haartman 1956). Similarly, raising or lowering the temperature of artificial eggs placed in the nests of Ringed Turtle-Doves and Savannah Sparrows decreases or increases the length of the incubation sessions, respectively (Franks 1967; Davis et al. 1984). The rate at which exposed eggs cool is faster than the rate at which they warm up under an incubating bird. More incubation is required after several short recesses than after one long recess. Herring Gull eggs, exposed for ten 6-minute periods, require a total warm-up time that is almost four times longer than that of eggs exposed for one 60-minute period (Drent 1975).

Experiments with Crested Mynas and European Starlings on Vancouver Island, British Columbia, demonstrate the effects of inadequate

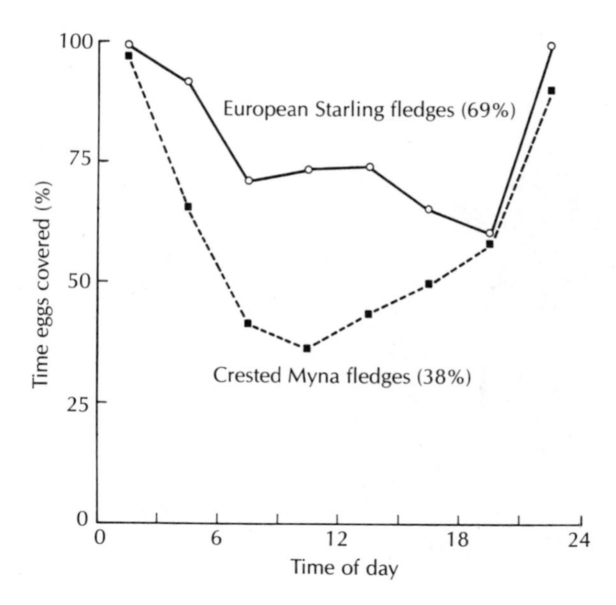

Figure 18–13 On Vancouver, introduced European Starlings are more attentive during incubation than introduced Crested Mynas; consequently, the starlings achieve greater fledging success and have expanded their range throughout North America whereas the mynas remain restricted to Vancouver. (Adapted from Drent 1972; Johnson 1971)

incubation behavior (Figure 18–13). Crested Mynas introduced to the area hatch and fledge young from only 38 percent of their eggs because they persist with an incubation rhythm that is suitable for the tropical climates of their native Hong Kong but unsuitable for the cool Vancouver climate (Johnson 1971). Unlike Great Tits, they do not regulate incubation time by air temperature. However, they hatch and fledge more young when their nest boxes are heated artificially. A low rate of reproduction is part of the reason Crested Mynas remain restricted to the vicinity of Vancouver. The related European Starling, in contrast, expanded its range rapidly after its introduction to North America. This starling is more attentive during incubation and therefore fledges 68 percent more young. Myna eggs that are incubated by starlings usually hatch, thus showing that the normal incubation of mynas, not the quality of their eggs, is at fault.

Moundbuilders use natural heat

Megapodes, or moundbuilders, are fowl-like birds of the Australasian region that use heat from the sun, volcanic steam, or decomposing vegetation to incubate their eggs. Reptilian as it may seem, this behavior appears to have evolved secondarily from normal avian incubation behavior. Some species, most notably the Mallee Fowl, regulate the temperatures of their nest heap with great sensitivity (Frith 1962). The Mallee Fowl hen lays her eggs in a large mound, up to 11 meters in diameter and 5 meters high, made of decaying vegetation and sand, not unlike a compost heap.

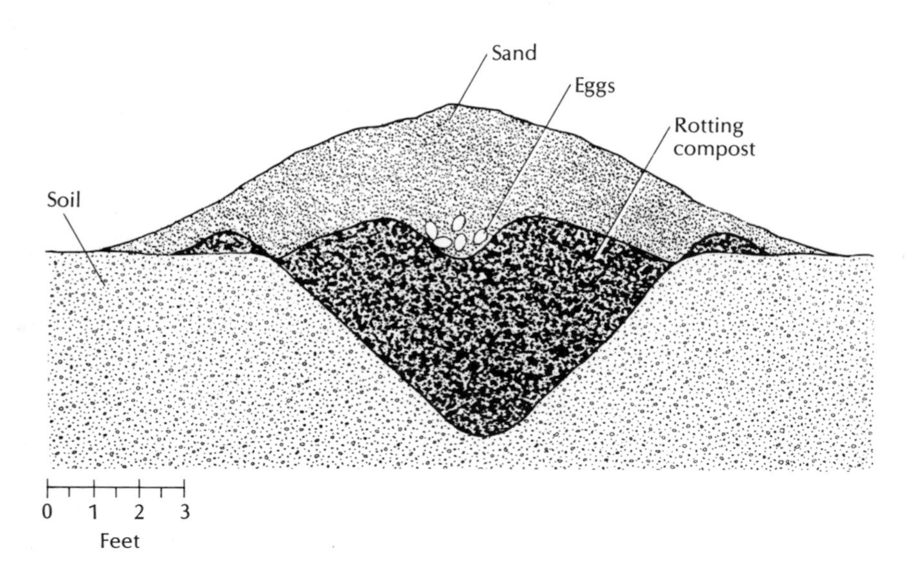

Figure 18–14 (Left) Moundbuilders do not incubate their eggs directly but regulate the temperatures inside their nests by varying the rate of natural heat loss and gain. (Right) Cross section of incubation mound with eggs. Underneath the egg chamber is a pit full of decaying vegetation. Sandy soil covers the eggs. (Photo courtesy J. Warham; drawing after Frith 1962, with permission of Scientific American)

The male, which tends the mound alone for 10 to 11 months a year, spends an average of five hours a day digging to manipulate the amount of material covering the eggs.

Incubation temperatures inside the mound remain at 32° to 35°C as external air temperatures range from 0° to 38°C. The male regularly checks the temperature inside the mound by testing the soil in his mouth. In the spring and summer, the problem facing an incubating male is one of cooling the mound, which he does by opening it to release accumulated heat and by replacing hot sand with cooler sand. Summer nests are made of deep pits, which protect the eggs from the hot sun. In the autumn the male faces the problem of heating the eggs because there is less sun and less decay. The pit for the eggs in autumn, therefore, is a shallow one that takes advantage of daytime solar heating. At night, the male adds extra insulation to seal in the heat (Figure 18–14).

Keeping eggs cool

Birds that nest in hot places face demanding incubation tasks of a different nature. The Two-banded Courser, which nests in the Kalahari Desert, shades its eggs from the desert sun when air temperatures exceed 36°C. Gray Gulls that nest in the deserts of northern Chile incubate their eggs at night when it is quite cold but shade them during the day when air temperatures are 38° to 39°C (Howell et al. 1974). The temperature in unprotected eggs of Sooty Terns on tropical islets rises to a lethal 44°C in only 15 minutes (Shea unpublished data).

Wetting the nest or eggs, which counteracts extreme heat with evaporative cooling, is a common practice among shorebirds, gulls, and terns (Grant 1982). Killdeer, for example, cool their eggs from a potentially lethal air temperature of 44° to 39°C by transferring water from their wetted belly feathers (Schardien and Jackson 1979). The Egyptian Plover, which nests on hot sandbars of the Nile River, cools its eggs by covering them with a thin layer of sand and then sprinkling water on top. The nest temperature stays near 37.5°C as a result (Howell 1979).

Heat and water problems impose stress on the parent trying to cool its eggs directly in a hot environment. To protect eggs from the hot sun, the incubating parent must absorb and dissipate enormous amounts of radiant energy without overheating itself. Sooty Terns dissipate heat by extending their legs fully, erecting their feathers, and panting (see Figure 6–6). The heat that its black back absorbs is removed by the breeze. The more sunlight that incubating Herring Gulls absorb, the more they must pant (Drent 1970). The stress on an individual's water balance is so great and the consequences of even temporary absences so severe that mates must take turns to provide continuous egg coverage.

Conservationists and sightseers should take the danger of exposure into consideration: The unwitting disturbance of seabird nesting colonies destroys embryos because disturbed parents leave their nests and expose their eggs to the sun.

Turning eggs

Observers of incubating birds see them rise periodically to peer sharply down at their eggs and then draw each egg backward with a sweeping motion of the bill, rearranging the clutch and turning the eggs. Parents rearrange the eggs so that those that have been on the outside of the clutch become more centrally situated, where the temperature is several degrees higher (a fact that is incorporated into the design of artificial incubators). Regular turning of eggs during early incubation prevents premature adhesion of the chorioallantois (page 336) to the inner shell membranes. Premature adhesion interferes with albumen uptake by the embryo and obstructs its ability to attain the tucking position essential for hatching.

Costs of incubation

Although incubation at normal temperatures consumes 16 to 25 percent of a bird's daily productive energy, incubating adults actually save energy. Studies of Red-winged Blackbirds and Willow Flycatchers demonstrate that the nest insulation and the favorable microclimate of the nest site cause the rate of heat loss to the environment to be less than the rate of heat loss away from the nest, even though the heat loss by an incubating bird is calculated to include the extra cost of heating the eggs (Walsberg and King 1978) (Figure 18–15). Because their foraging time is limited, incubating birds sometimes fast or depend on their mates for supplementary food. Female Snow Geese, for example, subsist on the reserves remaining after egg production. Inadequate reserves at this stage in the nesting cycle

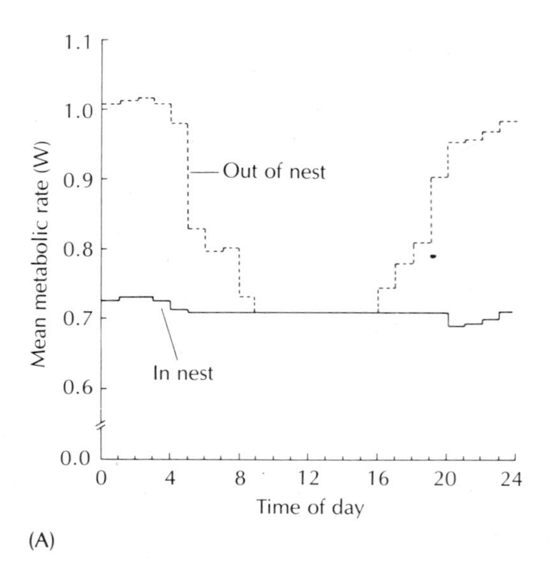

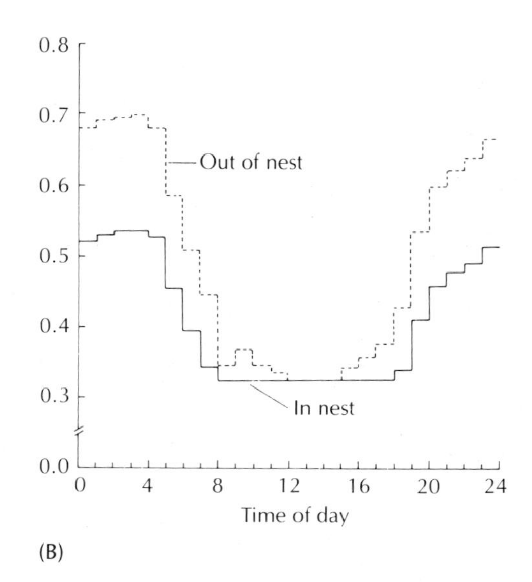

Figure 18–15 The energy expenditures of (A) a Red-winged Blackbird and (B) a Willow Flycatcher while perching near the nest are higher than those during incubation. (From Walsberg and King 1978)

cause some females to desert their eggs during incubation; some actually die of starvation (Ankney and MacInnes 1978). At the other extreme, male hornbills provide all their mates' food. Female Red Crossbills also receive all their food from their mate, which enables continuous incubation in the middle of winter. Female Yellow-billed Magpies, which forage 40 percent of the day while laying, limit themselves to foraging only 1.7 percent of the day while incubating. To compensate, the male's foraging time doubles to 78 percent of the day (Verbeek 1972).

Males feed their mates during the final days of incubation, seemingly more of an anticipation of their inherent need to feed hatchlings than a contribution to their mates' welfare (Verner and Willson 1969; Skutch 1976). Supplementary feeding at this time could be essential to mates that have exhausted their reserves during incubation or that need to build their own reserves for the demanding efforts that follow hatching. Males may call their mates off the nest to feed them or to escort them during a foraging recess. This habit is common among tyrant flycatchers, tanagers, vireos, wood warblers, and finches. An unexplored possibility is that males lead their mates to good feeding sites, especially if their food sources are unpredictable or local in distribution. Some males that do not accompany their mates on foraging recesses stay near the nest, apparently to guard it.

As mentioned earlier, birds abandon their eggs during incubation when foraging becomes difficult and their reserves dwindle to critical levels. When forced to choose between self-maintenance and caring for progeny, birds take care of themselves first. Desertion of eggs is fairly common, but in only a few cases have ornithologists correlated it with food availability. Eurasian Kestrels and Tawny (Wood) Owls readily desert their eggs when hunting becomes difficult (Cavé 1968; Southern 1970). Red-footed Boobies, Laysan Albatrosses, and Wood Pigeons desert their eggs if their mate fails to return on schedule to relieve them (Nelson 1969; Fisher 1971; Murton 1965).

In 54 percent of avian families, both sexes incubate (Van Tyne and Berger 1976). The female incubates alone in 25 percent of the remaining families, the male incubates alone in only 6 percent. The remaining 15 percent of avian families are not so simply characterized (Skutch 1976). Regular alternation of incubation shifts accomplishes nearly continuous coverage of the eggs in many groups, including penguins, woodpeckers, doves, trogons, hornbills, hoopoes, and antbirds. Alternating incubation shifts by mates may last for 1 or 2 hours; 12 hours when one sex incubates by day the other by night; 24 hours when each sex takes a day at a time; or several days, as in many pelagic seabirds. When changing the guard, Pied-billed Grebes touch bill tips lightly. Least Bitterns erect their crown feathers and rattle their bills. Some herons present a stick for the nest to their mates, and terns may offer a freshly caught fish, as the males do during courtship. Most small landbirds lack conspicuous relief ceremonies; they slip on and off the nest surreptiously to avoid detection by predators. In contrast, penguins have elaborate changeover rituals that facilitate individual recognition and reinforce the pair bond. The following observation is an example of one such ritual:

> As a Yellow-eyed Penguin approached his incubating partner, she broke into an "open yell." He ran up with arched back and beak to the ground. Then both put their heads together to perform a hearty welcome ceremony, in which a great volume of sound issued from their widely opened mouths as they faced each other, standing erect close together. After several less-intense displays of mutual affection and three repetitions of "welcome," the female resumed her position on the eggs, then rose to relinquish them to her mate. (Skutch 1976, p. 171, from Richdale 1951)

Incubation periods

The incubation period is the time required by embryos for development in a freshly laid egg that is given regular, normal attention by incubating parents. It is defined as the interval between the laying of the last egg of a clutch and the hatching of that egg (Drent 1975). Incubation periods vary from as little as 10 days in woodpeckers to 80 days in albatrosses and kiwis. Incubation periods are directly related to egg size and indirectly related to adult body weight (Heinroth 1922; Rahn and Ar 1974). Eggs of the same size but of different species may differ greatly in the amount of time they take to hatch. These variations are partially reflected in the baby bird's state of development upon hatching. The short incubation periods of woodpeckers, for example, relate to the undeveloped state of their young at hatching. The incubation periods of precocial birds are longer on the average than those of altricial birds (see Chapter 19). If the eggs fail to hatch on schedule, parent birds prolong the incubation period by 50 to 100 percent (Holcomb 1970).

In some places, such as deserts and at high altitudes, incubation periods may be limited by water loss from the egg. If more than 10 to 20 percent of the egg's initial weight is lost as a result of evaporation, the embryo could die. Short incubation periods are advantageous, therefore, where the rate of water loss is high. Incubation periods also reflect the probability of predation. Species that nest in holes have longer incubation periods, averaging 13.8 days, than species that nest in less safe, open sites, which have periods averaging 12 to 13 days (Skutch 1976).

Water loss is inevitable during incubation because of differences between the water-saturated interior atmosphere of the egg and its unsaturated external environment. Eggs lose from 10 to 23 percent of their weight during incubation, primarily as a result of water vapor loss. Water loss in 81 species of birds averages 15 percent of the original amount of water in the egg (Ar and Rahn 1980; Rahn and Ar 1980; Carey 1983). The net loss of water from the egg has both positive and negative effects. On the positive side, the space vacated inside the egg becomes the air cell at the blunt end of the egg. This is the source of air for a chick's first breath as it starts to break out of the egg. An adequate volume of air must be available for that critical inhalation. On the negative side, excessive water loss may fatally dehydrate the embryo. Hatchability of eggs, therefore, depends on the rate of water loss during incubation.

Different birds apparently achieve quite different rates of water loss by means that remain obscure. For example, Wedge-tailed Shearwater and chicken eggs lose 15 percent of their mass as water vapor despite the fact that the incubation period of the Wedge-tailed Shearwater is twice as long as that of the chicken (Ackerman et al. 1980), so the rate of water loss from the shearwater egg must be half that of the chicken egg. The eggs of seven species of terns vary considerably in their size, shell microstructure, and intrinsic rates of water loss (Rahn et al. 1976). The incubation periods of these terns range from 21 to 36 days, yet the eggs of all seven lose about 14 percent of their mass prior to hatching. The terns apparently regulate the

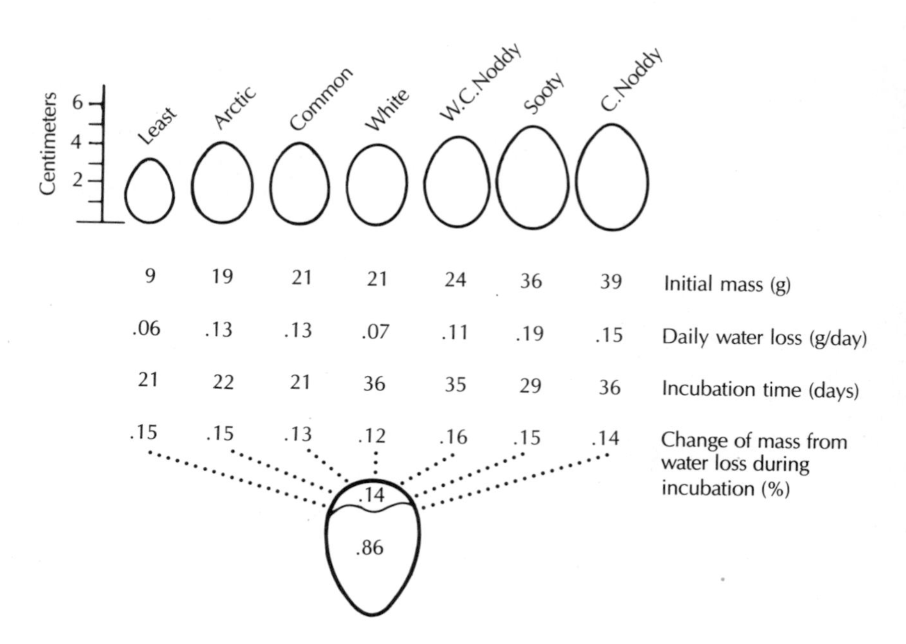

Figure 18–16 Water loss during incubation from tern eggs of various dimensions and masses. The eggs of these seven species differ in dimensions and initial masses and in the number of days they are incubated before hatching. Despite such differences, incubating terns control daily rates of water loss to limit total water loss to only about 14 percent of egg mass, which is represented here as the volume of the air cell. (After Rahn et al. 1976)

ventilation and relative humidity of the nest microclimate in ways that produce equal water loss (Figure 18–16).

Incubation ends when the young hatch, but the parents then enter a period of new efforts to brood and feed their young. The next chapter is devoted to this phase of avian reproduction.

Summary

Reproduction in birds requires the nurturing of eggs and young outside of the body. Nests, which provide a receptacle for eggs during incubation and for baby birds until they fledge, vary in construction from simple accumulations of sticks or scrapes in the earth to major architectural achievements. The woven, pensile nests and huge, apartmentlike, compound nests of certain weaver birds represent the pinnacles of nest construction. Nest materials may include specific plants with pharmacological properties, or feathers, hair, or spider webs. Particular methods of gathering nest materials and of constructing nests characterize each species and may have a genetic basis.

Nests have four primary purposes: protection from predators, provision of a microclimate suitable for egg incubation, cradles for dependent young, and roosting chambers for adults attending their eggs and young. Various forms of camouflage, inaccessible locations, and fortresslike structures reduce the vulnerability of nests to predators. Domed nests and cavity nests have many protective advantages. Birds also nest near stinging insects for protection, and are able to defend their nests or lure would-be predators away with great skill. The choice of safe nest sites may influence the evolution of other aspects of the morphology and behavior of a species.

The nest microclimate is governed in part by the thickness of insulation, which reduces cooling of eggs when the adult leaves the nest to feed. Nest location with respect to sun, shade, prevailing breezes, or sheltering objects has a major effect on incubation behavior. Burrows and cavity nests tend to buffer birds and their eggs from daily temperature cycles. Rigorous incubation schedules keep egg contents within the narrow limits of embryo temperature tolerances. Most birds transfer body heat directly to their eggs through the bare, flaccid, and highly vascularized brood or incubation patch. Moundbuilders, however, incubate their eggs by using the heat generated in heaps of decaying vegetation. In hot desert environments, cooling of eggs by means of midday shading or wetting may be necessary. Incubation not only exposes parents to temperature stress and predators but also reduces the time they have for feeding themselves. Male birds may relieve their mates by sharing incubation or may help provision them with food.

Further readings

Carey, C. 1983. Structure and function of avian eggs. Current Ornithology 1:69–103. *An excellent summary of recent work on the physiology of eggs.*

Collias, N.E., and E.C. Collias. 1984. Nest Building and Bird Behavior. Princeton, N.J.: Princeton University Press. *A comprehensive review of the natural history of avian nests.*

Drent, R.H. 1975. Incubation. Avian Biology 5:333–419. *An excellent summary of bird nests and incubation.*

Drent, R.H., and S. Daan. 1980. The prudent parent: Energetic adjustments in avian breeding. Ardea 68:225–252. *A well-illustrated summary of energetic aspects of parental care.*

Romanoff, A.L., and A.J. Romanoff. 1949. The Avian Egg. New York: John Wiley & Sons. *A classic reference.*

Skutch, A.F. 1976. Parent Birds and Their Young. Austin: University of Texas Press. *A readable account of the natural history of avian reproduction, with chapters on nests and incubation behavior.*

chapter 19

Young birds and their parents

Avian development begins with the initial proliferation of cells and differentiation of tissues that make up the embryo and ends with the learning of the complex behavioral skills of a capable young adult. Baby birds hatch from the egg, leave the nest, and migrate, often to distant places. They learn to feed, to fly, and to sing. They soon distinguish predators from prey and potential mates from potential rivals. In this chapter we follow the life of a baby bird from hatchling to fledgling. Baby birds vary in the way they develop after birth and in the degree to which they are dependent on parental care. The contrast between altricial and precocial modes of development is a central theme of this chapter. Included in these modes of development is the way fledglings leave the nest, a major event in their lives.

Hatching

Hatching is an extraordinary physical challenge. In its final stages of development, the folded and compact chick fills the limited space that was once occupied by yolk and albumen; the chick barely seems to fit inside the tight confines of the shell. The essential change of the prehatching position to that required to break out of the egg is called "tucking." The hatchling-to-be withdraws its head so that its bill passes between the body and its right wing. The tucking position increases the efficiency of pipping, or breaking the eggshell, and, thereby, the chances of hatching successfully (Brooks 1978). To hatch, the chick first punctures the membrane that encloses the air chamber at the large blunt end of the egg. Then the chick pecks awkwardly but regularly at the shell while rotating slowly in a counterclockwise direction by pivoting its legs. The power for the first feeble pecks comes from the hatching muscle, *M. complexus,* on the back of the neck. Bumping, as this feeble pecking is called, continues for one to two days, leaves a circular series of fractures on the eggshell, and finally results in penetration through the eggshell to the world outside (Figure 19–1).

A special structure on the tip of the bill, the egg tooth, helps the chick to break the shell. The hard, sharp-edged egg tooth is generally located just below the bill tip where the tip curves downward. The sheath of the egg tooth includes the lower mandible in loons, rails, bustards, pigeons, shorebirds, auks, hornbills, and woodpeckers. Moundbuilders have an egg tooth early in their development but lose it by hatching time; they kick rather than peck their way out of the egg. Egg teeth drop off the bills of most baby birds soon after hatching. Sandpipers and plovers drop them the first day, chickens and quail within two to three days, auks in two weeks to one month, and petrels in two to three weeks (Clark 1961). Passerines, on the other hand, gradually absorb the egg tooth (Parkes and Clark 1964). The hatching muscle also becomes vestigial once its task is accomplished (Figure 19–2).

Most birds chip a big hole out of the egg and finally shatter it with their body movements. Woodcocks and Willets, however, split the egg longitudinally, ripping open a seam rather than breaking it into pieces

Figure 19–1 Shortly before hatching, the chick shifts into the tucking position, breaks into the air chamber with its beak, and inflates its lungs for the first time. Prior to this event, the developing chick depends on oxygen exchanged through the capillary network of the chorioallantois. (After H. Rahn, et al. 1979, with permission of Scientific American)

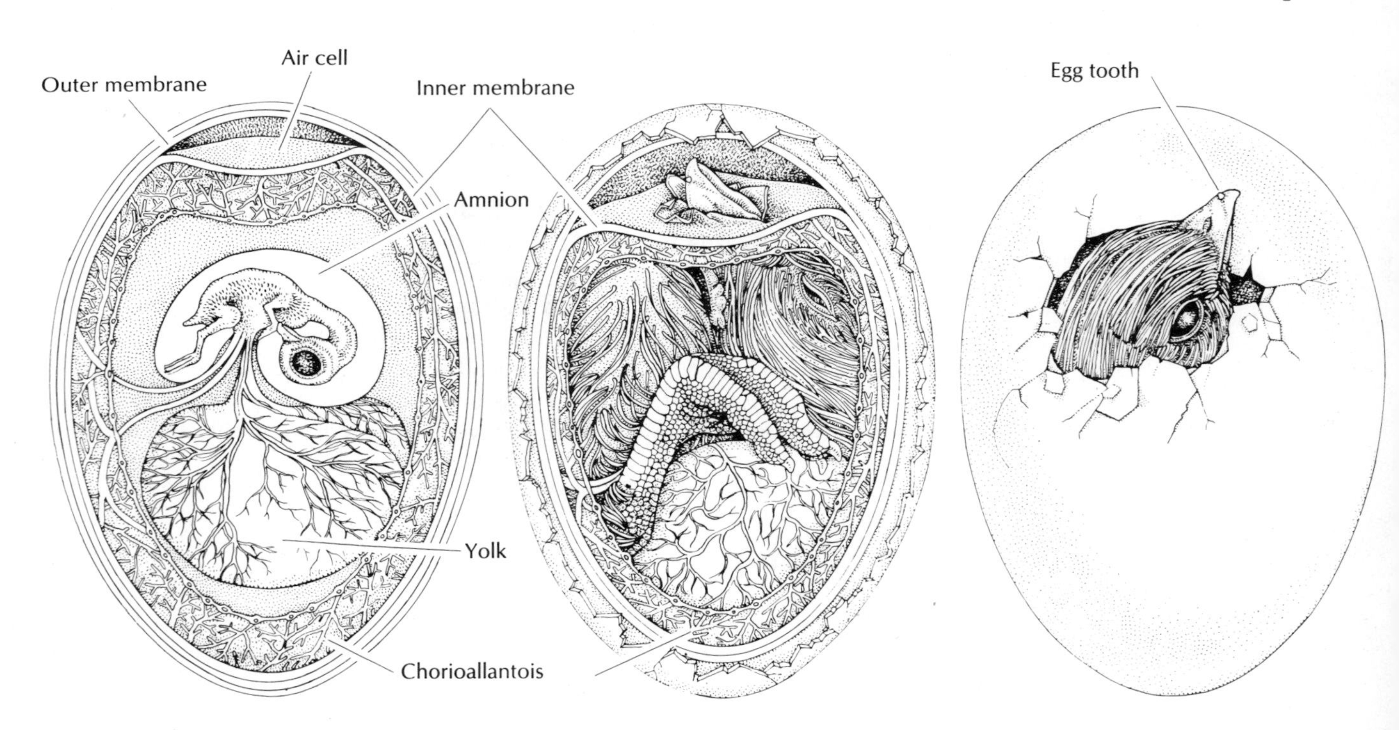

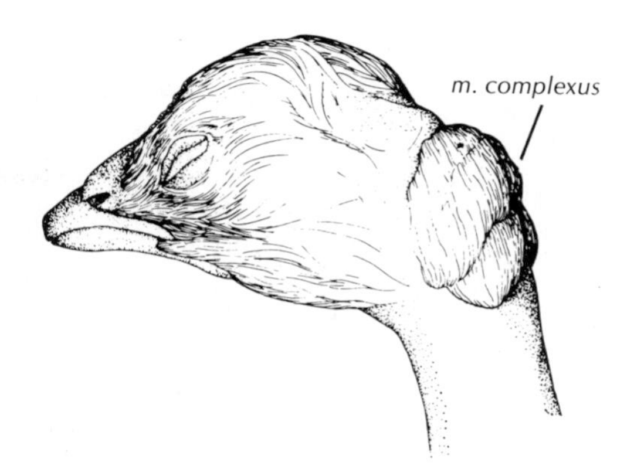

Figure 19–2 The hatching muscle, *M. complexus,* is a transient feature of chick anatomy that helps the chick to break out of the egg. (From Bock and Hikida 1968)

(Wetherbee and Bartlett 1962). Breaking out of their great, thick-shelled eggs is a major accomplishment for ostriches. After hours or even days of struggling, ostrich chicks virtually explode from their eggs, shattering the shell into many pieces (Sauer and Sauer 1966). Sometimes a parent may help crack the shell with its breastbone and pull the chick out by the head. Parents may help chicks that need assistance to hatch by enlarging the initial hole.

Synchronized hatching is especially important to waterfowl and quail, which move their large broods from the nest to safer sites soon after hatching. The 11 to 13 eggs that comprise the clutch of a Mallard duck, for example, all hatch within two hours despite having been fertilized and laid over a two-week period. Quail eggs that have been incubated beside one another hatch at the same time even though they are laid at 48-hour intervals (Johnson 1969). The stages of development of the eggs are brought closer together by differences in their rates of development (to within 30 hours on average), but coordinated adjustments during actual emergence are responsible for the final synchrony of hatching. Some eggs that are pipped early do not hatch for as long as 33 hours; sometimes eggs pipped later hatch in as little as 4 hours. Vocal communication among chicks within eggs enables them to synchronize the time of hatching. Older chicks about to hatch "click" slowly (1.5 to 60 times per second), accelerating the hatching of younger siblings. Conversely, younger chicks click rapidly (over 100 times per second), causing their older siblings to delay emergence (Vince 1969; Driver 1967; Forsythe 1971). Jarring of adjacent eggs by the first hatchling is the final signal; it stimulates nestmates to make their final moves and to escape 20 to 30 minutes later.

After baby birds hatch, they impose two pressing demands upon their parents: brooding and feeding. In their first week of life, baby birds cannot regulate their body temperatures or generate adequate heat. They need protection from cool air temperatures and from the hot sun. In cool climates, large broods need less brooding than small broods because decreased individual exposure, pooled metabolic heat, and the greater thermal inertia of their combined mass enables the chicks to keep each other warm (Clark 1983). Regardless of climate, single parents face potential conflicts between time for brooding on the nest and time for gathering food away from the nest. Within a week or so, however, the young have a greater tolerance for exposure, and a single parent is better able to gather the required food.

Altricial versus precocial modes of development

"Perhaps the single most striking feature of postnatal growth in birds is the dichotomy between precocial and altricial development" (Ricklefs 1983a, p. 3). Some hatchlings are helpless and must depend on their parents. Others are mobile and able to find their own food; a three-day-old Lesser Scaup duckling, for example, can dive, catch a minnow, and return to the surface. Even more precocious are the moundbuilders, whose chicks are completely independent and receive no parental care after they leave the egg. The terms *altricial* and *precocial* refer to the extremes of this spectrum of increasing maturity at hatching and decreasing dependence on parental care (Nice 1962; Skutch 1976) (Table 19–1).

Altricial birds are naked, blind, and virtually immobile when they hatch and thus are completely dependent on their parents (Figure 19–3). The helpless nestlings of altricial birds appear to have hatched prematurely; hummingbird hatchlings, for example, resemble large grubs. Altricial hatchlings have huge bellies and large viscera, which reflect their need for food and fast growth. In contrast, precocial chicks are well-developed little birds usually covered with fuzzy down. They can feed themselves, run about, and regulate their body temperatures shortly after they hatch. Their brains are quite large compared with those of altricial chicks.

Precocial birds such as quail and plovers lay larger eggs than do altricial birds of the same size such as European Starlings and Mourning Doves. The large eggs of precocial birds yield larger chicks that are either well advanced in their development at hatching or have large food stores that increase their chances of survival in the difficult first days out of the egg. The total energy requirements of the embryos of precocial birds are greater because their incubation periods are longer and their rates of metabolism higher. Their yolk reserve, which supplements their feeding for several days after hatching, is larger (Ricklefs 1974). Precocial birds lay eggs that are composed of 30 to 40 percent yolk; altricial birds' eggs are 15 to 27 percent yolk. The eggs of moundbuilders have extremely large yolks (62 percent of the egg mass).

Although we can classify most birds as altricial or precocial, a few species resist easy classification. We now recognize at least six categories of

Table 19–1 Comparison of Altricial and Precocial Modes of Development

Character	Altricial	Precocial
Eyes at hatching	Closed	Open
Down	Absent or sparse	Present
Mobility	Immobile	Mobile
Parental care	Essential	Minimal
Nourishment	Parents	Self-feeding
Egg size	Small (4–10%)*	Large (9–21%)*
Egg yolks	Small	Large
Brain size	Small (3%)*	Large (4–7%)*
Small intestine	Large (10.3–14.5%)*	Small (6.5–10.5%)*
Growth rate	Fast (3–4 times precocial rate)	Slow

* Percentage of adult weight.

(A)

(B)

(C)

(D)

(E)

(F)

Figure 19–3 Baby birds and their states of development at hatching. (A) Cedar Waxwing, altricial; (B) Ruby-throated Hummingbird, altricial; (C and D) Least Bittern, semialtricial; (E) Leach's Storm-Petrel, semiprecocial; (F) Whimbrel, precocial; note the egg tooth, the white structure at the tip of the bill, which the chick uses to break the eggshell. (Courtesy O. Pettingill/VIREO, A. Cruickshank/VIREO, W. Conway, and D. Hosking)

Figure 19–4 Development characteristics of baby birds, according to Margaret Nice's (1962) classification. (After Ricklefs 1983a)

Condition		Down	Eyes	Mobility	Parental nourishment	Parental attendance	Examples
Superprecocial		○	○	○	○	○	megapodes
Precocial	1	○	○	○	○	●	duck, shorebirds
	2	○	○	○	○	●	quail, grouse, murrelets
Subprecocial		○	○	○	◒	●	grebes, rails cranes, loons
Semiprecocial		○	○	◒	●	●	gull, terns, alcids, petrels, penguins
Semialtrical	1	○	○	●	●	●	herons, hawks
	2	○	●	●	●	●	owls
Atricial		●	●	●	●	●	passerines

○ precocial characters ● altricial characters

hatchlings (Nice 1962) based on primary criteria of mobility, open or closed eyes, presence or absence of down, and the nature of parental care (Box 19–1; Figure 19–4).

Each mode of development has distinct advantages, some of which reflect food habits (Nice 1962; Ricklefs 1983a). The food of most precocial birds — small invertebrates or, occasionally, seeds — can be procured by young chicks. Semiprecocial species and many altricial birds live on food

Box 19–1

Categories of hatchlings

Altricial: naked, blind, and helpless at hatching. Examples: songbirds, woodpeckers, hummingbirds, swifts, trogons, kingfishers, pigeons, parrots.

Semialtricial: stay in nest (nidicolous) though physically able to leave the nest within a few hours or the first day; fed and brooded by parents. Examples: goatsuckers, albatrosses, hawks, herons, and seriemas.

Semiprecocial: capable of temperature regulation; mobile but stay in the nest; fed by their parents. Examples: terns, auks, petrels, and penguins.

Subprecocial: leave the nest immediately and follow their parents; are fed directly by their parents. Examples: grebes, rails, cranes, guans, some pheasants, and loons.

Precocial: leave the nest immediately and follow their parents; pick up their own food soon after hatching though parents help to locate food. Examples: ostriches, pheasants, quail, murrelets, ducks, shorebirds, and kiwis.

Superprecocial: wholly independent. Examples: moundbuilders and Black-headed Ducks.

that must be located and captured with adult strength and skill. For these chicks, dependence on parents during a period of growth, maturation, and learning is essential before they can feed themselves. Semiprecocial chicks of gulls, terns, auks, and petrels are fed at the nest, but they can regulate their body temperature, thus freeing their parents to commute to distant feeding grounds (Ricklefs 1979a; Ricklefs et al. 1980). The precocial chicks of grebes and loons, which cannot dive and chase prey skillfully, are also fed by parents. But the young of these species must leave the nest to avoid predation and to free parents from costly flights between feeding areas and nests. Although it is tempting to suggest a simple evolutionary sequence from the precocial to the altricial condition, this cannot be done. Apparently, the altricial condition has evolved independently in many groups of birds; semialtricial or semiprecocial modes of development evolved secondarily from altricial modes of development (Ricklefs 1983a). The most reptilelike development patterns, seen in superprecocial moundbuilders and Black-headed Ducks, are breeding specializations rather than primitive states.

Temperature regulation

Homeothermy, the ability to generate metabolic heat (endothermy) and to maintain a high body temperature, is a great achievement of early development. Homeothermy releases a chick from its absolute dependence on parental brooding and enables it to face exposure without extreme vulnerability. Precocial chicks such as those of the Japanese Quail have some capability for thermoregulation when they hatch, but altricial chicks such as those of the European Starling do not. Precocial chicks quickly achieve 90 percent of their adult thermoregulatory capability. Even semiprecocial Western Gull and California Gull chicks can maintain high body temperatures rather well a day or two after hatching (Bartholomew and Dawson 1952; Behle and Goates 1957). Most small altricial passerines achieve homeothermy six or seven days after hatching.

Regulation of temperature by precocial and altricial chicks improves during development as a result of increased mass relative to surface area, improved insulation, increased metabolic heat production, and the development of nervous and endocrine system control. The doubling of weight improves the potential for thermoregulation by 26 percent, owing to the associated decrease in surface area relative to volume. The functional mass and thermal inertia of a chick also increase with brood size. As a result, altricial chicks in large broods achieve functional homeothermy earlier than those in small broods (Figure 19–5).

Skeletal muscle is the main source of heat production. The large leg muscles of a young chick are of primary importance in early thermogenesis. The growth and maturation of the pectoral muscles, however, cause the major improvements in thermoregulation. Early development of large pectoral muscles in the Willow Ptarmigan (Aulie 1976) and Leach's Storm-Petrel facilitates precocial heat production. The pectoral muscles of nestling Leach's Storm-Petrels mature early, by two weeks of age, even though the chicks do not fly for 9 to 10 weeks (Ricklefs et al. 1980). The ability of the chick to retain metabolic heat improves as its feather coat thickens. Oxygen consumption of nestling Great Tits and Pied Flycatchers, for example, increases by 25 and 15 percent, respectively, if they are shaved (Shi-

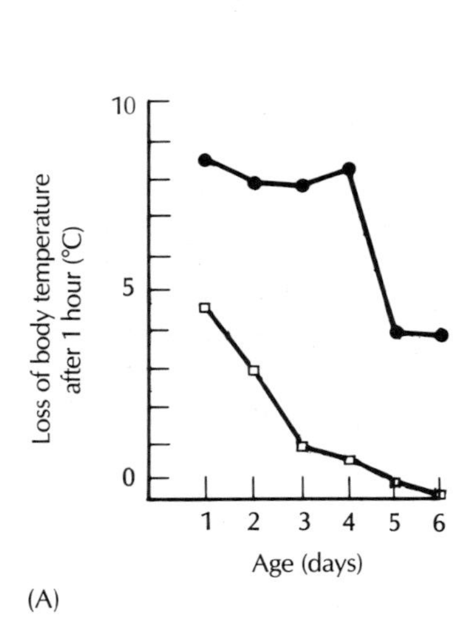

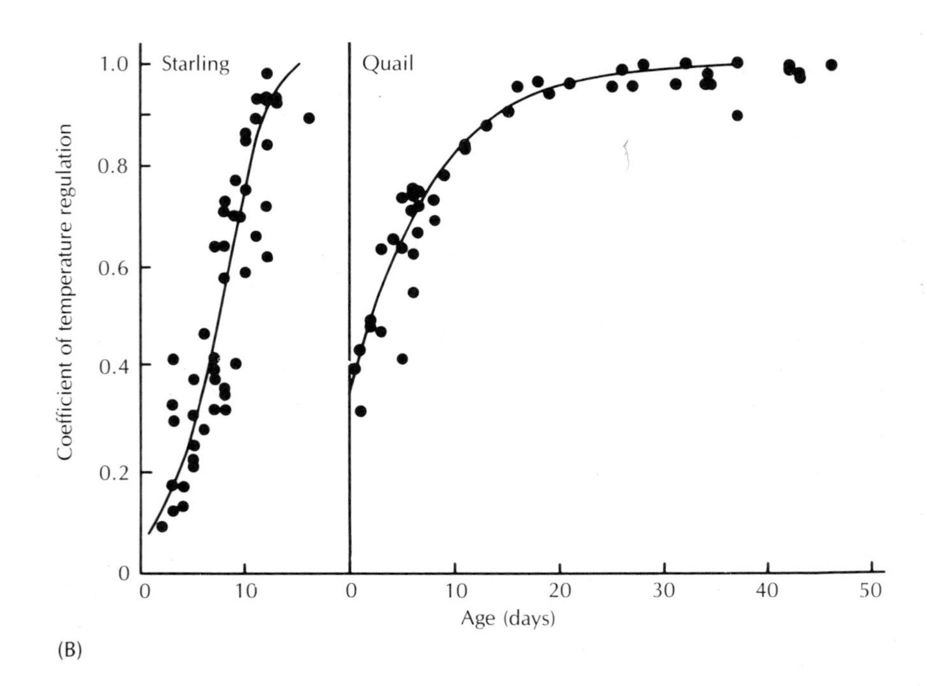

Figure 19–5 Development of homeothermy. (A) The ability of nestling European Starlings to maintain a body temperature of 39°C increases with age and presence of broodmates. The body temperature of a single nestling (solid circles) 1 to 4 days old drops 8° to 9°C after one hour of exposure to an ambient temperature of 20°C, whereas the body temperature of a 5-day-old nestling drops only 4°C under the same conditions. The presence of seven broodmates together (open circles) greatly reduces loss of body temperature. (B) Precocial chicks such as those of the Japanese Quail can maintain a high body temperature better on hatching than can European Starlings; this ability improves with age in both species. The coefficient of temperature regulation is the percentage of the difference between adult body temperature and air temperature at 20°C that a chick maintains after 30 minutes. (A, after Clark 1983; B, from Ricklefs 1979b)

lov 1973). Down insulation enhances the thermoregulation abilities of hatchling precocial birds.

Seabirds that nest in the hot sun need to lose heat. Young Least Terns and Sooty Terns can prevent overheating better than they can prevent chilling (Howell 1959; Howell and Bartholomew 1962). Laysan Albatrosses can also thermoregulate at an early age by dissipating excess heat from their large feet, which they expose to the breeze by leaning back on their ankles (Figure 19–6).

Figure 19–6 Young Laysan Albatrosses lose heat by leaning back and exposing their feet to the breeze. (Courtesy T.R. Howell and G.A. Bartholomew)

Growth rates

The growth of a baby bird during development follows an S-shaped or sigmoid curve (Figure 19–7). At first the chick grows slowly, then the growth rate accelerates and mass increases rapidly, and finally, growth decelerates as weight approaches the value (asymptote) of adult mass. The sigmoid curve enables us to compare species of differing masses and growth

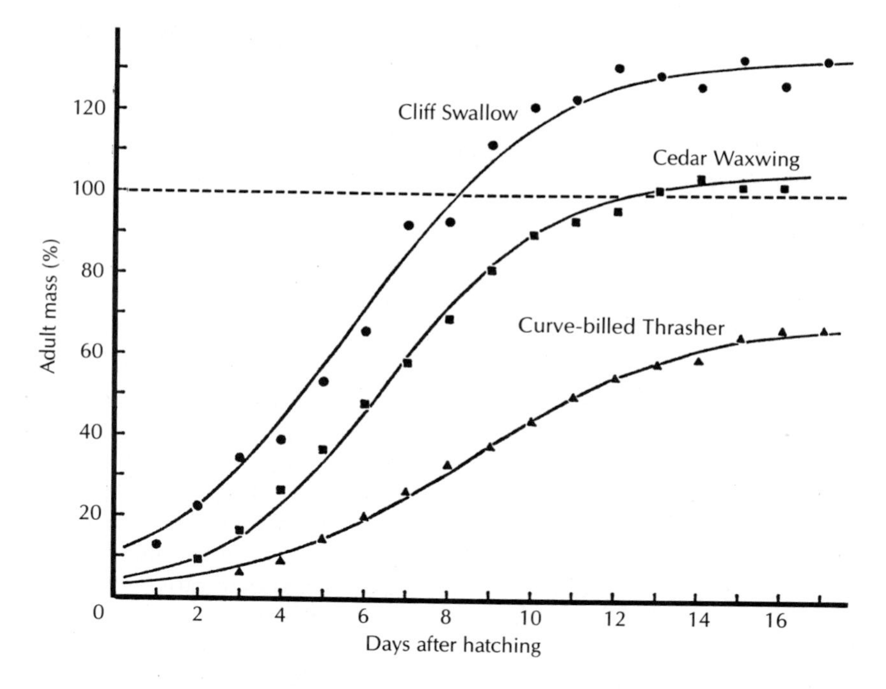

Figure 19–7 Nestling growth curves of three altricial birds, as determined from initial size, growth rate, and final asymptote (maximum value). Data are standardized to the maximum values of the growth curve in order to directly compare birds of different species size. (Adapted from Ricklefs 1968)

strategies because it is defined mathematically by only a few variables: initial size, growth rate, and final asymptote. Details of the relevant mathematical models and curve fitting are reviewed by Robert Ricklefs (1983b).

There are two major exceptions to the typical shape of the growth curve. The mass of young ground-feeding birds, such as doves and Curve-billed Thrashers, levels off below adult mass. Chicks gradually achieve full size after fledging when muscle and plumage development catch up with skeletal development. However, in aerial species, such as swallows, swifts, and pelagic seabirds, chick mass overshoots that of adults in the final stages of development and then declines as a result of metabolism of fat deposits or loss of water from maturing tissues, especially in the skin and at the bases of the feathers.

Some baby birds accumulate fat as insurance against poor food delivery by parents or as reserves for the days just after fledging when the chick is learning to feed for itself. Aerial passerines acquire more fat than other species; apparently it serves as an adaptation to the irregularity of their food supply (O'Connor 1977). The accumulation of fat is most striking in petrels; their obese chicks reach masses twice those of the adults. The chicks' lipid supplies act primarily as a reserve (Ricklefs et al. 1980). Young Oilbirds, which are raised on the oily lipid-rich fruits of palms and other tropical trees, also accumulate large lipid stores.

There is a 30-fold variance in the growth rates of chicks of the various species. Over half of the variation in growth rate relates directly to adult body weight: Growth rate decreases roughly as the cube root of body weight. The Wandering Albatross, one of the largest seabirds, has the longest known nestling period of any bird: 280 days. Hole nesters also grow more slowly than open nesters; pelagic seabirds (petrels) grow more slowly than inshore seabirds (gulls); and tropical landbirds grow more slowly than temperate landbirds. Rapid growth is the advantage of altricial development: Altricials grow three to four times as fast as precocials. The brevity of the altricial development period more than compensates for the chicks' vulnerability to predators and bad weather.

David Lack (1968) regarded growth rate as a balance between selection for rapid growth to escape predation and selection for slow growth to reduce food requirements: Parents might be able to rear more slow-growing offspring when food is limited. Ricklefs' (1983a) models of the energetics of growth suggest that these considerations are important to altricial birds but cannot explain the difference in growth rates between altricial and precocial birds.

The tissue allocation hypothesis is a unifying and simplifying explanation of the differences in growth rate among species (Ricklefs 1979a, b). The hypothesis states that variation in the growth rates of different species is directly related to relative precocity and to adult body proportions, rather than indirectly to food availability. Growth rate, it seems, is determined by a balance between the mature and embryonic functions of tissues. If growth and mature function of tissues (such as muscle contraction) are mutually exclusive, growth should slow as the individual matures. Species that mature early should grow slowly, and skeletal muscle growth is likely to play the greatest role in regulating development. This hypothesis predicts the observed relationship between growth rate and precocity.

Comparison of the altricial European Starling, the semiprecocial Common Tern, and the precocial Japanese Quail illustrates the interaction between precocity of tissue maturation and overall growth rate (Ricklefs 1979b) (Figure 19–8). Of these three species the starling grows fastest, the tern grows nearly as fast as the starling, and the quail grows relatively slowly. The rapid maturation of the quail's large leg muscles, essential for precocial locomotion, detracts from its potential growth rate. The tern also exhibits rapid leg development, but the material and energy needed for growth of its tiny legs are only minor investments relative to its overall growth. The starling puts energy into growth before tissue maturation.

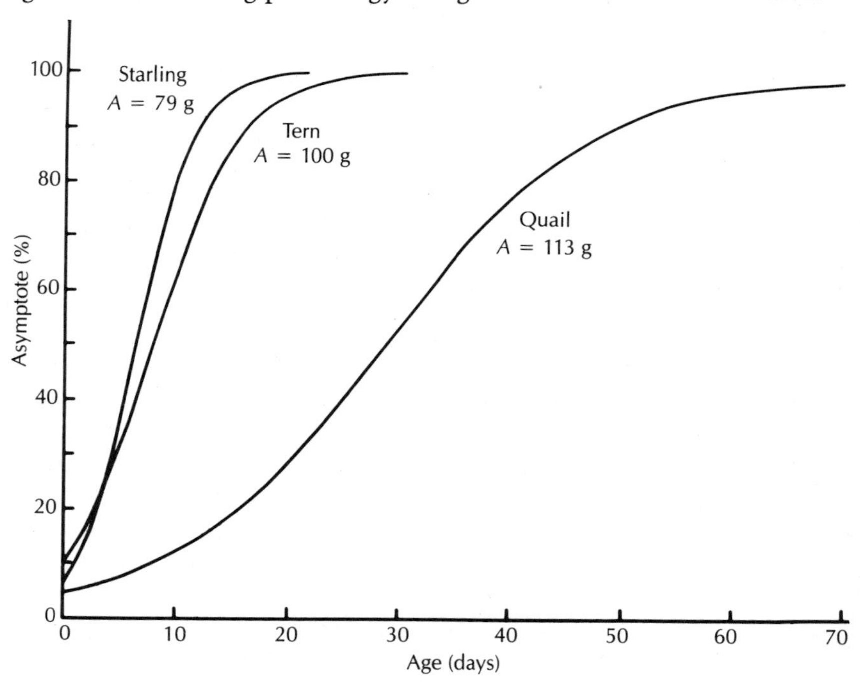

Figure 19–8 Growth curves for an altricial bird (European Starling), a semiprecocial bird (Common Tern), and a precocial bird (Japanese Quail). *A* is the bird's mass at the asymptote of the growth curve. (From Ricklefs 1979b)

Individuals of a species can also exhibit markedly different growth rates. Growth rates of individuals are affected by variations in quality and quantity of food, temporal pattern of feeding, and temperature, all of which vary according to locality, season, habitat, and weather. The fledging weights of Rhinoceros Auklets in Puget Sound vary from 339 grams in bad seasons to 521 grams in good seasons (Summers and Drent 1979). The effects of food supply on growth rate are perhaps best known in swifts and martins (Lack 1956; Bryant 1975, 1978a, b). The maturation of Common Swifts, for example, varies from 35 to 56 days, reflecting feeding conditions. Chicks of these swifts can survive up to 21 days of starvation by stopping growth and becoming hypothermic (Koskimies 1948). The nestling period of the Green Violet-Ear, a hummingbird of highland Mexico, ranges from 19 to 28 days depending on the amount of cold and rainy weather (Wagner 1945). Song Sparrows that are born early in the season stay 10 to 11 days in the nest, whereas those born later, when it is warmer and there is more food, stay in the nest only 8 to 9 days (Nice 1943).

The growth rate of Sooty Terns is slow compared with that of the related Common Tern (Ricklefs and White 1981). Sooty Terns take twice as long as Common Terns to make their first flight. The Common Tern initially uses its energy for growth and then later uses it for maintenance of full body size. The growth rate of Sooty Tern chicks declines slowly in relation to increasing maintenance requirements, and the balance of the two drains on energy result in a maximum of 30 percent less energy use than that of Common Terns. This difference may partly be due to food availability. The slower growth of Sooty Terns may also relate to the greater precocity of chicks in this tropical species, which nests in huge, dense colonies: Mobility, to escape aggression or predation, is advantageous in such an environment (Figure 19–9).

Growth may virtually stop if diet is inadequate. Domestic poultry chicks, for example, can remain at a physiological age of 10 days for months if their diets are deficient (McCance 1960; Dickerson and McCance 1960); with access to an adequate diet, normal growth resumes.

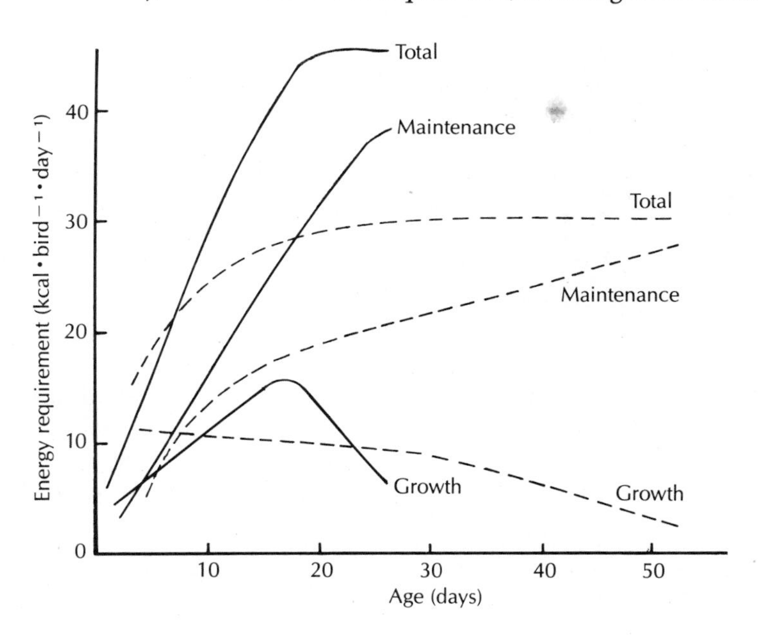

Figure 19–9 Energy requirements for growth of the Common Tern (solid lines) and Sooty Tern (dashed lines). (Adapted from Ricklefs and White 1981)

Brood size mediates the effects of food limitation on nestling growth rate (Mertens 1969; O'Connor 1975; Royama 1966b). Nestling growth often decreases with brood size, which suggests that parents cannot deliver enough food to all nestlings to insure maximum growth (Klomp 1970). Larger broods enhance the differences in growth rate between older and younger siblings of nestling Common House-Martins (Bryant 1978a, b). The predictable starvation of small chicks when food is insufficient tends to adjust brood size to food availability (Lack 1954). Brood reduction as a result of competition among nestmates for food may even be accomplished by vicious sibling rivalry.

Duration of parental care

Even after chicks leave the nest, most parents must still take care of them. Most small passerines, for example, tend their young for two to three weeks after they have left the nest (Nice 1943). Shearwaters and petrels desert their nestlings one to two weeks before they fledge, and megapode chicks are on their own from the start.

Prolonged parental care is typical of large birds. Young Bewick's Swans, for example, stay with their parents for one to two years, through several long-distance migrations. Boobies and terns feed their young for up to six months after they have fledged—until they have mastered the art of plunging after fish. In the tropics, where long apprenticeships also seem necessary to develop feeding skills, some young passerines stay with their parents for 10 to 23 weeks.

The chicks of quail and waterfowl leave the nest soon after hatching. Although they do not depend on their parents for delivery of food, they rely on them for food location and for protection. Gallinaceous chicks quickly learn what is and is not edible by pecking at objects shown to them by their parents. Parent ducks and geese also guide their chicks to food, with one exception: The Australian Pied Goose feeds its young from its bill tip to the chicks' bill tip (Kear 1963). Precocial chicks of quail and plovers require more parental care than is usually recognized (Safriel 1975; Walters 1984). In addition to brooding, protection from predators is a demanding effort that requires constant parental vigil. Male and female lapwing plovers, for example, face major time constraints while taking care of their mobile young (Walters 1982). They alternate "tending" behavior in order to feed.

Figure 19–10 Gray Gull shading its young from the intense desert sun. (Courtesy T.R. Howell)

Many species have evolved behaviors that enable parents to care for young away from the nest. These behaviors reduce the strain on parents and fulfill the nutritional and protective needs of young (Figure 19–10). Rails, coots, and gallinules brood their young away from the nest on special platforms that they build above the water. Grebes carry their young on their backs, often under their wings, diving and feeding relatively undisturbed. Sungrebes have special pouches under each wing for carrying new hatchlings (Alvarez el Toro 1971). Wood Ducks and other tree-nesting waterfowl have been known to carry their young to and from the ground in their bills or on their backs (Johnsgard and Kear 1968). Rails regularly pick up chicks with their bills and move them to safety (Turner 1924; Bent 1926).

A coucal, a kind of African cuckoo, has been observed carrying its young between its thighs when its nest was threatened by fire (Kilham 1957). Woodcocks and other shorebirds have been reported to carry their young, but these reports are discredited by woodcock experts (Tordoff 1984).

Feeding the young

How often do birds feed their young? The answer varies from once or twice daily for seabirds, swifts, and large raptors to once per minute for some small landbirds with large broods (Skutch 1976). Normal rates of food delivery by small and medium-sized landbirds average 4 to 12 times per hour. Trogons bring food to the nest once per hour, Bald Eagles 4 to 5 times per day, and Common Barn-Owls 10 times a night. Recorded extremes of rapid food delivery to large broods include 990 trips per day by the Great Tit and 491 trips per day by the House Wren. Food delivery rates vary according to the age of the young. Hatchlings require only small amounts of food, but as they develop, their appetites grow. The Eurasian Nuthatch, for example, delivers food 119 times per day when its young are 2 days old, and 353 times per day when they are 18 days old. Delivery rates slow to 270 per day just before the young fledge. Feeding young in the nest, especially large broods, can impose a major demand on the time and energy budgets of the parents. The Common Swift of Europe flies 1000 kilometers a day, scooping insects from the sky. The Pied Flycatcher, which brings food to the nest every two minutes, makes about 6200 feeding trips to nourish its young from hatchling stage to fledgling stage. It appears that, in general, parents must gather two to three times as much food as they need for themselves to cover the energy needs of their nestlings (Walsberg 1983).

The energy costs of reproduction in Phainopeplas have been especially well analyzed (Walsberg 1978) (Figure 19–11). This unusual species

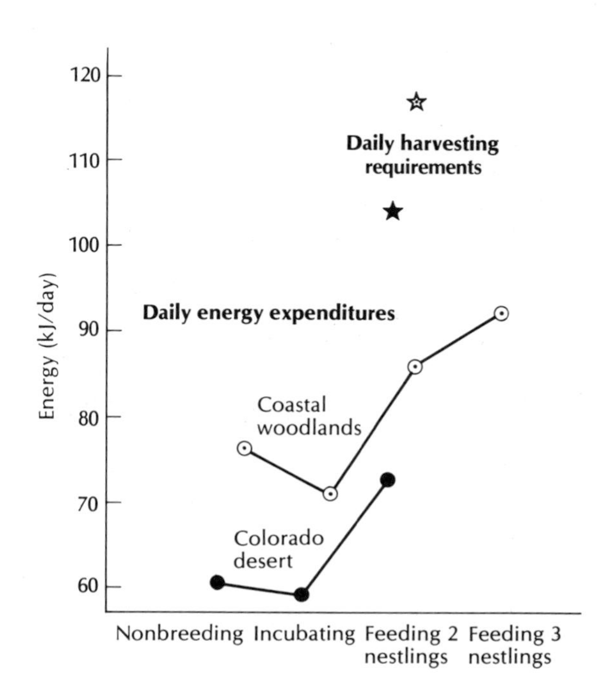

Figure 19–11 Energy costs of nestling care by Phainopeplas in two habitats, compared with the energy costs of its incubation and nonbreeding activity. Also shown (stars) is the amount of food energy an individual must harvest to feed both itself and two young. (After Walsberg 1978)

apparently breeds twice per year, once in the summer in the California coastal woodlands and once in the winter in the Arizona deserts. The daily energy expenditure of a breeding adult feeding two nestlings in the coastal woodlands is 12.5 percent more than the daily energy expenditures of nonbreeding birds in the woodlands, and 18.6 percent more than nonbreeding birds in the desert. A breeding Phainopepla must harvest 52 percent more food in the woodlands and 72 percent more food in the desert to meet its own requirements and those of its nestlings. Broods of three instead of the usual two increase adult daily energy expenditures by an additional 7 percent and the amount of food to be harvested by 18 percent.

In some birds, such as Great Tits in Holland (Kluijver 1950) and Scarlet-rumped Tanagers in Costa Rica (Skutch 1976), food delivery rates increase in proportion to the size of the brood. This is not always the case, however; four young do not necessarily require four times as much food as one. In fact, the rate of food delivery per nestling usually decreases as brood size increases (Moreau 1947; Skutch 1976), primarily because of the energy savings nestmates achieve by huddling together (Royama 1966b). The begging cries of young stimulate parents to deliver food to the nest. Experimental increases in the volume and continuity of begging cries at the nest have prompted greater activity. Lars von Haartman (1953), for example, hid extra young Pied Flycatchers behind the wall of a nest box. In response to their cries, the parents brought more food to the nest than was required for their nestlings.

Parents are also stimulated by the sight of their babies' gaping mouths. The chicks of some cavity-nesting species have brightly colored mouth markings to attract parental attention and serve as targets for food delivery. The bright yellow tongue of the Speckled Mousebird stands out against its satiny black mouth lining. Gouldian Finches of Australia have black spots inside their mouths and three opalescent green and blue spots in the corners. The arrangement of spots on the insides of the mouths of some African finches are unique to the species and apparently protect the young against competition from brood parasites (see Chapter 20). Most birds put food directly into a baby's gaping mouth, sometimes deeply into the digestive tract. Young hummingbirds receive an injection of nectar and insects through their mother's long, hypodermiclike bill (Figure 19–12). Regurgitation of a meal either directly into a nestling's mouth or onto the ground for the nestling to pick up is common among seabirds. Young penguins and pelicans plunge their heads deeply into their parents' gullets. Spoonbills and albatrosses cross their large bills over and under those of their young, like a pair of open scissors, so that the chicks' mouths are in position for food transfer.

Little Ringed Plovers, Common Ravens, anhingas, and Whale-headed Storks deliver water as well as food to their young. Young Namaqua Sandgrouse, which live in the hot Kalahari Desert, depend on their fathers to bring them water each day; the fathers visit waterholes as far as 80 kilometers from the nest and wet their belly feathers (Cade and Maclean 1967; Maclean 1968). Of the 25 to 40 milliliters of water soaked up each time a male visits the waterhole, the young at the nest glean only 10 to 18 milliliters.

Figure 19–12 Parent birds feeding young: (A) Anhinga young begging for food; (B) parent Anhinga feeding one of the young; (C) Ruby-throated Hummingbird nestlings begging for food; (D) parent hummingbird feeding one of the nestlings. (Courtesy A. Cruickshank/VIREO)

Energy and nutrition

Chicks require energy for maintenance, temperature regulation, activity, excretion, and growth. Growth accounts for a major fraction of total energy expenditures early in development—31 percent in the Common Tern, 46 percent in the Sooty Tern, and 56 percent in Leach's Storm-Petrel—but a rather small fraction in the later stages (see Ricklefs and White 1981; Ricklefs et al. 1980). Peak total energy expenditures occur late in development. The energy channeled into growth constitutes roughly 21 to 40 percent of a chick's energy budget for the entire developmental period. Energy may be less important in determining rates and patterns of development than is nutrition.

Production of new tissues during growth requires nutrients such as the essential amino acids and the sulfur-containing amino acids cysteine and methionine, which are used in feather production. To get the calcium they need, some chicks are fed fragments of teeth, bone, and eggshells as dietary supplements. The bone growth of Lapland Longspurs cannot depend on the meager 0.1-percent calcium (by dry weight) in the craneflies and sawflies they eat; their parents must also feed them lemming bones and teeth (Seastedt and Maclean 1977). Eurasian Wrynecks deliver eggshells, snail shells, and small bones to their young; and similarly, young Stanley Cranes often consume eggshells (Löhrl 1978; Walkinshaw 1963).

Baby birds require protein, especially in the early stages of their development. In many species of songbirds, parents supply mostly small, soft-bodied insects at first, gradually increasing the proportion of fruits and seeds. Protein-rich aquatic insects comprise 90 percent of the diet of baby American Black Ducks during the first five days after hatching but then drop to 43 percent of the diet (Reinecke 1979). The Resplendent Quetzal, primarily a fruit eater, feeds its young only insects for the first ten days (Skutch 1944).

Fruits do not usually provide an adequate diet for nestling growth (Morton 1973; White 1974; Foster 1978; Ricklefs 1976b). Tropical fruits tend to be deficient in some essential amino acids. The average ratio of protein (measured as a percentage of dry weight) to metabolizable energy (measured in kilocalories per gram) in tropical fruits is 1 : 6, whereas the average ratio in animal foods is 14 : 22. The chicks of Clay-colored Robins and Yellow-bellied Elaenias require protein/energy ratios of 6 : 9; thus they must supplement a diet of fruit with insects. Most tropical fruit-eating birds supplement their chick's fruit diet with insects. The chicks of Bearded Bellbirds and Oilbirds, which eat only fruits, grow half as fast as those of other tropical birds.

Pigeons, penguins, and flamingos feed their young nutritious esophageal fluids. Pigeon milk, the best known of these, is full of fat-laden cells sloughed off the epithelial lining of the crop. Pigeon crop milk often contains some small food fragments as well (Ziswiler and Farner 1972). Like the milk of marine mammals, this fluid is rich in protein (23 percent) and fat (10 percent). It also includes essential amino acids (Fisher 1972). Esophageal fluid is initially the sole source of nutrition for the chicks of Greater Flamingos. Flamingo milk has more fat and less protein than pigeon milk. The rich esophageal fluid on which Emperor Penguin chicks feed

Table 19–2 Composition (%) of Avian Esophageal Fluids

Bird	Protein	Lipid	Carbohydrates
Pigeon	23	10	0.0
Flamingo	8	18	0.2
Penguin	59	29	5.5

From Fisher 1972.

during their first week of life contributes to a doubling of their body weight. Penguin milk is rich in both fat and protein (Table 19–2).

Sibling rivalry

Vicious rivalry seems to be normal among the chicks of some birds (O'Connor 1978; Stinson 1979; Mock 1984a, b). Larger siblings often bully their nestmates and thereby get the first choice of food delivered by their parents. Occasionally, serious damage may result: Nestling Northern Flickers and Great Spotted Woodpeckers sometimes kill a sibling. Siblicide is standard in the nests of skuas, eagles, and some boobies. In the well-studied Verreaux's Eagle, for example, only once in 200 records did both siblings survive to the fledging stage (Brown et al. 1977; Gargett 1978). Detailed accounts of the deliberate killing of the younger eagle are recorded (Gargett 1978). The younger chick of two raised by Brown Boobies also usually dies as a result of abuse by its nestmate except in years of maximum food abundance. As a general rule, parents react passively to the destructive behavior of their offspring.

In part, siblicide seems preordained by the size differences that result from asynchronous hatching. In the South Polar Skua, the younger of two siblings has a good chance of surviving if it is nearly the same size as the older chick. If it is more than 8 grams lighter than its older nestmate, which is usually the case, the younger chick has a poor chance of survival (Spellerberg 1971; Procter 1975). Siblicidal behavior may be deeply rooted in species behavior: Experimental size matching of brood mates in the Lesser Spotted Eagle does not prevent the inevitable dominance of one over the other and eventual killing (Meyburg 1974).

In addition to the raptorial birds mentioned, sibling rivalry is a way of life in some colonial herons such as the Great Egret but not in others. In the Texas breeding colonies of the Great Egret studied by Douglas Mock (1984a), nestlings were often killed by an elder sibling, but he rarely saw siblicide in the Great Blue Heron. Why should two such similar species differ in this way? The type of food the parents bring their nestlings is part of the answer. Great Egrets bring small fish which are easily monopolized by the aggressive older sibling, whereas Great Blue Herons bring larger fish which cannot be easily monopolized. When experimentally cross-fostered in Great Egret nests, young Great Blue Herons adopted the siblicidal tactics typical of the egret, in response, it seems, to the opportunities presented by the smaller food. Surprisingly, the converse result did not take place: Great Egrets cross-fostered in Great Blue Heron nests did not become more toler-

Table 19–3 Effects of Cross-Fostering on the Relative Frequency of Siblicidal Mortality

	Great Egret Chicks			Great Blue Heron Chicks		
Parents	Number Alive by Day 25	Number of Siblicidal Deaths	Number of Other Deaths	Number Alive by Day 25	Number of Siblicidal Deaths	Number of Other Deaths
Natural	5	8	4	8	1	10
Foster	4	6	0	1	6	2

Presented as a comparison of the 1981–1982 foster broods with the 1979–1981 observed broods of three or four chicks (natural parents). The fates presented are for the youngest sibling in each brood.
From Mock 1984a.

ant of their nestmates. Sibling aggression in the Great Egret is a deep-seated behavior similar to that in raptors; it is not an optional behavior, as in the Great Blue Heron (Table 19–3).

Nest sanitation

Some birds exhibit little or no interest in nest sanitation. Nests of many pigeons, raptors, and carduelline finches, such as the House Finch, are well known for their filthy conditions. Many birds, however, are fastidious, regularly removing feces and other debris to prevent the nest from becoming a breeding ground for disease, insects, and parasites. Some insects are welcomed rather than ejected. The larvae of a particular moth species clean the nests of the Golden-shouldered Parrot of Australia, and a beetle provides sanitation services for Australian grassfinches. Some young birds instinctively eject liquid feces away from the nest, and others eliminate feces accurately through nest hole openings; for example, female hornbills, sealed in their nest holes, defecate through the narrow slit remaining in the mud-sealed opening.

Nest sanitation is made easier for most passerine bird and woodpeckers because their young excrete fecal sacs. Fecal sacs are packages of excrement surrounded by a gelatinous membrane. The parent can easily remove the sac from the nest and drop it away from the nest. Swallows and martins typically drop fecal sacs into nearby water. Similarly, the Superb Lyrebird takes fecal sacs to a nearby stream or, alternatively, scratches in the dirt and buries them. Incomplete digestion by nestlings leaves some residual food in their feces, which is often eaten by parents for nutrition as well as sanitation purposes. In one study, fecal sacs provided 10 percent of the daily energy requirements of adult White-crowned Sparrows (Morton 1979) (Figure 19–13). Consumption of fecal sacs is most frequent when the nestlings are young.

Nest sanitation may also contribute to safety because debris and feces in and around the nest may attract predators. Equally important is the removal of eggshells from the nest site. Eggshells betray the camouflage needed to escape predation and, therefore, are quickly removed after the young hatch. Parents may eat the shell, feed it to their chicks, or take it away from the nest for disposal. Niko Tinbergen (1963) demonstrated the cost of

Figure 19–13 Bearded Tit removing fecal sac from nest. (Courtesy E. Hosking)

leaving broken eggshells at the nest sites of Herring Gulls. The nests were robbed 65 percent of the time by crows if eggshells were not removed but only 22 percent of the time if they were removed.

Fledging from the nest

As a naked, blind hatchling is transformed into a feathered juvenile, the young bird approaches a pivotal event in its life: leaving the nest. Technically speaking, the nestling period is the interval between hatching and departure from the nest; the fledging period is the interval between hatching and flight (Skutch 1976). The nestling and fledging periods may be the same for altricial birds such as hummingbirds but different for subprecocial and precocial birds, which have short nestling and long fledging periods. We commonly refer to the moment of departure from the nest by altricial birds as fledging even though the young birds may only flutter and scramble about for a few days before their first flight. Departure from the nest increases vulnerability to predators and the weather. Unable to fly well, the baby bird cannot easily escape predators, and the mortality rate during this period is high. Once past the first dangerous days, however, the fledged chick is safer than a chick in a vulnerable nest. Fledglings respond to the warning calls of their parents by hiding or by staying still. Immobility combined with camouflaging plumage renders a chick extremely difficult to find.

Mobile young birds move with their parents closer to good feeding grounds, which reduces the strain on the parents. The initial journey away from the nest is often a heroic one. One brood of Wood Ducks, for example, jumped five to six yards to the ground from their nest in a tree cavity and then followed their mother down a bluff and across a railroad track before swimming three-quarters of a mile across the Mississippi River to feeding grounds in good bottomland (Leopold 1951). Precocial chicks that leave nests in tall trees or high cliffs must leap to the ground below, bouncing off soft earth if they are lucky or off jagged rocks if they are not. Torrent Ducks, for example, live in the dangerous waters of fast-flowing streams high in the South American Andes. To leave their nests in cliff crevices or holes above the streams, ducklings must plunge as much as 20 meters into the turbulent water of the rocky stream below. Only rarely do they hurt themselves. Their light weight, buoyancy, and downy cushioning protect them from severe impact.

Young seabirds that grow up on tiny cliff ledges overlooking the sea must also leap into space and flutter to the water below. Baby Xantus' Murrelets and Ancient Murrelets leave the nest shortly after hatching and swim rapidly out to sea; chicks only two days old have been found 15 miles from land. Similarly, the departure of young Common Murres from cliff ledges is a momentous event:

> Emerging in the evening from beneath their brooding parents, they walk around the ledge, preen, flap their incompletely feathered wings, and bob up and down. A parent also displays with its chick, bowing towards the chick and towards the sea, while it utters a long, high-pitched growl. Parent and chick bill and preen each other's head and neck. When the activity reaches a climax, the chick suddenly leaps from the ledge. Even if they strike jagged rocks [150 meters below], the young murres bounce off and are rarely injured. (Skutch 1976, p. 302)

No less dramatic is the emergence of a baby Mallee-Fowl from the compost heap where its egg was laid (see Chapter 18). The newly emerged, exhausted and weak hatchling must work its way up through several feet of sand and debris to the surface of the mound. This task takes 2 to 15 hours:

> Suddenly the back of its neck appears at the mound's surface. After the neck is free, the head quickly follows. The chick opens its eyes for the first time and rests briefly. Then it resumes its struggles, freeing one wing and then the other. Soon the whole body follows. Temporarily exhausted, the young Mallee-fowl may lie exposed on the surface for some time, an easy prey to predators; but more often it tumbles down the side of the mound and staggers to the nearest bush to collapse in the shade, where it recuperates its strength after such

> prolonged exertion. Its recovery is swift: within an hour it can run firmly; after two hours it runs very swiftly and can flutter above the ground for thirty to forty feet. Twenty-four hours after its escape from the mound, it flies strongly. (Skutch 1976, p. 234)

Long before they are ready to leave the nest or to fly, young birds develop essential strengths through exercise. Young pelicans jump up and down and flap their growing wings with increasingly effective strokes. Young hummingbirds grip nest fabric with their feet as they practice beating their new wings, anchoring themselves so as not to take off. When first airborne, the young bird responds to the new experience with astounding ability and control. When a young Osprey launches itself on its first flight over a northern lake, it wobbles and flaps uncertainly, loses altitude, and seems certain to splash into the lake. In the last possible moments, it flaps more effectively and gains altitude, climbing steadily until it is high above the lake. It then glides in circles and practices steering and control before disappearing over the horizon (Gill personal observation). Successfully launched from the nest, the young Osprey enters the local population of its species and may someday become a breeding adult.

Summary

A bird's first challenge is to break out of its shell. The egg tooth, a sharp-edged structure on the top of the bill, is a special feature for breaking the eggshell. Synchronized hatching of precocial young enables them to leave the nest together. Most nestlings fall into one of two categories: altricial or precocial. The former is a state of almost complete dependence on parents, the latter a state of relative independence. Ornithologists recognize the intermediate categories of semialtricial, semiprecocial, and subprecocial, plus an extreme category of superprecocial for young that are wholly independent when they hatch.

The growth of baby birds follows a sigmoid curve. There is a 30-fold variation in the growth rate of chicks of the various species. One school of thought regards growth rate as moderated by selection for slow growth to reduce food requirements. Another theory states that growth rates relate directly to precocity of development and adult body proportions. If growth and maturity of function are mutually exclusive, growth should slow as the individual matures. Growth rates of individuals of the same species may also vary considerably, owing to diet quality, food availability and reliability, and temperature. Brood size is also a factor here. Older, larger chicks have a greater chance of survival than smaller, younger siblings. Asynchronous hatching contributes to siblicide because it results in chicks of unequal size and strength. Food type is also a factor; food that can be monopolized, such as small fish, promotes brutal competition between siblings. Siblicide is not uncommon among birds.

Young birds show extraordinary skill and daring when they leave the nest. For example, flightless young of seabirds that nest on cliff ledges must leap great distances into turbulent waters below. Others, such as the

Osprey, master flight skills within minutes of launching themselves on their maiden flight, but most young birds are at risk to predators and rely on camouflaging plumage, immobility, and the warning calls of their parents for protection.

Further readings

Ricklefs, R.E. 1983. Avian postnatal development. Avian Biology 7:1–83. *A definitive review.*

Skutch, A.F. 1976. Parent Birds and Their Young. Austin: University of Texas Press. *A readable review of the natural history of avian reproduction, stressing the topics covered in this chapter.*

chapter 20

Brood parasites and cooperative breeders

The demands of parental care invite both cheating and cooperation. Brood parasitism and cooperative breeding lie at opposite ends of the spectrum of parental care practices among birds. Brood parasites are selfish cheaters whose evolution is consistent with Darwin's theory of natural selection. The apparent altruism of cooperative breeders, however, challenges the basic tenets of evolutionary theory. In this chapter we introduce avian brood parasites: their natural history, their mimicry of hosts, and their effects on their hosts. We then consider the evolution of the habit as an extension of intraspecific brood parasitism, which is widespread among birds. Later we focus on cases of cooperative breeding among birds and show how helpers increase their own prospects for reproduction by aiding their parents and stepparents. Beneath the appearance of cooperation lies a strong undercurrent of conflict and self interest.

Brood parasites

Cowbirds and cuckoos are the most familiar birds that relinquish care of their young to foster parents by laying their eggs in the nests of other birds (Figure 20–1). Roughly 1 percent of all bird species are brood parasites. This is a most unusual breeding strategy; a few insects but no mammals are brood parasites. Among birds, the practice has evolved independently at least seven times, in

1. cowbirds (Icteridae),
2. honeyguides (Indicatoridae),
3. Old World cuckoos (Cuculinae),
4. New World cuckoos (Neomorphinae),
5. whydahs and indigo-birds (Viduinae, Ploceidae),
6. the Cuckoo Weaver (Ploceidae), and
7. the Black-headed Duck (Anatidae).

In addition, brood parasitism is a covert practice among birds of the same species.

Parasitic birds take advantage of the parental care of other birds. By reducing their costs and risks, parasitic birds are able to lay more eggs each season. Parasitic cuckoos, for example, lay more and smaller eggs than nonparasitic cuckoos. By not putting all their eggs in one nest, brood parasites improve the chances that some of their offspring will escape predation (Payne 1977a). The absence of the trait of parental care reduces the advantages of monogamy in some brood parasites. Honeyguides and viduine finches, for example, lack a pair bond; every male has a principal calling site which females visit for purposes of mating. A male Yellow-rumped Honeyguide, which defends a prime feeding site of bee's honeycomb (page 310), may mate with as many as 18 females a day at the peak of the breeding season (Cronin and Sherman 1976). Dominant male Village Indigo-birds, which control the best territorial sites, typically mate with 3 or 4 females a day (Payne and Payne 1977). The mating systems of parasitic Old World cuckoos remain unknown.

Figure 20–1 Foster parent wagtail feeding a parasitic young Common Cuckoo. (Courtesy E. and D. Hosking)

Table 20–1 Host Specializations of African Honeyguides and Japanese Cuckoos

Brood Parasite	Primary Host(s)
African honeyguides	
Black-throated Honeyguide	Rollers, starlings, bee-eaters
Lesser Honeyguide	Large barbets, woodpeckers
Scaly-throated Honeyguide	Woodpeckers
Least Honeyguide	Tinkerbirds, small barbets
Cassin's Honeyguide	Rock sparrows
Wahlberg's Honeyguide	White-eyes, small warblers, flycatchers
Japanese cuckoos	
Common Cuckoo	Great Reed-Warbler, Bull-headed Shrike, Meadow Bunting
Oriental Cuckoo	Crowned Willow Warbler
Hodgson's Hawk Cuckoo	Chats
Little Cuckoo	Wren, Japanese Bush Warbler

From Lack 1963, 1968.

Mimicry and other adaptations of parasites

Obligatory brood parasites are highly specialized birds that use specific hosts (Table 20–1). The adaptations for brood parasitism include egg mimicry, nestling mimicry, host mimicry, and raptor mimicry. Protrusible cloacas, hard-shelled eggs, and deliberate destruction of host nests, host eggs, or host young also enhance a brood parasite's self-serving trade. Baby brood parasites are aggressive; it is not uncommon for hatchling cuckoos to shove the unhatched eggs of its host out of the nest. Baby honeyguides have special fanglike hooks at the ends of their bills for killing their nestmates (Figure 20–2).

Figure 20–2 Baby brood parasites dispose of their competitors: Hatchling Common Cuckoo (left) pushes the eggs of the host from the nest; a hatchling Black-throated Honeyguide (right) kills host nestlings with the hooklike tip of its bill. (Adapted from Lack 1968)

To minimize detection and destruction by the host, cuckoo eggs have come to resemble those of their primary hosts. Often only the heavier shell of the parasite's egg gives it away. The eggs of the Didric Cuckoo are so similar to those of its host, the Vitelline Masked Weaver, that one ornithologist resorted to chromosome analysis to distinguish them (Jensen 1980). The eggs of brood parasites are normally the same size or larger than those of their hosts. Brood parasites select hosts with same-sized or slightly smaller eggs to give their own young an advantage. The eggs of parasitic cuckoos are extraordinarily thick-shelled and resistant to cracking; females drop their eggs into deep nests, sometimes damaging the hosts' eggs rather than their own.

In Finland the blue eggs of the Common Cuckoo match those of its primary hosts, the Common Redstart and the Whinchat, whereas in Hungary the cuckoo lays greenish eggs with dark markings, similar to those of the Great Reed-Warbler. Common Cuckoos with particular egg types, or egg races, have either local or widespread geographical distributions. They usually have separate distributions that correspond to major habitats in Scandanavia and central Europe, which suggests that they have evolved in isolation. Nevertheless, in some parts of central Europe, up to four groups of cuckoos with distinctive eggs coexist. In this case, each egg race most likely evolved in a different region. The egg races then expanded until their distributions overlapped. No morphological and behavioral distinctions seem to be correlated with the egg races themselves. It is unlikely that coexisting egg races act as separate species; but the process of host selection by females of this species needs study (Payne 1977a). Possibly, females imprint on their hosts, which ensures that they will lay in nests in which their mimetic eggs best avoid detection (Figure 20–3).

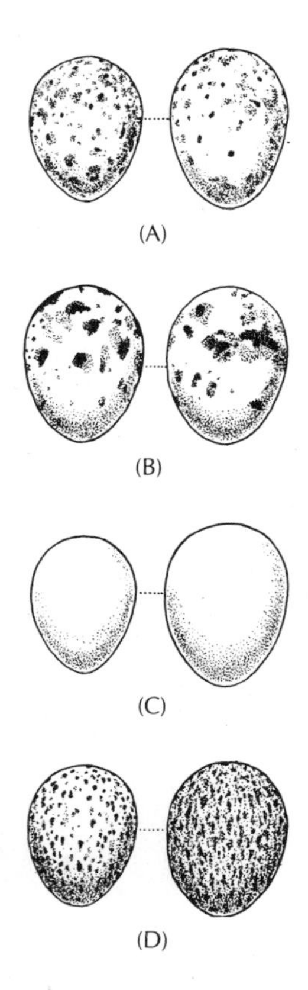

Figure 20–3 Common Cuckoo eggs (left) of particular types (egg races) closely match the color pattern of the eggs of their hosts (right). Identity of hosts: (A) Garden Warbler, (B) Great Reed-Warbler, (C) Common Redstart, (D) White Wagtail. (From Rensch 1947)

Prospective hosts try to drive brood parasites away. In India, the crow-black male Common Koel, a large Asian cuckoo, takes advantage of this reaction by drawing host crows from the vicinity of the nest so that its brown mate can slip in and deposit an egg. Male parasitic cuckoos have evolved hawklike plumage patterns, which incite mobbing behavior in potential hosts, drawing their attention away from the female parasite that is leaving an egg. The two Indian hawk-cuckoos, the Common Hawk-Cuckoo and the Large Hawk-Cuckoo, resemble the two sparrow hawks, the Shikra and Besra Sparrow Hawk, respectively. The Common Cuckoo resembles the Eurasian Sparrow Hawk and even flies like it. Host mimicry sometimes extends to nestlings. Feathered young Indian Common Koels, for example, look like nestling crows. Nestling whydahs have evolved mouth markings that mimic those of the nestlings of estrildine hosts. Young whydahs also imitate their nestmates by begging with their heads upside down (Nicolai 1964).

Like most birds, brood parasites lay eggs in clutches (Payne 1973a, 1976, 1977b). Female Brown-headed Cowbirds lay one to two dozen eggs (and in some regions more) per season in clutches of 2 to 5 eggs (Payne 1976). Brown-headed Cowbirds deposit their eggs at random; some host nests may have more than one cowbird egg (Mayfield 1965). African cuckoos of several species lay 16 to 25 eggs per season in clutches of 3 to 6 eggs, but they lay only one egg per nest. Very rarely do two females deposit an egg into one host nest. However, female indigo-birds occasionally put two or three eggs in each host nest. Even when they do so, a large number of their eggs become fledglings (Payne 1977b). As a rule the eggs of brood parasites require two to four days less incubation time than those of the host. This ensures earlier hatching and dominance by the young parasite. The Pied Cuckoo and Common Cuckoo incubate eggs in their oviduct for up to 18 hours prior to laying (Lack 1968; Liversidge 1971; Payne 1973a). Hatchling parasites also grow faster than nonparasites. Such advantages enable the young parasite to get most of the parental attention.

Effects of parasitism on the host

Parasitized nests rarely fledge young other than the parasite, so severely do brood parasites limit their host's breeding success. Parasitic adults or young may toss out or eat the host's eggs so that the host's clutch size appears normal. The young may also kill nestlings or cause the host to desert its nest. Feeding a large, insatiable parasite undoubtedly exhausts host parents and reduces their survivorship and ability to renest. Thus, the impact of brood parasitism can be substantial, especially in species with small populations. The Yellow-shouldered Blackbird, a species found only on Puerto Rico, is a case in point (Post and Wiley 1976, 1977). It has recently become a host for the parasitic Shiny Cowbird, which expanded rapidly throughout the West Indies from 1950 to 1970. Now the endangered Yellow-shouldered Blackbird breeds successfully only on tiny nearby islets where there are no Shiny Cowbirds.

Heavy parasitism by Brown-headed Cowbirds was responsible for the precipitous decline of the endangered Kirtland's Warbler in Michigan (Mayfield 1960). In 1957, the incidence of parasitism was high (about 55 percent); 75 percent of the nests examined between 1957 and 1971 were parasitized (Walkinshaw 1972). In just one decade, the number of singing male Kirtland Warbler's dropped from 502 (1961) to 201 (1971); parasitized nests produced nearly 40 percent fewer young than unparasitized nests. Control of cowbirds is now central to the management of the endangered Kirtland's Warbler populations.

Host responses to brood parasites

How well host birds accept the eggs of a brood parasite varies greatly among species. Stephen Rothstein (1975) placed artificial cowbird eggs in 640 nests of 30 species of North American birds. Twenty-three of these species usually accepted the eggs (meaning that they threw them out less than 40 percent of the time), whereas seven species usually rejected the different egg. "Rejectors" typically threw out the parasite egg as a natural extension of nest sanitation behavior. It appears that rejectors acquire true recognition of their own eggs through some form of imprinting that is less developed in "acceptor" species (Rothstein 1982).

Some birds, such as the Yellow Warbler, may respond to the discovery of a cowbird egg by deserting the nest or by burying the entire clutch in additional nest materials and laying a fresh clutch of eggs on top. A Yellow Warbler's method of rejection depends on the point in its laying cycle at which the cowbird adds the egg (Clark and Robertson 1981). Egg burial occurs most often when the warbler has just started its own clutch. This enables the warblers to renest without rebuilding the entire nest. Nests in

Table 20–2 Nesting Success of Parasitized Yellow Warblers

Nest Status	Number of Nests	Nest Success*
Parasitized		
Buried	13	0.78
Deserted	10	0.00
Accepted	12	0.53
Not parasitized	64	0.80

* After Clark and Robertson 1981. Average number of fledged young per egg laid, including buried eggs.

which eggs were buried have more reproductive success than nests in which cowbird eggs are accepted (Table 20–2).

In at least one case, young brood parasites are tolerated by their hosts because they may help their nestmates survive. In Panama, nestling Giant Cowbirds in Chestnut-headed Oropendola nests pluck bott fly larvae from nestmates (Smith 1968; Ricklefs 1979c). Bott flies cause a high level of mortality in oropendola nestlings; hence, adult oropendolas tolerate the presence of cowbirds and their eggs in colonies with serious infestations of bott flies but not in colonies with few bott flies. Cowbirds facing rejection have deceptive, mimetic eggs, whereas beneficial cowbirds in other colonies have plain, easily detected eggs (Figure 20–4).

Evolution of brood parasitism

The initial steps in the evolution of brood parasitism do not derive from a decay in normal breeding instincts or a hormonal imbalance. Rather, brood parasitism evolves because it enhances a female's reproductive output as long as chosen hosts accept the parasite's eggs (Lack 1968; Hamilton and Orians 1965). Intraspecific parasitism could be the first step in the evolution of obligatory brood parasitism. Intraspecific nest parasitism is known in at least 53 species of birds, of which 80 percent have precocial young (Yom-Tov 1980). The habit is most prevalent among waterfowl, but the list includes grebes, fowl, gulls, the ostrich, pigeons and doves, passerines, and cuckoos (other than the obligatory brood parasites). Most nests in some duck populations contain parasitic eggs (McCamant and Bolen 1979; Andersson and Ericksson 1982). European Starlings commonly lay eggs in nests other than their own (Yom-Tov et al. 1974). Ten percent of House Sparrow clutches in Australia contain eggs of other females (Manwell and Baker 1975). Both young females and those that have lost their nests may lay in the nests of other pairs. Intraspecific nest parasitism increases when there is a shortage of nest sites and when population density is high. For example, Cliff Swallows nesting in large, dense colonies in southwestern Nebraska regularly lay their eggs in each other's nests (Brown 1984). Careful daily monitors of the number of eggs in nests revealed that up to 24 percent of the nests in colonies of over 10 pairs of swallows receive eggs from others; many acts of parasitism that did not result in an increase of two

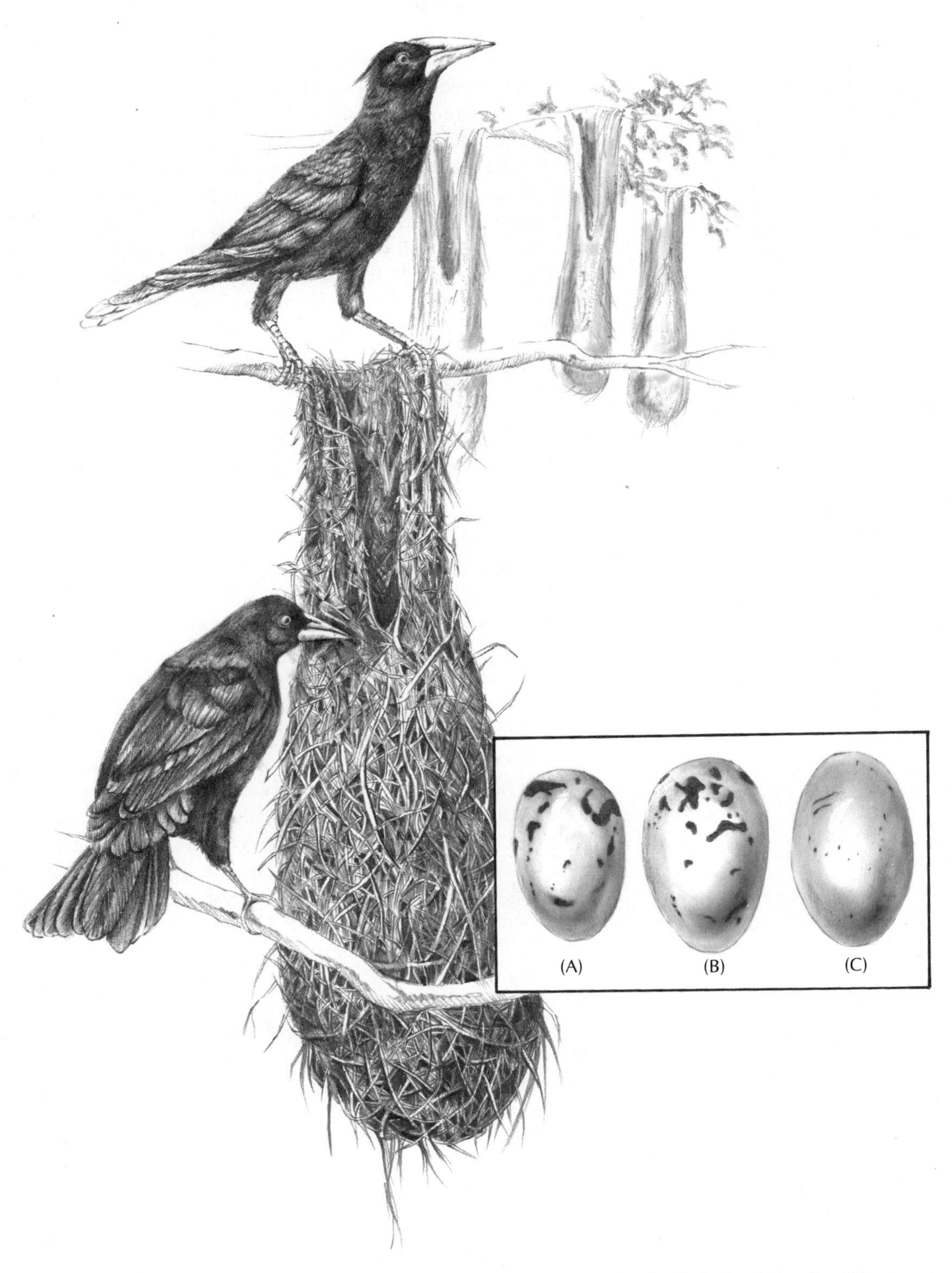

Figure 20–4 Chestnut-headed Oropendolas (above) are parasitized by Giant Cowbirds (below) that lay (B) mimetic and (C) nonmimetic eggs; (A) is an oropendola egg. (Courtesy Joel Ito and Chiron Press Inc.)

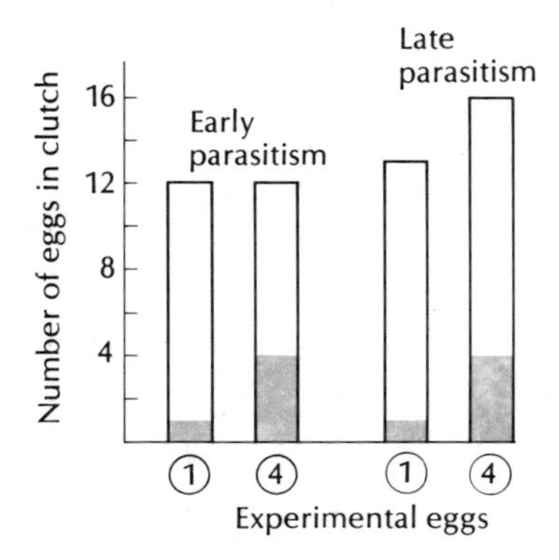

Figure 20–5 Female Common Goldeneyes respond to the experimental addition of eggs to their nests early in the laying cycle by laying fewer eggs. In this experiment, females laid significantly fewer eggs when four eggs were added than when only one egg was added (☐ denotes number of eggs added and ☐ denotes eggs laid by female). The final result in both cases was a clutch of 12 eggs for incubation. Female Common Goldeneyes laid their usual number of eggs when other eggs were added late in the laying cycle, resulting in larger clutch sizes. (After Andersson and Ericksson 1982)

eggs in the same nest on the same day escaped detection. Parasitic females quickly deposit an egg in a host nest when the host is away. In one instance a parasitic female swallow deposited her egg in only 15 seconds. Parasitism reduced the reproductive success of host females, which acted as though the parasitic egg were one of their own and laid fewer eggs.

Malte Andersson and Mats Eriksson (1982) studied the ways in which female Common Goldeneyes respond to intraspecific brood parasitism. The researchers observed that female goldeneyes accepted experimental parasitic eggs if they were added early in the laying period and then laid fewer eggs to reach a final clutch size, but when parasitic eggs were added late in the laying period, female goldeneyes did not adjust their own output and had to incubate unusually large clutches (Figure 20–5).

Facultative parasitism of the nests of related species with similar eggs probably is the next step in the evolution of brood parasitism. Ruddy Ducks and Redheads regularly lay additional eggs in the nests of other ducks (Weller 1959). Less well known is the fact that Black-billed Cuckoos and Yellow-billed Cuckoos of North America occasionally parasitize each other as well as other species, particularly when abundant food encourages the production of extra eggs (Nolan and Thompson 1975). Increasing specialization on another related species probably follows. In Argentina, the Screaming Cowbird parasitizes only the Bay-winged Cowbird, a nonparasitic cowbird that raises its own young in the old nests of other birds (Hudson 1920). An expanded repertoire of hosts is typical of more advanced obligatory brood parasites. Like the Brown-headed Cowbird of North America, the Shiny Cowbird of South America is an obligatory brood parasite of many small landbirds (Friedmann 1963).

Plumage, egg, mouth, and song mimicry may ultimately accompany host specialization. The evolution of host-specific whydahs has produced a variety of species adapted precisely to particular hosts, the estrilidine finches of Africa. These finches have distinctive mouth color markings: The mouth lining may be white, red, yellow, or blue; inside the mouth are black or violet spots; and the fleshy gapes at the corners of the mouth may be white, yellow, blue, or violet, elaborate or simple in morphology. Mimicry of mouth markings is important in these parasites because parents do not feed nestlings with the wrong markings (Nicolai 1974; Payne 1982a). Nestlings of the Paradise Whydah, for example, have one spot like nestlings of their host, the Green-winged Pytilia. Straw-tailed Whydah nestlings have three spots like nestlings of their host, the Purple Grenadier, and nestlings of the Village Indigobird have a ring of five spots like nestlings of their host, the Red-billed Waxbill. Both host and parasite of this latter pair of species also have the same mouth color pattern: white spots, blue corners, and yellow roof. Another pair of related species, the Jambandu Indigobird and its host, the Black-bellied Waxbill, share a mouth color pattern of blue spots, violet corners, purple roof, plus a pair of bright red spots. Color pattern may be a more important parental cue than the arrangement of the spots in these pairs of species (Figure 20–6).

How do brood parasites develop a sense of identity and association with other members of their own species? What little is known suggests that female Brown-headed Cowbirds rely largely on their genetic heritage. Females raised in isolation are fully responsive to songs of males of their own

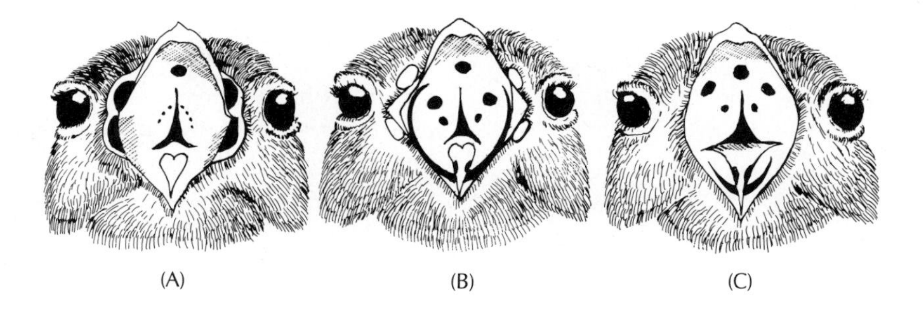

Figure 20–6 The mouth patterns of nestling parasitic whydahs match those of their hosts. (A) Paradise Whydah and Melba Finch; (B) Straw-tailed Whydah and Purple Grenadier; (C) Village Indigobird and Jameson's Fire-Finch. (After Nicolai 1974; Lack 1968)

species (King and West 1977): When they come into breeding condition and hear male songs, they solicit copulation. They do not respond to the songs of other species, even those of their foster parents. The calls of the Common Cuckoo are also inherited. Other parasites, however, imprint on the vocalizations of their foster parents. Young Village Indigobirds imitate their foster fathers' songs, including dialect variations (Payne 1982a). Thus, geographical patterns of song dialects correspond in host and parasite. The odd male Village Indigobird raised in the nest of a different host species acquires different songs, and the female Village Indigobird raised in such a nest becomes responsive to these different songs. Host vocalizations enable female indigobirds to recognize potential mates of their host heritage. In the short run, this ensures that young will have the host's mouth markings, which reduces chance of rejection. Host-specific lineages may then perpetuate themselves and, perhaps, evolve into separate species.

Cooperative breeding

Figure 20–7 The Florida Scrub Jay is one of the most thoroughly studied species of cooperatively breeding bird. (Courtesy A. Cruickshank/VIREO)

Brood parasitism evolves because it clearly favors the reproductive success of some individuals that exploit others. Advantages are not as obvious in cases of cooperative breeding, which occurs when individuals other than genetic parents help to incubate, feed, or protect young that are not their own. In Florida Scrub Jays, one of the best-known cases, about half of the breeding pairs have helpers, and the basic social unit is composed of a breeding pair with up to six helpers. Each group defends a territory throughout the year, with the helpers protecting and bringing food to the nestlings (Woolfenden and Fitzpatrick 1984). Thus the helpers are an integral part of the social system (Figure 20–7). The phenomenon of "helpers at the nest" was first reviewed by Alexander Skutch (1961). We now know of cooperative breeding in over 150 avian species (Emlen and Vehrencamp 1983). Cooperation is a major aspect of the breeding biology of birds, and one that has received intensive field study in the past decade. The apparent altruism of cooperative breeding challenges the basic tenets of evolution by natural selection. Charles Darwin himself offered the discovery of altruistic behavior as a way to disprove his theory. A century later, V.C. Wynne-Edwards (1962) shocked the establishment of evolutionary biologists when he concluded that individuals place the good of their populations or species above their individual well-being. In particular, helpers at the nest seemed to offer compelling cases of altruism.

Cooperation or opportunism?

Do helpers make positive contributions or do they interfere? Do helpers serve their own interests in some way or do they sacrifice their reproductive potential to help others? Answers to these questions could reconcile cooperative breeding with evolutionary theory. Two main avenues of reconciliation are possible. Helpers might be genetic relatives, or kin (but not parents), in which case they could propagate some of their own genes indirectly by helping parents or siblings. Kin selection is one route to understanding complex social behavior in ants, bees, and wasps, in which sterile castes help their mother produce sisters (Hamilton 1964; West-Eberhard 1975). Alternatively, help might be based on a principle of compensation. Mutualism or reciprocal altruism could be in an individual's best interest as long as cheating does not occur (Trivers 1971; West-Eberhard 1975; Axelrod and Hamilton 1981). We now recognize that individuals can enhance their own lifetime reproductive success through indirect means such as kin selection and reciprocal altruism as well as by direct means. Determining whether reproduction by means of helping kin contributes significantly to an individual's lifetime reproductive output is of primary importance (Vehrencamp 1979; Emlen 1984). Preliminary calculations suggest that indirect reproduction through kin contributes 30 to 50 percent of an individual's inclusive fitness (Vehrencamp 1979; Rowley 1981).

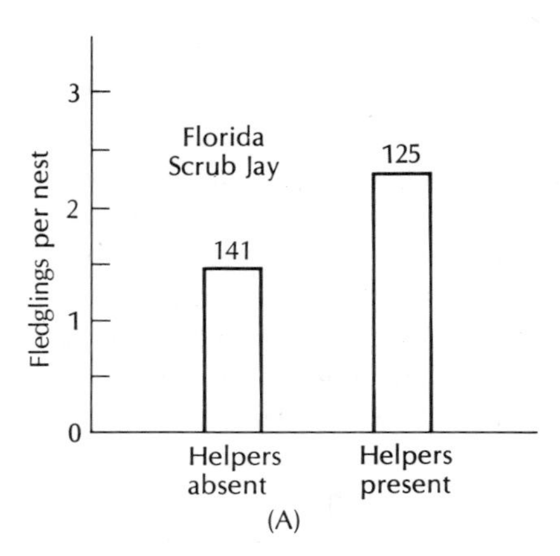

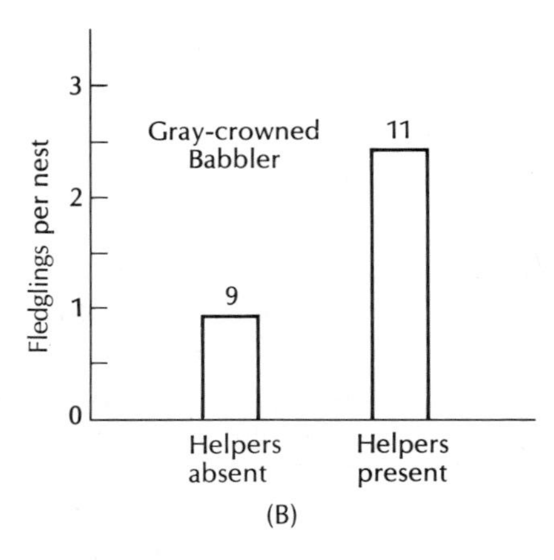

Figure 20–8 Groups with helpers fledge more young. (A) Groups of Florida Scrub Jays with helpers produce more fledglings per nest than do pairs without helpers. (B) Experimental removal of helpers from breeding groups of Gray-crowned Babblers reduces the average number of young fledged per nest. Number of pairs and pairs with helpers studied is indicated over each bar. (A, adapted from Woolfenden 1981; B, adapted from Brown et al. 1982)

Ian Rowley (1965) pioneered the study of the details of cooperative nesting in birds, specifically in the Australian Superb Blue Wren. His discovery that helpers are young from previous broods has been confirmed: Most helpers help genetic parents or stepparents to raise siblings or half-siblings. Of the 199 Florida Scrub Jays that Glen Woolfenden recorded as helping from 1969 to 1977, 118 (59 percent) helped both their mothers and their fathers, 49 (25 percent) helped one parent and one stepparent, and 32 (16 percent) helped distant kin as well as non-kin (Emlen 1978). Thus the functional social units are essentially extended families, and the production of genetically related siblings seems consistent with evolutionary theory.

The apparent importance of kin selection is weakened by observations that helpers continue to help unrelated stepparents after their own parents die. In the Gray-breasted Jay, helpers may be only distantly related to the nestlings they help feed because these jays regularly immigrate from other family units (Brown and Brown 1981a).

Most studies show that helpers truly help the parents in their social unit. In Florida Scrub Jays, for example, groups with helpers fledge more young per season than groups without helpers. In some cases the number of young fledged increases with the number of helpers. In a direct test of this question, Jerram Brown and his colleagues (1982) removed helpers from family units of the Gray-crowned Babbler. The average number of fledglings declined from 2.4 in the first broods (with helpers) to 0.8 in the second and third broods (without helpers). Therefore, each helper in this experiment increased the number of fledglings by 0.45 (Brown and Brown 1981b) (Figure 20–8).

The higher number of young produced with assistance results not so much from increased feeding and reduced starvation of nestlings as from relief of stress on the parents. Breeding Florida Scrub Jays with helpers survive more often (87 percent per year) than do those without helpers (80 percent per year) (Stallcup and Woolfenden 1978). Helpers also increase the parents' ability to start a second or third brood. Larger breeding groups of Gray-crowned Babblers renest sooner and start more clutches than do smaller breeding groups with few or no helpers (Brown and Brown 1981b).

The combined territories of helpers and parents presumably increase the survival rates of parents and young during periods of food shortage or unpredictability. Young that stay on familiar territory rather than disperse to unfamiliar areas have a better chance of surviving, which enhances the parents' long-term reproductive success. Parental tolerance of grown offspring on their natal territories, despite the fact that they may compete for resources, is a key step in the evolution of cooperative breeding systems (Brown 1974; Gaston 1978). Whether helpers reduce predation on nestlings or fledglings is not yet certain. This may be so in Florida Scrub Jays, which must fend off attacks by snakes (Woolfenden 1978). Larger groups may help birds detect predators earlier, as in the case of feeding flocks (Chapter 15) and may help birds defend their nests against predators, as in the case of breeding colonies. Robert Ricklefs (1980) has suggested that adults with helpers are safer at nest holes or covered nests because of the sentinel or "watchdog" behavior of the helpers. Reduced predation, however, is probably only a secondary benefit of helpers.

Why do helpers help rather than breed elsewhere on their own? In the role of a helper, a Gray-crowned Babbler rears an average of 0.46 fledglings compared with the average of 3.62 fledglings it could produce as a breeder, with help of its own (Brown and Brown 1981b). In fact, the per-capita reproduction of extended family units usually decreases with group size and is less than that of pairs without helpers. But such figures assume that helpers could breed on their own, and the growing conclusion is that young stay with their parents to help primarily because they cannot nest successfully on their own. Ecological constraints limit successful dispersal and reproduction of young birds entering the breeding population. Even though they may be physiologically capable of breeding on their own, young birds delay reproduction for several years until they achieve the required social status and control of territorial space.

Ecological factors

Achieving breeding status on an exclusive territory is difficult when occupied territories saturate the habitat. This situation is most common in species that reside in stable environments and that have specialized habitat requirements and high adult survivorship. Walter Koenig and Frank Pitelka (1981) suggested that the constraints on independent breeding are greatest when there is little marginal or secondary habitat for young individuals to occupy. Florida Scrub Jays, for example, are restricted to undisturbed oak-palmetto scrub habitat in central Florida, which exists only as small islands of scrub surrounded by other vegetation. In California and the southwestern United States, however, suitable Scrub Jay habitat is widespread; pairs of Scrub Jays in this region raise young without help.

Acorn Woodpeckers in central California are restricted to a habitat that can supply them with enough acorns to last the winter (Koenig 1981a,

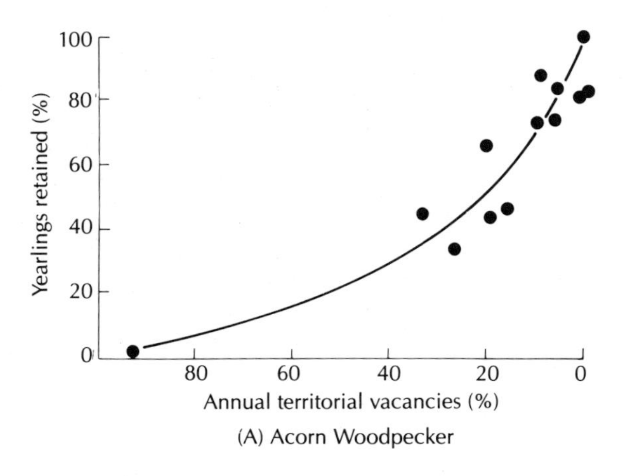

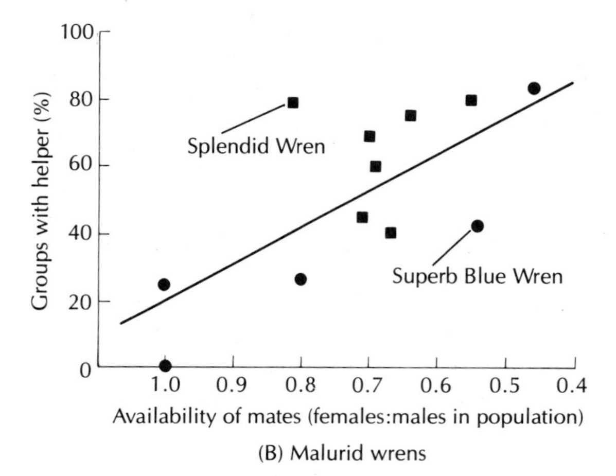

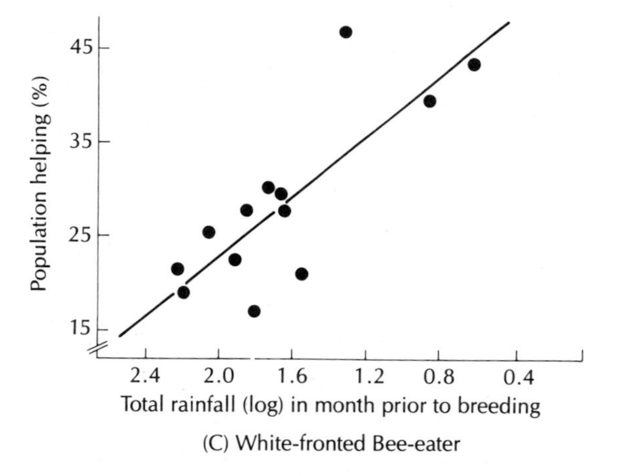

Figure 20–9 The retention of young as helpers may result from ecological constraints: (A) territory shortage, Acorn Woodpecker; (B) availability of female mates, malurid wrens; and (C) environmental harshness (lack of rain), the White-fronted Bee-eater. (Adapted from Emlen 1984)

b). This habitat is fully occupied by communal families that build and defend their granaries (large trees full of holes in which acorns are stored). Opportunities for young woodpeckers to establish themselves are scarce because communal families control key resources. Consequently, dispersal of young is limited. In Arizona and New Mexico, however, where Acorn Woodpecker habitats (with their resources) are scarce and of marginal quality, over-winter survival is less, dispersal is higher, and fewer groups

Table 20–3 Variation in Social Structure of Acorn Woodpeckers

Location	Dispersal (%)	Saturation (%)	Group Size	Units with Help (%)
Central California	51	100	5.1	70
Arizona and New Mexico	71	81	2.8	56
Southeastern Arizona	100	7	2.2	16

From Stacey and Bock 1978.

have helpers. The least favorable conditions occur in the Huachuca Mountains of southeastern Arizona where periodic scarcity of resources results in many vacant territories. Here, the young typically disperse, and few groups have helpers (Stacey and Bock 1978) (Figure 20–9) (Table 20–3).

Uli Reyer's (1980) studies of Pied Kingfishers on Lake Naivasha and Lake Victoria in Kenya shed more light on the importance of local ecology in the acceptance of helpers. Breeding male kingfishers' acceptance of unrelated male helpers is directly related to their need for help in delivering fish to their young. On Lake Victoria, where fishing is difficult, a single helper doubles the average fledging survival rate from 1.8 to 3.6 young per nest; on Lake Naivasha, where fishing is easier, helpers have less effect: The rate increases from 3.7 to 4.3 young per nest. Most breeding pairs on Lake Victoria have helpers, at least one of which is not their own progeny. On Lake Naivasha, however, few pairs have helpers and of those that do, almost all are their own young. These auxiliary male helpers increase their own chances of eventually pairing with a female by helping her raise the brood, and they also may recruit the young they helped raise as future helpers of their own (Ligon 1983).

How, then, do helpers finally achieve breeding status? Waiting for an opening is the first step. Female Florida Scrub Jays and Jungle Babblers monitor nearby groups and move quickly to replace females that disappear. Males, on the other hand, inherit breeding positions on their natal territories in relation to their age and status. The dominant (usually oldest) son replaces its (deceased) father, stepfather, or brother. Helpers may take over a separate portion of the family territory for their own breeding purposes. Defense of large group territories is typical of cooperatively breeding birds. Increased production of siblings is important in this respect; territories expand with group size. When the territory is large enough, one part is ceded to the oldest male, paired with a female recruited from outside the family unit.

Green Woodhoopoes, large hole-nesting birds of the African savannas, typically live in extended family groups of helpers (Ligon and Ligon 1978; Ligon 1981). Mortality among breeding adults is high, which creates openings for new breeders. However, young woodhoopoes usually cannot capture and hold a territory for themselves without help. Therefore pairs, usually an older and younger sibling or half-sibling, cooperate to secure new breeding space. David Ligon and Sandra Ligon (1983) suggest that woodhoopoes help to rear younger birds in order to form bonds with them. Later the younger bird will help its former helper take control of a quality territory. This initial cooperative effort leads to a long-term working partnership between siblings. The alliance is in the younger woodhoopoe's interest because it will eventually replace its partner as the breeding male of the new unit. The recruitment of younger siblings and the resettlement of sibling subgroups appear to be regular features of cooperative breeding systems (Emlen 1982a, 1984) (Figure 20–10).

Unpredictable or difficult breeding conditions are another ecological feature of cooperative breeding in some birds (Emlen 1982b). Many species that live in the dry forests of Africa and Australia breed cooperatively. Some are nomadic. Others, such as the White-fronted Bee-eater of East Africa, are resident and colonial. Nestling bee-eaters often starve when adequate rains and good supplies of insects fail to materialize. Helpers increase the rate of food delivery. Because space is available, the helpers

Figure 20–10 Scenario of cooperative partnerships and reciprocity in the Green Woodhoopoe, which suffers high and unpredictable mortality. (A–E) Offspring and their older helpers ultimately move together to take over a nearby breeding territory. (From Ligon and Ligon 1983)

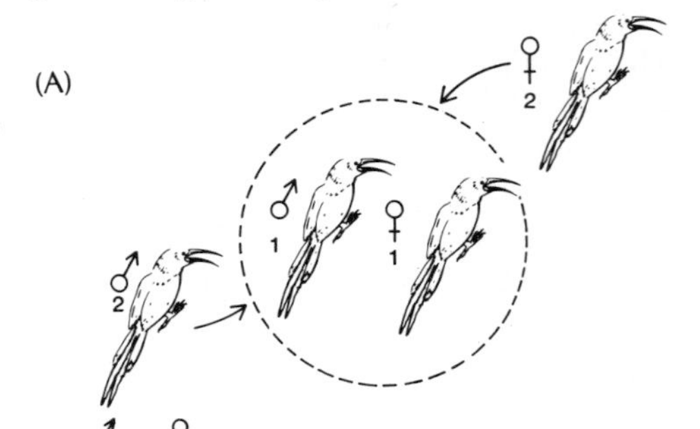

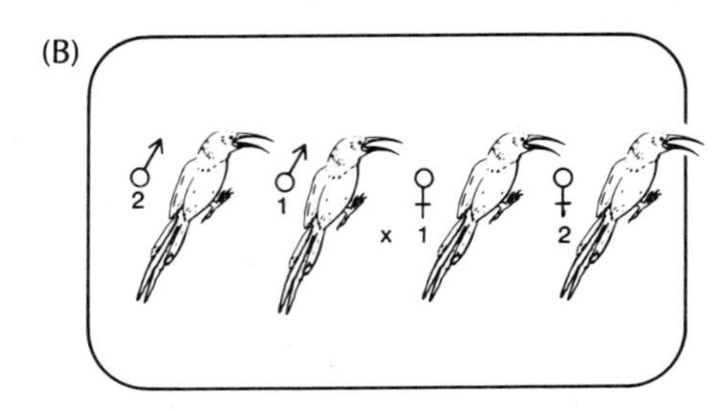

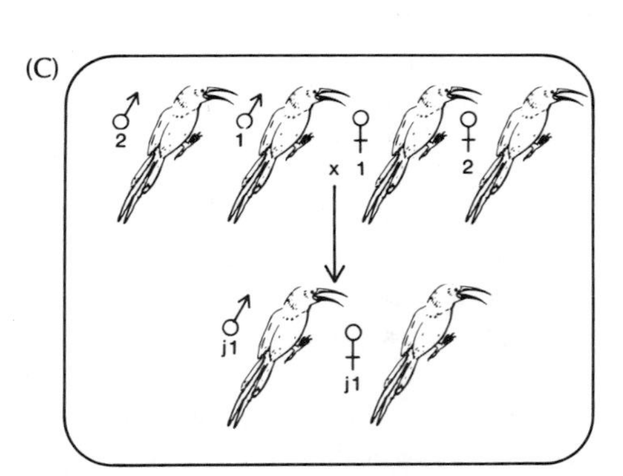

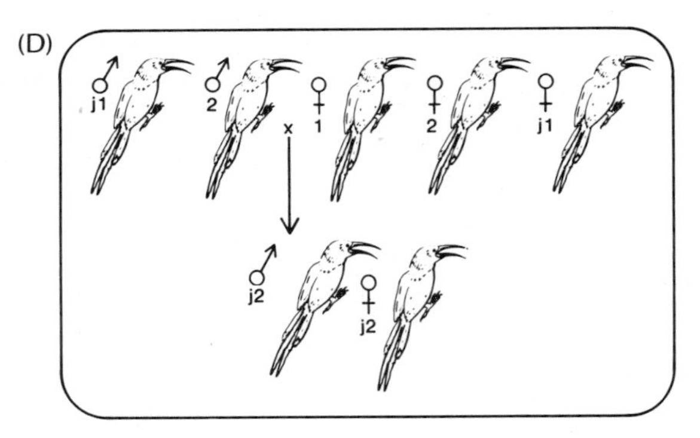

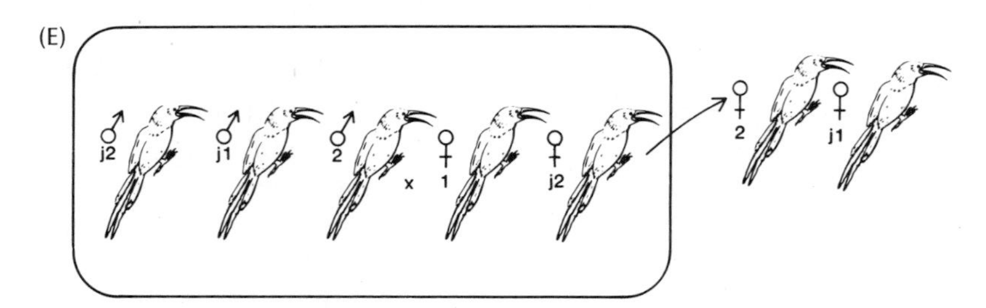

could start their own nests, but only in good years can they raise young successfully by themselves. Supporting the prediction that cooperative breeding should increase with harshness of environmental condition, group size increases with low rainfall and poor food availability in the month preceding the onset of breeding (see Figure 20–9C).

Ecological constraints set the stage for the conflicts that are a natural part of communal breeding. Behind the facade of a cooperative social order is a swirl of competition, strife, and harassment. Helpers may deliberately interfere with parental reproduction to increase turnover and thereby increase their own chances of breeding (Zahavi 1974). Conversely, adults may sabotage the initial breeding efforts of young to increase the incentives for the young to stay at the nest as helpers. Young helper males sometimes mate with their stepmothers (Emlen 1982a), and helper females sometimes slip an egg of their own into the parental clutch. In cases of communal nesting, females may destroy each other's eggs.

Groove-billed Anis, large black cuckoos of the New World tropics, form social units of one to four monogamous pairs. The unit's eggs are laid in a single nest, and all the individuals in the unit help incubate and feed the communal brood. The main advantage of communal nesting in this species is in sharing the high nocturnal predation risks during incubation and brooding, thereby improving individual survivorship (Vehrencamp 1978). Female anis compete among themselves to ensure the success of their respective contributions to the clutch. Because one nest cannot hold all of the eggs, the females throw each other's eggs out to make room for their own. Young subordinate females start laying first. The older females toss out some of these eggs to make room for their own eggs, which make up most of the clutch. Subordinate females counter these actions by increasing the total number of eggs added to large clutches, by prolonging the interval between eggs laid, and by producing a "late egg" as the clutch size nears completion. There are natural limits to the subordinate female's attempts because the last-born nestling is the smallest and most vulnerable member of the brood. It is unknown whether members of the ani's social units are genetic relatives.

Kinship

Kinship bonds do characterize some communal nesting species. Trios of the flightless Tasmanian Native-Hen, for example, consist of a pair of brothers and an unrelated female (Maynard Smith and Ridpath 1972). Both brothers copulate with their shared spouse and help care for their young. Acorn Woodpecker sisters, which lay eggs in the same communal nest, copulate promiscuously with the various males in the group (Koenig and Pitelka 1979). Participation in paternity is perhaps one incentive for the males to remain as helpers. The potential for complex social relationships may be greatest in colonial breeding birds with frequent predictable contacts with large numbers of individuals. The White-fronted Bee-eater is a case in point. These bee-eaters breed in large colonies in East Africa but function on a daily basis in clans of two to seven individuals that feed together and defend a group territory within 20 miles of the colony (Hegner et al. 1982). Members of a clan feed, roost, and may breed cooperatively.

Steve Emlen's studies (1981, 1982a, b) of color-marked White-fronted Bee-eaters in Kenya's Rift Valley near Lake Nakuru reveal that the

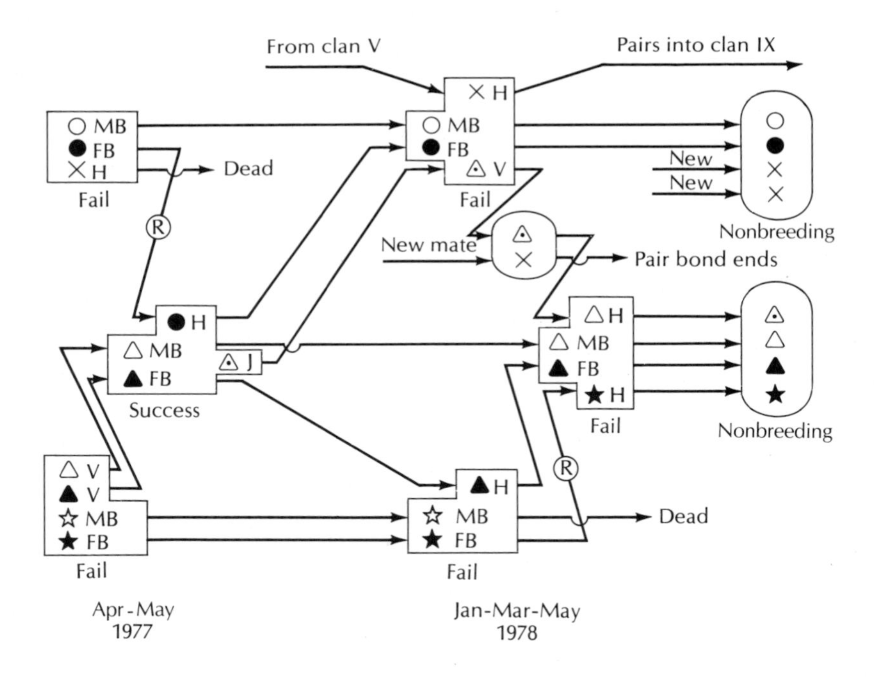

Figure 20–11 Clan relations in White-fronted Bee-eaters. Core members of the clan are identified individually by symbols (circles, triangles, stars), and connecting lines trace their social movements over time. Temporary associates are indicated by X. In 1977, the clan consisted of three monogamous pairs and their associates. Two of the pairs failed in their breeding attempts and one succeeded. Each box represents a breeding or roosting chamber in the colony (MB, male breeder; FB, female breeder; H, helper; V, visitor, i.e., a bee-eater that roosted in the chamber but did not help in the nesting effort; J, juvenile). Note the rearrangements of the associations within the clan, redirected helping by breeders whose own efforts failed, and the reciprocal helping between females represented by the closed star and closed triangle. (After Hegner et al. 1982; photo courtesy S. Emlen and N. Demong)

fabric of bee-eater society is a "mixture of openness and fluidity of group memberships on the one hand, with stability and fidelity of certain social bonds on the other" (Emlen 1981). Individuals appear to remember past associations. They leave groups to join other groups but return months or years later to roost or nest with old associates. Despite their flexibility and fluidity, personal relationships based on individual recognition and long-term memory could possibly provide a social basis for subtle forms of reciprocal altruism, social manipulation, and kinship responses, all extraordinary levels of social complexity far beyond what has been documented to date in bird groups (Figure 20–11).

The open cooperative breeding system of colonial bee-eaters such as the White-fronted Bee-eater seems to be adapted to the unpredictable environment of the Rift Valley. In some years pairs can breed successfully by themselves; in other years they cannot. Unlike closed cooperative breeding systems in saturated stable environments, where young cannot disperse and must compete with established individuals for breeding status, adult bee-eaters have less control over the breeding options of potential helpers. Emlen suggests that, as a result, potential breeders must sometimes allow helpers to share paternity or maternity of group clutches to attract their assistance. Complex social bonds result. Adult White-fronted Bee-eaters switch between breeding and helping more flexibly than in any closed (noncolonial) cooperative system known among birds.

Summary

Brood parasitism and cooperative breeding lie at opposite ends of the spectrum of parental care practices among birds. Brood parasites lay their eggs in the nests of other birds. By allowing foster parents to raise their young, brood parasites reduce their own costs and avoid the risk of losing all the young in a single nest. Cooperative breeders help other breeders (usually parents or stepparents) to raise young while waiting for an opportunity to breed themselves. Cowbirds and cuckoos are the most familiar brood parasites, but this practice has evolved along at least seven separate lines among birds. Adaptations for brood parasitism include egg color mimicry, nestling mimicry, host mimicry, raptor mimicry, egg size and hardness, and destruction of host eggs and young. A high incidence of brood parasitism is responsible for the decline of some host populations. Countermeasures evolved by hosts include egg recognition and nest abandonment. Intraspecific brood parasitism, the leaving of eggs in nests of other females of the same species, is proving to be quite common among birds. Facultative parasitism of other species is an intermediate step in the evolution of obligatory brood parasitism.

Although cooperative breeders may appear to act altruistically, they actually act in their own best interest. Cooperative breeding evolves under conditions of ecological constraint, such as lack of breeding territories, that prevent birds from breeding on their own. By helping to raise other broods, these birds enhance their own chances for breeding through inheritance of a territory or through other forms of territory acquisition. In some birds, helping seems to lead naturally to the recruitment of young allies to assist in territorial takeovers. Breeding pairs with helpers fledge more young than those without helpers, primarily because they suffer less stress and hence survive longer and are more likely to renest. Strife abounds in the social relations of cooperative breeders such as Groove-billed Anis, which throw each other's eggs out of their communal nests. White-fronted Bee-eaters are remarkable for the complex, cooperative social systems that help them adapt to the unpredictable environments of East Africa.

Further readings

Emlen, S.T. 1984. Cooperative breeding in birds and mammals. *In* Behavioral Ecology (J.R. Krebs, and N.B. Davies, eds.), 2nd ed., pp. 305–340. Sunderland, Mass.: Sinauer Assoc. *A major review of conceptual issues.*

Emlen, S.T., and S. Vehrencamp. 1983. Cooperative breeding strategies among birds. *In* Perspectives in Ornithology (A.H. Brush and G.A. Clark, eds.), pp. 93–120. New York: Oxford University Press. *A detailed review of the evolution of this behavior.*

Lack, D. 1968. Ecological Adaptations for Breeding in Birds. London: Methuen. *A classic.*

Skutch, A.F. 1976. Parent Birds and Their Young. Austin: University of Texas Press. *A readable account of the natural history of avian reproduction with good chapters on brood parasitism and cooperative breeding.*

Payne, R.B. 1977. The ecology of brood parasitism in birds. Ann. Rev. Ecol. Syst. 8:1–28. *An excellent and critical review of the literature.*

Woolfenden, G.E., and J.W. Fitzpatrick. 1984. Demography of a Cooperative Breeding Bird. Princeton, N.J.: Princeton University Press. *Details of an extraordinary long-term study.*

part VI

POPULATIONS

chapter 21

Demography

We have examined individual avian reproduction in terms of physiological process, behavior, energetic costs, and benefits. We now extrapolate from individual reproductive efforts to group life-history patterns. These patterns establish the dynamics of annual survival, fecundity, and ecology, on which basis we project overall population characteristics and population growth rates. A central issue in comparative avian demography — the study of variations in life-history patterns — concerns the evolution of clutch size. Under investigation are the ways in which clutch sizes maximize lifetime reproductive success and survival of parents and young and the reasons that clutch sizes vary among populations of the same species.

Life-history patterns

Some birds lay many eggs, others lay just one. Some birds live just a few years, others live for decades. Reproduction rates, on the one hand, and survival rates, on the other, combine to define an individual's contributions

Figure 21 – 1 Albatrosses are long-lived birds that raise only one offspring at a time. (Courtesy J. Warham)

to succeeding generations. These contributions drive natural selection. Several patterns characterize avian life histories. First, avian mortality rates are high in the first year of life and then drop to more or less constant lower levels among adults. Second, reproductive success and effort improve with age and experience. Third, long-lived species, such as albatrosses, penguins, and eagles, tend to low fecundity; short-lived species, such as songbirds and ducks, tend to higher fecundity (Figure 21 – 1). *Fecundity* is the number of young fledged per year. Finally, ecology molds life-history patterns in predictable ways. In the north and in arid environments, clutches are generally larger than those in the south and in mesic environments.

One of the central questions in comparative demography concerns the evolution of reproductive effort in relation to other demographic pa-

Table 21 – 1 Extremes of Avian Life History Patterns

Parameter	Albatross, Eagle	Small Passerine, Duck
Survival before breeding	Moderate (30% per year)	Low (15% per year)
Age of reproduction	Late (8 – 10 years)	Early (1 year)
Fecundity	Low (0.2 young per year)	Moderate (3 per year)
Adult mortality	Low (5% per year)	High (50% per year)

rameters. A bird's lifetime reproductive output reflects the age at which it first reproduces, the number of young it fledges each year, the survival of those young, and its own longevity as an adult. For individuals to maximize their contribution to the next generation they must hit upon optimal combinations of these four variables (Table 21 – 1).

Life tables

The extremes of avian life-history patterns can be illustrated by comparing a small, short-lived bird such as a sparrow with a large, long-lived bird such as a penguin. The sparrow starts breeding after one year and concentrates high reproductive achievements into a few consecutive years. The Yellow-eyed Penguin starts breeding after four years and spreads lower annual fecundity over a long span of years.

To study the life-history patterns of a particular kind of bird, the ornithologist follows the progress of a marked group of individuals, known as a cohort, from the time they fledge to the time the last one dies. The proportion of the cohort that survives each year defines the annual survivorship (s_x). The probability of survival to a particular age (l_x) is the product of the preceding annual survival rates. The number of young produced each year by adults in this cohort defines age-specific fecundity (b_x). The product $l_x b_x$ specifies an individual's expected annual fecundity, which is to say fecundity at a certain age discounted by the chance of dying before reaching that age (Figure 21 – 2).

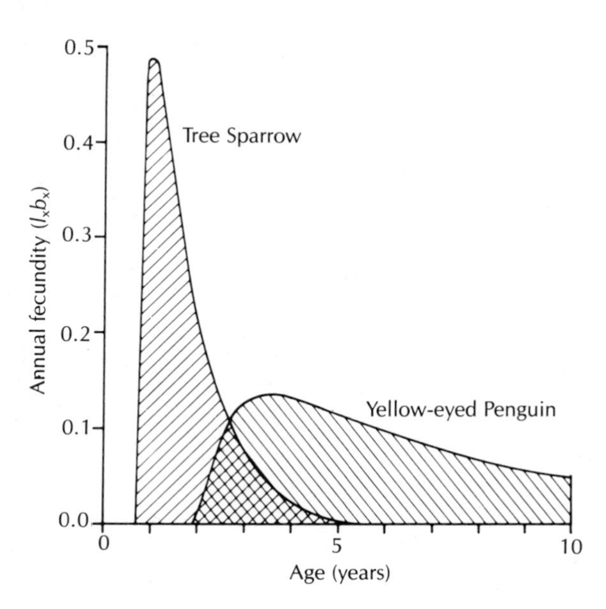

Figure 21 – 2 Reproductive efforts by an Eurasian Tree Sparrow and a Yellow-eyed Penguin, expressed in terms of expected annual fecundity $(l_x b_x)$, where l_x is the probability of survival to a particular age, and b_x is age-specific fecundity. The short-lived sparrow produces more young every year than does the long-lived penguin, but their total lifetime fecundities are roughly the same. (Adapted from Ricklefs 1973b)

Table 21–2 Life Table of the Eastern Screech-Owl in Ohio

Age (years)*	s_x	l_x	b_x†	l_xb_x	$x\lambda^{-x}l_xb_x$
0	0.305	1.000	0.00	0.000	0.000
1	0.594	0.305	1.04	0.317	0.317
2	0.632	0.181	1.30	0.235	0.470
3	0.667	0.115	1.30	0.150	0.450
4	0.750	0.086	1.30	0.112	0.448
5	0.750	0.064	1.30	0.083	0.415
6	0.750	0.048	1.30	0.062	0.372
7		0.036	1.30	0.047	0.329
Total				1.006	2.801

* Life table is arbitrarily truncated at 7 years, the age of the oldest individual recovered in the study.
† Fecundity (b_x) is the number of offspring reared, divided by two because only females are considered in the life table; assumes that sex ratio at fledging is 1 : 1.
From Ricklefs 1973.

Life tables summarize the vital population statistics of age-specific survivorship and age-specific fecundity under a given set of conditions (Table 21–2). From these statistics one can project individual life expectancy and family sizes of other individuals facing the same conditions. From life table data one can also project rates of population growth, which are the net result of fecundity and mortality. Life tables are usually based on the statistics of females because these are more reliably measured than those of males. It is easier to associate eggs in a nest with the female that laid them than with the male or males that might have fertilized them. Thus, age-specific survivorship is the proportion of females that survive to age x. Age-specific fecundity is the number of female offspring or eggs produced by a female of a particular age each year.

Life table data enable us to project future population trends (see Ricklefs 1973b for a review of the assumptions involved). The values of l_xb_x for each age category add up to R_0, which is the net reproductive rate, or the expected recruitment of new individuals into the population. If one female is replaced by one other during her lifetime, R_0 equals 1, and the population composed of many such individuals is stable in size. Larger values of R_0 are expected in growing populations and smaller values in declining populations. Thus, if $R_0 = 1.5$, the population will increase 50 percent in one generation. A value of 0.8 indicates a rapidly declining population. When life table data are compiled for a pair of mates rather than for single females, the effective values of R_0 are doubled; a pair that is simply replaced by another pair has an R_0 of 2.

The Eastern Screech-Owl, for example, lives about 7 years and fledges two to three young each year (see Table 21–2). In this case, annual survivorship (s_x) increases with age to a maximum of 75 percent per year. Screech-owls achieve full breeding potential by the age of 2 years, by which time each female produces an average of 1.3 offspring each year (b_x). The annual products of survivorship and fecundity (l_xb_x) sum to 1.006, which indicates simple lifetime replacement of a female by one daughter. Increases in either survivorship or fecundity cause an increase in lifetime reproduction and, hence, higher growth rates for the extended family or the population, whichever is the group of interest.

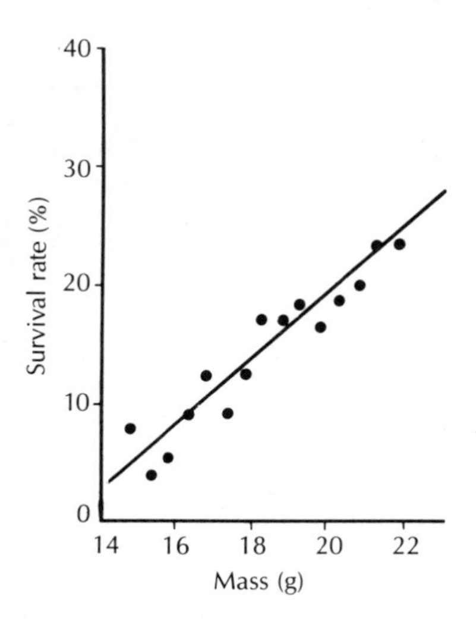

Figure 21 – 3 The probability of survival (and hence recapture by ornithologists) increases directly in relation to the mass a young Great Tit attains prior to leaving the nest. (Adapted from Perrins 1980)

Survival rates

Keeping in mind the overall patterns of avian demography and the relations between fecundity and survival as summarized by a life table, we now look in depth at the nature of survival and fecundity in birds.

A young bird's annual chance of survival from fledging to breeding age is about half that of an adult. Small landbirds are especially vulnerable in their first year. Mortality during the first few weeks out of the nest is particularly severe in some birds such as Eurasian Skylarks, only 22 percent of which survive the first 25 days (Delius 1965). Mortality of fledgling Black-capped Chickadees, on the other hand, is low: 82 to 97 percent may survive their first four weeks (Smith 1967). Generally speaking, the more advanced the young when they leave the nest, the greater are their chances of survival (Ricklefs 1973b). This is one of the advantages of longer nestling periods and of fast growth in altricial nestlings. A fledgling's chance of survival (measured by ornithologists in terms of future recaptures) increases in proportion to its mass at fledging (Figure 21 – 3). Food availability, the quality of parental care, the number of siblings competing for that care, and the timing of fledging are all important factors (Perrins 1980).

Some of the best field data of predation on young birds come from the systematic collection of the metal bands ("rings") on songbird carcasses that accumulated at the nests of Eurasian Sparrow Hawks that fed on the research populations of Great Tits and Blue Tits in Wytham Woods at Oxford (Perrins and Geer 1980). These avian predators took 922 ringed tits in 1976, 759 in 1977, and 1220 in 1978. Included were 18 to 34 percent of the Great Tit juveniles and 18 to 27 percent of the Blue Tit juveniles fledged each year.

Once birds reach adulthood, their survival rates increase and stay essentially constant. In the case of the Herring Gull, for example, only 60 percent survive the first year, but survival remains at about 90 percent each year thereafter (Paynter 1966) (Figure 21 – 4). Twenty percent of fledgling Great Tits survive their first year; survival rates climb to 48 percent in breeding females and 56 percent in males (Bulmer and Perrins 1973). Adult survival rates range from as little as 30 percent in Blue Tits and Song Sparrows to over 95 percent in Royal Albatrosses, Bald Eagles, and Atlantic

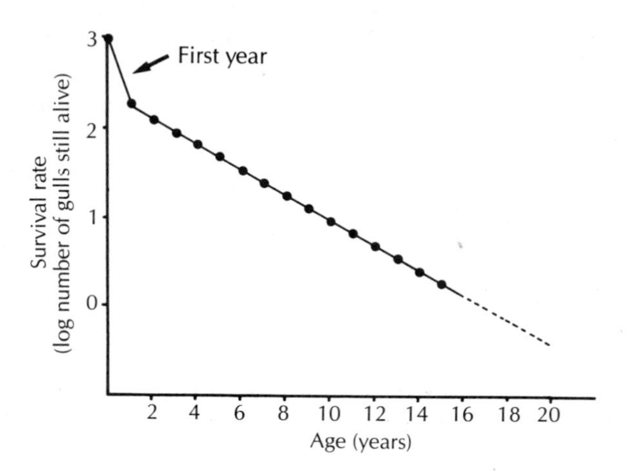

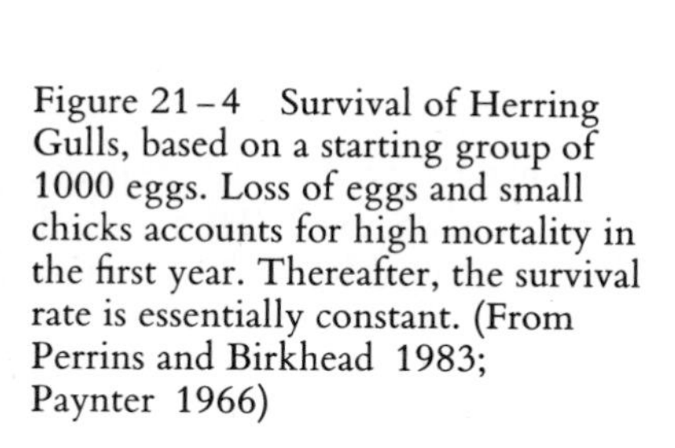
Figure 21 – 4 Survival of Herring Gulls, based on a starting group of 1000 eggs. Loss of eggs and small chicks accounts for high mortality in the first year. Thereafter, the survival rate is essentially constant. (From Perrins and Birkhead 1983; Paynter 1966)

Table 21 – 3 Annual Survival of Adults

	Survival (%/year)	Mean Further Life Expectancy (years)
Fowl	20 – 50	0.7 – 1.5
Small landbirds	30 – 65	0.9 – 2.4
Ducks	40 – 60	1.2 – 2.0
Raptors	50 – 96	1.5 – 24.5
Herons, gulls, waders	60 – 80	2.0 – 4.5
Seabirds	80 – 95	4.5 – 19.5
Tropical landbirds	50 – 90	1.5 – 9.5
Albatrosses	95	19.5

From Ricklefs 1973b; Welty 1982, Perrins and Birkhead 1983.

Puffins. In general, large species survive better than small species, tropical species survive better than temperate zone species, and seabirds survive better than landbirds. Survival rates vary greatly among similar kinds of birds, regionally within species, and between the sexes. Data on adult survival have been reviewed for gallinaceous birds (Hickey 1952, 1955), waders (Farner 1955; Boyd 1962), seabirds (Lack 1954; Ashmole 1971), and birds in general (Cody 1971; Haartman 1971; Ricklefs 1973b; Welty 1982) (Table 21 – 3).

Causes of mortality

The causes of mortality of adult birds are summarized by Joel Welty (1982). In general, the relative number of occurrences of the various causes is not very well documented, and for good reason. Most birds die inconspicuously of natural causes (Lincoln 1931). Among the most common causes of death are starvation, disease, predation, and climate. The numbers of deaths attributable directly or indirectly to human actions each year during the 1970s are staggering but are apparently minor in relation to the population level (Banks 1979). Human activities are responsible for roughly 270 million bird deaths every year in the continental United States (Table 21 – 4). This seemingly huge number is less than two percent of the 10 to 20 billion birds that inhabit the continental United States and appears to have no serious effect on the viability of any of the populations themselves, unlike human destruction of breeding habitat and interference with reproduction.

The human-related deaths of wild birds involve the following:

Nearly one-half of the annual deaths are the direct result of hunting and pest control.

Scientists collect a negligible number of birds for research purposes.

Collisions of various kinds constitute the vast majority of indirect human-related deaths. Assuming 15 bird deaths per road mile per year (Hodson and Snow 1965), a minimum of 57 million birds are killed by vehicles in the United States every year.

Large television towers kill about 1.2 million birds a year.

Table 21 – 4 Annual Human Related Mortality of Birds in the United States

Cause	Annual Mortality (millions)
Hunting	120
Pest control	2
Scientific research	0.02
Other direct	3.5
Pollution and poisoning	3.5
Collision	
Road kills	57
TV towers	1.2
Windows	80
Other indirect	3.5
Total	270.7

From Banks 1979; Klem 1979.

Miscellaneous accidents such as impact with golf balls, electrocution by transmission lines, and cat predation may amount to 3.5 million deaths a year.

Collisions with plate glass windows is perhaps the least recognized major cause of human-related mortality. Unable to distinguish between natural flight paths and the reflection of nearby vegetation in clear glass, millions of birds are stunned, injured, or killed on impact annually. Roughly half of the birds that collide with windows die of skull fractures and intracranial hemorrhaging. Systematic monitoring over one year registered 61 collisions at a house in Illinois and 47 collisions at a house in New York (Klem 1979). These samplings extrapolate to a minimum annual mortality from window collisions of 80 million songbirds throughout the United States, nearly two-thirds the annual harvest of waterfowl and gamebirds by hunters.

Longevity

Reflecting their high annual mortality rates, the life expectancy of most small birds is only 2 to 5 years. Large birds such as the Adelie Penguin live an average of 20 years. The average life span after an individual reaches adulthood is calculated from average survival rates by means of the formula

$$\text{mean life expectancy} = \frac{2 - m}{2m}$$

where m is the annual mortality rate, expressed as a fraction. Many individuals die younger, others live longer. The maximum ages recorded in wild birds average 10 to 20 years for various species of passerines and 20 to 30 years for waterbirds and raptors. Among the records are a 42-year-old Laysan Albatross, a 36-year-old Eurasian Oystercatcher, and a 34-year-old Great Frigatebird. As a rule, records of banded birds underestimate extremes because few birds keep their bands for life. Captive birds live much longer than their wild relatives (Flower 1938).

Fecundity

Fecundity, the number of young successfully raised, is a measure of an individual's reproductive success. Total lifetime fecundity depends on the age at which a bird starts to breed and on its lifespan. Annual fecundity reflects the length of the breeding season, the number of nesting attempts, and the success of each attempt, which is determined by losses to predation, starvation, or desertion, by the age and experience of the breeding individual, and by the number of eggs laid each time (clutch size).

Many birds, though by no means all, breed at the age of one year. Swifts breed at 2 years, parrots at 2 or 3 years, and raptors at 3 or more years. Waterbirds and seabirds generally take 4 or more years to breed for the first time. Large albatrosses and condors take 8 to 12 years. Among these species, there is a distinct correlation between longevity and age at first breeding.

Why should these birds delay breeding? Every extra year of nesting would seem to increase the chances of leaving some offspring; of all the variables that affect a bird's potential reproductive contribution to succeeding generations, early starting ages have the greatest impact. The age at first breeding controls mean generation time, which in turn drives the potential growth rate of a population. This effect can be seen in the formula for calculating r, the growth rate of a population:

$$r = \frac{\log_e (R_0)}{T}$$

where T is the mean generation time, and $\log_e (R_0)$ is the natural logarithm of the replacement rate R_0 from a life table. Doubling the value of R_0 causes population growth rate to increase by only 31 percent because of the discounting effect of logarithmic transformation. If the value of T is reduced by one-half, r increases by 100 percent. Thus, individuals that can breed in their first year should soon replace others that delay breeding for several years unless the costs of early reproduction are too severe.

Delayed maturity is a way of maximizing lifetime reproductive success in long-lived birds when reproduction enhances the risk of death. The factors favoring delayed maturity are well documented in Adelie Penguins (Ainley and DeMaster 1980; Ainley et al. 1983). First, breeding entails greater risk than not breeding: Survivorship of breeders (61 percent) is lower than survivorship of nonbreeders (78 percent). The greatest mortality occurs the first time young Adelie Penguins try to breed. Nearly 75 percent of 3-year-old females die during their first attempt to breed. Mortality then declines with age to 10 percent in 11-year-old breeding females. Offsetting the risks of initial reproduction in Adelie Penguins are improved prospects for raising young in subsequent attempts. Adelie Penguins that breed for the first time when they are 3 to 4 years old (and survive that effort) are less likely to lose their eggs or young in subsequent nesting seasons than those penguins that breed first at a later age. Whether these early starters are inherently better breeders or whether the early start somehow enhances subsequent breeding success is not known.

Three to four years seems to be the minimum possible age for reproduction in small penguins such as the Adelie. Three main factors are responsible for this age requirement. First, studies of another species, the Yellow-eyed Penguin, enable one to conclude that 2-year-old penguins are not usually reproductively mature; 65 percent of their eggs are infertile (Richdale 1957). Second, two to three years of experience seem to be essential for young penguins to develop the foraging efficiency that enables them to accumulate the large energy reserves required for long fasts while breeding. Third, at least one year of social experience is necessary to develop the behavioral skills required for successful pairing and defense of nest, eggs, and young. Given the increased risks of mortality associated with breeding even when well prepared, there is a clear advantage to the three- to four-year delay typical of this penguin.

Birds that breed for the first time produce fewer eggs and offspring than experienced birds, perhaps because smaller clutches reduce the costs and risks of reproduction. Black-legged Kittiwakes, for example, lay an average of 1.8 eggs their first year, compared with 2.4 eggs after their third year (Coulson and White 1961). European Starlings lay an average of 4.9 eggs per clutch in their first year and 5.9 eggs per clutch thereafter (Kluijver 1935). The probability that a female Yellow-eyed Penguin will lay a two-egg clutch rather than a one-egg clutch increases with her age though both clutch size and egg size decrease in very old females (Richdale 1957). In Snow Geese, clutch size increases with female age from an average of 3.3 eggs per clutch at 2 years of age to an average of 4.4 eggs per clutch at 5 years of age or older (Rockwell et al. 1983). This increase is partially due to the proportion of individuals breeding for the first time at each age. Most 2-year-olds are first-time breeders and lay fewer eggs than experienced breeders. In contrast, only 25 percent of the 4-year-old breeders are first-time breeders. As in the case of penguins, inexperienced female Snow Geese may be less able to accumulate the energy reserves required for egg production (see Chapter 17).

Older California Gulls produce more young then younger gulls (Pugesek 1981). The oldest gulls (12 to 18 years old) produce 1.5 young per year, the middle-aged gulls (7 to 9 years old) produce 0.80 young per year, and the youngest gulls (3 to 5 years old) produce 0.76 young per year. Older gulls feed their young more frequently, spend more time looking for food, and leave the nest less often than the younger members of the colony. It seems that older gulls invest heavily in reproduction during each season, even though doing so may decrease their chances for future reproduction. Young gulls invest less in their initial efforts (Figure 21–5).

Males of some short-lived songbirds do not acquire full adult breeding plumage in their first breeding season, even though they are capable of breeding. Thirty-one of the 105 sexually dimorphic passerines of North America, including Red-winged Blackbirds, Northern Orioles, Scarlet Tanagers, and American Redstarts, do not attain adult male plumage for a year (Rohwer et al. 1980). Such young males are less conspicuous and perhaps less vulnerable to predation. Also, because these young males resemble females more than they do rival males, they may avoid attack and eviction by older males and thereby occupy parts of established territories surreptitiously, establish some control in and access to that space, and ultimately gain priority in the use of the territory for breeding. In this way, perhaps, evolution has favored first-year males that postpone the acquisition of full breeding coloration. Otherwise, they would compete poorly with more experienced males. Evidence in support of this explanation remains,

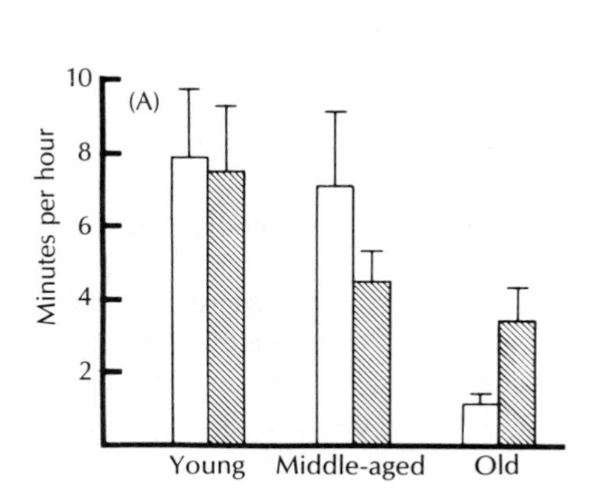

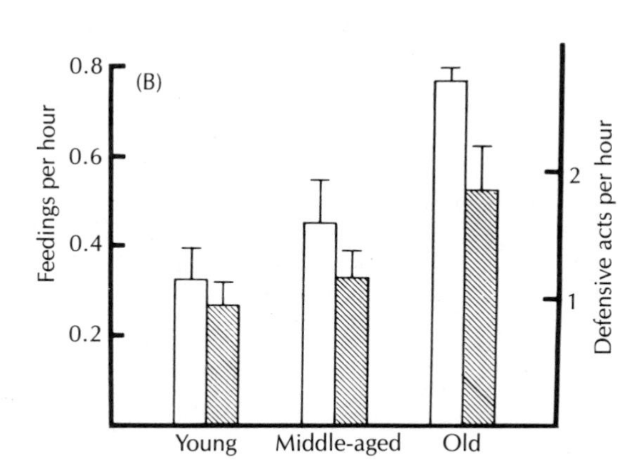

Figure 21–5 Older California Gulls put more effort into nesting than do young and middle-aged gulls. (A) During each hour, older gulls spend less time away from the nest and more time foraging (open bars denote time nest was unattended; shaded bars denote time that neither parent foraged). (B) Older gulls feed their young more often and defend their nest territories more often (open bars denote frequency of feeding; shaded bars denote frequency of defense). (From Pugesek 1981, © 1981 by the AAAS)

however, equivocal. Elizabeth Proctor-Gray and Richard Holmes (1981), for example, found that first-year male American Redstarts with female plumage establish territories, pair successfully, and fledge as many young as fully mature males with black-and-orange plumage. Although the young males in this study established themselves in habitats that were slightly different from those occupied by older males, no advantages to either plumage pattern could be demonstrated. The reasons for delayed maturity in sexually dimorphic birds such as these need more study.

Success in nesting

In general, nesting success increases in northern latitudes. Hole-nesting species and large species with hardy young tend to have relatively high rates of nesting success. The principal causes of failure in nesting are, in descending order, predation (see page 348), starvation, desertion, hatching failure, and adverse weather (Ricklefs 1969b). Starvation of young is most common in marsh- and field-nesting birds and in raptors. Desertion by parents occurs most often in hole-nesting birds. In colonial seabirds, separation from parents often results in the death of chicks, particularly in semiprecocial gulls and terns. Brood parasitism, nest-site competition, ectoparasites, and disease also contribute to nesting failure. Another cause of reproductive failure, reduction in fecundity as a result of inbreeding, is well documented in domestic animals and plants but has rarely been measured in birds. The Great Tit, however, has been observed to have an average nestling mortality of 28 percent among inbreeders and 16 percent among outbreeders. The viability of eggs decreased 7.5 percent for every 10 percent increase in the genetic relatedness of mates (Greenwood et al. 1978; Noordwijk and Scharloo 1981).

Multiple broods

The number of broods a pair can raise depends not only on the length of their nesting cycle but also on the length of the breeding season. Long nesting cycles or restricted breeding seasons such as those in temperate and arctic latitudes tend to preclude extra broods. Hence, many temperate and arctic birds attempt only one brood unless they are trying to replace one that was lost early (Lack 1966; Haartman 1971). Within species there may be a variation, due to geography, in numbers of broods attempted. The Common Song Thrush, for example, attempts only one brood in northern Europe (65° to 70°N) but attempts two to four broods in southern Europe (45° to 50°N) (Haartman 1971). Tropical birds generally attempt more broods than temperate birds, owing, in part, to prolonged breeding seasons. In addition, because losses of nests and young to predators are much higher in the tropics than in the temperate zone, renesting is often necessary to replace lost clutches. Two to four broods are not unusual in the tropics, and six are not rare. The White-bearded Manakin, for example, typically lays three to five clutches per season in Trinidad (Snow 1962).

Table 21 – 5 Interval between Broods of Temperate and Tropical Passerine Birds

	Interval (days)	
	Temperate	Tropical
After nest failure		
Mean	7.8	13.3
Range	3 – 11	6 – 24
After successful fledging		
Mean	8.2	25.8
Range	1 – 20	8 – 59

From Ricklefs 1969a.

Tropical birds require slightly longer than temperate species to renest, whether they fledge their first brood or whether they must replace a lost clutch or brood (Table 21 – 5). It is possible that tropical birds mobilize reserves more slowly and take longer to accumulate the additional energy needed for egg formation. The fact that tropical species lay on alternate days suggests that tropical food resources may be less certain than temperate zone resources. Also, tropical species, unlike temperate species, cannot renest soon after their young have left the nest because they care longer for their fledged young.

Some birds increase their seasonal productivity by overlapping successive clutches (Burley 1980). Overlapping small clutches can be a better way of increasing fecundity than enlarging a single clutch because it reduces stress by separating periods of peak parental care into several smaller peaks. Clutch overlap is practiced by Rock Doves, which are limited to two eggs per clutch because they cannot nourish more than two large young at a time with their pigeon milk (page 382). Proficiency in parental care is a prerequisite to the pigeons' management of overlapping clutches. The extent of overlap of clutches increases with a mated pair's combined experience as parents. In Goldcrests, which sometimes have two overlapping broods, the male builds the second nest alone (Haftorn 1978c) and assumes primary responsibility for the young in the first nest when the female begins to lay and incubate the second clutch. Faced with two broods of young of differing ages, the male puts his initial effort into the older first brood, for which the need is greatest, and then shifts his attention to the second brood after the first has achieved independence.

Effects of pesticides

Man's poisoning of the environment has had a devastating impact on the fecundity of birds. Accumulated pesticides, particularly DDT, not only kill birds directly but also interfere with eggshell production and cause nesting failure (Ratcliffe 1970; Cooke 1975). Eggs with shells 15 to 20 percent thinner than normal generally cause reproductive failure and population decline (Cooke 1975; Kiff et al. 1979). Reproduction of Bald Eagles in northwestern Ontario, for example, declined from an average of 1.26 young per breeding area (nest) in 1966 to a record low average of 0.46 in

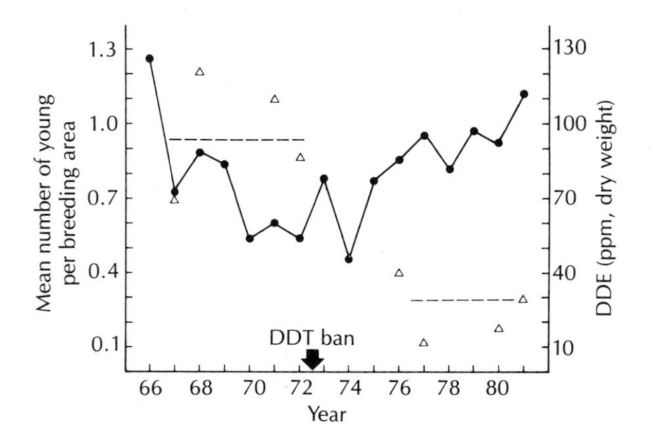

Figure 21–6 Reproduction in Bald Eagles (solid lines) improved following the ban on use of the pesticide DDT, which resulted in a drop in chemical residues (DDE) in eggs (triangles). Dashed lines represent weighted mean concentrations of DDE before and after the ban. (From Grier 1982, © 1981 by the AAAS)

1974, increasing to an average of 1.12 after DDT was banned (Grier 1982) (Figure 21–6).

The Brown Pelican, once one of the most familiar and abundant birds of the Gulf coast and west coast of North America, faced extinction in the 1960s because of widespread reproductive failure (Schreiber 1980b). Hydrocarbon pesticides in the marine food webs of coastal California and of coastal Louisiana and nearby Texas interfered with the production of normal eggshells, and the pelicans typically laid eggs with very thin or no shells. Fragile pelican eggs are easily broken by the weight of an incubating parent. The lack of reproduction in Brown Pelicans in California, where eggshell thinning was most severe, and the alarming disappearance of pelicans from Louisiana and Texas resulted in placement of this bird on the Endangered Species List in 1973. Recent reductions of pesticides in the environment have enabled Brown Pelicans to nest successfully in California once again.

Evolution of clutch size

The number of eggs a female bird lays is an essential and heritable component of fecundity (Klomp 1970; Perrins and Jones 1974; van Noordwijk et al. 1980; Findlay and Cooke 1983). Clutch size is an evolutionary adaptation molded by selection over many generations, but it is also sensitive to immediate environmental conditions (see Chapter 17). Passerines, and other small landbirds that feed their young, lay clutches of 2 to more than 12 eggs, the exact number of which varies among species and, within a species, with latitude, climate, age, and quality of territory. Waterfowl, pheasants, rails, and many other precocial birds have clutches of up to 20 eggs. Other birds have invariant clutch sizes: Precocial shorebirds typically lay 4 eggs, and oceanic pelecaniformes and procellariiformes lay only 1 egg. Hummingbirds and doves always lay 2 eggs.

Variations in clutch size pose intriguing questions. Why do some birds lay 2 eggs and others 20? Which particular number of eggs maximizes reproductive success for a particular species? In general, nutritional requirements for egg formation seem to limit clutch sizes of precocial birds whereas the feeding abilities of parents limit the clutch sizes of altricial birds. This

Table 21–6 Summary of Trends in Clutch Size

Correlate	Clutch Size: Small (2–3 eggs)	Large (4–6 eggs)
Latitude	Tropics	Temperate/arctic
Longitude	Eastern Europe	Western Europe
Altitude (temperate)	Lowlands	Highlands
Nest type	Vulnerable	Secure (hole)
Body size	Large species	Small species
Habitat	Maritime, island, and wet tropics	Continental, mainland, and arid tropics
Feeding rate	Pelagic seabirds	Inshore seabirds
Development mode	Altricial	Precocial

summary opens, rather than closes, further inquiry. No single topic has so occupied the attention of students of avian life-history patterns as has the evolution of clutch size. The literature summarizing conspicuous variations in clutch size (Table 21–6) is formidable, and the interpretations of the data are controversial.

Theoretically there should be an optimal clutch size, dictated by natural selection, that produces the maximum number of young that survive to sexual maturity, and this clutch size should prevail in local populations. However, the clutch size that is optimal for any given population of birds and the factors that are responsible for the evolution of a particular clutch size are questions that remain unresolved. The debate about the evolution of clutch sizes among birds centers on four major hypotheses:

1. Lack's hypothesis, which states that parental ability to care for nestlings dictates clutch size;

2. the tradeoff hypothesis, which states that future survival dictates maximum possible clutch sizes;

3. the predation hypothesis, which states that nest predation selects for smaller clutches because they are less conspicuous and minimize short-term losses; and

4. the seasonality hypothesis, which states that clutch size reflects the seasonal availability of resources relative to population size.

We now consider each of these hypotheses in detail.

Lack's hypothesis

The clutch size in birds is adjusted by natural selection to the maximum number of nestlings the parents can feed and nourish. This fundamental postulate by David Lack (1947b, 1948) has guided research for the last 40 years. Individuals that lay fewer eggs each year than they can raise are at a disadvantage. Observations, experiments, and theory (Royama 1969; Hussell 1972; Ricklefs 1977) all support this general concept, with some caveats and modifications.

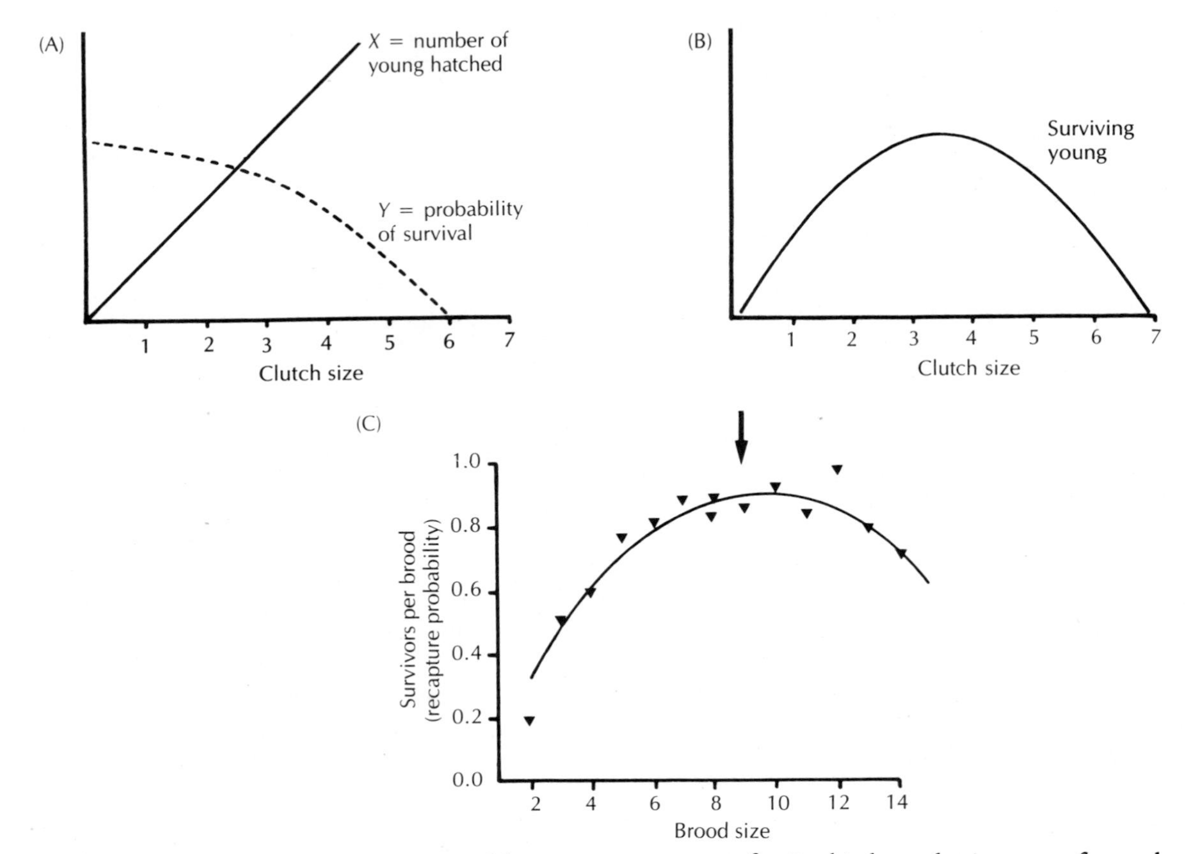

Figure 21–7 (A) Lack's hypothesis of optimal clutch size projects the maximum number of surviving young as a result of the balance between the number of young hatched *X* and the probability of survival *Y*. (B) Note the reduced survival owing to limited parental care of chicks in large broods. (C) In the Great Tit, broods of ten chicks are the most productive; the average clutch size in this species is about nine, good support for Lack's hypothesis. (From Perrins and Moss 1974)

The strongest support for Lack's hypothesis comes from observations of the relative success of various sizes of clutches and from experiments designed to test the ability of parents to feed extra young. For example, Christopher Perrins and D. Moss (1974) found that a clutch of 10 eggs produced the most surviving young Great Tits in Oxford (Figure 21–7). The probability of a chick's survival in a small brood (less than 10) is greater than in a large brood (10 and more) because the nestlings are better fed and heavier when they fledge, but the number of potential fledglings from small broods is necessarily low. Above a clutch size of 10, chicks tend to be underfed and to die. The average brood size in the population was 10, that is, the most productive, in 6 of 13 consecutive years but it was slightly lower than predicted in other years.

Lack's hypothesis seems to explain the average clutch size in a population and the individual variations. In southern Sweden, for example, female Black-billed Magpies lay clutches of optimal size (Högstedt 1980). Experimental additions of eggs increase losses to predation and starvation and thereby reduce a pair's reproductive output. Experimental removal of eggs simply cuts breeding potential (Table 21–7). Most of the variations in clutch size among females in any one year relate to differences in food availability on their territories. The mean clutch size in the population reflects the average quality of territory. Thus, no single, optimal clutch size exists, and the variations are adaptive.

The strengths and weaknesses of Lack's hypothesis can be seen in its application to the well-documented increases of clutch size with latitude. Lack (1947b) and others after him have tried to explain why the mean

Table 21–7 Reproductive Output by Black-billed Magpies with Natural and Altered Clutch Sizes

Initial Clutch Size	Experimental Clutch Size				
	4	5	6	7	8
5	0.3	0.7	0.5	0.3	0
6	1.7	1.9	2.8	0.8	1.2
7	3.5	2.3	3.1	3.6	2.4
8	2.5	3.5	3.5	4.3	4.5

After Högstedt 1980.

clutch size of many passerines, owls, hawks, herons, terns, gallinules, some galliformes, and some grebes increases with latitude. Lack suggested that longer daylength at high latitudes would enable birds to have larger broods. Birds nesting during the temperate summer have more time to find food for their young and themselves. However, as Gale Murray (1976) pointed out, owls, whose clutch sizes also increase with latitude, have less time, not more time, to feed. Increasing daylength is too simple an explanation for the geographical gradients in clutch sizes. In addition to the exception of the clutch sizes of owls, the effects of increasing daylength do not explain why clutch sizes increase with longitude from west to east in Europe, or with altitude in the temperate zone but not in the tropics (Table 21–6).

One problem with Lack's hypothesis is that the commonest clutch size observed in a population may be smaller than that which appears to be the most productive. This was the case for 7 of the 13 years that Great Tits were studied in Oxford, England (Perrins and Moss 1974). Another problem is that some species of birds can raise additional young, which are provided experimentally (Hussell 1972). Even some large seabirds such as gannets, which normally lay only a single egg, can raise two young when an extra egg is added to the nest (Nelson 1964). Thus, Lack's hypothesis does not explain why individuals that raise the maximum possible number of young every year do not dominate populations.

The tradeoff hypothesis

Another possible explanation for these observations seems to lie in the long-term costs of immediate reproductive effort and the tradeoffs inherent in the optimal allocation of energy (Williams 1966; Cody 1966; Charnov and Krebs 1974; Perrins and Moss 1974). The survival and total reproductive effort of a parent is an important variable. An individual's current reproductive effort takes into account its long-range reproductive interests. We have discussed this phenomenon with reference to age at first breeding. To the degree also that adult mortality increases with clutch size, due to the stress of caring for more young or to the heightened risk of predation, there may be advantages to reducing clutch size, especially in long-lived species. A distinction could be made, therefore, between the most productive clutch size, which maximizes number of young surviving to breed each year, and the optimal clutch size, which maximizes a parent's total reproductive contribution to the next generation.

There is some indirect evidence that larger broods are physiologically more stressful than small broods. Snow Buntings, Pied Flycatchers,

Table 21–8 Costs of Parental Care in Tree Swallows

Treatment	Mean Weight Loss (g) ± SE*	Survival (% return)
Control (no change)		
Yearling females	1.6 ± 0.6	75
Older females	1.3 ± 0.2	53
Enlarged broods (+2)		
Yearling females	1.9 ± 0.3	71
Older females	1.6 ± 0.2	64

* Over a 6-day period.
After DeSteven 1980.

and Common House-Martins that raise large broods sustain greater weight losses during the period of reproductive effort (Hussell 1972) than those that raise small broods. Eastern Bluebirds that raise large first broods are less likely to initiate a second brood (Pinkowski 1977). Such observations, however, have not been carried through to determine the long-term survival of the adults.

Contrary to the tradeoff hypothesis, it appears that adult survival may play only a minor role in shaping the life histories of short-lived bird species (Ricklefs 1977). Diane DeSteven (1980), for example, was unable to demonstrate an inverse relation between clutch size and adult survival in experiments that were specifically designed to test this hypothesis. Female Tree Swallows that raised experimentally enlarged broods did not lose significantly more weight or survive less well than females that raised normal-sized broods (Table 21–8). In addition, inexperienced yearling females were no more susceptible to stress during reproduction and brooding than older females. Thus, negative tradeoffs between clutch size and future reproductive effort (demographic optimization) do not help us understand the evolution of swallow clutch sizes.

The predation hypothesis

Predation is often postulated to be an important force in the evolution of clutch size (Perrins 1977). Increased predation could select for smaller clutches in at least three ways:

1. It takes longer to lay a large clutch than it does to lay a small one, which means that the eggs and chicks of a large clutch are at risk longer than those of a small clutch at the same site.

2. Larger broods of young in a nest are noisier and more conspicuous and therefore more likely to attract predators (Skutch 1949).

3. In the tropics, where most nests are lost to predators and where a pair of birds has the opportunity to renest several times, it may be advantageous to risk fewer eggs at a time. Mercedes Foster (1974) suggests that tropical birds can maximize their reproductive output by laying small clutches and by molting while they are breeding, which extends the breeding season and, therefore, renesting opportunities.

Greater safety from predators is probably the reason that hole-nesting birds lay larger clutches than open-nesting birds. Selection for small, inconspicuous nests with few eggs also seems to prevail in tropical forests and may even reduce the need for an active role by the male in some species (Lill 1974c; Snow 1978). Direct evidence for this strategy is scarce, however, except in the Great Tit and the Black-billed Magpie, for which nest predation does increase in relation to clutch size. Yet Ricklefs (1977) found no evidence that predation increases consistently as a function of clutch size in a way that would explain trends in clutch sizes. Furthermore, predation on nests and adult mortality during the breeding season accounts for only 10 to 25 percent of the variation in clutch size with latitude and seems to be too weak a force to counter the clear advantage of those larger clutches that escaped predation.

The number of fledged young that parents can attend or guard from predators may limit clutch size in some birds, particularly shorebirds, which lay no more than four eggs. Constraints of egg formation, incubation, and phylogenetic history have been suggested; or the demands of parental care of precocial hatchlings could be the limiting factor (Safriel 1975; Winkler and Walters 1983; Walters 1984; also see Chapter 19). Even though the parents of many shorebirds do not feed their precocial young, they brood and tend them actively and guard them from predators. Physical distance between parents and their mobile young increases with brood size and potentially sets an upper limit on brood size. Active tenders, those species that follow their young closely, have smaller clutches than inactive tenders, those species that monitor their young from a distance.

The seasonality hypothesis

The current hypothesis of perhaps the broadest application, the seasonality hypothesis, cites the seasonality of resources (Ashmole 1963b; Ricklefs 1980b). Resources that are available during the breeding season depend on demand, which in turn depends on population density, which is regulated by resource availability during the nonbreeding season. Seasonal variations control the amount of the resources available for breeding on a per capita basis (Figure 21–8).

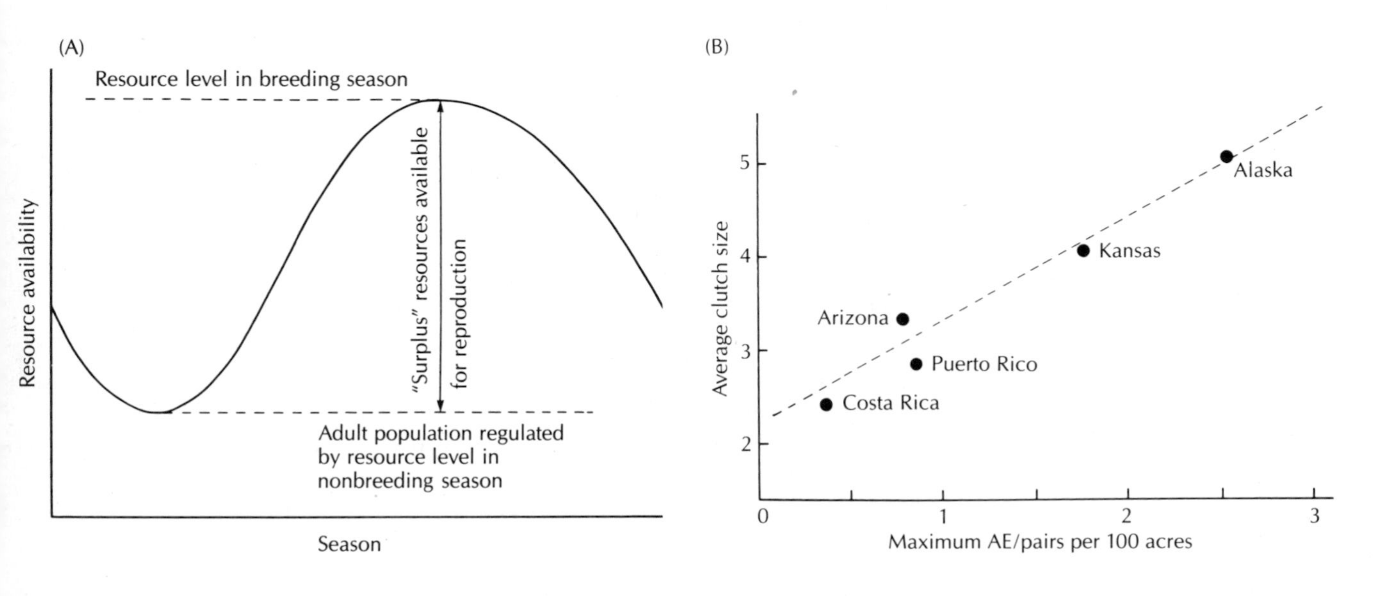

Figure 21–8 The seasonality hypothesis for geographical variation in clutch size. (A) Model of the seasonal increase in resources available for reproduction, measured in some months as the "surplus" above those resources that limit population size in the nonbreeding season; clutch size varies in relation to the ratio of the breeding season surplus to the adult population. (B) Clutch size increases with resource seasonality, measured as the ratio of maximum actual evapotranspiration (AE, an index of plant productivity) to the density of breeding pairs of birds. (From Ricklefs 1980b)

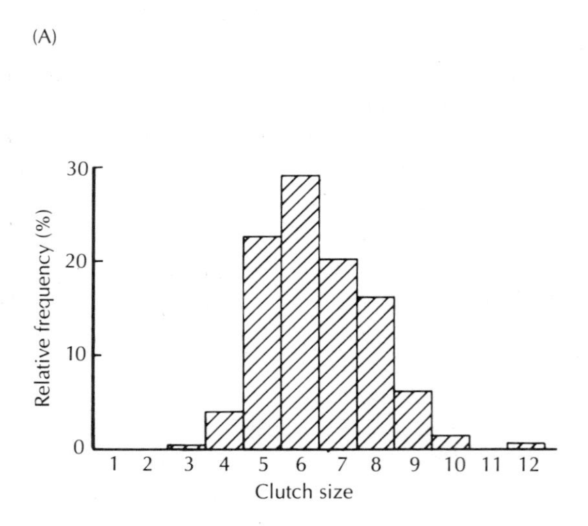

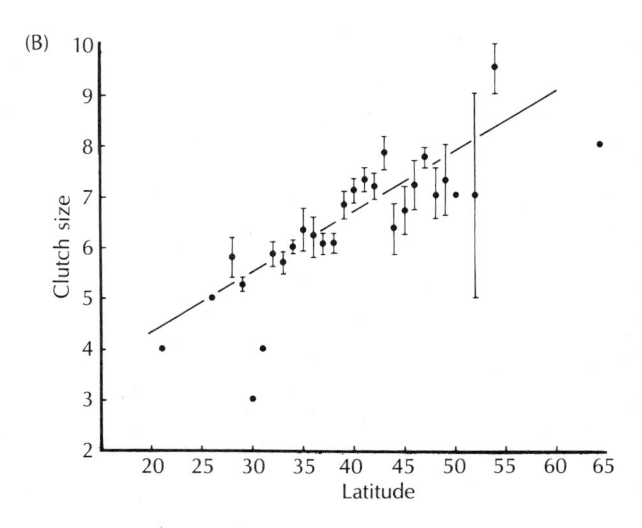

Figure 21–9 (A) The clutch sizes of Northern Flickers vary from 3 to 12 eggs. Relative frequency equals the percentage of total sample that had *x* number of eggs. (B) The increase in average clutch size with latitude supports the seasonality hypothesis. (From Koenig 1984)

Clutch sizes in some birds relate directly to seasonal increases in food production rather than to absolute level of production (Ricklefs 1980b). Birds of arid habitats in both Africa and Ecuador have larger clutches than those that live in less seasonal, humid habitats at the same latitude (Lack and Moreau 1965; Marchant 1960). The prediction that clutch sizes will be uniform in a region is also upheld. These results, as well as theoretical considerations, suggest that variation in the seasonability of resources experienced by a species is the ultimate cause of geographical variations in clutch size. In the broad context of the seasonal availability of resources, the immediate environment, such as territory quality, then molds the details of clutch size for each species.

The patterns of variation of clutch sizes of the Northern Flicker support the seasonality hypothesis (Koenig 1984). Clutches of this widespread North American woodpecker, which range from 3 to 12 eggs, increase by an average of 1 egg per 10 degrees of latitude. The variations in clutch size are directly correlated with the availability of resources per breeding woodpecker, which Walter Koenig has estimated as the ratio of local summer productivity (in terms of actual evapotranspiration, an index of plant productivity) to the breeding density of all woodpeckers. Local breeding densities of woodpeckers in turn are set by winter productivity, which apparently determines how many woodpeckers survive until the breeding season (Figure 21–9).

Brood reduction: Retrospective adjustment of brood size

One way that birds can cope with uncertainties about the maximum number of young they can raise in any particular year is to lay the number of eggs that might be successful in good years and then to sacrifice some of the eggs if necessary. "Brood reduction," as this behavior is called, protects parents against loss of the entire brood should conditions for raising young be poor

(O'Connor 1978). Parents can apportion their reproductive effort selectively among the offspring in a clutch in several ways. They may lay eggs of different sizes or nutritional qualities, start incubation before the last egg is laid, which promotes asynchronous hatching at the expense of the younger siblings, or selectively feed only some of the offspring. The smaller, weaker chicks of Herring Gulls and Common Terns that are laid third, for example, survive only when conditions are exceptionally good (Parsons 1970, 1975; Nisbet 1973). Older raptor nestlings often kill and eat their younger siblings when food is scarce (see Chapter 19).

The details of adaptive brood reduction by the Common Grackle are well documented (Howe 1978). Females that lay 2, 3, or 4 eggs per clutch apportion their reproductive investment differently from females that lay 5 or 6 eggs per clutch. Those that lay the larger clutches put more yolk and albumen into the last eggs and start incubation before they complete the clutch; but they rarely manage to raise the whole brood. The older siblings hatch first, are fed first, and grow faster. Although the extra egg provisions partially compensate for their disadvantages, the younger members of the brood starve if food is scarce. Females that lay the smaller clutches of 2, 3, or 4 eggs do not increase the size of the last eggs, hatch them all at the same time by waiting until the last egg is laid before they begin incubation, and usually raise the whole brood.

Brood reduction in Red-winged Blackbirds depends on the age of the breeding female and also affects the relative numbers of male and female offspring that are fledged (Table 21 – 9). Young females fledge more daughters than sons whereas old females fledge more sons than daughters. Although equal numbers of sons and daughters hatch in the broods of young females, starvation is common and sons starve more often than daughters. Young females lay poorly provisioned final eggs in the clutch, which causes those nestlings to be most vulnerable to starvation. Young females also tend to lack the experience required to feed their nestlings adequately. A sex bias exists in the probability of starvation because male offspring need more food than their sisters; they grow faster to a larger size and hence are more likely to starve. Older females, on the other hand, do not lay inferior final eggs, and they supply their young with better provisions. Hence their large, fast-growing sons are not likely to starve. Ornithologists do not know why older females hatch more sons than daughters, but apparently the female offspring are more likely to die as embryos as a result of unknown causes.

Table 21 – 9 Surviving Offspring of Female Age Classes in the Red-winged Blackbird

Surviving Offspring	Female Age Class		
	Young	Middle-aged	Old
Males	28	53	54
Females	50	66	34
Chi-square	5.65*	1.21	4.10*

* $P < 0.05$.
From Blank and Nolan 1983.

Summary

In this introduction to the evolution of life history traits of birds, we presented the basics of life table analysis of demographic patterns and looked in detail at the nature of survival and fecundity in birds and the typical patterns of avian life histories. A bird's lifetime reproductive output reflects the age at which it first reproduces, the number of young it fledges each year, the survival of those young, and its own longevity as an adult. In general, long-lived species tend not to breed until they are several years old and to produce few young each year. Short-lived species often breed when one year old and may produce many young each year. Reproductive success and effort usually improve with age and experience. Delayed maturity is a way of maximizing lifetime reproduction. Birds also increase the number of young produced by raising several broods sequentially in a season and in some cases by overlapping successive broods.

Most small birds live 2 to 5 years; large birds may live 20 to 40 years. Many young birds die during their first year as a result of predation and starvation. The survival rates of adults are much higher and remain the same each year. Human activities cause an estimated 270 million bird deaths each year, principally as a result of hunting and pest-control, but also from collisions with vehicles, television towers, and picture windows. Pesticides have devastated some bird populations by interfering with eggshell formation.

The evolution of clutch sizes among birds is a major research topic in ornithology. In general, birds raise as many young as they can with the food they are able to gather. There is little convincing evidence that birds raise fewer young each year as a way of increasing long-term survival. Risk of predation may favor small clutches in the tropics and in open nests as opposed to hole nests, but the role of predation in the evolution of clutch size remains uncertain. Geographical trends in average clutch size — large clutches in the north and in arid environments, small clutches in the south and in mesic environments — seem best explained by seasonal differences in food availability. Because it is difficult to predict the conditions that may favor a particular clutch size, some birds practice retrospective brood reduction, sacrificing surplus, disadvantaged young when necessary.

Further readings

Ainley, D.G., R.E. LeRosche, and W.J.L. Sladen. 1983. Breeding Biology of the Adelie Penguin. Berkeley: University of California Press. *A detailed long-term study of one species.*

Cody, M.L. 1971. Ecological aspects of reproduction. Avian Biology 1:20–55. *A review of early theory.*

Ricklefs, R.E. 1973. Fecundity, mortality, and avian demography. *In* Breeding Biology of Birds (D.S. Farner, ed.), pp. 366–435. Washington, D.C.: National Academy of Sciences *An excellent review of the literature.*

Ricklefs, R.E. 1983. Comparative avian demography. Current Ornithology 1:1–32. *A powerful summary of the theoretical foundations of this subject.*

chapter 22

Population size

The growth of populations is a geometric function of individual reproductive success: Potentially, a population can double every few years. But at some point, populations stop growing because their needs begin to exceed the availability of resources. A growing population may also become increasingly vulnerable to predation and disease. Determining the ecological, social, and competitive forces that limit the continued growth of bird populations is a central problem in the field of avian ecology. In this chapter we examine the phenomenon of population size. First, we establish that populations are dynamic, not static, entities. Then, extending the demographic discussion of population growth from the preceding chapter, we look at the growth potential of populations and explore the factors that act to limit population growth. The interactions among ecological and social factors are illustrated in the work done with the six-year population cycles of the Red Grouse of Scotland. The way in which fecundity and survival depend on local population density is illustrated by research in England and Holland on the Great Tit.

Populations are dynamic

Bird populations are rarely static, but fluctuate from year to year because of variations in breeding success and mortality. For example, populations of Pied Flycatchers, a hole-nesting European bird, fluctuate annually by 50 percent. The density of the Great Tit population in Marley wood near Oxford, England, varies threefold (Figure 22–1). Other populations of this species show similar fluctuations; Great Tit populations throughout Holland and England have grown gradually in average size since 1946 (Kluijver 1951; Lack 1966). The general increase is apparently due to a gradual amelioration of the European climate during this period whereas the annual fluctuations are caused by winter food limitations.

Avian abundance and distribution have undergone great changes in recent years. Many species, including the Brown-headed Cowbird and Glossy Ibis, for example, have gone from being rare to being abundant in the northeastern United States in the last 30 years. Others, such as the Red-headed Woodpecker and American Black Duck, have declined during this

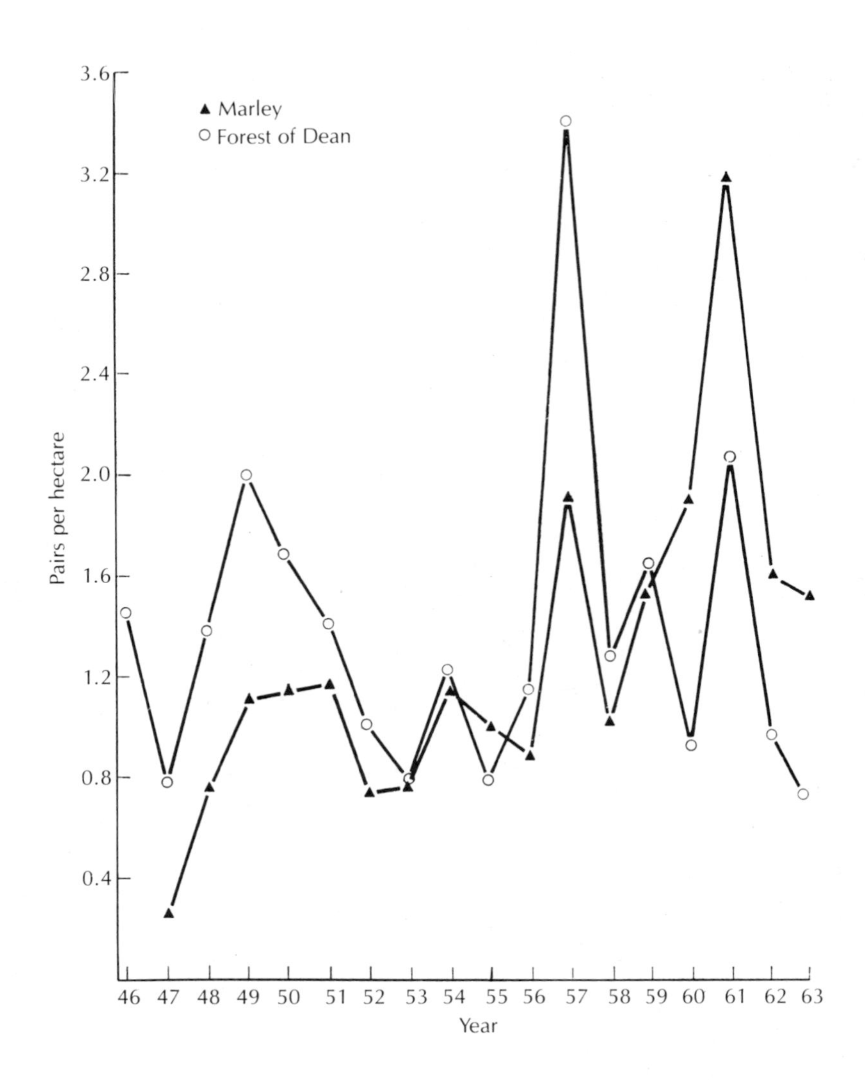

Figure 22–1 Annual fluctuations in the density of breeding pairs of the Great Tit in two British forests, Marley wood (triangles) and the Forest of Dean (circles). (From Lack 1966)

same period. In Finland during the past 100 years, 34 percent of the 233 breeding species have enlarged their geographical distribution or have become more numerous, 25 percent have declined in numbers, 22 percent have fluctuated, and 20 percent have been fairly stable (Haartman 1973; Järvinen 1980).

Population growth patterns

The typical growth pattern of a population in a new environment is sigmoid: As in a baby bird, the rate of growth increases slowly at first, accelerates, and then declines because of negative feedback that lowers reproduction and survival; finally the growth curve becomes asymptotic, gradually approaching the carrying capacity of the environment. The two key phases of this S-shaped growth curve are the period of maximum growth and the period of growth limitation. The changing growth rates derive from the demographic parameters of survival and fecundity reviewed in the previous chapter. Values of per capita replacement R_0 that are greater than 1 reflect a growing population, and values less than 1 reflect a declining population (Table 22–1). The instantaneous growth rate of a population is usually expressed as r, the logarithmic form of R_0. During the phase of accelerating growth, the rate of change in the number of individuals with time (dN/dt, from differential calculus) is the product of the instantaneous growth rate r and the population size N at time t. This is expressed mathematically as

$$\frac{dN}{dt} = rN$$

The growth potential of bird populations in this accelerated phase is tremendous. For example, the 120 European Starlings that were introduced into the United States in 1890 increased a millionfold in 50 years (Davis 1950). In general, large-bodied species with low reproductive rates have annual growth potentials of 10 to 30 percent, and small-bodied species with large brood sizes and high reproductive potentials have an annual growth potential of 50 to 100 percent in favorable years (Ricklefs 1973b).

Table 22–1 The Relationships between Exponential r and Geometric R_0 Growth Rates and Change in Population Size

	Growth Constant	
Change in Population Size	Exponential (r)	Geometric (R_0)
Decreasing	Less than 0	Between 0 and 1
Constant	0	1
Increasing	Greater than 0	Greater than 1

From Ricklefs 1979c.

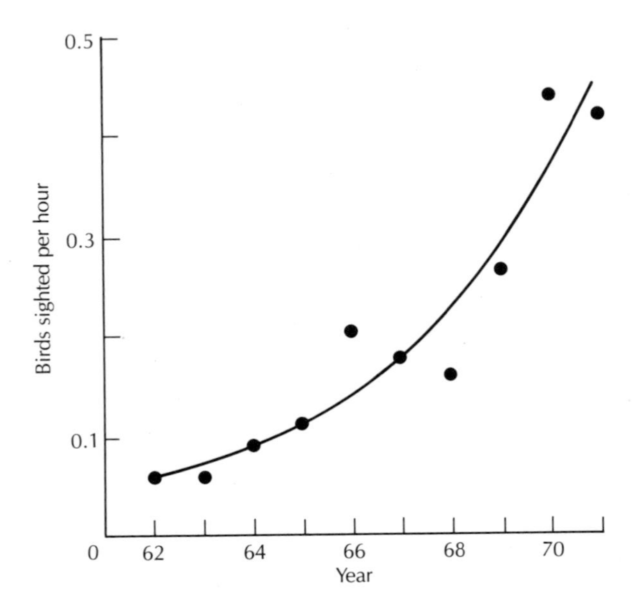

Figure 22–2 Exponential population growth of House Finches east of the Mississippi, based on annual Christmas counts. (Adapted from Bock and Lepthien 1976b)

Strong population growth enables species to recover quickly from short-term setbacks.

Newly established populations of the Cattle Egret and the House Finch have grown exponentially in recent years (Bock and Lepthien 1976a, b). The Cattle Egret colonized North America in the early 1950s and then spread dramatically, increasing 2000-fold from 1956 to 1971 (Bock and Lepthien 1976a). The calculated value of r was 0.21 for the 16-year period from 1956 to 1971 but was as high as 0.84 during the initial five years of this expansion. This very high value may be close to the species' maximum potential. Following the release of caged birds on Long Island in 1940, the eastern population of the House Finch grew rapidly. The average value of r for eastern House Finches from 1962 to 1971, which included some temporary declines, was 0.23, a figure nearly identical to the average value for the egret (Figure 22–2).

The carefully monitored growth of a population of Ring-necked Pheasants introduced in 1937 on Protection Island, Washington, illustrates the two phases of the population growth curve (Einarsen 1942, 1945; Ricklefs 1979c). In just five years the population multiplied from the initial 8 pheasants to a total of 1325 birds; R_0 was 2.8. The rate of increase dropped steadily as the population reached the environmental carrying capacity of the island. Similarly, the growth rate of the Cattle Egret population in North America appears to be dropping (Bock and Lepthein 1976a).

Population regulation

As the size of a growing population approaches the maximum (the environment's carrying capacity), its growth rate decelerates. The population then cycles, or fluctuates, close to an average population size. In the previous chapter on demography, we discussed causes of mortality and variations in individual fecundity. In this chapter we identify the major ecological forces that set limits on the population as a whole. But first we consider the role of population density as a force that helps maintain, on the average, a constant population size. The growth rate of a population may depend directly on the density of individuals, if mortality and reproductive failure increase with density (Lack 1954, 1966). On the other hand, annual variations in mortality and reproductive success can be unrelated to population density (Andrewartha and Birch 1954; Murray 1979). The limits of food supply can limit numbers in either a density-dependent or a density-independent way. For example, the proportion of individuals that starve may be independent of population size if starvation is due to a major ice storm or a sudden drought.

A distinction must be made between the terms *regulation* and *limitation,* both of which are used in reference to population sizes. Regulation implies that maintenance of the size of a population at some average depends on population density, which is determined by birth and death rates, whereas limitation refers to any ceiling on population growth. Demonstrating that fecundity or survival depends on population density, however, does not necessarily mean that density dependence is an example of population regulation. Density-dependent clutch size or adult survival, for example, may not affect population size if the death of juveniles during hurricanes each year severely limits the number of breeding pairs and, therefore, the population size. In the following sections, we consider the ecological forces that limit population sizes and the ways in which population density can influence these forces to create patterns of population regulation.

Factors that limit populations

Four ecological factors essentially set upper limits to bird populations: disease (including parasites), climate, food supply, and habitat. Diseases can devastate bird populations, as they apparently have done in Hawaii (Warner 1968), but the frequency of diseases and their impact on natural bird populations are not well known and deserve further study. The warming of the European climate, which was responsible for the gradual increase in Great Tit populations, has also favored the expansion of southern species into Scandinavia and forced the retreat of some northern species. The Thrush Nightingale occurred as far north as central Sweden in the 18th century, but it receded to southern Sweden as the climate turned colder in the 19th century. Owing to recent warming, it has expanded again during the 20th century to its earlier distribution.

Populations of migrant birds are subject to decrease on their wintering grounds. For example, the numbers of the Greater White-throat, a common British warbler, on its breeding grounds reflect the number of birds that are able to survive the winter in Africa (Winstanley et al. 1974; Batten and Marchant 1977). The breeding population of this species dropped 77 percent one year because of drought south of the Sahara. When wetter winters followed, the breeding population rebounded to its previous level (Figure 22–3).

Figure 22-3 The Greater White-throat, a European warbler, is a species whose numbers depend on the severity of conditions on African wintering grounds. (Courtesy E. Hosking)

Food limitation

The food supply is one of two principal ecological variables that seem to influence the population sizes of birds. The other variable is habitat availability. Food supply, which often depends on climatic conditions, can unquestionably limit population growth and influence population size (Lack 1954, 1966; Newton 1980). An extreme example is that of the millions of Peruvian seabirds that starve periodically when their main food, the anchovy, a small fish, disappears as a result of changes in surface water temperatures (Idyll 1973). The total population of cormorants, pelicans, and other seabirds went from 27 to 6 million birds in 1957 and 1958, increased to 17 million as food supplies returned, and then plummeted again to 4.3 million birds in 1965. Overall, the total number of seabirds during good years has been declining because overfishing depletes anchovy populations.

Most of the evidence of starvation among temperate-zone birds comes from losses of songbirds, waterfowl, and waders during hard winters (Ash 1957; Dobinson and Richards 1964; Trautman et al. 1939; Ogilvie 1967). The very cold winter of 1981-1982 was hard on British songbirds (O'Connor and Cawthorne 1982). During that winter, mortality rates in several species were 2 to 10 times the normal rate. Common Redshanks were unable to feed because their main food, shrimplike amphipods, remained deep in their burrows when intertidal areas froze. White Wagtails, searching for insects along frozen shorelines, could no longer find one every four seconds, the average rate needed for their subsistence.

Widespread food shortages can cause irruptions (mass dispersal) of populations, especially of populations of birds in the arctic and subarctic

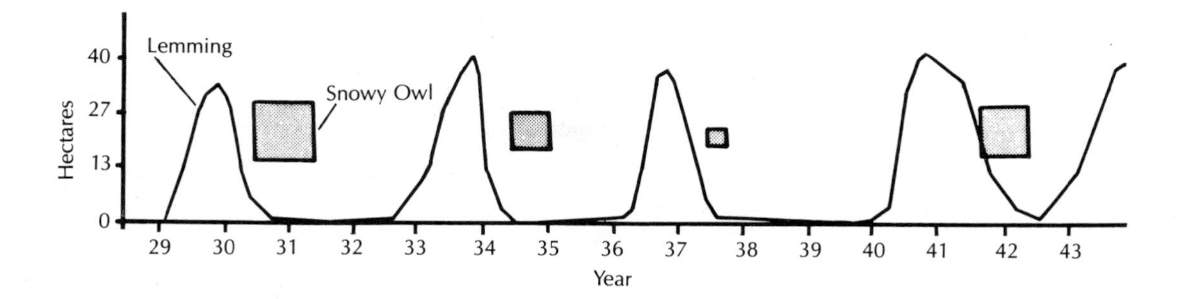

Figure 22–4 Irruptions of the Snowy Owl from the Arctic into Canada and the United States tend to occur in the years after severe decreases in the lemming population cycle. The sizes of the boxes indicate the relative sizes of the owl dispersal. (Adapted from Shelford 1945, Perrins and Birkhead 1983)

regions. One of the most familiar mass dispersals, the periodic southward invasions by hungry Snowy Owls, is linked to the cyclic scarcity of lemmings (Figure 22–4). Shelford's study ended in 1944, just before the great Snowy Owl invasion of 1945 to 1946, when over 14,000 Snowy Owls were counted in southeastern Canada and New England. Because they were away from their usual habitat, many of the owls were killed or died of starvation (Dorst 1962).

Irruptive invasions of dispersing populations of boreal seed-eating birds are dramatic ornithological events in both Europe and North America (Bock and Lepthien 1976c). During invasion years, flocks of northern finches appear along roadsides and at backyard feeders. Eight North American species, the Pine Siskin, the Red-breasted Nuthatch, the Red Crossbill, and White-winged Crossbill, the Purple Finch, Pine Grosbeak, the Evening Grosbeak, and the Common Redpoll, tend to invade during the same years (Figure 22–5). Invasion years, which are often the same in the New and Old World, correspond to years of poor boreal forest seed production.

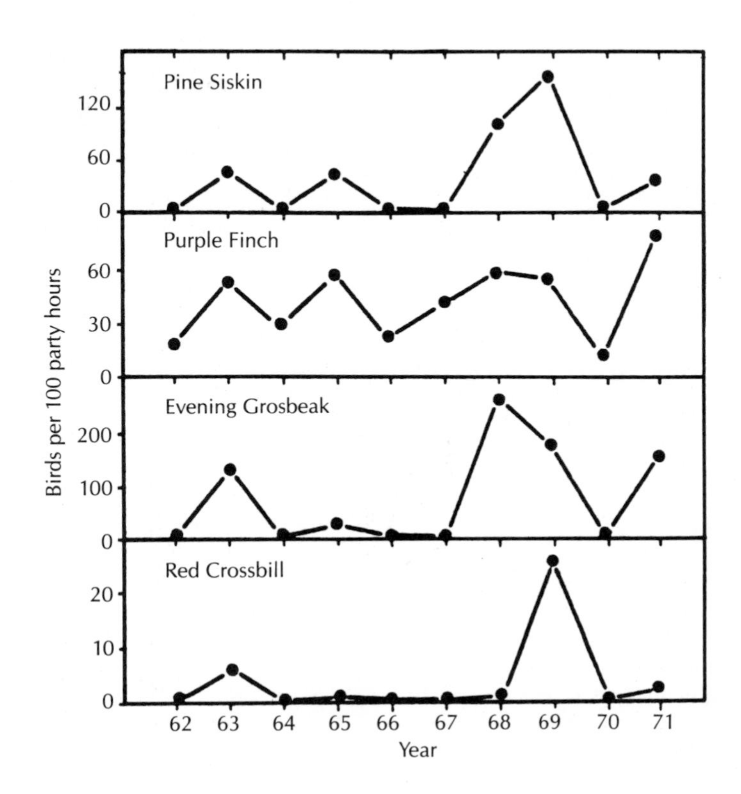

Figure 22–5 Annual variations in the winter abundance of four northern finches in the Chesapeake Bay region, based on Christmas count data. (Adapted from Bock 1980)

Table 22–2 Relation of Ground-finch Abundance to Seed Availability on Daphne Major in the Galapagos

	Year	
	1973 (wet)	1977 (dry)
Seeds		
Total number per m²	4821	295
Total volume (cm³) per m²	15	5
Finches		
Total number	1640	300
Biomass (kg)	26	6

From Grant and Grant 1980.

Detailed local studies of the correlation between food abundance and population are some of the best on this subject. Daphne Major, one of the small islands in the Galapagos archipelago, suffered severe drought in 1977, which caused a critical shortage of the seeds that are a staple for the resident ground-finches. When seed abundance, both in numbers and volume, plunged sharply compared with the number of seeds in 1973, a wet year, finch abundance plunged by a similar order of magnitude in both numbers and total biomass (Grant and Grant 1980) (Table 22–2). The effect of this event on average bill sizes in the population is described on page 134.

Some populations of sparrows that winter in southeastern Arizona are also limited by the availability of seeds, which is determined by the rainfall of the previous summer (Pulliam and Parker 1979; Dunning and Brown 1982). In poor years, the few wintering sparrows have eaten most of the seeds by the end of the winter, and populations of resident sparrows such as Grasshopper Sparrows are limited by supplies of winter food. Migrant sparrows such as Chipping Sparrows use the same grasslands as Grasshopper Sparrows when doing so is to their advantage. In good winters when food is abundant, some Chipping Sparrows stay in Arizona, and though many sparrows are present, they have little impact on seed abundance. In poor

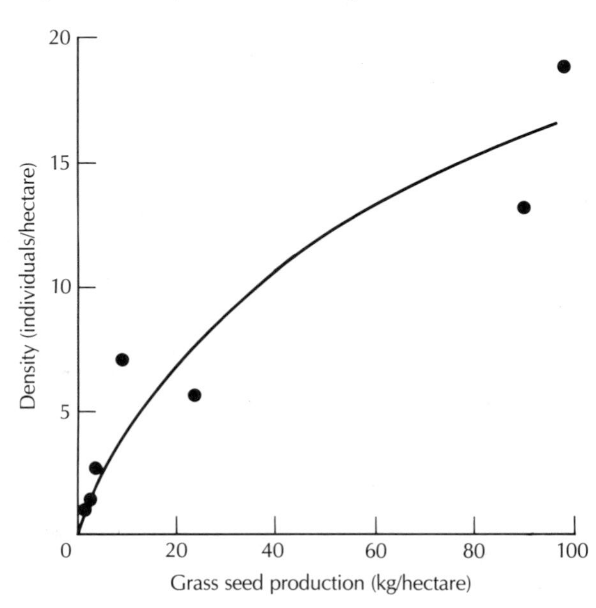

Figure 22–6 The density of sparrows wintering in southeastern Arizona increases with the abundance of the seeds that make up their food supply. In poor years, these sparrow populations may be limited by food supplies; in good years, food supply is not a limiting factor. (Adapted from Pulliam and Parker 1979)

winters, however, virtually all of the Chipping Sparrows move south to wintering grounds in Mexico, where the resulting high densities of sparrows can exhaust the usually abundant local food supplies (Figure 22–6). Raptor populations also reflect the availability of their food sources. For example, densities of breeding Peregrine Falcons in the Arctic are highest near large concentrations of seabirds. The populations of raptors such as the Northern Goshawk, which specializes on unstable populations of boreal prey, also fluctuate whereas Peregrine Falcons, which feed on more stable prey populations, have stable population densities and breeding rates (Newton 1979).

Habitat limitation

Habitat availability is the second major ecological variable that seems to influence population size. Rail and bittern populations, for example, are declining throughout the United States as marshlands are drained for industrial development. In recognition of this loss of habitat, the emphasis of many conservation efforts has shifted in recent years to the preservation of critical habitats, and wildlife refuges are being designed to fulfill the requirements of particular species. The essential resources provided by a particular habitat range from food to nest sites. The invasion of abandoned fields by various birds seems to be related to available nest cover. For some birds, the limited availability of nest holes clearly limits population size. Woodpeckers can dig their own nest holes, but other birds must either use abandoned woodpecker holes or dig their own in soft dead wood. In the managed forests of Britain and Europe, where dead trees and branches are routinely removed, the shortage of nest sites clearly limits the population densities of such species as the Great Tit and the Pied Flycatcher (Haartman 1951; Sternberg 1972). Larger birds such as the Eurasian Kestrel (Cavé 1968) and the Wood Duck (McLaughlin and Grice 1952) also face shortages of nest sites.

The full story of the impact of human activities on the availability of habitats for birds cannot yet be told. Some species benefit from human interference, but many do not. Widespread deforestation has aided species that inhabit open country but has hurt species that inhabit areas with large timber. Many species of the open grasslands of the midwestern United States have moved into the agricultural fields of the east. Birds that live in habitats with clearing and second growth, such as the Chestnut-sided Warbler, the American Robin, and the Indigo Bunting, were once rather scarce but are now abundant. Blackcaps and Dunnocks have greatly increased in abundance in Finland since 1956, when cattle were no longer permitted to graze in the forests. The growing appeal of birds to the general public has fostered the tending of winter feeding stations, which have increased the number of Great Tits that survive the winter in Europe and the number of Northern Cardinals, House Finches, and Evening Grosbeaks that survive the winter in North America.

The high densities of migrants on tropical wintering grounds make them especially vulnerable to the destruction of natural habitats. Clearing 1 hectare of forest in Mexico eliminates the same number of warblers as clearing 5 to 8 hectares in the United States (Terborgh 1980). Many migrants congregate in tropical highland areas, which is prime agricultural land. African migrants winter primarily in open woodlands, rather than the evergreen forests or rain forests, which are now being used for farming, pasturage, and firewood. Conservation of these tropical habitats may be essential to maintain the large numbers of migrants returning northward in the spring. In addition, the habitats of some birds have been poisoned by pesticides and pollutants, which have greatly diminished the populations of some species, especially those at the tops of food chains, where toxins tend to concentrate. For example, Peregrine Falcons and Ospreys in the eastern United States (Fyfe et al. 1976) and Ospreys and Common Sparrow Hawks in Britain were nearly exterminated as a result of such habitat poisoning (Newton 1979).

Social forces

Subtle social forces mediate the availability of habitat and, therefore, local population size. The spacing of territorial individuals in prime habitat may exclude some of them from the breeding population or force them to occupy secondary habitats where nesting is less successful and the risk of mortality is greater. Dispersal increases with population density. In one well-studied case, young male Great Tits dispersed as little as 354 meters (median) in years of low population density and up to 1017 meters (median) in years of high population density (Greenwood et al. 1979). Young male Great Tits disperse farther in populous years because more of them must look for unoccupied territories, which are scarce because of the high level of survival of established males. Young that fledge late in the season usually must disperse farther because young fledged earlier in the season occupy the nearest territorial openings.

The occupation of available habitat has three stages (Brown 1969):

1. Prime habitat is filled.
2. Surplus birds move to suboptimal habitat.
3. Any remaining birds wait for openings in any kind of habitat.

New males quickly replace established males that disappear or that are experimentally removed (Stewart and Aldrich 1951; Hensley and Cope 1951). Replacement males occupy inferior territories; others are floaters, which are unable to establish any territory (Smith 1978) (Figure 22–7).

Floaters either form flocks in areas that are not occupied by territorial breeders, as do Black-backed Magpies (Carrick 1963) and Dunlin (Holmes 1970), or they live singly on home ranges that overlap the breeding territories of established pairs. About 50 percent of a population of the Rufous-collared Sparrow was made up of nonterritorial floaters (Smith 1978). This tropical bunting, which is closely related to the White-crowned Sparrow of North America, defends territories and breeds throughout the year. Floaters, or members of the "underworld," live in well-defined, small home ranges; the ranges of young females were restricted to a single territory whereas the ranges of young males encompassed three to four established territories. Males and females of the underworld

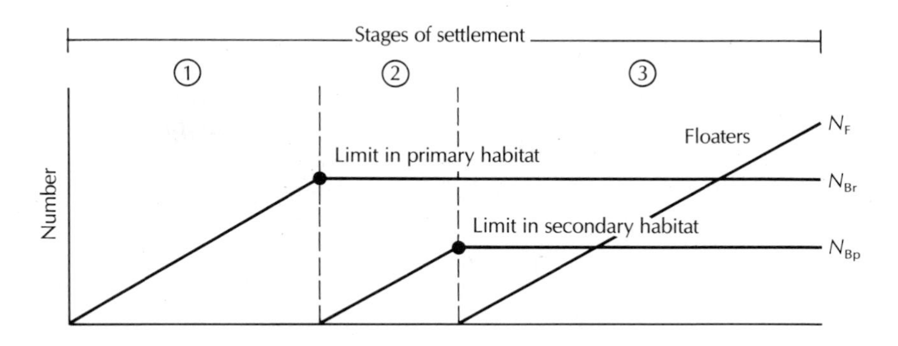

Figure 22–7 Stages of settlement in a local population of breeding birds. The first breeding birds to arrive in an area occupy primary habitat (stage 1). Individuals unable to establish territories in primary habitat settle in secondary, or poorer, habitat (stage 2). Floaters are individuals unable to establish territories because they arrive after all the breeding habitat is filled (stage 3). N_F = number of floaters; N_{Br} = number breeding in primary habitat; N_{Bp} = number breeding in secondary habitat. (Adapted from Brown 1969)

whose home ranges overlapped had well-defined, intrasexual dominance hierarchies. The dominant individuals of the appropriate sex filled the vacancies.

The dynamics of control and attempted takeover of limited territorial spaces are illustrated by Susan Smith's (1978) description of what happened when a territorial male Rufous-collared Sparrow (color banded RO) disappeared for nine days after capture and banding on August 10 (Figure 22-8):

> Less than one hour after his capture, two banded underworld males were courting his mate, GY, but she actively chased both throughout the day. Also, at least four neighbor male owners invaded the territory repeatedly and were driven out by GY. By August 15, one

Figure 22–8 Rufous-collared Sparrow

> of these, YO, had formed a stable pair with GY, and two other underworld males . . . had established small territories at each end of YO's former territory. Both actively courted YO's former mate, RRO, who, unlike GY, readily associated with both. On August 17 I saw RRO copulating with the one that sang more, RBO, and by August 18 they were established in her territory. Yet less than 24 hours later RO had returned and regained his territory and mate, and YO had reclaimed most of his old territory with RBO, holding a small corner, forming a trio of one female (RRO) and two males (YO and RBO). Five weeks later YO had regained all his territory, and RBO had rejoined the underworld. (Smith 1978, p. 577)

Population cycles of the Red Grouse

Food availability limits population directly by starvation and reduced fecundity or indirectly by causing territorial exclusion of surplus individuals, which subsequently starve or fail to breed. Social behavior mediates the effects of food limitation and habitat limitation. This process is well demonstrated in the population cycles of the Red Grouse of Scotland. Not all populations of a species are cyclic. For example, some populations of the Ruffed Grouse of North America peak locally every 10 years, but others do not. Nonetheless, many grouse (Tetraoninae; ptarmigans are a kind of grouse) undergo regular cyclic changes in population size (Watson and Moss 1979). Ten-year cycles characterize Rock Ptarmigan in Alaska and Iceland and Willow Ptarmigan in Newfoundland. In Scandinavia and northern Russia, these two ptarmigan have 3- to 4-year cycles, as do Hazel Grouse, Common Capercaillie, and Black Grouse.

Local populations of the Red Grouse of the Scottish moorlands fluctuate on a regular 6-year cycle from lows of 30 birds to highs of 120 birds per square kilometer. These cycles result from variations in the intensity or effectiveness of territorial behavior of aggressive individuals, whose social interactions are partly tied to the availability of food (Miller et al. 1970; Watson and Moss 1972, 1979, 1980). The leaves of alpine heathers are this grouse's primary food. The grouse select nutritious leaves, and in the spring the leaf quality affects maternal nutrition, egg quality, brood size, chick survival, and adult summer survival. Experimental additions of fertilizer to selected moorland plots increases heather growth and the quality of the leaves and, predictably, improves the breeding success and survival of the grouse on those plots (Miller et al. 1970) (Figure 22–9).

In the 1970s, a major decline of the Red Grouse population occurred despite an abundance of nutritious heather leaves. Changes in aggressiveness and territorial behavior at various phases of the population cycle were found to be responsible (Moss and Watson 1980; Watson and Moss 1980). The cocks that survive summers of high adult mortality are aggressive individuals that occupy larger-than-usual territories. This situation causes a decline in local population. The advantage of these aggressive cocks increases with population density; and as their numbers increase, the population declines because large numbers of less aggressive cocks are lost due to emigration and death. As the population declines further, however, aggressive cocks also lose their advantage. The decline then reverses as increasing

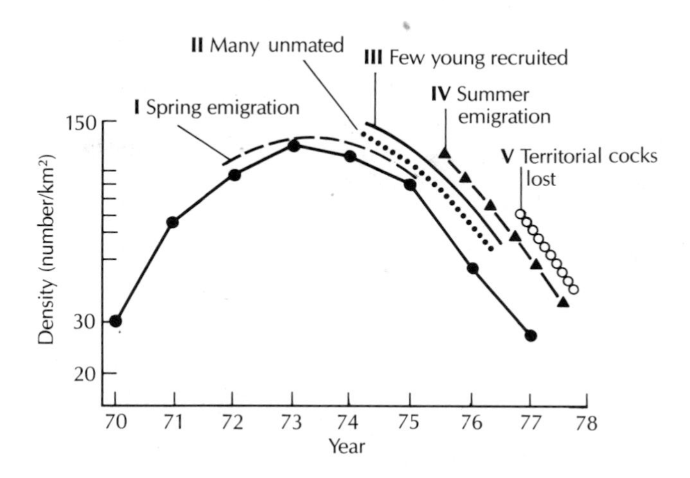

Figure 22–9 The density of Red Grouse cocks holding breeding territories cycles from lows of 30 birds per square kilometer to highs of over 100 birds. The solid line shows the density of cocks in the spring. The decline at high population densities is partly related to the control of large territories by aggressive males, which causes a series of effects (I-V). (Adapted from Watson and Moss 1980)

numbers of less aggressive cocks reestablish themselves on smaller territories at higher population densities, thus continuing the cycle.

Recruitment

Dispersal can limit a population, and conversely, recruitment of dispersing birds can increase a population. Recruits can comprise an important fraction of a population. The population of Atlantic Puffins on the Isle of Man, off eastern Scotland, for example, is increasing by 22 percent a year. Most of the additions are young birds from another population 80 kilometers away on the Farne Islands (Harris 1983). One can project the level of recruitment by direct observations of the recruitment of banded birds from the Farne Islands into the Isle of Man population and by life table calculations. Without immigration, the population of puffins on the Island of Man is growing by 5 percent a year, based on the observed average of 0.56 young fledged per pair per year and an annual survival rate of 83 percent.

Recruitment of young birds into the population every year reflects the number of young produced during the breeding season and, particularly, the proportion of those that survive their first six months of life. Mortality is great in Great Tit populations and may override large differences in the number of fledglings produced in various years. The yearlings recruited into the population are those that are produced locally and those that are immigrants from other places. In one study on an island off the coast of Holland, Great Tit recruits that were produced locally averaged about 20 percent (a range of 12 to 42 percent) of the breeding population each year. Recruitment of immigrants varied in relation to adult mortality over the winter (Van Balen 1980) (Figure 22–10). Providing winter food at bird feeders, starting in 1966, decreased adult mortality and, consequently, the immigration of young birds. These observations demonstrate once again that dispersal and recruitment patterns depend on the density of established adults. We mentioned at the beginning of this chapter that populations of the Great Tit generally fluctuate annually about an average size. Apparently, some form of population regulation is in effect.

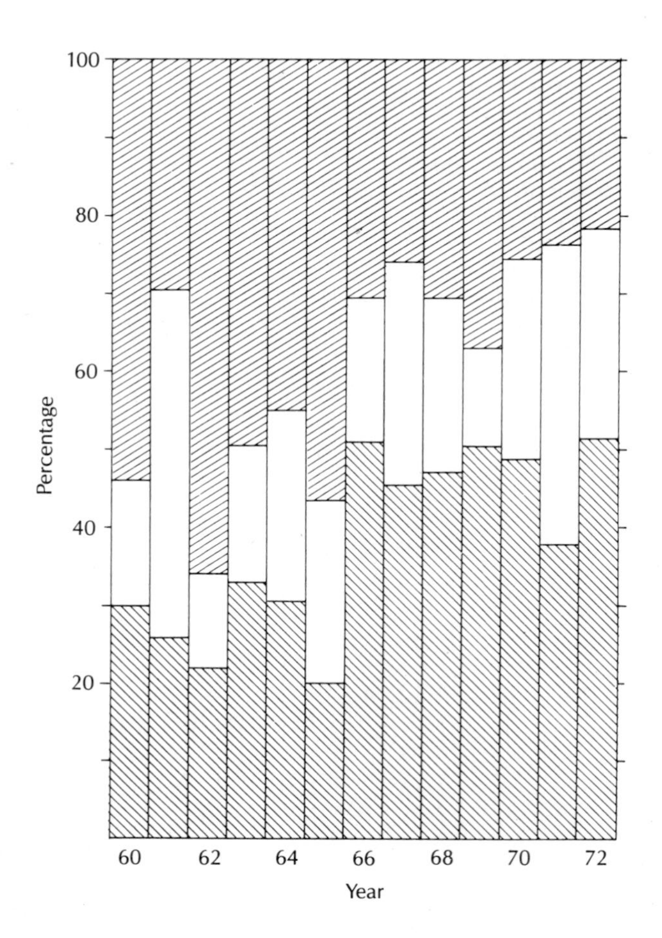

Figure 22–10 Composition of Great Tits breeding on the Hoge Veluwe, an island off the coast of Holland. Opportunities for immigration depend on the overwinter mortality of the resident adults. (Adapted from Van Balen 1980)

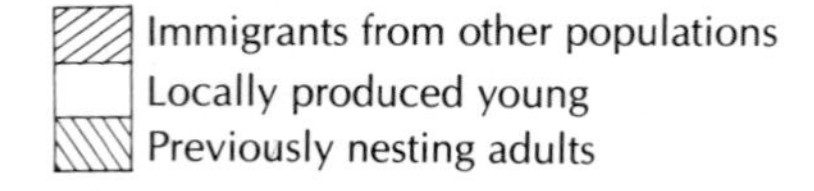

Regulation of Great Tit populations

Ornithologists have monitored populations of the Great Tit in Holland since 1912 and in England, especially Wytham woods near Oxford, since 1947. This species is quite sedentary and nests readily in boxes, especially in managed woodlands in which natural cavities are scarce. Inspection of the nest boxes facilitates accurate censuses of breeding pairs, clutch sizes, and young raised. Study of population regulation has been a primary goal of this research (Lack 1966; O'Connor 1980; Klomp 1980). The main finding is that population regulation in Great Tits is a density-dependent phenomenon, (Figure 22–11), which is evident mainly in the effects of food limitation on juvenile survival during the winter. Habitat limitation, territorial behavior, and dispersal also play mediating roles. Although fecundity and survival in the breeding season are density dependent, their impact on the overall population size is minor. Let us now examine these effects in detail.

Seasonal changes in Great Tit populations are influenced by reproduction, mortality, and migration. Each year, the population increases rapidly with the production of fledglings by a small number of breeding pairs. A steady decline follows because of the heavy mortality of young birds and the loss of some adults. The average percentage of juveniles that survive annually (22 percent, with a range of 4 to 40 percent) is less than half that of

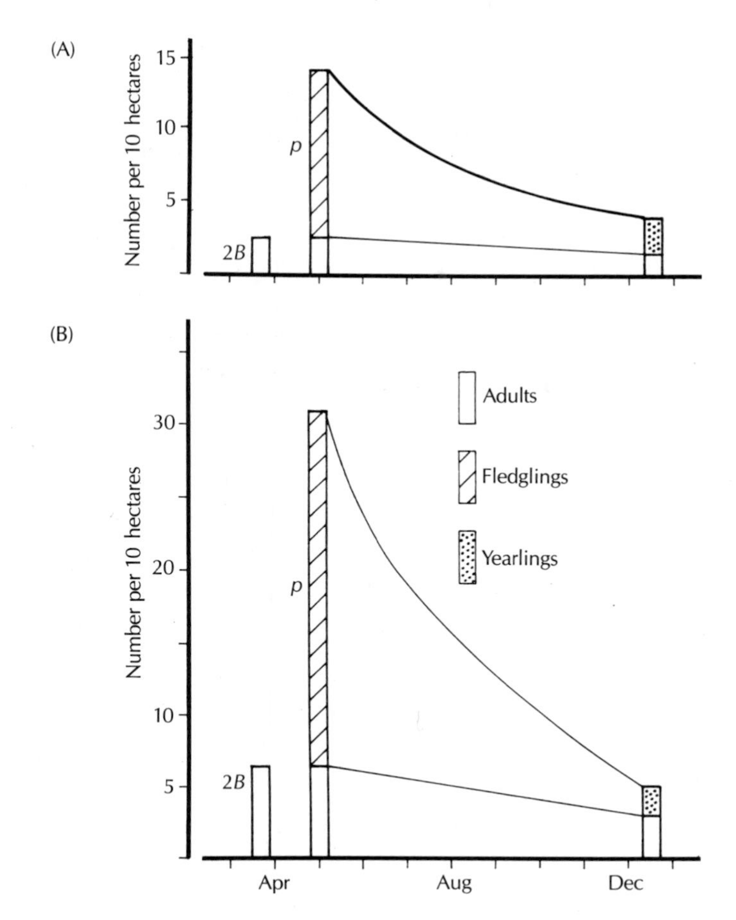

Figure 22–11 Great Tit production (*P*) and survival to the winter in years with (A) low and (B) high breeding densities in a model population. Nesting in the spring adds many young birds to the population, but most of these die by winter. An average of 9 and 7.5 young are produced per breeding pair (*B*) in low- and high-density years, respectively. (Adapted from Klomp 1980)

adults (48 percent of the females, 56 percent of the males, with a range of 40 to 81 percent).

Both the mean clutch size and the number of fledglings, or fecundity, depend on local population density (Perrins 1965). Great Tits lay fewer eggs when population density is high; 60 percent of the variation in annual mean clutch size is directly related to population density. Success in rearing nestlings also decreases as population density increases because of increased predation and because fewer females attempt second broods (Figure 22–12).

Survival of juvenile and adult Great Tits is also density dependent. Recoveries of banded individuals throughout Britain showed that females are less likely to survive in a high-density population year than in a low-density population year. Most telling were experiments by Kluijver (1966), who removed 60 percent of the eggs and nestlings some years, but not in others, from a population on an isolated Dutch island in the North Sea. Both juvenile and adult survival doubled. Juvenile survival rose from 11 to 20 percent and that of adults rose from 26 to 54 percent. Immigration and emigration did not affect these experimental results because it did not occur on the tiny island. In other experiments, the survival rate of juveniles in autumn and winter is positively correlated with the percentage of breeding birds removed in the summer.

Although annual variations in reproductive success, adult survival, and juvenile survival all potentially influence the density of the population

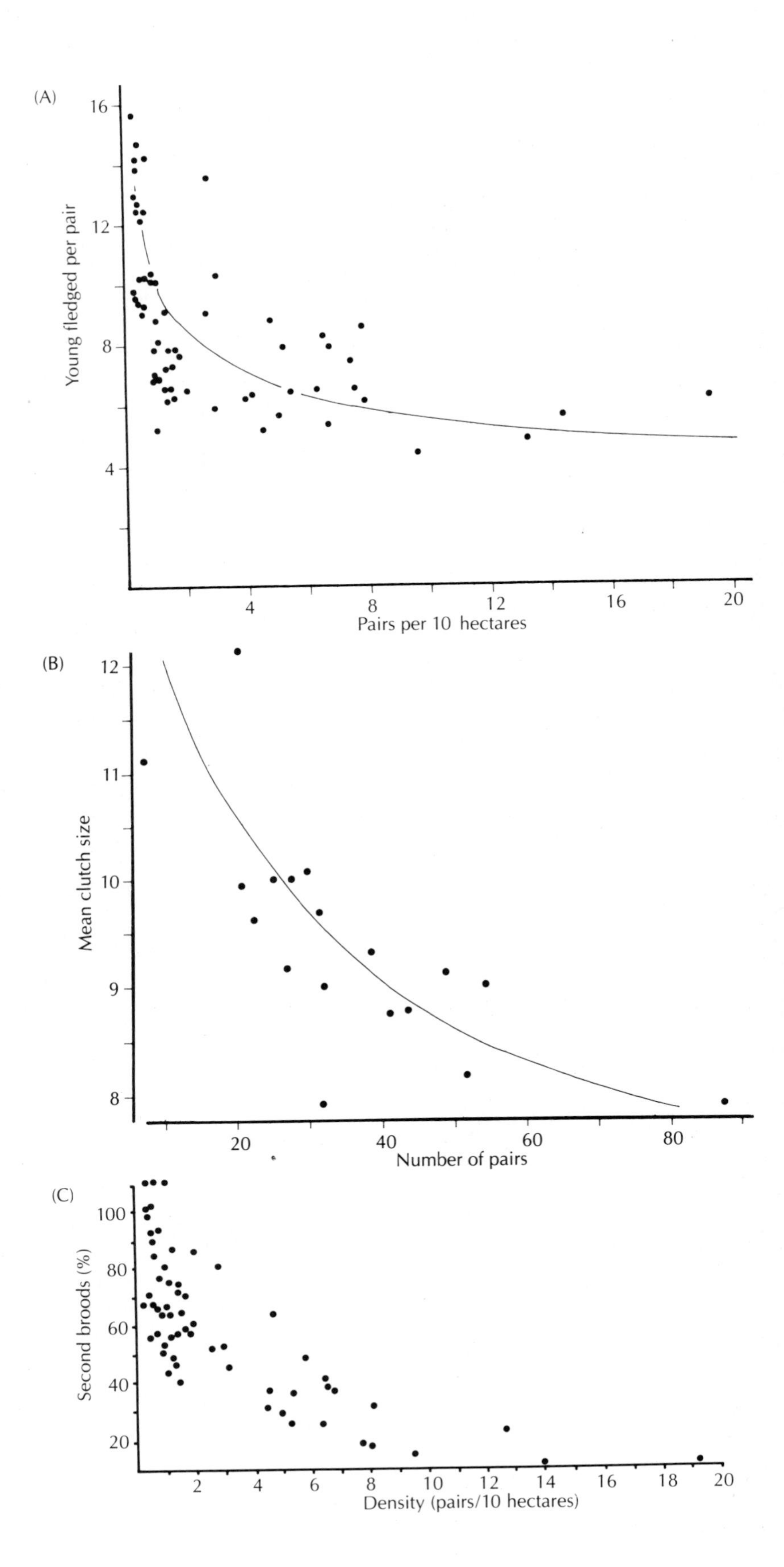

Figure 22–12 (A) Reduced fecundity at higher densities in the Great Tit reflects (B) smaller clutches at high population densities and (C) less frequent attempts to raise second broods. (Adapted from Klomp 1980; Kluijver 1951)

during the breeding season, survival outside the breeding season, particularly of juveniles, actually controls population size the following year (Klomp 1980). Winter food supplies, especially the seeds of beech trees, control juvenile survival in both Oxford and Holland. Young Great Tits in Holland, for example, depend on the beech mast from November to late February when other food sources are scarce. This essential reserve food supply is easily exhausted in poor crop years. Thus, annual variations in population size reflect the availability of winter food supplies. Beech mast crops vary greatly from year to year, partly because the production of flower buds, which determines the next year's food supply, cannot coincide with the production of many seeds. The production of flower buds is stimulated by high spring and summer temperatures throughout northwestern Europe.

The local population density of Great Tits also varies with habitat. Deciduous oak forests, for example, support 10 times as many breeding pairs as do pine forests. Mixed oak–pine forests support intermediate population densities. As the composition of trees in a local forest shifts from pine to oak, the density of tits increases (Figure 22–13). There is less food in pine forests than in oak forests, and, therefore, territories are larger in pine forests and more nestlings starve. The amplitude of annual population fluctuations is greater in thinly populated pine habitats than in densely populated deciduous woodland. The pine forest is a secondary or suboptimal habitat that is occupied by an overflow of Great Tits that are excluded from deciduous woods. Individuals from pine forest habitats quickly fill vacancies in the deciduous forest. Farmlands and other nonforest habitats act as suboptimal buffers to woodland populations, which are regulated by density-dependent factors.

The patterns of population regulation seen in the Great Tit have broad application to other birds. As a rule, bird populations seem to be limited by food scarcity during the nonbreeding season. Food abundance and scarcity fluctuate from year to year. Survival of juveniles in a year of scarcity has a major effect on recruitment levels the following year, and hence on population size.

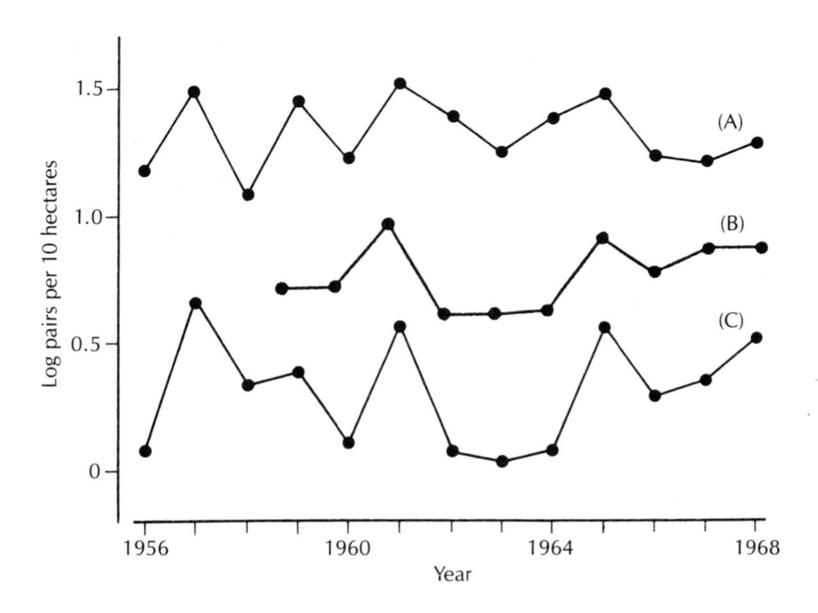

Figure 22–13 Densities of Great Tits breeding in (A) oak forests are higher than in (B) mixed oak-pine forests, and much higher than those in (C) pine forests. (Adapted from Klomp 1980).

Summary

Bird populations are dynamic entities, fluctuating from year to year as a result of variations in breeding success and mortality. Growth of a population in a new environment usually follows a pattern of a slow initial rate of increase, followed by accelerated growth rates, and finally, a decline in growth rates in response to factors that lower reproduction and survival. Established populations tend to stay close to a long-range average size. Growth rates may be directly dependent on population density.

Four ecological factors limit or set ceiling levels on the sizes of bird populations: disease and parasites, climate, food availability, and habitat. Social forces such as territoriality and aggressiveness also help to limit populations. The Red Grouse is an excellent case in point; population sizes are controlled directly by limited food sources and indirectly by limited territorial space.

Populations may grow as a result of recruitment. Population balance is dependent on dispersal and recruitment patterns, which, in turn, are often density dependent. Detailed studies of Great Tit populations in Holland and England illustrate the nature of density-dependent regulation of population size.

Further readings

Andrewartha, H.G., and L.C. Birch. 1954. The Distribution and Abundance of Animals. Chicago: University of Chicago Press. *A classic work on populations.*

Ashmole, N.P. 1971. Seabird ecology and the marine environment. Avian Biology 1:224–286. *An excellent review of the special problems of seabird ecology.*

Haartman, L. von. 1971. Population dynamics. Avian Biology 1:391–459. *A detailed review of European bird populations.*

Lack, D. 1954. The Natural Regulation of Animal Numbers. Oxford: Oxford University Press. *A classic.*

Lack, D. 1966. Population Studies of Birds. Oxford: Clarendon Press. *Another classic.*

Ricklefs, R.E. 1979. Ecology. New York: Chiron Press. *A well-written text that covers the fundamentals.*

chapter 23

Communities

The availability of resources such as food and nest holes determines not only the local population size of a species but also how many kinds of birds can coexist locally in a habitat. Such coexisting groups of species are called communities. The sizes, compositions, and diversity of bird communities vary conspicuously over the globe. Several hundred species coexist in a lowland tropical forest whereas less than 50 species coexist in a northern temperate forest; the sizes and diversity of communities increase regularly from temperate to tropical latitudes. The variety of species found on islands increases with island size but decreases with isolation from the mainland. There are conspicuous patterns of substitution of community members by similar species because similar species tend not to coexist but to replace each other geographically.

What determines the composition of local assemblages of bird species? What forces determine which species and how many species coexist? The answers to these basic questions lie in the study of the groups we call communities.

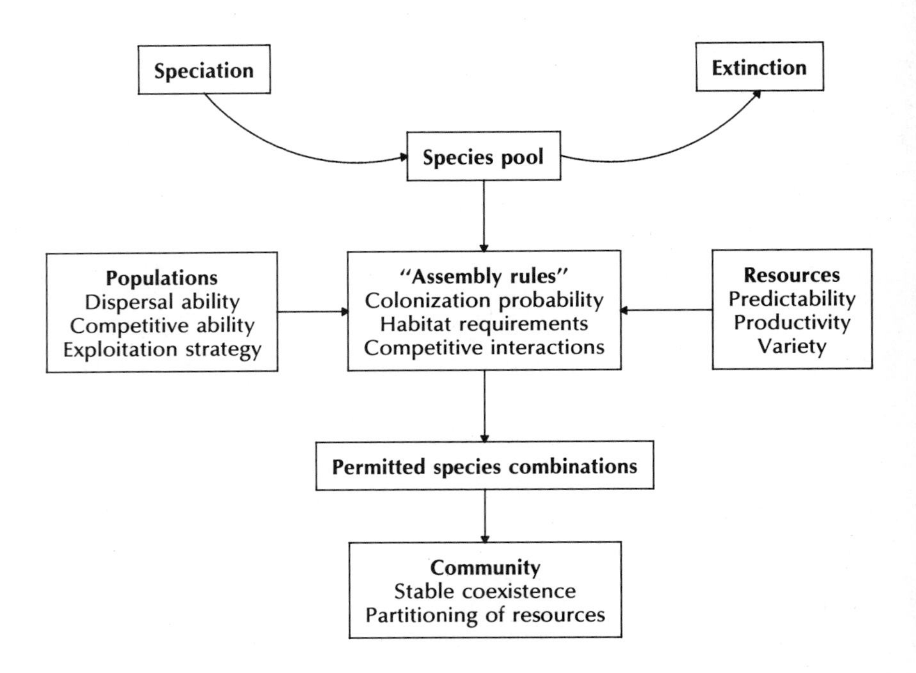

Figure 23–1 According to some ecologists, stable communities of coexisting bird species derive from a larger pool of species through the dynamics of population dispersal, colonization in relation to habitat or other resources, and sometimes by competitive resolution of unstable species combinations. (Adapted from Wiens 1983)

Evolution and resource availability play major roles in the formation of communities: Those species that coexist in a community change from epoch to epoch in accordance with evolution and from season to season in accordance with resource availability. Both long- and short-term changes greatly affect the species composition of communities. Resource availability and evolution give rise to a third force — competition — the interactions among species through which one species may directly or indirectly preclude another. Competition can select from a large pool of potential candidates for a community, those species best able to coexist (Figure 23–1). The relative importance of these forces in community ecology is a matter of continuing debate. Competition has long been assumed to be the main force in structuring bird communities, but we are now learning that irregular temporal changes may keep communities in a state of dynamic flux. Some communities, rather than maintaining stable, self-perpetuating sets of species that reach equilibrium in accord with theory, lack predictable structure and may simply be temporary and fortuitous associations of species.

Two polar views of communities exist today, which derive from early botanical thought. According to one view, communities are open systems in which each species arrays itself independently along environmental gradients according to its own ecological requirements (Gleason 1926, 1939). Thus open communities are fortuitous assemblages of noninteracting species. According to the other view, communities are closed, integrated sets of compatible species (Clements 1916, 1936). Closed communities include predictable sets of interacting species. These are extreme views, but evidence exists for both (Figure 23–2).

The birds that breed in upland hardwood stands in southern Wisconsin form open assemblages of species in habitats that range from open, dry deciduous forests dominated by Black Oak trees to denser, moist forests dominated by Sugar Maple trees (Bond 1957). Certain bird species, such as the Red-eyed Vireo, are more common in mesic forests; others, such as the

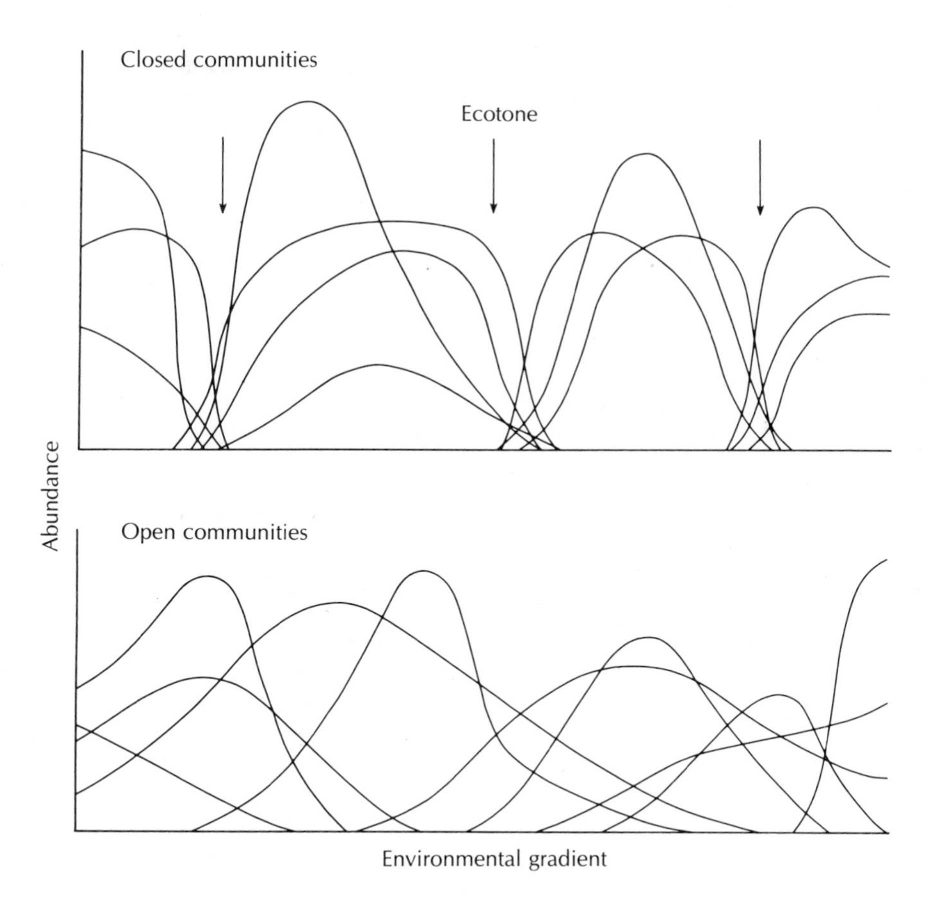

Figure 23–2 Open and closed communities are extreme forms of the continuum of possible community structures along environmental gradients, such as dry forest to wet forest. In open communities, species are arrayed independently in accord with their particular ecological needs; in closed communities, distinct sets of species occupy particular habitats with breaks at the interfaces between habitats, called ecotones. (From Ricklefs 1979a)

Black-capped Chickadee, are more common in xeric forests; and some, such as the American Redstart, are most common in forest types that are between these two. Each species has specific preferences or needs and chooses its habitat accordingly (Lack 1971; Lanyon 1981). These distributions suggest independent, ecologically related associations, not coincident relationships of coadapted species.

Nectar-feeding hummingbirds provide some of the best examples of closed communities based on competition for food (Wolf et al. 1976; Feinsinger and Colwell 1978). Species organize themselves in predictable sets around aggressive, dominant species that control access to preferred flowers (Table 23–1). In the highlands of Costa Rica, for example, the Fiery-throated Hummingbird and the Green Violet-ear determine which other species of hummingbirds are present.

We begin our review of bird communities by looking at the broad patterns of species diversity, which reflect spatial and temporal changes in

Table 23-1 Community Roles in Relation to Morphological Characteristics of Hummingbird Species

Role	Size	Bill Length
Low-reward trapliner	Small	Short/medium
Filcher	Small	Short
Generalist	Medium	Short/medium
Territorialist	Medium	Short/medium
High-reward trapliner	Large	Long
Marauder	Large	Medium

From Feinsinger and Colwell 1978.

the environment. The scale of these environmental changes plays a major role in this discussion. We then examine the phenomenon of competition from a theoretical standpoint. Competition is a corollary of density-dependent population regulation, discussed in the previous chapter. Here we examine two detailed examples of the effects of competition among species; the relationships among titmice in Europe and the evolution of Galapagos finches. Finally, we shall look at the patterns of geographical replacement of birds by similar species. The study of these patterns has fueled interest in the importance of competition in the community ecology of birds, but in truth such studies provide only weak evidence in support of the importance of competition in avian ecology.

Species diversity

The number of bird species in an area, or species richness, increases from the arctic to the tropics: Greenland has 56 breeding bird species, New York State about 135, Honduras over 550, and Colombia over 1300. The variety of species also increases from high to low altitudes. In Colombia, 47 species exist in the high paramo zone, 270 at temperate altitudes, over 480 at subtropical altitudes, and over 1000 in the tropical lowlands. Many more species are found in New World tropical forests than in comparable forests of the Old World (Pearson 1977). Why do these differences exist and, in particular, why are there so many more species in tropical communities than in the temperate zone communities?

Spatial components of diversity

The distributions of most birds are restricted globally and locally. Penguins are limited to the southern hemisphere, auks to the northern hemisphere, curassows to tropical South America, turacos to Africa, and the Dodo (once upon a time) to the island of Mauritius. Such boundaries may reflect physical restrictions: The flightless Dodo was limited to one oceanic island. More often, however, the limits of the distribution of a bird species, even on continents, reflect intrinsic limits of population growth, competitive replacement by another species, availability of resources, physiological tolerance, or some combination of these. Even on local levels, a county or shire perhaps, few birds fully use the variety of habitats available to them.

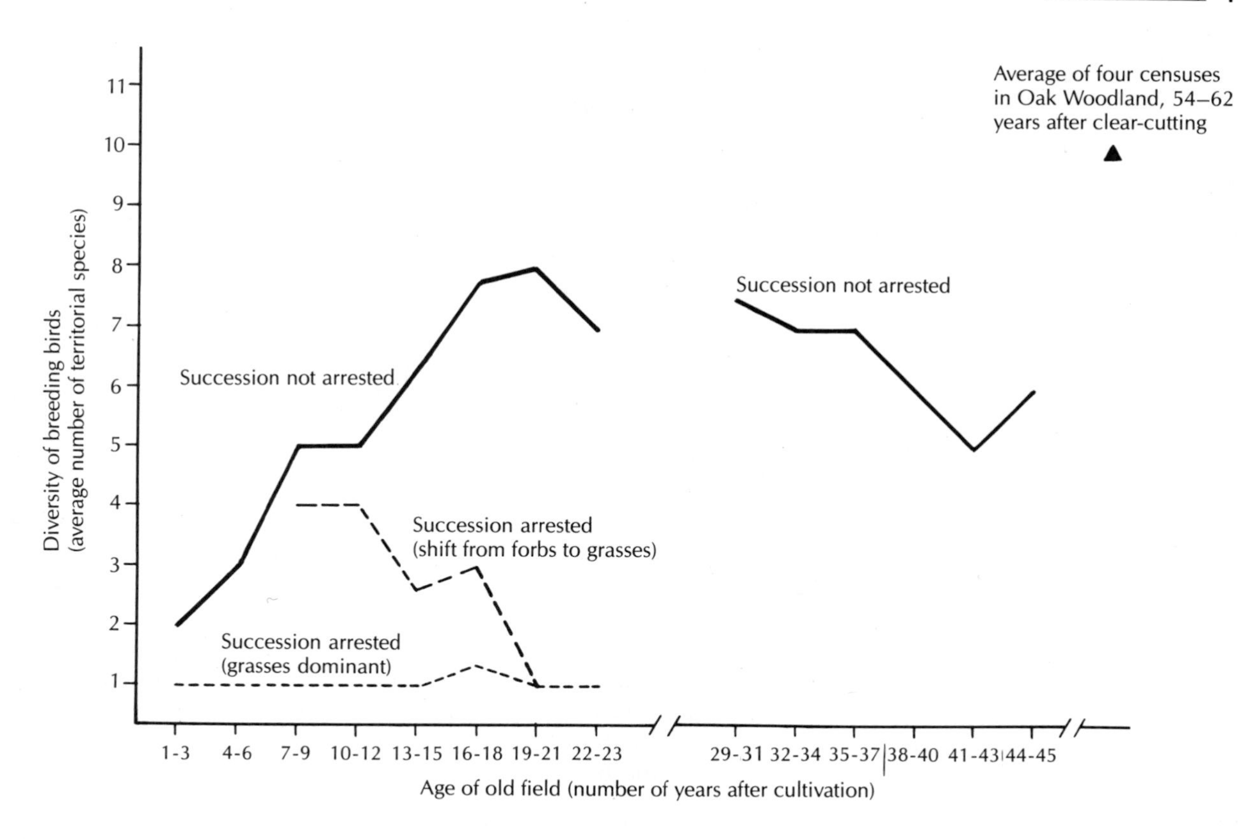

Figure 23–3 The diversity of species nesting in old fields increases with the age of the field and the successional change of the vegetation from grass, to bushes, and ultimately to trees. Successional change was arrested by cultivation. (From Lanyon 1981)

Each species usually has specific habitat preferences. Thus, we expect to find a Pileated Woodpecker in a forest with large trees rather than in the open prairie, where we look for a Burrowing Owl. More subtle differences in habitat preferences are seen in the preferences of birds that breed in upland hardwood forests and of birds that inhabit fields that are converting into shrublands and young forests.

In a 20-year study of birds associated with such field succession on Long Island, New York, Wesley Lanyon (1981) showed that nine species of birds successively established nesting territories in a sequence that corresponded to the availability of nest cover. Red-winged Blackbirds were the first to nest in a field, and Rufous-sided Towhees were the last. As the open field converts to shrubland and then to forest, the availability of nest-supporting vegetation and the amount of shade for nests determine the suitability of the habitat for breeding. In this example, the procession of species of overlapping tenures caused the average number of territorial species present to peak between 20 and 30 years after the field was last cultivated (Figure 23–3).

Habitat preferences and restricted distributions cause the variety of birds encountered in a geographical area to increase from local (alpha) diversity to regional (beta) diversity as the variety of habitats increases. Local diversity reflects the structural complexity of the habitat. The physical structure of habitats provides courtship and display stations, nest sites, protection from predators, and shelter from climatic stresses. Indirectly, it also provides a variety of prey. The vertical distribution of vegetation provides a rough index to the variety of foraging opportunities and, hence, the variety of species that can occupy a habitat (MacArthur and MacAr-

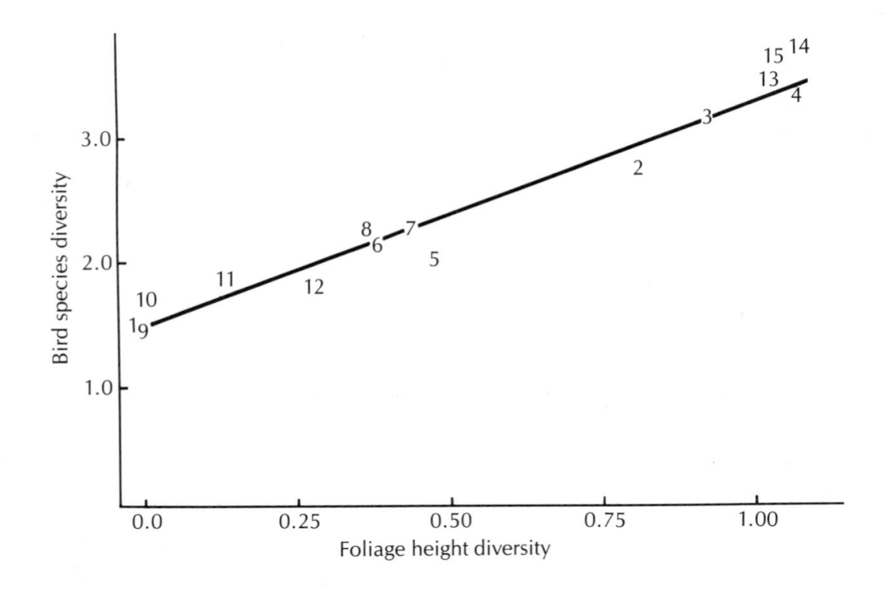

Figure 23–4 The local diversity and relative abundance of bird species is correlated with the relative height and diversity of the foliage, illustrated here for sites in Illinois (1–4), Texas (5–8), and Panama (9–15). (Adapted from Karr and Roth 1971)

thur 1961; Willson 1974) (Figure 23–4). Nearly 40 percent of the variance in foraging behavior among insectivorous birds in the Hubbard Brook forests of New Hampshire is attributable to variations in foliage height (Holmes et al. 1979).

However, the diversity of foliage height is too simplistic a measurement of habitat complexity and is of limited value to the study of community ecology (Lovejoy 1974; Karr and Roth 1971). Plant species and physical form, percentage of vegetative cover, and local variations in habitat structure all influence the local diversity of birds. Multivariate statistics enable community ecologists to integrate many variables into the definition of habitat complexity (James 1971). For example, a coordinate system with two independent axes summarizes the varied habitats on the slopes of the White Mountains of New Hampshire (Sabo 1980). One axis reflects elevation, the other reflects foliage types from coniferous to deciduous. The variety of bird species is greatest in low elevation habitats with conifers, but each species of wood warbler inhabits a preferred set of habitats (Figure 23–5). In a detailed statistical study of birds that inhabit the grasslands of the western United States, local species diversity increased with complexity of foliage heights in a habitat but not with spatial heterogeneity (Rotenberry and Wiens 1980a).

Distance to cover affects the variety of sparrows that can coexist in open grassland habitats (Pulliam and Mills 1977). The division of a habitat is not dependent on the kinds of seeds available or subtle differences in the composition of vegetation but on the distance that members of a species must fly to the nearest tree or shrub. Thus the habitat is divided into concentric rings of increasing distance from the nearest cover. For example, in southeastern Arizona, four species of sparrows inhabit open grasslands that have scattered mesquite trees, which provide some protection from predators such as Prairie Falcons. The Vesper Sparrow stays close to the mesquite trees (within 4 meters), the Savannah Sparrow feeds further out (4 to 16 meters), the Grasshopper Sparrow still further out (8 to 32 meters), and the Chestnut-collared Longspur feeds far from the trees in the most open grassland. The behavior of these species when flushed reflects the risks

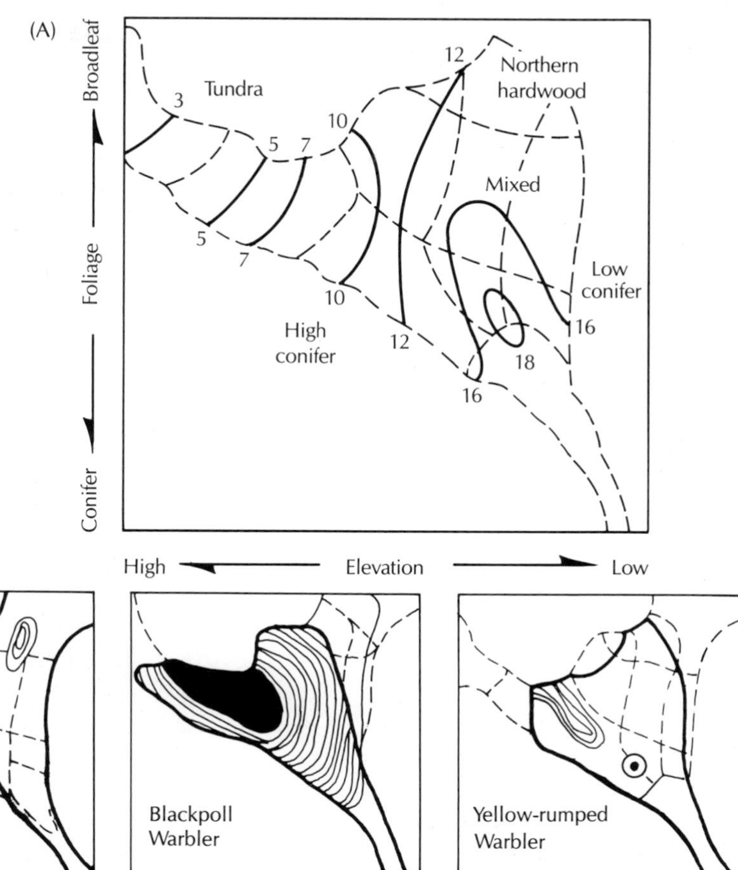

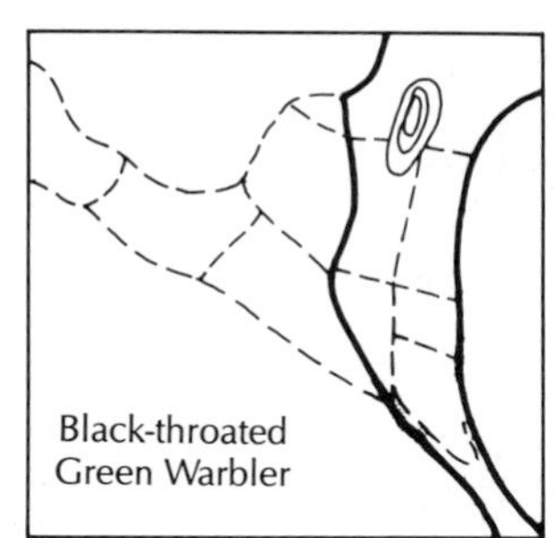

Figure 23-5 Model of bird species diversity (BSD) in relation to habitat in the White Mountains of New Hampshire. (A) Two variables, an elevation gradient and a foliage gradient, define the subalpine habitats outlined in dotted lines. Solid lines relate bird species diversity to habitat. Coniferous forest at moderately low elevations where some deciduous trees occur supports the greatest diversity. (B) Habitat use by four species of wood warblers on the same pair of gradients. Concentric rings indicate centers of highest density. (Adapted from Sabo 1980)

of flying increasing distances to cover. Vesper Sparrows fly quickly to nearby cover, Savannah Sparrows fly to an exposed perch the first time they are flushed and then to full cover if flushed again. Rather than face the risks of a longer flight, Grasshopper Sparrows usually drop back into the grass when flushed but fly for cover if repeatedly flushed. Longspurs, however, either crouch to the ground to hide or fly off in tight flocks that help thwart predators.

As we increase the number of distinct habitats sampled, the diversity of species we encounter also increases. For example, as we proceed from the west coast of the United States to the east coast, we find peaks in the number of species present in the Sierra, Rocky, and Appalachian Mountains, where markedly different habitats at various altitudes increase beta diversity. Peru and Colombia owe their extraordinary variety of birds in part to the topographic and ecological diversity of the Andes. On a more modest scale, the greater number of bird species on large islands compared with small ones is due in part to the greater variety of habitats (Hamilton and Rubinoff 1967).

Tropical diversity

More species of birds live in a given area in the tropics than in a comparable area in a temperate region; a 5-acre plot of forest in Panama has two-and-a-half times more species than a similar-sized area in Illinois. The density of species in an area increases as one proceeds south from North

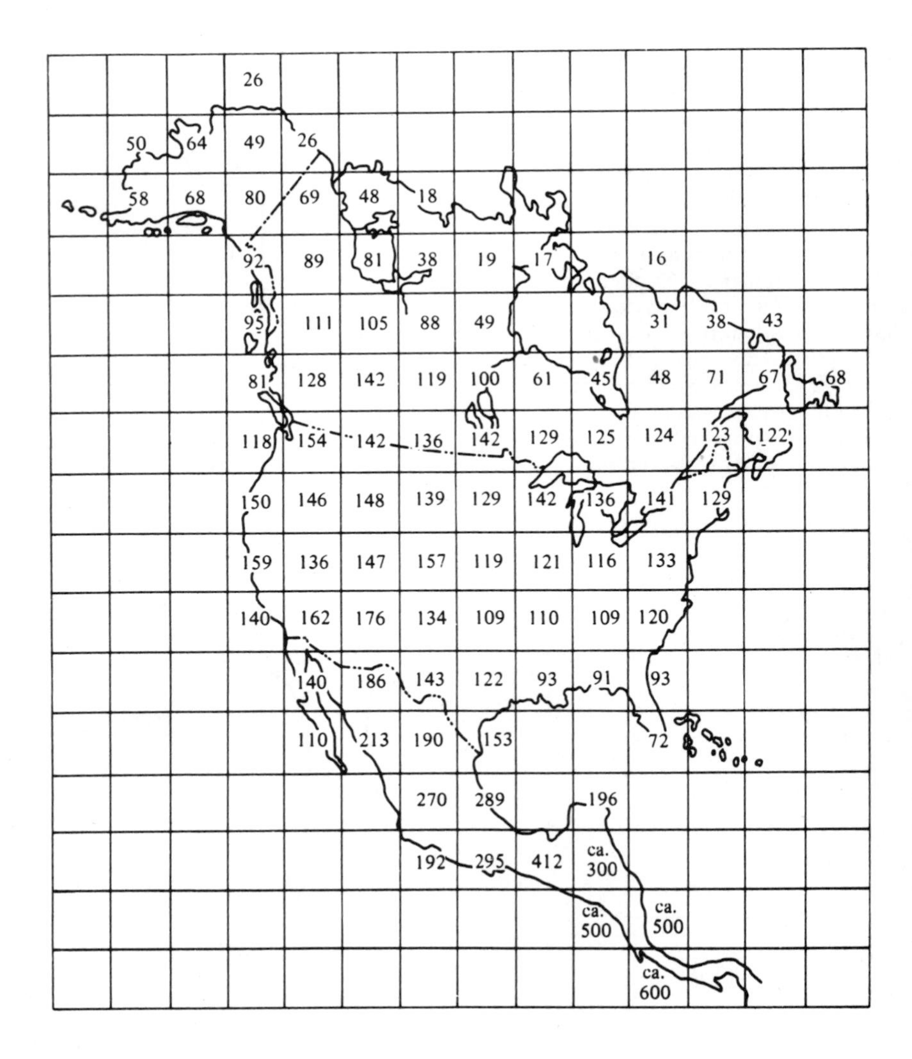

Figure 23–6 The number of land bird species that breed in geographical areas 400 miles square in North America decreases with increasing latitude. (From MacArthur 1969)

America through Central America and reaches its maximum in the tropical forests of western Amazonia. A 400-mile square section of the United States contains 120 to 150 species of breeding landbirds; but the same area in Central America contains 500 to 600 (MacArthur 1969) (Figure 23–6). In western Amazonia, over 1000 species can be found in such an area, and over 535 species can occur locally in a 100-hectare site (Terborgh personal communication).

The greater diversity of species in the tropics compared with the temperate zone is due in part to different and more varied food resources (Karr and James 1975; Ricklefs and Travis 1980). For example, groups of fruit-eating birds—toucans, hornbills, barbets, trogons, cotingas, manakins, broadbills, and turacos—expand the dimensions of tropical communities. Parrots, large and small, consume seeds, fruits, and nectars that are not available in northern forests. Hummingbirds and tanagers, of which only a few species live in the north, abound in New World tropical forests. Some families of strictly tropical birds—puffbirds, motmots, antbirds, wood hoopoes, rollers, and bee-eaters—depend on large insects and small reptiles that are not present in temperate ecosystems. Insect sizes are more diverse in the tropics than in the temperate zone habitats, and the diversity of bill sizes

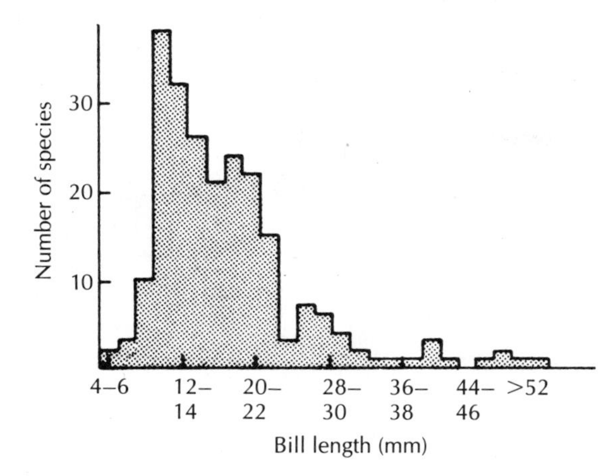

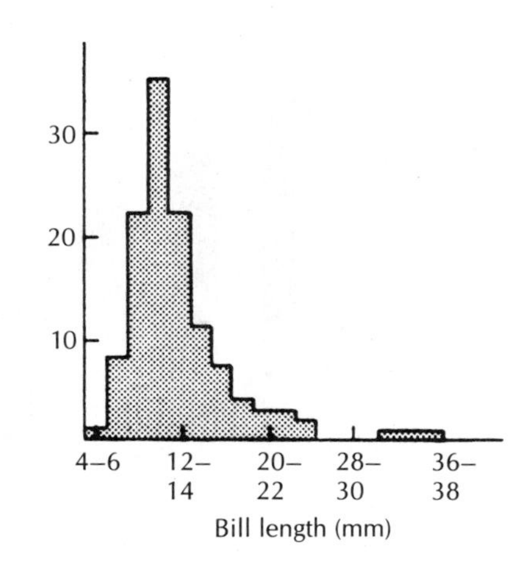

Figure 23–7 The addition of large-billed birds causes an increase in bird species diversity in the tropics. Shown here are the bill lengths of insectivorous birds that breed in (A) tropical latitudes (8° to 10°N) and (B) temperate latitudes (42° to 44°N). (From Schoener 1971)

of tropical birds increases accordingly (Schoener 1971) (Figure 23–7). Foraging specialists, such as ant followers and epiphyte probers, also add to tropical communities. More specialized species, large and small, can exist in benign climates than in cool climates.

Beta diversity greatly affects diversity in the tropics. The number of breeding birds increases more with area in the tropics than in the temperate zone, apparently because habitat heterogeneity is greater. This diversity is affected by mountainous topography and also by subtle variations among microhabitats. Tropical species tend to use a narrower range of habitats than temperate species (Karr 1971; Lovejoy 1974) and may also be more specialized in their foraging behavior (Terborgh and Weske 1969; Stiles 1978).

Diversity in tropical communities may reflect a long, stable history of species accumulation (Moreau 1966; Pianka 1966; MacArthur 1969; Mayr 1969). The diversity of avian communities tends to increase over time as speciation adds new specialized kinds of birds because food webs become more complex and mutualistic relationships increase. Ancient communities may be the most species-rich of all. Differences in tropical community diversity on different continents provide some evidence that certain communities accumulated members over a long period of time. For example, the forest fauna of Panama is richer than that of Africa, but African grasslands and savannas are richer than those of Panama. The scarcity of lowland forests during the Pleistocene period seems to have prevented the evolution of rich forest avifaunas in Africa (Moreau 1966; Karr 1976). The man-made grasslands in Panama are quite young (15,000 years) compared with the ancient natural grasslands and savannas of Africa. Thus grassland communities in Africa are species-rich whereas those in Central America are species-poor.

Temporal components of diversity

Local assemblages of species change in time as well as in space. Their composition fluctuates regularly with the season and irregularly with climate and resource availability. Disturbance due to deforestation, colonization, extinction, and fire also keeps many habitats in flux. The mobility of birds enables them to exploit variable environments opportunistically. Virtually all bird communities consist of both resident and nonresident species.

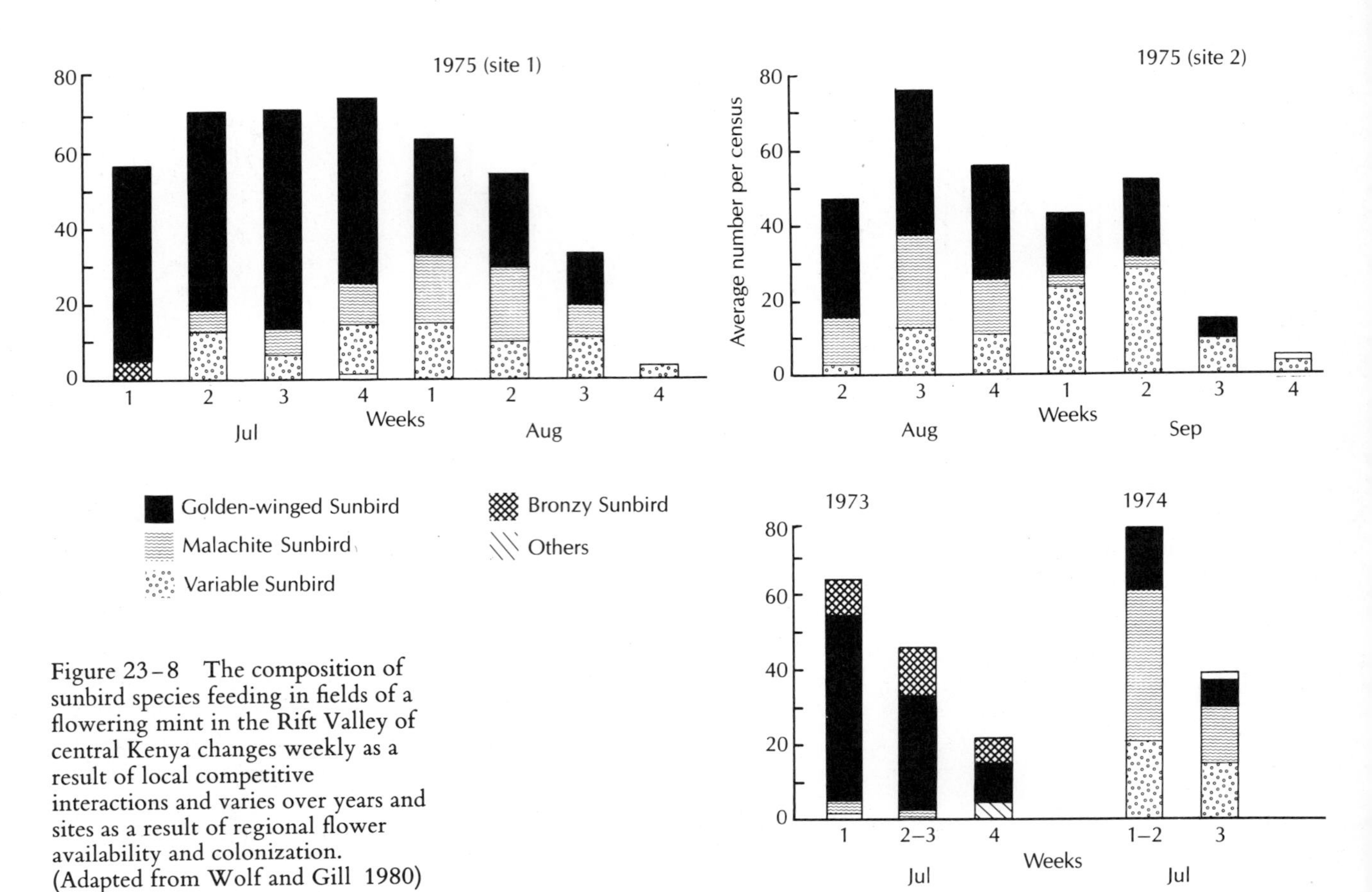

Figure 23–8 The composition of sunbird species feeding in fields of a flowering mint in the Rift Valley of central Kenya changes weekly as a result of local competitive interactions and varies over years and sites as a result of regional flower availability and colonization. (Adapted from Wolf and Gill 1980)

Nonresidents are seasonal specialists, which take advantage of predictable periods of regional food abundance. The mobility of birds and the evolution of the migratory habit have made nonresidency possible (see Chapter 13). Migrants are able to coexist temporarily with dominant or competitively superior residents, and they avoid permanent exclusion from a habitat by changing their community membership with the seasons.

Ephemeral resources attract opportunistic species. Temporary assemblages of highly mobile birds may last hours, weeks, or years. Aggregations of seabirds at a shoal of fish, for example, are brief and highly variable in species composition. Assemblages of sunbirds or hummingbirds at a flowering tree may last only a few weeks and are characterized by a high turnover of both individuals and species during that period (Figure 23–8).

The regional diversity of small, short-billed hummingbirds depends on their ability to circulate among locally blooming flowers (Feinsinger 1980). Only two short-billed species, the Copper-rumped Hummingbird and the Ruby-topaz Hummingbird, inhabit the small island of Tobago where they must coexist year-round in the principal nonforested habitat. Seven similar species coexist on the larger island of Trinidad where the habitats are more diverse, enabling more complicated seasonal patterns of local migration. The complex mosaics of locally seasonal habitats on the mountainsides of Costa Rica and Panama support even more species of short-billed hummingbirds than the comparatively homogeneous lowlands of these countries.

Temperate-zone migrant birds often consume superabundant or sporadically available foods on their tropical wintering grounds (Karr 1976). Large flycatchers such as the Eastern Kingbird, for example, move nomadically over long distances in the tropics, where they specialize on *Didymopanax* fruits (Morton 1980; Fitzpatrick 1980). On Barro Colorado Island in Panama, migrants concentrate along the lake shore and in disturbed scrub habitats but are uncommon in old forests (Willis 1980a). The preference of migrants for secondary growth in Africa, as well as in tropical America, contrasts with their behavior in the West Indies where they inhabit the forest (Terborgh and Faaborgh 1980). In Amazonia, small migrant flycatchers use early-succession habitats little used by permanent residents (Terborgh and Weske 1969). In most of Central America, however, migrants take over the main forest niches while resident flycatchers occupy peripheral niches. Migrants are so abundant in some Central American communities that they consume most of the resources and thereby limit the northward spread of some tropical flycatchers (Fitzpatrick 1980).

Seasonal residents comprise a major component of most bird communities. The influx of wintering migrants from the north triples the number of species found in the open pine forests of Grand Bahama Island and increases the density of individuals from 900 to 1600 per square kilometer (Emlen 1980) (Figure 23–9). Migrants comprise 1 out of every 16 birds (2 percent of the total biomass of birds) on Barro Colorado Island on an annual basis and up to 1 out of 7 birds (5 percent of the biomass) during the peak migration in October. In the tropical evergreen forests of western Mexico, the density of small foliage gleaners increases from an average of 1.7 to 64 per hectare with the arrival of the migrants (Hutto 1980). These extraordinary densities of migrant birds on their wintering grounds result from the compression of large populations into small areas. Migrant North American landbirds are compressed from 16 million square kilometers of breeding range into 2 million square kilometers of winter range in northern Central America and the West Indies.

The interactions between migrants and residents pose difficult and still unanswered questions about the spatial scale of community structure. Do local assemblages belong to a greater global community? To what degree do competitive interactions in one season influence community

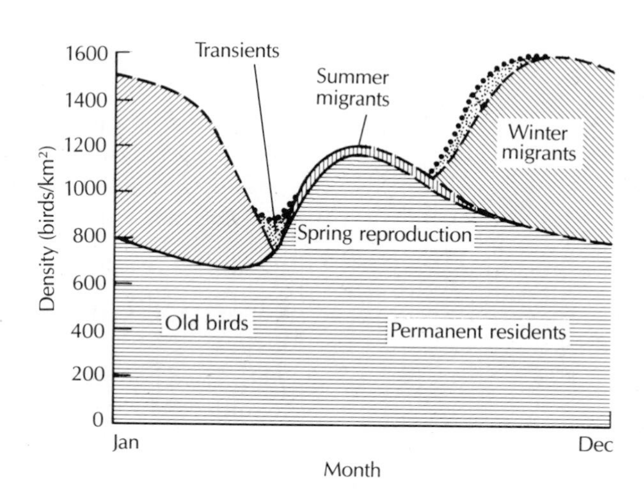

Figure 23–9 Model of seasonal composition of the pine forest bird community of Grand Bahama Island. Local numbers increase with the addition of young birds in the summer and also increase with the addition of wintering migrants, which leave in April. (Adapted from Emlen 1980)

structure in another season? In the extreme case, the local densities of migrants on the breeding grounds may be controlled primarily by interactions with a completely different set of species in another hemisphere in the opposite season.

Variable environments

Few environments are actually stable. Series of unpredictable wet and dry years or severe and benign winters are the norm worldwide. Pronounced year-to-year variations are typical even of tropical rainforests, once thought to be the most stable of ecosystems. Yearly variations in climate and food foster erratic fluctuations in bird populations such as those described in the previous chapter. Yearly variations in seed abundance, for example, have a major impact on the community structure of eight species of sparrows that winter in southeastern Arizona (Pulliam 1983). In years with poor seed crops, when the density of seeds fell below the minimum required by Chipping Sparrows (less than 1 kilogram per hectare), this species wintered elsewhere and was not a member of the sparrow community. Chipping Sparrows alone occupied woodland habitat when seed availability was just above this threshold (at 1 to 2 kilograms per hectare), but in better years they were joined by Brown Towhees. In some years of greatest seed abundance, White-crowned Sparrows and Vesper Sparrows joined them to form a community of four species.

Habitat disturbance, fluctuating resources, colonization, and extinction impose a series of short-term changes on avian communities. In addition, deforestation and rotation of farmlands have transformed woodland habitats and their avian communities into patchworks of chronically unstable and unsaturated habitats that invite opportunistic use. Island avifaunas, subject to frequent extinctions from climatic stress or human disturbance, also tend to be unstable (Abbott and Grant 1976). The composition of newly forming avian communities may be determined primarily by patterns of fortuitous colonization and extinction.

Fire, an extreme form of disturbance, naturally and periodically devastates chapparal communities in California and in the grasslands of both Africa and the western United States. Regrowth after a burn proceeds through regular patterns of plant succession and associated bird communities. Over one year, localized burns and recovery create an ever-changing mosaic of unstable habitats. The communities that occupy them are dynamic rather than self-perpetuating systems at equilibrium. Detailed analyses of shrub-steppe bird communities of the Great Basin of western North America reveal that these birds respond opportunistically to abundant, nonlimiting resources (Rotenberry and Wiens 1980b). Interspecific competition does not seem to be an important force in these dynamic communities.

Competition

The closed community concept dominates the modern study of bird communities (MacArthur 1971, 1972; Cody and Diamond 1975; Strong et al. 1984). In this view, stable combinations of species separate themselves from a pool of possible colonists and competitors and become a community.

Community efficiency, stability, and resistance to invasion by additional species increase with evolutionary adjustment among members of the community. Competition among species for limited resources is the single most important force that structures the closed communities most ornithologists believe are typical of birds. Competition occurs when use or defense of a resource by one individual reduces the availability of that resource to other individuals (Ricklefs 1979c). Interspecific competition occurs when individuals of coexisting species require some of the same limited resources; the use or defense of those resources by individuals of one species reduces the availability of resources to individuals of another species. Recall that competition among individuals of one species reduces the rate of population growth by limiting survival or reproduction. Competition among individuals of different species can have similar or more pronounced effects on the population growth of the species involved.

The competitive exclusion principle — a fundamental concept of ecology that is also called Gause's law after G.F. Gause, a pioneering Russian ecologist — states that two species with identical ecological requirements cannot coexist in the same environment. Firmly supporting this principle is the observation that one species usually replaces another similar species when the two are forced to share the same environment in a laboratory. Competition can be expressed in several ways:

1. Overt aggressive displacement of individuals, called interference competition;

2. direct reduction of the fecundity and survival of one species by another; or

3. interactions among several species, called diffuse competition.

Interspecific competition can take the form of direct aggressive exclusion of another species from vital resources. Large, dominant species of sunbirds and hummingbirds can exclude other species from the densest concentrations of flowers. Forced by dominant species to use other feeding grounds with fewer flowers, subordinate species quickly shift back to the best available feeding grounds when possible. Similarly, Golden-crowned Sparrows aggressively restrict juncos' use of foraging space near shrubs (Davis 1973). The juncos increase their use of sites closer to protective cover when Golden-crowns are removed experimentally but revert to infrequent use when the Golden-crowns return. Antbirds that gather at swarms of army ants exhibit similar behavior (see page 287). Most interspecific competition, however, is less conspicuous, more subtly depressing a species' survival or breeding success. Some of the best evidence of the effects of one species on the fecundity, survival, and population recruitment of another comes from research on Great Tits and Blue Tits (Dhondt and Eyckerman 1980). This research is an extension of the work on population regulation in the Great Tit, which is reviewed in the previous chapter.

Competition between Great Tits and Blue Tits

Competitive interactions should be most intense within "guilds" of similar species that are dependent on the same set of resources (Root 1967).

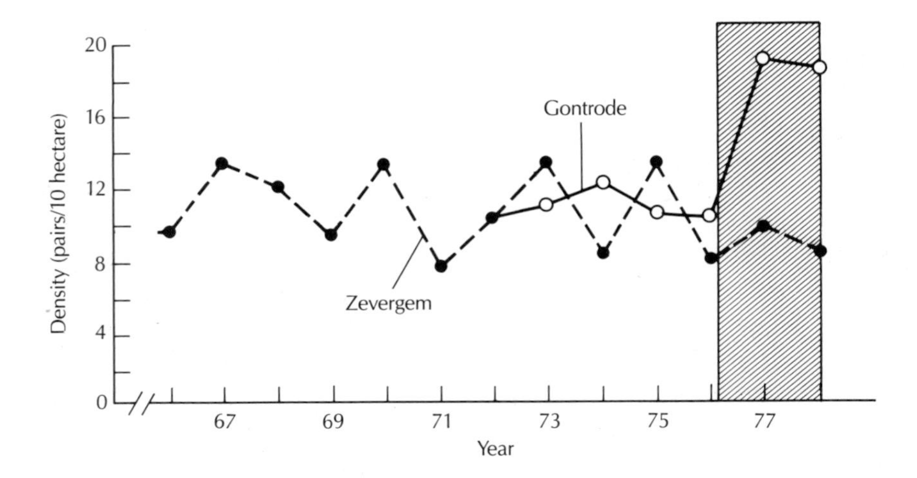

Figure 23–10 Experimental demonstration of interspecific competition. When Great Tits were excluded from nest boxes (from 1976 to 1978, striped area) more Blue Tits established themselves in the experimental area at Gontrode, Belgium (white circles) compared to a control area at Zevergem (black circles). (Adapted from Dhondt and Eyckerman 1980)

Local assemblages of titmice compose guilds that have been the focus of intense research on the role of interspecific competition in bird communities. Reproduction in Great Tits is sensitive to their population density, lessening as population density increases. High densities of Blue Tits in a woodlot also negatively affect the fecundity of Great Tits through competition for food. Blue Tits eat a variety of foods, the range of which encompasses those required by the Great Tit. High densities of Blue Tits reduce food availability and thereby reduce the reproductive output of Great Tits by increasing nestling mortality and by causing fewer Great Tits to have second broods. Reproduction of Blue Tits, however, is not density dependent and is affected only slightly or not at all by the presence of Great Tits (Dhondt 1977).

Despite this demonstration of interspecific competition at the level of fecundity, the population of Great Tits is not necessarily reduced by the presence of Blue Tits. The population density of the Great Tit is controlled instead by winter survival and recruitment of juveniles (see Chapter 22). Large numbers of Blue Tits may reduce the amplitude of annual fluctuations in the density of Great Tit populations, but not the average density. It is important to distinguish the effects of competition on demography and the effects of competition on either population density or on a species' use of the environment. Slight reductions in fecundity of the Great Tit due to competition with Blue Tits for food are of minimal consequence to the overall population level. The competitive interaction of greatest consequence is caused by Great Tits, which limit the number of Blue Tits in a woodlot by controlling the roost holes. When the number of Great Tits in a population dependent on man-made boxes for roosting (and also for nesting) is halved (by narrowing the nest entrances from 32 to 26 millimeters and thereby excluding the larger Great Tits), many more wandering juvenile Blue Tits settle in the woodlot in the autumn (Figure 23–10).

Ecological segregation

A corollary of the competitive exclusion principle is that the degree to which species potentially compete should relate directly to the extent that their use of limited resources overlaps. Detrimental ecological overlap may foster the evolution of ecological differences that reduce competition.

Table 23–2 Ecological Segregation of Titmice

	Europe	North America	Africa	Asiatic Mountains
Number of species	9	10	10	14
Number isolated by				
Range	3	13	7	2
Habitat	11.5	5	13	51
Feeding station	15.5	2	2	16
No contact	6	25	23	22
Percentage isolated by				
Habitat	32	11	30	56
Feeding station	43	18	4	4
Range (including no contact)	25	84	67	84

From Lack 1971.

Similar species that successfully coexist usually differ in some aspect of their ecology, suggesting the past resolution of competitive interactions. Observed ecological differences, therefore, may be the "ghosts of competition past" (Connell 1980).

Local separation by habitat and feeding stations is typical of titmice (Table 23–2) (Lack 1971). In Europe the Great Tit, Blue Tit, and Marsh Tit inhabit broadleaf forests. The Crested Tit and Coal Tit live primarily in coniferous forest used by the other three species only as a suboptimal habitat. The forest preferences of the Willow Tit vary geographically. The species that live together feed in different places: Great Tits on the ground, Marsh Tits on large branches, and Blue Tits in the smaller twigs. In the Himalayas, the world center of titmouse diversity, 14 species of titmice live in the same region but are isolated by habitat and altitude. Differences among European titmice in their feeding stations are associated with adaptive differences in body mass and beak size. Larger species feed at a lower level and on larger insects and harder seeds than the smaller species. Species that live in coniferous forest have longer and narrower beaks than those that live in broadleaf woods.

Each species of European tit has a counterpart in North America (Figure 23–11). However, only two of the North American species usually live together in the same habitat. In many areas, this may be a small chickadee that coexists with a large titmouse, which has different ecological requirements. Where two species of small chickadee occur together, they inhabit different habitats. In New England, the Boreal Chickadee inhabits dark conifer stands whereas the Black-capped Chickadee inhabits more open, mixed deciduous and conifer forest. On the west coast, coexisting Chestnut-backed Chickadees and Black-capped Chickadees also occupy habitats of similar differences.

If competition actually restricts a species, one would expect shifts in the distribution, habitat use, or foraging behavior of a species when it is not limited by a competitor. On the San Juan Islands of the Pacific Northwest, where there are no Black-capped Chickadees, the Chestnut-backed Chickadees inhabit broadleaf forests used elsewhere by the Black-caps. Shifts in habitat use in the absence of other species are well documented among European tits (Lack 1971). Marsh Tits inhabit pine plantations only in Denmark, where Willow Tits are absent from this habitat. On Gotland

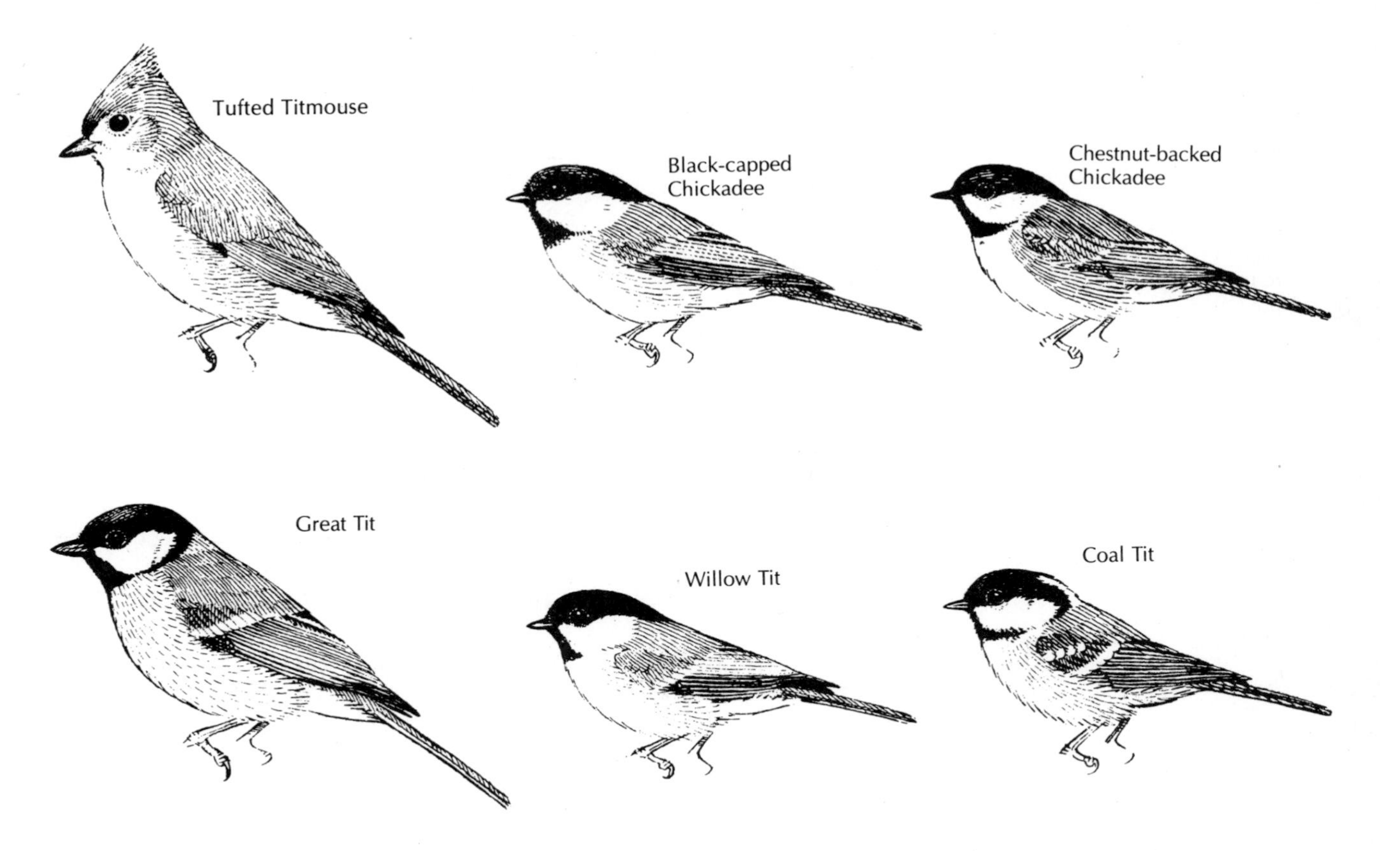

Figure 23–11 Many species of North American (top) and European (bottom) chickadees and titmice act as ecological equivalents. (From Lack 1971)

Island, 90 kilometers off the Swedish coast, both Blue and Great Tits occupy pine forests more than they do on the mainland, apparently because Willow, Marsh, and Crested Tits are absent. In Ireland, Coal Tits feed regularly in the understory of evergreen forests in the absence of the Marsh Tits, Willow Tits, and Crested Tits that normally preempt this niche.

Character displacement

Simple ecological displacements as a result of competition should theoretically lead to evolutionary reinforcement in the form of morphological character displacement (Brown and Wilson 1956). Despite intense interest in this phenomenon, there is little evidence for character displacement in birds (Grant 1972b, 1975). A possible exception, Darwin's finches of the Galapagos archipelago have been considered a classic example of the apparent role of competitive exclusion and character displacement (Lack 1947a; Grant 1981; Schluter and Grant 1984). The adaptive radiation of these finches has promulgated species with a variety of bill sizes that relate directly to seed sizes (Abbott et al. 1977). Ground-finches and Cactus-finches with distinctly different bill sizes inhabit every island. The differences in average bill size of coexisting species are consistent with the hypothesis of interspecific competition for food: Species with similar-sized bills replace each other on various islands, and the bills of various species are more alike when the species do not live together (Figure 23–12).

The evolution of the Large Cactus-Finch is an interesting case in point. On the two islands of Española and Genovesa, this species seems to

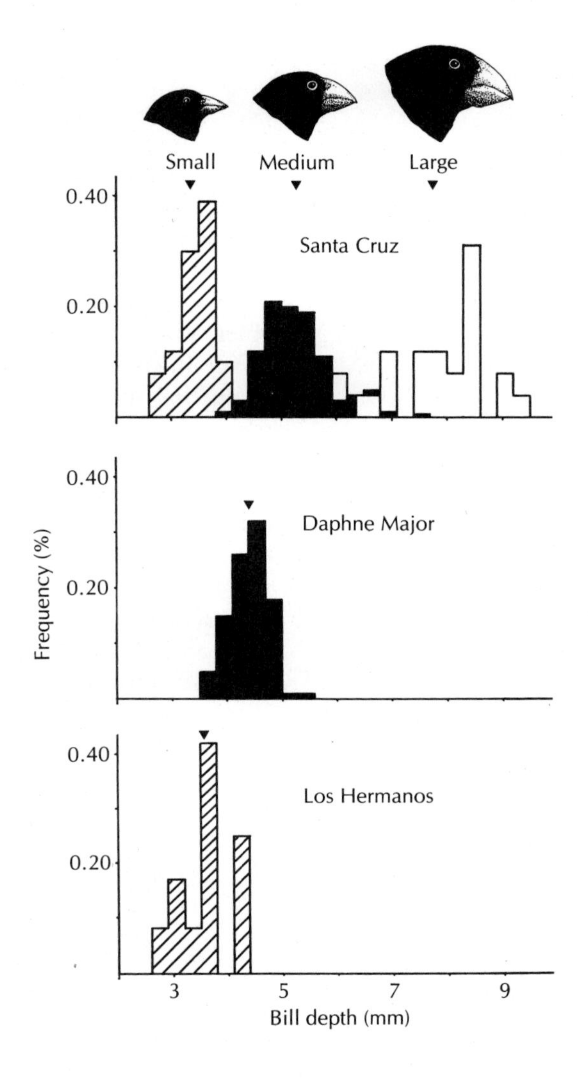

Figure 23-12 Three species of ground-finches coexisting on the Galapagos island of Santa Cruz have bills of different depths, which enable them to feed on different seeds. Only one species inhabits certain islands, such as Daphne Major and Los Hermanos. In the absence of other species, such solo populations evolve intermediate-sized bills. (Adapted from Grant 1986)

replace similar-sized species, the Medium Ground-Finch and the Small Cactus-Finch, which live on other islands of the archipelago. The Large Cactus-Finch lives with the Small Ground-Finch on the island of Española. On Genovesa Island, however, it lives with two other species, the Large Ground-Finch and a different small species, the Sharp-beaked Ground-Finch. The enhanced difference in bill size of Large Cactus-Finches on Genovesa versus Española is due to character displacement as a result of competition with the Large Ground-Finch. The broad feeding niche of the Large Cactus-Finch on Española is a composite of the narrower niches of the three species that are absent, with a concentration on the particular foods eaten by the Large Ground-Finch on Genovesa. Sharp-beaked Ground-Finches competitively exclude Small Ground-Finches from Genovesa. The absences of the Medium Ground-Finch and the Small Cactus-Finch appear to be the result of diffuse competition with the combination of species present rather than lack of suitable foods or failure to reach the islands. Immature Large Ground-Finches apparently fail to colonize Española successfully because they crack large seeds less efficiently than do resident adult Large Cactus-Finches.

Geographical replacement patterns

Active displacement of a bird species by a competitor in ways that demonstrate the competitive exclusion principle is rare. The Spanish Sparrow apparently displaced the Streaked Rock-Sparrow from towns in the Canary Islands during the last century (Lack 1971). Accompanying the dramatic expansion of urban and suburban habitats in Singapore, species of kingfisher, munia finch, tailorbird, and dove have replaced related species that were once common there. But these are exceptional, mostly anecdotal observations.

In contrast to the paucity of observations of direct displacement by competing species, much has been documented concerning geographical replacements that apparently resulted from past competition. These examples are largely responsible for the widespread conviction that competition is important in bird communities. Exclusions by past competition should be evident in patterns of geographical replacement in species too similar to coexist. Ornithologists who are documenting the distributions of species frequently discover that the ranges of closely related species are mutually exclusive, often involving a sharply delineated replacement of one species by another.

Patterns of species replacement are conspicuous on small, isolated islands. Each island in the West Indies, for example, is inhabited only by two

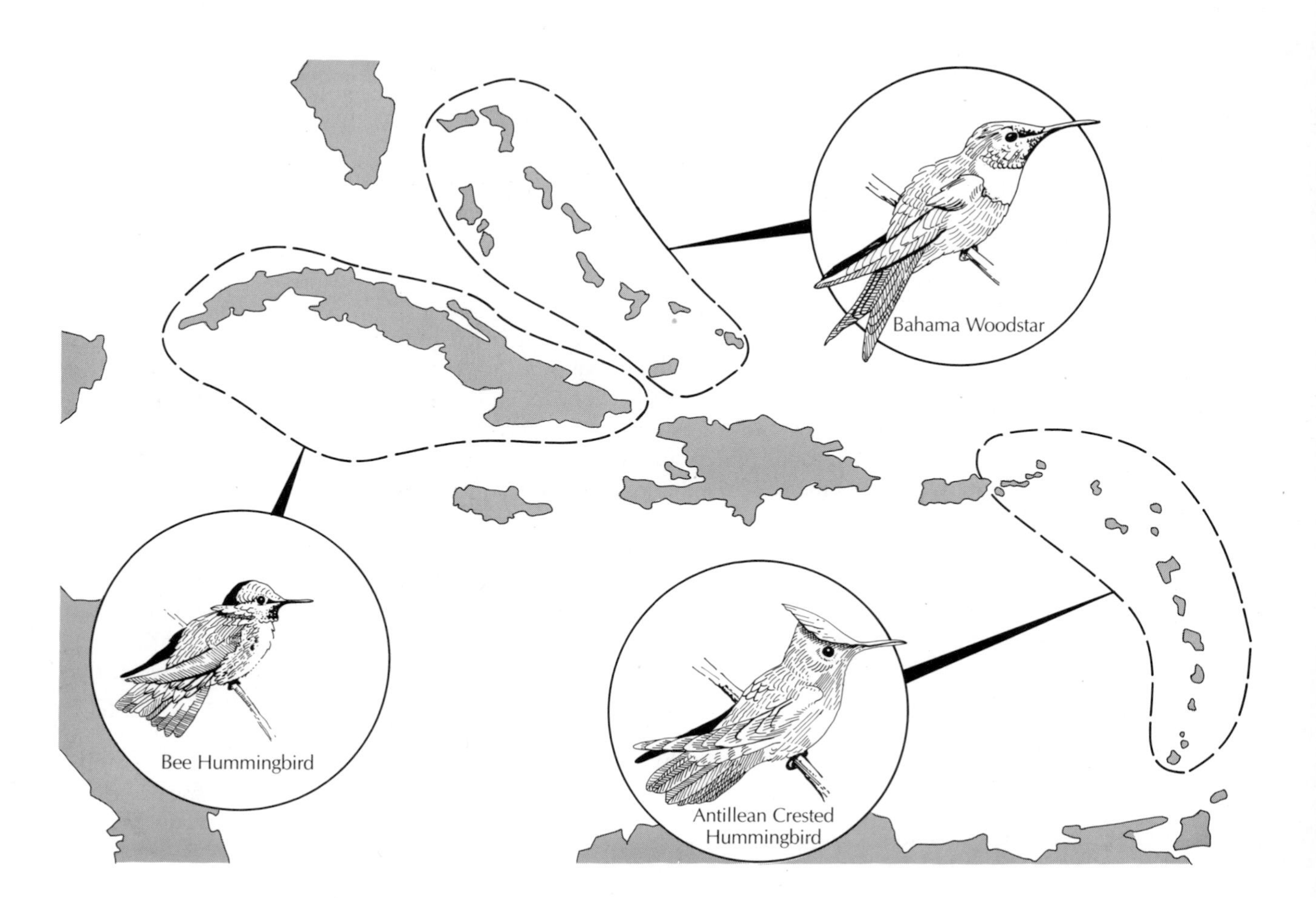

Figure 23–13 Geographical replacements of possible competitors: Different species of small hummingbirds do not coexist in the West Indies but have segregated distributions. (Adapted from Lack 1971)

or three hummingbird species though fifteen hummingbird species inhabit the region (Lack 1971) (Figure 23–13). Only two resident species, a small and a large one, inhabit low-lying islands. Mountainous islands are populated by three types of hummingbirds, a small, widespread species and two large ones, one in the lowlands and another in the highlands.

Abrupt replacement of one species by another at various altitudes in the Andes and in New Guinea is more evidence that competition from one species limits the distribution of another. Using data from the pristine, rugged eastern slopes of the Peruvian Andes, John Terborgh (1971) concluded that competitive exclusion accounted for about one-third of the distribution limits of the 261 species he studied; gradually changing conditions (such as temperature) accounted for about one-half of the limits; and shifts in habitat types accounted for less than one-fifth of the limits.

New Guinea birds with well-defined altitudinal distributions are also reported to replace each other abruptly at various elevations (Diamond 1973, 1975). The range of elevations occupied by a species seems to depend on the presence or absence of related species. For example, the Red-flanked Lorikeet is confined to low elevations in regions with a highland species, either the Red-spotted Lorikeet or the Red-chinned Lorikeet. In regions that the Red-flanked Lorikeet inhabits alone, however, it ascends to high elevations. Conversely, in regions where either of the other two lorikeets occurs alone, it lives at sea level (Figure 23–14).

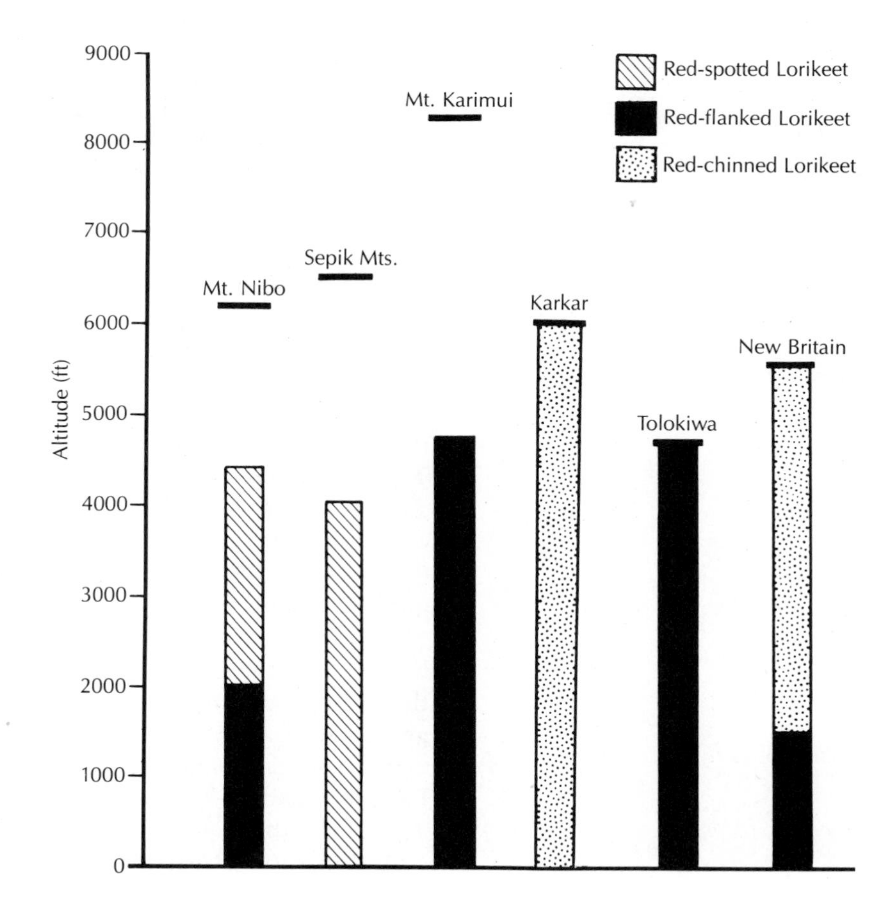

Figure 23–14 In regions where different species of lorikeets occur together on New Guinea they tend to occupy different elevations, but in the absence of others, one species occupies a broader range of elevations. (Adapted from Diamond 1975)

(A)

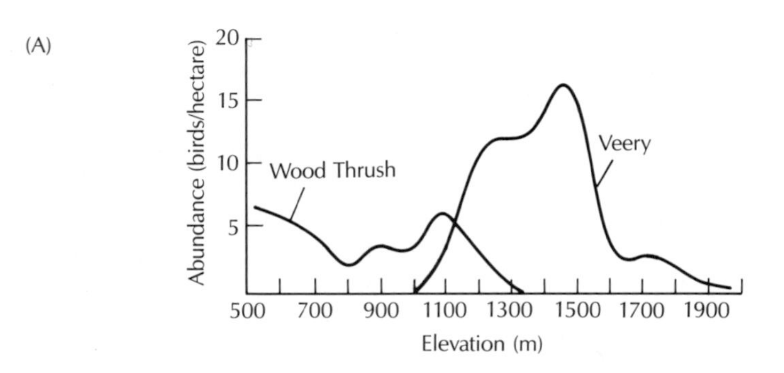

(B)

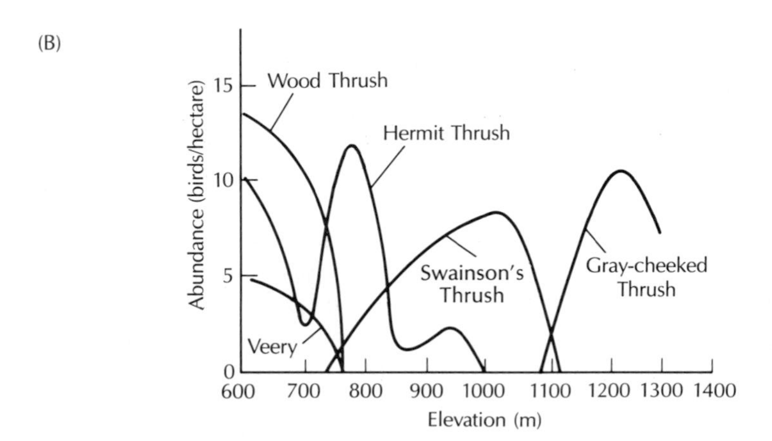

Figure 23–15 Altitudinal distributions and abundances of five species of thrushes: (A) The range of elevations occupied by Wood Thrushes and Veerys is greater in the Smoky Mountains of Tennessee and North Carolina than on (B) Mt. Mansfield, Vermont, where three additional species occupy the high altitudes. (Adapted from Noon 1981)

Similar patterns of altitudinal replacement are found in the mountains of eastern North America where up to five species of thrushes nest in the high mountain forests of New England (Noon 1981) (Figure 23–15). Veerys share low elevations with the Wood Thrush and are replaced at higher elevations by Swainson's Thrushes and Gray-cheeked Thrushes. In the Smoky Mountains, where Swainson's Thrushes and Gray-cheeked Thrushes are absent, the Veery shifts to higher elevations and overlaps only slightly with the Wood Thrush at low elevations. The conclusion that competition limits altitudinal distributions is based on the assumption that the various locations are identical except for the presence or absence of the purported competitors (Wiens 1983). This may or may not be true.

Two or more similar species that inhabit island archipelagos may have overlapping but mutually exclusive distributions in which only one of the species occurs on each island (Figure 23–16). The result is an irregular geographical array of species (Diamond 1975): "checkerboard distributions" of species combinations occur in the islands of the southwest Pacific, Indonesia, and off the coasts of Panama and New Guinea. Competitive exclusion is the most likely explanation of checkerboard distributions because it appears that colonists of both species have equal access to the islands.

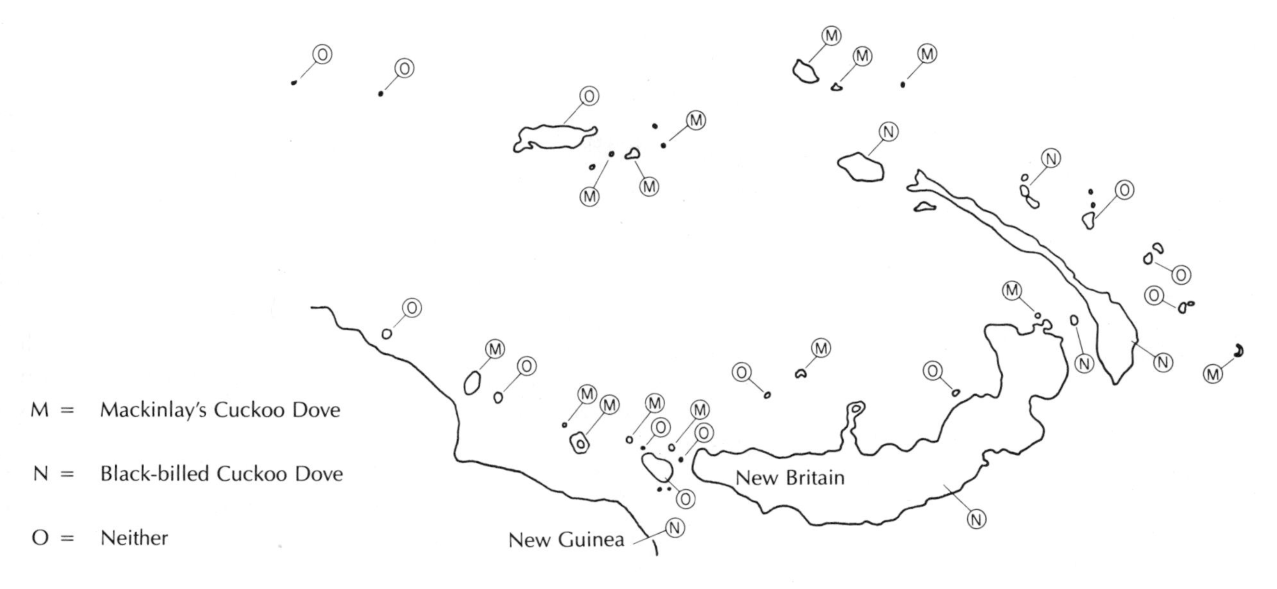

Figure 23–16 Checkerboard distribution of small cuckoo doves on small islands near New Guinea. Most islands have only one of the two widespread species, some islands have neither, and no island has both species. (From Diamond 1975)

Evidence of diffuse competition

Sometimes the presence, absence, or status of a given species relates not to any single other species but rather to combinations of other species (MacArthur 1972). Most of the Bismarck Islands, for example, are occupied by either the Black-billed Cuckoo Dove or the Mackinlay's Cuckoo Dove in checkerboard fashion. Neither of these two species, however, can coexist on small species-poor islands with a third species, the Pink-breasted Cuckoo Dove. They can, however, coexist with the Pink-breasted Cuckoo Dove on large islands *unless* the Long-tailed Pigeon is also present. The combination of Mackinlay's Cuckoo Dove, Pink-breasted Cuckoo Dove, and Long-tailed Pigeon never occurs. Diffuse competition potentially explains such complex mosaics of species distributions (Diamond 1975; see also Simberloff 1978; Diamond and Gilpin 1982).

It now appears that patterns of ecological isolation or replacement provide only weak evidence of structuring communities by competition (Newton 1980). Possibly the ecological and morphological differences we now see are due not directly to competitive displacement but only incidentally to past, unrelated events of evolutionary and geographical history. In no single case do we understand why a particular species of bird lives where it does rather than elsewhere. Species distribution is not a simple problem because present distributions reflect not only the confluence of proximate limiting forces but also the results of powerful historical forces, such as colonization and extinction, as well as ancient climatic changes and redistributions of habitats. The historical forces that determine the distributions of birds on a global scale are the topic of the next chapter.

Summary

The availability of resources such as food and nest holes determines not only the local population density of a given species but also the number and kinds of different species that can coexist in a given habitat. There are two major theories regarding the forces responsible for structuring avian communities. According to one theory, communities are open systems in which each species arrays itself independently along environmental gradients according to its specific ecological requirements. According to the other theory, communities are closed, integrated sets of mutually compatible species.

The local number of species increases from the arctic to the tropics. Species diversity is influenced by spatial components such as the availability of nesting and brooding sites, courtship and display stations, protection from predators, and protection from climatic stress. Habitat heterogeneity, a prime factor in the tropics, contributes to the species diversity of an area. Interactions between residents and migrants and variable environments also affect species diversity.

In the closed-community concept that dominates modern ornithology, competition is the key structuring force. Competition can be expressed in several ways:

1. Overt aggressive displacement of individuals, called interference competition;

2. direct reduction of the fecundity and survival of one species by another; or

3. interactions among several species, called diffuse competition.

Theoretically, the degree to which species compete should relate directly to overlapping use of shared resources. This corollary to the competition theory has been illustrated by tit populations. Another corollary of the competition theory is that ecological displacements as a result of competition should lead to evolutionary reinforcement in the form of morphological character displacements, of which the differing bill sizes of Galapagos Finches have been considered a classic example. Past competitive exclusion is generally evident in patterns of geographical replacement in species too similar to coexist. Replacement patterns are particularly well documented on the isolated islands of the West Indies and at different altitudes in the Andes. Diffuse competition, characterized by complex dynamics in a mosaic of species, may also play a role in structuring avian communities.

Further readings

MacArthur, R.H. 1972. Geographical Ecology. New York: Harper and Row. *An elegant summary of a dominant viewpoint.*

Ricklefs, R.E. 1979. Ecology. New York: Chiron Press. *A well-written text that covers the basics.*

Strong, D.R., Jr., D. Simberloff, L.G. Abele, and A.B. Thistle. 1984. Ecological Communities, Conceptual Issues and Evidence. Princeton: Princeton University Press. *A rich collection of advanced papers embracing polar views of current controversies in community ecology.*

Wiens, J. 1983. Avian community ecology: An iconoclastic view. *In* Perspectives in Ornithology (A.H. Brush and G.A. Clark, eds.), pp. 355–403. Oxford: Oxford University Press. *A presentation of an alternative view of avian communities.*

chapter 24

Geography

For over a century, biogeographers have divided the earth into six major faunal regions corresponding roughly to the major continental areas (Figure 24–1). Each faunal region has its characteristic birds, endemics found nowhere else as well as major adaptive radiations of more widespread taxa (Table 24–1). The birds that are endemic to Africa, or the Ethiopian region, for example, include ostriches, mousebirds, and turacos. The Australasian region has emus, honeyeaters, and birds-of-paradise. The Neotropical region has toucans, tanagers, and trumpeters. Waxwings and loons are restricted to the Nearctic and Palaearctic regions.

The history of avifaunas

Superficially, an *avifauna* is the set of species living in a particular geographical area. On a deeper level, avifaunas reflect a region's history because they are the grand result of millions of years of evolution, adaptive radiation, dispersal, and extinction of bird taxa (Mayr 1965). Invasions and fusions from other regions supplement adaptive radiations of ecological types

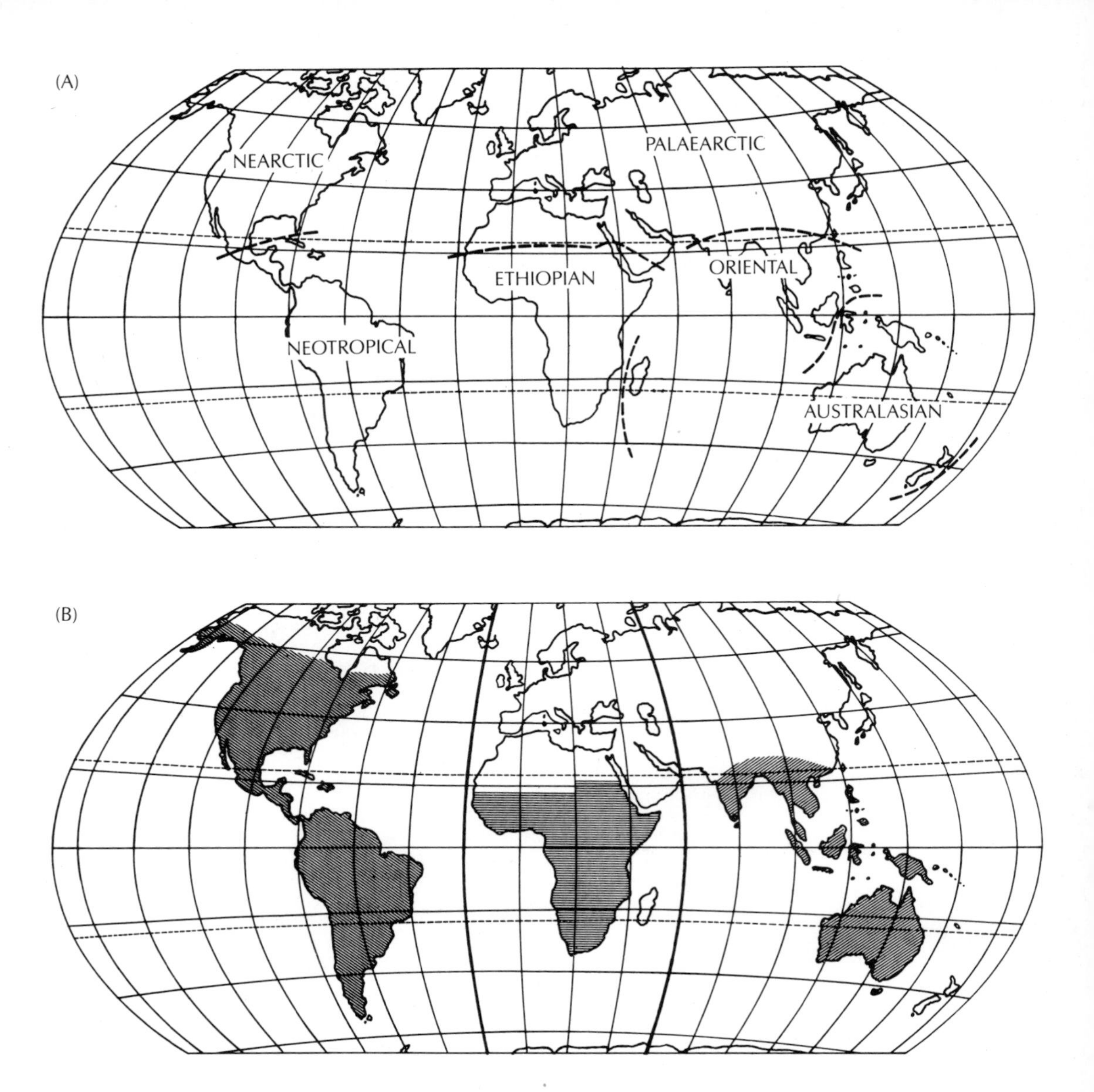

Figure 24 – 1 (A) The six major zoogeographical regions and (B) the distributions of three avian families: tyrant flycatchers, limited to the New World (Neotropical and Nearctic regions); turacos, limited to Africa south of the Sahara (Ethiopian region); and wood swallows, limited to the Oriental and Australasian regions. (From Thompson 1964)

within a region. The biogeographer's task is to distinguish these elements and to reconstruct the historical mosaics of modern avifaunas. Turnover, the addition and loss of species, controls the changing compositions of avifaunas. The time scales range from the long-term, gradual replacements of ancient taxa by more recently evolved taxa to short-term, yearly changes. An equilibrium in the number of species may be established if extinctions are balanced by new additions through colonization or speciation.

The early diversification of birds took place on a very different earth; neither the arrangement of the continental land masses nor the climates resembled those of today. Through much of the Tertiary period, 1 to 65 million years ago, the world's climates were warm from pole to pole; there was no striking polar gradient from frigid to hot as there is today. For example, during the late Eocene and early Oligocene epochs (see Table 2 – 1 for the geological time scale), subtropical to tropical climates with abundant precipitation and no frost prevailed in the far north of both North America and Eurasia (Cracraft 1973a). The floras of Great Britain and of western Europe in the early Eocene resembled the modern rain forests of southeast Asia. Tropical birds — trogons, parrots, hornbills, barbets, broadbills, and

Table 24–1 Avian Specialities of the Major Biogeographic Regions

Regions	Endemic Nonpasserine Families	Passerine Radiations
Nearctic and Palearctic (Holarctic): all of North America, Mexico, and the West Indies, plus all of Europe, northern Asia south to the Himalayas, and Africa north of the Sahara	Loons (Gaviidae)	Accentors (Prunellidae) [Palearctic]
	Auks (Alcidae)	Buntings (Emberizinae)
		Carduelline finches (Fringillidae)
		Wood warblers (Parulinae) [Nearctic]
		Old World warblers (Sylviidae) [Palearctic]
Neotropical: all of South America plus Central America south of the Isthmus of Tehauntepec, Mexico	Rheas (Rheidae)	Hummingbirds (Trochilidae)
	Tinamous (Tinamidae)	Tyrant flycatchers (Tyrannidae)
	Curassows (Cracidae)	Tanagers (Thraupinae)
	Trumpeters (Psophiidae)	Antbirds (Formicariidae)
	Sunbittern (Eurypygidae)	Ovenbirds (Furnariidae)
	Seriemas (Cariamidae)	Woodcreepers (Dendrocolaptidae)
	Limpkin (Aramidae)	Manakins (Pipridae)
	Oilbird (Steatornithidae)	Cotingas (Cotingidae)
	Hoatzin (Opisthocomidae)	
	Motmots (Momotidae)	
	Jacamars (Galbulidae)	
	Puffbirds (Bucconidae)	
	Toucans (Ramphastidae)	
Ethiopian: Africa south of the Sahara	Ostrich (Struthionidae)	Larks (Alaudidae)
	Secretarybird (Sagitariidae)	Sunbirds (Nectariniidae)
	Guineafowl (Numididae)	Weavers (Ploceidae)
	Roatelos (Mesoenatidae) (M)*	
	Turacos (Musophagidae)	
	Mousebirds (Coliidae)	
	Ground-rollers (Brachypteraciidae) (M)	
	Cuckoo-roller (Leptosomatidae) (M)	
	Woodhoopoes (Phoeniculidae)	
	Vanga shrikes (Vangidae) (M)	
Oriental: Southeast Asia from the Himalayas to northern Indonesia	None	Leafbirds (Irenidae)
		Pheasants (Phasianidae)
		Broadbills (Eurylaimidae)
		Pittas (Pittidae)
		Babblers (Timaliinae)
		Flowerpeckers (Dicaeidae)
Australasian: Australia and New Guinea from Lombok south plus the islands of the southwest Pacific	Emus (Dromacaeidae)	Birds-of-paradise (Paradisaeidae)
	Cassowaries (Casuariidae)	Whistlers (Pachycephalidae)
	Kiwis (Apterygidae)	Honeyeaters (Meliphagidae)
	Kagu (Rhyncochetidae)	Monarch flycatchers (Monarchidae)
	Cockatoos (Cacatuidae)	Australian warblers (Acanthizidae)
	Lories (Loriidae)	
	Owlet-frogmouths (Aegothelidae)	

* (M), Madagascar only.

mousebirds — once lived in central Europe. Alligators and large tortoises lived on Ellesmere Island, which is above the Arctic Circle. Joel Cracraft (1974b) postulates that the adaptive radiation of oscine passeriform birds took place primarily in Laurasia during the early Cenozoic era and then spread southward into Africa, Australasia, and North America. Passerines in each major area then underwent their own adaptive radiations.

The arrangements of continents and their connections have changed over the course of avian evolution. Plates of the earth's crust have been moving apart since the late Jurassic and Cretaceous periods. Much of the major reorganization during the Mesozoic era of the single great land mass known as Pangaea, with Laurasia in the north and Gondwanaland in the south, preceded the evolution of modern bird taxa. Nevertheless, modern

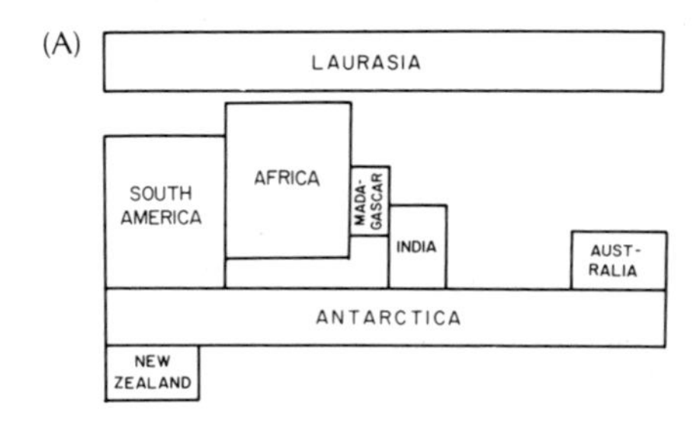

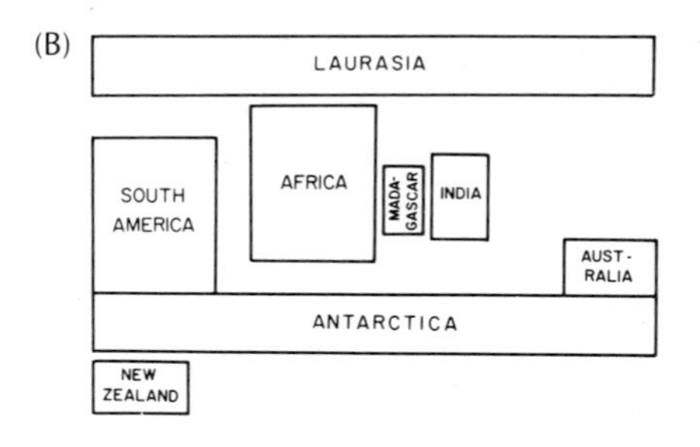

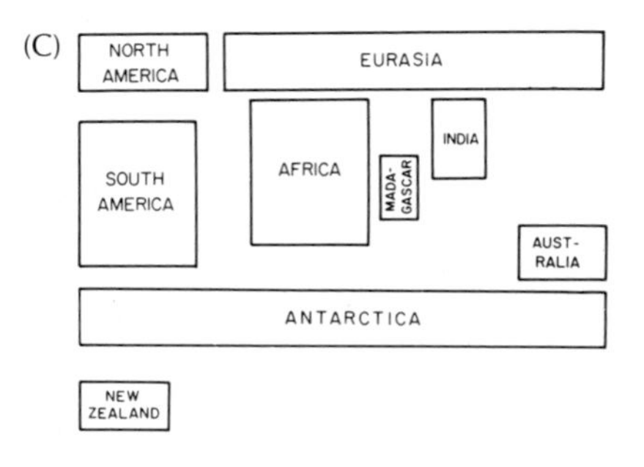

Figure 24–2 Schematic diagram of past configurations of the continents. (A) Once combined into two supercontinents, a northern Laurasia and a southern Gondwanaland, the continents drifted apart during the (B) Cretaceous periods and (C) Eocene epoch of the early Tertiary. into the present configuration (From Cracraft 1974b)

orders of birds were present during the next stages of continental drift. Brazil had separated from Africa by the late Cretaceous, and India had split from Antarctica and was moving north to collide with Asia. From the late Cretaceous period (65 to 70 million years ago) of the Mesozoic era and into the Paleocene epoch of the Cenozoic era, the northern land masses (North America, Europe, and Asia) were broadly connected as a single continent, called Laurasia. A continuous land connection existed between northeastern Asia (modern Siberia) and northwestern North America (modern Alaska) during much of the Tertiary period (Figure 24–2).

Throughout the Tertiary period, birds moved through the tropical-subtropical or warm temperate forests that covered Eurasia and North America. At first, movement was easy across a broad North Atlantic land bridge. After the separation of North America from Europe during the Eocene epoch, the Bering land bridge became the main corridor for faunal exchange in the Holarctic region. Wet tropical forests also covered most of South America during the Tertiary period. The rising of the Andes mountains in the late Tertiary period created a cooler, drier climate that favored the grasslands of Argentina and the coastal deserts of modern Chile and Peru. Even Antarctica supported lush temperate forests of southern beech, conifers, and palms during the Eocene epoch and into the Miocene epoch.

Coupled with favorable climates, the Gondwanaland association of the modern southern-hemisphere continents fostered the exchange of taxa among Africa, South America, and Australia. Dispersal between Australia and South America was once possible via the reasonable climates and forests of Antarctica. The flightless ratites may have taken advantage of such dispersal routes (Cracraft 1974a; see Chapter 3). Ancestral fowl—moundbuilders in Australia, cracids in South America, and guineafowl in Africa—appear to have originated in the main parts of Gondwanaland (Cracraft 1973a). Radiations of pheasants, partridge, New World quail, and grouse in North America came after the northward expansion of ancestral groups into Laurasia and the separation of Laurasia into North America and Eurasia in the Eocene. Representatives of these radiations on the continents of the northern hemisphere subsequently moved southward into Africa and the East Indies.

At the close of the Tertiary period the earth's climates cooled, especially in the polar regions, and became more strongly seasonal. During the Pleistocene epoch climatic changes and glaciers drastically altered the distributions of birds throughout the world. Alternating dry, cool and wet, warm climates split the geographical ranges of birds and promoted both speciation and extinction. Tropical birds became restricted to equatorial latitudes and, even there, were sometimes concentrated in limited refuges as the alpine zone extended to lower elevations and as grasslands expanded during dry, glacial periods. In Africa, for example, montane conditions extended to as low as 500 meters above sea level during glaciations, reducing the lowland forests to less than half of their area today (Moreau 1966). Later warm periods reversed this pattern, thereby restricting montane species to isolated mountaintops.

Remnant or relict populations are one of the major consequences of historical habitat changes. Ostriches, now restricted to Africa, once roamed throughout Asia. Todies, which are tiny, colorful relatives of kingfishers, are presently found only on the Greater Antilles of the West Indies but once

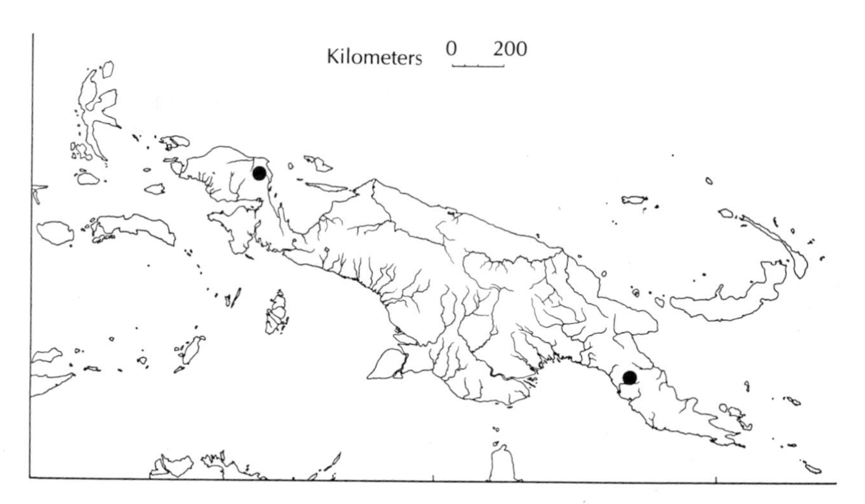

Figure 24–3 Disjunct distribution of the Obscure Berrypecker, which inhabits two sites (black circles) at the opposite ends of New Guinea. (Adapted from Diamond 1975)

lived also in Wyoming and France (Olson 1976, 1985). The extremely localized distributions of many birds of New Guinea are most likely a result of gradual extinction throughout most of the former range and of persistence of the species in a few favorable sites (Diamond 1973, 1975). Alan Feduccia and Storrs Olson (1982) suggest that the Australian lyrebirds and scrub birds and the South American tapaculos are closely related relicts of an early radiation of southern hemisphere suboscine birds that gave rise to the rest of the order Passeriformes.

Widely separated areas may share peculiar taxa as a result of historical range fragmentation. The African River Martin inhabits the Congo River basin, whereas the closely-related White-eyed River Martin inhabits only northern Thailand; no related species are found between these locations. The Scrub Jays of southern Florida are separated from other populations of this species in the southwestern United States by over 3000 kilometers. The Obscure Berrypecker now lives in only two localities at opposite ends of New Guinea (Figure 24–3).

The composition of avifaunas

The history of bird distribution can be viewed as a series of waves of adaptive radiations, moving north, south, east, and west, which were superimposed on older taxa, split by the movement of the continents, and fragmented, one by another, into complex mosaics of ancient, recent, and new elements. Most avifaunas are a mixture of elements of various ages and origins. Some, such as the major groups of South American birds and the Hawaiian honeycreepers, evolved locally in response to ecological opportunities, immune from regular invasion by members of other faunas. Other avifaunas consist of taxa of various ages that were derived from a series of invasions of species that evolved elsewhere, some long ago, some quite recently. Most of the birds of Madagascar came from Africa; a few (5 percent) came from India. The birds of New Zealand came primarily from Australia but also from Melanesia. Fifty-eight percent of the birds of the Venezuelan highlands came from distant highland avifaunas, especially those of the Andes; 42 percent came from the surrounding tropics.

Table 24–2 Composition of the Montane Avifauna of the Venezuelan Highlands

Species in endemic genera	2
Endemic species	23
Endemic species in widespread superspecies*	5
Total endemic species	30
Nonendemic species with only endemic subspecies	48
Nonendemic species with endemic subspecies and nonendemic subspecies	7
Nonendemic species without endemic subspecies on Pantepui	11
Total nonendemic species	66
Total species of birds	96

* A superspecies is a group of closely related species.
From Mayr 1965.

Environments that tend to be of recent origin, such as montane highlands, the Arctic, or major deserts, are apt to be colonized by a steady series of enterprising species, which then adapt to the new climatic rigors. The 96 montane species of the Venezuelan highlands (Pantepui), for example, range from older endemic forms, which are presumed to have arrived first, to recently arrived representatives of widespread species (Mayr and Phelps 1967) (Table 24–2).

New geographic or habitat connections permit fusions of isolated avifaunas (Mayr 1965). The birds of Central America, for example, include North and South American species. South American birds expanded into Central America with the northward movement of tropical rain forests in the late Pleistocene epoch. There was a reciprocal southward movement of some North American birds into the rain forests of South America. However, not many South American birds colonized the arid habitats in Central America because they were occupied by stable communities of well-adapted species that survived the Pleistocene (Table 24–3).

Origins of the birds of the West Indies

Unlike Central America, the West Indies (including both the Greater and Lesser Antilles) has been continuously colonized, despite its substantial open-water barriers (Bond 1948, 1963). Both ancient and modern contri-

Table 24–3 Faunal Origins of the Birds of Arid and Rain Forest Habitats of Central and South America

	Origin of Species (%)			
	Old World, North American	Pan-American	Expanding, and Secondarily, South American	Primarily South American
Tropical rain forest				
Central America	12.7	14.3	34.9	38.1
Amazonia forest	9.0	10.5	33.0	47.5
Arid habitats				
Middle America	49.1	15.1	27.3	8.5
Northern South America	29.5	13.6	45.5	11.4
Southern South America	23.1	17.9	7.7	51.3

From Mayr 1965.

butions are apparent among the birds of this avifauna. The original elements of the modern West Indies avifauna, the endemic genera, have come primarily from Central America. There are also species from South America, most of which are recent colonists. Thirty-three species of South American landbirds have entered the West Indies via the Lesser Antilles, fifteen of which remain on the fringes of the West Indies, not having moved past Grenada and St. Vincent, thirteen have moved to the middle of the central part of the Lesser Antilles, but only three have reached eastern Puerto Rico.

Of the South American families that spread northward during the Pleistocene epoch, the tyrant flycatchers have been one of the most aggressive colonists. Flycatchers of the genus *Myiarchus,* for example, have repeatedly invaded the West Indies (Lanyon 1967). One of them invaded Grenada from Venezuela, evolved into the Grenada Flycatcher and has spread no farther north. Three invasions of Jamaica from Central America account for the other six species now found in the West Indies (Figure 24–4).

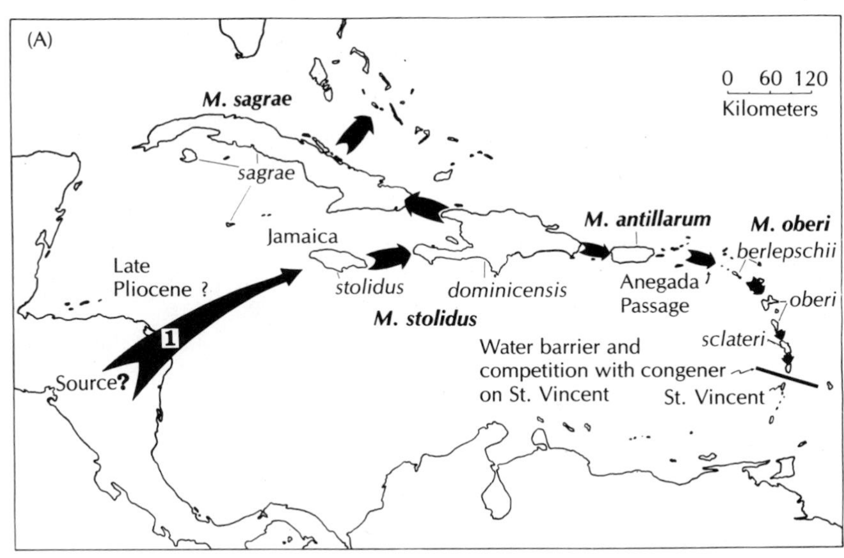

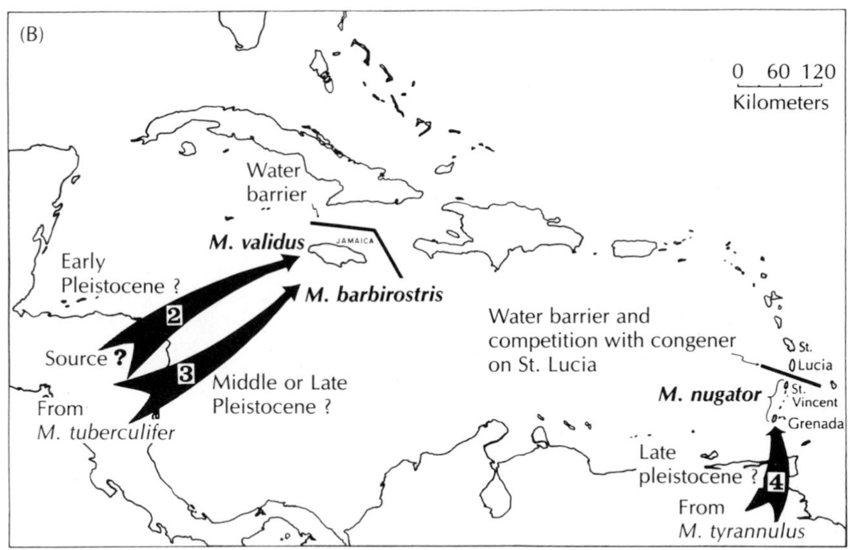

Figure 24–4 Four primary invasions account for flycatchers (genus *Myiarchus*) now found in the West Indies. (A) The first invasion of Jamaica from Central America was followed by the spread and differentiation of four related species: Stolid Flycatcher (*M. stolidus*); La Sagra's (*M. sagrae*); Puerto Rican Flycatcher (*M. antillarum*); and Lesser Antillean Flycatcher (*M. oberi*). (B) A second invasion by an unknown species and a third invasion by Dusky-capped Flycatcher (*M. tuberculifer*) from Central America were responsible for two endemic Jamaican species, Rufous-tailed Flycatcher (*M. validus*) and the Sad Flycatcher (*M. barbirostris*). An invasion of the southern Lesser Antilles from Venezuela by the Brown-crested Flycatcher (*M. tyrannulus*) was responsible for the evolution of Grenada Flycatcher (*M. nugator*) presently restricted to the southernmost Lesser Antilles. (From Lanyon 1967)

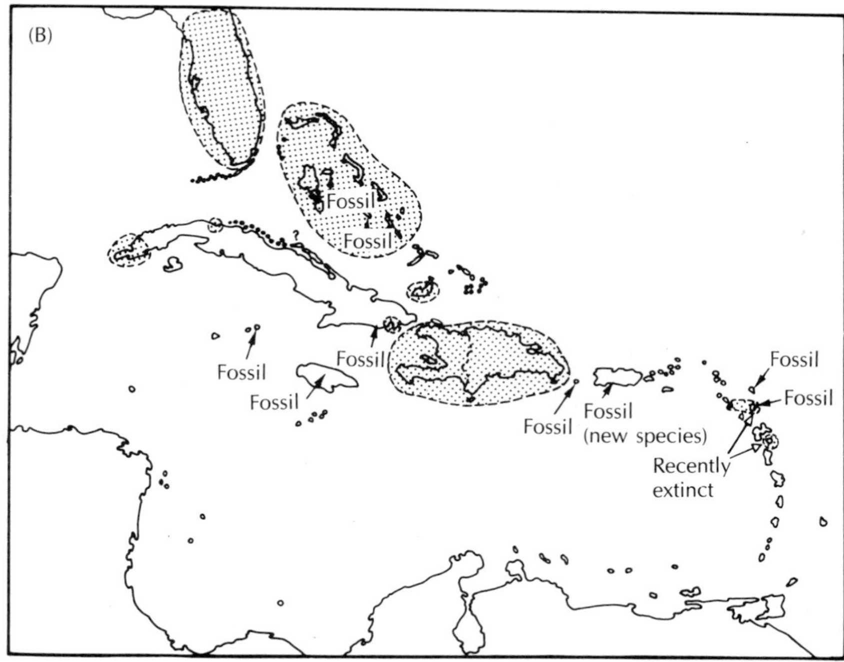

Figure 24–5 Fossils reveal that Burrowing Owls (A) were once more widely distributed in the West Indies. (Courtesy A. Cruickshank/VIREO) (B) Part of the current distribution is indicated by stippling. (From Pregill and Olson 1981)

Wesley Lanyon summarizes these invasions as follows:

> The oldest invasion, perhaps by an obscure species in the late Tertiary avifauna of Middle America, led to the development of *M. stolidus* on Jamaica and Hispaniola. Subsequent expansions from Hispaniola led to the evolution of *M. sagrae* and *M. antillarum* in the Greater Antilles and, more recently, to the polytypic *M. oberi* in the Lesser Antilles. . . . After the submergence of the Greater Antillean land masses in the early Pleistocene, a second invasion of Jamaica by a Middle American *Myiarchus,* now obscure, resulted in the endemic *M. validus.* A third invasion of Jamaica, probably in the mid- or late Pleistocene, by a representative of *M. tuberculifer* of Middle America, produced another endemic, *M. barbirostris.* (Lanyon 1967, p. 368)

The climates and principal habitats of the West Indies have changed in recent times. Dry climates prevailed in the West Indies during the Pleistocene epoch, enabling species in grasslands, savannahs, and xeric habitats to be widely distributed (Pregill and Olson 1981). Increases in annual rainfall during the last 10,000 years have caused many of these birds to become extinct. Today, only relict populations of Burrowing Owls exist (Figure 24–5). Fossils indicate they were once widespread in the West Indies. Caracaras, thickknees, and the Bahama Mockingbird of the dry country once were widely distributed in the West Indies, but they too exist now only as isolated relicts. The most diverse bird communities are those found in the older semiarid shrub communities on Gonave Island and on the islets of the northern Lesser Antilles.

Taxon cycles

Island species seem to progress through *taxon cycles,* the three phases of which — colonization, differentiation, and local extinction — can be likened to the life cycles of individuals: youth, maturity, and senescence (Simpson 1949; Darlington 1957; Ricklefs and Cox 1972). In the first stage of a taxon cycle, pioneering individuals of a species colonize a new area. The colonization of Jamaica by a species of *Myiarchus* flycatcher from Central America would represent this phase. In the second stage of the taxon cycle, the flycatcher would spread among neighboring islands, and the island populations would evolve differences in morphology, behavior, and ecology. Finally, in the third stage of the taxon cycle, some populations would fail and become extinct, creating gaps in the distribution of the forms derived from the original colonist (Figure 24–6).

Active dispersal is a trademark of bird behavior and one of the main reasons for colonization. The dynamics of colonization are most apparent on oceanic islands, such as the West Indies, which receive periodic arrivals of new individuals dispersing over water from larger source areas. Water barriers favor colonization by highly vagile species that travel in small groups. Bananaquits in the West Indies and white-eyes in the Indian and Pacific Oceans are superb colonists, or "supertramps" (Diamond 1974). Their high reproductive potential and extraordinary dispersal abilities

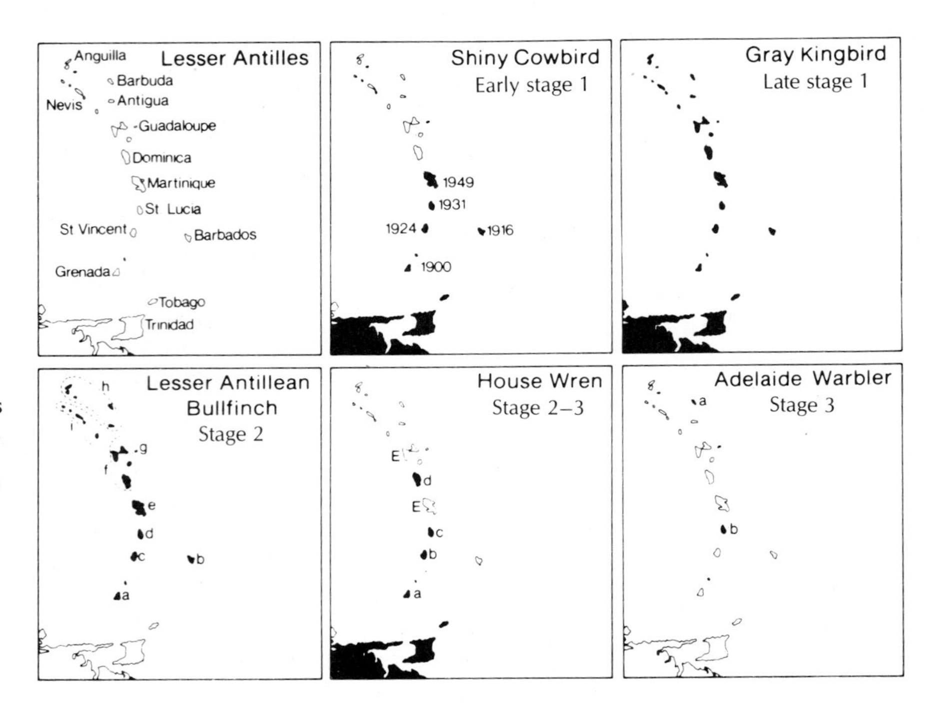

Figure 24–6 Stages of the taxon cycle in the Lesser Antilles, showing progressive colonization, differentiation, and extinction of a series of colonists. The Shiny Cowbird is a recent colonist that invaded the islands from the south during this century. The Gray Kingbird may also be a rather recent colonist; populations on islands distant from each other are all the same. Older colonists such as the Lesser Antillean Bullfinch and the House Wren have distinct subspecies (lowercase letters) on various islands. Some populations of the House Wren have become extinct in recent times (uppercase E). Adelaide's Warbler, an older endemic species, now persists only on the islands of Barbuda and St. Lucia. Intervening populations apparently have become extinct. (From Ricklefs and Cox 1972; Ricklefs 1976a)

enable them to be the most predictable first colonists of newly formed islands. Successful colonization of one island may be followed by colonization of adjacent islands and continued spread throughout a region.

The chances of establishing a population on a new island are influenced by a colonist's ecological flexibility and its ability to fit into the local community. Bananaquits and white-eyes, for example, are generalized opportunists, able to take advantage of local situations. They breed readily and repeatedly. Once established, their populations tend to thrive and grow rapidly in an environment with few specialized predators, competitors, diseases, or parasites. Population growth under such conditions of ecological release leads to large, dense populations and to use of a wider variety of habitats than is the case on the mainland. Resident birds of the Pearl Islands off western Panama, for example, achieve densities 20 to 40 percent higher than on the adjacent mainland. They also forage over a greater vertical range and use more habitats than their mainland counterparts (MacArthur et al. 1972). Comparison of the birds on small Caribbean islands such as St. Lucia and St. Kitts with the larger assemblages on Trinidad or mainland Panama reveals a fivefold increase in the relative abundance of birds of a species, corresponding to a sixfold decline in the number of species present (Cox and Ricklefs 1977) (Table 24–4). Both the average number of habitats used by a species and the density of each species in a particular habitat may double on an island.

Populations on different islands gradually diverge from each other as they adapt to unfilled local niches. Generalized colonists, such as white-eyes, may take over the specialized niches of species that are missing from island communities. The increased specialization of the first white-eyes to colonize an island then contributes to their ability, and to that of their descendants, to coexist with later arrivals. Unusually large white-eyes have evolved independently on 12 small Pacific islands that have few other

Table 24–4 Relative Abundance and Habitat Distribution of Birds in Five Tropical Localities*

Locality	Number of Species Observed (regional diversity)	Average Number of Species per Habitat (local diversity)	Habitats per Species	Relative Abundance per Species per Habitat (density)	Relative Abundance per Species	Relative Abundance of All Species
Panama	135	30.2	2.01	2.95	5.93	800
Trinidad	108	28.2	2.35	3.31	7.78	840
Jamaica	56	21.4	3.43	4.97	17.05	955
St. Lucia	33	15.2	4.15	5.77	23.95	790
St. Kitts	20	11.9	5.35	5.88	31.45	629

* Based on 10 counting periods in each of 9 habitats in each locality. The relative abundance of each species in each habitat is the number of counting periods in which the species was seen (maximum 10); this, times number of habitats, gives relative abundance per species; this, times number of species, gives relative abundance of all species together.
From Cox and Ricklefs 1977.

species, and these large white-eyes often coexist with one or more other white-eyes (Lack 1971). Ponape and Palau both have an extremely large white-eye (Large Ponape White-eye or Palau White-eye), the medium-sized Gray-brown White-eye, and the small Bridled White-eye (Figure 24–7). On Réunion Island in the Indian Ocean, where there is no nectar-feeding sunbird, the Olive White-eye has evolved into a specialized nectar-feeding species in both bill morphology and behavior (Gill 1971). It coexists there with a second, generalized white-eye, the Mascarene White-eye.

As the taxon cycle progresses, the established residents may become further restricted to specialized habitats because of competition with aggressive new colonists or changes in habitat distribution owing to climatic shifts (Pregill and Olson 1981). Local extinctions then fragment the distri-

(A)

(B)

Figure 24–7 White-eyes are excellent island colonists that often occupy unfilled niches on remote islands that have few other birds. Shown here are a typical species, the Bridled White-eye, and a large, thrushlike species, the Palau White-eye, found together on Palau in the Caroline Islands. (From Lack 1971)

bution of the species. Productivity and population size tends to decline as a result of increased susceptibility to parasites and predators. Declining population sizes and concomitant reductions in genetic variability increase the probability of extinction and relictual distribution. The House Wren, for example, has disappeared for no apparent reason from Guadeloupe and Martinique and is close to extinction on St. Vincent. Similarly, Adelaide's Warbler now exists on only two scattered islands of the Lesser Antilles.

The importance and direction of direct competition between new immigrants and endemics as a driving force in taxon cycles are uncertain and in need of detailed study. What may appear to be the competitive replacement of older endemic species by new colonists may only be coincidence. For example, the endemic Socorro Dove was replaced on Socorro Island, off Baja California, by Mourning Doves in the 1970s. It turns out that the extinction of the Socorro Dove and the colonization by the Mourning Dove were independent events related to the establishment of a garrison of the Mexican army, whose cats ate the tame endemic doves and whose new wells provided the drinking water required by the colonizing Mourning Doves (Jehl and Parkes 1983).

Equilibrium theory

Although island avifaunas may change identity, the number of species present reflects a balance between losses due to extinction and gains due to immigration (MacArthur and Wilson 1967). The resulting number of species present at any time is the equilibrium species number, which is the point of intersection between the extinction curve and the immigration curve. The rate of extinction increases with the number of species on an island simply because of the increased number of candidates for extinction, but the rate of extinction also probably increases in communities with large numbers of species because of the effects of diffuse competition (page 465) and the presence of many species with small populations, which are prone to extinction. The rate of immigration, on the other hand, falls as the number of species on the island increases, for two reasons: Additions of new species from source areas become fewer, and it becomes more difficult for a new species to colonize an island on which resources have been preempted by earlier immigrants (Figure 24–8).

Different rates of extinction and colonization are found on islands of different sizes and degrees of isolation. Colonization of islands close to the mainland is more extensive than colonization of remote islands. The probability of extinction is greater on small islands with small populations of resident bird species than on large islands with large populations of many resident bird species. Thus, small, isolated islands have the lowest number of equilibrium species because of infrequent immigration and high extinction levels, and large islands near other land masses (mainland or another large island) have frequent immigration and infrequent extinction and, therefore, large numbers of equilibrium species. The number of equilibrium species for large land-bridge islands, which were once part of a mainland with a full complement of species, is much greater (often three times) than that for large oceanic islands. Oceanic islands do not begin with a full complement

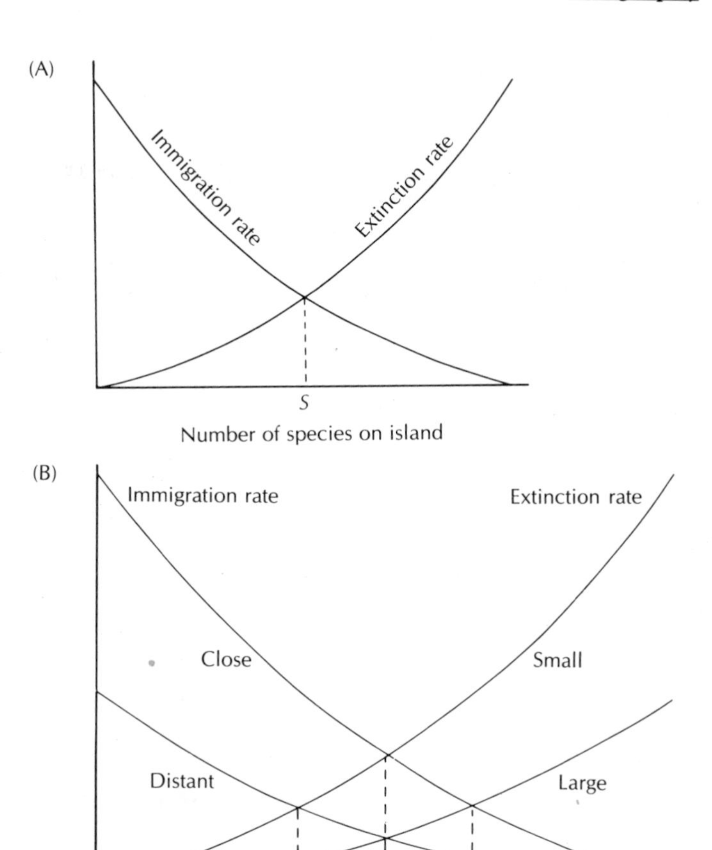

Figure 24–8 (A) The number of species found on an island reflects the balance between the rates of immigration (colonization) and extinction. (B) Immigration rates on islands that are distant from source areas are less than rates on islands close to source areas. Extinction rates on large islands are less than rates on small islands. Extinction rates increase as the number of species present on an island increases. The point of intersection of the two curves for any particular island defines the expected equilibrium number of species *(S)*. (From Ricklefs 1976a)

of mainland species and depend solely on colonists that cross the seas. Whether a true equilibrium is established as a result of the balance between immigration and extinction (as envisioned by Robert MacArthur and Edward Wilson) is not yet certain (Lynch and Johnson 1974; Abbott and Grant 1976; Simberloff 1976). Nevertheless, the observed relations between the number of species and island size seem to be in general accord with the model.

When plotted logarithmically, the direct relationship between number of species and island size is a straight line described by the equation $S = bA^z$, where S is the number of species on the island, A is island area, and z is the slope of the relationship, or rate of change of species number in relation to island area (Preston 1962; MacArthur and Wilson 1967; Ricklefs and Cox 1972). Values of z, a measure of the dependence of extinction rate on island size, are usually 0.2 to 0.4 for various sets of islands. Islands close to source areas have lower values of z than distant islands because frequent immigration masks regular extinctions on the smaller islands, increasing the total number of species on them relative to the number of species present on a large island where extinctions occur less frequently. The change of species number in relation to woodlot area on the mainland has an even lower slope ($z = 0.12$ to 0.17) because of the rapid replacement of faltering local populations by species of surrounding areas. For similar reasons, the relationships between species and area for highly mobile, aerial

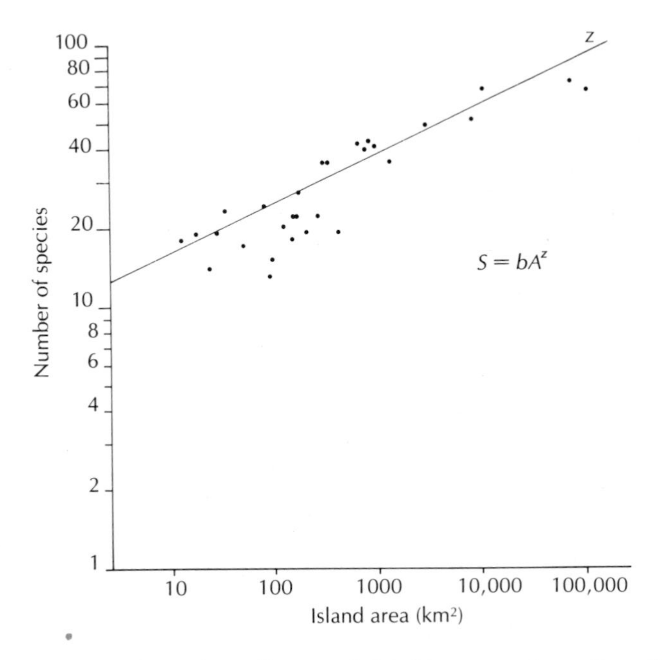

Figure 24-9 The number of species found on islands increases in direct relation to island area. The slope z of this relationship represents the degree to which extinction rate varies with island size, assuming that immigration is constant (z is 0.22 on this graph plotted for the West Indies). (From Ricklefs and Cox 1972)

feeders such as swifts and swallows exhibit lower z values than those for sedentary birds such as woodpeckers (Figure 24-9).

Species turnover

The turnover of species is a fundamental of geographical ecology. Species are subject to evolutionary change, multiplication through speciation, and extinction. Avifaunas, therefore, are dynamic in their composition and reflect the changes in the species that compose them. Island avifaunas are subject to conspicuous species turnover. Turnover rates on small offshore islands such as the Channel Islands off the coast of California are roughly 1 to 20 percent a year (Diamond 1980). Annual censuses of birds on a small British island, Calf of Man, revealed that Northern Wheatears and Stonechats repeatedly became extinct on the island, only to be replaced within a short time by new immigrants. Thus, a high turnover will not be detected if an island is censused infrequently because regular immigrations may obscure regular extinctions. Jared Diamond (1969, 1980) estimates that censuses spaced more than a decade apart will substantially underestimate turnover (Figure 24-10).

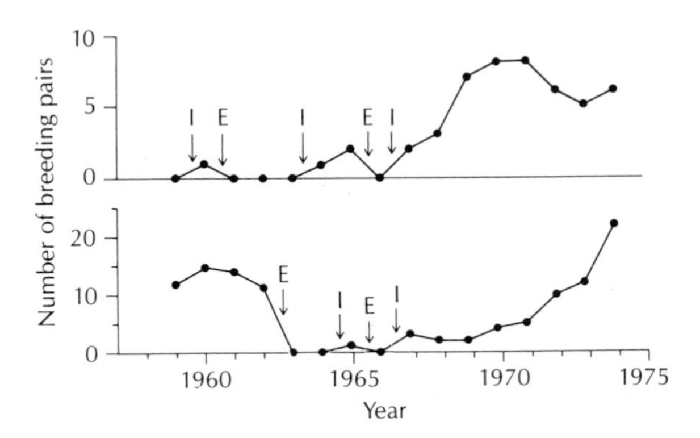

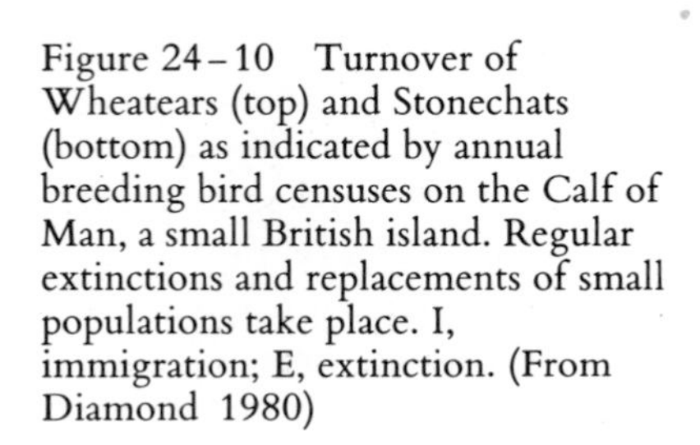
Figure 24-10 Turnover of Wheatears (top) and Stonechats (bottom) as indicated by annual breeding bird censuses on the Calf of Man, a small British island. Regular extinctions and replacements of small populations take place. I, immigration; E, extinction. (From Diamond 1980)

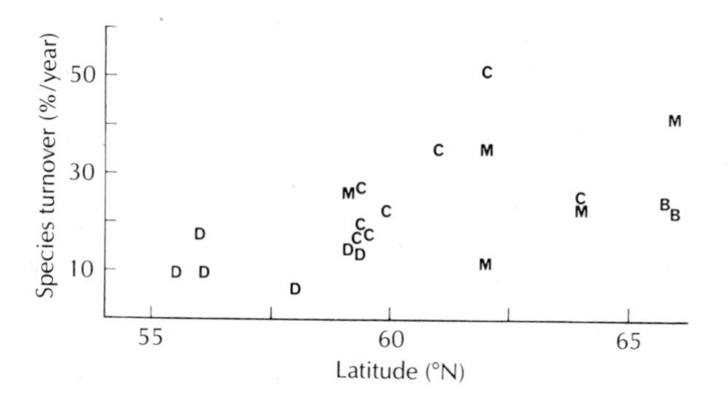

Figure 24–11 Species turnover, which occurs in continental bird communities as well as in island communities, increases with latitude in Swedish forests. All habitats are forests: B, mountain birch; C, coniferous; D, deciduous; M, mixed deciduous-coniferous. (From Järvinen 1980)

Rates of turnover vary among birds of different ecological types. Specialized populations of aerial feeders such as swifts and swallows are quite susceptible to local extinction, especially on small islands. The mobility of such birds means that their replacement by new immigrants is likely; hence their turnover is high. More sedentary birds, such as woodpeckers, reach oceanic islands infrequently and experience low turnover rates. Turnover is typical of continental as well as of island bird communities. The number of species present in northern Scandinavian bird communities, for example, varies by 15 percent per year (Järvinen 1980). Farther south, in the rich deciduous forests of Birdsong Valley, Sweden, annual turnover is about 10 percent. In both regions, extinction and immigration of small numbers of rare species (one to two pairs total) is the primary cause of annual variations in community diversity. Higher turnover in the north is caused by the greater unpredictability of breeding season weather and the greater proportion of rare species, which are more likely to become extinct (Figure 24–11).

Extinction can result simply from the inevitable fluctuations in the size of small populations. The probability of extinction, therefore, depends on population size and the area containing the population (MacArthur and Wilson 1967). Small islands lose species more frequently than large ones, an effect that is seen most clearly in the analysis of land-bridge islands such as Trinidad. Land-bridge islands have lost species steadily since they were isolated by rising sea levels at the end of the Pleistocene epoch (10,000 years ago), and small land-bridge islands have lost a greater proportion of their initial populations of birds than have large land-bridge islands of comparable age (Table 24–5).

Table 24–5 Present and Probable Past Landbird Faunas of Five Major Land-bridge Islands

Island	Area (miles²)	Number of Species: Original	Number of Species: Present	Number of Species: Extinct	Species Extinct (%)
Fernando Po	2,036	360	128	232	84
Trinidad	4,834	350	220	130	37
Hainan	33,710	198	123	75	38
Ceylon	65,688	239	171	68	28
Tasmania	67,978	180	88	92	51

From Terborgh and Winter 1980.

Barro Colorado became a land-bridge island in 1914 when Lake Gatun was formed in Panama. Frank Chapman found 208 species of birds there when he surveyed it for the first time in 1920. By 1970, 45 of these had disappeared (Willis 1974). Many of the losses reflected the regrowth of forests on abandoned farmland and the disappearance of second-growth species. Eighteen forest species also disappeared; 15 of these were either large for their trophic class or were species that forage or nest on the ground. The Ocellated Antbird, for example, declined steadily and is now gone. Mammals, such as peccaries and coatimundies, which now thrive on Barro Colorado Island in the absence of larger predators, destroy large numbers of the ground nests.

Habitat islands, which also lose species because of isolation, may be natural, such as bog and mountaintop forest habitats, or they may be the result of man's agricultural activities, such as are woodlots surrounded by fields or housing developments. As on land-bridge islands, the loss of species from these habitat islands decreases with area. When isolated by the growth of sugarcane and coffee plantations in the last century, subtropical woodlots in southern Brazil supported about 220 bird species (Willis 1980b). Today a large, isolated woodlot (1400 hectares) continues to support 202 species whereas a medium-sized woodlot (250 hectares) and a small woodlot (21 hectares) have only 146 species and 93 species, respectively. The birds lost from the largest plots were mostly large species found in low densities, such as eagles, macaws, parrots, toucans, tinamous, a wood-quail, a pigeon, and a fruit-crow. The birds most likely to disappear from the small woodlots were primarily large, canopy fruit-eating birds and ground, insect-eating birds. The birds lost from the small woodlot also included many microhabitat specialists, which were replaced by birds that thrived in edge and second growth.

Extinction is inevitable, but expanding human populations have hastened the process. The full extent of extinctions caused by human activities worldwide is much greater than is generally appreciated. Losses of island birds due primarily to man account for 90 percent of animal extinctions during historical times. The slaughter of the Dodo and other birds on the Mascarene Islands in the late 1600s is a classic example of the eradication of vulnerable island birds. Human influences are responsible for 19 cases of extinction (and 59 cases of immigration) in the avifaunas of the satellite islands of Australia and New Zealand during the last century (Abbott and Grant 1976). The vulnerability of island bird populations relates in part to their tendency to lose their resistance to mainland diseases. Lowland populations of the Hawaiian Honeycreeper, for example, were destroyed by bird pox and malaria when mosquitos that carried these diseases were accidentally introduced in the early 1800s (Warner 1968). In addition, long before Captain Cook brought European civilization and mosquitos to the islands, the early Polynesians destroyed most of the lowland forests on the Hawaiian islands after reaching them, roughly 1500 years ago. As a result of this destruction, they eliminated at least 39 species of landbirds, including 7 geese, 2 flightless ibis, 3 owls, 7 flightless rails, and 15 species of honeycreepers (Olson and James 1982).

Birds now face a global transformation of their habitats. In the past, climatic changes were responsible for habitat transformation, but today this has become man's role. All biologists, including ornithologists, are distressed by the rapid destruction of the natural habitats of the world, which includes the replacement of virgin rainforest by pastures and coffee or banana plantations, the conversion of rich grasslands into overgrazed desert scrub, the draining of wetlands, and the urban destruction of biological habitats in general. Many of the species of birds in the world today will become extinct in 50 to 100 years, perhaps sooner in tropical regions.

Populations of the most common birds number in the hundreds of millions. A large population, however, is no guarantee of continued existence. Legendary were the estimated two billion Passenger Pigeons that flew over colonial America. Now they are extinct because of systematic slaughter for the human food markets (Blockstein and Tordoff 1985). A host of species that once existed in large numbers now hover at the brink of extinction (King 1981). California Condors, which now only live in zoos, are the remnants of a once-thriving radiation of giant vultures that depended on the large mammals of North America, most of which are now extinct. Many of the world's rarest birds persist in small numbers on remote islands. Forty Seychelles Magpie-Robins remain on the islet of La Digue in the Seychelles Islands. Only 10 Kakapo, a flightless nocturnal parrot, remain on the south island of New Zealand; 9 of these are males. The future of this bird depends on the 30 individuals recently discovered on nearby tiny Stewart Island. Ivory-billed Woodpeckers disappeared from North America with the destruction of swamp forests in the southern United States. Today, just a few pairs of this magnificent bird exist in eastern Cuba.

Some species will certainly adapt to man's new environments. Canada Geese are thriving along polluted streams and now perch on chimneys, and American Crows, known for 200 years as shy, rural birds in the United States, are invading suburban backyards and city parks. This relatively recent behavior is similar to that exhibited by House Crows and Common Jackdaws in Asia and Europe centuries ago. Equally impressive is the growing dominance of introduced species throughout the world. The commonest birds on Hawaii and Puerto Rico are exotic species. Rock Doves, European Starlings, and House Sparrows are human associates with prospects of global success. A redistribution of those birds of the world that are able to coexist with human societies is pending. The future of most of the other species will depend increasingly on the reserves of natural habitat we can set aside for them.

Summary

The diversity of birds is comprised of both local assemblages of species, called communities, and geographical assemblages of species, called avifaunas. The six major avifaunas of the world are the Nearctic (North America), Neotropical (Central and South America), Palearctic (Europe and Asia), Ethiopian (Africa), Oriental (southeast Asia), and Australasian (Australia and New Guinea). Each region has its characteristic birds.

Avifaunas are the grand result of millions of years of evolution, adaptive radiation, dispersal, and extinction of avian taxa with varied ecological roles. Throughout the history of avian evolution, neither the world's climates nor the arrangement of the continents were as we know them today. Many modern birds occupy only remnants of their original distributions. Turnover, the addition and loss of species, drives the changing compositions of avifaunas. The composition of island avifaunas reflects not only ancient history but an ongoing cycle of colonization and extinction, the frequency of which depends on the isolation and the size of the island. Small, isolated islands have the smallest equilibrium number of species whereas large islands near continental source areas have the highest equilibrium number of species. Increasingly, mainland forests are reduced to small, island fragments subject to loss of species. Extinction of bird species worldwide is now reaching unprecedented rates as a result of human-related transformation of habitats.

Further readings

Cracraft, J. 1973. Continental drift, paleoclimatology, and the evolution and biogeography of birds. J. Zool. Lond 169:455–545. *A detailed review of the modern perspective.*

Darlington, P.J., Jr. 1957. Zoogeography: The Geographical Distribution of Animals. New York: John Wiley & Sons. *A classic.*

MacArthur, R.H. 1972. Geographical Ecology. New York: Harper and Row. *A review of the interface between community ecology and biogeography.*

Vuilleumier, F. 1975. Zoogeography. Avian Biology 5:421–495. *A detailed review of the classic perspective.*

chapter 25

Speciation

The evolutionary legacy of the earliest Mesozoic birds is believed to be 100,000 species, of which 9021 are now with us. Behind this legacy lies the process of speciation, the multiplication of species through the irreversible separation of one species into two or more. New species evolve as a result of the genetic divergence of isolated populations. Geographical separation of populations provides the opportunity for speciation to occur. The process of speciation involves genetic transformation and, ultimately, reproductive isolation. Although the general patterns of geographical speciation in birds are well known, the precise mechanisms are not. Still to be resolved is the relative importance of ecological and social adaptation as well as the timing and nature of the genetic changes related to them. Whatever their origin, new genetic identities spread and become established in the population as a whole. The chance that a new mutation or arrangement of genes will become established depends on the size, historical origin, and structure of the population in which it takes place.

Population structure

The movement of a young bird from the site where it hatches to the site where it breeds, called dispersal, is the key determinant of population structure. Insufficiency of habitat and social interactions foster dispersal. Natal dispersal is the permanent movement of young birds from their birth sites to their own breeding location (Figure 25–1). Many birds tend to stay near their birthplaces, a habit called philopatry. Colonial seabirds, such as albatrosses, gulls, and terns, return to their natal colonies, and songbirds such as the Great Tit, the Pied Flycatcher, and the Song Sparrow stay within a few kilometers of their natal territory. A few individuals of philopatric species disperse widely. Generally speaking, females of a species disperse farther than males, a pattern that is most extreme in communal breeders such as the Arabian Babbler and Florida Scrub Jay, in which young females disperse and join new family groups while young males wait at home to inherit a portion of their family plot (see Chapter 20). Waterfowl, however, behave in the opposite way. Females return to their natal marsh or colony, and males disperse widely, following whichever female chose them on the wintering grounds. In the case of the Snow Goose, approximately 50 percent of all birds breeding for the first time at the La Perouse Bay colony are immigrant males from other colonies (Cooke et al. 1975).

Dispersal distance of a species defines the size of a local population, or *deme*, in which gene exchange is theoretically a random process and the laws of population genetics are manifest. The number of individuals in a deme is called the effective population size. The evolutionary potential of small demes differs from that of large demes in that small demes evolve

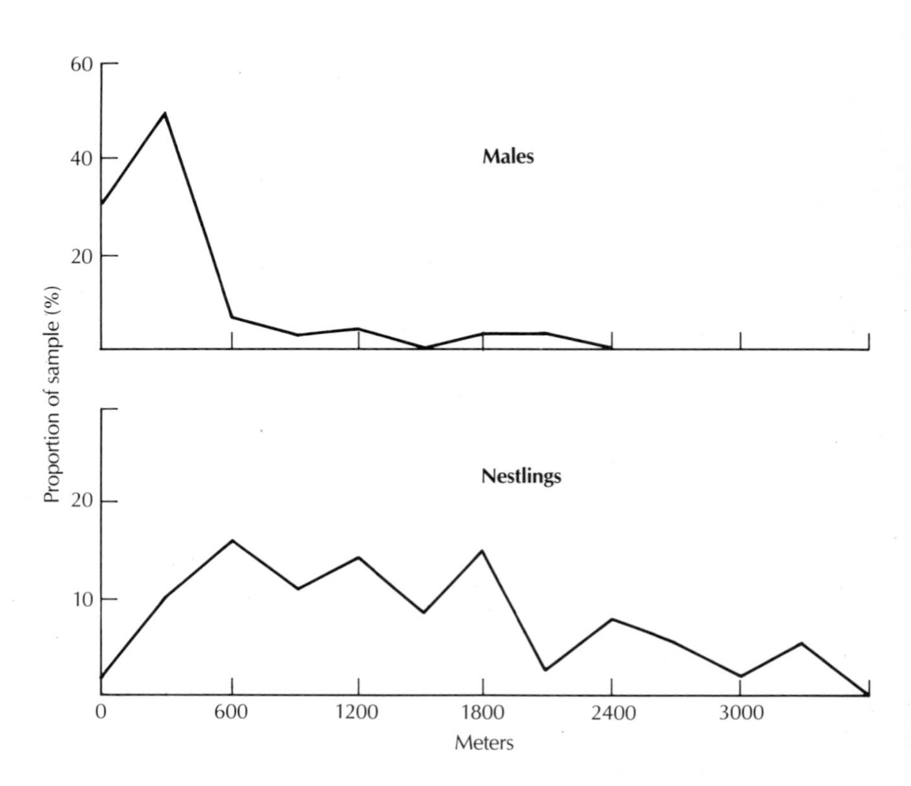

Figure 25–1 Dispersal of adult male (top) and nestling (bottom) House Wrens. Most adult males disperse over a small area whereas young wrens disperse more widely. (Adapted from Barrowclough 1978)

faster and in directions that sometimes can be dictated by gene loss or by genes that are passed to the next generation by chance, independent of their adaptive value. Effective population size decreases as average dispersal distance decreases and as patterns of dispersal or pairing relationships among individuals become nonrandom. The effective sizes of populations of some birds may decrease because of

1. fragmentation of populations into small islands, which limits dispersal;
2. founding of new populations by a small number of colonists;
3. confinement of populations to isolated colonies on long, narrow coastlines; or
4. existence of nonmonogamous breeding systems.

Ornithologists now estimate that noncolonial passerine birds disperse roughly one kilometer per year, with a range of 350 to 1700 meters per year (Barrowclough 1980a). This means that the effective population sizes of such birds are quite large, roughly 175 to 7700 individuals, and that evolutionary change tends to be slow and adaptive except in very small, isolated groups of birds. It follows also that bird speciation usually results from gradual adaptive divergence of large, fragmented populations or from rapid genetic reorganization in small, founder populations (Barrowclough 1983).

With this general introduction to population structure and the modes of speciation in birds, we now look at the phenomenon of geographical speciation, which proceeds through four main stages:

1. fragmentation and isolation of populations;
2. genetic divergence of the isolates;
3. reuniting or secondary contact of divergent populations; and
4. resolution of the requirements for coexistence, including both reproductive isolation and ecological isolation.

We begin our review of geographical speciation with the phenomenon of geographical variation in birds and the classic concept of subspecies (Figure 25–2).

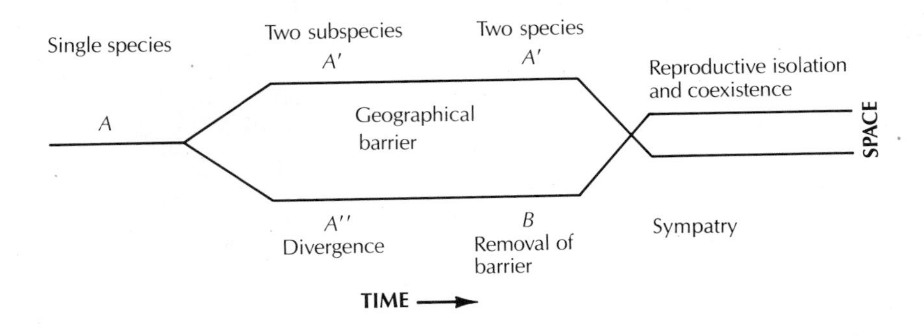

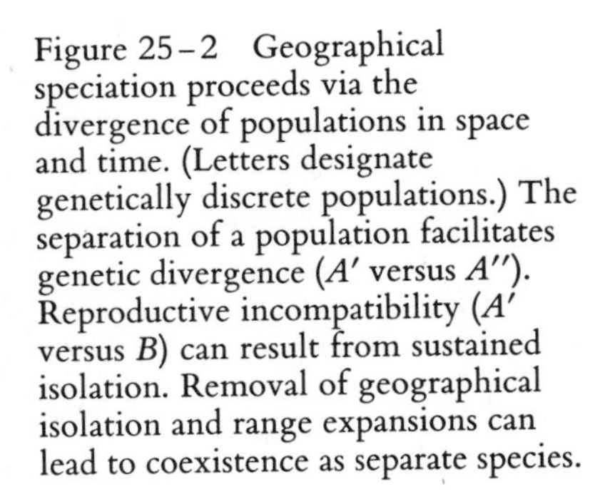
Figure 25–2 Geographical speciation proceeds via the divergence of populations in space and time. (Letters designate genetically discrete populations.) The separation of a population facilitates genetic divergence (*A′* versus *A″*). Reproductive incompatibility (*A′* versus *B*) can result from sustained isolation. Removal of geographical isolation and range expansions can lead to coexistence as separate species.

Figure 25–3 Geographical variation in the Fox Sparrow. Local populations diverge in bill dimensions as they adapt to local environments. (Adapted from Zink 1986)

Evolution of geographical variation

Gradual adaptive divergence of large populations is one of the primary modes of speciation in birds. Geographical speciation starts with the evolution of geographical variations among populations. Geographical variation in size and color is characteristic of many birds. One-third of the species of North American birds show conspicuous geographical variation. Some, such as the Song Sparrow, have many recognizably different populations, or "subspecies," whereas the Cedar Waxwing and Bay-breasted Warbler have few or none. Historically, a subspecies is said to exist when 75 percent of the individuals in a population are distinguishable, usually by their plumage coloration and size characteristics, from other populations of the same species. The concept of subspecies played an important role in early studies of geographical variation in birds, but it is now under growing threat of obsolescence (Ford 1974; Barrowclough 1982; Gill 1982; Mayr 1982), primarily because thorough study using large samples from throughout a species range reveals that the patterns of geographical variation in color and size are far more complex than was previously thought. Furthermore, some of the variations that were used to diagnose subspecies may not have genetic bases, which means that subspecies may not be meaningful evolutionary units.

Genetic variation can come about because different attributes are favored in different environments. For example, populations of Steller's Jays on the west coast of the United States have evolved crests of various lengths in relation to the openness of the vegetation in their habitats (Brown 1964a) (see Figure 9–8). The early differences that evolve between populations are subtle. Populations of Fox Sparrows in the western United States, for example, differ slightly in bill dimensions, which presumably reflect slight differences in diet (Zink 1986) (Figure 25–3). The Song Sparrow exhibits striking geographical variations among its populations. In the Pacific northwest they are blackish whereas in the deserts of California they are pale brown. In Ohio, Song Sparrows are medium-sized and brownish whereas in the Aleutian Islands they are large, about the size of a thrush.

Climatic adaptation is a conspicuous feature of geographical divergence. Geographical trends in the body sizes of permanent residents relate precisely to climatic gradients that include both temperature and humidity (James 1970). Downy Woodpeckers, for example, are smallest in hot, humid climates such as the Mississippi Valley, where the potential for heat loss is lowest and where small individuals with a large surface area relative to volume are favored. Conversely, cool, dry air increases heat loss and favors larger bodies (Figure 25–4).

Geographical differences among populations can evolve rapidly. Within 100 years of their introduction to both New Zealand and the United States, House Sparrows living in various environments in these countries have evolved visible differences in body color and size (Johnston and Selander 1964, 1971; Selander and Johnston 1967; Baker 1980). As in the Song Sparrows, desert dwellers are paler than those living in humid climates. Body size and sexual dimorphism increase with latitude. Direct evidence of the advantage of larger size in colder climates was obtained by Herman Bumpus, who compared the measurements of House Sparrows

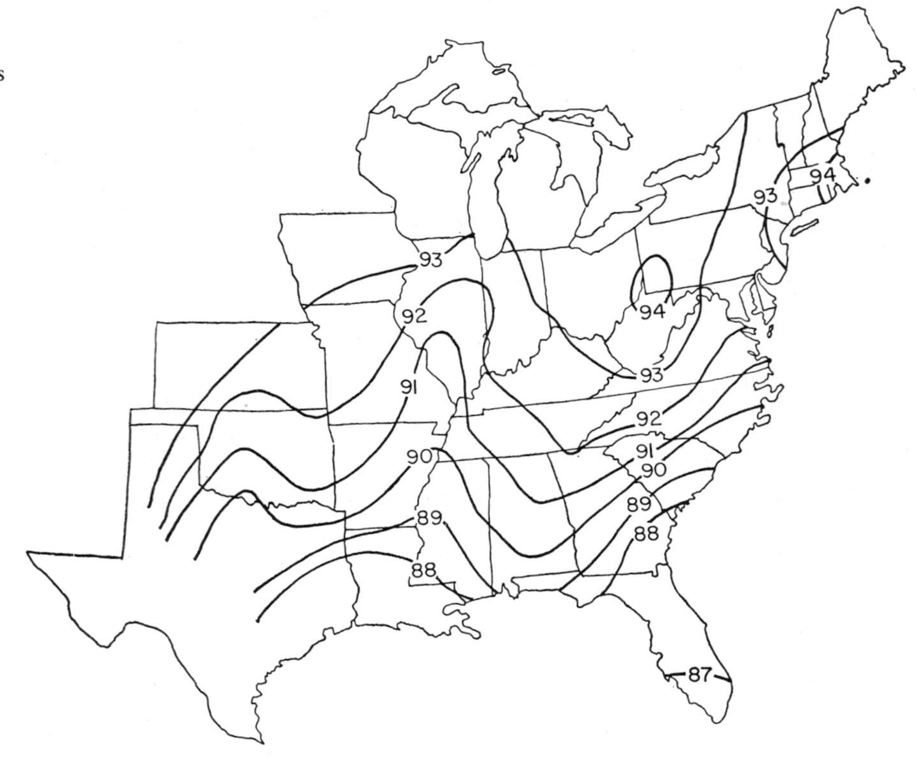

Figure 25–4 Size variation in Downy Woodpeckers. Body size increases to the north, but individuals in the warm, humid Mississippi Valley and coastal areas are small compared with those at other localities at similar latitudes. Numbers on map indicate average wing lengths in millimeters. (From James 1970)

that survived a severe winter storm in Rhode Island in 1898 with sparrows that did not (Grant 1972a; O'Donald 1973). Female House Sparrows suffered greater mortality than males, which are larger and can live longer without food.

When we relate such episodes of natural selection to the evolution of the geographical variation in body size, we assume that size and color variation are genetically controlled, and that they are not directly affected by the environment. This is a reasonable assumption, but it is not a simple or certain one.

To elucidate early reports that climate can directly affect feather pigmentation and size of exposed body parts (Beebe 1907; Allee and Lutherman 1940), experiments are needed to determine the ways in which heritage and environment interact to produce patterns of geographical variation. In a pioneering study, Frances James (1983) demonstrated the effect of environment on size in Red-winged Blackbirds, which vary geographically in bill shape and in wing length relative to leg (tarsus) length. Some of this variation is attributable directly to the environment. When eggs were transplanted from nests of one population to nests of another, morphologically distinct population, the dimensions of fostered chicks resembled those of their foster parents. Red-wings that were transplanted from the Everglades to Tallahassee, Florida, grew shorter, thicker bills, similar to those of Tallahassee Red-wings. Colorado Red-wings that were transplanted to Minnesota developed longer wings and toes (Figure 25–5).

To understand the environment's effect on the evolution of geographical variation, we must first understand the nature of heritability.

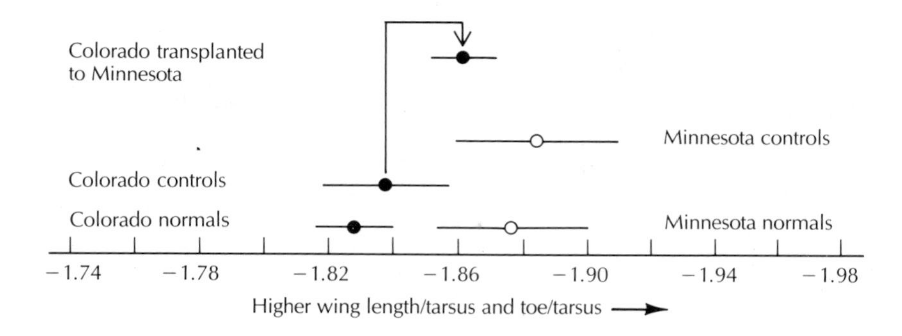

Figure 25–5 Environmental influence on the dimensions of nestling Red-winged Blackbirds. When transplanted to nests in Minnesota, eggs from nests in Colorado yielded nestlings that were shaped more like Minnesota Red-wings than the controls in Colorado, which were those nestlings transplanted to new nests in the same locality. Nestling shape is here defined in terms of a discriminant function that relates wing length to size of the legs and feet. (From James 1983)

Traits such as body size, bill depth, and bone dimensions are controlled by many genes, but as we have seen, these traits are also directly affected by the environment. The proportion of total observed variability that is controlled by genes is called the heritability H of a character, which has a maximum value of 1.0. One-half to most of the size variation observed in species of birds has a genetic basis. Body weights of chickens are moderately heritable ($H = 0.53$) whereas feathering traits, breast angle, body depth, keel length, and shank pigmentation have lower heritabilities ($H = 0.25$ to 0.40) (Kinney 1969). The few recent studies of character heritability in wild birds indicate moderate to high heritabilities: 0.43 to 0.95 (Boag and Grant 1978; Smith and Zach 1979; Smith and Dhondt 1980; Garnett 1981). Because the potential for influence by natural selection increases with the heritability of a trait, we can be confident of the potential for evolutionary change. Nevertheless, some direct environmental influence is probable unless $H = 1.0$.

Gene flow and clines

The evolution of geographical differences among natural bird populations depends on the relative strength of two opposing forces: the degree to which natural selection favors one genetic attribute over another and the rate of genetic blending as a result of gene flow from interbreeding with individuals from other populations. Theoretically, adaptations that confer a slight relative advantage are powerful evolutionary forces, even if the advantage is only a one percent difference in the number of surviving offspring. However, strong gene flow can override weak selection and can prevent the evolution of differences among populations.

Clines, which are gradients of character change, such as in body size or feather pigmentation, are the expression of the opposing actions of divergent selection and blending gene flow in contiguous populations. Clinal variation is especially conspicuous in birds that have color phases, which usually have a simple genetic basis. For example, the relative num-

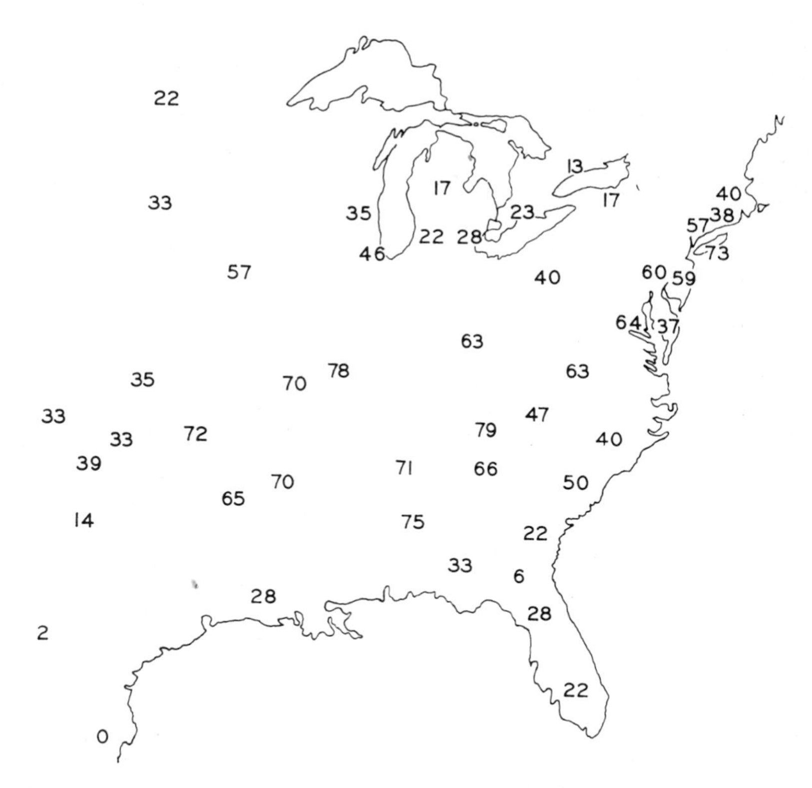

Figure 25–6 The proportions of red-phase Eastern Screech-Owls found in local populations decline from high values of 70 to 80 percent in the center of the range of this species to 30 percent or less at the edges of the range. (From Owen 1963)

bers of red- and gray-phase Eastern Screech-Owls and of dark and light phase jaegers, which are predaceous gull-like birds that breed in the arctic, change systematically with locality. In the case of the screech-owls, local populations change from mostly red-phase owls in Tennessee to mostly gray-phase owls in Maine (Owen 1963) (Figure 25–6). We do not know what favors gray-phase screech-owls in the north, but studies of another species, the Ruffed Grouse, may be illuminating (Gullion and Marshall 1968). This grouse also has red and gray color phases, and the proportion of gray-phase individuals in local populations increases toward the north. Gray-phase grouse survive better than red-phase grouse in gray-colored coniferous and aspen-birch forests and where there is deep winter snow, which allows grouse to roost at night in burrows rather than in trees. As the type of forest changes from hardwoods in the south to conifers in the north, and as the depth of winter snow increases, so does the advantage of the gray-phase grouse. Quite likely, this advantage relates to protective coloration and exposure to nocturnal predators such as the Great Horned Owl.

Clines may be either static or dynamic in character. Static clines are those in which an equilibrium between selection and gene flow has been established, and no future changes in the composition of the populations are expected. Dynamic clines change with time as a result of slow diffusion of neutral traits or as a result of a selective advantage of one trait over its alternatives. An example of a dynamic cline is found in Bananaquits, a small warblerlike bird that feeds on nectar and fruit on the Caribbean island of Grenada (Wunderle 1981, 1983). The yellow-and-black color form of Bananaquits prevails throughout most of the Caribbean, but until the early

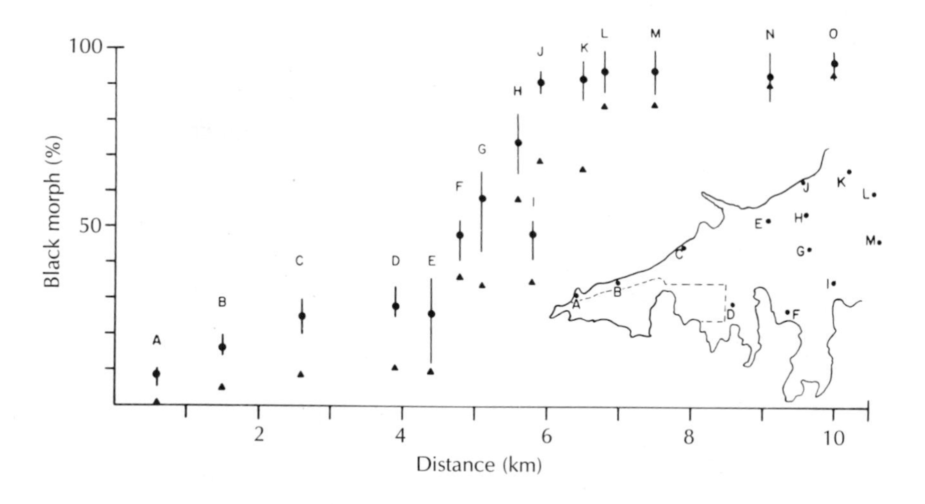

Figure 25–7 Yellow-and-black Bananaquits are common only in southwestern Grenada, the region of their recent colonization of this island. Black-morph birds occupy the rest of the island. Shown here are means and ranges of the proportions of the black morph at sampling sites 1974–1978, graphed in terms of distance from the westernmost point of the island. Triangles indicate the results of resampling the localities in 1981. The declines in the proportions of the black morph to the east suggest expansion of the yellow-and-black morph at the expense of the black. (From Wunderle 1983)

1900s, a black color form of this species inhabited Grenada. About 70 years ago, yellow-and-black Bananaquits colonized the arid southwestern corner of Grenada, where they replaced the black Bananaquits. The relative numbers of the two color forms change rapidly in favor of black Bananaquits as one proceeds north and east. Yellow-and-black Bananaquits are advancing eastward at a rate of roughly 400 meters per year, mixing with and then replacing black Bananaquits. This rate of advance suggests that yellow-and-black Bananaquits have a 17 percent selective advantage over black Bananaquits and will soon replace them throughout the island (Figure 25–7).

Assortative mating

So far, our discussion of population structure in birds has been based on the assumption that individuals pair at random. However, preferential pairing of like types, or assortative mating, violates this premise and can strongly affect clinal variation. One of the best studied cases of assortative mating is that of the color phases of the Snow Goose of the Canadian Arctic (Cooke 1978; Geramita et al. 1982; Rockwell et al. 1985). An east-west cline exists in the proportions of dark-phase ("blue") geese in various colonies. Dark-phase geese are most common in eastern breeding colonies; white-phase geese are most common in western breeding colonies.

Snow Geese pair assortatively with geese of their own color phase (Table 25–1). At the La Perouse Bay colony in northern Manitoba only 15

Table 25–1 Mate Selection by Snow Geese at La Perouse Bay

Parent Color	Mate Color		
	White	Blue	Total
White pair	191	37	228
Mixed pair	37	27	64
Blue pair	9	57	66

After Cooke 1978.

to 18 percent of all pairs were mixed, much less than the 35 to 41 percent expected if pairing were random with respect to color phase. Absolute assortative mating could eventually serve to isolate reproductively two forms and thus be tantamount to speciation without geographical isolation. In the case of the Snow Goose, however, the frequency of mixed pairings is too high for this to occur. One of the effects of assortative mating is to stabilize the proportions of dark- and light-phase birds in a colony despite the fact that gene flow among colonies with different proportions of phases is extremely high. The exceptions to the assortative mating, on the other hand, contribute to gradual changes in the phase compositions of populations.

The difference between the white and the dark phases of this species has a simple genetic basis: The dark phase is controlled by a single dominant allele, the white phase by alternative recessive alleles in homozygous condition. Heterozygotes are close to the dark phase in appearance, with more white on the underparts. The color phases do not differ even slightly in fecundity or survivorship.

Family color is the force behind mate preference; young geese choose mates of the same color as their families, principally that of their parents. Regardless of what color phase they themselves may be, geese raised by white-phase parents choose white-phase mates, those raised by dark-phase parents choose dark-phase mates, and those raised by mixed pairs choose mates of either color phase in proportion to availability on the wintering grounds where pair formation takes place. The exceptions, cases of mate choices of phases that are opposite to the parents, reflect some influence of sibling color in families where some young geese are not the same color as their parents. Offspring that are different in color from their parents are primarily white-phase offspring that are produced by heterozygote dark-phase parents (as a result of segregation of the recessive alleles, analogous to the production of blue-eyed human children by brown-eyed parents). Offspring that are different in color from their parents also result from intraspecific brood parasitism, which occurs when females of one color phase deposit their eggs in the nests of females of another color phase.

The proportions of white- and dark-phase geese are converging slowly towards equality throughout the arctic, in part because of changes in wintering ground ecology. Once upon a time, dark-phase geese, which predominated in the eastern Arctic, wintered on the Texas coast, and white-phase colonies from the western arctic wintered on the Mississippi delta. Pair choices were therefore limited to the color phase that corresponded to the family color. In recent times, however, human disturbance, marsh management, and the cultivation of rice have increased the contact between white and dark phases on the wintering grounds, which increases the frequency of mixed pairings by young geese of mixed family exposure. The result is a continuing increase in numbers of heterozygous offspring, mixed family experience, and mixed pairings.

Local scale variation

Sometimes bird populations evolve differences on a local scale. One of the most extreme cases of microgeographic variation occurs within the confines of small, rugged Réunion Island in the western Indian Ocean, the home of the Mascarene White-eye, an Old World ecological equivalent of the Bananaquit (see Chapter 24). The complex patterns of local geographic variation in the Mascarene White-eye exemplify different aspects of geographical variation, including climatic adaptation and clinal variation of color phases (Gill 1973).

The Mascarene White-eye has a brown color phase and a gray color phase. Unlike Snow Geese, these color phases pair randomly. The proportions of gray-phase white-eyes in local populations increase with altitude from none in coastal populations to over 90 percent in populations above 2000 meters. Whereas the proportions of color phases change gradually with latitude in the screech-owl and the grouse, the clines in the white-eye change rapidly over distances of only a few kilometers on steep mountainsides. The advantage of gray-phase white-eyes at high altitudes is not known, but it must be at least 1 percent to maintain such steep clines.

Independent of the proportions of the two color phases, brown-phase white-eyes vary strikingly in color and size at different localities on the island. The variations involve three distinct populations, each with altitudinal clines in size and pigmentation. Occupying the coastal regions of Réunion Island are three distinct populations of brown-phase white-eyes: a gray-headed form, a brown-headed form, and a gray-crowned, brown-naped form. These three forms apparently evolved as small, isolated populations on different parts of the island before the arrival of humans. Expansion followed widespread cutting of the forests. The ranges of each form now abut one another at major river beds and at a lava flow, and thus remain partially isolated. The gray-crowned, brown-naped form appears to be a result of hybridization of the other two forms (Figure 25–8).

Representatives of brown-phase white-eyes are larger and darker at higher elevations and in the cold, wet interior of the island. In contrast to the brown-headed birds of the coast, which are small and pale, with white bellies, those of the central highlands, only 30 kilometers away, are six to eight percent larger and dark, with rufous-brown or charcoal-gray bellies.

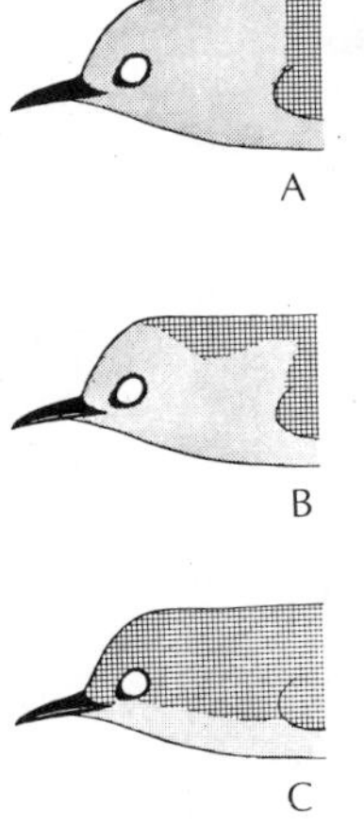

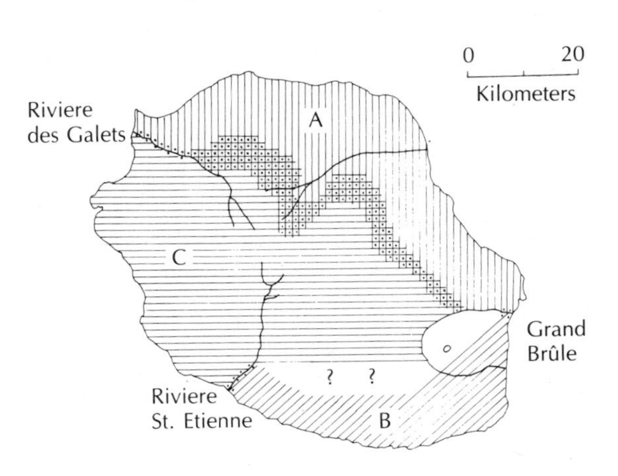

Figure 25–8 Three distinct populations of the Mascarene White-eye evolved on Réunion Island in the Indian Ocean. Zones of contact and hybridization are indicated by dots. Population B, which is isolated coastally from the other two populations by the Riviére St. Etienne and lava flows of Grand Brûlé, probably originated as a result of hybridization between populations A and C. (From Gill 1973)

This aspect of geographical variation in the Mascarene White-eye is similar to the variations recently evolved in House Sparrows in North America and is characteristic of geographical variation in birds in general. However, the short distances between distinct forms of the white-eye are unusual.

Geographical isolation

It is believed that most species of birds evolve as geographical isolates, or in allopatry, under conditions of minimal gene flow. Recall from our earlier discussion that, because the demes of birds tend to be rather large, the primary modes of speciation seem to be through slow adaptive divergence of fragments of large populations or through rapid genetic reorganization in small founder populations. Historical isolation by various kinds of barriers makes these modes possible.

Bird populations become isolated in two principal ways:

1. Individuals may colonize an oceanic island that is well separated from their main population on the mainland or other islands, as the Bananaquits and Mascarene White-eyes have done. Classical examples of major evolutionary change come from remote islands such as the Galápagos and Hawaiian archipelagos. The birds on the Channel Islands off the coast of southern California are also well differentiated. Small satellite islands off the coast of New Guinea have distinct populations of kingfishers. Islands of special habitats, such as oases in the desert, may set a similar stage for speciation for the populations that occupy them.

2. Fragmentation of habitats that were once continuous may also fragment a population of birds. Historical contractions of habitats have isolated remnants of once-widespread species. The dry climates of the Pleistocene epoch shrank the great continuous Amazonian rain forests into much smaller areas surrounded by grasslands. Restricted to these forest refugia, the toucans, manakins, and flycatchers were among the many kinds of birds that underwent speciation (Haffer 1969, 1974) (Figure 25–9).

The patterns of divergence among fragmented populations are not easily predicted. All populations within a species may evolve diagnostic distinctions, or some populations may change while others retain the original characteristics. Of particular interest in this regard are the "leapfrog" patterns of geographical variation among birds in the Andes of South America (Remsen 1984). In the case of a small tanager with a big name, the Superciliaried Hemispingus, for example, two identical populations are separated by a third distinct population. The simplest explanation for this pattern is that only the central fragment of the three evolved new features. Over 20 percent of the species and superspecies of birds in the Andes with at least three differentiated populations exhibit similar patterns of geographical variation. In the other cases, the peripheral populations seem to have changed.

Historical fragmentation of major habitats may fragment the ranges of whole communities that have been associated with a particular habitat. Pairs of sister species, or *vicariants,* may then evolve concordantly in a variety of taxa that were separated at the same time. For example, Joel Cracraft

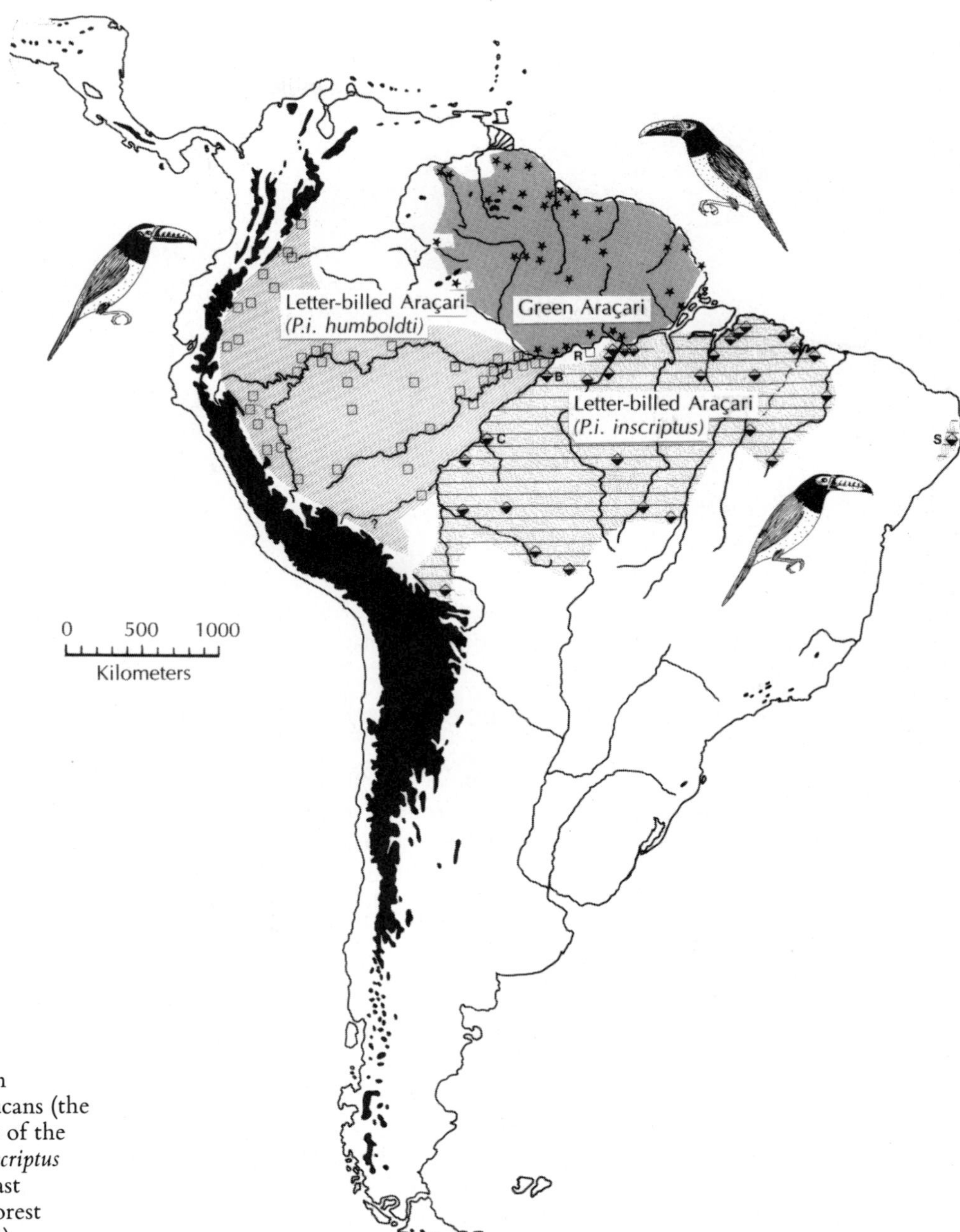

Figure 25-9 The ranges in Amazonia of three small toucans (the Green Araçari and two races of the Letter-billed Araçari *P. i. inscriptus* and *P. i. humboldti*) reflect past isolation in refuges of wet forest habitats. (From Haffer 1974)

(1982a) postulates ten specific fractures, or vicariance barriers, in the recent history of the Australian avifauna. Among these was a major separation of the northern and eastern avifaunas from the central and southern avifaunas. The birds that occupy wet habitats on the southern edge of Australia also split from those of the central arid region. Later the central arid avifaunas and southern mesic avifaunas split again. The Pleistocene deterioration of climates caused the two northern avifaunas to split repeatedly at different sites. Accompanying these separations was the divergence of sister taxa, such as the species and subspecies of Australian robins (Figure 25-10).

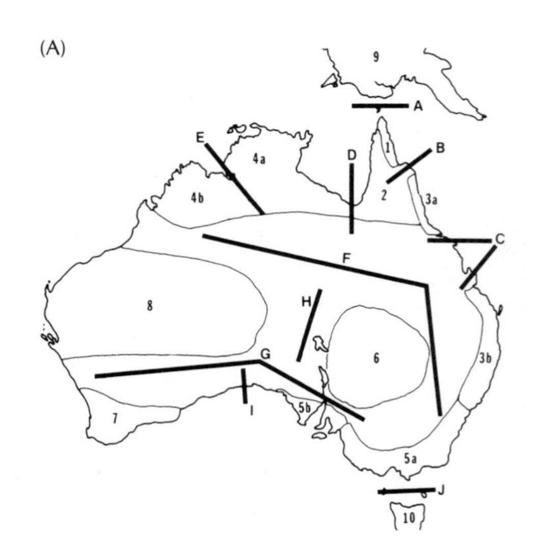

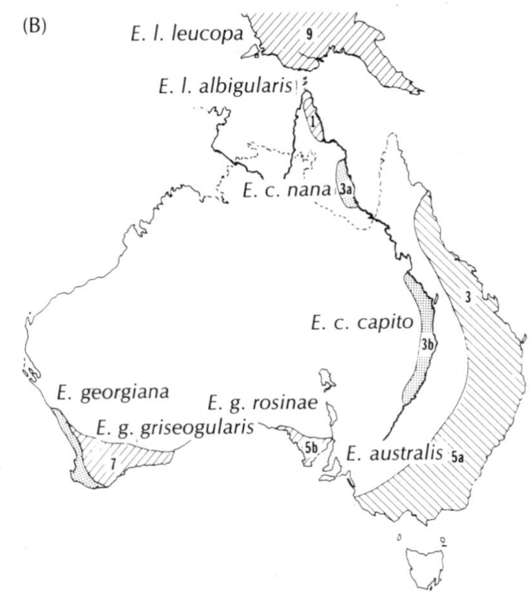

Figure 25 – 10 Fragmentations of Australian habitats were responsible for the speciation patterns on that continent. (A) Ten primary ecological barriers (A – J) separate the geographical regions of endemism of Australian birds (1 – 10). (B) Distributions of Australian robins *(Eopsaltria)* in relation to the major areas of endemism of Australian birds. The east coast appears in a double image to show the distributions of the two species that occur together there. (From Cracraft 1982a)

These sister taxa came back into contact when their ranges expanded during favorable climatic periods.

Secondary contact and hybridization

Secondary contact, the reuniting of previously isolated populations, tests the reproductive compatibility of populations. Members of sister taxa face the new options of pairing with dissimilar individuals. The possible consequences of these matings range from free interbreeding and blending of the attributes of the sister taxa through limited hybridization to strict reproductive isolation and conformity to the definition of biological species.

We find an example of free interbreeding without any penalty of inferior hybrid offspring in populations of the Yellow-rumped Warbler (Hubbard 1969; Barrowclough 1980b). Separated by glaciers during the Wisconsin glaciation, eastern and western populations evolved into the distinctive Myrtle Warbler and Audubon's Warbler (Figure 25 – 11). The two divergent populations came into secondary contact about 7500 years ago when the forests reunited as the glaciers retreated. Myrtle Warblers and Audubon's Warblers now interbreed freely in mountain passes of the Canadian Rockies. Movement of Myrtle Warbler genes westward from sites of hybridization and of Audubon's Warbler genes eastward has produced a zone of clinal intergradation that extends far beyond the mountain passes where hybridization takes place. The 150-kilometer width of the zone is close to the theoretical prediction for a dynamic cline involving neutral characters that has been widening slowly but steadily for the past 7500 years. Only the restriction of contact between the two populations to the high mountain passes of Alberta prevents a more rapid blurring of the differences between them.

As this example demonstrates, there is blending, or introgression, of the attributes of two populations that interbreed freely; characteristics may penetrate two introgressing populations at different rates, reflecting the respective advantages of the characteristics in the new environments. The

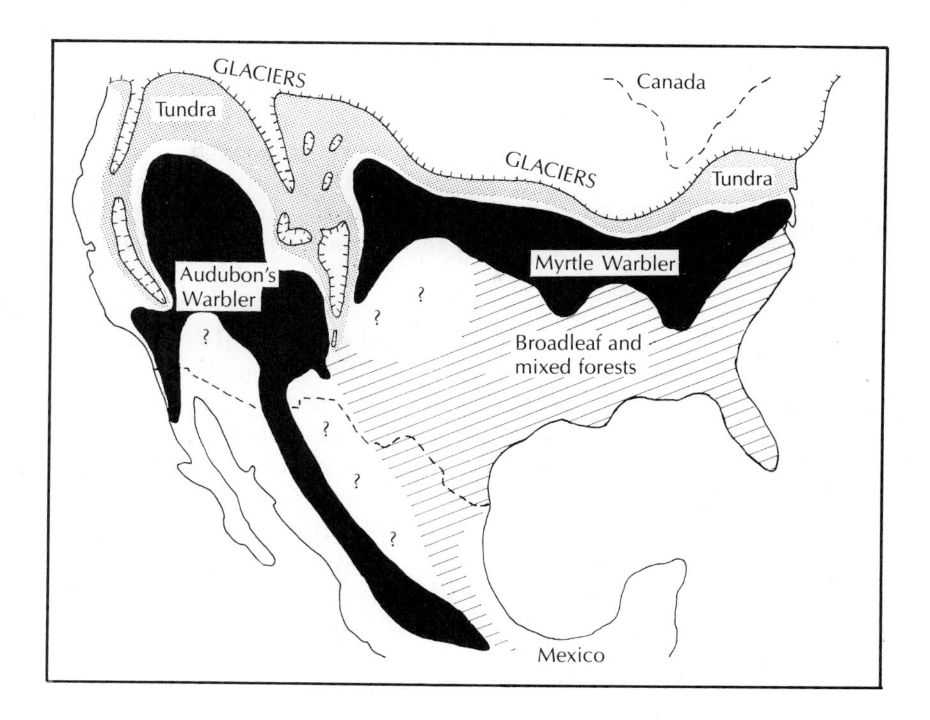

Figure 25–11 Model of the distributions of the eastern Myrtle Warbler and the western Audubon's Warbler separated during the Pleistocene epoch (Wisconsin glaciation), the time when they may have diverged from a common ancestor. (From Hubbard 1969)

width of the hybrid zone may enlarge with time, as in the case of the warblers, and a new population of intermediate characters may result. The Gilded Flicker of the southwestern United States is probably of hybrid origin (Short 1965) as is one population of Mascarene White-eyes (Figure 25–8). Some local populations of towhees in Mexico consist only of hybrids with characteristics of both the Rufous-sided Towhee and the Collared Towhee (Sibley 1954). Quite possibly, hybridization is responsible for the modern characteristics of other bird species (Short 1972).

Hybridization may sometimes continue unabated for centuries, as in the warblers. Alternatively, the hybrid zone between the Hooded Crow and the Carrion Crow of Europe appears to be stable in width and location. These two crows interbreed freely where their ranges abut, creating a variable hybrid population. The precise location of the narrow hybrid zone has shifted slightly over the years, but early records indicate that it has been there for at least 500 years (Mayr 1963). Such long-term stability is rarely documented for avian hybrid zones.

Hybridization may take place on initial contact of two species and then stop. One such case involves a distant relative of the Mascarene White-eye, the Gray-backed White-eye of Australia (Gill 1970). This species has colonized Norfolk Island at least three times, most recently in 1904. Shortly after the third invasion, some of the Gray-backed White-eyes hybridized with the descendants of the previous invasion, which in the interim had evolved into the larger Slender-billed White-eye. But hybridization did not continue; the two white-eyes now coexist as distinct species on Norfolk Island without interbreeding.

Hybridization does not normally continue between populations with genetic, behavioral, or ecological incompatibilities because hybridization would produce inferior offspring. Like mules, the hybrids of female horses and male donkeys, almost all the hybrids of the Eastern Meadowlark and the Western Meadowlark are sterile (Lanyon 1979). Although they appear normal and healthy, 90 percent of the eggs produced by hybrids that are paired in captivity with an Eastern Meadowlark or a Western Meadowlark are infertile.

Hybrids may be able to produce viable sperm or fertile eggs, but the final test of fertility comes later when the gene combinations carried by these gametes must function correctly during early development of the fertilized zygote. Blocks of genes of one parental species recombine with the genes of the other species for the first time during meiosis and gamete formation in the F1 hybrid. Incompatible gene combinations may then disrupt the delicate process of embryo development in F2 offspring. F2 breakdown, as this phenomenon is called, is a common result of hybridization. For example, in junglefowl a large proportion of the eggs produced by female F1 hybrids are fertile, but few of the chicks hatch (Morejohn 1968). The embryos perish before hatching as a result of developmental failure caused by incompatible sets of genes.

Hybrid inferiority may also be evident in intermediate plumage or displays that render the hybrid less effective in courtship. Hybrid crosses of Anna's Hummingbirds and Costa's Hummingbirds, for example, are intermediate in many details of plumage as well as in the circular courtship flight displays characteristic of these species (Wells et al. 1978). Similarly, a hybrid cross of the Sharp-tailed Grouse and the Greater Prairie-Chicken was unable to perform bobs, bows, and foot stomps correctly and mated infrequently as a result (Evans 1966). Hybrid crosses of the Blue-winged Warbler and the Golden-winged Warbler take a few days longer than nonhybrids to secure mates (Ficken and Ficken 1968).

Production of inferior hybrid offspring may select for reduced frequency of hybridization as a result of improved discrimination by both species. Perhaps this is the reason why hybridization did not continue in the two species of white-eyes on Norfolk Island. One intriguing report of greater discrimination and reduced hybridization concerns the Baltimore and Bullock's orioles of the northern United States. In the 1950s the two kinds of orioles hybridized extensively near Crook, Colorado (Sibley and Short 1964). This evidence led to a recommendation that the two orioles be considered representatives of the same species, the Northern Oriole (AOU Checklist, Sixth ed. 1983). Twenty years later, however, hybrids were much less common. Parental forms outnumbered hybrids in the population, and most of them chose mates of their own species (Corbin and Sibley 1977). This observation of reduced hybridization pertains only to the orioles at Crook; in western Kansas they continue to hybridize freely (Rising 1983) (Figure 25–12).

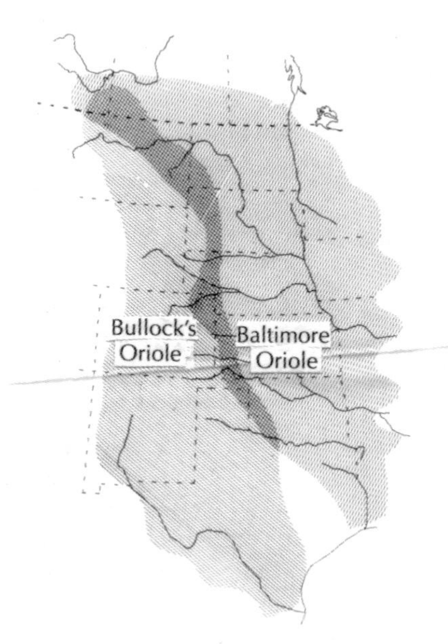

Figure 25–12 The eastern Baltimore Oriole and western Bullock's Oriole interbreed in a narrow zone of overlap in the Great Plains. (From Rising 1983)

As noted earlier, the Baltimore Oriole and Bullock's Oriole have been combined into a single species, the Northern Oriole. The Myrtle Warbler and Audubon's Warbler, which interbreed in the mountain passes of the Canadian Rockies, have also been combined into a single species, the Yellow-rumped Warbler. Decisions to lump or split species are efforts to bring existing information into accord with the biological species concept, which states

> Species are groups of interbreeding natural populations that are reproductively isolated from other such groups. (Mayr 1970, p. 28)

Species have characteristic sizes, shapes, and colors, as well as ecological niches and geographical ranges. Species may interact ecologically, but they cannot exchange genes or novel genetic-based adaptations. Species evolve independently of each other because, by definition, they are reproductively isolated from one another. The criteria involved in the definition of biological species are principles of population biology and the reproductive compatibility of individuals. The essential question from an evolutionary point of view is whether differences that have evolved between two populations are potentially subject to blending or are irreversible.

The proportions of hybrid versus parental phenotypes in a zone of overlap can serve as a criterion for judging whether or not two populations are the same species. If no hybrids are present, reproductive isolation is manifest and species status has been achieved. If hybrids appear in low frequencies, interspecific pairings are infrequent or else hybrids must be less viable than the parental forms. Thus, taxonomic distinction as species is justified. When hybrids are fairly common and blend with parental types, however, the taxonomic decision becomes more difficult. Lester Short (1969) proposed that we adopt the conservative rule that as long as parental phenotypes constitute five percent each (ten percent total) of a population, there is sufficient reproductive isolation in effect to satisfy the biological species concept.

Samples of individuals from a series of localities along a transect through zones of overlap and hybridization provide evidence of the amount of hybridization and thereby enable us to make taxonomic decisions. A variety of bird species found throughout eastern North America is replaced by closely related sister taxa in the western part of the continent. In the Great Plains alone, there are 14 such pairs of species, and of these, 11 engage in hybridization (Rising 1983). Farther north are additional cases of replacement with hybridization, such as the contact between Myrtle and Audubon warblers. As one proceeds east to west through the zones of hybridization of these taxa, the first samples include only the eastern representative of the pair, the samples from the hybrid zone consist of intermediate and variable phenotypes, and finally, the composition switches to include only the western representative. In the case of Northern Flickers (the eastern Yellow-shafted Flicker and the western Red-shafted Flicker), and in the case of Yellow-rumped Warblers (the eastern Myrtle Warbler and the western Audubon's Warbler), most individuals at certain localities in the hybrid zone are intermediate in appearance, which indicates a lack of reproductive isolation. The two flicker populations and the two warbler populations thus are now lumped into a single species, which is consistent with the biological species concept.

In contrast, hybridization is limited between Lazuli Buntings and Indigo Buntings and between Rose-breasted Grosbeaks and Black-headed Grosbeaks (Anderson and Daugherty 1974; Rising 1983). Hybrid grosbeaks, for example, never constitute more than 37 percent of local populations in South Dakota. Although no evidence of assortative mating was found, hybrid female grosbeaks lay smaller clutches (Anderson and Daugherty 1974) than pure females. In the buntings, mixed matings and hybrid buntings are both quite uncommon, and little introgression is evident. Hybrids tend to be excluded from optimal habitats and may have reduced viability (Emlen et al. 1975; Kroodsma 1975). Many pure buntings and grosbeaks persist despite some hybridization, which suggests greater reproductive isolation than is operating between the flickers or the warblers. The two grosbeaks and the two buntings act as biological species despite limited hybridization.

Studies of hybridizing populations aid the application of the biological species concept to decisions about the status of sister taxa, but the vast majority of recently isolated and divergent populations do not come into contact. They remain geographically separated, forcing ornithologists to guess what might happen if contact should be established in the future. This limitation and other concerns about practical application prompt some ornithologists to question the modern utility of the biological species concept and to recommend new approaches to the study of speciation and geographic variation in birds (Zink and Remsen 1986, McKitrick and Zink 1987).

Ecology of speciation

Divergence and reproductive isolation are the primary ingredients in the speciation process; ecological isolation is another ingredient (see Chapter 23). Incomplete reproductive isolation fosters hybridization and genetic fusion of divergent populations. Incomplete ecological isolation leads to competition and possible geographical replacement. An expanding species tends to displace the other, at least initially. Closely related, ecologically incompatible species tend to have contiguous, or parapatric, distributions. Together such species form a superspecies. Each species, called an allospecies, is reproductively isolated but is ecologically too similar to coexist with the other. In time, ecological differences may evolve and permit expansion of one species into the geographical range of the other. An overlapping distribution of reproductively isolated and ecologically compatible species is the final stage of the speciation process.

If sister taxa have evolved ecological differences while isolated, they may not compete at all when they first come into contact, and coexistence is automatic. If they have not diverged ecologically, however, they tend to compete, which may lead to the extinction of one by the other or to open confrontation. The Eastern Meadowlark and the Western Meadowlark, for example, defend mutually exclusive territories where their ranges overlap (Lanyon 1957). Dusky Flycatchers and Gray Flycatchers also defend mutually exclusive territories in areas of recent secondary contact (Johnson 1963). Such similar species may at first defend mutually exclusive territories, but with time one of the competitors is likely to exclude the other from optimal habitats, restricting it to suboptimal habitats (Murray 1971). The weaker species either adapts to that habitat or it remains vulnerable to local extinction (Jaeger 1980). Alternatively, the subordinate species may cease trying to defend territories against the dominant species and adopt a less conspicuous, nonterritorial breeding behavior. Both solutions could promote coexistence in one region. Coexistence of LeConte's Sparrows and Sharp-tailed Sparrows in the marshes of North Dakota seems to be based on just this kind of resolution of their ecological interactions (Murray 1969, 1971). LeConte's Sparrows are territorial; Sharp-tailed

Sparrows are not. Territorial male LeConte's Sparrows attack Sharp-tails when they first arrive in the spring, but the Sharp-tails remain, nonaggressively sharing both marsh and song perches with the aggressive species.

Geographical replacement

The case of Blue-winged Warblers and Golden-winged Warblers brings together both the hybridization and the ecological factors of speciation. The Blue-winged Warbler was once an uncommon species of the south central United States. Its sister species, the Golden-winged Warbler, occurred farther north and at higher altitudes in the Appalachians; the ranges of the two species were well separated (Short 1963). Both warblers benefitted from the clearing of forests and the increases in second growth throughout the northeastern United States that occurred in the mid-1800s. The Blue-winged Warbler expanded northward into the range of the Golden-winged Warbler (Figure 25–13).

The two warblers have strikingly different color patterns. Blue-wings are bright yellow with white wing bars and a narrow black line through the eye. Golden-wings are gray above, white below, with yellow wing bars and crown, and bold, black patches on the throat and the eyes. The contrasting facial color patterns of the two species have a simple genetic basis, analogous to the color phases of Bananaquits. The plain throat and narrow black eye line of the Blue-wing is dominant to the black throat and black eye patch of the Golden-wing. Other plumage color characteristics are controlled by several additive genes. First-generation hybrids, called Brewster's Warblers, inherit the genetically dominant Blue-wing face pattern but are intermediate between the two parental species in other aspects of plumage coloration. These fertile hybrids produce viable offspring when they mate with either Blue-wings, Golden-wings, or other hybrids. Introgressive blending of plumage colors coupled to either facial color pattern

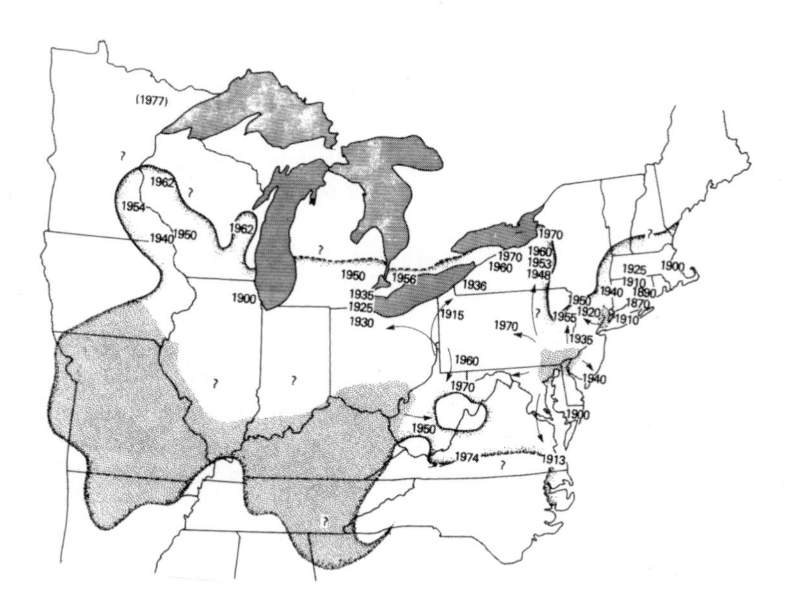

Figure 25–13 Distribution and spread of the Blue-winged Warbler in the last century. Dates on the map indicate when Blue-wings first established themselves at that locality. Range boundaries and arrows indicating patterns of spread are hypotheses based on historical information. Stippled area indicates the range of Blue-wings in the 1800s. Stippled area is approximate. (From Gill 1980)

produces a variety of hybrid types, including Lawrence's Warbler, which looks like a yellow Blue-wing with a bold, black Golden-winged Warbler facial pattern.

Blue-winged Warblers generally replace Golden-winged Warblers within 50 years of local contact (Gill 1980). The Golden-winged Warbler faces widespread extinction as a result. Blue-winged Warblers use a broad range of habitats, including those of the Golden-winged Warbler, which usually inhabits early stage succession habitats (Confer and Knapp 1981). These two warblers apparently have not achieved the ecological compatibility needed for coexistence. Hybridization may hasten the extinction of the rarer Golden-wing species, but ecological forces apparently will be the major determinant of its fate.

The steady replacement of Golden-wings by Blue-wings causes a predictable shift in the composition of local assemblages of these warblers. At first only Golden-wings are present, with perhaps an occasional pioneering Blue-wing or an odd hybrid. As Blue-wings move in, one finds balanced proportions of the two extreme forms plus an assortment of intermediate hybrids. Finally, as the number of Golden-wings declines and Blue-wings continue to increase, the distribution shifts in favor of a majority of Blue-wings plus various hybrids that are the remnants of previous hybridization. In the last phase, only Blue-wings remain, perhaps with a few individuals of mixed blood. The same dynamic may be at work in Baltimore and Bullock's Orioles in the western Great Plains. Baltimore Orioles are expanding their range westward in northern Nebraska at the expense of the Bullock's Oriole. Elsewhere, Black-crested Titmice are expanding their range in Texas at the expense of Tufted Titmice and are doing so with a moving front of local hybridization. Likewise, Indigo Buntings are now replacing Lazuli Buntings in northern Nebraska and in western Kansas (Rising 1983).

Behavior and speciation

We return now to the phenomenon of speciation in birds in terms of the broad questions posed at the beginning of this chapter. Which features catalyze speciation? The population structure based on large demes that are connected by much gene flow is less susceptible to speciation than are the small, partially isolated demes characteristic of less mobile vertebrates. Yet birds overcome this apparent handicap and speciate prolifically.

The behavioral attributes of birds, particularly their capacity for new behavior and its cultural transmission, may be extraordinary advantages (Wyles et al. 1983; West-Eberhard 1983). Behavior, rather than the environment, can be the driving force of evolutionary change when individuals exploit the environment in new ways, followed by rapid spread of the new habit through the population by cultural transmission, followed finally by the evolution of anatomical traits that enhance the effectiveness of individuals that practice the new habit. The final step of the anatomical or physiological reinforcement of an adaptive shift is well illustrated in the evolution of island birds (see Chapter 24). New behaviors that ultimately spawn anatomical change are more likely to arise in large populations of individuals that have the intelligence to develop such behavior. Such populations, as well as cultural transmission of new habits, are characteristic avian traits (see Chapter 11).

It is appropriate that the birds that inspired Darwin's thinking about the origin of species, the finches of the Galapagos Islands, provide perhaps the best modern examples of the speciation process in birds. Despite the fact that these finches live on small oceanic islands, they have moderately large populations united by gene flow (Barrowclough 1983). Regular interisland exchange of individuals at low frequencies sustains the genetic coherence of the populations. Speciation apparently takes place when rapid genetic change in small colonizing founder populations is followed by rapid population growth and adaptive divergence. Rather little overall genetic divergence has occurred among these finches. Nevertheless, their evolutionary history is marked by several episodes of strong selection for changes in plumage or skeletal characteristics. The drought of 1976, for example, resulted in severe natural selection of bill sizes (page 130). These finches are a prime example of adaptive radiation of bill sizes, associated feeding habits, and behavioral innovations (see Chapter 7). Periodic and stringent sorting of individuals from such populations with new behaviors and anatomical variations would promote the evolution of new species of finches and is precisely the process we envision for speciation in birds generally.

New behavior, with its subsequent cultural transmission, is one catalyst in avian speciation. Social selection appears to be another (West-Eberhard 1983). Attributes associated with communication and competition among individuals for mates, space, or access to food may ensure that some individuals survive or reproduce while others do not (see discussion of sexual selection in Chapter 14). Attributes that are favored in social competition, such as the extraordinary display plumages of male birds-of-paradise, can be included in a process of evolutionary elaboration in habitats that lack severe limits. Such attributes may not be adaptive in other contexts and, in fact, may prove to be a liability when they render the bird conspicuous to predators or desirable to humans: Thousands of birds-of-paradise are killed every year by New Guinea tribesmen for head ornaments. Songs also are subject to elaboration through vocal contests. The interesting feature of this process with respect to speciation is that, through ritualization, these same attributes often enable pair formation, species recognition, and reproductive isolation. Thus, social selection in the form of contests, or group coherence coupled with behavioral flexibility, could be a force in the speciation process. Charles Darwin (1871) recognized the importance of this force, but only very recently have ornithologists started to appreciate its implications.

Summary

New species evolve from the genetic divergence of isolated populations. Speciation in birds usually proceeds, ornithologists believe, via the gradual divergence of large isolated populations adapting to different environments or via a rapid reorganization of some of the genes of birds in small, isolated populations. The evolution of geographical differences among natural bird populations depends on the relative strength of two opposing forces: the intensity of natural selection favoring one genetic attribute over another and the rate of genetic blending as a result of interbreeding of individuals from different locations.

The classification of species is guided by the biological species concept, which states that a species comprises a population or set of populations that are capable of successfully interbreeding under natural conditions. Secondary contact, the reuniting of previously isolated populations, tests the ability of populations to interbreed. Once considered separate species, the Audubon's Warbler and the Myrtle Warbler interbreed freely where they come into contact in the Canadian Rockies. On this evidence they are now considered to be populations of the same species. Many other species that interbreed do not produce viable hybrids. Hybrids resulting from interspecific breeding have one of several selective disadvantages: They may be unable to produce viable sperm and eggs, they may have incompatible blocks of genes that disrupt early embryo development, or they may suffer from intermediate plumage and display behaviors that reduce their reproductive success.

Birds are prone to rapid speciation. The capacities of birds to develop new behaviors and to learn new behaviors from each other may contribute to this process. Behavior, rather than the environment, can be the driving force of evolutionary change if a new behavior is followed by the evolution of new anatomical traits that support the behavior.

Further readings

Barrowclough, G. 1983. Biochemical studies of microevolutionary processes. *In* Perspectives in Ornithology (A.H. Brush and G.A. Clark, Jr., eds.), pp. 223–261. Oxford: Oxford University Press. *An excellent review of new methodologies and their meaning.*

Cracraft, J. 1983. Species concepts and speciation analysis. Current Ornithology 1:159–188. *A provocative challenge to the classic viewpoint.*

Mayr, E. 1970. Populations, Species, and Evolution. Cambridge, Mass.: Belknap Press. *A classic by the leading proponent of geographical speciation in birds.*

Selander, R.K. 1971. Systematics and speciation in birds. Avian Biology 1:57–147. *A detailed review of the old literature on avian speciation.*

West-Eberhard, M.J. 1983. Sexual selection, social competition, and speciation. Quarterly Review of Biology 58:155–183. *A vital new perspective.*

Appendixes

1
2
3
4
5
♂
COE

appendix I

Birds of the world

This appendix introduces the major groups, or taxonomic orders, of extant birds of the world. Included for each order are

1. illustrations by James Coe of some representative species;

2. summaries of general characteristics and distinguishing features;

3. component families with their geographical distribution, plus the number of genera and species in each; and

4. current thought about evolutionary relations.

The first three chapters of this book introduced the characteristics of birds and their evolution, as well as the principles of classification and phylogeny. This introductory material should be reviewed before proceeding to the technical presentations of this appendix.

As emphasized in Phylogeny (Chapter 3), no classification of birds can be presented as gospel; rather, classification serves as a framework for discussion and future discovery. In most cases the relationships among orders are unknown, and, therefore, the sequence in which orders are listed is based on tradition. For the most part, I use the highly regarded, traditional classification of Robert W. Storer (1971b). I also have tried to reflect recent advances, particularly the genetic analyses of evolutionary relationships using biochemical techniques. In some cases, I adopt a more conservative presentation than Storer's. For example, I treat sandgrouse (Pteroclidi-

Figure A–1 Ratites and tinamous: (1) Elegant Crested Tinamou (Tinamidae); (2) Australian Cassowary (Casuariidae); (3) Brown Kiwi (Apterygidae); (4) Greater Rhea (Rheidae); (5) Ostrich (Struthionidae). Accompanying text is on page 511.

formes) and turacos (Musophagiformes) as separate orders, as does Karl Voous (1973), to await resolution of their phylogenetic affinities. I also treat many groups of passerine birds as separate families, so that in the future, they can be grouped properly in accordance with new evidence. For example, I treat the current subfamilies of Muscicapidae as separate families (e.g., Old World warblers as Sylviidae, thrushes as Turdidae). Further, I group the families of oscine passerine birds as recommended by Charles Sibley and Jon Ahlquist (1985) based on comparisons of DNA. Even though the relationships among passerine birds suggested by Sibley and Ahlquist are still fresh and under discussion, I believe them to be more firmly based than those of any previous classification. By and large, I have not incorporated Sibley and Ahlquist's (1985c) and Sibley et al.'s (1988) analyses of relationships among nonpasserine birds for two reasons: These analyses are decidedly more controversial than their analyses of passerine birds, and the work was still in progress when I wrote the final manuscript of this appendix. Nor have I adopted the controversial nomenclature and hierarchical structure of Sibley and Ahlquist's classification; instead, I preserve familiar traditional names standardized at the family level.

As a guide to the recent technical literature and the decisions involved in my classification, I include taxonomic comments on each order of nonpasserine birds and on major subgroups of passerine birds. These comments may be of particular interest to advanced students and to faculty who teach courses that focus on avian systematics. A more detailed analysis of the evidence supporting avian relationships is beyond the scope of this book. I have not included technical anatomical diagnoses unless adequate background is provided in the main body of the book. I could, for example, have mentioned that the thigh muscle formula of the Caprimulgiformes was AYZ except for the Steatornithidae, for which the thigh muscle formula is AZ; but this fact would mean little without a detailed presentation of the thigh muscles and their configurations. Sibley and Ahlquist (1970, 1972) provide excellent summaries of the technical diagnoses available in the historical literature.

Tallies of species and genera are taken directly from Walter Bock and John Farrand (1980), with a few exceptions as noted.

Birds: Class Aves

Birds are tetrapod vertebrates with feathers. Feathers alone distinguish birds from all other vertebrates. Feathers evolved from scales, and birds have reptilelike scales on their legs and feet. Although the skeleton of birds retains many reptilian features, it is modified for flight. The bones and feathers of the forelimb, particularly, are transformed into wings capable of powerful, flapping flight. Birds maintain a high body temperature through endothermic heat production, and have a large, four-chambered heart that supports the demands of sustained activity and high metabolism.

The Class Aves includes two superorders: (1) the Palaeognathae, which includes the ratites and tinamous, and (2) the Neognathae, which includes all other modern birds (Cracraft 1986). The complete classification of living birds is a hierarchical arrangement of 30 orders, 174 families, 2044 genera, and 9021 species.

Ratites and tinamous
Superorder Palaeognathae

The ratites and tinamous are an ancient group of relict families that share a unique configuration of bones between the nasal passages (the palaeognathous palate) and, except for the tinamous, are flightless birds with reduced wing bones and a sternum that lacks a keel, the bony plate that serves as an anchor for flight muscles in most birds. Tinamous have a reduced keel.

The larger ratites, the Ostrich, rheas, emus, and cassowaries, are long-necked birds with strong, muscular legs adapted for running. All are terrestrial, and the number of toes is reduced to two in the Ostrich and to three in the rheas, emus, and cassowaries. The kiwis and tinamous, the smaller members of the group, still have four toes.

In the Ostrich, rheas, emus, and cassowaries, the bill is relatively flat and the nostrils are oval; in the kiwis the bill is long and slightly decurved, and the nostrils are located at the tip; in the tinamous the bill is more chickenlike, and the nostrils are covered with a fleshy cere.

Most members of the superorder have distinct feather tracts, but the Ostrich has feathers distributed continuously over its body. The plumage is lax in all groups except the tinamous, and hairlike in the kiwis. An aftershaft is strongly developed in emus and cassowaries, absent in the Ostrich, rheas, and kiwis, and small to well-developed in tinamous. Kiwis lay enormous eggs and tinamous lay glossy, heavily pigmented, often brightly colored eggs. Young ratites and tinamous are clad in down at hatching.

Order	Family	Members	Distribution	Genera	Species
Tinamiformes	Tinamidae	Tinamous	Neotropics	9	47
Rheiformes	Rheidae	Rheas	Neotropics	2	2
Struthioniformes	Struthionidae	Ostriches	Africa	1	1
Casuariiformes	Dromiceidae	Emus	Australia	1	2
	Casuariidae	Cassowaries	Australia, New Guinea	1	3
Dinornithiformes	Apterygidae	Kiwis	New Zealand	1	3

Taxonomic Comments The monophyly of the ratites now seems well established by morphological, biochemical, and chromosomal evidence (see Bock 1963; Parkes and Clark 1966; Sibley and Frelin 1972; Cracraft 1974a; Prager et al. 1976; de Boer 1980; Sibley and Ahlquist 1981a; Stapel et al. 1984).

Grouping the ratites together in the Superorder Palaeognathae reflects their common ancestry and distinguishes them from other modern birds. Classifying the ratites in five separate orders emphasizes the fact that each of these groups is ancient and highly differentiated.

1
2
3
4
5
COE

Figure A-2 Grebes: (1) Eared Grebe; (2) Horned Grebe; (3) Great Crested Grebe; (4) Pied-billed Grebe; (5) Western Grebe.

Grebes
Superorder Neognathae
Order Podicipediformes

Grebes are medium-sized, foot-propelled diving birds with stocky bodies, slender necks, and small heads. The toes are lobed and the tarsi are laterally compressed so that they offer little resistance when drawn forward through the water. The legs are located far back on the body, making it difficult for the birds to move on land but giving them a powerful forward thrust during dives. The claws on the toes are unusual in being flattened like finger nails.

The plumage of grebes is dense and satiny and is waterproofed by secretions from a feathered oil gland above the base of the tail. The tail feathers are reduced, so that grebes appear tailless, a feature that contributes to their stocky appearance.

In most species, the bill is slender and pointed, but in the Pied-billed Grebe of the New World the bill is stouter, an adaptation to this bird's diet of hard-shelled crustaceans.

Like many birds that are highly adapted to diving, grebes have difficulty taking off from the water, and, with rapidly beating wings, they must taxi across the surface before becoming airborne.

The young are clad in fine down at hatching; in most species, the down is handsomely striped in black and white, but in the Western and Clark's grebes of North America, the down is plain gray.

Most grebes build floating platforms of aquatic weeds on which they mate and lay their eggs. They carry their young on their back, even when diving. Grebes eat their own feathers, which then trap fish bones in the stomach, holding the bones for prolonged digestion or regurgitation. Elaborate courtship displays include fancy dives and rises out of the water, and prolonged rushes side by side.

Family	Members	Distribution	Genera	Species
Podicipedidae	Grebes	Worldwide	6	21

Taxonomic Comments The affinities of the grebes remain uncertain (Storer 1960, 1986), though traditionally and, probably, incorrectly, they have been placed next to loons. Species relationships are reviewed by Storer (1963).

1
2
3
4
CoE

Penguins
Superorder Neognathae
Order Sphenisciformes

Figure A–3 Penguins: (1) Chinstrap Penguins; (2) Rockhopper Penguin; (3) Cape Penguin; (4) King Penguin, juvenile (left), adult (right).

The penguins form a distinctive order of flightless, marine diving birds of the southern oceans. The most striking adaptation of these birds involves the wings. The bones of the wing are flattened and somewhat fused so it cannot be folded as in other birds; the result is a very efficient flipper, which is the principal means of locomotion under water. Because the wing is still used for locomotion, the keel of the sternum is well developed unlike the sternum of the flightless ratites. At the base of the wing is a complex network of blood vessels, a vascular rete, in which cooled blood returning to the body from the wing absorbs heat from blood flowing outward into the wing. The result is that heat is retained in the body and not lost during the inevitable chilling of the wings.

The legs are short, and the stout, webbed feet are located far back on the body, giving the birds an upright posture when they are on land; unlike many other diving birds, penguins can walk easily onshore.

The plumage of penguins is dense and waterproof; the feathers are distributed continuously over the body rather than in discrete tracts as in most birds. The body is further insulated from cold water by a heavy layer of fat beneath the skin.

The bill of most species is somewhat laterally compressed, but in the largest penguins, the King and Emperor, the bill is long and slender. Most penguins are clad in black and white, and often have distinctive head patterns. The young are clad in dense down at hatching. The order is confined to the Southern Hemisphere, but reaches the Equator in the Galapagos Islands.

Family	Members	Distribution	Genera	Species
Spheniscidae	Penguins	Southern oceans	6	18

Taxonomic Comments Penguins evolved from flying procellariiform ancestors (Simpson 1946; Sibley and Ahlquist 1972; Ho et al. 1976). The presence of tubular nostrils in both fossil penguins and in the Fairy Penguin, considered the most primitive living member of the order, supports this relationship. The two groups also have similar bill-fencing courtship displays and methods of feeding their young.

1
2
3
4
COE

Tube-nosed seabirds
Superorder Neognathae
Order Procellariiformes

Figure A–4 Tube-nosed seabirds: (1) Wandering Albatross (Diomedeidae); (2) Cape Petrel (Procellariidae); (3) Wilson's Storm-Petrel (Hydrobatidae); (4) Short-tailed Shearwater (Procellariidae).

The Procellariiformes consist of several families of pelagic birds, most of which seek food by flying over the surface of the sea. They range in size from the six-inch Least Storm-Petrel to the huge Wandering Albatross, with a wingspan of nearly 12 feet. All have tubular nostrils and a distinctly hooked bill. The plumage is dense and waterproof and beneath the outer feathering is a dense coat of down. In the roof of the orbit is a large gland that concentrates and excretes salt in drops from the bill. Most tube-noses have rather long wings, held stiffly and used for soaring and planing over the waves, but the storm-petrels flutter close to the surface on shorter wings, and the diving-petrels of the southern hemisphere have very short wings, which they use for locomotion under water. Diving petrels are convergent in morphology and ecology with the unrelated auklets of the northern hemisphere. Tube-nosed seabirds are clad in black, white, brown or gray, and show little color except in the bill and feet of some species. The young are densely downy. Tube-noses have a well-developed sense of smell, which some species use to find their nest burrows at night. Both parents and young squirt foul-smelling stomach oil at intruders. The order is distributed throughout the oceans and seas of the world, but a majority of genera and species are found in the southern hemisphere.

Family	Members	Distribution	Genera	Species
Diomedeidae	Albatrosses	North Pacific, southern oceans	2	13
Procellariidae	Shearwaters, petrels, fulmars	All oceans	12	66
Hydrobatidae	Storm-petrels	All oceans	8	21
Pelecanoididae	Diving-petrels	Southern oceans	1	4

Taxonomic Comments The Procellariiformes and Sphenisciformes are closely related (Simpson 1946; Ho et al. 1976). In turn, skeletal similarities between albatrosses and frigatebirds, the most primitive members of their respective orders, link the Procellariiformes and Pelecaniformes (Cracraft 1981).

♂
♂
1
2
3
4
5
6
COE

Pelicans and allies
Superorder Neognathae
Order Pelecaniformes

Figure A–5 Pelicans and allies: (1) Red-footed Booby (Sulidae); (2) Magnificent Frigatebird (Fregatidae); (3) White-tailed Tropicbird (Phaethontidae); (4) American Anhinga (Anhingidae); (5) Brown Pelican (Pelecanidae); (6) Guanay Cormorant (Phalacrocoracidae).

The varied Order Pelecaniformes comprises several groups of large or medium-sized, aquatic birds that eat fish or squid. They differ from all other birds in having totipalmate feet, with all four toes joined by webs. Most have a more or less distensible pouch of bare skin between the branches of the lower mandible; this gular pouch is absent in the tropicbirds and most highly developed in the pelicans, which use it to capture fish, and in the frigatebirds, in which it is inflated by males and used in courtship displays. In the cormorants, boobies, and gannets the external nostrils are closed, and the birds breathe through the mouth. The bill is hooked in cormorants and frigatebirds, pointed and spear-shaped in tropicbirds, gannets, boobies, and snakebirds, and long and pouched in the pelicans. In all except the tropicbirds, the young are without down at hatching. The plumage of many members of the order is black and white, and others are clad in somber black or dark brown. The feet and bare areas of facial skin may be brightly colored. The distribution of the order is worldwide.

The diversity of the Pelecaniformes reflects the various means the birds employ to obtain fish or squid. In the boobies, gannets, and tropicbirds, prey is caught in dives from the air. Cormorants and snakebirds pursue fish under water, cormorants seizing their prey with their hooked bills, and snakebirds spearing it. Pelicans dive and scoop fish up in their pouched bills, allowing water to drain out before swallowing their catch. Frigatebirds snatch food from other fish-eating birds. Nearly all members of the order nest in colonies, those of the gannets numbering in hundreds or thousands of pairs. One to three eggs are laid on bare ground or among rocks (tropicbirds, some boobies and cormorants) or in a nest of sticks, reeds, seaweed, or guano. The young are helpless at hatching and are tended by both parents. Tropicbirds, frigatebirds, boobies, and gannets are exclusively marine, cormorants and pelicans inhabit both salt or fresh water, and snakebirds are confined to fresh water.

Family	Members	Distribution	Genera	Species
Phaethontidae	Tropicbirds	Pantropical	1	3
Fregatidae	Frigatebirds	Pantropical	1	5
Sulidae	Boobies and gannets	All oceans	3	9
Phalacrocoracidae	Cormorants	All continents and oceans	3	33
Anhingidae	Snakebirds, or anhingas	Pantropical	1	4
Pelecanidae	Pelicans	All continents	1	8

Taxonomic Comments The Pelecaniformes are thought to be related to the Procellariiformes and also to the Ciconiiformes (Cracraft 1984; Sibley and Ahlquist 1985a). They are generally grouped into three suborders: the Phaethontes (tropicbirds), the Fregatae (frigatebirds), and the Pelecani (boobies, gannets, cormorants, snakebirds, and pelicans).

♂
1
♂
2
3
♂
4
5
6
COE

Waterfowl
Superorder Neognathae
Order Anseriformes

Figure A–6 Waterfowl: (1) Musk Duck (Anatidae); (2) Smew (Anatidae); (3) Black-necked Swan (Anatidae); (4) Mallard (Anatidae); (5) Pied Goose (Anatidae); (6) Horned Screamer (Anhimidae).

The Order Anseriformes contains two rather distinctive families that share few external features. In both, the aftershaft is reduced or absent, the oil gland is feathered, and the young are clad in down at hatching. Internally, they share characters of the skull, sternum, and syrinx. The ducks, geese, and swans are a diverse group of mainly aquatic birds that have webbed feet with a hind toe that is somewhat elevated, a flattened, blunt-tipped bill covered with a thin layer of skin and bearing a nail at the tip of the maxilla and fine lamellae along the margins of the maxilla and mandible, pointed wings, and a dense coat of firm, waterproof feathers, in distinct tracts, with a layer of down beneath; males have a penis. Many of these birds are colorful, and the sexes are often patterned differently.

Screamers have long, slender toes that bear only rudimentary webs, a hind toe on the same level as the front toes, a short, slightly hooked bill, rounded wings, and a continuous coat of feathers. They have stout spurs at the bend of the wings. The males do not have a penis, and, unusual among birds, screamers lack uncinate processes, bones that strengthen the rib cage. They are patterned in grays and browns, and the sexes are alike. The most peculiar feature of the screamers is a skin filled with small bubbles of air about a quarter of an inch thick, which produce a crackling sound when pressed; the function of this layer of air bubbles is unknown.

Waterfowl vary in ecology from terrestrial grazers to deepwater divers to agile riders of ocean surfs and mountain streams. Bill form varies with diet from those with lateral lamellae for straining microscopic food from mud, to strong, broad bills for wrenching molluscs from rock moorings, to narrow bills with sharp tooth-like serrations for capturing fish, to short, blunt bills for grazing field plants.

Family	Members	Distribution	Genera	Species
Anhimidae	Screamers	South America	2	3
Anatidae	Swans, geese, and ducks	Worldwide	43	147

Taxonomic Comments Waterfowl traditionally have been regarded as relatives of the Galliformes (Johnsgard 1968; Prager and Wilson 1976). An Eocene fossil, *Presbyornis,* intermediate between ducks and shorebirds, suggests that the relationships of the Anseriformes may lie with the Charadriiformes instead (Olson and Feduccia 1980b).

Rather than retaining primitive, galliformlike features, as was once thought, the turkeylike screamers may be highly specialized waterfowl (Olson and Feduccia 1980b). The Pied Goose of Australia, however, retains primitive characters and represents an early lineage (Delacour and Mayr 1945).

See Livezey (1986) for a modern and comprehensive taxonomic revision of waterfowl.

Flamingos
Superorder Neognathae
Order Phoenicopteriformes

Figure A–7 Flamingos: Greater Flamingos.

The flamingos are distinguished by their long necks and legs with webbed feet, generally pink coloration, and the highly specialized structure of their bills, which are bent downward in the middle. The margins bear long lamellae for filtering small organisms out of mud or water. The maxilla fits within the mandible, and during feeding the bill is placed in the water with the maxilla downward; the tongue is thick and fleshy and is used to circulate water between the lamellae. The young are clad in whitish down at hatching, and their bills gradually assume the specialized adult shape. Flamingos breed in dense colonies and construct tall cone-shaped nests of mud.

Family	Members	Distribution	Genera	Species
Phoenicopteridae	Flamingos	All continents except Australia	3	6

Taxonomic Comments The relationships of the flamingos remain controversial. Traditional debate has been over the question of whether they are closest to the Ciconiiformes or to the Anseriformes (see Sibley and Ahlquist 1972). Studies of DNA support a relationship with the Ciconiiformes (Sibley and Ahlquist 1985a). Olson and Feduccia (1980a) contend that the flamingos are related to the stilts (Recurvirostridae) and recommend placing them in the Charadriiformes.

1
2
3
4
5
6
COE

Figure A-8 Herons, storks, and allies: (1) Eurasian Spoonbill (Threskiornithidae); (2) White Stork (Ciconiidae); (3) Sacred Ibis (Threskiornithidae); (4) Great Blue Heron (Ardeidae); (5) Hammerhead (Scopidae); (6) Whale-headed Stork (Balaenicipitidae).

Herons, storks, and allies
Superorder Neognathae
Order Ciconiiformes

Most Ciconiiformes are long-legged, long-necked birds that wade in shallow water or, in some cases, feed on open ground. The various families differ in the shape of the bill: herons have rather long, spear-shaped bills; storks have bills that are usually straight and sharp, but sometimes have a slight curve at the tip; ibises have long, curved bills; spoonbills have long bills with a broad, flattened, spoon-shaped tip; the Whale-headed Stork has a massive, bulbous bill with a hooked tip; and the Hammerhead has a more slender bill, also somewhat hooked. Aside from these obvious external features, the families that make up this order share few features. The herons have powder-downs, feathers that disintegrate and are used to condition the rest of the plumage, a comblike margin on the claw of the middle toe, and a modification of the vertebrae of the neck that permits these birds to fold the neck into an S-shaped curve. The Whale-headed Stork also has a tract of powder-downs and a shorter neck than most other members of the order. The Hammerhead has a slender crest, which roughly matches the length of the bill and is responsible for the species' English name. Ibises and spoonbills, despite the difference in the shape of their bills, show their relationship in a number of features, the most noticeable of which is a pair of grooves that extend from the nostrils to the tip of the bill. The storks lack a syrinx, and are well known for clattering their bills. Members of the order range in size from the least bitterns *(Ixobrychus)*, about a foot tall, to the Goliath Heron of Africa and the marabou storks *(Leptoptilos)*, the latter attaining a length of five feet.

Family	Members	Distribution	Genera	Species
Ardeidae	Herons, bitterns	Worldwide	16	62
Balaenicipitidae	Whale-headed Stork	Africa	1	1
Scopidae	Hammerhead	Africa, Madagascar	1	1
Threskiornithidae	Ibises, spoonbills	Pantropical; a few in temperate regions	20	33
Ciconiidae	Storks	Pantropical; temperate Eurasia	6	17

Taxonomic Comments The Ciconiiformes may be a polyphyletic group. The Whale-headed Stork and Hammerhead may belong with the Pelecaniformes (Cottam 1957; Sibley and Ahlquist 1985a), though Cracraft (1986) disagrees. Herons may belong to the Gruiformes (Olson 1978a). Storks may be related to the New World vultures (Ligon 1967; Sibley and Ahlquist 1985a).

1
2
3
4
5
6

Figure A–9 Birds of prey:
(1) Cooper's Hawk (Accipitridae);
(2) Osprey (Pandionidae);
(3) Harpy Eagle (Accipitridae);
(4) Peregrine Falcon (Falconidae);
(5) Andean Condor (Cathartidae);
(6) Secretary Bird (Sagittariidae).

Birds of prey
Superorder Neognathae
Order Falconiformes

Most members of the Order Falconiformes are diurnal, raptorial birds with short, strongly hooked bills with a fleshy cere containing the nostrils, and sharp, curved talons. The legs are generally short, but are very long in the largely terrestrial Secretarybird of Africa, which also has blunt claws. The wings vary considerably in shape: broad and rounded in eagles and many hawks, long and narrow in harriers, and pointed in falcons and many kites. The tails of diurnal raptors may be broad and fan-shaped, or long and narrow; in the Secretarybird, the two central tail feathers are greatly elongated. The Osprey is distinguished from other members of the order by having a reversible outer toe and sharp spicules on the underside of the foot, both adaptations for holding slippery fish. The typical falcons *(Falco)* have a bony tubercle in the nostril. In most families in the order, females are larger than males. The New World vultures differ from other Falconiformes in having perforate nostrils and in lacking a syrinx; like the Old World vultures, which are members of the Accipitridae, the New World vultures have bare heads. Falconiform birds are patterned in black, white, brown, rufous, and gray, and some members of the order have color phases, often including an all-black form. They range in size from the Philippine Falconet six inches long, to the Andean Condor the largest flying bird in the world, with a length of over four feet and a wingspan of nine and a half feet.

Family	Members	Distribution	Genera	Species
Cathartidae	New World vultures	North and South America	5	7
Pandionidae	Ospreys	Worldwide	1	1
Accipitridae	Hawks, eagles, kites	Worldwide	64	217
Sagittariidae	Secretarybird	Africa	1	1
Falconidae	Falcons, caracaras	Worldwide	10	62

Taxonomic Comments The monophyly of the Falconiformes is not certain, and the relationships of its member families are still debated (Jollie 1953; Cracraft 1981). New World vultures may be related to storks (Ligon 1967; Sibley and Ahlquist 1985a).

1
2
♂
3
♂
4
5
♂
CoE

Fowllike birds
Superorder Neognathae
Order Galliformes

Figure A–10 Fowllike birds: (1) Lady Amherst Pheasant (Phasianidae); (2) Red Junglefowl (Phasianidae); (3) Great Curassow (Cracidae); (4) Vulturine Guineafowl (Numididae); (5) Sage Grouse (Phasianidae).

The Galliformes are medium-to-large terrestrial birds with short, rounded wings, a well-developed keel, and sturdy legs with four toes. In the Phasianidae and Numididae the hind toe is elevated and not in contact with the ground, but in the Megapodiidae and Cracidae, the hind toe is on the same level as the ground. The bill is short and more or less conical in most species, with an arched culmen and with the tip of the maxilla overlapping the mandible. All members of the order have a large, muscular gizzard, a well-developed aftershaft, and large intestinal caeca. Many members of the Phasianidae have spurs on the tarsus, and in the grouse, the tarsi, and sometimes the toes, are feathered. The turkeys have bare heads, ornamented with wattles. Most galliform birds are cryptically colored, patterned in black, gray, and brown, but among the true pheasants, the males and sometimes the females are clad in reds, yellows, silver, and other bright colors; in the peacocks, large members of the pheasant group, the upper tail coverts are greatly lengthened and bear large, iridescent eyespots; these great trains are erected during display. Intricate patterns are also found in some of the New World quails and in some of the small quails of the Old World. In most members of the order, the young are clad in down at hatching, but in the Megapodiidae, the young are fully feathered and capable of flight when they emerge from their mound nests.

Family	Members	Distribution	Genera	Species
Cracidae	Curassows, guans, chachalacas	Neotropics	8	44
Megapodiidae	Moundbuilders	Australasian region	7	12
Numididae	Guineafowl	Africa	5	7
Phasianidae	Pheasants, quail, grouse, turkeys	Nearly worldwide	58	205

Taxonomic Comments The Galliformes are usually considered to be most closely related to the Anseriformes and secondarily to the Falconiformes (Cracraft 1981; Olson and Feduccia 1980b; Sibley and Ahlquist 1985a).

The Hoatzin (Opisthocomidae), often placed in the Galliformes, has been assigned to the Cuculiformes (Sibley and Ahlquist 1973; but see Brush 1979; Cracraft 1981).

1
2
3
4
♂
5
COE

Cranes, rails, and allies
Superorder Neognathae
Order Gruiformes

Figure A–11 Cranes, rails, and allies: (1) Sunbittern (Eurypygidae); (2) Crowned Crane (Gruidae); (3) Water Rail (Rallidae); (4) Gray-winged Trumpeter (Psophiidae); (5) Great Bustard (Otididae).

The Order Gruiformes is an old, widely dispersed, and diverse group of small-to-large birds with few unifying characters. No member of the order has a crop, and most share certain skeletal and palatal features. An oil gland is present in most families but absent in the bustards. Most are terrestrial, aquatic, or marsh-dwelling birds with strong, unwebbed or only slightly webbed toes, though the aquatic sungrebes and the coots (Family Rallidae) have lobed toes. The condition of the hind toe varies: it is lacking, for example, in the terrestrial bustards and the buttonquails, elevated in rails, trumpeters, and roatelos of Madagascar, the seriemas of South America, and the cranes, and on the same level as the front toes in the Sunbittern and the Limpkin. The nostrils are pervious in rails, sungrebes, the Sunbittern, cranes, seriemas, and bustards, and covered with an operculum in the Kagu. The young of all Gruiformes are downy, and leave the nest soon after hatching.

Except for the large, stately cranes and the heavy (up to 18 kg) bustards, most gruiform birds are secretive and little known. Rails, which comprise the majority of species in the order, are heard more often than they are seen.

Family	Members	Distribution	Genera	Species
Rallidae	Rails, coots	Worldwide	53	142
Heliornithidae	Sungrebes	Pantropical	3	3
Rhynochetidae	Kagu	New Caledonia	1	1
Eurypygidae	Sunbittern	Neotropics	1	1
Mesoenatidae	Roatelos	Madagascar	2	3
Turnicidae	Button-quail	Tropical and warm temperate parts of Old World	2	14
Gruidae	Cranes	All continents, except South America	4	15
Aramidae	Limpkin	Neotropics	1	1
Psophiidae	Trumpeters	South America	1	3
Cariamidae	Seriemas	South America	2	2
Otididae	Bustards	Old World	11	24

Taxonomic Comments The boundary between the Gruiformes and the Charadriiformes, the most closely related order, is vague (Cracraft 1973b). The jacanas are intermediate, now placed in the Charadriiformes. The Plains-wanderer (Pedionomidae), once included in this order, belongs in the Charadriiformes (Olson and Steadman 1981; Sibley and Ahlquist 1985a).

Cracraft (1973b, 1981) reviewed relationships among the gruiform families. Skeletal characters link the trumpeters, Sunbittern, and Kagu. DNA analysis suggests that the Kagu of New Caledonia is most closely allied to the seriemas of South America (Sibley and Ahlquist 1985a). The relationships of the roatelos and buttonquail remain obscure.

1
2
3
4
5
6
7
COE

Shorebirds, gulls, and allies
Superorder Neognathae
Order Charadriiformes

Figure A–12 Shorebirds, gulls, and allies: (1) Pheasant-tailed Jacana (Jacanidae); (2) Snowy Sheathbill (Chionididae); (3) Eurasian Woodcock (Scolopacidae); (4) Atlantic Puffin (Alcidae); (5) Blacksmith Plover (Charadriidae); (6) Ring-billed Gull (Laridae); (7) Black Skimmer (Rynchopidae).

The 18 families and more than 300 species contained in the Order Charadriiformes are all waterbirds or birds, like the terrestrial lapwings, that are clearly derived from waterbirds. The group is united by various characteristics of the skull, vertebral column, and syrinx, but it is so varied that it is divided into three suborders that bear little outward resemblance to one another.

Members of the Suborder Charadrii, which includes the sandpipers, plovers, and other birds that are collectively known as "shorebirds" or (in Britain) "waders," are small to medium-sized birds with slender, probing bills and rather long legs. The feet are webbed in only a few species, and the hind toe is well developed in all but a small number of species. Many are cryptically colored, but some are boldly patterned. In other ways, they vary greatly; they make up about two-thirds of the order, and are the most varied of the three suborders.

The Suborder Lari is comprised of long-winged, rather short-legged, generally web-footed birds, often with a large salt-excreting gland located in the orbit above the eye, and with the hind toe small or lacking. These birds are usually clad in white, gray and white, or black and white. The gulls have a stout, somewhat hooked bill, the jaegers have a more strongly-hooked bill, the bills of terns are slender and sharply pointed, and the bill of the skimmer is uniquely modified, with the lower mandible bladelike and longer than the upper.

The auks, murres, and puffins make up the Suborder Alcae by themselves. They are small to medium-sized, stocky marine birds, usually black and white, with webbed feet, no hind toe, and dense, waterproof plumage. In most members of this group, the feet are located so far back on the body that the birds have a distinctive upright stance and resemble the penguins of the Southern Hemisphere. Their bills are quite varied in shape, ranging from the flattened, triangular bills of the puffins to the slender bills of murres and guillemots. All modern species of auks retain the ability to fly, but one flightless species, the Great Auk of the North Atlantic, became extinct in 1844, after years of persecution by seafarers.

The three suborders reflect three distinct foraging lifestyles among these largely aquatic birds. The Charadrii feed by wading in shallow water or along the edge of the water, using their bills to probe in mud or sand or to pluck prey items from the surface of the ground. All of the Charadriiformes that feed on open ground away from water are members of this suborder. The Suborder Lari consists of birds that feed in flight, scavenging along shores in the case of gulls, diving for fish that swim near the surface in the case of terns, and "skimming" over the surface snapping up small fish in the case of skimmers. The jaegers and skuas are pirates, stealing food from other members of the suborder, or preying on small rodents and nestlings of other birds. Many members of this suborder are colonial nesters.

Although the Charadrii feed mainly at the edge of the water, and the Lari primarily on the surface, the Suborder Alcae consists of diving birds that forage below the surface, pursuing small fish and other marine animals by swimming rapidly with their short, paddlelike wings. Most species are colonial nesters, though many nest in burrows so that colonies are not as conspicuous as those of gulls and terns.

Family	Members	Distribution	Genera	Species
Jacanidae	Jacanas	Pantropical	6	8
Rostratulidae	Painted-snipes	Nearly pantropical	2	2
Scolopacidae	Woodcocks, snipes, sandpipers, phalaropes, turnstones	Worldwide	23	86
Dromadidae	Crab Plover	Indian Ocean	1	1
Chionididae	Sheathbills	Subantarctic	1	2
Pluvianellidae	Magellanic Plover	Patagonia	1	1
Pedionomidae	Plains-wanderer	Australia	1	1
Thinocoridae	Seedsnipes	Temperate South America	2	4
Burhinidae	Thick-knees	Worldwide, excluding North America	2	9
Haematopodidae	Oystercatchers	Worldwide	1	7
Ibidorhynchidae	Ibisbill	Asia	1	1
Recurvirostridae	Avocets, stilts	All continents	3	10
Glareolidae	Pratincoles and coursers	Warm Old World	5	16
Charadriidae	Plovers, lapwings	Worldwide	8	64
Laridae	Gulls, terns	Worldwide	4	88
Stercorariidae	Jaegers, skuas	Polar regions, all	1	5
Rynchopidae	Skimmers	Americas, Africa, southeastern Asia	1	3
Alcidae	Auks, murres, puffins	Northern oceans	13	23

Taxonomic Comments The Order Charadriiformes is considered closely related to the Gruiformes and possibly to the Columbiformes through the sandgrouse. Sibley and Ahlquist (1985a) include sandgrouse as typical members of the Charadriiformes (but see Pteroclidiformes).

Several important revisions of charadriiform birds have appeared in recent years (Jehl 1968; Strauch 1978; Stegmann 1978; Gochfeld et al. 1984). It is now believed that the Scolopacidae split off early in the evolution of the order. Jacanas and the painted-snipes are related. The relations of the sheathbills and seedsnipes remain uncertain. Other papers that discuss enigmatic genera include: *Pluvianellus* (Jehl 1975); *Aechmorhynchus, Prosobonia,* and *Phegornis* (Zusi and Jehl 1970); and *Pedionomus* (Olson and Steadman 1981; Sibley and Ahlquist 1985a).

Figure A-13 Loons: Common Loon (Gaviidae).

Loons
Superorder Neognathae
Order Gaviiformes

Loons are a homogenous group of large, foot-propelled diving birds with spear-shaped bills, stocky necks, fusiform bodies, webbed feet, and laterally compressed tarsi. The legs are positioned far back on the body as in many other diving birds, and the birds have difficulty moving on land. Loons have different breeding and nonbreeding plumages; the breeding plumages are boldly patterned in black and white or with gray or chestnut, and the winter plumages are dark brownish or gray above and white below. The young are downy at hatching and wear a second coat of down before acquiring adult feathers.

Family	Members	Distribution	Genera	Species
Gaviidae	Loons	Holarctic	1	5

Taxonomic Comments Debate continues over the relationships of loons (Cracraft 1982b). They have been associated with other foot-propelled diving birds, such as the grebes, because of similarities in hind-limb anatomy; these are now believed to be due to convergence (Storer 1971a). Instead, loons probably evolved from gull-like ancestors (Storer 1956). Two species of primitive loons *(Colymboides),* known as fossils from the late Oligocene, are not as highly specialized for diving as modern loons and resemble gulls in skeletal characters such as the canals of the hypotarsus and the structure of the coracoid.

Figure A–14 Sandgrouse: Chestnut-bellied Sandgrouse (Pteroclididae).

Sandgrouse
Superorder Neognathae
Order Pteroclidiformes

Sandgrouse are stocky, medium-sized pigeonlike birds with small heads, short feathered tarsi, a large crop, and chickenlike bills. The hind toe is elevated in *Pterocles* and absent in *Syrrhaptes.* Some species have long, pointed central tail feathers. Unlike the pigeons and doves, they lack a fleshy cere and have a well-developed oil gland and an aftershaft; although they have a crop, they do not produce "pigeon's milk." The birds are clad in brown, tan, buff, and rufous, and often have breast bands. The sexes differ somewhat in color, with the females being more cryptically patterned than the males. The young are downy at hatching, another feature in which these birds differ from the Columbiformes.

Clouds of sandgrouse visit remote desert waterholes at dawn and dusk to drink. Some species carry water in soaked breast feathers for many miles to their young. Contrary to early reports, sandgrouse, unlike pigeons, raise their heads to swallow water (Maclean 1968).

Family	Members	Distribution	Genera	Species
Pteroclididae	Sandgrouse	Old World	2	16

Taxonomic Comments The relationships of sandgrouse spark vigorous debate. They seem intermediate between the Charadriiformes and Columbiformes, but are probably closer to the latter, according to some (Zusi 1986; Maclean 1967, 1969; Fjeldså 1976). Sibley and Ahlquist (1985a) consider sandgrouse to be typical shorebirds, closer to the Charadriidae than are the Scolopacidae. Following Voous (1973), I shall treat the sandgrouse as a separate order until their relationships can be resolved.

1
2
3
4
5
COE

Pigeons and doves
Superorder Neognathae
Order Columbiformes

Figure A-15 Pigeons and doves: (1) Mourning Dove (Columbidae); (2) Wood Pigeon (Columbidae); (3) Superb Fruit Dove (Columbidae); (4) Dodo (Columbidae); (5) Tooth-billed Pigeon (Columbidae).

Pigeons and doves are small, medium-sized, or large, plump birds with small heads, short legs covered with small, reticulate scales, and a fleshy cere at the base of the bill. The plumage is dense and is easily detached from the thin skin; there is little down under the outer feathering. The tail is commonly fan-shaped but may be long and narrow or pointed. Most have a muscular gizzard and all have a large crop, the lining of which secretes a substance known as "pigeon's milk," which is fed to nestlings. The oil gland is small or absent, and there is no aftershaft. A large group, pigeons and doves exhibit a wide variety of colors and patterns, with soft, pastel grays and buff predominating; many species have patches of softly iridescent feathers on the sides of the neck, and others have dark bands on the nape. Some, such as the large crowned pigeons *(Goura)* of New Guinea, bear elaborate crests. The young are nearly naked at hatching. Pigeons are unusual among birds in that they can drink by sucking water rather than tilting the head back to swallow it mouthful by mouthful. The extinct Dodo of Mauritius Island and the Solitaire of Rodriguez Island in the western Indian Ocean were large, flightless pigeons (Storer 1970).

Family	Members	Distribution	Genera	Species
Columbidae	Pigeons, doves	Worldwide	42	303

Taxonomic Comments This is a distinct order, possibly related to the Charadriiformes through the sandgrouse (Pteroclidiformes). The sequence of these orders is mainly based on tradition, not strong evidence of genealogy.

1
2
3
4
COE

Figure A–16 Parrots: (1) Greater Sulphur-crested Cockatoo (Cacatuidae); (2) Rainbow Lorikeet (Loriidae); (3) Chestnut-fronted Macaw (Psittacidae); (4) Kea (Psittacidae).

Parrots
Superorder Neognathae
Order Psittaciformes

Parrots are a well-defined group of small to medium-sized birds with stout, hooked bills, in which the maxilla is movable, being attached by a hingelike articulation to the skull. There is a fleshy cere. The tongue is fleshy and some species in the Australasian region have brush-tipped tongues for feeding on nectar. The neck and legs are short, and the toes are zygodactyl, adapted for perching and climbing; the scales on the legs and toes are small and granular. Parrots hold and manipulate food with their feet, and use their bill as well as their feet in climbing. Parrots tend to be gregarious, vocal birds and can be destructive. The raptorlike Kea of New Zealand pulls large nails from buildings, rips automobile upholstery, and kills sick sheep to feed on kidney fat. The wings are generally pointed and the keel is well-developed except in the flightless Owl Parrot of New Zealand. The plumage is sparse, and some species, most notably the cockatoos, are crested. The tail is most commonly fan-shaped, but may be long and pointed. Many parrots are green or largely green, but others, especially in the Australasian region, are clad in a variety of brilliant colors as well as solid black. The young are naked at hatching in some species but covered with down in other species.

Family	Members	Distribution	Genera	Species
Psittacidae	Parrots, macaws	Pantropical, a few temperate	64	268
Cacatuidae	Cockatoos	Australia	6	18
Loriidae	Lories	Australasian region, Pacific	11	54

Taxonomic Comments The parrots are so distinctive that their affinities remain obscure. They may be distantly related either to the pigeons and doves or to the cuckoos. See Forshaw (1978) for a recent review of the parrots of the world.

COE

Mousebirds
Superorder Neognathae
Order Coliiformes

Figure A–17 Mousebirds: Speckled Mousebird (Coliidae).

The mousebirds, or colies, are a distinctive group of small, crested, African birds with dense, gray or brown plumage, and long, pointed tails. The first and fourth toes of the pamprodactyl foot are reversible, so that all four toes can be directed forward.

Mousebirds inhabit savannah, woodland edge, and brushy country throughout Africa south of the Sahara. They travel in small, tight flocks during most of the year. They often hang upside down from branches, and can scurry about in bushes like mice, a habit that explains their name. They feed mainly on fruit, buds, and flowers, and at times may damage crops.

Family	Members	Distribution	Genera	Species
Coliidae	Mousebirds	Africa	1	6

Taxonomic Comments Although they have no obvious allies, the mousebirds may be distantly related to parrots, which they resemble in the structure of the palate, heart, pelvis, intestines, and oil gland (Berman and Raikow 1982). Sibley and Ahlquist (1985a) suggest that mousebirds are part of a complex that includes owls, nightjars, and turacos.

Turacos
Superorder Neognathae
Order Musophagiformes

Figure A–18 Turacos: Ross' Turaco (Musophagidae).

The African turacos are medium-sized, long-tailed, chiefly arboreal birds with a foot on which the outer toe is reversible, but not permanently reversed, so that the toes are not truly zygodactyl as they are in the cuckoos. The bill is short and chickenlike but with the edges serrated, and most species are crested and have patches of bare facial skin. A few are gray or clad in blue or purple, but most species are green, with a patch of bright red concealed in the flight feathers of the wing. Both of these colors are due to pigments that are unique to the Musophagidae: the green pigment is called turacoverdin and the red pigment is turacin.

Family	Members	Distribution	Genera	Species
Musophagidae	Turacos, Plantaineaters	Africa	5	18

Taxonomic Comments The turacos are a distinctive group not clearly related to any other order (Voous 1973; Sibley and Ahlquist 1985a). In the past they have been associated with the Cuculiformes and by some with the Galliformes.

1
2
3
COE

Cuckoos
Superorder Neognathae
Order Cuculiformes

Figure A–19 Cuckoos: (1) Common Cuckoo (Cuculidae); (2) Smooth-billed Ani (Cuculidae); (3) Greater Roadrunner (Cuculidae); see also Figure 3–12 for Hoatzin (Opisthocomidae).

Cuckoos are small to medium-sized, slender, usually long-tailed birds with a zygodactyl foot. Terrestrial species have sturdy legs, and arboreal species somewhat weaker ones, but in all, the foot is well adapted for perching. The bill is usually slender and slightly decurved but is laterally compressed in the anis, chickenlike in the coucals, and almost toucanlike in the Channel-billed Cuckoo. Many species have a colorful fleshy eye-ring, and some are crested. There are ten primaries. The plumage colors of cuckoos range from streaked or solid brown to solid black, or brilliant, metallic green in the emerald cuckoos *(Chrysococcyx)*. About 50 species are brood parasites, some of which produce eggs that mimic the eggs of their hosts. Anis of the New World tropics have communal nests.

The Hoatzin is a distinctive bird, slender, with rounded wings, and with the keel shortened to accommodate a greatly enlarged crop (Figure 3–12). The flight muscles are weakly developed, perhaps to make room for the large crop, so the Hoatzin can fly only short distances. It has a long loose crest and ten primaries. The species' most peculiar feature is found in young birds. At hatching, the young have two functional claws on the second and third digits of the wing; for a few days, the young are capable of clambering about among the branches near the nest. After a few days, the claws regress, and the wing develops like that of any other bird.

Family	Members	Distribution	Genera	Species
Cuculidae	Cuckoos	Worldwide	38	129
Opisthocomidae	Hoatzin	South America	1	1

Taxonomic Comments The Cuculiformes may be distantly related to the Psittaciformes. The Hoatzin is tentatively included here on the basis of biochemical evidence (Sibley and Ahlquist 1973, 1985a; see Brush 1979; Cracraft 1981 for alternative view).

1
2
3
4
COE

Figure A–20 Owls: (1) Elf Owl (Strigidae); (2) Northern Hawk-Owl (Strigidae); (3) Northern Eagle-Owl (Strigidae); (4) Common Barn-Owl (Tytonidae).

Owls
Superorder Neognathae
Order Strigiformes

The owls are a group of mainly nocturnal birds of prey with large, rounded heads; their eyes face forward and are surrounded by large facial disks of feathers that concentrate sound and greatly increase the hearing ability of these birds. In some species, the outer, bony portions of the two ears are differently shaped, creating a stereophonic effect that enables the birds to sense the precise location of a sound made by a prey animal. The eyes themselves are strengthened by bony plates like those of other birds, but in owls these plates form a lengthened cylinder that provides telescopic vision, capable of detecting even very scant amounts of light. The bill is long and hooked, and bears a cere but because it is directed downward, with its base covered by bristles, it appears short. The legs, and often the toes, are feathered, and the outer toe can be reversed. The plumage is soft, enabling the birds to fly silently. The barn-owls differ from typical owls of the Family Strigidae in a number of internal characters, among them a longer and narrower skull and a furcula fused to the sternum; the legs are longer than in most typical owls. Nearly all owls are cryptically colored; even the Snowy Owl, almost entirely white, is camouflaged when perched on snow. In both families, the young are clad in whitish down at hatching.

Family	Members	Distribution	Genera	Species
Tytonidae	Barn-owls	Worldwide	2	11
Strigidae	Owls	Worldwide	28	135

Taxonomic Comments Owls may be related to the diurnal birds of prey (Falconiformes, especially the Falconidae), or to the nightjars and their allies (Caprimulgiformes), but both possibilities remain controversial (Jollie 1976–1977; Cracraft 1981). Sibley and Ahlquist (1985a) consider owls and nightjars to be close relatives.

1
2
3
COE

Nightjars and allies
Superorder Neognathae
Order Caprimulgiformes

Figure A-21 Nightjars and allies: (1) Common Nighthawk (Caprimulgidae); (2) Tawny Frogmouth (Podargidae); (3) European Nightjar (Caprimulgidae).

The Caprimulgiformes are nocturnal or crepuscular birds with soft and cryptically patterned plumage and short legs. Most have a small, weak bill with a very large mouth opening, enabling them to capture insects on the wing; the bill is often surrounded by long bristles, which aid in catching prey and protect the eyes from damage in an aerial encounter with an insect. The frogmouths have heavier bills but still have the wide gape of other members of the order. The Oilbird has a stronger, more hooked bill, with a smaller mouth opening. The cave-dwelling Oilbird, which feeds on oil-rich fruits, is one of the few birds known to navigate by means of echolocation. It also is unusual in having a well-developed sense of smell. The true nightjars have small, weak feet, but those of other families whose members dwell in trees have stronger toes, well suited to perching.

Family	Members	Distribution	Genera	Species
Steatornithidae	Oilbirds	Northern South America	1	1
Podargidae	Frogmouths	Oriental and Australasian regions	2	13
Aegothelidae	Owlet-frogmouths	Australasian region	1	8
Nyctibiidae	Potoos	Neotropics	1	6
Caprimulgidae	Nightjars	Worldwide	19	77

Taxonomic Comments The Caprimulgiformes may be related to the owls, though both the Apodiformes and Trogoniformes are also possibilities. The Oilbird resembles owls in some anatomical characters but has egg-white proteins like those of nightjars and also resembles nightjars in characters of the skull, humerus, and sternum (Sibley and Ahlquist 1972; Cracraft 1981).

Nightjars of the Caribbean genus *Siphonorhis* may be primitive members of the order, related to the owlet-frogmouths of Australasia (Olson 1978b).

1
2
3
4
COE

Swifts and hummingbirds
Superorder Neognathae
Order Apodiformes

Figure A-22 Swifts and hummingbirds: (1) White-collared Swift (Apodidae); (2) Streamer-tail Hummingbird (Trochilidae); (3) Tufted Coquette (Trochilidae); (4) Black-chinned Hummingbird (Trochilidae).

Swifts and hummingbirds are small or very small birds with tiny feet, extremely short humeri, long bones in the outer portion of the wing, ten long, sturdy primaries, and short secondaries. These are all adaptations for special flight; the Apodiformes are perhaps the most accomplished fliers of all birds. In the swifts and crested swifts, the bill is short and the gape very broad to aid in capturing insects in flight. In the hummingbirds the gape is small and the bill is slender and quite variable in shape, adapted to feeding at flowers of diverse structures. All hummingbirds have a long and extensile tongue for reaching nectar. The nostrils are rounded and exposed in swifts and crested swifts, and slitlike and covered by an operculum in the hummingbirds. In the true swifts (Apodidae), there is a small claw on the hand; this is absent in the crested swifts and hummingbirds. A crop is lacking in adult Apodiformes, but is present in nestling hummingbirds. An oil gland is present and unfeathered. The swifts tend to be dull-colored or patterned in blackish and white, whereas most hummingbirds are brilliantly iridescent and males usually have brightly colored throat patches or crests. The hummingbirds include the smallest bird in the world, the Cuban Bee Hummingbird two and a quarter inches from bill tip to tail tip. The young are blind and helpless at hatching.

Family	Members	Distribution	Genera	Species
Hemiprocnidae	Crested swifts	Southern Asia to Solomon Islands	1	4
Apodidae	Swifts	Worldwide	18	83
Trochilidae	Hummingbirds	New World	116	341

Taxonomic Comments The Apodiformes are related to the Caprimulgiformes and perhaps to the Passeriformes.

Although usually placed in the same order, the relationship of the swifts and hummingbirds is uncertain. Their similar, specialized wing morphology may reflect common ancestry but could be due to convergence (Cohn 1968; Zusi and Bentz 1982, 1984). They share a unique form of the enzyme malate dehydrogenase (Kitto and Wilson 1966) and are classified together by most ornithologists.

♂
♂
1
2

Trogons
Superorder Neognathae
Order Trogoniformes

Figure A–23 Trogons: (1) Resplendent Quetzal (Trogonidae); (2) Red-headed Trogon (Trogonidae).

Trogons are a small group of tropical, forest-dwelling, frugivorous birds with short bills, strongly arched culmens, and serrated edges on the upper mandibles. The feet, small and weak, differ from those of all other birds in having heterodactyl feet, with the first and second toes directed backward, and the third and fourth toes directed forward; all other birds with two toes in front and two behind have the first and fourth directed backward. Trogons have dense, lax plumage and a well-developed aftershaft; the feathers tear loose from the thin skin very easily. The wings are short and rounded, and there are ten primaries on each. The tail is long, usually squared at the tip. The upper tail coverts are much longer than the tail in the quetzals of the New World tropics. Most species have iridescent green plumage; many have bright red or yellow underparts. The red pigment is unusual because it is unstable, fading quickly in museum specimens, and is maintained in life only as a result of feather replacement during molt. The young are naked and blind at hatching.

Family	Members	Distribution	Genera	Species
Trogonidae	Trogons, quetzals	Pantropical, except Australasia	8	37

Taxonomic Comments The trogons are an enigmatic group, probably related to the Coraciiformes (Maurer and Raikow 1981).

1
2
3
4
5
6

Rollers and allies
Superorder Neognathae
Order Coraciiformes

Figure A-24 Roller and allies: (1) Puerto Rican Tody (Todidae); (2) European Beeeater (Meropidae); (3) Lilac-breasted Roller (Coraciidae); (4) Turquoise-browed Motmot (Momotidae); (5) Indian Pied Hornbill (Bucerotidae); (6) Pied Kingfisher (Alcedinidae).

The rollers and their allies are small to medium-sized, stocky birds with large heads and small feet. The group is diverse, and its unifying characters largely involve the structure of the palate and leg muscles, and fusion of the toes (syndactyly). In addition to their varied patterns and often brilliant colors, coraciiform families differ greatly in the shape of the bill, ranging from the slender bills of bee-eaters, wood-hoopoes and the Hoopoe and the sturdier bills of kingfishers and rollers, to the huge and ornamented bills of the larger hornbills; the most distinctive bill is that of the Shovel-billed Kingfisher of New Guinea, whose bill is shaped like a horse's hoof and is used to dig for earthworms in the forest floor. The kingfishers, todies, motmots, bee-eaters, and, to a lesser extent, the other members of the order, have the anterior toes fused at the base. The number of primaries is eleven in most kingfishers, the motmots, and some bee-eaters, and ten in the other families. Features peculiar to individual families include eyelashes and a large, bony casque above the bill in many hornbills, a long, erectile crest in the Hoopoe, and spatulate tips to the central tail feathers in motmots.

Family	Members	Distribution	Genera	Species
Alcedinidae	Kingfishers	Worldwide	14	91
Todidae	Todies	Greater Antilles	1	5
Momotidae	Motmots	Neotropics	6	9
Meropidae	Bee-eaters	Warm Old World	3	24
Coraciidae	Rollers	Warm Old World	2	11
Brachypteraciidae	Groundrollers	Madagascar	3	5
Leptosomatidae	Cuckoo Roller	Madagascar	1	1
Upupidae	Hoopoe	Africa, Madagascar, warm Eurasia	1	1
Phoeniculidae	Wood-hoopoes	Africa	2	8
Bucerotidae	Hornbills	Tropical Africa and Asia; East Indies	12	45

Taxonomic Comments The characteristics of the Coraciiformes link them to the Piciformes and the Passeriformes, but the Cuculiformes, Psittaciformes, and even the Caprimulgiformes have been considered possible relatives.

Whether this order is truly monophyletic is a matter of debate (Cracraft 1971, 1981; Feduccia 1977, 1980; Maurer and Raikow 1981). The Hoopoe and wood-hoopoes are relatively closely related, as are the motmots and todies, and probably also the kingfishers and bee-eaters. The rollers, and particularly the ground-rollers and Cuckoo Roller of the Malagasy region, probably represent an ancient lineage.

1
2
3
4
5
6
COE

Figure A–25 Woodpeckers and allies: (1) Great Spotted Woodpecker (Picidae); (2) Black-throated Honeyguide (Indicatoridae); (3) Double-toothed Barbet (Capitonidae); (4) White-chinned Jacamar (Galbulidae); (5) White-eared Puffbird (Bucconidae); (6) Keel-billed Toucan (Ramphastidae).

Woodpeckers and allies
Superorder Neognathae
Order Piciformes

The Piciformes are a varied group of six families, of which the woodpeckers are the largest. All have zygodactyl feet and a unique arrangement of tendons in the toes. Most have an aftershaft. The families differ most obviously in the shape of the bill: it is a strong, sturdy, and chisel-like in woodpeckers, wrynecks, and piculets; long, slender, and sharply pointed in jacamars; stout, anteriorly compressed, and somewhat decurved at the tip in puffbirds; stout, more or less conical, and surrounded by bristles in barbets; small and slightly hooked in honeyguides; and very large and inflated in toucans. There are usually ten primaries but only nine in honeyguides and 13 in some woodpeckers. The true woodpeckers (Subfamily Picinae) have stiffened tail feathers that serve as a prop when the bird is clinging to a tree. Color and pattern vary, from the dull greens of some honeyguides and the black-and-white patterns of woodpeckers to the brilliant reds, blues, and yellows of many barbets and toucans. The sexes are generally similar, but among the woodpeckers, the males often have patches of red on the head that are reduced or lacking in females. All species nest in holes, or cavities. The young are naked and blind at hatching. The diversity of feeding specializations is a feature of this order. Woodpeckers hitch up vertical tree trunks and chip wood and bark with their chisel-like bills. Long, barb-tipped tongues enable them to extract woodboring insects from tiny crevices. Jacamars catch insects on the wing. Puffbirds pluck large caterpillars from tropical foliage. Barbets and toucans consume much fruit, but toucans also snatch eggs and young from the nests of other birds. Some honeyguides eat beeswax. Honeyguides are also specialized brood parasites that lay their eggs in the nests of woodpeckers, barbets, and other birds.

Family	Members	Distribution	Genera	Species
Galbulidae	Jacamars	Neotropics	5	17
Bucconidae	Puffbirds	Neotropics	7	32
Capitonidae	Barbets	Pantropical, except Australasia	13	81
Indicatoridae	Honeyguides	Africa, tropical Asia	4	16
Ramphastidae	Toucans	Neotropics	6	33
Picidae	Woodpeckers, wrynecks, piculets	Worldwide except Australasia	27	204

Taxonomic Comments The Piciformes may be related to the Coraciiformes.

Whether the Piciformes is a monophyletic group is currently a matter of debate (Swierczewski and Raikow 1981; Simpson and Cracraft 1981; Olson 1983; Raikow and Cracraft 1983). The main issue is whether the jacamars and puffbirds are perhaps closer to the Coraciiformes than to other members of the Piciformes. Toucans evolved from barbetlike ancestors, but the origins of honeyguides and woodpeckers are uncertain.

1
2
3
4
5
COE

Perching birds
Superorder Neognathae
Order Passeriformes

Perching birds constitute more than half of the species of birds of the world. They represent a diverse, species-rich, monophyletic order of mostly small landbirds with distinctive feet, oil glands, sperm, bony palates, and a reduced number of neck vertebrae. Many species rub ants (with noxious fluids) on their feathers for protection against ectoparasites. The metabolism of perching birds is higher than that of other birds of comparable size. Differences in the anatomy of the syrinx distinguish two suborders of perching birds, the suboscines (Tyranni) and the oscines (Passeres).

Passerine birds have repeatedly evolved into certain basic ecological forms: flycatchers, thrushes, warblers, creepers, and seedeaters. Convergence in ecology and morphology has confused their family classification. Biochemistry helps to deal with the problem of pervasive convergence and allows more successful grouping of these landbirds by their true affinities. Karl Voous (1985) reviews the familial taxonomy of perching birds.

Perching birds — suboscines
Suborder Tyranni

Figure A–26 Perching birds — suboscines: (1) Scissor-tailed Flycatcher (Tyrannidae); (2) Guianan Cock-of-the-Rock (Cotingidae); (3) Banded Pitta (Pittidae); (4) Barred Woodcreeper (Dendrocolaptidae); (5) Great Antshrike (Formicariidae).

These are passeriform birds with few unifying characters other than arrangements of the syringeal muscles that are simpler than those distinguishing the oscine passeriform birds. Traditionally, suboscine families include broadbills and pittas of Africa and Asia, asities of Madagascar, and the New Zealand wrens. Broadbills are sluggish, tropical, fruit-eating birds that are traditionally separated from other suboscine birds on the basis of distinct foot tendons (though see Olson 1971). Pittas are secretive, brightly colored, ground birds with long legs and short tails. The little known asities include two sluggish, fruit-eating species and two nectar-feeding species that closely resemble sunbirds (Nectariniidae). Both DNA and morphological evidence suggest that the New Zealand wrens (Xenicidae) may be the most ancient passerine lineage (Sibley et al. 1982; Raikow 1984).

The suboscines also include two major radiations of South American birds: tyrant-flycatchers, cotingas, and manakins in one case, and woodcreepers, ovenbirds, antbirds, and tapaculos in the other. Although their body forms vary greatly with a diversity of feeding styles, tyrant-flycatchers are distinguished by cranial, syringeal, and tarsal characters. Cotingas are fairly large tropical fruit-eating birds with broad bills, rounded wings, and short legs. Manakins are small, stocky, tropical fruit-eating birds with short wings and tail and broad bills. Some cotingas and manakins are brightly colored birds with elaborate courtship behavior. Woodcreepers, ovenbirds, antbirds, and tapaculos include a variety of small to medium-sized, insect-eating birds of tropical forests. The woodpeckerlike woodcreepers typically have powerful feet with sharp claws and stiff, bracing tails. Ovenbirds are a varied group of small, brown birds. Antbirds and tapaculos, which mostly have thick, often hook-tipped bills and short, rounded wings, reside in deeply shaded thick vegetation.

Family	Members	Distribution	Genera	Species
Xenicidae	New Zealand wrens	New Zealand	2	4
Pittidae	Pittas	Old World tropics	1	24
Eurylaimidae	Broadbills	Africa, southeastern Asia	8	14
Philepittidae	Asities, False-sunbirds	Madagascar	2	4
Dendrocolaptidae	Woodhewers	Neotropical	13	52
Furnariidae	Ovenbirds	Neotropical	34	218
Formicariidae	Antbirds, gnateaters	Neotropical	52	240
Rhinocryptidae	Tapaculos	Neotropical	12	30
Cotingidae	Cotingas	Neotropical	27	79
Oxyruncidae	Sharpbill	Neotropical	1	1
Phytotomidae	Plantcutters	South America	1	3
Pipridae	Manakins	Neotropical	17	52
Tyrannidae	Tyrant-fly-catchers	New World	111	376

Taxonomic Comments The Old World suboscines are more closely related to one another than to New World suboscines (Feduccia 1974; Olson 1971; Sibley and Ahlquist 1985c).

Voous (1985) suggests that the term suboscines be restricted to the lyrebirds (Menuridae) and scrubbirds (Atrichornithidae), which have been moved to the suborder Passeres, close to bowerbirds.

Sharpbill and plantcutters may be related to Cotingidae (see Sibley et al. 1984).

Sibley and Ahlquist (1985c) propose that there are two distinct groups of antbirds, the typical antbirds (e.g., *Thamnophilus*), and the ground antbirds (e.g., *Formicarius*) together with gnateaters and tapaculos.

Perching birds — oscines
Suborder Passeres

Oscine families are distinguished by their complex vocal musculature. Diverse in habits and form, they constitute the majority of passerine birds and include over 4000 species of finches, warblers, larks, swallows, chickadees, crows, birds-of-paradise, orioles, nuthatches, shrikes, vireos, bulbuls, babblers, wrens, mockingbirds, thrushes, pipits, sunbirds, weavers, buntings, blackbirds, tanagers, and many other songbirds.

Sibley and Ahlquist (1985b) group the oscine passerines by the similarities of their nuclear DNA, as follows: (1) Crow relatives, which include most of the families of Australasian birds as well as crows, jays, drongos, shrikes, vangas, birds-of-paradise, and also vireos; (2) thrush relatives, which include waxwings, starlings, mimic thrushes, plus Old World flycatchers and chats; (3) Old World insecteaters, which include swallows, titmice, nuthatches, white-eyes, babblers, and Old World warblers (4) weaver relatives, which include crossbills, weavers, sunbirds and larks, plus the New World nine-primaried oscines, which include wood warblers, tanagers, blackbirds, and buntings. Here, I present the families of oscine birds in these major groups, separating the nine-primaried oscines for convenience.

Perching birds — crow relatives

This group includes most families of Australasian passerine birds as well as crows, jays, drongos, shrikes, vangas, birds-of-paradise, and vireos. Many of the Australasian families were once squeezed into Asian bird families, which they resembled in ecology and morphology. Now, they are thought to be convergent on those Asian groups and closely related instead to one another or to a limited number of non-Australian groups.

Family	Members	Distribution	Genera	Species
Climacteridae	Australasian treecreepers	Australia, New Guinea	1	6
Menuridae	Lyrebirds	Australia	1	2
Atrichornithidae	Scrubbirds	Australia	1	2
Ptilonorhynchidae	Bowerbirds	Australasia	8	18
Maluridae	Fairy Wrens	Australia	6	30
Acanthizidae	Australian warblers	Australasia	16	75
Meliphagidae	Honeyeaters	Australasia	39	172
Callaeidae	Wattlebirds	New Zealand	3	3
Eopsaltriidae	Australasian robins	Australasia	10	34
Pachycephalidae	Whistlers	Australasia	12	49
Orthonychidae	Log-runners	Australasia	1	2
Pomatostomatidae	Pseudo-babblers	Australasia	1	4
Cinclosomatidae	Quail-thrushes, whip-birds	Australia	8	15

(continued)

1
2
3
♂
4
5
♂
6

Figure A-27 Perching birds—crow relatives: (1) Fork-tailed Drongo (Dicruridae); (2) Eurasian Jay (Corvidae); (3) Common Fiscal Shrike (Laniidae); (4) New Holland Honeyeater (Meliphagidae); (5) Blue Wren (Maluridae); (6) Spotted Quail-thrush (Cinclosomatidae).

Family	Members	Distribution	Genera	Species
Corcoracidae	Australian Chough, Apostlebird	Australia	2	2
Monarchidae	Monarch flycatchers	Southeastern Asia, Australasia	19	132
Dicruridae	Drongos	Old World tropics	2	20
Corvidae	Crows, jays, magpies	Worldwide	26	106
Paradisaeidae	Birds-of-paradise	Australasia	20	42
Artamidae	Wood-swallows	Oriental, Australasia	1	10
Cracticidae	Bellmagpies, piping-crows, *Pityriasis*	Australasia	4	11
Grallinidae	Mudnest builders	Australia, New Guinea	1	2
Oriolidae	Old World orioles	Warm Old World	2	25
Campephagidae	Cuckoo-shrikes	Old World tropics	9	70
Irenidae	Fairy-bluebirds, leafbirds	Oriental, East Indies	3	14
Vireonidae	Vireos, peppershrikes, shrike-vireos	New World	4	43
Laniidae	Shrikes	North America, Africa, Eurasia	11	73
Vangidae	Vangas, Coral-billed Nuthatch	Madagascar	9	13

Taxonomic Comments Lyrebirds and scrubbirds are now recognized to be Passeres, perhaps related to bowerbirds (Olson 1971; Sibley 1974; Feduccia 1975). Together with the Climacteridae, these families may be a distinct superfamily (Sibley and Ahlquist 1985b).

Bowerbirds and birds-of-paradise are not closely related (Sibley and Ahlquist 1980); *Turnagra* of New Zealand may be related to one or the other of these two groups (Olson et al. 1983);

Vireos apparently belong to this primarily Old World assemblage (Sibley and Ahlquist 1982b).

Laniidae includes shrikes, bush shrikes (Malaconotinae), helmet shrikes and puffback shrikes. See Raikow et al. (1980) for myological relationships among shrike subfamilies. DNA indicates that *Pityriasis* belongs to the Cracticidae (Ahlquist et al. 1984, though see also Raikow et al. 1980).

1
2
3
4
5

Figure A-28 Perching birds—thrush relatives: (1) Northern Mockingbird (Mimidae); (2) Pied Flycatcher (Muscicapidae); (3) Bohemian Waxwing (Bombycillidae); (4) Brahminy Starling (Sturnidae); (5) MistleThrush (Turdidae).

Perching birds—thrush relatives

This group includes waxwings and their relatives, dippers and thrushes, starlings, mockingbirds and their relatives, plus Old World flycatchers and chats. The Australasian robins, including Drymodes (Eopsaltriidae), formerly placed in or close to the Turdidae, are now excluded. Chats were previously thought to be close to large thrushes but are now thought to be "ground-dwelling" Old World flycatchers.

Family	Members	Distribution	Genera	Species
Bombycillidae	Waxwings, silky flycatchers, *Hypocolius*	Holarctic	5	8
Dulidae	Palmchat	Hispaniola	1	1
Cinclidae	Dippers	Holarctic, South America	1	5
Turdidae	Thrushes	Worldwide	38	275
Mimidae	Mimic thrushes	New World	13	31
Sturnidae	Starlings, mynas	Old World	26	111
Muscicapidae	Old World flycatchers and chats	Eurasia	24	153

Taxonomic Comments Mimic thrushes and starlings may be related (Sibley and Ahlquist 1984).

1
2
3
4
5
6

Perching birds — Old World insecteaters

Figure A–29 Perching birds—insecteaters: (1) Kikuyu White-eye (Zosteropidae); (2) Barn Swallow (Hirundinidae); (3) Sardinian Warbler (Sylviidae); (4) Long-tailed Tit (Aegithalidae); (5) White-breasted Nuthatch (Sittidae); (6) White-crested Laughing-Thrush (Timaliidae).

This group includes many Eurasian bird families. Included are some of our most familiar birds, such as nuthatches, titmice, wrens and swallows, as well as the babblers, and Old World warblers, which have a few representatives in North America, e.g. the Wrentit is a babbler, kinglets are related to Old World warblers. Excluded from these families are convergent Australasian birds, most of which are classified in their own families as crow relatives.

Family	Members	Distribution	Genera	Species
Sittidae	Nuthatches	Widespread, not subsaharan Africa or South America	2	23
Certhiidae	Creepers	Holarctic, Africa, India	2	6
Troglodytidae	Wrens	New World; one species Holarctic	14	60
Aegithalidae	Long-tailed Tits	Eurasia, western North America	3	7
Hirundinidae	Swallows, martins	Worldwide	20	80
Pycnonotidae	Bulbuls	Africa, southern Asia	15	123
Zosteropidae	White-eyes	Old World tropics and subtropics	11	83
Sylviidae	Old World warblers, kinglets, gnatcatchers	Mostly Old World	66	361
Timaliidae	Babblers, Wrentit	Old World Tropics; one in North America	44	257
Rhabdornithidae	Philippine creepers	Philippine Islands	1	2
Paridae	Titmice	Eurasia, Africa, North America	4	47
Remizidae	Penduline-tits, Verdins	Eurasia, Africa, western North America	4	10

Taxonomic Comments Timaliidae and Sylviidae may belong together with ecological types crossing traditional taxonomic boundaries (Sibley and Ahlquist 1983). See Ames (1975) for a taxonomic discussion of syringeal morphology of these taxa.

♂
♂
1
2
♂
3
4
♂
5
COE

Figure A-30 Perching birds—weaver relatives: (1) Village Weaver (Ploceidae); (2) Beautiful Sunbird (Nectariniidae); (3) Pine Grosbeak (Fringillidae); (4) Skylark (Alaudidae); (5) Alpine Accentor (Prunellidae).

Perching birds—weaver relatives

Primarily birds of the Old World, this group includes ground birds such as larks, sparrows, finches, wagtails and accentors, plus the ubiquitous African weavers, the nectar-feeding sunbirds and flowerpeckers, the carduelline finches, including crossbills and siskins, and the Hawaiian honeycreepers. It also includes the nine-primaried oscines according to Sibley et al. (1988). Seed-eating passerines (sparrows, buntings, weavers, waxbills, siskins, finches, grosbeaks) are assigned to at least five different families. The commonly used English names of seed-eating birds, such as finch or sparrow, do not relate in a simple way to this classification, but reflect traditional, conflicting uses.

Family	Members	Distribution	Genera	Species
Alaudidae	Larks	Worldwide, except most of South America	15	78
Passeridae	Old World sparrows, rock sparrows	Old World	8	37
Estrildidae	Waxbills, manakins, grassfinches	Old World tropics	28	127
Motacillidae	Wagtails, pipits	Nearly worldwide	5	54
Prunellidae	Accentors	Eurasia	1	12
Ploceidae	Weavers	Eurasia, Africa	10	107
Promeropidae	Sugarbirds	South Africa	1	2
Dicaeidae	Flowerpeckers	Oriental	6	50
Nectariniidae	Sunbirds	Old World tropics	5	117
Fringillidae	Siskins, crossbills, and allies	Worldwide, except Australasia	20	122
Drepanididae	Hawaiian honeycreepers	Hawaiian Islands	10	23

Taxonomic Comments The Passeridae is characterized by unique character state of the preglossale (Bock and Morony 1971). Sunbirds and flowerpeckers are close to the weavers and might be merged into a single family (Sibley and Ahlquist 1981b). Hawaiian honeycreepers evolved from cardueline finches (Beecher 1953; Raikow 1977; Sibley and Ahlquist 1982a).

1
♂
2
♂
3
4
♂
5

Perching birds — nine-primaried oscines

Figure A-31 Perching birds — nine-primaried oscines: (1) Yellow-rumped Cacique (Icteridae); (2) Magnolia Warbler (Parulidae); (3) Rose-breasted Grosbeak (Emberizidae); (4) Magpie Tanager (Emberizidae); (5) Lapland Longspur (Emberizidae)

This is a diverse, colorful group of primarily New World birds that includes the wood warblers, tanagers, and blackbirds, as well as the buntings and New World "sparrows." Many species migrate from the tropics to the United States and Canada to nest in June and July.

Family	Members	Distribution	Genera	Species
Parulidae	New World warblers	New World	28	126
Emberizidae	Tanagers, buntings, New World sparrows, cardinal-grosbeaks, Plush-capped Finch	New World, Eurasia	134	560
Icteridae	Troupials, meadowlarks, New World blackbirds	New World	23	95

Taxonomic Comments The sixth edition of the AOU *Checklist of North American Birds* now recognizes the Emberizidae as a broadly inclusive family that encompasses these three taxa as subfamilies.

appendix II

Bird names

English	Scientific
Accentor, Alpine	*Prunella collaris*
Albatross, Black-footed	*Diomedea nigripes*
Albatross, Laysan	*Diomedea immutabilis*
Albatross, Royal	*Diomedea irrorata*
Albatross, Wandering	*Diomedea exulans*
Amakihi, Common	*Hemignathus virens*
Anhinga, American	*Anhinga anhinga*
Ani, Groove-billed	*Crotophaga sulcirostris*
Ani, Smooth-billed	*Crotophaga ani*
Antbird, Bicolored	*Gymnopithys leucaspis*
Antbird, Ocellated	*Phaenostictus mcleannani*
Antbird, Spotted	*Hylophylax naevioides*
Antbird, White-plumed	*Pithys albifrons*
Antshrike, Barred	*Thamnophilus doliatus*
Antshrike, Bluish-slate	*Thamnomanes schistogynus*
Antshrike, Dusky-throated	*Thamnomanes ardesiacus*
Antshrike, Great	*Taraba major*
Antwren, Dot-winged	*Microrhopias quixensis*
Araçari, Curl-crested	*Pteroglossus beauharnaesii*
Araçari, Green	*Pteroglossus viridis*
Araçari, Letter-billed	*Pteroglossus inscriptus*
Auk, Great	*Pinguinus impennis*
Auk, Razorbill	*Alca torda*
Auklet, Least	*Aethia pusilla*
Auklet, Rhinoceros	*Cerorhinca monocerata*
Babbler, Arabian	*Turdoides squamiceps*
Babbler, Common	*Turdoides caudatus*
Babbler, Gray-crowned	*Pomatostomus temporalis*

English	Scientific
Babbler, Jungle	*Turdoides striatus*
Bananaquit	*Coereba flaveola*
Barbet, Double-toothed	*Lybius bidentatus*
Barn-Owl, Common	*Tyto alba*
Bateleur	*Terathopius ecaudatus*
Beardless-Tyrannulet, Northern	*Camptostoma imberbe*
Bee-eater, Carmine	*Merops nubicus*
Bee-eater, European	*Merops apiaster*
Bee-eater, Rainbow	*Merops ornatus*
Bee-eater, White-fronted	*Merops bullockoides*
Bellbird, Bearded	*Procnias averano*
Bellbird, Crested	*Oreoica gutturalis*
Bellbird, Three-wattled	*Procnias tricarunculata*
Berrypecker, Obscure	*Melanocharis arfakiana*
Bird-of-Paradise, Count Raggi's	*Paradisaea raggiana*
Bird-of-Paradise, King	*Cicinnurus regius*
Bird-of-Paradise, Lesser	*Paradisaea minor*
Bird-of-Paradise, Magnificent	*Diphyllodes magnificus*
Bird-of-Paradise, Red	*Paradisaea rubra*
Bird-of-Paradise, Superb	*Lophorina superba*
Bird-of-Paradise, Twelve-wired	*Seleucidis melanoleuca*
Bishop, Red	*Euplectes orix*
Bittern, American	*Botaurus lentiginosus*
Bittern, Least	*Ixobrychus exilis*
Blackbird, Brewer's	*Euphagus cyanocephalus*
Blackbird, Eurasian	*Turdus merula*
Blackbird, Red-winged	*Agelaius phoeniceus*
Blackbird, Tricolored	*Agelaius tricolor*
Blackbird, Yellow-headed	*Xanthocephalus xanthocephalus*
Blackbird, Yellow-shouldered	*Agelaius xanthomus*
Blackcap; see Warbler	
Bluebird, Eastern	*Sialia sialis*
Bluebird, Mountain	*Sialia currucoides*
Bluethroat	*Luscinia svecica*
Bobolink	*Dolichonyx oryzivorus*
Bobwhite, Northern	*Colinus virginianus*
Booby, Blue-footed	*Sula nebouxii*
Booby, Brown	*Sula leucogaster*
Booby, Masked	*Sula dactylatra*
Booby, Red-footed	*Sula sula*
Boubou, Tropical	*Laniarius aethiopicus*
Bowerbird, Archbold's	*Archboldia papuensis*

English	Scientific
Bowerbird, Gardener; see Bowerbird, Striped	
Bowerbird, Golden	*Xanthomelus aureus*
Bowerbird, Great Gray	*Chlamydera nuchalis*
Bowerbird, MacGregor's	*Amblyornis macgregoriae*
Bowerbird, Regent	*Xanthomelus chrysocephalus*
Bowerbird, Satin	*Ptilonorhynchus violaceus*
Bowerbird, Spotted	*Chlamydera maculata*
Bowerbird, Striped	*Amblyornis subalaris*
Bowerbird, Tooth-billed	*Scenopocetes dentirostris*
Bowerbird, Yellow-breasted	*Chlamydera lauterbachi*
Bowerbird, Yellow-fronted	*Amblyornis flavifrons*
Brambling	*Fringilla montifringilla*
Brant	*Branta bernicla*
Brush-Finch, Yellow-throated	*Atlapetes gutturalis*
Budgerigar	*Melopsittacus undulatus*
Bulbul, Yellow-vented	*Pycnonotus goiavier*
Bullfinch, Eurasian	*Pyrrhula pyrrhula*
Bullfinch, Lesser Antillean	*Loxigilla noctis*
Bunting, Indigo	*Passerina cyanea*
Bunting, Lark	*Calamospiza melanocorys*
Bunting, Lazuli	*Passerina amoena*
Bunting, Meadow	*Emberiza cioides*
Bunting, Ortolan	*Emberiza hortulana*
Bunting, Painted	*Passerina ciris*
Bunting, Snow	*Plectrophenax nivalis*
Bush-Tanager, Sooty-capped	*Chlorospingus pileatus*
Bush-Warbler, Japanese	*Cettia diphone*
Bustard, Great	*Otis tarda*
Buzzard, Common	*Buteo buteo*
Cacique, Scarlet-rumped	*Cacicus uropygialis*
Cacique, Yellow-rumped	*Cacicus cela*
Camaroptera, Gray-backed	*Camaroptera brevicaudata*
Canary, Common	*Serinus canaria*
Capercaillie, Common	*Tetrao urogallus*
Cardinal, Northern	*Cardinalis cardinalis*
Cassowary, Australian	*Casuarius casuarius*
Catbird, Gray	*Dumetella carolinensis*
Chaffinch, Canary Islands	*Fringilla teydea*
Chaffinch, Common	*Fringilla coelebs*
Chat, River	*Chaimarrornis leucocephalus*
Chickadee, Black-capped	*Parus atricapillus*
Chickadee, Boreal	*Parus hudsonicus*
Chickadee, Chestnut-backed	*Parus rufescens*
Chicken, Domestic	*Gallus gallus*

English	Scientific
Chiffchaff; see Warbler	
Cock-of-the-Rock, Andean	*Rupicola peruviana*
Cock-of-the-Rock, Guianan	*Rupicola rupicola*
Cockatoo, Gr. Sulphur-crested	*Cacatua galerita*
Cockatoo, Palm	*Probosciger atterimus*
Condor, Andean	*Vultur gryphus*
Condor, California	*Gymnogyps californianus*
Conebill, Blue-backed	*Conirostrum sitticolor*
Conebill, White-browed	*Conirostrum ferrugineiventre*
Coot, Eurasian	*Fulica atra*
Coot, Horned	*Fulica cornuta*
Coquette Tufted; see Hummingbird	
Cormorant, Common	*Phalacrocorax aristotelis*
Cormorant, Flightless	*Nannopterum harrisi*
Cormorant, Great	*Phalacrocorax carbo*
Cormorant, Guanay	*Phalacrocorax bougainvillii*
Cormorant, Olivaceous	*Phalacrocorax olivaceus*
Cormorant, Pelagic	*Phalacrocorax pelagicus*
Cotinga, Pompadour	*Xipholena punicea*
Courser, Two-banded	*Rhinoptilus africanus*
Courser, Violet-tipped	*Hemerodromus chalcopterus*
Cowbird, Bay-winged	*Molothrus badius*
Cowbird, Bronzed	*Molothrus aeneus*
Cowbird, Brown-headed	*Molothrus ater*
Cowbird, Giant	*Scaphidura oryzivora*
Cowbird, Screaming	*Molothrus rufoaxillaris*
Cowbird, Shiny	*Molothrus bonariensis*
Crane, Crowned	*Balearica pavonina*
Crane, Paradise	*Anthropoides paradisea*
Crane, Whooping	*Grus americana*
Creeper, Brown	*Certhia americana*
Creeper, Common	*Certhia familiaris*
Crossbill, Red	*Loxia curvirostra*
Crossbill, White-winged	*Loxia leucoptera*
Crow, American	*Corvus brachyrhynchos*
Crow, Carrion	*Corvus corone cornix*
Crow, Fish	*Corvus ossifragus*
Crow, Hooded	*Corvus corone corone*
Crow, House	*Corvus splendens*
Crow, Northwestern	*Corvus caurinus*
Cuckoo, Black-billed	*Coccyzus erythropthalmus*
Cuckoo, Channel-billed	*Scythrops novaehollandiae*
Cuckoo, Common Hawk	*Cuculus varius*
Cuckoo, Common	*Cuculus canorus*
Cuckoo, Didric	*Chrysococcyx caprius*
Cuckoo, Drongo	*Surniculus lugubris*

English	Scientific
Cuckoo, Hodgson's Hawk	*Cuculus fugax*
Cuckoo, Horsfield's Bronze	*Chalcites basalis*
Cuckoo, Large Hawk	*Cuculus sparverioides*
Cuckoo, Little	*Cuculus poliocephalus*
Cuckoo, Oriental	*Cuculus saturatus*
Cuckoo, Pied	*Clamator jacobinus*
Cuckoo, Shining Bronze	*Chalcites lucidus*
Cuckoo, Yellow-billed	*Coccyzus americanus*
Curlew, Bristle-thighed	*Numenius tahitiensis*
Curlew, Eskimo	*Numenius borealis*
Currasow, Great	*Crax rubra*
Dickcissel	*Spiza americana*
Dikkop, Water	*Burhinus vermiculatus*
Diuca-Finch, White-winged	*Diuca speculifera*
Diving Petrel, Common	*Pelecanoides urinatrix*
Dodo	*Raphus cucullatus*
Dove, Black-billed Cuckoo	*Macropygia nigrirostris*
Dove, Brown's Cuckoo	*Reinwardtoena browni*
Dove, Eared	*Zenaida auriculata*
Dove, Great Cuckoo	*Reinwardtoena reinwardtsi*
Dove, Mackinlay's Cuckoo	*Macropygia mackinlayi*
Dove, Mourning	*Zenaida macroura*
Dove, Pink-breasted Cuckoo	*Macropygia amboinensis*
Dove, Ring-necked	*Streptopelia capicola*
Dove, Rock	*Columba livia*
Dove, Socorro	*Zenaida graysoni*
Dove, Superb Fruit	*Ptilinopus perousii*
Dove, Turtle	*Streptopelia turtur*
Dove, White-tipped	*Leptotila verreauxi*
Dove, White-winged	*Zenaida asiatica*
Dovekie	*Alle alle*
Drongo, Black	*Dicrurus macrocercus*
Drongo, Fork-tailed	*Dicrurus adsimilis*
Duck, American Black	*Anas rubripes*
Duck, Black-headed	*Heteronetta atricapilla*
Duck, Blue	*Hymenolaimus malacorhynchos*
Duck, Lesser Scaup	*Aythya affinis*
Duck, Mallard	*Anas platyrhynchos*
Duck, Mandarin	*Aix galericulata*
Duck, Muscovy	*Cairina moschata*
Duck, Musk	*Biziura lobata*
Duck, Northern Shoveler	*Anas clypeata*
Duck, Northern Pintail	*Anas acuta*
Duck, Redhead	*Aythya americana*
Duck, Ruddy	*Oxyura jamaicensis*

English	Scientific
Duck, Smew	*Mergellus albellus*
Duck, Torrent	*Merganetta armata*
Duck, Wood	*Aix sponsa*
Dunlin; see Sandpiper	
Dunnock	*Prunella modularis*
Eagle, Bald	*Haliaeetus leucocephalus*
Eagle, Crowned	*Stephanoaetus coronatus*
Eagle, Golden	*Aquila chrysaeotos*
Eagle, Greater Spotted	*Aquila clanga*
Eagle, Harpy	*Harpia harpyja*
Eagle, Imperial	*Aquila heliaca*
Eagle, Lesser Spotted	*Aquila pomarina*
Eagle, Solitary	*Harpyhaliaetus solitarius*
Eagle, Verreaux's	*Aquila verreauxi*
Eagle, White-tailed	*Haliaeetus albicilla*
Egret, Cattle	*Bubulcus ibis*
Egret, Great	*Casmerodius albus*
Egret, Reddish	*Egretta rufescens*
Egret, Snowy	*Egretta thula*
Eider, Common	*Somateria mollissima*
Eider, Spectacled	*Somateria fischeri*
Elaenia, Lesser	*Elaenia chiriquensis*
Elaenia, Yellow-bellied	*Elaenia flavogaster*
Euphonia, Thick-billed	*Euphonia laniirostris*
Euphonia, Violaceous	*Euphonia violacea*
Falcon, Eleanora's	*Falco eleanorae*
Falcon, Peregrine	*Falco peregrinus*
Falcon, Prairie	*Falco mexicanus*
Falcon, Sooty	*Falco concolor*
Falconet, Philippine	*Microhierax erythrogonys*
Falconet, Spot-winged	*Spiziapteryx circumcinctus*
Fieldfare	*Turdus pilaris*
Finch, Cactus	*Geospiza scandens*
Finch, Cassin's	*Carpodacus cassinii*
Finch, Gouldian	*Chloebia gouldiae*
Finch, House	*Carpodacus mexicanus*
Finch, Large Cactus	*Geospiza conirostris*
Finch, Melba; see Pytilia, Green-winged	
Finch, Plush-capped	*Catamblyrhynchus diadema*
Finch, Purple	*Carpodacus purpureus*
Finch, Rosy	*Leucosticte arctoa*
Finch, Woodpecker	*Camarhynchus pallidus*
Finch, Yellow-thighed	*Pselliophorus tibialis*
Finch, Zebra	*Poephila guttata*
Finch; see also Fire-Finch and Ground-Finch	
Fire-Finch, Jameson's	*Lagonosticta rhodopareia*
Firecrest	*Regulus ignicapillus*

English	Scientific
Flamingo, Chilean	*Phoenicopterus chilensis*
Flamingo, Greater	*Phoenicopterus ruber*
Flamingo, Lesser	*Phoeniconaias minor*
Flicker, Andean	*Colaptes rupicola*
Flicker, Gilded	*Colaptes auratus chrysoides*
Flicker, Northern	*Colaptes auratus*
Flicker, Red-shafted	*Colaptes auratus cafer*
Flicker, Yellow-shafted	*Colaptes auratus auratus*
Flower-piercer, Masked	*Diglossa cyanea*
Flowerpecker, Black-sided	*Dicaeum celebicum*
Flycatcher, Alder	*Empidonax alnorum*
Flycatcher, Brown-crested	*Myiarchus tyrannulus*
Flycatcher, Collared	*Ficedula albicollis*
Flycatcher, Crowned Slaty	*Empidonomous aurantioatro-cristatus*
Flycatcher, Dusky	*Empidonax oberholseri*
Flycatcher, Dusky-capped	*Myiarchus tuberculifer*
Flycatcher, Gray	*Empidonax wrightii*
Flycatcher, Grenada	*Myiarchus nugator*
Flycatcher, La Sagra's	*Myiarchus sagrae*
Flycatcher, Least	*Empidonax minimus*
Flycatcher, Lesser Antillean	*Myiarchus oberi*
Flycatcher, Ochre-bellied	*Mionectes oleagineus*
Flycatcher, Pied	*Ficedula hypoleuca*
Flycatcher, Puerto Rican	*Myiarchus antillarum*
Flycatcher, Rufous-tailed	*Myiarchus validus*
Flycatcher, Rusty-margined	*Myiozetetes cayanensis*
Flycatcher, Sad	*Myiarchus barbirostris*
Flycatcher, Scissor-tailed	*Tyrannus forficatus*
Flycatcher, Spotted	*Muscicapa striata*
Flycatcher, Stolid	*Myiarchus stolidus*
Flycatcher, Sulphur-rumped	*Myiobius barbatus*
Flycatcher, Willow	*Empidonax traillii*
Fowl, Mallee	*Leipoa ocellata*
Fregatebird, Lesser	*Fregata ariel*
Frigatebird, Ascension	*Fregata aquila*
Frigatebird, Great	*Fregata minor*
Frigatebird, Magnificent	*Fregata magnificens*
Frogmouth, Javan	*Batrachostomus javensis*
Frogmouth, Tawny	*Podargus strigoides*
Fulmar, Northern	*Fulmarus glacialis*
Gallinule, Purple	*Porphyrula martinica*
Gannet, Northern	*Sula bassanus*
Gerygone, Black-throated	*Gerygone palpebrosa*
Giant-Petrel, Antarctic	*Macronectes giganteus*
Gnatcatcher, Blue-gray	*Polioptila caerulea*
Gnateater, Chestnut-belted	*Conopophaga aurita*
Godwit, Bar-tailed	*Limosa lapponica*
Godwit, Hudsonian	*Limosa haemastica*

English	Scientific
Goldcrest	*Regulus regulus*
Golden-Plover, Lesser	*Pluvialis dominica*
Goldeneye, Common	*Bucephala clangula*
Goldfinch, American	*Carduelis tristis*
Goldfinch, European	*Carduelis carduelis*
Gonolek, Black-headed	*Laniarius barbarus*
Goose, Canada	*Branta canadensis*
Goose, Greylag	*Anser anser*
Goose, Kelp	*Chloephaga hybrida*
Goose, Pied	*Anseranas semipalmata*
Goose, Ross'	*Chen rossii*
Goose, Snow	*Chen caerulescens*
Goshawk, Northern	*Accipiter gentilis*
Grackle, Boat-tailed	*Quiscalus major*
Grackle, Common	*Quiscalus quiscula*
Grackle, Great-tailed	*Quiscalus mexicanus*
Grasshopper-Warbler, Pale	*Locustella naevia*
Grebe, Atitlan	*Podilymbus gigas*
Grebe, Clark's	*Aechmophorus clarkii*
Grebe, Eared	*Podiceps nigricollis*
Grebe, Great Crested	*Podiceps cristatus*
Grebe, Horned	*Podiceps auritus*
Grebe, Pied-billed	*Podilymbus podiceps*
Grebe, Short-winged	*Rollandia micropterum*
Grebe, Western	*Aechmophorus occidentalis*
Greenfinch, European	*Carduelis chloris*
Grenadier, Purple	*Uraeginthus ianthinogaster*
Grosbeak, Black-headed	*Pheucticus melanocephalus*
Grosbeak, Evening	*Coccothraustes vespertinus*
Grosbeak, Pine	*Pinicola enucleator*
Grosbeak, Rose-breasted	*Pheucticus ludovicianus*
Ground-Dove, Blue	*Claravis pretiosa*
Ground-Dove, Ruddy	*Columbina talpacoti*
Ground-Finch, Large	*Geospiza magnirostris*
Ground-Finch, Medium	*Geospiza fortis*
Ground-Finch, Sharp-beaked	*Geospiza difficilis*
Ground-Finch, Small	*Geospiza fuliginosa*
Grouse, Black	*Tetrao tetrix*
Grouse, Hazel	*Tetrastes bonasia*
Grouse, Red; see Ptarmigan, Willow	
Grouse, Ruffed	*Bonasa umbellus*
Grouse, Sage	*Centrocercus urophasianus*
Grouse, Sharp-tailed	*Tympanuchus phasianellus*
Grouse, Spruce	*Dendragapus canadensis*
Guillemot, Black	*Cepphus grylle*
Guineafowl, Helmeted	*Numida meleagris*
Guineafowl, Vulturine	*Acryllium vulturinum*

English	Scientific
Gull, Black-billed	*Larus bulleri*
Gull, California	*Larus californicus*
Gull, Common Black-headed	*Larus ridibundus*
Gull, Glaucous	*Larus hyperboreus*
Gull, Glaucous-winged	*Larus glaucescens*
Gull, Gray	*Larus modestus*
Gull, Great Black-backed	*Larus marinus*
Gull, Herring	*Larus argentatus*
Gull, Iceland	*Larus glaucoides*
Gull, Laughing	*Larus atricilla*
Gull, Ring-billed	*Larus delawarensis*
Gull, Sabine's	*Xema sabini*
Gull, Swallow-tailed	*Creagrus furcatus*
Gull, Thayer's	*Larus thayeri*
Gull, Western	*Larus occidentalis*
Hammerhead	*Scopus umbretta*
Hawfinch	*Coccothraustes coccothraustes*
Hawk, Besra Sparrow	*Accipiter virgatus*
Hawk, Broad-winged	*Buteo platypterus*
Hawk, Cooper's	*Accipiter cooperi*
Hawk, Eurasian Sparrow	*Accipiter nisus*
Hawk, Galapagos	*Buteo galapagoensis*
Hawk, Harris'	*Parabuteo unicinctus*
Hawk, Red-tailed	*Buteo jamaicensis*
Hawk, Rough-legged	*Buteo lagopus*
Hawk, Sharp-shinned	*Accipiter striatus*
Hawk, Shikra	*Accipiter badius*
Hawk, Swainson's	*Buteo swainsoni*
Hawk, Tiny	*Accipiter superciliosus*
Hemipode, Andalusian	*Turnix sylvatica*
Hemispingus, Black-eared	*Hemispingus melanotis*
Hemispingus, Superciliaried	*Hemispingus superciliaris*
Hermit, Long-tailed	*Phaethornis superciliosus*
Hermit, Reddish	*Phaethornis ruber*
Heron, Boat-billed	*Cochlearius cochlearius*
Heron, Goliath	*Ardea goliath*
Heron, Gray	*Ardea cinerea*
Heron, Great Blue	*Ardea herodias*
Heron, Green-backed	*Butorides striatus*
Heron, Little Blue	*Egretta caerulea*
Heron, Purple	*Ardea purpurea*
Heron; see also Night-Heron	
Hillstar, Andean	*Oreotrochilus estella*
Hoatzin	*Opisthocomus hoazin*
Honey-Buzzard	*Pernis apivorus*
Honeyeater, New Holland	*Phylidornis novaehollandiae*
Honeyguide, Black-throated	*Indicator indicator*

English	Scientific
Honeyguide, Cassin's	*Prodotiscus insignis*
Honeyguide, Least	*Indicator exilis*
Honeyguide, Lesser	*Indicator minor*
Honeyguide, Lyre-tailed	*Melichneutes robostus*
Honeyguide, Scaly-throated	*Indicator variegatus*
Honeyguide, Yellow-rumped	*Indicator xanthonotus*
Honeyguide, Wahlberg's	*Prodotiscus, regulus*
Hoopoe	*Upupa epops*
Hornbill, Ground	*Bucorvus leadbeateri*
Hornbill, Indian Pied	*Anthracoceros albirostris*
Hornero, Rufous	*Furnarius rufus*
House-Martin, Common	*Delichon urbica*
Huia	*Heteralocha acutirostris*
Hummingbird, Anna's	*Calypte anna*
Hummingbird, Antillean Crested	*Orthorhynchus cristatus*
Hummingbird, Bahama Woodstar	*Calliphlox evelynae*
Hummingbird, Bee	*Mellisuga helenae*
Hummingbird, Black-chinned	*Archilochus alexandri*
Hummingbird, Broad-tailed	*Selasphorus platycercus*
Hummingbird, Calliope	*Stellula calliope*
Hummingbird, Copper-rumped	*Amazilia tobaci*
Hummingbird, Costa's	*Calypte costae*
Hummingbird, Fiery-throated	*Panterpe insignis*
Hummingbird, Giant	*Patagona gigas*
Hummingbird, Magnificent	*Eugenes fulgens*
Hummingbird, Puerto Rican Emerald	*Chlorostilbon maugaeus*
Hummingbird, Purple-throated Carib	*Eulampis jugularis*
Hummingbird, Ruby-throated	*Archilochus colubris*
Hummingbird, Ruby-topaz	*Chrysolampis mosquitus*
Hummingbird, Rufous	*Selasphorus rufus*
Hummingbird, Rufous-tailed	*Amazilia tzacatl*
Hummingbird, Streamer-tail	*Trochilus polytmus*
Hummingbird, Tufted Coquette	*Lophornis ornata*
Hummingbird, Vervain	*Mellisuga minima*
Hummingbird, Volcano	*Selasphorus flammula*
Ibis, Buff-necked	*Theristicus caudatus*
Ibis, Glossy	*Plegadis falcinellus*

English	Scientific
Ibis, Sacred	*Threskiornis aethiopicus*
Ibisbill	*Ibidorhyncha struthersii*
Indigobird, Village	*Vidua chalybeata*
Jacamar, White-chinned	*Galbula tombacea*
Jacana, Lesser	*Microparra capensis*
Jacana, Northern	*Jacana spinosa*
Jacana, Pheasant-tailed	*Hydrophasianus chirurgus*
Jackdaw, Common	*Corvus monedula*
Jaeger, Pomarine	*Stercorarius pomarinus*
Jay, Blue	*Cyanocitta cristata*
Jay, Eurasian	*Garrulus glandarius*
Jay, Florida Scrub	*Aphelocoma c. coerulescens*
Jay, Gray	*Perisoreus canadensis*
Jay, Gray-breasted	*Aphelocoma ultramarina*
Jay, Green	*Cyanocorax yncas*
Jay, Pinyon	*Gymnorhinus cyanocephalus*
Jay, Scrub	*Aphelocoma coerulescens*
Jay, Steller's	*Cyanocitta stelleri*
Junco, Dark-eyed	*Junco hyemalis*
Junco, Yellow-eyed	*Junco phaeonotus*
Junglefowl, Red	*Gallus gallus*
Kagu	*Rhynochetos jubatus*
Kakapo	*Strigops habroptilus*
Kea	*Nestor notabilis*
Kestrel, American	*Falco sparverius*
Kestrel, Eurasian	*Falco tinnunculus*
Kestrel, Mauritius	*Falco punctatus*
Killdeer	*Charadrius vociferus*
Kingbird, Eastern	*Tyrannus tyrannus*
Kingbird, Gray	*Tyrannus dominicensis*
Kingfisher, Belted	*Ceryle alcyon*
Kingfisher, Pied	*Ceryle rudis*
Kingfisher, Shovel-billed	*Clytoceyx rex*
Kinglet, Golden-crowned	*Regulus satrapa*
Kiskadee, Great	*Pitangus sulphuratus*
Kittiwake, Black-legged	*Rissa tridactyla*
Kiwi, Brown	*Apteryx australis*
Knot, Red	*Calidris canutus*
Koel, Common	*Eudynamys scolopacea*
Lark, Horned	*Eremophila alpestris*
Laughing-thrush, White-crested	*Garrulax leucolophus*
Longclaw, Yellow-throated	*Macronyx croceus*
Longspur, Chestnut-colored	*Calcarius ornatus*
Longspur, Lapland	*Calcarius lapponicus*
Loon, Common	*Gavia immer*
Lorikeet, Rainbow	*Trichoglossus haematod*
Lorikeet, Red-chinned	*Charmosyna rubrigularis*
Lorikeet, Red-flanked	*Charmosyna placentis*

English	Scientific
Lorikeet, Red-spotted	*Charmosyna rubronotata*
Lovebird, Masked	*Agapornis personata*
Lovebird, Peach-faced	*Agapornis roseicollis*
Lyrebird, Superb	*Menura superba*
Macaw, Chestnut-fronted	*Ara severa*
Macaw, Scarlet	*Ara macao*
Magpie, Black-backed	*Gymnorhina tibicen*
Magpie, Black-billed	*Pica pica*
Magpie, Green	*Cissa chinensis*
Magpie, Yellow-billed	*Pica nuttalli*
Magpie-Robin, Seychelles	*Copsychus sechellarum*
Malimbe, Cassin's	*Malimbus cassini*
Mallard; see Duck	
Mallee-Fowl	*Leipoa ocellata*
Manakin, Blue-backed	*Chiroxiphia pareola*
Manakin, Club-winged	*Allocotopterus deliciosus*
Manakin, Red-capped	*Pipra mentalis*
Manakin, Swallow-tailed	*Chiroxiphia caudata*
Manakin, White-bearded	*Manacus manacus*
Martin, African River	*Pseudochelidon eurystomina*
Martin, Purple	*Progne subis*
Martin, White-eyed River	*Pseudochelidon sirintarae*
Martin, see also House-Martin	
Meadowlark, Eastern	*Sturnella magna*
Meadowlark, Western	*Sturnella neglecta*
Merganser, Hooded	*Lophodytes cucullatus*
Merlin	*Falco columbarius*
Mockingbird, Bahama	*Mimus gundlachii*
Mockingbird, Northern	*Mimus polyglottos*
Motmot, Blue-crowned	*Momotus momota*
Motmot, Blue-throated	*Aspatha gularis*
Motmot, Turquoise-browed	*Eumomota superciliosa*
Mountain-Tanager, Black-cheeked	*Anisognathus melanogenys*
Mountain-Tanager, Blue-winged	*Anisognathus flavinuchus*
Mountain-Tanager, Buff-breasted	*Dubusia taeniata*
Mountain-Tanager, Chestnut-bellied	*Dubusia castaneoventris*
Mountain-Tanager, Hooded	*Buthraupis montana*
Mountain-Tanager, Masked	*Buthraupis wetmorei*
Mousebird, Speckled	*Colius striatus*
Munia, White-rumped	*Lonchura striata*
Murre, Common	*Uria aalge*
Murre, Thick-billed	*Uria lomvia*
Murrelet, Ancient	*Synthliboramphus antiquus*
Murrelet, Xantus'	*Synthliboramphus hypoleucus*

English	Scientific
Myna, Crested	*Acridotheres cristatellus*
Myna, Hill	*Gracula religiosa*
Native-Hen, Tasmanian	*Tribonyx mortierii*
Night-Heron, Black-crowned	*Nycticorax nycticorax*
Nighthawk, Common	*Chordeiles minor*
Nighthawk, Lesser	*Chordeiles acutipennis*
Nightingale, Thrush	*Erithacus luscinia*
Nightjar, European	*Caprimulgus europaeus*
Nightjar, Pennant-winged	*Semeiophorus vexillarius*
Nightjar, Standard-wing	*Macrodipteryx longipennis*
Noddy, Black	*Anous tenuirostris*
Noddy, Blue-gray	*Procelsterna cerulea*
Noddy, Brown	*Anous stolidus*
Nunbird, White-fronted	*Monasa morphoeus*
Nutcracker, Clark's	*Nucifraga columbiana*
Nutcracker, Eurasian	*Nucifraga caryocatactes*
Nuthatch, Brown-headed	*Sitta pusilla*
Nuthatch, Eurasian	*Sitta europaea*
Nuthatch, Persian	*Sitta tephronota*
Nuthatch, Pygmy	*Sitta pygmaea*
Nuthatch, Red-breasted	*Sitta canadensis*
Nuthatch, Rock	*Sitta neumayer*
Nuthatch, White-breasted	*Sitta carolinensis*
Oilbird	*Steatornis caripensis*
Oriole, Baltimore	*Icterus galbula galbula*
Oriole, Bullock's	*Icterus galbula bullocki*
Oriole, Hooded	*Icterus cucullatus*
Oriole, Northern	*Icterus galbula*
Oriole, Orchard	*Icterus spurius*
Oropendola, Chestnut-headed	*Psarocolius wagleri*
Oropendola, Crested	*Psarocolius decumanus*
Osprey	*Pandion haliaetus*
Ostrich	*Struthio camelus*
Ovenbird	*Seiurus aurocapillus*
Owl, Barn	*Tyto alba*
Owl, Barred	*Strix varia*
Owl, Burrowing	*Athene cunicularia*
Owl, Common Scops	*Otus scops*
Owl, Eastern Screech	*Otus asio*
Owl, Elf	*Micrathene whitneyi*
Owl, Flammulated	*Otus flammeolus*
Owl, Great Horned	*Bubo virginianus*
Owl, Little	*Athene noctua*
Owl, Long-eared	*Asio otus*
Owl, Northern Hawk	*Surnia ulula*
Owl, Northern Eagle	*Bubo bubo*
Owl, Snowy	*Nyctea scandiaca*

English	Scientific
Owl, Tawny	*Strix aluco*
Owl, Whiskered Screech	*Otus trichopsis*
Owlet-Nightjar, Australian	*Aegotheles cristatus*
Oystercatcher, Eurasian	*Haematopus ostralegus*
Palmchat	*Dulus dominicus*
Parakeet, Blossom-headed	*Psittacula roseata*
Parakeet, Monk	*Myiopsitta monachus*
Parakeet, Orange-fronted	*Aratinga canicularis*
Parakeet, Plum-headed	*Psittacula cyanocephala*
Parrot, Bourke's	*Neophema bourkii*
Parrot, Golden-shouldered	*Psephotus chrysopterygius*
Parrot, Gray	*Psittacus erithacus*
Parrot, Owl: see Kakapo	
Partridge, Red-legged	*Alectoris rufa*
Peafowl, Common	*Pavo cristatus*
Pelican, Brown	*Pelecanus occidentalis*
Penguin, Adelie	*Pygoscelis adeliae*
Penguin, Cape	*Spheniscus demersus*
Penguin, Chinstrap	*Pygocelis antarctica*
Penguin, Emperor	*Aptenodytes forsteri*
Penguin, Erect-crested	*Eudyptes sclateri*
Penguin, Humboldt	*Spheniscus humboldti*
Penguin, King	*Aptenodytes patagonicus*
Penguin, Little Fairy	*Eudyptula minor*
Penguin, Rockhopper	*Eudyptes crestatus*
Penguin, Yellow-eyed	*Megadyptes antipodes*
Petrel, Cape	*Daption capense*
Petrel, Phoenix	*Pterodroma alba*
Petrel, Snow	*Pagodroma nivea*
Petrel; see also Giant-Petrel and Storm-Petrel	
Pewee, Tropical	*Contopus cinereus*
Phainopepla	*Phainopepla nitens*
Pheasant, Blood	*Ithaginis cruentus*
Pheasant, Golden	*Chrysolophus pictus*
Pheasant, Great Argus	*Argusianus argus*
Pheasant, Himalayan Monal	*Lophophorus impejanus*
Pheasant, Lady Amherst	*Chrysolophus amherstiae*
Pheasant, Ring-necked	*Phasianus colchicus*
Phoebe, Black	*Sayornis nigricans*
Phoebe, Eastern	*Sayornis phoebe*
Phoebe, Say's	*Sayornis saya*
Pigeon, Black-chequered Homing	*Columbia livia*
Pigeon, Crested	*Ocyphaps lophotes*
Pigeon, Long-tailed	*Reinwardtoena crassirostris*
Pigeon, Nicobar	*Caloenas nicobarica*
Pigeon, Passenger	*Ectopistes migratorius*
Pigeon, Tooth-billed	*Didunculus strigirostris*
Pigeon, Wood	*Columba palumbus*
Pitta, Banded	*Pitta guajana*
Plover, Black-bellied	*Pluvialis squatarola*
Plover, Blacksmith	*Vanellus armatus*
Plover, Common Ringed	*Charadrius hiaticula*
Plover, Crab	*Dromas ardeola*
Plover, Egyptian	*Pluvianus aegyptius*
Plover, Little Ringed	*Charadrius dubius*
Plover, Mountain	*Charadrius montanus*
Plover, Wilson's	*Charadrius wilsonia*
Plover; see also Golden-Plover	
Poorwill, Common	*Phalaenoptilus nuttallii*
Potoo, Common	*Nyctibius griseus*
Prairie-Chicken, Greater	*Tympanuchus cupido*
Prion, Fairy	*Pachyptila turtur*
Ptarmigan, Rock	*Lagopus mutus*
Ptarmigan, White-Tailed	*Lagopus leucurus*
Ptarmigan, Willow	*Lagopus lagopus*
Puffbird, White-eared	*Nystacalus chacuru*
Puffin, Atlantic	*Fratercula arctica*
Pytilia, Green-winged	*Pytilia melba*
Quail, California	*Callipepla californica*
Quail, European	*Coturnix coturnix*
Quail, Gambel's	*Callipepla gambelii*
Quail, Japanese	*Coturnix japonica*
Quail, Painted	*Excalfactoria chinensis*
Quail-thrush, Spotted	*Cinclosoma punctatum*
Quelea, Red-billed	*Quelea quelea*
Quetzal, Resplendent	*Pharomachrus mocinno*
Rail, King	*Rallus elegans*
Rail, Water	*Rallus aquaticus*
Raven, Brown-necked	*Corvus ruficollis*
Raven, Common	*Corvus corax*
Razorbill; see Auk	
Redpoll, Common	*Carduelis flammea*
Redshank, Common	*Tringa totanus*
Redstart, American	*Setophaga ruticilla*
Redstart, Collared	*Myioborus torquatus*
Redstart, Common	*Phoenicurus phoenicurus*
Redstart, Slate-throated	*Myioborus miniatus*
Reed-Warbler, European	*Acrocephalus scirpaceus*
Reed-Warbler, Great	*Acrocephalus arundinaceus*
Rhea, Greater	*Rhea americana*
Riflebird, Magnificent	*Ptiloris magnificus*
Roadrunner, Greater	*Geococcyx californianus*
Robin, American	*Turdus migratorius*
Robin, Clay-colored	*Turdus grayi*
Robin, European	*Erithacus rubecula*

English	Scientific
Roller, Broad-billed	*Eurystomus glaucurus*
Roller, Cuckoo	*Leptosomus discolor*
Roller, Lilac-breasted	*Coracias caudata*
Rook, Eurasian	*Corvus frugilegus*
Rosy-Finch, Gray-crowned	*Leucosticte tephrocotis*
Ruff	*Philomachus pugnax*
Rush-Tyrant, Many-colored	*Tachuris rubrigastra*
Sanderling	*Calidris alba*
Sandgrouse, Chestnut-bellied	*Pterocles exustus*
Sandgrouse, Namaqua	*Pterocles namaqua*
Sandpiper, Buff-breasted	*Tryngites subruficollis*
Sandpiper, Curlew	*Calidris ferruginea*
Sandpiper, Dunlin	*Calidris alpina*
Sandpiper, Green	*Tringa ocrophus*
Sandpiper, Pectoral	*Calidris melanotos*
Sandpiper, Sharp-tailed	*Calidris acuminata*
Sandpiper, Spotted	*Actitis macularia*
Sandpiper, Western	*Calidris mauri*
Sandpiper, White-rumped	*Calidris fuscicollis*
Sandpiper, Wood	*Tringa glareola*
Screamer, Horned	*Anhima cornuta*
Screamer, Northern	*Chauna chavaria*
Secretary Bird	*Sagittarius serpentarius*
Seedeater, Dull-colored	*Sporophila obscura*
Serin, Common	*Serinus serinus*
Shearwater, Audubon's	*Puffinus lherminieri*
Shearwater, Christmas	*Puffinus nativitatis*
Shearwater, Greater	*Puffinus gravis*
Shearwater, Manx	*Puffinus puffinus*
Shearwater, Short-tailed	*Puffinus tenuirostris*
Shearwater, Sooty	*Puffinus griseus*
Shearwater, Wedge-tailed	*Puffinus pacificus*
Sheathbill, Snowy	*Chionis alba*
Shikra; see Hawk	
Shoveler, Northern; see Duck	
Shrike, Bull-headed	*Lanius bucephalus*
Shrike, Common Fiscal	*Lanius collaris*
Shrike, Loggerhead	*Lanius ludovicianus*
Siskin, Pine	*Carduelis pinus*
Sittella, Orange-winged	*Neositta chrysoptera*
Skimmer, Black	*Rynchops niger*
Skua, Great	*Catharacta skua*
Skua, South Polar	*Catharacta maccormicki*
Skylark	*Alauda arvensis*
Smew; see Duck	
Snipe, Common	*Gallinago gallinago*
Snipe, Greater Painted	*Rostratula benghalensis*

English	Scientific
Solitaire	*Pezophaps solitarius*
Sparrow, American Tree	*Spizella arborea*
Sparrow, Chipping	*Spizella passerina*
Sparrow, Clay-colored	*Spizella pallida*
Sparrow, Eurasian Tree	*Passer montanus*
Sparrow, Field	*Spizella pusilla*
Sparrow, Fox	*Passerella iliaca*
Sparrow, Gambel's White-crowned	*Zonotrichia leucophrys gambelli*
Sparrow, Golden-crowned	*Zonotrichia atricapilla*
Sparrow, Grasshopper	*Ammodramus savannarum*
Sparrow, Harris'	*Zonotrichia querula*
Sparrow, House	*Passer domesticus*
Sparrow, Ipswich	*Passerculus sandwichensis princeps*
Sparrow, Lark	*Chondestes grammacus*
Sparrow, LeConte's	*Ammodramus leconteii*
Sparrow, Rock	*Petronia petronia*
Sparrow, Rufous-collared	*Zonotrichia capensis*
Sparrow, Sage	*Amphispiza belli*
Sparrow, Savannah	*Passerculus sandwichensis*
Sparrow, Seaside	*Ammodramus maritimus*
Sparrow, Sharp-tailed	*Ammodramus caudacutus*
Sparrow, Song	*Melospiza melodia*
Sparrow, Spanish	*Passer hispaniolensis*
Sparrow, Swamp	*Melospiza georgiana*
Sparrow, Vesper	*Pooecetes gramineus*
Sparrow, White-crowned	*Zonotrichia leucophrys*
Sparrow, White-throated	*Zonotrichia albicollis*
Spatule-tail, Marvelous	*Loddigesia mirabilis*
Spiderhunter, Little	*Arachnothera longirostra*
Spoonbill, Eurasian	*Platalea leucorodia*
Starling, Brahminy	*Sturnus pagodarum*
Starling, European	*Sturnus vulgaris*
Starling, Grosbeak	*Scissirostrum dubium*
Stilt, Banded	*Cladorhynchus leucocephalus*
Stint, Little	*Calidris minuta*
Stint, Temminck's	*Calidris temminckii*
Stonechat	*Saxicola torquata*
Stork, Marabou	*Leptoptilos crumeniferus*
Stork, Whale-headed	*Balaeniceps rex*
Stork, White	*Ciconia ciconia*
Stork, White-bellied	*Ciconia abdimii*
Stork, Wood	*Mycteria americana*
Storm-Petrel, Band-rumped	*Oceanodroma castro*
Storm-Petrel, Leach's	*Oceanodroma leucorhoa*
Storm-Petrel, Least	*Oceanodroma microsoma*
Storm-Petrel, White-faced	*Pelagodroma marina*

English	Scientific
Storm-Petrel, White-throated	*Nesofregetta albigularis*
Storm-Petrel, Wilson's	*Oceanites oceanicus*
Streamer-tail, see Hummingbird	
Sunbird, Beautiful	*Nectarinia pulchella*
Sunbird, Bronzy	*Nectarinia kilimensis*
Sunbird, Golden-winged	*Nectarinia reichenowi*
Sunbird, Malachite	*Nectarinia famosa*
Sunbird, Variable	*Nectarinia venusta*
Sunbittern	*Eurypyga helias*
Swallow, Bank	*Riparia riparia*
Swallow, Barn	*Hirundo rustica*
Swallow, Cliff	*Hirundo pyrrhonota*
Swallow, Northern Rough-winged	*Stelgidopteryx serripennis*
Swallow, Southern Rough-winged	*Stelgidopteryx ruficollis*
Swallow, Tree	*Tachycineta bicolor*
Swan, Bewick's	*Cygnus bewickii*
Swan, Black-necked	*Cygnus melanocoryphus*
Swan, Mute	*Cygnus olor*
Swan, Tundra	*Cygnus columbianus*
Swift, Black	*Cypseloides niger*
Swift, Common	*Apus apus*
Swift, White-collared	*Streptoprocne collaris*
Swiftlet, Edible-nest	*Collocalia fuciphaga*
Tanager, Blue-and-Black	*Tangara vassorii*
Tanager, Blue-capped	*Thraupis cyanocephala*
Tanager, Blue-gray	*Thraupis episcopus*
Tanager, Golden-collared	*Iridosornis jelskii*
Tanager, Magpie	*Cissopis leveriana*
Tanager, Palm	*Thraupis palmarum*
Tanager, Plain-colored	*Tangara inornata*
Tanager, Scarlet	*Piranga olivacea*
Tanager, Scarlet-rumped	*Ramphocelus passerinii*
Tanager, Silver-throated	*Tangara icterocephala*
Teal, Blue-winged	*Anas discors*
Teal, Cinnamon	*Anas cyanoptera*
Teal, Speckled	*Anas flavirostris*
Tern, Antarctic	*Sterna vittata*
Tern, Arctic	*Sterna paradisaea*
Tern, Caspian	*Sterna caspia*
Tern, Common	*Sterna hirundo*
Tern, Gray-backed	*Sterna lunata*
Tern, Crested	*Sterna bergii*
Tern, Inca	*Larosterna inca*
Tern, Least	*Sterna antillarum*
Tern, Little	*Sterna albifrons*
Tern, Royal	*Sterna maxima*
Tern, Sooty	*Sterna fuscata*
Tern, White	*Gygis alba*
Thornbird, Rufous-fronted	*Phacellodomus rufifrons*
Thrasher, Brown	*Toxostoma rufum*
Thrasher, California	*Toxostoma redivivum*
Thrasher, Curve-billed	*Toxostoma curvirostre*
Thrasher, Long-billed	*Toxostoma longirostre*
Thrush, Common Song	*Turdus philomelos*
Thrush, Gray-cheeked	*Catharus minimus*
Thrush, Hermit	*Catharus guttatus*
Thrush, Island	*Turdus poliocephalus*
Thrush, Mistle	*Turdus viscivorus*
Thrush, Olive-backed; see Thrush, Swainson's	
Thrush, Swainson's	*Catharus ustulatus*
Thrush, Wood	*Hylocichla mustelina*
Tinamou, Elegant Crested	*Eudromia elegans*
Tinamou, Great	*Tinamus major*
Tinamou, Little	*Crypturellus soui*
Tit, Bearded	*Panurus biarmicus*
Tit, Blue	*Parus caeruleus*
Tit, Coal	*Parus ater*
Tit, Crested	*Parus cristatus*
Tit, Great	*Parus major*
Tit, Long-tailed	*Aegithalos caudatus*
Tit, Marsh	*Parus palustris*
Tit, Siberian	*Parus cinctus*
Tit, Willow	*Parus montanus*
Titmouse, Black-crested	*Parus bicolor atricristatus*
Titmouse, Tufted	*Parus bicolor bicolor*
Tody, Puerto Rican	*Todus mexicanus*
Toucan, Keel-billed	*Ramphastos sulfuratus*
Towhee, Abert's	*Pipilo aberti*
Towhee, Brown	*Pipilo fuscus*
Towhee, Collared	*Pipilo ocai*
Towhee, Rufous-sided	*Pipilo erythrophthalmus*
Triller, Polynesian	*Lalage maculosa*
Triller, Varied	*Lalage leucomela*
Trogon, Collared	*Trogon collaris*
Trogon, Diard's	*Harpactes diardii*
Trogon, Red-headed	*Harpactes erythrocephalus*
Trogon, Violaceous	*Trogon violaceus*
Tropicbird, Red-tailed	*Phaethon rubricauda*
Tropicbird, White-tailed	*Phaethon lepturus*
Trumpetbird	*Phonygammus keraudrenii*
Trumpeter, Gray-winged	*Psophia crepitans*
Turaco, Hartlaub's	*Turaco hartlaubi*
Turaco, Ross'	*Musophaga rossae*

English	Scientific
Turkey, Wild	*Meleagris gallopavo*
Turnstone, Ruddy	*Arenaria interpres*
Turtle-Dove, Ringed	*Streptopelia risoria*
Veery	*Catharus fuscescens*
Verdin	*Auriparus flaviceps*
Violet-Ear, Green	*Colibri thalassinus*
Vireo, Bell's	*Vireo bellii*
Vireo, Red-eyed	*Vireo olivaceus*
Vireo, Warbling	*Vireo gilvus*
Vireo, Yellow-throated	*Vireo flavifrons*
Vulture, Bearded	*Gypaetus barbatus*
Vulture, Black	*Coragyps atratus*
Vulture, Egyptian	*Neophron percnopterus*
Vulture, Griffon	*Gyps fulvus*
Vulture, Turkey	*Cathartes aura*
Wagtail, White	*Motacilla alba*
Wagtail, Yellow	*Motacilla flava*
Warbler, Adelaide's	*Dendroica adelaidae*
Warbler, Arctic	*Phylloscopus borealis*
Warbler, Arrow-headed	*Dendroica pharetra*
Warbler, Audubon's	*Dendroica coronata auduboni*
Warbler, Bay-breasted	*Dendroica castanea*
Warbler, Black-cheeked	*Basileuterus melanogenys*
Warbler, Blackburnian	*Dendroica fusca*
Warbler, Blackcap	*Sylvia atricapilla*
Warbler, Blackpoll	*Dendroica striata*
Warbler, Black-throated Green	*Dendroica virens*
Warbler, Blue-winged	*Vermivora pinus*
Warbler, Chestnut-sided	*Dendroica pensylvanica*
Warbler, Chiffchaff	*Phylloscopus collybita*
Warbler, Crowned Willow	*Phylloscopus occipitalis*
Warbler, Dartford	*Sylvia undata*
Warbler, European Marsh	*Acrocephalus palustris*
Warbler, Garden	*Sylvia borin*
Warbler, Golden-winged	*Vermivora chrysoptera*
Warbler, Greater Whitethroat	*Sylvia communis*
Warbler, Japanese Bush	*Cettia diphone*
Warbler, Kirtland's	*Dendroica kirtlandii*
Warbler, Magnolia	*Dendroica magnolia*
Warbler, Marmora's	*Sylvia sarda*
Warbler, Myrtle	*Dendroica coronata coronata*
Warbler, Nashville	*Vermivora ruficapilla*
Warbler, Orange-crowned	*Vermivora celata*
Warbler, Palm	*Dendroica palmarum*
Warbler, Prairie	*Dendroica discolor*
Warbler, Sardinian	*Sylvia melanocephala*
Warbler, Subalpine	*Sylvia cantillans*
Warbler, Tennessee	*Vermivora peregrina*
Warbler, Willow	*Phylloscopus trochilus*
Warbler, Worm-eating	*Helmitheros vermivorus*
Warbler, Yellow	*Dendroica petechia*
Warbler, Yellow-rumped	*Dendroica coronata*
Waterthrush, Northern	*Seiurus noveboracensis*
Wattlebird, Spine-cheeked	*Anthochaera rufogularis*
Waxbill, Black-bellied	*Lagonosticta rara*
Waxbill, Red-billed	*Lagonosticta senegala*
Waxwing, Bohemian	*Bombycilla garrulus*
Waxwing, Cedar	*Bombycilla cedrorum*
Weaver, Buffalo	*Bubalornis albirostris*
Weaver, Cuckoo	*Anomalospiza imberbis*
Weaver, Forest	*Ploceus bicolor*
Weaver, Golden-backed	*Ploceus jacksoni*
Weaver, Orange	*Ploceus aurantius*
Weaver, Social	*Philetairus socius*
Weaver, Village	*Ploceus cucullatus*
Weaver, Vitelline Masked	*Ploceus velatus*
Wheatear, Northern	*Oenanthe oenanthe*
Whimbrel	*Numenius phaeopus*
Whinchat	*Saxicola rubetra*
Whip-poor-will	*Caprimulgus vociferus*
White-eye, Bridled	*Zosterops conspicillata*
White-eye, Gray-backed	*Zosterops lateralis*
White-eye, Gray-brown	*Zosterops cinerea*
White-eye, Large Ponape	*Rukia longirostris*
White-eye, Mascarene	*Zosterops borbonica*
White-eye, Kikuyu	*Zosterops poliogaster*
White-eye, Palau	*Rukia palauensis*
White-eye, Reunion Olive	*Zosterops olivacea*
White-eye, Slender-billed	*Zosterops tenuirostris*
Whitethroat, see Warbler	
Whydah, Paradise	*Vidua paradisaea*
Whydah, Straw-tailed	*Vidua fischeri*
Widowbird, Long-tailed	*Euplectes progne*
Willet	*Catoptrophorus semipalmatus*
Wood-Partridge, Crested	*Rollulus rouloul*
Woodcock, American	*Scolopax minor*
Woodcock, Eurasian	*Scolopax rusticola*
Woodcreeper, Barred	*Dendrocolaptes certhia*
Woodcreeper, White-chinned	*Dendrocincla merula*
Woodhoopoe, Green	*Phoeniculus purpureus*
Woodpecker, Acorn	*Melanerpes formicivorus*
Woodpecker, Downy	*Picoides pubescens*
Woodpecker, Gila	*Melanerpes uropygialis*
Woodpecker, Great Spotted	*Dendrocopos major*

English	Scientific
Woodpecker, Hairy	*Picoides villosus*
Woodpecker, Ivory-billed	*Campephilus principalis*
Woodpecker, Lewis'	*Melanerpes lewis*
Woodpecker, Pileated	*Dryocopus pileatus*
Woodpecker, Red-bellied	*Melanerpes carolinus*
Woodpecker, Red-headed	*Melanerpes erythrocephalus*
Woodpecker, White-headed	*Picoides albolarvatus*
Wren, Bewick's	*Thryomanes bewickii*
Wren, Blue	*Malurus cyaneus*
Wren, Cactus	*Campylorhynchus brunneicapillus*
Wren, Canyon	*Catherpes mexicanus*
Wren, Carolina	*Thryothorus ludovicianus*
Wren, House	*Troglodytes aedon*
Wren, Marsh	*Cistothorus palustris*
Wren, Rock	*Salpinctes obsoletus*
Wren, Sedge	*Cistothorus platensis*
Wren, Splendid	*Malurus splendens*
Wren, Winter	*Troglodytes troglodytes*
Wrenthrush	*Zeledonia coronata*
Wrentit	*Chamaea fasciata*
Wryneck, Eurasian	*Jynx torquilla*
Yellowhammer	*Emberiza citrinella*
Yellowlegs, Lesser	*Tringa flavipes*
Yellowthroat, Common	*Geothlypis trichas*

Bibliography

Abbott, I., and P.R. Grant. 1976. Nonequilibrial bird faunas on islands. Am. Nat. 110: 507–528.

Abbott, I., L.K. Abbott, and P.R. Grant. 1977. Comparative ecology of Galápagos ground finches (*Geospiza* Gould): evaluation of the importance of floristic diversity and interspecific competition. Ecol. Monogr. 47: 151–184.

Abbott, I., L.K. Abbott, and P.R. Grant. 1975. Seed selection and handling ability of four species of Darwin's finches. Condor 77: 332–335.

Ackerman, R.A., G.C. Whittow, C.V. Paganelli, and T.N. Pettit. 1980. Oxygen consumption, gas exchange, and growth of embryonic Wedge-tailed Shearwaters *(Puffinus pacificus chlororhynchus)*. Physiol. Zool. 53: 210–221.

Ahlquist, J.E., F.H. Sheldon, and C.G. Sibley. 1984. The relationships of the Bornean Bristlehead *(Pityriasis gymnocephala)* and the Black-collared Thrush *(Chlamydochaera jefferyi)*. J. Ornithol. 125: 129–140.

Ainley, D.G., and D.P. DeMaster. 1980. Survival and mortality in a population of Adelie Penguins. Ecology 61: 522–530.

Ainley, D.G., R.E. LeRosche, and W.J.L. Sladen. 1983. Breeding Biology of the Adelie Penguin. Berkeley: University of California Press.

Åkerman, B. 1966ab. Behavioural effects of electrical stimulation in the forebrain of the pigeon, I and II. Behaviour 26: 323–338, 339–350.

Alcock, J. 1984. Animal Behavior (third ed.). Sunderland, MA: Sinauer Assoc. Inc.

Allee, W.C. 1938. The Social Life of Animals. New York: Norton.

Allee, W.C., and C.Z. Lutherman. 1940. An experimental study of certain effects of temperature on differential growth of pullets. Ecology 21: 29–33.

Alvarez del Toro, M. 1971. On the biology of the American Finfoot in southern Mexico. Living Bird 10: 79–88.

Amadon, D. 1980. Varying proportions between young and old raptors. Proc. IV Pan-Afr. Ornithol. Congr.: 327–331.

American Ornithologists' Union. 1983. Check-list of North American Birds (sixth ed.). Lawrence, KS: American Ornithologists' Union.

Ames, P.L. 1971. The morphology of the syrinx in passerine birds. Peabody Mus. Nat. Hist. Yale Univ. Bull. No. 37.

Ames, P.L. 1975. The application of syringeal morphology to the classification of the Old World insect-eaters. Bonn. Zool. Beitr. 26: 107–134.

Anderson, B.W., and R.J. Daugherty. 1974. Characteristics and reproductive biology of grosbeaks *(Pheucticus)* in the hybrid zone in South Dakota. Wilson Bull. 86: 1–11.

Anderson, D.W., J.R. Jehl Jr., R.W. Risebrough, L.A. Woods Jr., L.R. Deweese, and W.G. Edgecomb. 1975. Brown Pelicans: Improved reproduction off the southern California coast. Science (Wash., D.C.) 190: 806–808.

Andersson, M. 1976. Social behaviour and communication in the Great Skua. Behaviour 58: 40–77.

Andersson, M. 1978. Optimal egg shape in waders. Ornis Fenn. 55: 105–109.

Andersson, M. 1982. Female choice selects for extreme tail length in a widowbird. Nature (Lond.) 299: 818–820.

Andersson, M., and M.O.G. Eriksson. 1982. Nest parasitism in Goldeneyes *Bucephala clangula*: some evolutionary aspects. Am. Nat. 120: 1–16.

Andrewartha, H.G., and L.C. Birch. 1954. The Distribution and Abundance of Animals. Chicago: University of Chicago Press.

Ankney, C.D., and D.M. Scott. 1980. Changes in nutrient reserves and diet of breeding Brown-headed Cowbirds. Auk 97: 684–696.

Ankney, C.D., and C.D. MacInnes. 1978. Nutrient reserves and reproductive performance of female Lesser Snow Geese. Auk 95: 459–471.

Ar, A., and H. Rahn. 1980. Water in the avian egg: Overall budget of incubation. Am. Zool. 20: 373–384.

Armstrong, E.A. 1942. Bird Display. Cambridge: Cambridge University Press.

Armstrong, E.A. 1947. Bird Display and Behaviour. New York: Oxford University Press.

Armstrong, E.A. 1963. A Study of Bird Song. New York: Academic Press.

Arnold, A.P. 1980. Anatomical and electrophysiological studies of sexual dimorphism in a passerine vocal control system. Acta XVII Congr. Int. Ornithol.: 648–652.

Arnold, A.P. 1982. Neural control of passerine song. *In* Acoustic Communication in Birds, Vol. 1, D.E. Kroodsma and E.H. Miller, Eds.: 75–94. New York: Academic Press.

Arnold, A.P., and A. Saltiel. 1979. Sexual difference in pattern of hormone accumulation in the brain of a songbird. Science (Wash., D.C.) 205: 702–705.

Aschoff, J. 1955. Jahresperiodite der Fortpflanzung beim Warmblütern. Stud. Gen. 8: 742–776.

Aschoff, J. 1980. Biological clocks in birds. Acta XVII Congr. Int. Ornithol.: 113–136.

Ash, J.S. 1957. Post-mortem examinations of birds found dead during the cold spells of 1954 and 1956. Bird Study 4: 159–166.

Ashkenazie, S., and U.N. Safriel. 1979. Time-energy budget of the Semipalmated Sandpiper *(Calidris pusilla)* at Barrow, Alaska. Ecology 60: 783–799.

Ashmole, N.P. 1963a. The biology of the Wideawake or Sooty Tern *Sterna fuscata* on Ascension Island. Ibis 103b: 297–364.

Ashmole, N.P. 1963b. The regulation of numbers of tropical oceanic birds. Ibis 103b: 458–473.

Ashmole, N.P. 1965. Adaptive variation in the breeding regime of a tropical sea bird. Proc. Natl. Acad. Sci. U.S.A. 53: 311–318.

Ashmole, N.P. 1968. Breeding and molt in the White Tern *(Gygis alba)* on Christmas Island, Pacific Ocean. Condor 70: 35–55.

Ashmole, N.P. 1971. Seabird ecology and the marine environment. *In* Avian Biology, Vol. 1, D.S. Farner, J.R. King, and K.C. Parkes, Eds.: 223–286. New York: Academic Press.

Ashmole, N.P., and H.S. Tovar. 1968. Prolonged parental care in Royal Terns and other birds. Auk 85: 90–100.

Aulie, A. 1976. The pectoral muscles and the development of thermoregulation in chicks of Willow Ptarmigan *(Lagopus lagopus)*. Comp. Biochem. Physiol. 53A: 343–346.

Austin, G.T. 1976. Behavioral adaptations of the Verdin to the desert. Auk 93: 245–262.

Austin, O.L., Jr., and A. Singer. 1985. Families of Birds (new, revised ed.). New York: Golden Press.

Axelrod, R., and W.D. Hamilton. 1981. The evolution of cooperation. Science (Wash., D.C.) 211: 1390–1396.

Bagg, A.M., W.W.H. Gunn, D.S. Miller, J.T. Nichols, W. Smith, and F.P. Wolfarth. 1950. Barometric pressure patterns and spring bird migration. Wilson Bull. 62: 5–19.

Bailey, R.E. 1952. The incubation patch of passerine birds. Condor 54: 121–136.

Baker, A.J. 1980. Morphometric differentiation in New Zealand populations of the House Sparrow *(Passer domesticus)*. Evolution 34: 638–653.

Baker, M.C. 1975. Song dialects and genetic differences in white-crowned sparrows *(Zonotrichia leucophrys)*. Evolution 29: 226–241.

Baker, M.C. 1982. Genetic population structure and vocal dialects in *Zonotrichia* (Emberizidae). *In* Acoustic Communication in Birds, Vol. 2, D.E. Kroodsma and E.H. Miller, Eds.: 209–235. New York and London: Academic Press.

Baker, M.C., and S.F. Fox. 1978. Dominance, survival, and enzyme polymorphism in Dark-eyed Juncos, *Junco hyemalis.* Evolution 32: 697–711.

Bakker, R.T. 1975. Dinosaur renaissance. Sci. Am. 232(4): 58–78.

Balda, R.P. 1980. Recovery of cached seeds by a captive *Nucifraga caryocatactes.* Z. Tierpsychol. 52: 331–346.

Balph, M.H., D.F. Balph, and H.C. Romesburg. 1979. Social status signaling in winter flocking birds: An examination of a current hypothesis. Auk 96: 78–93.

Bancroft, G.T., and G.E. Woolfenden. 1982. The molt of Scrub Jays and Blue Jays in Florida. Ornithol. Monogr. No. 29.

Bang, B.G. 1971. Functional anatomy of the olfactory system in 23 orders of birds. Acta Anat. Suppl. 58: 1–76.

Bang, B.G., and S. Cobb. 1968. The size of the olfactory bulb in 108 species of birds. Auk 85: 55–61.

Banks, R.C. 1979. Human related mortality of birds in the United States. U.S. Fish Wildl. Serv. Spec. Sci. Rep. Wildl. No. 215.

Barash, D.P. 1977. Sociobiology of rape in Mallards *(Anas platyrhynchos):* Responses of the mated male. Science (Wash. D.C.) 197: 788–789.

Barraud, E.M. 1961. The development of behaviour in some young passerines. Bird Study 8: 111–118.

Barrowclough, G.F. 1978. Sampling bias in dispersal studies based on finite area. Bird-Banding 49: 333–341.

Barrowclough, G.F. 1980a. Gene flow, effective population sizes, and genetic variance components in birds. Evolution 34: 789–798.

Barrowclough, G.F. 1980b. Genetic and phenotypic differentiation in a wood warbler (genus *Dendroica*) hybrid zone. Auk 97: 655–668.

Barrowclough, G.F. 1982. Geographic variation, predictiveness, and subspecies. Auk 99: 601–603.

Barrowclough, G.F. 1983. Biochemical studies of microevolutionary processes. *In* Perspectives in Ornithology, A.H. Brush and G.A. Clark Jr., Eds.: 223–261. Cambridge: Cambridge University Press.

Barrowclough, G.F., and K.W. Corbin. 1978. Genetic variation and differentiation in the Parulidae. Auk 95: 691–702.

Barrowclough, G.F., K.W. Corbin, and R.M. Zink. 1981. Genetic differentiation in the Procellariiformes. Comp. Biochem. Physiol. 69B: 629–632.

Bartholomew, G.A. 1942. The fishing activities of Double-crested Cormorants on San Francisco Bay. Condor 44: 13–21.

Bartholomew, G.A. 1982. Body temperature and energy metabolism. *In* Animal Physiology: Principles and adaptations (fourth ed.), M.S. Gordon et al., Eds.: 333–406. New York: Macmillan.

Bartholomew, G.A., and T.J. Cade. 1963. The water economy of landbirds. Auk 80: 504–539.

Bartholomew, G.A., and C.H. Trost. 1970. Temperature regulation in the Speckled Mousebird, *Colius striatus.* Condor 72: 141–146.

Bateman, A.J. 1948. Intra-sexual selection in *Drosophila.* Heredity 2: 349–368.

Bateson, P.P.G. 1976. Specificity and the origins of behavior. *In* Advances in the Study of Behavior, Vol. 6, J. Rosenblatt, R.A. Hinde, and C. Beer, Eds.: 1–20. New York: Academic Press.

Batten, L.A., and J.H. Marchant. 1977. Bird population changes for the years 1974-75. Bird Study 24: 55–61.

Baxter, M., and M.D. Trotter. 1969. The effect of fatty materials extracted from keratins on the growth of fungi, with particular reference to the free fatty acid content. Sabouraudia 7: 199–206.

Baylis, J.R. 1982. Avian vocal mimicry: Its function and evolution. *In* Acoustic Communication in Birds, Vol. 2., D.E. Kroodsma and E.H. Miller, Eds.: 51–83. New York: Academic Press.

Becker, P.H. 1976. Artkennzeichnende Gesangsmerkmale bei Winter- und Sommergoldhänchen *(Regulus regulus, R. ignicapillus).* Z. Tierpsychol. 42: 411–437.

Becker, P.H. 1982. The coding of species-specific characteristics in bird sounds. *In* Acoustic Communication in Birds, Vol. 1, D.E. Kroodsma and E.H. Miller, Eds.: 213–252. New York: Academic Press.

Becking, J.H. 1975. The ultrastructure of the avian eggshell. Ibis 117: 143–151.

Beebe, W. 1907. Geographic variation in birds, with special reference to the effects of humidity. Zoologica 1: 1–41.

Beecher, M.D. 1982. Signature systems and kin recognition. Am. Zool. 22: 477–490.

Beecher, W.J. 1953. A phylogeny of the oscines. Auk 70: 270–333.

Beehler, B. 1983. Lek behavior of the Lesser Bird of Paradise. Auk 100: 992–995.

Beehler, B. 1985. Adaptive significance of monogamy in the Trumpet Manucode *Manucodia keraudrenii* (Aves: Paradisaeidae). Ornithol. Monog. 37: 83–99.

Beehler, B., and S.G. Pruett-Jones. 1983. Display dispersion and diet of birds of paradise: a comparison of nine species. Behav. Ecol. Sociobiol. 13: 229–238.

Beer, C.G. 1970. Individual recognition of voice in the social behaviour of birds. Adv. Study Behav. 3: 27–74.

Beer, C.G. 1979. Vocal communication between Laughing Gull parents and chicks. Behaviour 70: 118–146.

Behle, W.H., and W.A. Goates. 1957. Breeding biology of the California Gull. Condor 59: 235–246.

Bellairs, A. d'A., and C.R. Jenkin. 1960. The skeleton of birds. *In* Biology and Comparative Physiology of Birds, Vol. 1, A.J. Marshall, Ed.: 241–300. New York: Academic Press.

Bellairs, R. 1960. Development of birds. *In* Biology and Comparative Physiology of Birds, Vol. 1, A.J. Marshall, Ed.: 127–188. New York: Academic Press.

Bellrose, F.C. 1967. Orientation in waterfowl migration. Proc. Annu. Biol. Colloq. (Oreg. State Univ.) 27: 73–79.

Benedict, F.G., W. Landauer, and E.L. Fox. 1932. The physiology of normal and Frizzle fowl, with special reference to the basal metabolism. Storrs (Conn.) Agr. Expt. Sta. Bull. 177: 12–101.

Bennett, A.F. 1980. The metabolic foundations of vertebrate behavior. Bioscience 30: 452–456.

Bent, A.C. 1926. Life histories of North American marsh birds. U.S. Natl. Mus. Bull. No. 135.

Bent, A.C. 1939. Life histories of North American woodpeckers. U.S. Natl. Mus. Bull. No. 174.

Berger, A.J. 1957. On the anatomy and relationships of *Fregilupus varius,* an extinct starling from the Mascarene Islands. Bull. Am. Mus. Nat. Hist. 113: 225–272.

Berger, M., and J.S. Hart. 1974. Physiology and energetics of flight. *In* Avian Biology, Vol. 4, D.S. Farner, J.R. King, and K.C. Parkes, Eds.: 415–477. New York: Academic Press.

Berman, S.L., and R.J. Raikow. 1982. The hindlimb musculature of the mousebirds (Coliiformes). Auk 99: 41–57.

Berthold, P. 1975. Migration: Control and metabolic physiology. *In* Avian Biology, Vol. 5, D.S. Farner, J.R. King, and K.C. Parkes, Eds.: 77–128. New York: Academic Press.

Berthold, P., 1978. Circannuale Rhythmik: Freilaufende selbsterregte Periodik mit lebenslanger Wirksamkeit bei Vögeln. Naturwissenschaften 65: 546–547.

Berthold, P., and U. Querner. 1981. Genetic basis of migratory behavior in European warblers. Science (Wash., D.C.) 212: 77–79.

Bertram, B.C.R. 1980. Vigilance and group size in ostriches. Anim. Behav. 28: 278–286.

Best, L.B. 1978. Field Sparrow reproductive success and nesting ecology. Auk 95: 9–22.

Biebach, H. 1983. Genetic determination of partial migration in the European Robin *(Erithacus rubecula).* Auk 100: 601–606.

Bitterman, M.E. 1965. Phyletic differences in learning. Am. Psychol. 20: 396–410.

Blank, J.L., and V. Nolan Jr. 1983. Offspring sex ratio in red-winged blackbirds is dependent on maternal age. Proc. Natl. Acad. Sci. U.S.A. 80: 6141–6145.

Bledsoe, A.H. 1988. A phylogenetic analysis of postcranial skeletal characters of the ratite birds. Annals Carnegie Museum 57: 73–90.

Blockstein, D.E., and H.B. Tordoff. 1985. Gone forever—a contemporary look at the extinction of the Passenger Pigeon. Am. Birds 39: 845–851.

Boag, P.T., and P.R. Grant. 1978. Heritability of external morphology in Darwin's finches. Nature (Lond.) 274: 793–794.

Boag, P.T., and P.R. Grant. 1981. Intense natural selection in a population of Darwin's finches (Geospizinae) in the Galapagos. Science (Wash., D.C.) 214: 82–85.

Board, R.G., and G. Love. 1980. Magnesium distribution in avian eggshells. Comp. Biochem. Physiol. 66A: 667–672.

Board, R.G., and V.D. Scott. 1980. Porosity of the avian eggshell. Am. Zool. 20: 339–349.

Bock, C.E. 1980. Winter bird population trends: scientific evaluation of Christmas Bird count data. Atl. Nat. 33: 28–37.

Bock, C.E., and L.W. Lepthien. 1976a. Population growth in the Cattle Egret. Auk 93: 164–166.

Bock, C.E., and L.W. Lepthien. 1976b. Growth in the eastern House Finch population. 1962–1971. Am. Birds 30: 791–792.

Bock, C.E., and L.W. Lepthien. 1976c. Synchronous eruptions of boreal seed-eating birds. Am. Nat. 110: 559–571.

Bock, W.J. 1961. Salivary glands in the gray jays *(Perisoreus)*. Auk 78: 355–365.

Bock, W.J. 1963. The cranial evidence for ratite affinities. Proc. XIII Int. Ornithol. Congr.: 39–54.

Bock, W.J. 1965. The role of adaptive mechanisms in the origin of higher levels of organization. Syst. Zool. 14: 272–287.

Bock, W.J. 1966. An approach to the functional analysis of bill shape. Auk 83: 10–51.

Bock, W.J. 1973. Philosophical foundations of classical evolutionary classification. Syst. Zool. 22: 375–392.

Bock, W.J., and J. Farrand Jr. 1980. The number of species and genera of recent birds: a contribution to comparative systematics. Am. Mus. Novit. No. 2703.

Bock, W.J., and R.S. Hikida. 1968. An analysis of twitch and tonus fibers in the hatching muscle. Condor 70: 211–222.

Bock, W.J., and W.D. Miller. 1959. The scansorial foot of the woodpeckers, with comments on the evolution of perching and climbing feet in birds. Am. Mus. Novit. No. 1931.

Bock, W.J., and J. Morony. 1971. The preglossale of *Passer* (Aves)—a skeletal neomorph. Am. Zool. 11: 705.

Boer, M.H. den. 1971. A colour-polymorphism in caterpillars of *Bupalus piniarius* (L.) (Lepidoptera: Geometridae). Neth. J. Zool. 21: 61–116.

Böker, H. 1929. Flugvermögen und Kropf bei *Opisthocomus cristatus* und *Strigops habroptilus*. Jahrbuch für Morphologie und Mikroskopische Anatomie, Erste Abteilung: Gegenbauers Morphologische Jahrbuch, Vol. 63 (2), E. Göppert, Ed.: 152–207. Leipzig: Akademische Verlagsgesellschaft.

Bolze, G. 1968. Anordnung und Bau der Herbstschen Körperchen in Limicolenschnäbeln im Zusammenhang mit der Nahrungsfindung. Zool. Anz. 181: 313–355.

Bond, J. 1948. Origin of the bird fauna of the West Indies. Wilson Bull. 60:207–229.

Bond, J. 1963. Derivation of the Antillean avifauna. Proc. Acad. Nat. Sci. Phila. 115: 79–98.

Bond, R.R. 1957. Ecological distribution of breeding birds in the upland forests of southern Wisconsin. Ecol. Monogr. 27: 351–384.

Borgia, G. 1985. Bower quality, number of decorations and mating success of male Satin Bowerbirds *(Ptilonorhynchus violaceus)*: An experimental analysis. Anim. Behav. 33: 266–271.

Borgia, G. 1986. Sexual selection in bowerbirds. Sci. Amer. 254 (6): 92–100.

Borgia, G., and M.A. Gore. 1986. Sexual competition by feather stealing in the Satin Bowerbird *(Ptilonorhynchus violaceus)*. Anim. Behav. 34: 727–738.

Borgia, G., S.G. Pruett-Jones, and M.A. Pruett-Jones. 1985. The evolution of bower-building and the assessment of male quality. Z. Tierpsychol. 67: 225–236.

Borror, D.J., and C.R. Reese. 1956. Vocal gymnastics in Wood Thrush songs. Ohio J. Sci. 56: 177–182.

Boswall, J. 1977. Tool-using by birds and related behavior. Avic. Mag. 83: 88–97, 146–159, 220–228.

Boswall, J. 1983. Tool-using and related behaviour in birds: More notes. Avic. Mag. 89: 94–108.

Boughey, M.J., and N.S. Thompson. 1976. Species specificity and individual variation in the songs of the Brown Thrasher *(Toxostoma rufum)* and Catbird *(Dumetella carolinensis)*. Behaviour 57: 64–90.

Bourne, W.R.P. 1955. The birds of the Cape Verde Islands. Ibis 97: 508–556.

Bowman, R.I., and S.L. Billeb. 1965. Blood-eating in a Galápagos finch. Living Bird 4: 29–44.

Boyd, H. 1962. Mortality and fertility of European Charadrii. Ibis 104: 368–387.

Brackenbury, J.H. 1982. The structural basis of voice production and its relationship to sound characteristics. *In* Acoustic Communication in Birds, Vol. 1, D.E. Kroodsma and E.H. Miller, Eds.: 53–73. New York: Academic Press.

Bradbury, J.W. 1981. The evolution of leks. *In* Natural Selection and Social Behavior: Recent Research and New Theory, R.D. Alexander and D.W. Tinkle, Eds.: 138–169. New York: Chiron Press.

Bradbury, J.W., and R. Gibson. 1980. Leks and mate choice. *In* Mate Choice, P.P.G. Bateson, Ed.: 109–183. Cambridge: Cambridge University Press.

Brockway, B.F. 1964. Social influences on reproductive physiology and ethology of budgerigars *(Melopsittacus undulatus)*. Anim. Behav. 12: 493–501.

Brodkorb, P. 1955. Number of feathers and weights of various systems in a Bald Eagle. Wilson Bull. 67: 142.

Brodkorb, P. 1971. Origin and evolution of birds. *In* Avian Biology, Vol. 1, D.S. Farner, J.R. King, and K.C. Parkes, Eds.: 19–55. New York: Academic Press.

Brooks, W.S. 1978. Avian prehatching behavior: Functional aspects of the tucking pattern. Condor 80: 442–444.

Broom, R. 1913. On the South-African pseudosuchian *Euparkeria* and allied genera. Proc. Zool. Soc. Lond. 1913: 619–633.

Brosset, A. 1971. L'"imprinting", chez les Colombidés—étude des modifications comportementales au cours du vieillissement. Z. Tierpsychol. 29: 279–300.

Brower, L.P., B.S. Alpert, and S.C. Glazier. 1970. Observational learning in the feeding behavior of Blue Jays (*Cyanocitta cristata* Oberholser, Fam. Corvidae). Am. Zool. 10: 475–476.

Brown, C.R. 1984. Laying eggs in a neighbor's nest: benefit and cost of colonial nesting in swallows. Science (Wash., D.C.) 224: 518–519.

Brown, J.L. 1964a. The integration of agonistic behavior in the Steller's Jay, *Cyanocitta stelleri* (Gmelin). Univ. Calif. Publ. Zool. No. 60.

Brown, J.L. 1964b. The evolution of diversity in avian territorial systems. Wilson Bull. 76: 160–169.

Brown, J.L. 1969. Territorial behaviour and population regulation in birds: A review and re-evaluation. Wilson Bull. 81: 293–329.

Brown, J.L. 1974. Alternate routes to sociality in jays—with a theory for the evolution of altruism and communal breeding. Am. Zool. 14: 63–80.

Brown, J.L. 1975. The Evolution of Behavior. New York: Norton.

Brown, J.L., and E.R. Brown. 1981a. Extended family system in a communal bird. Science (Wash., D.C.) 211: 959–960.

Brown, J.L., and E.R. Brown. 1981b. Kin selection and individual selection in babblers. *In* Natural Selection and Social Behavior: Recent Research and New Theory, R.D. Alexander and D.W. Tinkle, Eds.: 244–256. New York: Chiron Press.

Brown, J.L., E.R. Brown, S.D. Brown, and D.D. Dow. 1982. Helpers: Effects of experimental removal on reproductive success. Science (Wash., D.C.) 215: 421–422.

Brown, J.L., and G.H. Orians. 1970. Spacing patterns in mobile animals. Annu. Rev. Ecol. Syst. 1: 239–262.

Brown, L.H., V. Gargett, and P. Steyn. 1977. Breeding success in some African eagles related to theories about sibling aggression and its effects. Ostrich 48: 65–71.

Brown, L.H., and E.K. Urban. 1969. The breeding biology of the Great White Pelican *Pelecanus onocrotalus roseus* at Lake Shala, Ethiopia. Ibis 111: 199–237.

Brown, R.G.B. 1962. The aggressive and distraction behavior of the Western Sandpiper *Ereunetes mauri.* Ibis 104: 1–12.

Brown, W.L., Jr., and E.O. Wilson. 1956. Character displacement. Syst. Zool. 5: 49–64.

Brush, A.H. 1967. Pigmentation in the Scarlet Tanager, *Piranga olivacea.* Condor 69: 549–559.

Brush, A.H. 1969. On the nature of "cotingin." Condor 71: 431–433.

Brush, A.H. 1972. Review of structure and spectral reflectance of green and blue feathers of the Rose-faced Lovebird *(Agapornis roseicollis),* by J. Dyke, 1971, Biol. Skr. 18: 1–67. Auk 89: 679–681.

Brush, A.H. 1979. Comparison of egg-white proteins: effect of electrophoretic conditions. Biochem. Syst. Ecol. 7: 155–165.

Brush, A.H. 1989. The evolution of feathers: a novel approach. *In* Avian Biology, Vol. 9, in press.

Bryant, D.M. 1975. Breeding biology of House Martins *Delichon urbica* in relation to aerial insect abundance. Ibis 117: 180–216.

Bryant, D.M. 1978a. Environmental influences on growth and survival of nestling House Martins *Delichon urbica.* Ibis 120: 271–283.

Bryant, D.M. 1978b. Establishment of weight hierarchies in the broods of House Martins *Delichon urbica.* Ibis 120: 16–26.

Buckley, F.G., and P.A. Buckley. 1974. Comparative feeding ecology of wintering adult and juvenile Royal Terns (Aves: Laridae, Sterninae). Ecology 55: 1053–1063.

Buckley, P.A., and F.G. Buckley. 1977. Hexagonal packing of Royal Tern nests. Auk 94:36–43.

Bulmer, M.G., and C.M. Perrins. 1973. Mortality in the Great Tit *Parus major.* Ibis 115: 277–281.

Burger, J. 1981. On becoming independent in Herring Gulls: parent-young conflict. Am. Nat. 117: 444–456.

Burley, N. 1980. Clutch overlap and clutch size: alternative and complementary reproductive tactics. Am. Nat. 115: 223–246.

Burley, N. 1981. Sex ratio manipulation and selection for attractiveness. Science (Wash., D.C.) 211: 721–722.

Burley, N., and N. Moran. 1979. The significance of age and reproductive experience in the mate preferences of feral pigeons, *Columba livia.* Anim. Behav. 27: 686–698.

Burtt, E.H., Jr. 1979. Tips on wings and other things. *In* The Behavioral Significance of Color, E.H. Burtt Jr., Ed.: 75–110. New York: Garland STPM Press.

Burtt, E.H., Jr., and J.P. Hailman. 1978. Head-scratching among North American wood-warblers (Parulidae). Ibis 120: 153–170.

Buskirk, W.H. 1976. Social systems in a tropical forest avifauna. Am. Nat. 110: 293–310.

Buskirk, W.H. 1980. Influence of meteorological patterns and trans-gulf migration on the calendars of latitudinal migrants. *In* Migrant Birds in the Neotropics, A. Keast and E.S. Morton, Eds.: 485–491. Washington, DC: Smithsonian Institution Press.

Buxton, J. 1950. The Redstart. London: Collins.

Cade, T.J. 1953. Sub-nival feeding of the Redpoll in interior Alaska: A possible adaptation to the northern winter. Condor 55: 43–44.

Cade, T.J., and G.L. Maclean. 1967. Transport of water by adult sandgrouse to their young. Condor 69: 323–343.

Calder, W.A. 1971. Temperature relationships and nesting of the Calliope Hummingbird. Condor 73: 314–321.

Calder, W.A. 1973. An estimate of the heat balance of a nesting hummingbird in a chilling climate. Comp. Biochem. Physiol. 46A: 291–300.

Calder, W.A. 1974. Consequences of body size for avian energetics. *In* Avian Energetics: 86–151. Publ. Nuttall Ornithol. Club No. 15.

Calder, W.A. 1979. The Kiwi and egg design: Evolution as a package deal. Bioscience 29: 461–467.

Calder, W.A., and J.R. King. 1974. Thermal and caloric relations of birds. *In* Avian Biology, Vol. 4, D.S. Farner, J.R. King, and K.C. Parkes, Eds.: 259–413. New York: Academic Press.

Campbell, B., and E. Lack. 1985. A Dictionary of Birds. Calton: T & AD Poyser.

Canady, R.A., D.E. Kroodsma, and F. Nottebohm. 1984. Population differences in complexity of a learned skill are correlated with the brain space involved. Proc. Nat. Acad. Sci. 81: 6232–6234.

Caple, G., R.P. Balda, and W.R. Willis. 1983. The physics of leaping animals and the evolution of preflight. Am. Nat. 121: 455–476.

Caple, G., R.P. Balda, and W.R. Willis. 1984. Flap about flight. Anim. Kingdom 87: 33–38.

Caraco, T. 1979. Time budgeting and group size: A test of theory. Ecology 60: 618–627.

Caraco, T. 1982. Aspects of risk-aversion in foraging White-crowned Sparrows. Anim. Behav. 30: 719–727.

Caraco, T., S. Martindale, and T.S. Whittam. 1980a. An empirical demonstration of risk-sensitive foraging preferences. Anim. Behav. 28: 820–830.

Caraco, T., S. Martindale, and H.R. Pulliam. 1980b. Avian flocking in the presence of a predator. Nature (Lond.) 285: 400–401.

Carey, C. 1980. Adaptation of the avian egg to high altitude. Am. Zool. 20: 449–459.

Carey, C. 1983. Structure and function of avian eggs. *In* Current Ornithology, Vol. 1, R.F. Johnston, Ed.: 69–103. New York: Plenum Press.

Carey, C., W.R. Dawson, L.C. Maxwell, and J.A. Faulkner. 1983. Seasonal acclimatization to temperature in carduelline finches. II. Changes in body composition and mass in relation to season and acute cold stress. J. Comp. Physiol. 125: 101–103.

Carey, M., and V. Nolan Jr. 1975. Polygyny in Indigo Buntings: A hypothesis tested. Science (Wash., D.C.) 190: 1296–1297.

Carlson, A., and J. Moreno. 1982. The loading effect in central place foraging Wheatears (*Oenanthe oenanthe* L.). Behav. Ecol. Sociobiol. 11: 173–183.

Carrick, R. 1963. Ecological significance of territory in the Australian Magpie, *Gymnorhina tibicen.* Proc. XIII Int. Ornithol. Congr.: 740–753.

Caryl, P.G. 1979. Communication by agonistic displays: What can games theory contribute to ethology? Behaviour 68: 136–169.

Cavé, A.J. 1968. The breeding of the Kestrel, *Falco tinnunculus* L., in the reclaimed area Oostelijk Flevoland. Neth. J. Zool. 18: 313–407.

Chandra-Bose, D.A., and J.C. George. 1964. Studies on the structure and physiology of the flight muscles of birds, 12: Observations on the structure of the pectoralis and supracoracoideus of the Ruby-throated Hummingbird. Pavo 2: 111–114.

Chaplin, S.B. 1974. Daily energetics of the black-capped chickadee, *Parus atricapillus,* in winter. J. Comp. Physiol. 89: 321–330.

Chapman, F.M. 1935. The courtship of Gould's Manakin *(Manacus vitellinus vitellinus)* on Barro Colorado Island, Canal Zone. Bull. Am. Mus. Nat. Hist. 68: 471–525.

Chappuis, C. 1971. Un exemple de l'influence du milieu sur les émissions vocales des oiseaux: l'évolution des chants en forêt équatoriale. Terre Vie 25: 183–202.

Charnov, E.L. 1976. Optimal foraging: The marginal value theorem. Theor. Popul. Biol. 9: 129–136.

Chen, D.-M., J.S. Collins, and T.H. Goldsmith. 1984. The ultraviolet receptor of bird retinas. Science (Wash., D.C.) 225: 337–340.

Clara, M. 1925. Ueber den Bau des Schnabels der Waldschnepfe. Z. Mikrosk.-Anat. Forsch. (Leipz.) 3: 1–108.

Clark, G.A., Jr. 1961. Occurrence and timing of egg teeth in birds. Wilson Bull. 73: 268–278.

Clark, G.A., Jr. 1964a. Ontogeny and evolution in the megapodes (Aves: Galliformes). Postilla No. 78.

Clark, G.A., Jr. 1964b. Life histories and the evolution of megapodes. Living Bird 3: 149–167.

Clark, K.L., and R.J. Robertson. 1981. Cowbird parasitism and evolution of anti-parasite strategies in the Yellow Warbler. Wilson Bull. 93: 249–258.

Clark, L. 1983. The development of effective homeothermy and endothermy by nestling starlings. Comp. Biochem. Physiol. 73A: 253–260.

Clark, L., and J.R. Mason. 1985. Use of nest material as insecticidal and anti-pathogenic agents by the European Starling. Oecologia (Berl.) 67: 169–176.

Clements, F.E. 1916. Plant succession: an analysis of the development of vegetation. Carnegie Inst. Wash. Publ. No. 242.

Clements, F.E. 1936. Nature and structure of the climax. J. Ecol. 24: 252–284.

Clements, J. 1981. Birds of the World: A Checklist. New York: Facts on File, Inc.

Clench, M.H. 1978. Tracheal elongation in birds-of-paradise. Condor 80: 423–430.

Clench, M.H., and O.L. Austin Jr. 1986. Birds: Passeriformes (perching birds). Encyclopaedia Britannica (15th ed.), Macropaedia, Vol. 15: 95–108. Chicago: Encyclopaedia Britannica, Inc.

Cochran, W.W., G.G. Montgomery, and R.R. Graber. 1967. Migratory flights of *Hylocichla* thrushes in spring: A radiotelemetry study. Living Bird 6: 213–225.

Cody, M.L. 1966. A general theory of clutch size. Evolution 20: 174–184.

Cody, M.L. 1971. Ecological aspects of reproduction. *In* Avian Biology, Vol. 1, D.S. Farner, J.R. King, and K.C. Parkes, Eds.: 461–512. New York: Academic Press.

Cody, M.L., and J.M. Diamond, Eds. 1975. Ecology and Evolution of Communities. Cambridge: Harvard University Press.

Cohn, J.M.W. 1968. The convergent flight mechanism of swifts (Apodi) and hummingbirds (Trochili) (Aves). Ph.D. dissertation, University of Michigan, Ann Arbor, Michigan.

Collias, N.E. 1952. The development of social behavior in birds. Auk 69: 127–159.

Collias, N.E., and E.C. Collias. 1964. Evolution of nest-building in the weaverbirds (Ploceidae). Univ. Calif. Publ. Zool. No. 73.

Collias, N.E., and E.C. Collias. 1971. Some observations on behavioral energetics in the Village Weaverbird, I: Comparison of colonies from two subspecies in nature. Auk 88: 124–133.

Collias, N.E., and E.C. Collias. 1984. Nest Building and Bird Behavior. Princeton: Princeton University Press.

Confer, J.L., and K. Knapp. 1981. Golden-winged Warblers and Blue-winged Warblers: The relative success of a habitat specialist and a generalist. Auk 98: 108–114.

Connell, J.H. 1980. Diversity and the coevolution of competitors, or the ghost of competition past. Oikos 35: 131–138.

Conrads, K. 1966. Der Egge-Dialekt des Buchfinken *(Fringilla coelebs):* ein Beitrag zur geographischen Gesangvariation. Vogelwelt 87: 176–182.

Contino, F. 1968. Observations on the nesting of *Sporophila obscura* in association with wasps. Auk 85: 137–138.

Cooke, A.S. 1975. Pesticides and eggshell formation. Symp. Zool. Soc. Lond. 35: 339–361.

Cooke, F. 1978. Early learning and its effect on population structure. Studies of a wild population of Snow Geese. Z. Tierpsychol. 46: 344–358.

Cooke, F., C.S. Findlay, R.F. Rockwell, and J.A. Smith. 1985. Life history studies of the Lesser Snow Goose *(Anser caerulescens caerulescens)*, III: The selective value of plumage polymorphism: net fecundity. Evolution 39: 165–177.

Cooke, F., C.D. MacInnes, and J.P. Prevett. 1975. Gene flow between breeding populations of Lesser Snow Geese. Auk 92: 493–510.

Corbin, K.W., and C.G. Sibley. 1977. Rapid evolution in orioles of the genus *Icterus.* Condor 79: 335–342.

Cortopassi, A.J., and L.R. Mewaldt. 1965. The circumannual distribution of White-crowned Sparrows. Bird-Banding 36: 141–169.

Cott, H.B. 1940. Adaptive Coloration in Animals. London: Methuen.

Cottam, P.A. 1957. The pelecaniform characters of the skeleton of the Shoe-bill Stork, *Balaeniceps rex.* Bull. Br. Mus. (Nat. Hist.) Zool. 5: 49–72.

Cox, G.W. 1968. The role of competition in the evolution of migration. Evolution 22: 180–192.

Cox, G.W. 1985. The evolution of avian migration systems between temperate and tropical regions of the New World. Am. Nat. 126: 451–474.

Cox, G.W., and R.E. Ricklefs. 1977. Species diversity and ecological release in Caribbean land bird faunas. Oikos 28: 113–122.

Cracraft, J. 1971. The relationships and evolution of the rollers: Families Coraciidae, Brachypteraciidae, and Leptosomatidae. Auk 88: 723–752.

Cracraft, J. 1973a. Continental drift, paleoclimatology, and the evolution and biogeography of birds. J. Zool. (Lond.) 169: 455–545.

Cracraft, J. 1973b. Systematics and evolution of the Gruiformes (Class Aves), 3: Phylogeny of the suborder Grues. Bull. Am. Mus. Nat. Hist. 151: 1–128.

Cracraft, J. 1974a. Phylogeny and evolution of the ratite birds. Ibis 116: 494–521.

Cracraft, J. 1974b. Continental drift and vertebrate distribution. Annu. Rev. Ecol. Syst. 5: 215–261.

Cracraft, J. 1981. Toward a phylogenetic classification of the recent birds of the world (Class Aves). Auk 98: 681–714.

Cracraft, J. 1982a. Geographic differentiation, cladistics, and vicariance biogeography: Reconstructing the tempo and mode of evolution. Am. Zool. 22: 411–424.

Cracraft, J. 1982b. Phylogenetic relationships and monophyly of loons, grebes, and hesperornithiform birds, with comments on the early history of birds. Syst. Zool. 31: 35–56.

Cracraft, J. 1985. Monophyly and phylogenetic relationships of the Pelecaniformes: A numerical cladistic analysis. Auk 102: 834–853.

Cracraft, J. 1986. The origin and early diversification of birds. Paleobiology 12: 383–399.

Craig, J.V., L.L. Ortman, and A.M. Guhl. 1965. Genetic selection for social dominance ability in chickens. Anim. Behav. 13: 114–131.

Cronin, E.W., Jr., and P.W. Sherman. 1976. A resource-based mating system: The Orange-rumped Honeyguide. Living Bird 15: 5–32.

Crook, J.H. 1964. The evolution of social organization and visual communication in the weaver birds (Ploceinae). Behav. Suppl. No. 10.

Crowe, T.M., and P.C. Withers. 1979. Brain temperature regulation in Helmeted Guineafowl. S. Afr. J. Sci. 75: 362–365.

Cullen, E. 1957. Adaptations in the kittiwake to cliff-nesting. Ibis 99: 275–302.

Cullen, J.M. 1966. Reduction of ambiguity through ritualization. Philos. Trans. R. Soc. Lond. B Biol. Sci. 251: 363–374.

Curio, E. 1959. Beiträge zur Populationsökologie des Trauerschnäppers (*Ficedula h. hypoleuca* Pallas). Zool. Jahrb. Abt. Syst. Oekol. Geogr. Tiere 87: 185–230.

Curio, E., U. Ernst, and W. Vieth. 1978. The adaptive significance of avian mobbing, II: Cultural transmission of enemy recognition in blackbirds: Effectiveness and some constraints. Z. Tierpsychol. 48: 184–202.

Dane, B., C. Walcott, and W.H. Drury. 1959. The form and duration of the display actions of the Goldeneye *(Bucephala clangula)*. Behaviour 14: 265–281.

Darlington, P.J. 1957. Zoogeography: The Geographical Distribution of Animals. New York: Wiley.

Darwin, C. 1859. On the Origin of Species by Means of Natural Selection. London: J. Murray.

Darwin, C. 1871. The Descent of Man, and Selection in Relation to Sex. New York: D. Appleton and Company.

Davidson, N.C. 1983. Identification of refuelling sites by studies of weight changes and fat deposition. *In* Shorebirds and Large Waterbirds Conservation, P.R. Evans, H. Hafner, and P.L'Hermite, Eds.: 68–78. Brussels: Commission of the European Communities.

Davies, N.B. 1976a. Parental care and the transition to independent feeding in the young spotted flycatcher *(Muscicapa striata)*. Behaviour 59: 280–295.

Davies, N.B. 1976b. Food, flocking, and territorial behavior of the pied wagtail (*Motacilla alba yarrellii* Gould) in winter. J Anim. Ecol. 45: 235–253.

Davies, N.B. 1977. Prey selection and social behaviour in wagtails (Aves: Motacillidae). J. Anim. Ecol. 46:37–57.

Davies, N.B. 1978a. Ecological questions about territorial behaviour. *In* Behavioural Ecology: An Evolutionary Approach, J.R. Krebs and N.B. Davies, Eds.: 317–350. Sunderland, MA: Sinauer Associates, Inc.

Davies, N.B., 1978b. Parental meanness and offspring independence: An experiment with hand-reared Great Tits *Parus major*. Ibis 120: 509–514.

Davies, N.B., and R.E. Green. 1976. The development and ecological significance of feeding techniques in the Reed warbler *(Acrocephalus scirpaceus)*. Anim. Behav. 24: 213–229.

Davis, D.E. 1950. The growth of Starling, *Sturnus vulgaris*, populations. Auk 67: 460–465.

Davis, J. 1973. Habitat preferences and competition of wintering juncos and Golden-crowned Sparrows. Ecology 54: 174–180.

Davis, S.D., J.B. Williams, W.J. Adams, and S.L. Brown. 1984. The effect of egg temperature on attentiveness in the Belding's Savannah Sparrow. Auk 101: 556–566.

Dawkins, M. 1971. Perceptual changes in chicks: another look at the "search image" concept. Anim. Behav. 19: 566–574.

Dawson, W.R., and C. Carey. 1976. Seasonal acclimatization to temperature in carduelline finches. J. Comp. Physiol. 112: 317–333.

Dawson, W.R., R.L. Marsh, and M.E. Yacoe. 1983. Metabolic adjustments of small passerine birds for migration and cold. Am. J. Physiol. 245: R755–R767.

DeBenedictis, P.A. 1966. The bill-brace feeding behavior of the Galapagos finch *Geospiza conirostris*. Condor 68: 206–208.

DeBenedictis, P.A., F.B. Gill, F.R. Hainsworth, G.H. Pyke, and L.L. Wolf. 1978. Optimal meal size in hummingbirds. Am. Nat. 112: 301–316.

de Boer, L.E.M. 1980. Do the chromosomes of the kiwi provide evidence for a monophyletic origin of the ratites? Nature (Lond.) 287: 84–85.

De Schauensee, R.M. 1982. A guide to the birds of South America, with new addenda. Acad. Nat. Sci., Philadelphia (reprinted by Pan Am. Section, ICBP).

De Steven, D. 1980. Clutch size, breeding success, and parental survival in the Tree Swallow *(Iridoprocne bicolor)*. Evolution 34: 278–291.

Delacour, J., and E. Mayr. 1945. The family Anatidae. Wilson Bull. 57: 1–55.

Delacour, J., and E. Mayr. 1946. Supplementary notes on the family Anatidae. Wilson Bull. 58: 104–110.

Delius, J.D. 1965. A population study of Skylarks *Alauda arvensis.* Ibis 107: 466–492.

Dement'ev, G.P., and V.D. Il'ichev. 1963. Die äussere Ohr der Greivogel. Falke 10: 123–125, 158–164, 187–191.

De Sante, D.F. 1983. Annual variability in the abundance of migrant landbirds on southeast Farallon Island, California. Auk 100: 826–852.

Desselberger, H. 1931. Der Verdauungskanal der Dicaeiden nach Gestalt und Funktion. J. Ornithol. 79: 353–370.

Dhondt, A.A. 1977. Interspecific competition between Great and Blue Tit. Nature (Lond.) 268: 521–523.

Dhondt, A.A., and R. Eyckerman. 1980. Competition and the regulation of numbers in Great and Blue Tit. Ardea 68: 121–132.

Diamond, J.M. 1969. Avifaunal equilibria and species turnover rates on the Channel Islands of California. Proc. Natl. Acad. Sci. U.S.A. 64: 57–63.

Diamond, J.M. 1973. Distributional ecology of New Guinea birds. Science (Wash., D.C.) 179: 759–769.

Diamond, J.M. 1974. Colonization of exploded volcanic islands by birds: the supertramp strategy. Science (Wash., D.C.) 184: 803–806.

Diamond, J.M. 1975. Assembly of species communities. *In* Ecology and Evolution of Communities, M.L. Cody and J.M. Diamond, Eds.: 342–444. Cambridge, MA: Harvard University Press.

Diamond, J.M. 1980. Species turnover in island bird communities. Acta XVII Congr. Int. Ornithol.: 777–782.

Diamond, J.M. 1982. Mimicry of friarbirds by orioles. Auk 99: 187–196.

Diamond, J.M. 1983. Taxonomy by nucleotides. Nature (Lond.) 305: 17–18.

Diamond, J.M., and M.E. Gilpin. 1982. Examination of the "null" model of Connor and Simberloff for species co-occurrences on islands. Oecologia (Berl.) 52: 64–74.

Dickerson, J.W.T., and R.A. McCance. 1960. Severe undernutrition in growing and adult animals, III: Aian skeletal muscle. Br. J. Nutr. 14: 331–338.

Dilger, W. 1962. The behavior of lovebirds. Sci. Am. 206(1): 88–98.

Dobinson, H.M., and A.J. Richards. 1964. The effects of the severe winter of 1962/63 on birds in Britain. Br. Birds 57: 373–434.

Dobkin, D.S. 1979. Functional and evolutionary relationships of vocal copying phenomena in birds. Z. Tierpsychol. 50: 348–363.

Dobson, C.W., and R.E. Lemon. 1975. Re-examination of monotony threshold hypothesis in bird song. Nature (Lond.) 257: 126–128.

Dobson, C.W., and R.E. Lemon. 1977. Markovian versus rhomboidal patterning in the song of Swainson's Thrush. Behaviour 62: 277–297.

Docters van Leeuwen, W.M. 1954. On the biology of some Javanese Loranthaceae and the role birds play in their life-historie. Beaufortia 4: 105–207.

Dolnik, V.R., and V.M. Gavrilov. 1979. Bioenergetics of molt in the Chaffinch *(Fringilla coelebs).* Auk 96: 253–264.

Dooling, R.J. 1982. Auditory perception in birds. *In* Acoustic Communication in Birds, Vol. 1, D.E. Kroodsma and E.H. Miller, Eds.: 95–130. New York: Academic Press.

Dorst, J. 1962. The Migrations of Birds. Boston: Houghton Mifflin.

Dow, D.D. 1965. The role of saliva in food storage by the Gray Jay. Auk 82: 139–154.

Downhower, J.F. 1976. Darwin's finches and the evolution of sexual dimorphism in body size. Nature (Lond.) 263: 558–563.

Downing, R.L. 1959. Significance of ground nesting by Mourning Doves in northwestern Oklahoma. J. Wildl. Manage. 23: 117–118.

Dowsett-Lemaire, F. 1979. The imitative range of the song of the Marsh Warbler *Acrocephalus palustris,* with special reference to imitations of African birds. Ibis 121: 453–468.

Drent, R.H. 1970. Functional aspects of incubation in the Herring Gull. Behav. Suppl. 17: 1–132.

Drent, R.H. 1972. Adaptive aspects of the physiology of incubation. Proc. XV Int. Ornithol. Congr.: 255–280.

Drent, R.H. 1975. Incubation. *In* Avian Biology, Vol. 5, D.S. Farner, J.R. King, and K.C. Parkes, Eds.: 333–420. New York: Academic Press.

Drent, R.H. and S. Daan. 1980. The prudent parent: Energetic adjustments in avian breeding. Ardea 68: 225–252.

Driver, P.M. 1967. Notes on the clicking of avian egg-young, with comments on its mechanism and function. Ibis 109: 434–437.

Drobney, R.D. 1980. Reproductive bioenergetics of Wood Ducks. Auk 97: 480–490.

Duke, G.E., Ciganek, J.G., and O.A. Evanson. 1973. Food consumption and energy, water, and nitrogen budgets in captive Great-Horned Owls *(Bubo virginianus)*. Comp. Biochem. Physiol. 44A: 283–292.

Dunning, J.B., Jr., and J.H. Brown. 1982. Summer rainfall and winter sparrow densities: a test of the food limitation hypothesis. Auk 99: 123–129.

Durrer, H., and W. Villiger. 1966. Schillerfarben der Trogoniden. J. Ornithol. 107: 1–26.

Dwight, J., Jr. 1907. Sequence in moults and plumages, with an explanation of plumage-cycles. Proc. Fourth Int. Ornithol. Congr.: 513–518.

Dyck, J. 1971. Structure and spectral reflectance of green and blue feathers of the Rose-faced Lovebird *(Agapornis roseicollis)*. Biol. Srk. No. 18 (2).

Dyrcz, A. 1977. Polygamy and breeding success among Great Reed Warblers *Acrocephalus arundinaceus* at Milicz, Poland. Ibis 119: 73–77.

East, M. 1981. Aspects of courtship and parental care of the European Robin *Erithacus rubecula.* Ornis Scand. 12: 230–239.

Edwards, E.P. 1974. A Coded List of Birds of the World. Sweet Briar, VA: Ernest P. Edwards.

Einarsen, A.S. 1942. Specific results from Ring-necked Pheasant studies in the Pacific northwest. Trans. N. Am. Wildl. Nat. Resour. Conf. 7: 130–145.

Einarsen, A.S. 1945. Some factors affecting Ring-necked Pheasant population density. Murrelet 26: 39–44.

Ekman, J. 1979. Coherence, composition and territories of winter social groups of the Willow Tit *Parus montanus* and the Crested Tit *P. cristatus.* Ornis Scand. 10: 56–68.

Emlen, J.T. 1980. Interactions of migrant and resident land birds in Florida and Bahama pinelands. *In* Migrant Birds in the Neotropics, A. Keast and E.S. Morton, Eds.: 133–144. Washington, DC: Smithsonian Institution Press.

Emlen, J.T. and R.L. Penney. 1964. Distance navigation in the Adelie Penguin. Ibis 106: 417–431.

Emlen, S.T. 1967a. Migratory orientation in the Indigo Bunting, *Passerina cyanea.* Part I: Evidence for use of celestial cues. Auk 84: 309–342.

Emlen, S.T. 1967b. Migratory orientation in the Indigo Bunting, *Passerina cyanea,* Part II: Mechanism of celestial orientation. Auk 84: 463–489.

Emlen, S.T. 1969. Bird migration: Influence of physiological state upon celestial orientation. Science (Wash., D.C.) 165: 716–718.

Emlen, S.T. 1970. Celestial rotation: Its importance in the development of migratory orientation. Science (Wash., D.C.) 170: 1198–1201.

Emlen, S.T. 1972. An experimental analysis of the parameters of bird song eliciting species recognition. Behaviour 41: 130–171.

Emlen, S.T. 1975a. Migration: Orientation and navigation. *In* Avian Biology, Vol. 5, D.S. Farner, J.R. King, and K.C. Parkes, Eds.: 129–219. New York: Academic Press.

Emlen, S.T. 1975b. The stellar-orientation system of a migratory bird. Sci. Am. 233(2): 102–111.

Emlen, S.T. 1978. The evolution of cooperative breeding in birds. *In* Behavioural Ecology: An Evolutionary Approach, J.R. Krebs and N.B. Davies, Eds.: 245–281. Sunderland, MA: Sinauer Associates, Inc.

Emlen, S.T. 1981. Altruism, kinship, and reciprocity in the White-fronted Bee-eater. *In* Natural Selection and Social Behavior: Recent Research and New Theory, R.D. Alexander and D.W. Tinkle, Eds.: 217–230. New York: Chiron Press.

Emlen, S.T. 1982a. The evolution of helping. II: The role of behavioral conflict. Am. Nat. 119: 40–53.

Emlen, S.T. 1982b. The evolution of helping. I: An ecological constraints model. Am. Nat. 119: 29–39.

Emlen, S.T. 1984. Cooperative breeding in birds and mammals. *In* Behavioural Ecology: An Evolutionary Approach (second ed.), J.R. Krebs and N.B. Davies, Eds.: 305–339. Sunderland, MA: Sinauer Associates, Inc.

Emlen, S.T., and H.W. Ambrose III. 1970. Feeding interactions of Snowy Egrets and Red-breasted Mergansers. Auk 87: 164–165.

Emlen, S.T., and N.J. Demong. 1975. Adaptive significance of synchronized breeding in a colonial bird: a new hypothesis. Science (Wash., D.C.) 188: 1029–1031.

Emlen, S.T., and J.T. Emlen. 1966. A technique for recording migratory orientation of captive birds. Auk 83: 361–367.

Emlen, S.T., and L.W. Oring. 1977. Ecology, sexual selection, and the evolution of mating systems. Science (Wash., D.C.) 197: 215–223.

Emlen, S.T., J.D. Rising, and W.L. Thompson. 1975. A behavioral and morphological study of sympatry in the Indigo and Lazuli Buntings of the Great Plains. Wilson Bull. 87: 145–179.

Emlen, S.T., and S.L. Vehrencamp. 1983. Cooperative breeding strategies among birds. *In* Perspectives in Ornithology, A.H. Brush and G.A. Clark Jr., Eds.: 93–120. Cambridge: Cambridge University Press.

Erickson, C.J., and P.G. Zenone. 1978. Courtship differences in male Ring Doves: avoidance of cuckoldry? Science (Wash., D.C.) 192: 1353–1354.

Ettinger, A.O., and J.R. King. 1980. Time and energy budgets of the Willow Flycatcher *(Empidonax traillii)* during the breeding season. Auk 97: 533–546.

Evans, K. 1966. Observations on a hybrid between the Sharp-tailed Grouse and the Greater Prairie Chicken. Auk 83: 128–129.

Evans, H.E. 1969. Anatomy of the budgerigar. *In* Diseases of Cage and Aviary Birds, M.L. Petrak, Ed.: 45–112. Philadelphia: Lea and Febiger.

Ewald, P.W. 1980. Energetics of resource defense: an experimental approach. Acta XVII Congr. Int. Ornithol.: 1093–1099.

Fagen, R. 1981. Animal Play Behavior. New York: Oxford University Press.

Falls, J.B. 1982. Individual recognition by sounds in birds. *In* Acoustic Communication in Birds, Vol. 2, D.E. Kroodsma and E.H. Miller, Eds.: 237–278. New York: Academic Press.

Farabaugh, S.M. 1982. The ecological and social significance of duetting. *In* Acoustic Communication in Birds, Vol. 2, D.E. Kroodsma and E.H. Miller, Eds.: 85–124. New York: Academic Press.

Farner, D.S. 1955. Birdbanding in the study of population dynamics. *In* Recent Studies in Avian Biology, A. Wolfson, Ed.: 397–449. Urbana, IL: University of Illinois Press.

Farner, D.S. 1964. The photoperiodic control of reproductive cycles in birds. Am. Sci. 52: 137–156.

Farner, D.S. 1967. The control of avian reproductive cycles. Proc. XIV Int. Ornithol. Congr.: 107–133.

Farner, D.S. 1970. Some glimpses of comparative avian physiology. Fed. Proc. 29: 1649–1663.

Farner, D.S. 1980a. Endogenous periodic functions in the control of reproductive cycles. *In* Biological Rhythms in Birds: Neural and Endocrine Aspects, Y. Tanabe et al., Eds.: 123–138. Tokyo: Japan Sci. Soc. Press.

Farner, D.S. 1980b. Evolution of the control of reproductive cycles in birds. *In* Hormones, Adaptation and Evolution, S. Ishii et al., Eds.: 185–191. Tokyo: Japan Sci. Soc. Press.

Farner, D.S. 1980c. The regulation of the annual cycle of the White-crowned Sparrow, *Zonotrichia leucophrys gambelii.* Acta XVII Congr. Int. Ornithol.: 71–82.

Farner, D.S., and R.A. Lewis. 1971. Photoperiodism and reproductive cycles in birds. Photophysiology 6: 325–370.

Farner, D.S., and L.R. Mewaldt. 1952. The relative roles of photoperiod and temperature in gonadal recrudescence in male *Zonotrichia leucophrys gambelii.* Anat. Rec. 113: 612–613.

Feduccia, A. 1974. Morphology of the bony stapes in New and Old World suboscines: New evidence for common ancestry. Auk 91: 427–429.

Feduccia, A. 1975. Morphology of the bony stapes in the Menuridae and Acanthisittidae: evidence for oscine affinities. Wilson Bull. 87: 418–420.

Feduccia, A. 1977. A model for the evolution of perching birds. Syst. Zool. 26: 19–31.

Feduccia, A. 1979. Comments on the phylogeny of perching birds. Proc. Biol. Soc. Wash. 92: 689–696.

Feduccia, A. 1980. The Age of Birds. Cambridge: Harvard University Press.

Feduccia, A., and S.L. Olson. 1982. Morphological similarities between the Menurae and the Rhinocryptidae, relict passerine birds of the southern hemisphere. Smithson. Contrib. Zool. No. 366.

Feduccia, A., and H.B. Tordoff. 1979. Feathers of *Archaeopteryx:* Asymmetric vanes indicate aerodynamic function. Science (Wash., D.C.) 203: 1021–1022.

Feinsinger, P. 1980. Asynchronous migration patterns and the coexistence of tropical hummingbirds. *In* Migrant Birds in the Neotropics, A. Keast and E.S. Morton, Eds.: 411–419. Washington, DC: Smithsonian Institution Press.

Feinsinger, P., and R.K. Colwell. 1978. Community organization among neotropical nectar-feeding birds. Am. Zool. 18: 779–795.

Fenna, L., and D.A. Boag. 1974. Adaptive significance of the caeca in Japanese Quail and Spruce Grouse (Galliformes). Can. J. Zool. 52: 1577–1584.

Ferns, P.N. 1978. Individual differences in the head and neck plumage of Ruddy Turnstones *(Arenaria interpres)* during the breeding season. Auk 95: 753–755.

Ficken, M.S. 1977. Avian play. Auk 94: 573–582.

Ficken, M.S., and R.W. Ficken. 1968. Courtship of Blue-winged Warblers, Golden-winged Warblers, and their hybrids. Wilson Bull. 80: 161–172.

Findlay, C.S., and F. Cooke. 1983. Genetic and environmental components of clutch size variance in a wild population of Lesser Snow Goose *(Anser caerulescens caerulescens).* Evolution 37: 724–734.

Fisher, C.D., E. Lindgren, and W.R. Dawson. 1972. Drinking patterns and behavior of Australian desert birds in relation to their ecology and abundance. Condor 74: 111–136.

Fisher, H.I. 1971. The Laysan Albatross: its incubation, hatching, and associated behaviors. Living Bird. 10: 19–78.

Fisher, H. 1972. The nutrition of birds. *In* Avian Biology, Vol. 2, D.S. Farner, J.R. King, and K.C. Parkes, Eds.: 431–469. New York: Academic Press.

Fisher, J., and R.A. Hinde. 1949. The opening of milk bottles by birds. Br. Birds 42: 347–357.

Fisher, R.A. 1930. The Genetical Theory of Natural Selection. Oxford: Clarendon Press.

Fitzpatrick, J.W. 1980. Wintering of North American tyrant flycatchers in the Neotropics. *In* Migrant Birds in the Neotropics, A. Keast and E.S. Morton, Eds.: 67–78. Washington, DC: Smithsonian Institution Press.

Fjeldså, J. 1976. The systematic affinities of sandgrouses, Pteroclididae. Vidensk. Medd. Dan. Naturhist. Foren. 139: 179–243.

Flower, S.S. 1938. Further notes on the duration of life in animals, IV: Birds. Proc. Zool. Soc. Lond. Series A, 108: 195–235.

Fogden, M.P.L. 1972. The seasonality and population dynamics of equatorial forest birds in Sarawak. Ibis 114: 307–342.

Fogden, M.P.L. and P.M. Fogden. 1979. The role of fat and protein reserves in the annual cycle of the Grey-backed Camaroptera in Uganda (Aves: Sylviidae). J. Zool. (Lond.) 189: 233–258.

Ford, J. 1974. Concepts of subspecies and hybrid zones, and their application in Australian ornithology. Emu 74: 113–123.

Ford, N.L. 1983. Variation in mate fidelity in monogamous birds. *In* Current Ornithology, Vol. 1, R.F. Johnston, Ed.: 329–356. New York: Plenum Press.

Forshaw, J.M. 1978. Parrots of the World (second ed.). Melbourne: Lansdowne Eds.

Forsythe, D.M. 1971. Clicking in the egg-young of the Long-billed Curlew. Wilson Bull. 83: 441–442.

Foster, M.S. 1974. A model to explain molt-breeding overlap and clutch size in some tropical birds. Evolution 28: 182–190.

Foster, M.S. 1975. The overlap of molting and breeding in some tropical birds. Condor 77: 304–314.

Foster, M.S. 1977. Odd couples in manakins: A study of social organization and cooperative breeding in *Chiroxiphia linearis.* Am. Nat. 111: 845–853.

Foster, M.S. 1978. Total frugivory in tropical passerines: A reappraisal. Trop. Ecol. 19: 131–154.

Foster, M.S. 1981. Cooperative behavior and social organization of the Swallow-tailed Manakin *(Chiroxiphia caudata).* Behav. Ecol. Sociobiol. 9: 167–177.

Foster, M.S. 1983. Disruption, dispersion, and dominance in lek-breeding birds. Am. Nat. 122: 53–72.

Fox, G.A. 1976. Eggshell quality: its ecological and physiological significance in a DDE-contaminated Common Tern population. Wilson Bull. 88: 459–477.

Franks, E.C. 1967. The response of incubating Ringed Turtle Doves *(Streptopelia risoria)* to manipulated egg temperatures. Condor 69: 268–276.

Fretwell, S. 1968. Habitat distribution and survival in the Field Sparrow *(Spizella pusilla).* Bird-Banding 39: 293–306.

Fretwell, S. 1980. Evolution of migration in relation to factors regulating bird numbers. *In* Migrant Birds in the Neotropics, A. Keast and E.S. Morton, Eds.: 517–527. Washington, DC: Smithsonian Institution Press.

Friedmann, H. 1963. Host relations of the parasitic cowbirds. U.S. Natl. Mus. Bull. 233.

Friedmann, H. and J. Kern. 1956. The problem of cerophagy or wax-eating in the Honey-guides. Q. Rev. Biol. 31: 19–30.

Frith, H.J. 1962. The Mallee-Fowl. Sydney: Angus and Robertson.

Frith, H.J. 1959. Incubator birds. Sci. Am. 201 (2): 52–58.

Fyfe, R.W., S.A. Temple, and T.J. Cade. 1976. The 1975 North American Peregrine Falcon survey. Can. Field-Nat. 90: 228–273.

Fürbringer, M. 1888. Bijdragen tot de Dierkunde. Vol. 15. Untersuchungen zur Morphologie und Systematik der Vögel. Amsterdam: Tj. Van Holkema.

Gadow, H. 1892. On the classification of birds. Proc. Zool. Soc. Lond. 1892: 229–256.

Gadow, H. 1893. Dr. H.G. Bronn's Klassen und Ordnungen des Thier-Reichs. Vol. 6, Pt. 4. Vogel, II: Systematischer Theil. Leipzig: C.F. Winter.

Gardner, L.L. 1925. The adaptive modifications and the taxonomic value of the tongue in birds. Proc. U.S. Natl. Mus. 67 (19): 1–49.

Gargett, V. 1978. Sibling aggression in the Black Eagle in the Matopos, Rhodesia. Ostrich 49: 57–63.

Garnett, M.C. 1981. Body size, its heritability and influence on juvenile survival among Great Tits, *Parus major.* Ibis 123: 31–41.

Garrod, A.H. 1876. Notes on the anatomy of *Plotus anhinga.* Proc. Zool. Soc. Lond. 1876: 335–345.

Gasaway, W.C. 1976a. Seasonal variation in diet, volatile fatty acid production and size of the cecum of Rock Ptarmigan. Comp. Biochem. Physiol. 53A: 109–114.

Gasaway, W.C. 1976b. Volatile fatty acids and metabolizable energy derived from cecal fermentation in the Willow Ptarmigan. Comp. Biochem. Physiol. 53A: 115–121.

Gaston, A.J. 1978. The evolution of group territorial behavior and cooperative breeding. Am. Nat. 112: 1091–1100.

Gaunt, A.S., and M.K. Wells. 1973. Models of syringeal mechanisms. Am. Zool. 13: 1227–1247.

Gaunt, A.S., and S.L.L. Gaunt. 1985. Syringeal structure and avian phonation. *In* Current Ornithology, Vol. 2., R.F. Johnston, Ed.: 213–245. New York: Plenum.

Gauthreaux, S.A., Jr. 1971. A radar and direct visual study of passerine spring migration in southern Louisiana. Auk 88: 343–365.

Gauthreaux, S.A., Jr. 1972. Behavioral responses of migrating birds to daylight and darkness: a radar and direct visual study. Wilson Bull. 84: 136–148.

Gauthreaux, S.A., Jr. 1982. The ecology and evolution of avian migration systems. *In* Avian Biology, Vol. 6, D.S. Farner, J.R. King, and K.C. Parkes, Eds.: 93–168. New York: Academic Press.

George, F.W., J.F. Noble, and J.D. Wilson. 1981. Female feathering in Sebright cocks is due to conversion of testosterone to estradiol in skin. Science (Wash., D.C.) 213: 557–559.

George, J.C., and A.J. Berger. 1966. Avian Myology. New York and London: Academic Press.

Geramita, J.M., F. Cooke, and R.F. Rockwell. 1982. Assortative mating and gene flow in the Lesser Snow Goose: a modelling approach. Theor. Popul. Biol. 22: 177–203.

Gessaman, J.A. 1972. Bioenergetics of the Snowy Owl *(Nyctea scandiaca)*. Arct. Alp. Res. 4: 223–238.

Gibb, J. 1954. Feeding ecology of tits, with notes on Treecreeper and Goldcrest. Ibis 96: 513–543.

Gibb, J. 1956. Food, feeding habits and territory of the Rock Pipit *Anthus spinoletta.* Ibis 98: 506–530.

Gibb, J.A. 1960. Populations of tits and goldcrests and their food supply in pine plantations. Ibis 102: 163–208.

Gill, F.B. 1970. Hybridization in Norfolk Island white-eyes *(Zosterops)*. Condor 72: 481–482.

Gill, F.B. 1971. Ecology and evolution of the sympatric Mascarene white-eyes, *Zosterops borbonica* and *Zosterops olivacea.* Auk 88: 35–60.

Gill, F.B. 1973. Intra-island variation in the Mascarene White-eye, *Zosterops borbonica.* Ornithol. Monogr. No. 12.

Gill, F.B. 1980. Historical aspects of hybridization between Blue-winged and Golden-winged Warblers. Auk 97: 1–18.

Gill, F.B. 1982. Might there be a resurrection of the subspecies? Auk 99:598–599.

Gill, F.B. 1985. Hummingbird flight speeds. Auk 102: 97–101.

Gill, F.B., and L.L. Wolf. 1975. Economics of feeding territoriality in the Golden-winged Sunbird. Ecology 56: 333–345.

Gill, F.B., and L.L. Wolf. 1977. Nonrandom foraging by sunbirds in a patchy environment. Ecology 58: 1284–1296.

Gill, F.B., and L.L. Wolf. 1978. Comparative foraging efficiencies of some montane sunbirds in Kenya. Condor 80: 391–400.

Gill, F.B., and L.L. Wolf. 1979. Nectar loss by Golden-winged Sunbirds to competitors. Auk 96: 448–461.

Gilliard, E.T. 1956. Bower ornamentation versus plumage characters in bowerbirds. Auk 73: 450–451.

Gilliard, E.T. 1958. Living Birds of the World. Garden City: Doubleday.

Gilliard, E.T. 1969. Birds of paradise and Bowerbirds. Garden City: Natural History Press.

Gleason, H.A. 1926. The individualistic concept of the plant association. Bull. Torrey Bot. Club 53: 7–26.

Gleason, H.A. 1939. The individualistic concept of the plant association. Am. Midl. Nat. 21: 92–110.

Glue, D.E. 1971. Ringing recovery circumstances of small birds of prey. Bird Study 18: 137–146.

Gochfeld, M., J. Burger, and J.R. Jehl Jr., 1984. The classification of the shorebirds of the world. *In* Behavior of Marine Animals, Vol. 5, Shorebirds: Breeding Behavior and Populations, J. Burger and B.L. Olla, Eds.: 1–15. New York: Plenum Press.

Goldsmith, T.H. 1980. Hummingbirds see near ultraviolet light. Science (Wash., D.C.) 207: 786–788.

Goldstein, D.L. 1984. The thermal environment and its constraint on activity of desert quail in summer. Auk 101: 542–550.

Goss-Custard, J.D. 1975. Beach Feast. Birds (Sept/Oct): 23–26.

Goss-Custard, J.D. 1977a. Optimal foraging and the size selection of worms by Redshank, *Tringa totanus,* in the field. Anim. Behav. 25: 10–29.

Goss-Custard, J.D. 1977b. Predator responses and prey mortality in Redshank, *Tringa totanus* (L.), and a preferred prey, *Corophium volutator* (Pallas). J. Anim. Ecol. 46: 21–35.

Goss-Custard, J.D. 1977c. The ecology of the Wash, III: Density-related behaviour and the possible effects of loss of feeding grounds on wading birds (Charadrii). J. Appl. Ecol. 14: 721–739.

Goss-Custard, J.D., R.A. Jenyon, R.E. Jones, P.E. Newberry, and R. le B. Williams. 1977. The ecology of the Wash, II: Seasonal variation in the feeding conditions of wading birds (Charadrii). J. Appl. Ecol. 14: 701–719.

Gottlieb, G. 1968. Prenatal behavior of birds. Q. Rev. Biol. 43: 148–174.

Gottlieb, G. 1971. Development of Species Identification in Birds. Chicago: University of Chicago Press.

Gould, S. 1985. A clock of evolution. Nat. Hist. 94(4): 12–25.

Gowaty, P.A. 1981. Aggression of breeding Eastern Bluebirds *(Sialia sialis)* toward their mates and models of intra- and interspecific intruders. Anim. Behav. 29: 1013–1027.

Grant, G.S. 1982. Avian incubation: Egg temperature, nest humidity, and behavioral thermoregulation in a hot environment. Ornithol. Monogr. No. 30.

Grant, P.R. 1968. Polyhedral territories of animals. Am. Nat. 102: 75–80.

Grant, P.R. 1972a. Centripetal selection and the House Sparrow. Syst. Zool. 21: 23–30.

Grant, P.R. 1972b. Convergent and divergent character displacement. Biol. J. Linn. Soc. 4: 39–68.

Grant, P.R. 1975. The classical case of character displacement. Evol. Biol. 8: 237–337.

Grant, P.R. 1981. Speciation and adaptive radiation of Darwin's finches. Am. Sci. 69: 653–663.

Grant, P.R. 1986. Ecology and Evolution of Darwin's Finches. Princeton, NJ: Princeton University Press.

Grant, P.R., and B.R. Grant. 1980. Annual variation in finch numbers, foraging and food supply on Isla Daphne Major, Galápagos. Oecologia (Berl.) 46: 55–62.

Grassé, P.-P. (Ed.). 1950. Traité de Zoologie. Vol. XV. Oiseaux. Paris: Masson et Cie.

Grau, C.R. 1976. Ring structure of avian egg yolk. Poult. Sci. 55: 1418–1422.

Grau, C.R. 1982. Egg formation in Fiordland Crested Penguins *(Eudyptes pachyrhynchus).* Condor 84: 172–177.

Green, C. 1972. Use of tool by Orange-winged Sitella. Emu 72: 185–186.

Greenberg, R. 1983. The role of neophobia in determining the degree of foraging specialization of some migrant warblers. Am. Nat. 122: 444–453.

Greenberg, R. 1984. Neophobia in the foraging-site selection of a neotropical migrant bird: An experimental study. Proc. Natl. Acad. Sci. U.S.A. 81: 3778–3780.

Greenberg, R., and J. Gradwohl. 1983. Sexual roles in the Dot-winged Antwren *(Microrhopias quixensis),* a tropical forest passerine. Auk 100: 920–925.

Greenewalt, C.H. 1960a. Hummingbirds. Garden City, NY: Doubleday.

Greenewalt, C.H. 1960b. The wings of insects and birds as mechanical oscillators. Proc. Am. Philos. Soc. 104: 605–611.

Greenewalt, C.H. 1968. Bird Song: Acoustics and Physiology. Washington, DC: Smithsonian Institution Press.

Greenewalt, C.H. 1969. How birds sing. Sci. Am. 221(5): 126–139.

Greenewalt, C.H. 1975. The flight of birds. Trans. Am. Philos. Soc. New Series 65(4): 1–67.

Greenewalt, C.H., Brandt, W. and D.D. Friel. 1960. Iridescent colors of hummingbird feathers. J. Opt. Soc. Am. 50: 1005–1013.

Greenlaw, J.S. 1969. The importance of food in the breeding system of the Rufous-sided Towhee, *Pipilo erythrophthalamus* (L.). Ph.D. dissertation, Rutger's University.

Greenwood, P.J., P.H. Harvey, and C.M. Perrins. 1978. Inbreeding and dispersal in the Great Tit. Nature (Lond.) 271: 52–54.

Greenwood, P.J., P.H. Harvey, and C.M. Perrins. 1979. The role of dispersal in the Great Tit *(Parus major):* the causes, consequences and heritability of natal dispersal. J. Anim. Ecol. 48: 123–142.

Gregg, K., S.D. Wilton, D.A.D. Parry, and G.E. Rogers. 1984. A comparison of genomic coding sequences for feather and scale keratins: structural and evolutionary implications. EMBO J. 3: 175–178.

Gregg, K., S.D. Wilton, G.E. Rogers, and P.I. Malloy. 1983. Avian keratin genes: Organization and evolutionary relationships. *In* Manipulation and expression of genes in Eukaryotes, P. Nagley, A.W. Linnane, W.J. Peacock, and J.A. Pateman, Eds.: 65–72. Sydney: Academic Press.

Grier, J.W. 1982. Ban of DDT and subsequent recovery of reproduction in Bald Eagles. Science (Wash., D.C.) 218: 1232–1235.

Griffin, D.R. 1974. Bird Migration. New York: Dover.

Grubb, T.C., Jr. 1972. Smell and foraging in shearwaters and petrels. Nature (Lond.) 237: 404–405.

Grubb, T.C., Jr. 1974. Olfactory navigation to the nesting burrow in Leach's Petrel *(Oceanodroma leucorrhoa).* Anim. Behav. 22: 192–202.

Grubb, B., J.M. Colacino, and K. Schmidt-Nielsen. 1978. Cerebral blood flow in birds: effect of hypoxia. Am. J. Physiol. 234(3): H230–H234.

Grubb, B., J.H. Jones, and K. Schmidt-Nielsen. 1979. Avian cerebral blood flow: Influence of the Bohr effect on oxygen supply. Am. J. Physiol. 236(5): H744–H749.

Guhl, A.M. 1968. Social inertia and social stability in chickens. Anim. Behav. 16: 219–232.

Gullion, G.W., and W.H. Marshall. 1968. Survival of Ruffed Grouse in a boreal forest. Living Bird 7: 117–167.

Gurney, M.E., and M. Konishi. 1980. Hormone-induced sexual differentiation of brain and behavior in Zebra Finches. Science (Wash., D.C.) 208: 1380–1383.

Gwinner, E. 1966. Ueber einige Bewegungsspiele des Kolkraben (*Corvus corax* L.). Z. Tierpsychol. 23: 28–36.

Gwinner, E. 1975. Circadian and circannual rhythms in birds. *In* Avian Biology, Vol. 5, D.S. Farner, J.R. King, and K.C. Parkes, Eds.: 221–285. New York: Academic Press.

Gwinner, E. 1977. Circannual rhythms in bird migration. Annu. Rev. Ecol. Syst. 8: 381–405.

Gwinner, E. 1980. Relationship between circadian activity patterns and gonadal function: Evidence for internal coincidence? Acta XVII Congr. Int. Ornithol.: 409–416.

Haartman, L. von. 1951. Der Trauerfliegenschnäpper, II: Populationsprobleme. Acta Zool. Fenn. 67: 1–60.

Haartman, L. von. 1953. Was reizt den Trauerfliegenschnäpper *(Muscicapa hypoleuca)* zu füttern? Vogelwarte 16: 157–164.

Haartman, L. von. 1956. Der Einfluss der Temperatur auf den Brutrhythmus experimentell nachgewiesen. Ornis Fenn. 33: 100–107.

Haartman, L. von. 1957. Adaptation in hole-nesting birds. Evolution 11: 339–347.

Haartman, L. von. 1958. The incubation rhythm of the female Pied Flycatcher *(Ficedula hypoleuca)* in the presence and absence of the male. Ornis Fenn. 35: 71–76.

Haartman, L. von. 1969. Nest-site and evolution of polygamy in European passerine birds. Ornis Fenn. 46: 1–12.

Haartman, L. von. 1971. Population dynamics. *In* Avian Biology, Vol. 1, D.S. Farner, J.R. King, and K.C. Parkes, Eds.: 391–459. New York: Academic Press.

Haartman, L. von. 1973. Changes in the breeding bird fauna of north Europe. *In* Breeding Biology of Birds, D.S. Farner, Ed.: 448–481. Washington, DC: National Academy of Sciences.

Häcker, V. 1900. Der Gesang der Vögel. Jena: Gustav Fischer.

Haffer, J. 1969. Speciation in Amazonian forest birds. Science (Wash., D.C.) 165: 131–137.

Haffer, J. 1974. Avian speciation in tropical South America. Publ. Nuttall Ornithol. Club No. 14.

Haftorn, S. 1956. Contribution to the food biology of tits especially about storing of surplus food, IV: A comparative analysis of *Parus atricapillus* L., *P. cristatus* L., and *P. ater* L. K. Nor. Vidensk. Selsk. Skr. (N.S.) 4:1–54.

Haftorn, S. 1978a. Egg-laying and regulation of egg temperature during incubation in the Goldcrest *Regulus regulus.* Ornis Scand. 9: 2–21.

Haftorn, S. 1978b. Energetics of incubation by the Goldcrest *Regulus regulus* in relation to ambient air temperatures and the geographical distribution of the species. Ornis Scand. 9: 22–30.

Haftorn, S. 1978c. Cooperation between the male and female Goldcrest *Regulus regulus* when rearing overlapping double broods. Ornis Scand. 9: 124–129.

Hailman, J.P. 1967. The ontogeny of an instinct. Behav. Suppl. No. 15.

Hailman, J.P. 1969. How an instinct is learned. Sci. Am. 221(6): 98–106.

Hailman, J.P. 1977. Optical Signals: Animal Communication and Light. Bloomington: Indiana University Press.

Hainsworth, F.R. 1974. Food quality and foraging efficiency: The efficiency of sugar assimilation by hummingbirds. J. Comp. Physiol. 88: 425–431.

Hainsworth, F.R., and L.L. Wolf. 1970. Regulation of oxygen consumption and body temperature during torpor in a hummingbird, *Eulampis jugularis.* Science (Wash., D.C.) 168: 368–369.

Hamilton, T.H., and I. Rubinoff. 1967. On predicting insular variation in endemism and sympatry for the Darwin Finches in the Galápagos Archipelago. Am. Nat. 101: 161–171.

Hamilton, W.D. 1964. The genetical evolution of social behaviour I, II. J. Theor. Biol. 7: 1–52.

Hamilton, W.D. 1971. Geometry for the selfish herd. J. Theor. Biol. 31: 295–311.

Hamilton, W.D., and M. Zuk. 1982. Heritable true fitness and bright birds: a role for parasites? Science (Wash., D.C.) 218: 384–387.

Hamilton, W.J., III. 1965. Sun-oriented display of the Anna's Hummingbird. Wilson Bull. 77: 38–44.

Hamilton, W.J., III. 1973. Life's Color Code. New York: McGraw-Hill.

Hamilton, W.J., III and F.H. Heppner. 1967. Radiant solar energy and the function of black homeotherm pigmentation: an hypothesis. Science (Wash., D.C.) 155: 196–197.

Hamilton, W.J., III, and G.H. Orians. 1965. Evolution of brood parasitism in altricial birds. Condor 67: 361–382.

Hann, H.W. 1937. Life history of the Oven-bird in southern Michigan. Wilson Bull. 49: 145–237.

Harris, M.P. 1969. The biology of storm petrels in the Galápagos Islands. Proc. Calif. Acad. Sci. 37: 95–166.

Harris, M.P. 1970. Abnormal migration and hybridization of *Larus argentatus* and *L. fuscus* after interspecies fostering experiments. Ibis 112: 488–498.

Harris, M.P. 1983. Biology and survival of the immature Puffin *Fratercula arctica.* Ibis 125: 56–73.

Harrison, C. 1975. A Field Guide to the Nests, Eggs, and Nestlings of European Birds. New York: Quadrangle, The New York Times Book Company.

Hartshorne, C. 1973. Born to Sing. An Interpretation and World Survey of Bird Song. Bloomington: University of Indiana Press.

Hays, H. 1972. Polyandry in the Spotted Sandpiper. Living Bird 11: 43–57.

Hegner, R.E., S.T. Emlen, and N.J. Demong. 1982. Spatial organization of the White-fronted Bee-eater. Nature (Lond.) 298: 264–266.

Heilmann, G. 1927. The Origin of Birds. New York: D. Appleton and Company.

Heinroth, O. 1922. Die Beziehungen zwischen Vogelgewicht, Eigewicht, Gelegegewicht und Brutdauer. J. Ornithol. 70: 172–285.

Henley, C., A. Feduccia, and D.P. Costello. 1978. Oscine spermatozoa: A light- and electron-microscopy study. Condor 80: 41–48.

Hensley, M.M., and J.B. Cope. 1951. Further data on removal and repopulation of the breeding birds in a spruce–fir forest community. Auk 68: 483–493.

Heppner, F. 1970. The metabolic significance of differential absorption of radiant energy by black and white birds. Condor 72: 50–59.

Herrick, F.H. 1932. Daily life of the American Eagle: Early phase. Auk 49: 307–323.

Hess, E.H. 1959a. Imprinting. Science (Wash., D.C.) 130: 133–141.

Hess, E.H. 1959b. Two conditions limiting critical age for imprinting. J. Comp. Physiol. Psychol. 52: 515–518.

Hess, E.H. 1973. Imprinting. New York: Van Nostrand Reinhold.

Hickey, J.J. 1952. Survival studies of banded birds. U.S. Fish Wildl. Serv. Spec. Sci. Rep. Wildl. 15: 1–177.

Hickey, J.J. 1955. Some American population research on gallinaceous birds. *In* Recent Studies in Avian Biology, A. Wolfson, Ed.: 326–396. Urbana: University of Illinois Press.

Higginson, T.W. 1863. Outdoor Papers. Boston: Lee and Shepard.

Hinde, R.A. 1956. The biological significance of territories of birds. Ibis 98: 340–369.

Hinde, R.A. 1970. Animal Behaviour (second ed.). New York: McGraw-Hill.

Hindwood, K.A. 1959. The nesting of birds in the nests of social insects. Emu 59: 1–36.

Ho, C. Y.-K., E.M. Prager, A.C. Wilson, D.T. Osuga, and R.E. Feeney. 1976. Penguin evolution: protein comparisons demonstrate phylogenetic relationship to flying aquatic birds. J. Mol. Evol. 8: 271–282.

Hodson, N.L., and D.W. Snow. 1965. The road deaths enquiry, 1960-61. Bird Study 12: 90–99.

Hoffman, K. 1954. Versuche zu der im Richtungsfinden der Vögel enthaltenen Zeitschätzung. Z. Tierpsychol. 11: 453–475.

Hogan-Warburg, A.J. 1966. Social behavior of the Ruff, *Philomachus pugnax* (L.). Ardea 54: 109–229.

Hogstad, O. 1967. Seasonal fluctuation in bird populations within a forest area near Oslo (southern Norway) in 1966-67. Nytt Mag. Zool. (Oslo). 15: 81–96.

Högstedt, G. 1980. Evolution of clutch size in birds: adaptive variation in relation to territory quality. Science (Wash., D.C.) 210: 1148–1150.

Höhn, E.O. 1961. Endocrine glands, thymus, and pineal body. *In* Biology and Comparative Physiology of Birds, Vol. 2, A.J. Marshall, Ed.: 87–114. New York and London: Academic Press.

Holcomb, L.C. 1970. Prolonged incubation behaviour of Red-winged Blackbird incubating several egg sizes. Behaviour 36: 74–83.

Holm, C.H. 1973. Breeding sex ratios, territoriality, and reproductive success in the red-winged blackbird *(Agelaius phoeniceus)*. Ecology 54: 356–365.

Holmes, R.T. 1970. Differences in population density, territoriality, and food supply of Dunlin on arctic and subarctic tundra. *In* Animal Populations in Relation to Their Food Resources, A. Watson, Ed.: 303–319. Oxford and Edinburgh: Blackwell Scientific Publications.

Holmes, R.T., R.E. Bonney Jr., and S.W. Pacala. 1979. Guild structure of the Hubbard Brook bird community: a multivariate approach. Ecology 60: 512–520.

Hoogland, J.L., and P.W. Sherman. 1976. Advantages and disadvantages of Bank Swallow *(Riparia riparia)* coloniality. Ecol. Monogr. 46: 33–58.

Horn, H.S. 1968. The adaptive significance of colonial nesting in the Brewer's Blackbird *(Euphagus cyanocephalus)*. Ecology 49: 682–694.

Houde, P., and S.L. Olson. 1981. Paleognathous carinate birds from the early Tertiary of North America. Science (Wash., D.C.) 214: 1236–1237.

Howard, H.E. 1920. Territory in Bird Life. London: John Murray.

Howard, R.D. 1974. The influence of sexual selection and interspecific competition on mockingbird song *(Mimus polyglottos)*. Evolution 28: 428–438.

Howard, R., and A. Moore. 1980. A Complete Checklist of the Birds of the World. Oxford: Oxford University Press.

Howe, H.F. 1978. Initial investment, clutch size, and brood reduction in the Common Grackle *(Quiscalus quiscula* L.). Ecology 59: 1109–1122.

Howell, T.R. 1959. A field study of temperature regulation in young Least Terns and Common Nighthawks. Wilson Bull. 71: 19–32.

Howell, T.R. 1979. Breeding biology of the Egyptian Plover, *Pluvianus aegyptius.* Univ. Calif. Publ. Zool. No. 113.

Howell, T.R., B. Araya, and W.R. Millie. 1974. Breeding biology of the Gray Gull, *Larus modestus.* Univ. Calif. Publ. Zool. No. 104.

Howell, T.R., and G.A. Bartholomew. 1962. Temperature regulation in the Sooty Tern *Sterna fuscata.* Ibis 104: 98–105.

Hubbard, J.P. 1969. The relationships and evolution of the *Dendroica coronata* complex. Auk 86: 393–432.

Hudson, W.H. 1920. Birds of La Plata. London: J.M. Dent and Sons Ltd.
Hudson, G.I., and P.J. Lanzilloti. 1964. Muscles of the pectoral limb in galliform birds. Amer. Midl. Natur. 71: 1–113.
Humphrey, P.S., and K.C. Parkes. 1959. An approach to the study of molts and plumages. Auk 76: 1–31.
Hunt, J.H. 1971. A field study of the Wrenthrush, *Zeledonia coronata.* Auk 88: 1–20.
Hussell, D.J.T. 1969. Weight loss of birds during nocturnal migration. Auk 86: 75–83.
Hussell, D.J.T. 1972. Factors affecting clutch size in Arctic passerines. Ecol. Monogr. 42: 317–364.
Hussell, D.J.T., and A.B. Lambert. 1980. New estimates of weight loss in birds during nocturnal migration. Auk 97: 547–558.
Hutchison, L.V., and B.M. Wenzel. 1980. Olfactory guidance in foraging by Procellariiformes. Condor 82: 314–319.
Hutto, R.L. 1980. Winter habitat distribution of migratory land birds in western Mexico, with special reference to small foliage-gleaning insectivores. *In* Migrant Birds in the Neotropics, A. Keast and E.S. Morton, Eds.: 181–204. Washington, DC: Smithsonian Institution Press.
Huxley, T.H. 1867. On the classification of birds and on the taxonomic value of the modifications of certain of the cranial bones observable in that class. Proc. Zool. Soc. Lond. 1867: 415–472.
Huxley, T.H. 1868. On the animals which are most nearly intermediate between birds and reptiles. Ann. Mag. Nat. Hist. 4th Series 2: 66–75.

Idyll, C.P. 1973. The Anchovy crisis. Sci. Am. 228 (6): 23–29.
Il'ichev, V.D. 1961. Morphological and functional details of the external ear in crepuscular and nocturnal birds. Dokl. Biol. Sci. (Engl. Transl. Dokl. Akad. Nauk SSSR) 137: 253–256.
Immelmann, K. 1967. Periodische Vorgänge in der Fortpflanzung tierischer Organismen. Stud. Gen. 20: 15–33.
Immelmann, K. 1969. Ueber den Einfluß frühkindlicher Erfahrungen auf die geschlechtliche Objektfixierung bei Estrildiden. Z. Tierpsychol. 26: 677–691.
Immelmann, K. 1970. Zur ökologischen Bedeutung prägungsbedingter Isolationsmechanismen. Verhandlungen der Deutschen Zoologischen Gesellschaft, 64. Jahresversammlung: 304–313.
Immelmann, K. 1971. Ecological aspects of periodic reproduction. *In* Avian Biology, Vol. 1, D.S. Farner, J.R. King, and K.C. Parkes, Eds.: 341–389. New York: Academic Press.
Immelmann, K. 1975. Ecological significance of imprinting and early learning. Annu. Rev. Ecol. Syst. 15–37.
Ingolfsson, A. 1969. Behaviour of gulls robbing eiders. Bird Study 16: 45–52.
Ingram, W.J. 1907. On the display of the King Bird-of-Paradise. Ibis 1 (Ninth Series): 225–229.
Isleib, M.E. "P". 1979. Migratory shorebird populations on the Copper River Delta and eastern Prince William Sound, Alaska. *In* Studies in Avian Biology, Vol. 2: Shorebirds in Marine Environments, F.A. Pitelka, Ed.: 125–129. Los Angeles: Cooper Ornithological Society.

Jacob, J., and V. Ziswiler. 1982. The uropygial gland. *In* Avian Biology, Vol. 6, D.S. Farner, J.R. King, and K.C. Parkes, Eds.: 199–324. New York: Academic Press.
Jaeger, R.G. 1980. Density-dependent and density-independent causes of extinction of a salamander population. Evolution 34: 617–621.
James, F.C. 1970. Geographic size variation in birds and its relationship to climate. Ecology 51: 365–390.

James, F.C. 1971. Ordinations of habitat relationships among breeding birds. Wilson Bull. 83: 215–236.

James, F.C. 1983. Environmental component of morphological differentiation in birds. Science (Wash., D.C.) 221: 184–186.

Järvinen, O. 1979. Geographical gradients of stability in European land bird communities. Oecologia (Berl.) 38: 51–69.

Järvinen, O. 1980. Dynamics of North European bird communities. Acta XVII Congr. Int. Ornithol.: 770–776.

Jehl, J.R. 1968. Relationships in the Charadrii (shorebirds): A taxonomic study based on color patterns of the downy young. Mem. San Diego Soc. Nat. Hist. 3: 1–54.

Jehl, J.R., Jr. 1975. *Pluvianellus socialis:* Biology, ecology, and relationships of an enigmatic Patagonian shorebird. Trans. San Diego Soc. Nat. Hist. 18: 25–74.

Jehl, J.R., Jr., and K.C. Parkes. 1983. "Replacements" of landbird species on Socorro Island, Mexico. Auk 100: 551–559.

Jenkin, P.M. 1957. The filter-feeding and food of flamingoes (Phoenicopteri). Philos. Trans. R. Soc. Lond. B Biol. Sci. 240: 401–493.

Jenni, D.A. 1974. Evolution of polyandry in birds. Am. Zool. 14: 129–144.

Jenni, D.A., and G. Collier. 1972. Polyandry in the American Jacana *(Jacana spinosa)*. Auk 89: 743–765.

Jensen, R.A.C. 1980. Cuckoo egg identification by chromosome analysis. Proc. IV Pan-Afr. Ornithol. Congr: 23–25.

Johnsgard, P.A. 1967. Animal Behavior. Dubuque, IA: Wm. C. Brown.

Johnsgard, P.A. 1975. Waterfowl of North America. Bloomington: Indiana University Press.

Johnsgard, P.A., and J. Kear. 1968. A review of parental carrying of young by waterfowl. Living Bird 7: 89–102.

Johnson, N.K. 1963. Biosystematics of sibling species of flycatchers in the *Empidonax hammondii-oberholseri-wrightii* complex. Univ. Calif. Publ. Zool. 66: 79–238.

Johnson, R.A. 1969. Hatching behavior of the Bobwhite. Wilson Bull. 81: 79–86.

Johnson, R.F., Jr., and N.F. Sloan. 1978. White Pelican production and survival of young at Chase Lake National Wildlife Refuge, North Dakota. Wilson Bull. 90: 346–352.

Johnson, S.R. 1971. Thermal adaptability of Sturnidae introduced into North America. Unpubl. thesis, University of British Columbia.

Johnston, R.F., and R.K. Selander. 1964. House Sparrows: Rapid evolution of races in North America. Science (Wash., D.C.) 144: 548–550.

Johnston, R.F., and R.K. Selander. 1971. Evolution in the house sparrow, II: Adaptive differentiation in North American populations. Evolution. 25: 1–28.

Jollie, M. 1953. Are the Falconiformes a monophyletic group? Ibis 95: 369–371.

Jollie, M. 1976–77. A contribution to the morphology and phylogeny of the Falconiformes. Evol. Theory 1: 285–298, 2: 115–300, 3: 1–141.

Jones, D.R., and K. Johansen. 1972. The blood vascular system of birds. *In* Avian Biology, Vol. 2, D.S. Farner, J.R. King, and K.C. Parkes, Eds.: 157–285. New York: Academic Press.

Jones, P.J., and P. Ward. 1976. The level of reserve protein as the proximate factor controlling the timing of breeding and clutch-size in the Red-billed Quelea *Quelea quelea.* Ibis 118: 547–574.

Jones, P.J., and P. Ward. 1979. A physiological basis for colony desertion by Red-billed Queleas *(Quelea quelea)*. J. Zool. (Lond.) 189: 1–19.

Jouventin, P., M. Guillotin, and A. Cornet. 1979. Le chant du Manchot empereur et sa signification adaptative. Behaviour 70: 231–250.

Kahl, M.P., Jr. 1963. Thermoregulation in the Wood Stork with special reference to the role of the legs. Physiol. Zool. 36: 141–151.

Kamil, A.C. 1978. Systematic foraging by a nectar-feeding bird, the Amakihi *(Loxops virens)*. J. Comp. Physiol. Psychol. 92: 388–396.

Kanwisher, J.W., G. Gabrielsen, and N. Kanwisher. 1981. Free and forced diving in birds. Science (Wash., D.C.) 211: 717–719.

Karr, J.R. 1971. Structure of avian communities in selected Panama and Illinois habitats. Ecol. Monogr. 41: 207–233.

Karr, J.R. 1976. Within- and between-habitat avian diversity in African and Neotropical lowland habitats. Ecol. Monogr. 46: 457–481.

Karr, J.R., and F.C. James. 1975. Eco-morphological configurations and convergent evolution in species and communities. *In* Ecology and Evolution of Communities, M.L. Cody and J.M. Diamond, Eds.: 258–291. Cambridge: Harvard University Press.

Karr, J.R., and R.R. Roth. 1971. Vegetation structure and avian diversity in several New World areas. Am. Nat. 105: 423–435.

Kear, J. 1963. Parental feeding in the Magpie Goose. Ibis 105: 428.

Keast, A. 1972. Ecological opportunities and dominant families, as illustrated by the Neotropical Tyrannidae (Aves). Evol. Biol. 5: 229–277.

Keast, A., and E.S. Morton, Eds. 1980. Migrant Birds in the Neotropics. Washington, D.C.: Smithsonian Institution Press.

Keeton, W.T. 1971. Magnets interfere with pigeon homing. Proc. Natl. Acad. Sci. U.S.A. 68: 102–106.

Keeton, W.T. 1972. Effects of magnets on pigeon homing. NASA SP 262: 579–594.

Keeton, W.T. 1974. The mystery of pigeon homing. Sci. Am. 231 (6): 96–107.

Keeton, W.T., and A. Gobert. 1970. Orientation by untrained pigeons requires the sun. Proc. Natl. Acad. Sci. U.S.A. 65: 853–856.

Keeton, W.T., M.L. Kreithen, and K.L. Hermayer. 1977. Orientation by pigeons deprived of olfaction by nasal tubes. J. Comp. Physiol. 114: 289–299.

Kendeigh, S.C. 1949. Effect of temperature and season on energy resources of the English Sparrow. Auk 66: 113–127.

Kendeigh, S.C. 1952. Parental care and its evolution in birds. Ill. Biol. Monogr. No. 22.

Kendeigh, S.C. 1976. Latitudinal trends in the metabolic adjustments of the House Sparrow. Ecology 57: 509–519.

Kendeigh, S.C., V.R. Dol'nik, and V.M. Gavrilov. 1977. Avian energetics. *In* Granivorous Birds in Ecosystems, J. Pinowski and S.C. Kendeigh, Eds.: 127–204. Cambridge: Cambridge University Press.

Kenward, R.E. 1978. Hawks and doves: factors affecting success and selection in goshawk attacks on wood pigeons. J. Anim. Ecol. 47: 449–460.

Ketterson, E.D. 1977. Male Prairie Warbler dies during courtship. Auk 94: 393.

Ketterson, E.D., and V. Nolan, Jr. 1982. The role of migration and winter mortality in the life history of a temperate-zone migrant, the Dark-eyed Junco, as determined from demographic analyses of winter populations. Auk 99: 243–259.

Ketterson, E.D., and V. Nolan, Jr. 1983. The evolution of differential bird migration. *In* Current Ornithology, Vol. 1, R.F. Johnston, Ed.: 357–402. New York: Plenum Press.

Kiff, L.F., D.B. Peakall, and S.R. Wilbur. 1979. Recent changes in California Condor eggshells. Condor 81: 166–172.

Kilham, L. 1957. Egg-carrying by the Whip-poor-will. Wilson Bull. 69: 113.

King, A.P., and M.J. West. 1977. Species identification in the North American cowbird: appropriate responses to abnormal song. Science (Wash., D.C.) 195: 1002–1004.

King, J.R. 1972a. Adaptive periodic fat storage by birds. Proc. XV Int. Ornithol. Congr. 200–217.

King, J.R. 1973. Energetics of reproduction in birds. *In* Breeding Biology of Birds, D.S. Farner, Ed.: 78–107. Washington, D.C.: National Academy of Sciences.

King, J.R. 1974. Seasonal allocation of time and energy resources in birds. *In* Avian Energetics, 4–85. Publ. Nuttall Ornithol. Club No. 15.

King, J.R., and D.A. Farner. 1965. Studies of fat deposition in migratory birds. Ann. N.Y. Acad. Sci. 131: 422–440.

King, S.C., and C.R. Henderson. 1954. Variance components in heritability studies. Poult. Sci. 33: 147–154.

King, W.B. 1981. Red Data Book, Vol. 2: Aves (third rev. ed.). Morges, Switzerland: International Union for Conservation of Nature and Natural Resources.

Kinney, T.B., Jr. 1969. A summary of reported estimates of heritabilities and of genetic and phenotypic correlations for traits of chickens. U.S. Dep. Agric. Agric. Handb. 363.

Kinsky, F.C. 1971. The consistent presence of paired ovaries in the Kiwi *(Apteryx)* with some discussion of this condition in other birds. J. Ornithol. 112: 334–357.

Kitto, G.B., and A.C. Wilson. 1966. Evolution of malate dehydrogenase in birds. Science (Wash., D.C.) 153: 1408–1410.

Klem, D.A. 1979. Biology of collisions between birds and windows. Ph.D. thesis, Southern Illinois University.

Klomp, H. 1970. The determination of clutch-size in birds. Ardea 58: 1–124.

Klomp, H. 1980. Fluctuations and stability in Great Tit populations. Ardea 68: 205–224.

Klopfer, P.H. 1959. Social interactions in discrimination learning with special reference to feeding behaviour in birds. Behaviour 14: 282–299.

Klopfer, P. 1963. Behavioral aspects of habitat selection: the role of early experience. Wilson Bull. 75: 15–22.

Kluijver, H.N. 1935. Waarnemingen over de levenswijze van den Spreeuw (*Sturnus v. vulgaris* L.) met behulp van geringde individuen. Ardea 24: 133–166.

Kluijver, H.N. 1950. Daily routines of the Great Tit, *Parus m. major* L. Ardea 38: 99–135.

Kluijver, H.N. 1951. The population ecology of the Great Tit, *Parus m. major* L. Ardea 39: 1–135.

Kluijver, H.N. 1966. Regulation of a bird population. Ostrich Suppl. 6: 389–396.

Knorr, O.A. 1957. Communal roosting of the Pygmy Nuthatch. Condor 59: 398.

Knudsen, E.I. 1980. Sound localization on the neuronal level. Acta XVII Congr. Int. Ornithol.: 718–723.

Knudsen, E.I. 1981. The hearing of the Barn Owl. Sci. Am. 245 (6): 112–125.

Knudsen, E.I., and M. Konishi. 1978. A neural map of auditory space in the owl. Science (Wash., D.C.) 200: 795–797.

Kodric-Brown, A., and J.H. Brown. 1978. Influence of economics, interspecific competition, and sexual dimorphism on territoriality in migrant hummingbirds. Ecology 59: 285–296.

Koehler, O. 1950. The ability of birds to "count". Bull. Anim. Behav. 9: 41–45.

Koenig, W.D. 1981a. Space competition in the Acorn Woodpecker: power struggles in a cooperative breeder. Anim. Behav. 29: 396–409.

Koenig, W.D. 1981b. Reproductive success, group size, and the evolution of cooperative breeding in the Acorn Woodpecker. Am. Nat. 117: 421–443.

Koenig, W.D. 1984. Geographic variation in clutch size in the Northern Flicker *(Colaptes auratus):* support for Ashmole's hypothesis. Auk 101: 698–706.

Koenig, W.D., and F.A. Pitelka. 1979. Relatedness and inbreeding avoidance: counterploys in the communally nesting Acorn Woodpecker. Science (Wash., D.C.) 206: 1103–1105.

Koenig, W.D., and F.A. Pitelka. 1981. Ecological factors and kin selection in the evolution of cooperative breeding in birds. *In* Natural Selection and Social Behavior: Recent Research and New Theory, R.D. Alexander and D.W. Tinkle, Eds.: 261–280. New York: Chiron Press.

Konishi, M. 1963. The role of auditory feedback in the vocal behavior of the domestic fowl. Z. Tierpsychol. 20: 349–367.

Konishi, M., and E.I. Knudsen. 1979. The Oilbird: hearing and echolocation. Science (Wash., D.C.) 204: 425–427.

Konishi, M., and F. Nottebohm. 1969. Experimental studies in the ontogeny of avian vocalisations. *In* Bird Vocalisations, R.A. Hinde, Ed.: 29–48. London: Cambridge University Press.

Kontogiannis, J.E. 1968. Effect of temperature and exercise on energy intake and body weight of the White-throated Sparrow *Zonotrichia albicollis.* Physiol. Zool. 41: 54–64.

Korhonen, K. 1981. Temperature in the nocturnal shelters of the Redpoll (*Acanthis flammea* L.) and the Siberian Tit (*Parus cinctus* Budd.) in winter. Ann. Zool. Fenn. 18: 165–168.

Koskimies, J. 1948. On temperature regulation and metabolism in the Swift, *Micropus a. apus* L., during fasting. Experientia (Basel) 4: 274–276.

Koskimies, J. 1950. The life of the swift, *Micropus apus* (L.), in relation to the weather. Ann. Acad. Sci. Fenn. Ser. A IV Biol. No. 15.

Kramer, G. 1950. Orientierte Zugaktivität gekäfigter Singvögel. Naturwissenschaften 37: 188.

Kramer, G. 1951. Eine neue Methode zur Erforschung der Zugorientierung und die bisher damit erzielten Ergenbnisse. Proc. X Int. Ornithol. Congr. 269–280.

Kramer, G. 1952. Experiments on bird orientation. Ibis 94: 265–285.

Krebs, J.R. 1970. The efficiency of courtship feeding in the Blue Tit *Parus caeruleus.* Ibis 112: 108–110.

Krebs, J.R. 1973. Social learning and the significance of mixed-species flocks of chickadees (*Parus* spp.). Can. J. Zool. 51: 1275–1288.

Krebs, J.R. 1974. Colonial nesting and social feeding as strategies for exploiting food resources in the Great Blue Heron *(Ardea herodias).* Behaviour 51: 99–134.

Krebs, J.R. 1977. The significance of song repertoires: The Beau Geste hypothesis. Anim. Behav. 25: 475–478.

Krebs, J.R. 1978. Optimal foraging: Decision rules for predators. *In* Behavioural Ecology: An Evolutionary Approach, J.R. Krebs and N.B. Davies, Eds.: 23–63. Sunderland, MA: Sinauer Associates, Inc.

Krebs, J.R., and C.J. Barnard. 1980. Comments on the function of flocking in birds. Acta XVII Congr. Int. Ornithol.: 795–799.

Krebs, J.R., and R. Dawkins. 1984. Animal signals: mind-reading and manipulation. *In* Behavioural Ecology: An Evolutionary Approach (second ed.), J.R. Krebs and N.B. Davies, Eds.: 380–402. Sunderland, MA: Sinauer Associates, Inc.

Krebs, J.R., J.T. Erichsen, M.I. Webber, and E.L. Charnov. 1977. Optimal prey selection in the Great Tit *(Parus major).* Anim. Behav. 25: 30–38.

Krebs, J.R., and D.E. Kroodsma. 1980. Repertoires and geographical variation in bird song. Adv. Study Behav. 11: 143–177.

Krebs, J.R., M.H. MacRoberts, and J.M. Cullen. 1972. Flocking and feeding in the Great Tit *Parus major*—an experimental study. Ibis 114: 507–530.

Kreithen, M.L. 1979. The sensory world of the homing pigeon. *In* Neural Mechanisms of Behavior in the Pigeon, A.M. Granda and J.H. Maxwell, Eds.: 21–33. New York: Plenum Press.

Kreithen, M.L., and T. Eisner. 1978. Ultraviolet light detection by the homing pigeon. Nature (Lond.) 272: 347–348.

Kreithen, M.L., and W.T. Keeton. 1974a. Detection of changes in atmospheric pressure by the homing pigeon, *Columba livia.* J. Comp. Physiol. 89: 73–82.

Kreithen, M.L., and W.T. Keeton. 1974b. Detection of polarized light by the homing pigeon, *Columba livia.* J. Comp. Physiol. 89: 83–92.

Kreithen, M.L., and D.B. Quine. 1979. Infrasound detection by the homing pigeon: a behavioral audiogram. J. Comp. Physiol. A 129: 1–4.

Kroodsma, D.E. 1976. Reproductive development in a female songbird: differential stimulation by quality of male song. Science (Wash., D.C.) 192: 574–575.

Kroodsma, D.E. 1977. Correlates of song organisation among North American wrens. Am. Nat. 111: 995–1008.

Kroodsma, D.E. 1978. Continuity and versatility in bird song: support for the monotony threshold hypothesis. Nature (Lond.) 274: 681–683.

Kroodsma, D.E. 1979. Vocal dueling among male Marsh Wrens: Evidence for ritualized expressions of dominance/subordinance. Auk 96: 506–515.

Kroodsma, D.E. 1980. Winter Wren singing behavior: A pinnacle of song complexity. Condor 82: 357–365.

Kroodsma, D.E. 1982a. Song repertoires: Problems in their definition and use. *In* Acoustic Communication in Birds, Vol. 2, D.E. Kroodsma and E.H. Miller, Eds.: 125–146. New York: Academic Press.

Kroodsma, D.E. 1982b. Learning and ontogeny of sound signals in birds. *In* Acoustic Communication in Birds, Vol. 2, D.E. Kroodsma and E.H. Miller, Eds.: 1–23. New York: Academic Press.

Kroodsma, D.E. 1984. Songs of the Alder Flycatcher *(Empidonax alnorum)* and Willow Flycatcher *(Empidonax traillii)* are innate. Auk 101: 13–24.

Kroodsma, D.E., and R. Pickert. 1980. Environmentally dependent sensitive periods for avian vocal learning. Nature (Lond.) 288: 477–479.

Kroodsma, R.L. 1975. Hybridization in buntings *(Passerina)* in North Dakota and eastern Montana. Auk 92: 66–80.

Kruijt, J.P., and J.A. Hogan. 1967. Social behavior on the lek in Black Grouse, *Lyrurus tetrix tetrix* (L.). Ardea 55: 203–240.

Kruijt, J.P., G.J. de Vos, and I. Bossema. 1972. The arena system of Black Grouse. Proc. XV Int. Ornithol. Congr.: 399–423.

Kruuk, H. 1964. Predators and anti-predator behavior of the black-headed gull (*Larus ridibundus* L.). Behav. Suppl. 11: 1–130.

Künzi, W. 1918. Versuch einer systematischen Morphologie des Gehirns der Vögel. Rev. Suisse Zool. 26: 17–112.

Lack, D. 1947a. Darwin's Finches. Cambridge: Cambridge University Press.

Lack, D. 1947b. The significance of clutch-size, I-II. Ibis 89: 302–352.

Lack, D. 1948. The significance of clutch-size, III. Ibis 90: 25–45.

Lack, D. 1954. The Natural Regulation of Animal Numbers. Oxford: Clarendon Press.

Lack, D. 1956. Swifts in a Tower. London: Methuen.

Lack, D. 1958. The significance of the colour of turdine eggs. Ibis 100: 145–166.

Lack, D. 1963. Cuckoo hosts in England. (With an appendix on cuckoo hosts in Japan by T. Royama.) Bird Study 10: 185–202.

Lack, D. 1966. Population Studies of Birds. Oxford: Clarendon Press.

Lack, D. 1968. Ecological Adaptations for Breeding in Birds. London: Methuen.

Lack, D. 1971. Ecological Isolation in Birds. Cambridge: Harvard University Press.

Lack, D., and R.E. Moreau. 1965. Clutch-size in tropical passerine birds of forest and savanna. Oiseau Rev. Fr. Ornithol. 35 (No. Special): 76–89.

Lande, R. 1981. Models of speciation by sexual selection on polygenic traits. Proc. Natl. Acad. Sci. U.S.A. 78: 3721–3725.

Lanyon, W.E. 1957. The comparative biology of meadowlarks *(Sturnella)* in Wisconsin. Publ. Nuttall Ornithol. Club No. 1.

Lanyon, W.E. 1960. Vocal characters and avian systematics. *In* Bird Vocalisations, R.A. Hinde, Ed.: 291–310. Cambridge: Cambridge University Press.

Lanyon, W.E. 1967. Revision and probable evolution of the *Myiarchus* flycatchers of the West Indies. Bull. Am. Mus. Nat. Hist. 136: 329–370.

Lanyon, W.E. 1978. Revision of the *Myiarchus* flycatchers of South America. Bull. Am. Mus. Nat. Hist. 161: 427–628.

Lanyon, W.E. 1979. Hybrid sterility in meadowlarks. Nature (Lond.) 279: 557–558.

Lanyon, W.E. 1981. Breeding birds and old field succession on fallow Long Island farmland. Bull. Am. Mus. Nat. Hist. 168: 1–60.

Lasiewski, R.C. 1962. The energetics of migrating hummingbirds. Condor 64: 324.

Lasiewski, R.C. 1972. Respiratory function in birds. *In* Avian Biology, Vol. 2, D.S. Farner, J.R. King, and K.C. Parkes, Eds.: 287–342. New York: Academic Press.

Lasiewski, R.C., W.W. Weathers, and M.H. Bernstein. 1967. Physiological responses of the giant hummingbird, *Patagona gigas.* Comp. Biochem. Physiol. 23: 797–813.

Lawick-Goodall, J. van. 1968. Tool-using bird: the Egyptian Vulture. Natl. Geogr. Mag. 133: 630–641.

Laybourne, R.C. 1967. Bilateral gynandrism in an Evening Grosbeak. Auk 84: 267–272.

Lehrman, D.S. 1953. A critique of Konrad Lorenz's theory of instinctive behavior. Q. Rev. Biol. 28: 337–363.

Lehrman, D.S. 1961. Gonadal hormones and parental behavior in birds and infrahuman mammals. *In* Sex and Internal Secretions, Vol. 2, W.C. Young, Ed.: 1268–1382. Baltimore: Williams and Wilkins.

Lein, M.R. 1978. Song variation in a population of Chestnut-sided Warblers *(Dendroica pensylvanica):* its nature and suggested significance. Can. J. Zool. 56: 1266–1283.

Lemon, R.E. 1977. Bird song: an acoustic flag. Bioscience 27: 402–408.

Lemon, R.E. and C. Chatfield. 1971. Organization of song in cardinals. Anim. Behav. 19: 1–17.

Leopold, F. 1951. A study of nesting Wood Ducks in Iowa. Condor 53: 209–220.

Ligon, J.D. 1967. Relationships of the cathartid vultures. Occas. Pap. Mus. Zool. Univ. Mich. No. 651.

Ligon, J.D. 1970. Still more responses of the Poor-will to low temperatures. Condor 72: 496–498.

Ligon, J.D. 1974. Green cones of the piñon pine stimulate late summer breeding in the piñon jay. Nature (Lond.) 250: 80–82.

Ligon, J.D. 1981. Demographic patterns and communal breeding in the Green Woodhoopoe, *Phoeniculus purpureus. In* Natural Selection and Social Behavior: Recent Research and New Theory, R.D. Alexander and D.W. Tinkle, Eds.: 231–243. New York: Chiron Press.

Ligon, J.D. 1983. Cooperation and reciprocity in avian social systems. Am. Nat. 121: 366–384.

Ligon, J.D., and S.H. Ligon. 1978. The communal social system of the Green Woodhoopoe in Kenya. Living Bird 17: 159–197.

Ligon, J.D., and S.H. Ligon. 1983. Reciprocity in the Green Woodhoopoe *(Phoeniculus purpureus).* Anim. Behav. 31: 480–489.

Lill, A. 1974a. Sexual behavior of the lek-forming White-bearded Manakin (*Manacus manacus trinitatis* Hartert). Z. Tierpyschol. 36: 1–36.

Lill, A. 1974b. Social organization and space utilization in the lek-forming White-bearded Manakin, *M. manacus trinitatis* Hartert. Z. Tierpsychol. 36: 513–530.

Lill, A. 1974c. The evolution of clutch size and male "chauvinism" in the White-bearded Manakin. Living Bird 13: 211–231.

Lincoln, F.C. 1931. Some causes of mortality among birds. Auk 48: 538–546.

Lind, E. 1964. Nistzeitliche Gesselligkeit der Mehlschwalbe, *Delichon u. urbica* (L.). Ann. Zool. Fenn. 1: 7–43.

Linnaeus, C. 1758. Systema Naturae. Regnum Animale (tenth ed., tomus I). Holminae: L. Salvii.

Liversidge, R. 1971. The biology of the Jacobin Cuckoo *Clamator jacobinus.* Proc. Third Pan-Afr. Ornithol. Congr., Ostrich Suppl. 8: 117–137.

Livezy, B.C. 1986. A phylogenetic analysis of recent anseriform genera using morphological characters. Auk 103: 737–754.

Lockley, R.M. 1953. Puffins. London: J.M. Dent and Sons.

Lofts, B., and R.K. Murton. 1973. Reproduction in birds. *In* Avian Biology, Vol. 3, D.S. Farner, J.R. King, and K.C. Parkes, Eds.: 1–107. New York: Academic Press.

Löhrl, H. 1978. Beiträge zur Ethologie und Gewichtsentwicklung beim Wendehals *Jynx torquilla.* Ornithol. Beob. 75: 193–201.

Lorenz, K. 1937. The companion in the bird's world. Auk 54: 245–273.

Lorenz, K. 1941. Vergleichende Bewegungsstudien an Anatinen. J. Ornithol. 89: 194–294 (see translations Avic. Mag. 1951-53.)

Lorenz, K. 1951-1953. Comparative studies on the behaviour of the Anatinae. Avic. Mag. 57: 157–182; 58: 8–17, 61–72, 86–94, 172–184; 59: 24–34, 80–91.

Lorenz, K. 1961. Phylogenetische Anpassung und adaptive Modifikation des Verhaltens. Z. Tierpsychol. 18: 139–187.

Lorenz, K. 1965. Evolution and Modification of Behavior. Chicago: University of Chicago Press.

Lorenz, K. 1969. Innate bases of learning. *In* On the Biology of Learning, K.H. Pribram, Ed.: 13–93. New York: Harcourt Brace and World.

Lovejoy, T.E. 1974. Bird diversity and abundance in Amazon forest communities. Living Bird 13: 127–191.

Lovell, H.B. 1958. Baiting of fish by a Green Heron. Wilson Bull. 70: 280–281.

Lucas, A.M., and P.R. Stettenheim. 1972. Avian Anatomy: Integument. U.S. Dep. Agric. Agric. Handb. 362.

Lumley, William Faithful (Ed.). 1985. Fulton's Book of Pigeons (new, revised, enlarged ed.). London: Cassell and Company, Ltd.

Lustick, S. 1969. Bird energetics: effects of artificial radiation. Science (Wash., D.C.) 163: 387–390.

Lustick, S. 1970. Energy requirements of molt in cowbirds. Auk 87: 742–746.

Lynch, J.F., and N.K. Johnson. 1974. Turnover and equilibria in insular avifaunas, with special reference to the California Channel Islands. Condor 76: 370–384.

MacArthur, R.H. 1969. Patterns of communities in the tropics. Biol. J. Linn. Soc. 1: 19–30.

MacArthur, R.H. 1971. Patterns of terrestrial bird communities. *In* Avian Biology, Vol. 1, D.S. Farner, J.R. King, and K.C. Parkes, Eds.: 189–221. New York: Academic Press.

MacArthur, R.H. 1972. Geographical Ecology. New York: Harper and Row.

MacArthur, R.H., J.M. Diamond, and J.R. Karr. 1972. Density compensation in island faunas. Ecology 53: 330–342.

MacArthur, R.H., and J.W. MacArthur. 1961. On bird species diversity. Ecology 42: 594–598.

MacArthur, R.H., and E.O. Wilson. 1967. The Theory of Island Biogeography. Princeton: Princeton University Press.

McFarland, D.J., and E. Baher. 1968. Feathers affecting feather posture in the Barbary Dove. Anim. Behav. 16: 171–177.

Maclean, G.L. 1967. Die systematische Stellung der Flughühner (Pteroclididae). J. Ornithol. 108: 203–217.

Maclean, G.L. 1968. Field studies on the sandgrouse of the Kalahari desert. Living Bird 7: 209–235.

Maclean, G.L. 1969. The sandgrouse—doves or plovers? J. Ornithol. 110: 104–107.

MacRoberts, M.H., and B.R. MacRoberts. 1976. Social organization and behavior of the Acorn Woodpeckers in central coastal California. Ornithol. Monogr. No. 21.

Maher, W.J. 1962. Breeding biology of the Snow Petrel near Cape Hallett, Antarctica. Condor 64: 488–499.

Maher, W.J. 1970. The Pomarine Jaeger as a brown lemming predator in northern Alaska. Wilson Bull. 82: 130–157.

Manwell, C., and C.M.A. Baker. 1975. Molecular genetics of avian proteins, XIII: Protein polymorphism in three species of Australian passerines. Aust. J. Biol. Sci. 28: 545–557.

Marchant, S. 1960. The breeding of some S.W. Ecuadorian birds. Ibis 102: 349–382, 584–599.

Marler, P. 1955a. Characteristics of some animal calls. Nature (Lond.) 176: 6–8.

Marler, P. 1956. The voice of the Chaffinch and its function as a language. Ibis 98: 231–261.

Marler, P. 1967. Animal communication signals. Science (Wash., D.C.) 157: 769–774.

Marler, P. 1969. Tonal quality of bird sounds. *In* Bird Vocalisations, R.A. Hinde, Ed.: 5–18. Cambridge: Cambridge University Press.

Marler, P. 1981. Birdsong: the acquisition of a learned motor skill. Trends Neurosci. 4: 88–94.

Marler, P., and W.J. Hamilton III. 1966. Mechanisms of Animal Behavior. New York: John Wiley and Sons.

Marler, P., and P. Mundinger. 1971. Vocal learning in birds. *In* The Ontogeny of Vertebrate Behavior, H. Moltz, Ed.: 389–450. New York and London: Academic Press.

Marler, P., and S. Peters. 1977. Selective vocal learning in a sparrow. Science (Wash., D.C.) 198: 519–521.

Marler, P., and S. Peters. 1981. Sparrows learn adult song and more from memory. Science (Wash., D.C.) 213: 780–782.

Marler, P., and S. Peters. 1982a. Subsong and plastic song: their role in the vocal learning process. *In* Acoustic Communication in Birds, Vol. 2, D.E. Kroodsma and E.H. Miller, Eds.: 25–50. New York: Academic Press.

Marler, P., and S. Peters. 1982b. Structural changes in song ontogeny in the Swamp Sparrow, *Melospiza georgiana.* Auk 99: 446–458.

Marler, P., and M. Tamura. 1962. Song "dialects" in three populations of White-crowned Sparrows. Condor 64: 368–377.

Marsh, O.C. 1877. Introduction and succession of vertebrate life in America. Am. J. Sci. (third ser.) 14: 337–378.

Marsh, R.L., and W.R. Dawson. 1982. Substrate metabolism in seasonally acclimatized American goldfinches. Am. J. Physiol. 242 (Regulatory Intregrative Comp. Physiol. 11): R563–R569.

Marshall, A.J. 1954. Bowerbirds. Cambridge: Oxford University Press.

Marshall, A.J. 1961. Reproduction. *In* Biology and Comparative Physiology of Birds, Vol. 2., A.J. Marshall, Ed.: 169–213. New York and London: Academic Press.

Martella, M.B., and E.H. Bucher. 1984. Nesting of the Spot-winged Falconet in Monk Parakeet nests. Auk 101: 614–615.

Martin, L.D. 1983. The origin and early radiation of birds. *In* Perspectives in Ornithology, A.H. Brush and G.A. Clark Jr., Eds.: 291–338. Oxford: Oxford University Press.

Martinez, M.M. 1983. Nidification de *Hirundo rustica erythrogaster* (Boddaert) en la Argentina (Aves, Hirundinidae). Neotropica (La Plata) 29: 83–86.

Matthews, G.V.T. 1951. The experimental investigation of navigation in homing pigeons. J. Exp. Biol. 28: 508–536.

Matthews, G.V.T. 1953. Sun navigation in homing pigeons. J. Exp. Biol. 30: 243–267.

Matthews, G.V.T. 1968. Bird Navigation (second ed.). London: Cambridge University Press.

Maurer, D.R., and R.J. Raikow. 1981. Appendicular myology, phylogeny, and classification of the avian order Coraciiformes (including Trogoniformes). Ann. Carnegie Mus. 50: 417–434.

Mayer, L., S. Lustick, and B. Battersby. 1982. The importance of cavity roosting and hypothermia to the energy balance of the winter acclimatized Carolina Chickadee. Int. J. Biometeorol. 26: 231–238.

Mayfield, H.F. 1960. The Kirtland's Warbler. Bloomfield Hills, MI: Cranbrook Institute of Science.

Mayfield, H.F. 1965. Chance distribution of cowbird eggs. Condor 67: 257–263.

Maynard Smith, J., and M.G. Ridpath. 1972. Wife sharing in the Tasmanian native hen, *Tribonyx mortierii:* a case of kin selection? Am. Nat. 106: 447–452.

Mayr, E. 1926. Die Ausbreitung des Girlitz (*Serinus canaria serinus* L.). J. Ornithol. 74: 571–671.

Mayr, E. 1963. Animal Species and Evolution. Cambridge, MA: The Belknap Press of Harvard University Press.

Mayr, E. 1965. What is a fauna? Zool. Jahrb. Abt. Syst. Oekol. Geogr. Tiere 92: 473–486.

Mayr, E. 1969. Bird speciation in the tropics. Biol. J. Linn. Soc. 1: 1–17.

Mayr, E. 1970. Populations, Species, and Evolution. Cambridge, MA: The Belknap Press of Harvard University Press.

Mayr, E. 1982. Of what use are subspecies? Auk 99: 593–595.

Mayr, E., and D. Amadon. 1951. A classification of recent birds. Am. Mus. Novit. No. 1496.

Mayr, E., and W.H. Phelps Jr. 1967. The origin of the bird fauna of the South Venezuelan highlands. Bull. Am. Mus. Nat. Hist. 136: 269–328.

Mazzeo, R. 1953. Homing of the Manx Shearwater. Auk 70: 200–201.

McCamant, R.E., and E.G. Bolen. 1979. A 12-year study of nest box utilization by Black-bellied Whistling Ducks. J. Wildl. Manage. 43: 936–943.

McCance, R.A. 1960. Severe undernutrition in growing and adult animals, I: production and general effects. Br. J. Nutr. 14: 59–73.

McFarlane, R.W. 1963. The taxonomic significance of avian sperm. Proc. XIII Int. Ornithol. Congr. 91–102.

McKinney, F., J. Barrett, and S.R. Derrickson. 1978. Rape among Mallards. Science (Wash., D.C.) 201: 281–282.

McKitrick, M.C., and R.M. Zink. 1987. Species concepts in ornithology. Condor 90: 1–14.

McLaughlin, C.L., and D. Grice. 1952. The effectiveness of large-scale erection of Wood Duck boxes as a management procedure. Trans. N. Am. Wildl. Nat. Resour. Conf. 17: 242–259.

McLelland, J. 1975. Aves digestive system. *In* Sisson and Grossman's The anatomy of the domestic animals, Vol. 2, R. Getty, Ed. 5th ed.: 1857–1882. Philadelphia: Saunders.

McLelland, J. 1979. Digestive system. *In* Form and Function in Birds, A.S. King and J. McLelland, Eds.: 69–181. New York and London: Academic Press.

Meanley, B., and J.S. Webb. 1965. Nationwide population estimates of blackbirds and Starlings. Atl. Nat. 20: 189–191.

Medway, Lord, and J.D. Pye. 1977. Echolocation and the systematics of swiftlets. *In* Evolutionary Ecology, B. Stonehouse and C. Perrins, Eds.: 225–238. Baltimore, London, Tokyo: University Park Press.

Meinertzhagen, R. 1964. Grit. *In* A New Dictionary of Birds, A.L. Thomson, Ed., New York: McGraw-Hill.

Merkel, F.W. 1971. Orientation behavior of birds in Kramer cages under different physical cues. Ann. N.Y. Acad. Sci. 188: 283–294.

Merkel, F.W., and W. Wiltschko. 1965. Magnetismus und Richtungsfinden zugunruhiger Rotkehlchen *(Erithacus rubecula)*. Vogelwarte 23: 71–77.

Mertens, J.A.L. 1969. The influence of brood size on the energy metabolism and water loss of nestling Great Tits *Parus major major.* Ibis 111: 11–16.

Mewaldt, L.R. 1964. California sparrows return from displacement to Maryland. Science (Wash., D.C.) 146: 941–942.

Mewaldt, L.R., and J.R. King. 1978. Latitudinal variation of postnuptial molt in Pacific coast White-crowned Sparrows. Auk 95: 168–179.

Meyburg, B. 1974. Sibling aggression and mortality among nestling eagles. Ibis 116: 224–228.

Michener, M.C., and C. Walcott. 1967. Homing of single pigeons—analysis of tracks. J. Exp. Biol. 47: 99–131.

Miller, A.H. 1959. Response to experimental light increments by Andean Sparrows from an equatorial area. Condor 61: 344–347.

Miller, A.H. 1962. Bimodal occurrence of breeding in an equatorial sparrow. Proc. Natl. Acad. Sci. U.S.A. 48: 396–400.

Miller, D.B. 1977. Two-voice phenomenon in birds: Further evidence. Auk 94: 567–572.

Miller, G.R., A. Watson, and D. Jenkins. 1970. Responses of Red Grouse populations to experimental improvement of their food. Symp. Br. Ecol. Soc. 10: 323–334.

Millikan, G.C., and R.I. Bowman. 1967. Observations on Galápagos tool-using finches in captivity. Living Bird 6: 23–41.

Minton, C.D.T. 1968. Pairing and breeding of Mute Swans. Wildfowl 19: 41–60.

Miskimen, M. 1951. Sound production in passerine birds. Auk 68: 493–504.

Mock, D.W. 1975. Social behavior of the Boat-billed Heron. Living Bird 14: 185–214.

Mock, D.W. 1976. Pair-formation displays of the Great Blue Heron. Wilson Bull. 88: 185–230.

Mock, D.W. 1978. Pair-formation displays of the Great Egret. Condor 80: 159–172.

Mock, D.W. 1980. Communication strategies of Great Blue Herons and Great Egrets. Behaviour 72: 156–170.

Mock, D.W. 1983. On the study of avian mating systems. *In* Perspectives in Ornithology, A.H. Brush and G.A. Clark Jr., Eds.: 55–91. Cambridge: Cambridge University Press.

Mock, D.W. 1984a. Siblicidal aggression and resource monopolization in birds. Science (Wash., D.C.) 225: 731–733.

Mock, D.W. 1984b. Infanticide, siblicide, and avian nestling mortality. *In* Infanticide: Comparative and Evolutionary Perspectives. G. Hausfater and S. B. Hrdy, Eds.: 3–30. New York: Aldine Publ. Co.

Moore, F.R. 1977. Geomagnetic disturbance and the orientation of nocturnally migrating birds. Science (Wash., D.C.) 196: 682–684.

Moore, F.R. 1978. Interspecific aggression: toward whom should a mockingbird be aggressive? Behav. Ecol. Sociobiol. 3: 173–176.

Moreau, R.E. 1947. Relations between number in brood, feeding-rate and nestling period in nine species of birds in Tanganyika Territory. J. Anim. Ecol. 16: 205–209.

Moreau, R.E. 1950. The breeding seasons of African birds, I: landbirds. Ibis 92: 223–267.

Moreau, R.E. 1961. Problems of Mediterranean-Saharan migration. Ibis 103a: 373–427; 580–623.

Moreau, R.E. 1966. The Bird Faunas of Africa and Its Islands. London: Academic Press.

Moreau, R.E. 1972. The Palaearctic-African Bird Migration Systems. London: Academic Press.

Morejohn, G.V. 1968. Breakdown of isolation mechanisms in two species of captive junglefowl (*Gallus gallus* and *Gallus sonneratii*). Evolution 22: 576–582.

Moreno, J. 1984a. Parental care of fledged young, division of labor, and the development of foraging techniques in the Northern Wheatear (*Oenanthe oenanthe* L.) Auk 101: 741–752.

Moreno, J. 1984b. Search strategies of Wheatears *(Oenanthe oenanthe)* and Stonechats *(Saxicola torquata)*: adaptive variation in perch height, search time, sally distance and inter-perch move length. J. Anim. Ecol. 53: 147–159.

Morony, J.J., Jr., W.J. Bock, and J. Farrand Jr. 1975. Reference List of the Birds of the World. New York: American Museum of Natural History.

Morowitz, H.J. 1968. Energy Flow in Biology. New York: Academic Press.

Morris, D. 1956. The feather postures of birds and the problem of the origin of social signals. Behaviour 9: 75–113.

Morse, D.H. 1970. Ecological aspects of some mixed-species foraging flocks of birds. Ecol. Monogr. 40: 119–168.

Morse, D.H. 1975. Ecological aspects of adaptive radiation in birds. Biol. Rev. Camb. Philos. Soc. 50: 167–214.

Morse, D.H. 1978. Structure and foraging patterns of tit flocks in an English woodland. Ibis 120: 298–312.

Morse, D.H. 1980. Behavioral Mechanisms in Ecology. Cambridge, MA: Harvard University Press.

Morton, E.S. 1973. On the evolutionary advantages and disadvantages of fruit eating in tropical birds. Am. Nat. 107: 8–22.

Morton, E.S. 1975. Ecological sources of selection on avian sounds. Am. Nat. 109: 17–34.

Morton, E.S. 1976. Vocal mimicry in the Thick-billed Euphonia. Wilson Bull. 88: 485–487.

Morton, E.S. 1980. Adaptations to seasonal changes by migrant land birds in the Panama Canal Zone. *In* Migrant Birds in the Neotropics, A. Keast and E.S. Morton, Eds.: 421–436. Washington, D.C.: Smithsonian Institution Press.

Morton, E.S. 1982. Grading, discreteness, redundancy, and motivation-structural rules. *In* Acoustic Communication in Birds, Vol. 1, D.E. Kroodsma and E.H. Miller, Eds.: 183–212. New York and London: Academic Press.

Morton, M.L. 1979. Fecal sac ingestion in the Mountain White-crowned Sparrow. Condor 81: 72–77.

Morton, M.L., and L.R. Mewaldt. 1962. Some effects of castration on a migratory sparrow *(Zonotrichia atricapilla)*. Physiol. Zool. 35: 237–247.

Moseley, L.J. 1979. Individual auditory recognition in the Least Tern *(Sterna albifrons)*. Auk 96: 31–39.

Moss, R. 1972. Food selection by Red Grouse (*Lagopus lagopus scoticus* (Lath.)) in relation to chemical composition. J. Anim. Ecol. 41: 411–428.

Moss, R., and A. Watson. 1980. Inherent changes in the aggressive behavior of a fluctuating Red Grouse *Lagopus lagopus scoticus* population. Ardea 68: 113–119.

Moss, R., G.R. Miller, and S.E. Allen. 1972. Selection of heather by captive red grouse in relation to the age of the plant. J. Appl. Ecol. 9: 771–781.

Moynihan, M. 1955. Some aspects of reproductive behavior in the Black-headed Gull (*Larus ridibundus ridibundus* L.) and related species. Behav. Suppl. 4: 1–201.

Moynihan, M. 1962. The organization and probable evolution of some mixed species flocks of neotropical birds. Smithson. Misc. Collect. 143: 1–140.

Moynihan, M. 1968. Social mimicry: Character convergence versus character displacement. Evolution 22: 315–331.

Mugaas, J.N., and J.R. King. 1981. Annual variation of daily energy expenditure by the Black-billed Magpie: a study of thermal and behavioral energetics. Stud. Avian Biol. No. 5.

Mulligan, J.A. 1966. Singing behavior and its development in the Song Sparrow, *Melospiza melodia.* Univ. Calif. Publ. Zool. No. 81.

Mundinger, P.C. 1982. Microgeographic and macrogeographic variation in the acquired vocalizations of birds. *In* Acoustic Communication in Birds, Vol. 2, D.E. Kroodsma and E.H. Miller, Eds.: 147–208. New York: Academic Press.

Munn, C.A., and J.W. Terborgh. 1979. Multi-species territoriality in Neotropical foraging flocks. Condor 81: 338–347.

Murray, B.G., Jr. 1969. A comparative study of the LeConte's and Sharp-tailed Sparrows. Auk 86: 199–231.

Murray, B.G., Jr. 1971. The ecological consequences of interspecific territorial behavior in birds. Ecology 52: 414–423.

Murray, B.G., Jr. 1979. Population Dynamics. New York: Academic Press.

Murray, G.A. 1976. Geographic variation in the clutch sizes of seven owl species. Auk 93: 602–613.

Murton, R.K. 1965. The Wood Pigeon. London: Collins.

Murton, R.K. 1967. The significance of endocrine stress in population control. Ibis 109: 622–623.

Murton, R.K. 1971a. The significance of a specific search image in the feeding behaviour of the wood-pigeon. Behaviour 40: 10–42.

Murton, R.K. 1971b. Why do some bird species feed in flocks? Ibis 113: 534–536.

Murton, R.K., A.J. Isaacson, N.J. Westwood. 1971. The significance of gregarious feeding behavior and adrenal stress in a population of wood pigeons *Columba palumbus.* J. Zool. (Lond.) 165: 53–84.

Myers, J.P. 1980. The Pampas shorebird community: Interactions between breeding and nonbreeding members. *In* Migrant Birds in the Neotropics, A. Keast and E.S. Morton, Eds.: 37–49. Washington, DC: Smithsonian Institution Press.

Myers, J.P. 1981a. Cross-seasonal interactions in the evolution of sandpiper social systems. Behav. Ecol. Sociobiol. 8: 195–202.

Myers, J.P. 1981b. A test of three hypotheses for latitudinal segregation of the sexes in wintering birds. Can. J. Zool. 59: 1527–1534.

Myers, J.P., P.G. Connors, and F.A. Pitelka. 1979. Territory size in wintering Sanderlings: the effects of prey abundance and intruder density. Auk 96: 551–561.

Myers, J.P., J.L. Maron, and M. Sallaberry. 1985. Going to extremes: why do Sanderlings migrate to the Neotropics? *In* Neotropical Ornithology, P.A. Buckley, M.S. Foster, E.S. Morton, R.S. Ridgely, and F.G. Buckley, Eds. (Ornithol. Monogr. No. 36): 520–535. Washington, DC: The American Ornithologists' Union.

Nathusius, W. von. 1869. Ueber die Hüllen, welche den Dotter des Vogeleies umgeben. Z. Wiss. Zool. 18: 225–270.

Nelson, J.B. 1964. Factors influencing clutch-size and chick growth in the North Atlantic Gannet *Sula bassana.* Ibis 106: 63–77.

Nelson, J.B. 1966. The breeding biology of the Gannet, *Sula bassana,* on the Bass Rock, Scotland. Ibis 108: 584–626.

Nelson, J.B. 1969. The breeding ecology of the Red-footed Booby in the Galapagos. J. Anim. Ecol. 38: 181–198.

Newton, I. 1972. Finches. London: Collins.

Newton, I. 1979. Population Ecology of Raptors. Berkhamsted, England: T. and A.D. Poyser, Ltd.

Newton, I. 1980. The role of food in limiting bird numbers. Ardea 68: 11–30.

Nice, M.M. 1943. Studies in the life history of the Song Sparrow, II. Trans. Linn. Soc. N.Y. No. 6.

Nice, M.M. 1957. Nesting success in altricial birds. Auk 74: 305–321.

Nice, M.M. 1962. Development of behavior in precocial birds. Trans. Linn. Soc. N.Y. 8: 1–211.

Nicolai, J. 1964. Der Brutparasitismus der Viduinae als ethologisches Problem: Prägungsphaenomene als Faktoren der Rassen- und Artbildung. Z. Tierpsychol. 21(2): 129–204.

Nicolai, J. 1974. Mimicry in parasitic birds. Sci. Am. 231(4): 92–98.

Nisbet, I.C.T. 1973. Courtship-feeding, egg-size and breeding success in Common Terns. Nature (Lond.) 241: 141–142.

Nishiyama, H. 1955. Studies of the accessory reproductive organs in the cock. Reprinted from J. Fac. Agric. Kyushu Univ. 10 (3).

Nitzsch, C.L. 1867. Pterylography (trans. W.S. Dallas, P.L. Sclater, Eds.). London: Robert Hardwicke.

Nolan, V., Jr. 1963. Reproductive success of birds in a deciduous scrub habitat. Ecology 44: 305–313.

Nolan, V., Jr., and C.F. Thompson. 1975. The occurrence and significance of anomalous reproductive activities in two North American non-parasitic cuckoos *Coccyzus* spp. Ibis 117: 496–503.

Noon, B.R. 1981. The distribution of an avian guild along a temperate elevational gradient: the importance and expression of competition. Ecol. Monogr. 51: 105–124.

Noordwijk, A.J. van and W. Scharloo. 1981. Inbreeding in an island population of the Great Tit. Evolution 35: 674–688.

Noordwijk, A.J. van, J.H. Van Balen, and W. Scharloo. 1980. Heritability of ecologically important traits in the Great Tit. Ardea 68: 193–203.

Norberg, R.Å. 1977. Occurrence and independent evolution of bilateral ear asymmetry in owls and implications on owl taxonomy. Philos. Trans. R. Soc. Lond. B Biol. Sci. 280: 375–408.

Norton-Griffiths, M. 1969. The organization, control and development of parental feeding in the oystercatcher *(Haematopus ostralegus)*. Behaviour 34: 55–114.

Nottebohm, F. 1967. The role of sensory feedback in the development of avian vocalizations. Proc. XIV Int. Ornithol. Congr.: 265–280.

Nottebohm, F. 1971. Neural lateralization of vocal control in a passerine bird, I: song. J. Exp. Zool. 177: 229–262.

Nottebohm, F. 1972. Neural lateralization of vocal control in a passerine bird, II: subsong, calls, and a theory of vocal learning. J. Exp. Zool. 179: 35–49.

Nottebohm, F. 1975. Vocal behavior in birds. *In* Avian Biology, Vol. 5, D.S. Farner, J.R. King, and K.C. Parkes, Eds.: 287–332. New York: Academic Press.

Nottebohm, F. 1980. Neural pathways for song control: A good place to study sexual dimorphism, hormonal influences, hemispheric dominance and learning. Acta XVII Congr. Int. Ornithol.: 642–647.

Nottebohm, F. 1981. A brain for all seasons: Cyclical anatomical changes in song control nuclei of the Canary brain. Science (Wash., D.C.) 214: 1368–1370.

Nottebohm, F., and M.E. Nottebohm. 1971. Vocalizations and breeding behaviour of surgically deafened Ring Doves *(Streptopelia risoria)*. Anim. Behav. 19: 313–327.

Nottebohm, F., and M.E. Nottebohm. 1978. Relationship between song repertoire and age in the canary *Serinus canarius*. Z. Tierpsychol. 46: 298–305.

Nottebohm, F., S. Kasparian and C. Pandazis. 1981. Brain space for a learned task. Brain Res. 213: 99–110.

O'Connor, R.J. 1975. The influence of brood size upon metabolic rate and body temperature in nestling Blue Tits *Parus caeruleus* and House Sparrows *Passer domesticus*. J. Zool. (Lond.) 175: 391–403.

O'Connor, R.J. 1977. Growth strategies in nestling passerines. Living Bird 16: 209–238.

O'Connor, R.J. 1978. Brood reduction in birds: Selection for fratricide, infanticide and suicide? Anim. Behav. 26: 79–96.

O'Connor, R.J. 1980. Pattern and process in Great Tit *(Parus major)* populations in Britain. Ardea 68: 165–183.

O'Connor, R.J. and A. Cawthorne. 1982. How Britain's birds survived the winter. New Sci. 93: 786–788.

O'Donald, P. 1973. A further analysis of Bumpus' data: The intensity of natural selection. Evolution 27: 398–404.

Odum, E.P., and C.E. Connell. 1956. Lipid levels in migrating birds. Science (Wash., D.C.) 123: 892–894.

Odum, E.P., C.E. Connell, and H.L. Stoddard. 1961. Flight energy and estimated flight ranges of some migratory birds. Auk 78: 515–527.

Odum, E.P., and J.D. Perkinson, Jr. 1951. Relation of lipid metabolism to migration in birds: Seasonal variation in body lipids of the migratory White-throated Sparrow. Physiol. Zool. 24: 216–230.

Ogilvie, M.A. 1967. Population changes and mortality of the Mute Swan in Britain. Rep. Wildfowl Trust 18: 64–73.

Ohmart, R.D., and R.C. Lasiewski. 1971. Roadrunners: Energy conservation by hypothermia and absorption of sunlight. Science (Wash., D.C.) 172: 67–69.

Oliphant, L.W. 1983. First observations of brown fat in birds. Condor 85: 350–354.

Olsen, M.W. 1960. Performance record of a parthenogenetic turkey male. Science (Wash., D.C.) 132: 1661.

Olson, S.L. 1971. Taxonomic comments on the Eurylaimidae. Ibis 113: 507–516.

Olson, S.L. 1973. Evolution of the rails of the South Atlantic islands (Aves: Rallidae). Smithson. Contrib. Zool. No. 152.

Olson, S.L. 1976. Oligocene fossils bearing on the origins of the Todidae and the Momotidae (Aves: Coraciiformes). Smithson. Contrib. Paleobiol. 27: 111–119.

Olson, S.L. 1978a. Multiple origins of the Ciconiiformes. Proc. Colonial Waterbird Group 1978: 165–170.

Olson, S.L. 1978b. A paleontological perspective of West Indian birds and mammals. *In* Zoogeography in the Caribbean, Spec. Publ. Acad. Nat. Sci. Phila. No. 13. F.B. Gill, Ed.: 99–117.

Olson, S.L. 1983. Evidence for a polyphyletic origin of the Piciformes. Auk 100: 126–133.

Olson, S.L. 1985. The fossil record of birds. *In* Avian Biology, Vol. 5, D.S. Farner, J.R. King, and K.C. Parkes, Eds.: 79–238. New York: Academic Press.

Olson, S.L., and A. Feduccia. 1979. Flight capability and the pectoral girdle of *Archaeopteryx.* Nature (Lond.) 278 (5701): 247–248.

Olson, S.L., and A. Feduccia. 1980a. Relationships and evolution of flamingos. Smithson. Contrib. Zool. No. 316.

Olson, S.L., and A. Feduccia. 1980b. *Presbyornis* and the origin of the Anseriformes (Aves: Charadriomorphae). Smithson. Contrib. Zool. No. 323.

Olson, S.L., and H.F. James. 1982. Fossil birds from the Hawaiian Islands: evidence for wholesale extinction by man before western contact. Science (Wash., D.C.) 217: 633–635.

Olson, S.L., K.C. Parkes, M.H. Clench, and S.R. Borecky. 1983. The affinities of the New Zealand passerine genus *Turnagra.* Notornis 30: 319–336.

Olson, S.L., and D.W. Steadman. 1981. The relationships of the Pedionomidae (Aves: Charadriiformes). Smithson. Contrib. Zool. No. 337.

Orians, G.H. 1969a. Age and hunting success in the Brown Pelican *(Pelecanus occidentalis).* Anim. Behav. 17: 316–319.

Orians, G.H. 1969b. On the evolution of mating systems in birds and mammals. Am. Nat. 103: 589–603.

Orians, G.H., and G.M. Christman. 1968. A comparative study of the behavior of Red-winged, Tri-colored, and Yellow-headed Blackbirds. Univ. Calif. Publ. Zool. 84: 1–81.

Orians, G.H., and N.E. Pearson. 1979. On the theory of central place foraging. *In* Analysis of Ecological Systems, D.J. Horn, G.R. Stairs and R. Mitchell, Eds.: 155–177. Columbus: Ohio State University Press.

Orians, G.H., and M.F. Willson. 1964. Interspecific territories of birds. Ecology 45: 736–745.

Oring, L.W. 1982. Avian mating systems. *In* Avian Biology, Vol. 6, D.S. Farner, J.R. King, and K.C. Parkes, Eds.: 1–92. New York: Academic Press.

Oring, L.W., and M.L. Knudson. 1972. Monogamy and polyandry in the Spotted Sandpiper. Living Bird 11: 59–73.

Österlöf, S. 1966. Kungsfågelns *(Regulus regulus)* flyttning. Vår Fågelvärld 25: 49–56.

Ostrom, J.H. 1974. *Archaeopteryx* and the origin of flight. Q. Rev. Biol. 49: 27–47.

Ostrom, J.H. 1975. The origin of birds. Annu. Rev. Earth Planet Sci. 3: 55–77.

Ostrom, J.H. 1976a. *Archaeopteryx* and the origin of birds. Biol. J. Linn. Soc. 8: 91–182.

Ostrom, J.H. 1976b. Some hypothetical stages in the evolution of avian flight. Smithson. Contrib. Paleobiol. 27: 1–21.

Ostrom, J.H. 1979. Bird flight: how did it begin? Am. Sci. 67: 46–56.

Owen, D.F. 1963. Polymorphism in the Screech Owl in eastern North America. Wilson Bull. 75: 183–190.

Packard, G.C., and M.J. Packard. 1980. Evolution of the cleidoic egg among reptilian antecedents of birds. Am. Zool. 20: 351–362.

Page, G., and D.F. Whitacre. 1975. Raptor predation on wintering shorebirds. Condor 77: 73–83.

Papi, F., L. Fiore, V. Fiaschi, and S. Benvenuti. 1971. The influence of olfactory nerve section on the homing capacity of carrier pigeons. Monit. Zool. Ital. (N.S.) 5: 265–267.

Papi, F., L. Fiore, V. Fiaschi, and S. Benvenuti. 1972. Olfaction and homing in pigeons. Monit. Zool. Ital. (N.S.) 6: 85–95.

Parkes, K.C. 1966. Speculations on the origin of feathers. Living Bird 5: 77–86.

Parkes, K.C. 1972. Tail molt in the family Icteridae. Proc. XV Int. Ornithol. Congr.: 674.

Parkes, K.C., and G.A. Clark Jr. 1966. An additional character linking ratites and tinamous, and an interpretation of their monophyly. Condor 68: 459–471.

Parrish, J.W., J.A. Ptacek, and K.L. Will. 1984. The detection of near-ultraviolet light by nonmigratory and migratory birds. Auk 101: 53–58.

Parsons, J. 1970. Relationship between egg size and post-hatching chick mortality in the Herring Gull *(Larus argentatus)*. Nature (Lond.) 228: 1221–1222.

Parsons, J. 1975. Asynchronous hatching and chick mortality in the Herring Gull *Larus argentatus.* Ibis 117: 517–520.

Payne, R.B. 1967. Interspecific communication signals in parasitic birds. Am. Nat. 101: 363–376.

Payne, R.B. 1969a. Overlap of breeding and molting schedules in a collection of African birds. Condor 71: 140–145.

Payne, R.B. 1969b. Breeding seasons and reproductive physiology of Tricolored Blackbirds and Red-winged Blackbirds. Univ. Calif. Publ. Zool. 90: 1–115.

Payne, R.B. 1972. Mechanism and control of molt. *In* Avian Biology, Vol. 2, D.S. Farner, J.R. King, and K.C. Parkes, Eds.: 103–155. New York: Academic Press.

Payne, R.B. 1973a. Individual laying histories and the clutch size and numbers of eggs of parasitic cuckoos. Condor 75: 414–438.

Payne, R.B. 1973b. Behavior, mimetic songs and song dialects, and relationships of the parasitic indigobirds *(Vidua)* of Africa. Ornithol. Monogr. No. 11.

Payne, R.B. 1976. The clutch size and numbers of eggs of Brown-headed Cowbirds: effects of latitude and breeding season. Condor 78: 337–342.

Payne, R.B. 1977a. The ecology of brood parasitism in birds. Annu. Rev. Ecol. Syst. 8: 1–28.

Payne, R.B. 1977b. Clutch size, egg size, and the consequences of single vs. multiple parasitism in parasitic finches. Ecology 58: 500–513.

Payne, R.B. 1979. Sexual selection and intersexual differences in variance of breeding success. Am. Nat. 114: 447–452.

Payne, R.B. 1981a. Population structure and social behavior: models for testing the ecological significance of song dialects in birds. *In* Natural Selection and Social Behavior: Recent Research and New Theory, R.D. Alexander and D.W. Tinkle, Eds.: 108–120. New York: Chiron Press.

Payne, R.B. 1981b. Song learning and social interaction in Indigo Buntings. Anim. Behav. 29: 688–697.

Payne, R.B. 1982a. Species limits in the indigobirds (Ploceidae, *Vidua*) of West Africa: mouth mimicry, song mimicry, and description of new species. Misc. Publ. Mus. Zool. Univ. Mich. No. 162.

Payne, R.B. 1982b. Ecological consequences of song matching: Breeding success and intraspecific song mimicry in Indigo Buntings. Ecology 63: 401–411.

Payne, R.B. 1983. The social context of song mimicry: Song-matching dialects in Indigo Buntings *(Passerina cyanea)*. Anim. Behav. 31: 788–805.

Payne, R.B., and K. Payne. 1977. Social organization and mating success in local song populations of Village Indigobirds, *Vidua chalybeata*. Z. Tierpsychol. 45: 113–173.

Payne, R.B., W.L. Thompson, K.L. Fiala, and L.L. Sweany. 1981. Local song traditions in Indigo Buntings: Cultural transmission of behavior patterns across generations. Behaviour 77: 199–221.

Payne, R.S. 1971. Acoustic location of prey by Barn Owls *(Tyto alba)*. J. Exp. Biol. 54: 535–573.

Payne, R.S., and W.H. Drury Jr. 1958. *Tyto alba,* II: Marksman of the darkness. Nat. Hist. 67: 316–323.

Paynter, R.A., Jr. 1966. A new attempt to construct life tables for Kent Island Herring Gulls. Bull. Mus. Comp. Zool. Harv. Univ. 133: 489–528.

Pearson, D.L. 1977. A pantropical comparison of bird community structure on six lowland forest sites. Condor 79: 232–244.

Pearson, O.P. 1950. The metabolism of hummingbirds. Condor 52: 145–152.

Pearson, R. 1972. The Avian Brain. London: Academic Press.

Peek, F.W. 1972. An experimental study of the territorial function of vocal and visual display in the male Red-winged Blackbird *(Agelaius phoeniceus)*. Anim. Behav. 20: 112–118.

Pendergast, B.A., and D.A. Boag. 1971. Nutritional aspects of the diet of Spruce Grouse in central Alberta. Condor 73: 437–443.

Pennycuick, C.J. 1960. The physical basis of astronavigation in birds: Theoretical considerations. J. Exp. Biol. 37: 573–593.

Pennycuick, C.J. 1969. The mechanics of bird migration. Ibis 111: 525–556.

Pennycuick, C.J. 1975. Mechanics of flight. *In* Avian Biology, Vol. 5, D.S. Farner, J.R. King, and K.C. Parkes, Eds.: 1–75. New York: Academic Press.

Perdeck, A.C. 1958. Two types of orientation in migrating Starlings *Sturnus vulgaris* L., and Chaffinches *Fringilla coelebs* L., as revealed by displacement experiments. Ardea 46: 1–37.

Perdeck, A.C. 1967. Orientation of Starlings after displacement to Spain. Ardea 55: 194–202.

Perek, M., and F. Sulman. 1945. The basal metabolic rate in molting and laying hens. Endocrinology 36: 240–243.

Pernkopf, E., and J. Lehner. 1937. Vorderdarm. Vergleichende Beschreibung des Vorderdarmes bei den einzelnen Klassen der Kranioten. *In* Handbuch der Vergleichenden Anatomie der Wirbeltiere, Vol. 3, L. Bolk, E. Göppert, E. Kallius, and W. Lubosch, Eds.: 349–476. Berlin and Vienna: Urban and Schwarzenberg.

Perrins, C.M. 1965. Population fluctuations and clutch-size in the Great Tit, *Parus major* L. J. Anim. Ecol. 34: 601–647.

Perrins, C.M. 1970. The timing of birds' breeding seasons. Ibis 112: 242–255.

Perrins, C.M. 1977. The role of predation in the evolution of clutch size. *In* Evolutionary Ecology, B. Stonehouse and C. Perrins, Eds.: 181–191. Baltimore, London, Tokyo: University Park Press.

Perrins, C.M. 1980. Survival of young Great Tits, *Parus major.* Acta XVII Congr. Int. Ornithol.: 159–174.

Perrins, C.M., and T.R. Birkhead. 1983. Avian Ecology. New York: Chapman and Hall.

Perrins, C.M., and T.A. Geer. 1980. The effect of Sparrowhawks on tit populations. Ardea 68: 133–142.

Perrins, C.M., and P.J. Jones. 1974. The inheritance of clutch size in the Great Tit (*Parus major* L.). Condor 76: 225–229.

Perrins, C.M., and A.L. Middleton. 1985. The Encyclopedia of Birds. Oxford: Equinox (Oxford) Ltd.

Perrins, C.M., and D. Moss. 1974. Reproductive rates in the Great Tit. J. Anim. Ecol. 44: 695–706.

Peterson, R.T. 1980. A Field Guide to the Birds. Boston: Houghton Mifflin.

Petrinovich, L., T. Patterson, and L.F. Baptista. 1981. Song dialects as barriers to dispersal: a re-evaluation. Evolution 35: 180–188.

Pettingill, O.S. 1984. Ornithology (fourth ed.). New York: Academic Press.

Phillips, R.E. and O.M. Youngren. 1981. Effects of denervation of the tracheo-syringeal muscles on frequency control in vocalizations in chicks. Auk 98: 299–306.

Pianka, E.R. 1966. Latitudinal gradients in species diversity: A review of concepts. Am. Nat. 100: 33–46.

Pietrewicz, A.T., and A.C. Kamil. 1979. Search image formation in the Blue Jay *(Cyanocitta cristata).* Science (Wash., D.C.) 204: 1332–1333.

Pinkowski, B.C. 1977. Breeding adaptations in the Eastern Bluebird. Condor 79: 289–302.

Pinowski, J. 1968. Fecundity, mortality, numbers and biomass dynamics of a population of the Tree Sparrow (*Passer m. montanus* L.). Ekol. Pol. Ser. A No. 16 (1).

Pitelka, F.A. 1958. Timing of molt in Steller Jays of the Queen Charlotte Islands, British Columbia. Condor 60: 38–49.

Pitelka, F.A., R.T. Holmes, and S.F. MacLean Jr. 1974. Ecology and evolution of social organization in Arctic sandpipers. Am. Zool. 14: 185–204.

Pitelka, F.A., P.Q. Tomich, and G.W. Treichel. 1955. Ecological relations of jaegers and owls as lemming predators near Barrow, Alaska. Ecol. Monogr. 25: 85–117.

Pleszczynska, W.K. 1978. Microgeographic prediction of polygyny in the Lark Bunting. Science (Wash., D.C.) 201: 935–937.

Pleszczynska, W.K., and R.I.C. Hansell. 1980. Polygyny and decision theory: Testing of a model in Lark Buntings *(Calamospiza melanocorys).* Am. Nat. 116: 821–830.

Pohlman, A.G. 1921. The position and functional interpretation of the elastic ligaments in the middle-ear region of *Gallus.* J. Morphol. 35: 229–262.

Portmann, A. 1961. Sensory organs: Skin, taste and olfaction. *In* Biology and Comparative Physiology of Birds, Vol. 2, A.J. Marshall, Ed.: 37–48. New York and London: Academic Press.

Portmann, A., and W. Stingelin. 1961. The central nervous system. *In* Biology and

Comparative Physiology of Birds, Vol. 2, A.J. Marshall, Ed.: 1–36. New York and London: Academic Press.

Post, W., and F. Enders. 1970. The occurrence of Mallophaga on two bird species occupying the same habitat. Ibis 112: 539–540.

Post, W., and J.W. Wiley. 1976. The Yellow-shouldered Blackbird—present and future. Am. Birds 30: 13–20.

Post, W., and J.W. Wiley. 1977. Reproductive interactions of the Shiny Cowbird and the Yellow-shouldered Blackbird. Condor 79: 176–184.

Potter, R.K., G.A. Kopp, and H.C. Green. 1947. Visible Speech. New York: D. Van Nostrand, Co. (reprinted 1966 by Dover Publ. Inc., New York).

Powell, G.V.N. 1974. Experimental analysis of the social value of flocking by Starlings *(Sturnus vulgaris)* in relation to predation and foraging. Anim. Behav. 22: 501–505.

Powell, G.V.N. 1979. Structure and dynamics of interspecific flocks in a neotropical mid-elevation forest. Auk 96: 375–390.

Powell, G.V.N. 1985. Sociobiology and adaptive significance of interspecific foraging flocks in the Neotropics. Ornithol. Monogr. No. 36: 713–732.

Prager, E.M., A.C. Wilson, D.T. Osuga, and R.E. Feeney. 1976. Evolution of flightless land birds on southern continents: Transferrin comparison shows monophyletic origin of ratites. J. Mol. Evol. 8: 283–294.

Prager, E.M., and A.C. Wilson. 1976. Congruency of phylogenies derived from different proteins. J. Mol. Evol. 9: 45–57.

Prange, H.D., and K. Schmidt-Nielsen. 1970. The metabolic cost of swimming in ducks. J. Exp. Biol. 53: 763–777.

Pregill, G.K., and S.L. Olson. 1981. Zoogeography of West Indian vertebrates in relation to Pleistocene climatic cycles. Annu. Rev. Ecol. Syst. 12: 75–98.

Preston, F.W. 1962. The canonical distribution of commonness and rarity. Ecology 43: 185–215, 410–432.

Price, T.D. 1984. Sexual selection on body size, territory and plumage variables in a population of Darwin's finches. Evolution 38: 327–341.

Procter, D.L.C. 1975. The problem of chick loss in the South Polar Skua *Catharacta maccormicki.* Ibis 117: 452–459.

Proctor-Gray, E., and R.T. Holmes. 1981. Adaptive significance of delayed attainment of plumage in male American Redstarts: Tests of two hypotheses. Evolution 35: 742–751.

Prys-Jones, O.E. 1973. Interactions between gulls and eiders in St. Andrews Bay, Fife. Bird Study 20: 311–313.

Pugesek, B.H. 1981. Increased reproductive effort with age in the California Gull *(Larus californicus).* Science (Wash. D.C.) 212: 822–823.

Pugh, G.J.F., and M.D. Evans. 1970. Keratinophilic fungi associated with birds, II: Physiological studies. Trans. Br. Mycol. Soc. 54: 241–250.

Pulliam, H.R. 1983. Ecological community theory and the coexistence of sparrows. Ecology 64: 45–52.

Pulliam, H.R. and G.S. Mills. 1977. The use of space by wintering sparrows. Ecology 58: 1393–1399.

Pulliam, H.R., and T.H. Parker. 1979. Population regulation of sparrows. Fortschr. Zool. 25: 137–147.

Pumphrey, R.J. 1961. Sensory organs: Hearing. *In* Biology and Comparative Physiology of Birds, Vol. 2, A.J. Marshall, Ed.: 69–86. New York and London: Academic Press.

Rahn, H., and A. Ar. 1974. The avian egg: Incubation time and water loss. Condor 76: 147–152.

Rahn, H., and A. Ar. 1980. Gas exchange of the avian egg: Time, structure, and function. Am. Zool. 20: 477–484.

Rahn, H., A. Ar, and C.V. Paganelli. 1979. How bird eggs breathe. Sci. Amer. 240(2):46–55.

Rahn, H., T. Ledoux, C.V. Paganelli, and A.H. Smith. 1982. Changes in eggshell conductance after transfer of hens from an altitude of 3800 m to 1200 m. J. Appl. Physiol. 53: 1429–1431.

Rahn, H., C.V. Paganelli, I.C.T. Nisbet, and G.C. Whittow. 1976. Regulation of incubation water loss in eggs of seven species of terns. Physiol. Zool. 49: 245–259.

Raikow, R.J. 1976. The origin and evolution of the Hawaiian honeycreepers (Drepanididae). Living Bird 15: 75–117.

Raikow, R.J. 1978. Appendicular myology and relationships of the New World nine-primaried oscines (Aves: Passeriformes). Bull. Carnegie Mus. Nat. Hist. No. 7.

Raikow, R.J. 1982. Monophyly of the Passeriformes: Test of a phylogenetic hypothesis. Auk 99: 431–445.

Raikow, R.J. 1984. Hindlimb myology and phylogenetic position of the New Zealand wrens. Am. Zool. 24: 446.

Raikow, R.J., S.R. Borecky, and S.L. Berman. 1979. The evolutionary reappearance of a lost ancestral muscle in the bowerbird assemblage. Condor 81: 203–206.

Raikow, R.J., and J. Cracraft. 1983. Monophyly of the Piciformes: A reply to Olson. Auk 100: 134–138.

Raikow, R.J., P.J. Polumbo, and S.R. Borecky. 1980. Appendicular myology and relationships of the shrikes (Aves: Passeriformes: Laniidae). Ann. Carnegie Mus. 49: 131–152.

Rand, A.L. 1954. Social feeding behavior of birds. Fieldiana Zool. 36: 1–71.

Rappole, J.H., E.S. Morton, T.E. Lovejoy III, and J.L. Ruos. 1983. Nearctic Avian Migrants in the Neotropics. Washington, D.C.: U.S. Department of the Interior, Fish and Wildlife Service.

Rappole, J.H., and D.W. Warner. 1980. Ecological aspects of migrant bird behavior in Veracruz, Mexico. *In* Migrant Birds in the Neotropics, A. Keast and E.S. Morton, Eds.: 353–393. Washington, DC: Smithsonian Institution Press.

Ratcliff, D.A. 1970. Changes attributable to pesticides in egg breakage frequency and eggshell thickness in some British birds. J. Appl. Ecol. 7: 67–107.

Raveling, D.G. 1979. The annual cycle of body composition of Canada Geese with special reference to control of reproduction. Auk 96: 234–252.

Rayner, J.M.V. 1985. Flight, speeds of. *In* A Dictionary of Birds, B. Campbell and E. Lack, Eds.: 224–226. Staffordshire, England: T. and A.D. Poyser, Ltd.

Raynor, G.S. 1956. Meteorological variables and the northward movement of nocturnal landbird migrants. Auk 73: 153–175.

Recher, H.F., and J.A. Recher. 1969. Comparative foraging efficiency of adult and immature Little Blue Herons *(Florida caerulea)*. Anim. Behav. 17: 320–322.

Regal, P.J. 1975. The evolutionary origin of feathers. Q. Rev. Biol. 50: 35–66.

Regal, P.J. 1977. Ecology and evolution of flowering plant dominance. Science (Wash., D.C.) 196: 622–629.

Reinecke, K.J. 1979. Feeding ecology and development of juvenile Black Ducks in Maine. Auk 96: 737–745.

Remsen, J.V., Jr. 1984. High incidence of "leapfrog" pattern of geographic variation in Andean birds: Implications for the speciation process. Science (Wash. D.C.) 224: 171–173.

Rensch, B. 1947. Neure probleme der Abstammungslehne. Stuttgart: Ferdinand Enke Verlag.

Reyer, H.-U. 1980. Flexible helper structure as an ecological adaptation in the Pied Kingfisher (*Ceryle rudis rudis* L.). Behav. Ecol. Sociobiol. 6: 219–227.

Rhijn, J.G. van. 1973. Behavioural dimorphism in male Ruffs, *Philomachus pugnax* (L.). Behaviour 47: 153–229.

Rice W.R. 1982. Acoustical location of prey by the Marsh Hawk: Adaptation to concealed prey. Auk 99: 403–413.

Richards, D.G. 1981. Estimation of distance of singing conspecifics by the Carolina Wren. Auk 98: 127–133.
Richardson, W.J. 1978. Timing and amount of bird migration in relation to weather: A review. Oikos 30: 224–272.
Richdale, L.E. 1951. Sexual Behavior of Penguins. Lawrence: University of Kansas Press.
Richdale, L.E. 1957. A Population Study of Penguins. Oxford: Clarendon Press.
Ricklefs, R.E. 1968. Patterns of growth in birds. Ibis 110: 419–451.
Ricklefs, R.E. 1969a. The nesting cycle of songbirds in tropical and temperate regions. Living Bird 8: 165–175.
Ricklefs, R.E. 1969b. An analysis of nesting mortality in birds. Smithson. Contrib. Zool. 9: 1–48.
Ricklefs, R.E. 1973a. Patterns of growth in birds, II: Growth rate and mode of development. Ibis 115: 177–201.
Ricklefs, R.E. 1973b. Fecundity, mortality, and avian demography. *In* Breeding Biology of Birds, D.S. Farner, Ed.: 366–435. Washington, DC: National Academy of Sciences.
Ricklefs, R.E. 1974. Energetics of reproduction in birds. *In* Avian Energetics. Publ. Nuttall. Ornithol. Soc. No. 15: 152–292.
Ricklefs, R.E. 1975. Dwarf eggs laid by a Starling. Bird-Banding 46: 169.
Ricklefs, R.E. 1976a. The Economy of Nature. New York: Chiron Press.
Ricklefs, R.E. 1976b. Growth rates of birds in the humid New World tropics. Ibis 118: 179–207.
Ricklefs, R.E. 1977. A note on the evolution of clutch size in altricial birds. *In* Evolutionary Ecology, B. Stonehouse and C. Perrins, Eds.: 193–214. London: Macmillan.
Ricklefs, R.E. 1979a. Adaptation, constraint, and compromise in avian postnatal development. Biol. Rev. Camb. Philos. Soc. 54: 269–290.
Ricklefs, R.E. 1979b. Patterns of growth in birds. V. A comparative study of development in the Starling, Common Tern, and Japanese Quail. Auk 96: 10–30.
Ricklefs, R.E. 1979c. Ecology (second ed.). New York: Chiron Press.
Ricklefs, R.E. 1980a. To the editor. Condor 82: 476–477.
Ricklefs, R.E. 1980b. Geographical variation in clutch size among passerine birds: Ashmole's hypothesis. Auk 97: 38–49.
Ricklefs, R.E. 1980c. 'Watch-dog' behaviour observed at the nest of a cooperative breeding bird, the Rufous-margined Flycatcher *Myiozetetes cayanensis.* Ibis 122: 116–118.
Ricklefs, R.E. 1983a. Avian postnatal development. *In* Avian Biology, Vol. 7, D.S. Farner, J.R. King, and K.C. Parkes, Eds.: 1–83. New York: Academic Press.
Ricklefs, R.E. 1983b. Comparative avian demography. *In* Current Ornithology, Vol. 1, R.F. Johnston, Ed.: 1–32. New York: Plenum Press.
Ricklefs, R.E., and G.W. Cox. 1972. Taxon cycles in the West Indian avifauna. Am. Nat. 106: 195–219.
Ricklefs, R.E., and F.R. Hainsworth. 1969. Temperature regulation in nestling Cactus Wrens: The nest environment. Condor 71: 32–37.
Ricklefs, R.E., and J. Travis. 1980. A morphological approach to the study of avian community organization. Auk 97: 321–338.
Ricklefs, R.E., and S.C. White. 1981. Growth and energetics of chicks of the Sooty Tern *(Sterna fuscata)* and Common Tern *(S. hirundo).* Auk 98: 361–378.
Ricklefs, R.E., S. White, and J. Cullen. 1980. Postnatal development of Leach's Storm-Petrel. Auk 97: 768–781.
Ridpath, M.G. 1972. The Tasmanian Native Hen, *Tribonyx mortierii.* Commonwealth Scientific and Industrial Research Organization. Wildlife Research 17: 53–90; 91–118.
Ripley, S.D. 1957. Notes on the Horned Coot, *Fulica cornuta* Bonaparte. Postilla No. 30.

Rising, J.D. 1983. The Great Plains hybrid zones. *In* Current Ornithology, Vol. 1, R.F. Johnston, Ed.: 131–157. New York: Plenum Press.

Roberts, B. 1934. Notes on the birds of central and south-east Iceland with special reference to food-habits. Ibis 13: 239–264.

Roby, D. 1986. Diet and reproduction in high latitudes by plankton-feeding seabirds. Ph.D. dissertation, University of Pennsylvania, Philadelphia.

Rochon-Duvigneaud, A. 1950. Les yeux et la vision. *In* Oiseaux, Traité de Zoologie, P.-P. Grassé, Ed., Vol. 15: 221–242 Paris: Masson et Cie.

Rockwell, R.F., C.S. Findlay, and F. Cooke. 1983. Life history studies of the Lesser Snow Goose *(Anser caerulescens caerulescens),* I: The influence of age and time on fecundity. Oecologia (Berl.) 56: 318–322.

Rockwell, R.F., C.S. Findlay, F. Cooke, and J.A. Smith. 1985. Life history studies of the Lesser Snow Goose *(Anser caerulescens caerulescens),* IV: The selective value of plumage polymorphism: Net viability, the timing of maturation, and breeding propensity. Evolution 39: 178–189.

Rohwer, S. 1977. Status signaling in Harris Sparrows: Some experiments in deception. Behaviour 61: 107–129.

Rohwer, S. 1982. The evolution of reliable and unreliable badges of fighting ability. Am. Zool. 22: 531–546.

Rohwer, S., S.D. Fretwell, and D.M. Niles. 1980. Delayed maturation in passerine plumages and the deceptive acquisition of resources. Am. Nat. 115: 400–437.

Romanoff, A.L., and A.J. Romanoff. 1949. The Avian Egg. New York: John Wiley.

Romer, A.S. 1955. The Vertebrate Body (second ed.). Philadelphia: W.B. Saunders Co.

Root, R.B. 1967. The niche exploitation pattern of the Blue-gray Gnatcatcher. Ecol. Monogr. 37: 317–350.

Rosowoski, J.J., and J.C. Saunders. 1980. Sound transmission through the avian interaural pathways. J. Comp. Physiol. A 136: 183–190.

Rotenberry, J.T., and J.A.Wiens. 1980a. Habitat structure, patchiness, and avian communities in North American steppe vegetation: A multivariate analysis. Ecology 61: 1228–1250.

Rotenberry, J.T., and J.A. Wiens. 1980b. Temporal variation in habitat structure and shrubsteppe bird dynamics. Oecologia (Berl.) 47: 1–9.

Rothstein, S.I. 1975. An experimental and teleonomic investigation of avian brood parasitism. Condor 77: 250–271.

Rothstein, S.I. 1982. Successes and failures in avian egg and nestling recognition with comments on the utility of optimality reasoning. Am. Zool. 22: 547–560.

Roudybush, T.E., C.R. Grau, M.R. Petersen, D.G. Ainley, K.V. Hirsch, A.P. Gilman, and S.M. Patten. 1979. Yolk formation in some charadriiform birds. Condor 81: 293–298.

Rowan, W. 1929. Experiments in bird migration, I: Manipulation of the reproductive cycle: Seasonal histological changes in the gonads. Proc. Boston Soc. Nat. Hist. 39: 151–208.

Rowley, I. 1965. The life history of the Superb Blue Wren, *Malurus cyaneus.* Emu 64: 251–297.

Rowley, I. 1981. The communal way of life in the Splendid Wren, *Malurus splendens.* Z. Tierpsychol. 55: 228–267.

Royama, T. 1963. Cuckoo hosts in Japan. Bird Study 10: 201–202.

Royama, T. 1966a. A re-interpretation of courtship feeding. Bird Study 13: 116–129.

Royama, T. 1966b. Factors governing feeding rate, food requirement and brood size of nestling Great Tits *Parus major.* Ibis 108: 313–347.

Royama, T. 1969. A model for the global variation of clutch size in birds. Oikos 20: 562–567.

Rüppell, G. 1977. Bird Flight. New York: Van Nostrand Reinhold Co.

Sabo, S.R. 1980. Niche and habitat relations in subalpine bird communities of the White Mountains of New Hampshire. Ecol. Monogr. 50: 241–259.

Safriel, U.N. 1975. On the significance of clutch size in nidifugous birds. Ecology 56: 703–708.

Sagitov, A.K. 1964. The vestibular apparatus and the degree of mobility of gallinaceous birds. Tr. Samark. Gos. Univ. 137: 5–38.

Salt, G.W. 1964. Respiratory evaporation in birds. Biol. Rev. Camb. Philos. Soc. 39: 113–136.

Samson, F.B. 1976. Territory, breeding density, and fall departure in Cassin's Finch. Auk 93: 477–497.

Sargent, T.D. 1965. The role of experience in the nest building of the Zebra Finch. Auk 82: 48–61.

Sauer, E.G.F. 1957. Die Sternenorientierung nächtlich ziehender Grasmücken *(Sylvia atricapilla, borin und curruca)*. Z. Tierpsychol. 14: 29–70.

Sauer, E.G.F. 1958. Celestial navigation by birds. Sci. Am. 199: 42–47.

Sauer, E.G.F. and E.M. Sauer. 1966. The behavior and ecology of the South Africa Ostrich. Living Bird 5: 45–75.

Saunders, A.A. 1959. Forty years of spring migration in southern Connecticut. Wilson Bull. 71: 208–219.

Schardien, B.J., and J.A. Jackson. 1979. Belly-soaking as a thermoregulatory mechanism in nesting Killdeers. Auk 96: 604–606.

Schenkel, R. 1956. Zur Deutung der Balzleistungen einiger Phasianiden und Tetraoniden. Ornithol. Beob. 53: 182–201.

Schjeldrup-Ebbe, T. 1935. Social behavior of birds. *In* A Handbook of Social Psychology, C.A. Murchison, Ed.: 947–973. Worcester, MA: Clark University Press.

Schlichte, H.-J. 1973. Untersuchungen über die Bedeutung optischer Parameter für das Heimkehrverhalten der Brieftaube. Z. Tierpsychol. 32: 257–280.

Schlichte, H.-J., and K. Schmidt-Koenig. 1971. Zum Heimfindevermögen der Brieftaube bei erschwerter optischer Wahrnehmung. Naturwissenschaften 58: 329–330.

Schluter, D., and P.R. Grant. 1984. Determinants of morphological patterns in communities of Darwin's finches. Am. Nat. 123: 175–196.

Schmidt-Nielsen, K. 1983. Animal Physiology: Adaptation and Environment (third ed.). Cambridge: Cambridge University Press.

Schnell, G.D. 1965. Recording the flight-speed of birds by Doppler radar. Living Bird 4: 79–87.

Schnell, G.D., and J.J. Hellack. 1979. Bird flight speeds in nature: Optimized or a compromise? Am. Nat. 113: 53–66.

Schnell, G.D., B.L. Woods, and B.J. Ploger. 1983. Brown Pelican foraging success and kleptoparasitism by Laughing Gulls. Auk 100: 636–644.

Schoener, T.W. 1968. Sizes of feeding territories among birds. Ecology 49: 123–141.

Schoener, T.W. 1971. Large-billed insectivorous birds: A precipitous diversity gradient. Condor 73: 154–161.

Scholander, P.F., R. Hock, V. Walters, F. Johnson, and L. Irving. 1950. Heat regulation in some arctic and tropical mammals and birds. Biol. Bull. (Woods Hole) 99: 237–258.

Schönwetter, M. 1960–1980. Handbuch der Oologie. Berlin: Akademie-Verlag.

Schreiber, R.W. 1980a. Nesting chronology of the Eastern Brown Pelican. Auk 97: 491–508.

Schreiber, R.W. 1980b. The Brown Pelican: An endangered species? Bioscience 30: 742–747.

Schreiber, R.W., and N.P. Ashmole. 1970. Sea-bird breeding seasons on Christmas Island, Pacific Ocean. Ibis 112: 363–394.

Schutz, F. 1965. Sexuelle Prägung bei Anatiden. Z. Tierpsychol. 22: 50–103.

Schutz, F. 1971. Prägung des Sexualverhaltens von Enten und Gänsen durch Sozialeindrücke während der Jugendphase. J. Neuro-visc. Rel., Suppl. 10: 339–357.

Schüz, E. 1971. Gundriss der Vogelzugskunde. Berlin: Paul Parey.

Schwartz, P. 1964. The Northern Waterthrush in Venezuela. Living Bird 3: 169–184.

Schwartzkopff, J. 1968. Structure and function of the ear and of the auditory brain area in birds. *In* Hearing Mechanisms in Vertebrates, A.V.S. De Reuck and J. Knight, Eds.: 41–59. London: Churchill.

Schwartzkopff, J. 1973. Mechanoreception. *In* Avian Biology, Vol. 3, D.S. Farner, J.R. King, and K.C. Parkes, Eds.: 417–477. New York: Academic Press.

Scott, D.M., and C.D. Ankney. 1980. Fecundity of the Brown-headed Cowbird in southern Ontario. Auk 97: 677–683.

Searcy, W.A., and P. Marler. 1981. A test for responsiveness to song structure and programming in female sparrows. Science (Wash., D.C.) 213: 926–928.

Searcy, W.A., P. Marler, and S. Peters. 1981. Species song discrimination in adult female song and swamp sparrows. Anim. Behav. 29: 997–1003.

Searcy, W.A., and K. Yasukawa. 1983. Sexual selection and Red-winged Blackbirds. Am. Sci. 71: 166–174.

Seastedt, T.R., and S.F. Maclean Jr. 1977. Calcium supplements in the diet of nestling Lapland Longspurs *(Calcarius lapponicus)* near Barrow, Alaska. Ibis 119: 531–533.

Seebohm, H. 1885. A history of British birds with coloured figures of their eggs. London.

Selander, R.K., and R.F. Johnston. 1967. Evolution in the House Sparrow, I: Intrapopulation variation in North America. Condor 69: 217–258.

Selander, R.K., and L.L. Kuich. 1963. Hormonal control and development of the incubation patch in icterids, with notes on behavior of cowbirds. Condor 65: 73–90.

Senner, S.E. 1979. An evaluation of the Copper River Delta as critical habitat for migrating shorebirds. *In* Studies in Avian Biology, Vol. 2, F.A. Pitelka, Ed.: 131–145. Lawrence, KS: Cooper Ornithological Society.

Shelford, V.E. 1945. The relation of Snowy Owl migration to the abundance of the Collared Lemming. Auk 62: 592–596.

Shepard, J.M. 1975. Factors influencing female choice in the lek mating system of the Ruff. Living Bird 14: 87–111.

Shilov, I.A. 1973. Heat Regulation in Birds. An Ecological-physiological Outline. New Delhi: Amerind Publishing Co.

Shoemaker, V.H. 1972. Osmoregulation and excretion in birds. *In* Avian Biology, Vol. 2, D.S. Farner, J.R. King, and K.C. Parkes, Eds.: 527–574. New York: Academic Press.

Short, L.L., Jr. 1963. Hybridization in the wood warblers *Vermivora pinus* and *V. chrysoptera.* Proc. XIII Int. Ornithol. Congr.: 147–160.

Short, L.L., Jr. 1965. Hybridization in the flickers *(Colaptes)* of North America. Bull. Am. Mus. Nat. Hist. 129: 307–428.

Short, L.L., Jr. 1969. Taxonomic aspects of avian hybridization. Auk 86: 84–105.

Short, L.L., Jr. 1972. Hybridization, taxonomy and avian evolution. Ann. Mo. Bot. Gard. 59: 447–453.

Short, L.L., Jr. 1986. Birds: Piciformes (woodpeckers, barbets, honeyguides, toucans). *In* Encyclopaedia Britannica (15th ed.), Macropaedia, Vol. 15: 90–95. Chicago: Encyclopaedia Britannica, Inc.

Sibley, C.G. 1954. Hybridization in the red-eyed towhees of Mexico. Evolution 8: 252–290.

Sibley, C.G. 1968. The relationships of the "wren-thrush", *Zeledonia coronata* Ridgway. Postilla No. 125.

Sibley, C.G. 1974. The relationships of the lyrebirds. Emu 74: 65–79.

Sibley, C.G. 1970. A comparative study of the egg-white proteins of passerine birds. Peabody Mus. Nat. Hist. Yale Univ. Bull. No. 32.

Sibley, C.G., and J.E. Ahlquist. 1972. A comparative study of the egg white proteins of non-passerine birds. Peabody Mus. Nat. Hist. Yale Univ. Bull. No. 39.

Sibley, C.G., and J.E. Ahlquist. 1973. The relationships of the Hoatzin. Auk 90: 1–13.

Sibley, C.G., and J.E. Ahlquist. 1980. The relationships of the "primitive insect eaters" (Aves: Passeriformes) as indicated by DNA × DNA hybridization. Acta XVII Int. Ornithol.: 1215–1220.

Sibley, C.G., and J.E. Ahlquist. 1981a. The phylogeny and relationships of the ratite birds as indicated by DNA-DNA hybridization. Proc. Second Int. Congr. Syst. Evol. Biol.: 301–335.

Sibley, C.G., and J.E. Ahlquist. 1981b. The relationships of the wagtails and pipits (Motacillidae) as indicated by DNA-DNA hybridization. Oiseau Rev. Fr. Ornithol. 51: 189–199.

Sibley, C.G., and J.E. Ahlquist. 1982a. The relationships of the Hawaiian Honeycreepers (Drepaninini) as indicated by DNA-DNA hybridization. Auk 99: 130–140.

Sibley, C.G., and J.E. Ahlquist. 1982b. The relationships of the vireos (Vireoninae) as indicated by DNA-DNA hybridization. Wilson Bull. 94: 114–128.

Sibley, C.G., and J.E. Ahlquist. 1983. Phylogeny and classification of birds based on the data of DNA-DNA hybridization. *In* Current Ornithology, Vol. 1, R.F. Johnston, Ed.: 245–292. New York: Plenum Press.

Sibley, C.G., and J.E. Ahlquist. 1984. The relationships of the starlings (Sturnidae: Sturnini) and the mockingbirds (Sturnidae: Mimini). Auk 101: 230–243.

Sibley, C.G., and J.E. Ahlquist. 1985a. The relationships of some groups of African birds, based on comparisons of the genetic material, DNA. *In* Proceedings of the International Symposium on African Vertebrates: Systematics, Phylogeny and Evolutionary Biology, K.-L. Schuchmann, Ed.: 115–161. Bonn: Zoologisches Forschungsinstitut und Museum Alexander Koenig.

Sibley, C.G., and J.E. Ahlquist. 1985b. The phylogeny and classification of the Australo-papuan passerine birds. Emu 85: 1–13.

Sibley, C.G., and J.E. Ahlquist. 1985c. The phylogeny and classification of the passerine birds, based on comparisons of the genetic material, DNA. *In* Proc. XVIII Int. Orn. Congr. Moscow (1982). V.N. Ilyichev and V.M. Gavrilov, Eds.: 83–121. Moscow: Navka Publ.

Sibley, C.G., J.E. Ahlquist, and B.L. Monroe Jr. 1988. A classification of the living birds of the world based on DNA-DNA hybridization studies. Auk 105: 409–423.

Sibley, C.G., and C. Frelin. 1972. The egg white protein evidence for ratite affinities. Ibis 114: 377–387.

Sibley, C.G., S.M. Lanyon, and J.E. Ahlquist. 1984. The relationships of the Sharpbill *(Oxyruncus cristatus)*. Condor 86: 48–52.

Sibley, C.G., and L.L. Short Jr. 1964. Hybridization in the orioles of the Great Plains. Condor 66: 130–150.

Sibley, C.G., G.R. Williams, and J.E. Ahlquist. 1982. The relationships of the New Zealand Wrens (Acanthisittidae) as indicated by DNA-DNA hybridization. Notornis 29: 113–130.

Sick, H. 1964. Hoatzin. *In* A New Dictionary of Birds, A.L. Thomson, Ed.: 369–371. New York: McGraw-Hill.

Sick, H. 1967. Courtship behavior in the manakins (Pipridae): A review. Living Bird. 6: 5–22.

Sillman, A.J. 1973. Avian vision. *In* Avian Biology, Vol. 3, D. S. Farner, J.R. King, and K.C. Parkes, Eds.: 349–387. New York: Academic Press.

Simberloff, D. 1976. Species turnover and equilibrium island biogeography. Science (Wash., D.C.) 194: 572–578.

Simberloff, D. 1978. Using island biogeographic distributions to determine if colonization is stochastic. Am. Nat. 112: 713–726.

Simmons, K.E.L. 1952. The nature of the predator-reactions of breeding birds. Behaviour 4: 161–171.

Simmons, K.E.L. 1955. The nature of the predator-reactions of waders towards humans; with special reference to the role of the aggressive- , escape- , and brooding-drives. Behaviour 8: 130–173.

Simmons, K.E.L. 1957. The taxonomic significance of the head-scratching methods of birds. Ibis 99: 178–181.

Simmons, K.E.L. 1964. Feather maintenance. *In* A New Dictionary of Birds, A.L. Thomson, Ed.: 278–286. New York: McGraw-Hill.

Simpson, G.G. 1946. Fossil penguins. Bull. Am. Mus. Nat. Hist. 87: 1–100.

Simpson, G.G. 1949. The Meaning of Evolution. New Haven: Yale University Press.

Simpson, S.F., and J. Cracraft. 1981. The phylogenetic relationships of the Piciformes (Class Aves). Auk 98: 481–494.

Skowron, C., and M. Kern. 1980. The insulation in nests of selected North American songbirds. Auk 97: 816–824.

Skutch. A.F. 1944. Life history of the Quetzal. Condor 46: 213–235.

Skutch, A.F. 1949. Do tropical birds rear as many young as they can nourish? Ibis 91: 430–455.

Skutch, A.F. 1960. Life histories of Central American birds, II. Pacific Coast Avifauna No. 34.

Skutch, A.F. 1961. Helpers among birds. Condor 63: 198–226.

Skutch, A.F. 1976. Parent Birds and Their Young. Austin: University of Texas Press.

Smith, J.N.M. 1974. The food searching behaviour of two European thrushes, II: The adaptiveness of the search patterns. Behaviour 49: 1–61.

Smith, J.N.M., and A.A. Dhondt. 1980. Experimental confirmation of heritable morphological variation in a natural population of Song Sparrows. Evolution 34: 1155–1158.

Smith, J.N.M., and H.P.A. Sweatman. 1974. Food-searching behavior of titmice in patchy environments. Ecology 55: 1216–1232.

Smith, J.N.M., and R. Zach. 1979. Heritability of some morphological characters in a Song Sparrow population. Evolution 33: 460–467.

Smith, N.G. 1966. Evolution of some Arctic gulls *(Larus)*: An experimental study of isolating mechanisms. Ornithol. Monogr. No. 4.

Smith, N.G. 1968. The advantage of being parasitized. Nature (Lond.) 219: 690–694.

Smith, N.G. 1980. Hawk and vulture migrations in the Neotropics. *In* Migrant Birds in the Neotropics, A. Keast and E.S. Morton, Eds.: 51–65. Washington, DC: Smithsonian Institution Press.

Smith, S.M. 1967. Seasonal changes in the survival of the Black-capped Chickadee. Condor 69: 344–359.

Smith, S.M. 1972. The ontogeny of impaling behaviour in the Loggerhead Shrike, *Lanius ludovicianus* L. Behaviour 42: 232–247.

Smith, S.M. 1975. Innate recognition of coral snake pattern by a possible avian predator. Science (Wash., D.C.) 187: 759–760.

Smith, S.M. 1977. Coral-snake pattern recognition and stimulus generalisation by naive great kiskadees (Aves: Tyrannidae). Nature (Lond.) 265: 535–536.

Smith, S.M. 1978. The "underworld" in a territorial sparrow: Adaptive strategy for floaters. Am. Nat. 112: 571–582.

Smith, S.M. 1983. The ontogeny of avian behavior. *In* Avian Biology, Vol. 7, D.S. Farner, J.R. King, and K.C. Parkes, Eds.: 85–160. New York: Academic Press.

Smith, W.J. 1969. Messages of vertebrate communication. Science (Wash., D.C.) 165: 145–150.

Smith, W.J. 1977. The Behavior of Communicating. Cambridge: Harvard University Press.

Smith, W.J., Pawlukiewicz, J. and S.T. Smith. 1978. Kinds of activities correlated with singing patterns of the Yellow-throated Vireo. Anim. Behav. 26: 862–884.

Smith, W.K., S.W. Roberts, and P.C. Miller. 1974. Calculating the nocturnal energy expenditure of an incubating Anna's Hummingbird. Condor 76: 176–183.

Snow, B.K., and D.W. Snow. 1979. The Ochre-bellied Flycatcher and the evolution of lek behavior. Condor 81: 286–292.

Snow, D.W. 1956. The annual mortality of the Blue Tit in different parts of its range. Br. Birds 49: 174–177.

Snow, D.W. 1962. A field study of the Black and White Manakin, *Manacus manacus,* in Trinidad. Zoologica (N.Y.) 47: 65–104.

Snow, D.W. 1978. The nest as a factor determining clutch-size in tropical birds. J. Ornithol. 119: 227–230.

Snow, D.W. 1976. The Web of Adaptation. New York: Quadrangle/NY Times Books.

Snow, D.W., and B.K. Snow. 1964. Breeding seasons and annual cycles of Trinidad land-birds. Zoologica (N.Y.) 49: 1–39.

Southern, H.N. 1970. The natural control of a population of Tawny Owls *(Strix aluco).* J. Zool. (Lond.) 162: 197–285.

Southern, W.E. 1971. Gull orientation by magnetic cues: A hypothesis revisited. Ann. N.Y. Acad. Sci. 188: 295–311.

Southern, W.E. 1972. Influence of disturbances in the earth's magnetic field on Ring-billed Gull orientation. Condor 74: 102–105.

Southwick, E.E., and D.M. Gates. 1975. Energetics of occupied hummingbird nests. *In* Perspectives of biophysical ecology, D.M. Gates and R.B. Schmerl, Eds.: 417–430. New York: Springer Verlag.

Spellerberg, I.F. 1971. Aspects of McCormick Skua breeding biology. Ibis 113: 357–363.

Spitzer, G. 1972. Jahreszeitliche Aspekte der Biologie der Bartmeise *(Panurus biarmicus).* J. Ornithol. 113: 241–275.

Stacey, P.B., and C.E. Bock. 1978. Social plasticity in the Acorn Woodpecker. Science (Wash., D.C.) 202: 1298–1300.

Stager, K.E. 1964. The role of olfaction in food location by the Turkey Vulture *(Cathartes aura).* Los Ang. Cty. Mus. Contrib. Sci. No. 81.

Stager, K.E. 1967. Avian olfaction. Am. Zool. 7: 415–419.

Stallcup, J.A., and G.E. Woolfenden. 1978. Family status and contributions to breeding by Florida Scrub Jays. Anim. Behav. 26: 1144–1156.

Stapel, S.O., J.A.M. Leunissen, M. Versteeg, J. Wattel, and W.W. de Jong. 1984. Ratites as oldest offshoot of avian stem—evidence from alpha-crystallin A sequences. Nature (Lond.) 311: 257–259.

Stegmann, B.C. 1978. Relationships of the superorders Alectoromorphae and Charadriomorphae (Aves): A comparative study of the avian hand. Publ. Nuttall Ornithol. Club No. 17.

Sternberg, H. 1972. The origin and age composition of newly formed populations of Pied Flycatchers *(Ficedula hypoleuca).* Proc. XV Int. Ornithol. Congr.: 690–691.

Stettenheim, P. 1973. The bristles of birds. Living Bird 12: 201–234.

Stettenheim, P. 1976. Structural adaptations in feathers. Proc. 16th Int. Ornithol. Congr.: 385–401.

Stettner, L.J., and K.A. Matyniak. 1968. The brain of birds. Sci. Am. 218: 64–76.

Stewart, R.E., and J.W. Aldrich. 1951. Removal and repopulation of breeding birds in a spruce-fir forest community. Auk 68: 471–482.

Stiles, E.W. 1978. Avian communities in temperate and tropical alder forests. Condor 80: 276–284.

Stiles, F.G. 1971. Time, energy, and territoriality of the Anna Hummingbird *(Calypte anna).* Science (Wash., D.C.) 173: 818–821.

Stiles, F.G. 1978. Possible specialization for hummingbird-hunting in the Tiny Hawk. Auk 95: 550–553.

Stiles, F.G. 1980. Evolutionary implications of habitat relations between permanent and winter resident landbirds in Costa Rica. *In* Migrant Birds in the Neotropics, A. Keast and E.S. Morton, Eds.: 421–435. Washington, DC: Smithsonian Institution Press.

Stinson, C.H. 1979. On the selective advantage of fratricide in raptors. Evolution 33: 1219–1225.

Stokes, A.W. 1960. Nest-site selection and courtship behaviour of the Blue Tit *Parus caeruleus.* Ibis 102: 507–519.

Stonehouse, B. 1960. The King Penguin *Aptenodytes patagonica* of South Georgia, I: Breeding behaviour and development. Sci. Rep. Falkland Is. Depend. Surv. No. 23: 1–81.

Storer, J.H. 1948. The flight of birds. Cranbrook Inst. Sci. Bull. No. 28.

Storer, R.W. 1956. The fossil loon, *Colymbus minutus.* Condor 58: 413–426.

Storer, R.W. 1960. Evolution in the diving birds. Proc. XII Int. Ornithol. Congr.: 694–707.

Storer, R.W. 1963. Courtship and mating behavior and the phylogeny of the grebes. Proc. XIII Int. Ornithol. Congr.: 562–569.

Storer, R.W. 1967. The patterns of downy grebes. Condor 69: 469–478.

Storer, R.W. 1970. Independent evolution of the Dodo and the Solitaire. Auk 87: 369–370.

Storer, R.W. 1971a. Adaptive radiation of birds. *In* Avian Biology, Vol. 1, D.S. Farner, J.R. King, and K.C. Parkes, Eds.: 149–188. New York: Academic Press.

Storer, R.W. 1971b. Classification of birds. *In* Avian Biology, Vol. 1, D.S. Farner, J.R. King, and K.C. Parkes, Eds.: 1–18. New York: Academic Press.

Storer, R.W. 1986. Birds: Podicipediformes (grebes). *In* Encyclopaedia Britannica (15th ed.) Macropaedia, Vol. 15: 14–16. Chicago: Encyclopaedia Britannica, Inc.

Strauch, J.G., Jr. 1978. The phylogeny of the Charadriiformes (Aves): A new estimate using the method of character compatibility analysis. Trans. Zool. Soc. Lond. 34: 269–345.

Stresemann, E. 1927-1934. Aves. *In* Handbuch Zoologie, Vol. VII B, W. Kükenthal, Ed.: 729–853. Berlin: W. de Gruyter.

Stresemann, E. 1959. The status of avian systematics and its unsolved problems. Auk 76: 269–280.

Stresemann, E. 1967. Inheritance and adaptation in moult. Proc. XIV Int. Ornithol. Congr.: 75–80.

Stresemann, E., and V. Stresemann. 1966. Die Mauser der Vögel. J. Ornithol., Sonderheft 107.

Strong, D.R., Jr., D. Simberloff, L.G. Abele, and A.B. Thistle. 1984. Ecological Communities. Conceptual Issues and the Evidence. Princeton, NJ: Princeton University Press.

Sturkie, P.D. 1976. Avian Physiology (third ed.). New York: Springer-Verlag.

Sulkava, S. 1969. On small birds spending the night in the snow. Aquilo Ser Zool. 7: 33–37.

Summers, K.R., and R.H. Drent. 1979. Breeding biology and twinning experiments of Rhinoceros Auklets on Cleland Island, British Columbia. Murrelet 60: 16–22.

Summers, R.W., and M. Waltner. 1979. Seasonal variations in the mass of waders in southern Africa, with special reference to migration. Ostrich 50: 21–37.

Swennen, C. 1968. Nest protection of Eiderducks and Shovelers by means of faeces. Ardea 56: 248–258.

Swierczewski, E.V., and R.J. Raikow. 1981. Hind limb morphology, phylogeny, and classification of the Piciformes. Auk 98: 466–480.

Sy, M. 1936. Funktionell-anatomische Untersuchungen am Vogelflügel. J. Ornithol. 84: 199–296.

Szumowski, P., and M. Theret. 1965. Causes possibles de la faible fertilité des oies et des difficultés de son amelioration. Recl. Méd. Vét Ec. Alfort 141: 583.

Talesara, G.L., and G. Goldspink. 1978. A combined histochemical and biochemical study of myofibrillar ATPase in pectoral, leg and cardiac muscle of several species of bird. Histochem. J. 10: 695–710.

Tarsitano, S., and M.K. Hecht. 1980. A reconsideration of the reptilian relationships of *Archaeopteryx*. Zool. J. Linn. Soc. 69: 149–182.

Tasker, C.R., and J.A. Mills. 1981. A functional analysis of courtship feeding in the Red-billed Gull, *Larus novaehollandiae scopulinus.* Behaviour 77: 222–241.

Taylor, T.G. 1970. How an eggshell is made. Sci. Am. 222 (3): 88–95.

Temple, S.A. 1977. The status and conservation of endemic kestrels on Indian Ocean Islands. *In* Proceedings of the ICBP World Conference on Birds of Prey, R.D. Chancellor, Ed.: 74–92. London: International Council for Bird Preservation.

Temple, S.A. 1978. Manipulating behavioral patterns of endangered birds: a potential management technique. *In* Endangered Birds: Management Techniques for Preserving Threatened Species, S.A. Temple, Ed.: 435–446. Madison: University of Wisconsin Press.

Terborgh, J.W. 1971. Distribution on environmental gradients: Theory and a preliminary interpretation of distributional patterns in the avifauna of the Cordillera Vilcabamba, Peru. Ecology 52: 23–40.

Terborgh, J.W. 1980. The conservation status of Neotropical migrants: Present and future. *In* Migrant Birds in the Neotropics, A. Keast and E.S. Morton, Eds.: 21–30. Washington, DC: Smithsonian Institution Press.

Terborgh, J.W., and J. Faaborgh. 1980. Factors affecting the distribution and abundance of North American migrants in the eastern Caribbean region. *In* Migrant Birds in the Neotropics, A. Keast and E.S. Morton, Eds.: 145–155. Washington, DC: Smithsonian Institution Press.

Terborgh, J.W., and J.S. Weske. 1969. Colonization of secondary habitats by Peruvian birds. Ecology 50: 765–782.

Terborgh, J.W., and B. Winter. 1980. Some causes of extinction. *In* Conservation Biology: An Evolutionary and Ecological Perspective, M.E. Soulé and B.A. Wilcox, Eds.: 119–133. Sunderland, MA: Sinauer Associates, Inc.

Thayer, G.H. 1909. Concealing-coloration in the Animal Kingdom. New York: Macmillan.

Thielcke, G. 1961. Stammegeschichte und geographische Variation des Gesanges unserer Baumläufer *(Certhia familiaris* L. und *C. brachydactyla* Brehm). Z. Tierpsychol. 18: 188–204.

Thielcke, G. 1969. Geographic variation in bird vocalizations. *In* Bird Vocalisations, R.A. Hinde, Ed.: 311–339. Cambridge: Cambridge University Press.

Thompson, A.L. (Ed.). 1964. A New Dictionary of Birds. New York: McGraw-Hill.

Thompson, W.A., I. Vertinsky, and J.R. Krebs. 1974. The survival value of flocking in birds: A simulation model. J. Anim. Ecol. 43: 785–820.

Thorpe, W.H. 1958. The learning of song patterns by birds, with especial reference to the song of the Chaffinch *Fringilla coelebs.* Ibis 100: 535–570.

Thorpe, W.H. 1961. Bird-Song. London: Cambridge University Press.

Thorpe, W.H. 1963. Antiphonal singing in birds as evidence for avian auditory reaction time. Nature (Lond.) 197: 774–776.

Thorpe, W.H. 1973. Duet-singing birds. Sci. Am. 229 (2): 70–79.

Thorpe, W.H., and M.E.W. North. 1966. Vocal imitation in the tropical Bou-bou Shrike *Laniarus aethiopicus major* as a means of establishing and maintaining social bonds. Ibis 108: 432–435.

Tinbergen, N. 1951. The Study of Instinct. London: Oxford University Press.

Tinbergen, N. 1952. "Derived" activities; their causation, biological significance, origin, and emancipation during evolution. Q. Rev. Biol. 27: 1–32.

Tinbergen, N. 1959. Comparative studies of the behaviour of gulls (Laridae): A progress report. Behaviour 15: 1–70.

Tinbergen, N. 1963. The shell menace. Nat. Hist. 72 (7): 28–35.

Tinbergen, N., M. Impekoven, and D. Franck. 1967. An experiment on spacing-out as a defence against predation. Behaviour 28: 307–321.

Tinbergen, N., and A.C. Perdeck. 1950. On the stimulus situation releasing the begging response in the newly hatched Herring Gull chick (*Larus argentatus argentatus* Pont.). Behaviour 3: 1–39.

Tompa, F.S. 1964. Factors determining the numbers of Song Sparrows, *Melospiza melodia* (Wilson), on Mandarte Island, B.C., Canada. Acta Zool. Fenn. No. 109.

Tordoff, H.B. 1984. Do woodcock carry their young? Loon 56 (2): 81–82.

Trail, P.W. 1985. Courtship disruption modifies mate choice in a lek-breeding bird. Science (Wash., D.C.) 227: 778–780.

Trautman, M.B., W.E. Bills, and E.L. Wickliff. 1939. Winter losses from starvation and exposure of waterfowl and upland game birds in Ohio and other northern states. Wilson Bull. 51: 86–104.

Traylor, M.A., Jr., 1977. A classification of the tyrant flycatchers (Tyrannidae). Bull. Mus. Comp. Zool. 148: 129–184.

Trivers, R.L. 1971. The evolution of reciprocal altruism. Q. Rev. Biol. 46: 35–57.

Truslow, F.K. 1967. Egg-carrying by the Pileated Woodpecker. Living Bird 6: 227–236.

Tschanz, B. 1959. Zur Brutbiologie der Trottellumme (*Uria aalge aalge* Pont.). Behaviour 14: 1–100.

Tschanz, B. 1968. Trottellummen. Z. Tierpsychol., Beiheft 4: 1–103.

Tschanz, B., P. Ingold, and H. Lengacher. 1969. Eiform und Bruterfolg bei Trottellummen *Uria aalge aalge* Pont. Ornithol. Beob. 66: 25–42.

Tucker, V. 1968. Respiratory exchange and evaporative water loss in the flying Budgerigar. J. Exp. Biol. 48: 67–87.

Tucker, V.A. 1969. The energetics of bird flight. Sci. Am. 220 (5): 70–78.

Tullett, S.G., and R.G. Board. 1977. Determinants of avian eggshell porosity. J. Zool. (Lond.) 183: 203–211.

Turner, E.L. 1924. Broadland birds. London.

Tyler, C., and K. Simkiss. 1959. A study of the egg shells of ratite birds. Proc. Zool. Soc. Lond. 133: 201–243.

Van Balen, J.H. 1980. Population fluctuations of the Great Tit and feeding conditions in winter. Ardea 68: 143–164.

Van Iersel, J.J.A., and A.C.A. Bol. 1958. Preening of two tern species: A study on displacement activities. Behaviour 13: 1–88.

Van Tets, G.F. 1965. A comparative study of some social communication patterns in the Pelecaniformes. Ornithol. Monogr. No. 2.

Van Tyne, J., and A.J. Berger. 1976. Fundamentals of Ornithology (second ed.). New York: John Wiley.

Vanden Berge, J.C. 1975. Aves myology. *In* Sisson and Grossman's The anatomy of the domestic animals (fifth ed.), Vol 2., R. Getty with editorial coordination and completion by C.A. Rosenbaum, N.G. Ghoshal, and D. Hillman: 1802–1848. Philadelphia: Saunders.

Vander Wall, S.B. 1982. An experimental analysis of cache recovery in Clark's Nutcracker. Anim. Behav. 30: 84–94.

Vehrencamp, S.L. 1978. The adaptive significance of communal nesting in Groove-billed Anis *(Crotophaga sulcirostris)*. Behav. Ecol. Sociobiol. 4: 1–33.

Vehrencamp, S.L. 1979. The roles of individual, kin, and group selection in the evolution of sociality. *In* Handbook of Behavioural Neurobiology, Vol. 3, P. Marler and J.G. Vandenbergh, Eds.: 351–394. New York: Plenum Press.

Vehrencamp, S.L. and J.W. Bradbury. 1984. Mating systems and ecology. *In* Behavioral Ecology: An Evolutionary Approach (second ed.), J.R. Krebs and N.B. Davies, Eds.: 251–278. Sunderland, MA: Sinauer Associates, Inc.

Verbeek, N.A.M. 1972. Daily and annual time budget of the Yellow-billed Magpie. Auk 89: 567–582.

Verbeek, N.A.M. 1973. The exploitation system of the Yellow-billed Magpie. Univ. Calif. Publ. Zool. 99: 1–58.

Verner, J. 1964. Evolution of polygamy in the Long-billed Marsh Wren. Evolution 18: 252–261.

Verner, J., and G.H. Engelson. 1970. Territories, multiple nest building, and polygyny in the Long-billed Marsh Wren. Auk 87: 557–567.

Verner, J., and M.F. Willson. 1966. The influence of habitats on mating systems of North American passerine birds. Ecology 47: 143–147.

Verner, J., and M.F. Willson. 1969. Mating systems, sexual dimorphism, and the role of male North American passerine birds in the nesting cycle. Ornithol. Monogr. No. 9.

Vernon, C.J. 1973. Vocal imitation by southern African birds. Ostrich 44: 23–30.

Vince, M.A. 1969. How quail embryos communicate. Ibis 111: 441.

Voous, K.H. 1957. Studies on the Fauna of Curaçao and other Caribbean Islands, Vol. 7, The Birds of Aruba, Curaçao and Bonaire. The Hague: Martinus Nijhoff.

Voous, K.H. 1973. List of recent Holarctic bird species: Non-passerines. Ibis 115: 612–638.

Voous, K.H. 1985. Passeriformes. *In* A Dictionary of Birds. B. Campbell and E. Lack, Eds.: 440–441. Calton: T & AD Poyser.

Vuilleumier, F. 1975. Zoogeography. *In* Avian Biology, Vol. 5, D.S. Farner, J.R. King, and K.C. Parkes, Eds.: 421–496. New York: Academic Press.

Wagner, H.O. 1945. Notes on the life history of the Mexican Violet-ear. Wilson Bull. 57: 165–187.

Walcott, C., and R.P. Green. 1974. Orientation of homing pigeons altered by a change in the direction of an applied magnetic field. Science (Wash., D.C.) 184: 180–182.

Walcott, C., J.L. Gould, and J.L. Kirschvink. 1979. Pigeons have magnets. Science (Wash., D.C.) 205: 1027–1028.

Walkinshaw, L.H. 1963. Some life history studies of the Stanley Crane. Proc. XIII Int. Ornithol. Congr.: 344–353.

Walkinshaw, L.H. 1972. Kirtland's Warbler—Endangered. Am. Birds 26: 3–9.

Wallace, A.R. 1874. Migration of Birds. Nature (Lond.) 10: 459.

Walls, G.L. 1942. The Vertebrate Eye and Its Adaptive Radiation. Cranbrook Inst. Sci. Bull. No. 19. Bloomfield Hills, MI: Cranbrook Institute of Science.

Walsberg, G.E. 1975. Digestive adaptations of *Phainopepla nitens* associated with the eating of mistletoe berries. Condor 77: 169–174.

Walsberg, G.E. 1978. Brood size and the use of time and energy by the Phainopepla. Ecology 59: 147–153.

Walsberg, G.E. 1983. Avian ecological energetics. *In* Avian Biology, Vol. 7, D.S. Farner, J.R. King, and K.C. Parkes, Eds.: 161–220. New York: Academic Press.

Walsberg, G.E., and J.R. King. 1978. The energetic consequences of incubation for two passerine species. Auk 95: 644–655.

Walter, H. 1979. Eleonora's Falcon: Adaptations to Prey and Habitat in a Social Raptor. Chicago: University of Chicago Press.

Walters, J.R. 1982. Parental behavior in lapwings (Charadriidae) and its relationships with clutch sizes and mating systems. Evolution 36: 1030–1040.

Walters, J.R. 1984. The evolution of parental behavior and clutch size in shorebirds. *In* Behavior of Marine Animals, Vol. 5, J. Burger and B.L. Olla, Eds.: 243–287. New York: Plenum Press.

Wangensteen, O.D., D. Wilson, and H. Rahn. 1970. Diffusion of gases across the shell of the hen's egg. Respir. Physiol. 11: 16–30.

Ward, P. 1965. The breeding biology of the Black-faced Dioch *Quelea quelea* in Nigeria. Ibis 107: 326–349.

Ward, P. 1969. The annual cycle of the Yellow-vented Bulbul *Pycnonotus goiavier* in a humid equatorial environment. J. Zool. (Lond.) 157: 24–45.

Ward, P., and A. Zahavi. 1973. The importance of certain assemblages of birds as "information-centres" for food-finding. Ibis 115: 517–534.

Warner, R.E. 1968. The role of introduced diseases in the extinction of the endemic Hawaiian avifauna. Condor 70: 101–120.

Wasserman, F.E. 1977. Mate attraction function of song in the White-throated Sparrow. Condor 79: 125–127.

Waterman, A.J. 1977. The integumentary system. *In* Chordate Structure and Function, (second ed.), A. Kluge, Ed. New York: Macmillan.

Watson, A., and R. Moss. 1972. A current model of population dynamics in Red Grouse. Proc. XV Int. Ornithol. Congr.: 134–149.

Watson, A., and R. Moss. 1979. Population cycles in the Tetraonidae. Ornis Fenn. 56: 87–109.

Watson, A., and R. Moss. 1980. Advances in our understanding of the population dynamics of Red Grouse from a recent fluctuation in numbers. Ardea 68: 103–111.

Watson, G.E. 1963. The mechanism of feather replacement during natural molt. Auk 80: 486–495.

Weathers, W.W. 1979. Climatic adaptation in avian standard metabolic rate. Oecologia (Berl.) 42: 81–89.

Weathers, W.W., W.A. Buttemer, A.M. Hayworth, and K.A. Nagy. 1984. An evaluation of time-budget estimates of daily energy expenditure in birds. Auk 101: 459–472.

Weathers, W.W., and K.A. Nagy. 1980. Simultaneous doubly labeled water ($^3HH^{18}O$) and time budget estimates of daily energy expenditure in *Phainopepla nitens.* Auk 97: 861–867.

Weeden, J.S. 1965. Territorial behavior of the Tree Sparrow. Condor 67: 193–209.

Wegge, P. 1980. Distorted sex ratio among small broods in a declining capercaille population. Ornis Scand. 11: 106–109.

Weller, M.W. 1959. Parasitic egg laying in the Redhead *(Aythya americana)* and other North American Anatidae. Ecol. Monogr. 29: 333–365.

Wells, S., R.A. Bradley, and L.F. Baptista. 1978. Hybridization in *Calypte* hummingbirds. Auk 95: 537–549.

Welsh, D.A. 1975. Savannah Sparrow breeding and territoriality on a Nova Scotia dune beach. Auk 92: 235–251.

Welty, J.C. 1982. The Life of Birds (third ed.). Philadelphia: Saunders.

Wenzel, B.M. 1973. Chemoreception. *In* Avian Biology, Vol. 3, D.S. Farner, J.R. King, and K.C. Parkes, Eds.: 389–415. New York: Academic Press.

Werner, C.F. 1958. Der Canaliculus (Aquaeductus) cochleae und seine Beziehungen zu den Kanälen des IX. und X. Hirnnerven bei den Vögeln. Zool. Jahr. Abt. Anat. Ontog. Tiere 77: 1–8.

West, M.J., and A.P. King. 1980. Enriching cowbird song by social deprivation. J. Comp. Physiol. Psychol. 94: 263–270.

West, M.J., A.P. King, and D.H. Eastzer. 1981. The cowbird: Reflections on development from an unlikely source. Am. Sci. 69: 56–66.

West Eberhard, M.J. 1975. The evolution of social behavior by kin selection. Q. Rev. Biol. 50: 1–33.

West Eberhard, M.J. 1983. Sexual selection, social competition and speciation. Q. Rev. Biol. 58: 155–183.

Wetherbee, D.K., and L.M. Bartlett. 1962. Egg teeth and shell rupture of the American Woodcock. Auk 79: 117.

Wetmore, A. 1936. The number of contour feathers in passeriform and related birds. Auk 53: 159–169.

Wetmore, A. 1960. A classification for the birds of the world. Smithson. Misc. Collect. 139 (11).

White, F.N., G.A. Bartholomew, and J.L. Kinney. 1978. Physiological and ecological correlates of tunnel nesting in the European Bee-eater, *Merops apiaster.* Physiol. Zool. 51: 140–154.

White, F.N., G.A. Bartholomew, and T.R. Howell. 1975. The thermal significance of the nest of the Sociable Weaver *Philetairus socius:* Winter observations. Ibis 117: 171–179.

White, F.N., and J.L. Kinney. 1974. Avian incubation. Science (Wash. D.C.) 186: 107–115.

White, S.C. 1974. Ecological aspects of growth and nutrition in tropical fruit-eating birds. Ph.D. dissertation, University of Pennsylvania, Philadelphia, PA.

White, S.J. 1971. Selective responsiveness by the Gannet *(Sula bassana)* to played-back calls. Anim. Behav. 19: 125–131.

Whittow, G.C., P.D. Sturkie, and G. Stein Jr. 1964. Cardiovascular changes associated with thermal polypnea in the chicken. Am. J. Physiol. 207: 1349–1353.

Wiens, J.A. 1983. Avian community ecology: An iconoclastic view. *In* Perspectives in Ornithology, A.H. Brush and G.A. Clark Jr., Eds.: 355–403. Cambridge: Cambridge University Press.

Wiens, J.A., and J.T. Rotenberry. 1980. Bird community structure in cold shrub deserts: competition or chaos? Acta XVII Congr. Int. Ornithol.: 1063–1070.

Wiley, R.H. 1974. Evolution of social organization and life history patterns among grouse (Aves: Tetraonidae). Q. Rev. Biol. 49: 201–227.

Wiley, R.H., and D.G. Richards. 1982. Adaptations for acoustic communication in birds: Sound transmission and signal detection. *In* Acoustic Communication in Birds, Vol. 1, D.E. Kroodsma and E.H. Miller, Eds.: 132–181. New York and London: Academic Press.

Williams, G.C. 1966. Natural selection, the costs of reproduction, and a refinement of Lack's principle. Am. Nat. 100: 687–690.

Williams, H.W., A.W. Stokes, and J.C. Wallen. 1968. The food call and display of the Bobwhite Quail *(Colinus virginianus).* Auk 85: 464–476.

Williams, J.B., and K.A. Nagy. 1984. Daily energy expenditure of Savannah Sparrows: Comparison of time-energy budget and doubly-labeled water estimates. Auk 101: 221–229.

Williams, T.C., and J.M. Williams. 1978. An oceanic mass migration of land birds. Sci. Am. 239 (4): 166–176.

Willis, E.O. 1967. The behavior of bicolored antbirds. Univ. Calif. Publ. Zool. 79: 1–132.

Willis, E.O. 1972. The behavior of Spotted Antbirds. Ornithol. Monogr. No. 10.

Willis, E.O. 1974. Populations and local extinctions of birds on Barro Colorado Island, Panama. Ecol. Monogr. 44: 153–169.

Willis, E.O. 1980a. Ecological roles of migratory and resident birds on Barro Colorado Island, Panama. *In* Migrant Birds in the Neotropics, A. Keast and E.S. Morton, Eds.: 205–225. Washington, DC: Smithsonian Institution Press.

Willis, E.O. 1980b. Species reduction in remanescent woodlots in southern Brazil. Acta XVII Congr. Int. Ornithol.: 783–786.

Willis, E.O., and Y. Oniki. 1978. Birds and army ants. Annu. Rev. Ecol. Syst. 9: 243–263.

Willis, E.O., D. Wechsler, and Y. Oniki. 1978. On behavior and nesting of McConnell's Flycatcher *(Pipromorpha macconnelli)*: Does female rejection lead to male promiscuity? Auk 95: 1–8.

Willson, M.F. 1971. Seed selection in some North American finches. Condor 73: 415–429.

Willson, M.F. 1974. Avian community organization and habitat structure. Ecology 55: 1017–1029.

Willson, M.F., and J.C. Harmeson. 1973. Seed preferences and digestive efficiency of Cardinals and Song Sparrows. Condor 75: 225–234.

Wilson, B.W. 1980. Birds. New York: W.H. Freeman and Company.

Wilson, A.C., S.S. Carlson, and T.J. White. 1977. Biochemical evolution. Ann. Rev. Biochem. 46: 573–639.

Wilson, E.O. 1975. Sociobiology. Cambridge, MA: Belknap Press

Wiltschko, W. 1982. The migratory orientation of Garden Warblers, *Sylvia borin*. *In* Avian Navigation, F. Papi and H.G. Wallraff, Eds.: 50–58. Berlin: Springer-Verlag.

Wiltschko, R., D. Nohr, and W. Wiltschko. 1981. Pigeons with a deficient sun compass use the magnetic compass. Science (Wash., D.C.) 214: 343–345.

Wimberger, P.H. 1984. The use of green plant material in bird nests to avoid ectoparasites. Auk 101: 615–618.

Wingfield, J.C. 1984. Androgens and mating systems: Testosterone-induced polygyny in normally monogamous birds. Auk 101: 665–671.

Wingfield, J.C., and D.S. Farner. 1978. The endocrinology of a natural breeding population of the White-crowned Sparrow *(Zonotrichia leucophrys pugetensis)*. Physiol. Zool. 51: 188–205.

Winkler, D.W., and J.R. Walters. 1983. The determination of clutch size in precocial birds. *In* Current Ornithology, Vol. 1, R.F. Johnston, Ed.: 33–68. New York: Plenum Press.

Winstanley, D., R. Spencer, and K. Williamson. 1974. Where have all the whitethroats gone? Bird Study 21: 1–14.

Winter, P. 1963. Vergleichende qualitative und quantitative Untersuchungen an der Hörbahr von Vögeln. Z. Morphol. Oekol. Tiere. 52: 365–400.

Winterbottom, J.M. 1971. Priest's Eggs of Southern African Birds. Johannesburg: Winchester Press.

Winternitz, B.L. 1976. Temporal change and habitat preference of some montane breeding birds. Condor 78: 383–393.

Witschi, E. 1935. Seasonal sex characters in birds and their hormonal control. Wilson Bull. 47: 177–188.

Wittenberger, J.F. 1981. Animal Social Behavior. Boston: Duxbury Press.

Wolf, L.L. 1969. Breeding and molting periods in a Costa Rican population of the Andean Sparrow. Condor 71: 212–219.

Wolf, L.L. 1975. Energy intake and expenditures in a nectar-feeding sunbird. Ecology 56: 92–104.

Wolf, L.L. 1978. Aggressive social organization in nectarivorous birds. Am. Zool. 18: 765–778.

Wolf, L.L., and F.B. Gill. 1980. Resource gradients and community organization of nectarivorous birds. Acta XVII Congr. Int. Ornithol.: 1105–1113.

Wolf, L.L., and F.R. Hainsworth. 1971. Time and energy budgets of territorial hummingbirds. Ecology 52: 980–988.

Wolf, L.L., F.R. Hainsworth, and F.B. Gill. 1975. Foraging efficiencies and time budgets in nectar-feeding birds. Ecology 56: 117–128.

Wolf, L.L., F.G. Stiles, and F.R. Hainsworth. 1976. Ecological organization of a tropical, highland hummingbird community. J. Anim. Ecol. 45: 349–379.

Wolfson, A. 1942. Regulation of spring migration in juncos. Condor 44: 237–263.

Wolfson, A. 1954. Sperm storage at lower-than-body temperature outside the body cavity in some passerine birds. Science (Wash., D.C.) 120: 68–71.

Wooller, R.D. 1978. Individual vocal recognition in the Kittiwake Gull, *Rissa tridactyla* (L.). Z. Tierpsychol. 48: 68–86.

Woolfenden, G.E. 1978. Growth and survival of young Florida Scrub Jays. Wilson Bull. 90: 1–18.

Woolfenden, G.E. 1981. Selfish behavior by Florida Scrub Jay helpers. *In* Natural Selection and Social Behavior: Recent Research and New Theory, R.D. Alexander and D.W. Tinkle, Eds.: 257–260. New York: Chiron Press.

Woolfenden, G.E., and J.W. Fitzpatrick. 1978. The inheritance of territory in group-breeding birds. Bioscience 28: 104–108.

Woolfenden, G.E., and J.W. Fitzpatrick. 1984. The Florida Scrub Jay. Princeton, NJ: Princeton University Press.

Wunderle, J.M., Jr. 1981. An analysis of a morph ratio cline in the Bananaquit *(Coereba flaveola)* on Grenada, West Indies. Evolution 35: 333–344.

Wunderle, J.M., Jr. 1983. A shift in the morph ratio cline in the Bananaquit of Grenada, West Indies. Condor 85: 365–367.

Wyles, J.S., J.G. Kunkel, and A.C. Wilson. 1983. Birds, behavior, and anatomical evolution. Proc. Natl. Acad. Sci. U.S.A. 80: 4394–4397.

Wynne-Edwards, V.C. 1962. Animal Dispersion in Relation to Social Behavior. Edinburgh: Oliver and Boyd.

Yasukawa, K. 1981. Song repertoires in the Red-winged Blackbird *(Agelaius phoeniceus)*: A test of the Beau Geste hypothesis. Anim. Behav. 29: 114–125.

Yeagley, H.L. 1947. A preliminary study of a physical basis of bird navigation. J. Appl. Phys. 18: 1035–1063.

Yodlowski, M.L., M.L. Kreithen, and W.T. Keeton. 1977. Detection of atmospheric infrasound by homing pigeons. Nature (Lond.) 265: 725–726.

Yokoyama, K., and D.S. Farner. 1978. Induction of *Zugunruhe* by photostimulation of encephalic receptors in White-crowned Sparrows. Science (Wash., D.C.) 201: 76–79.

Yom-Tov, Y. 1974. The effect of food and predation on breeding density and success, clutch size and laying date of the crow (*Corvus corone* L.). J. Anim. Ecol. 43: 479–498.

Yom-Tov, Y. 1980. Intraspecific nest parasitism in birds. Biol. Rev. Camb. Philos. Soc. 55: 93–108.

Yom-Tov, Y., G.M. Dunnet, and A. Anderson. 1974. Intraspecific nest parasitism in the Starling *Sturnus vulgaris.* Ibis 116: 87–90.

Zach, R. 1979. Shell dropping, decision making and optimal foraging in Northwestern Crows. Behaviour 68: 106–117.

Zach, R., and J.B. Falls. 1977. Influences of capturing prey on subsequent search in the ovenbird (Aves: Parulidae). Can. J. Zool. 55: 1958–1969.

Zahavi, A. 1971a. The function of pre-roost gatherings and communal roosts. Ibis 113: 106–109.

Zahavi, A. 1971b. The social behavior of the White Wagtail *Motacilla alba alba* wintering in Israel. Ibis 113: 203–211.

Zahavi, A. 1974. Communal nesting by the Arabian Babbler: A case of individual selection. Ibis 116: 84–87.

Zahavi, A. 1977. Reliability in communication systems and the evolution of altruism. *In* Evolutionary Ecology, B. Stonehouse and C. Perrins, Eds.: 253–259. Baltimore, London, Tokyo: University Park Press.

Zahavi, A. 1979. Parasitism and nest predation in parasitic cuckoos. Am. Nat. 113: 157–159.

Zahavi, A. 1980. Ritualisation and the evolution of movement signals. Behaviour 72: 77–81.

Zenone, P.G., M.E. Sims, and C.J. Erickson. 1979. Male Ring Dove behavior and the defense of genetic paternity. Am. Nat. 114: 615–626.

Ziegler, H.P. 1964. Displacement activity and motivational theory: A case study in the history of ethology. Psychol. Bull. 61: 362–376.

Zimmer, J.T. 1926. Catalogue of the Edward E. Ayer ornithological library. Field Mus. Nat. Hist. Publ. Zool. Ser. No. 16.

Zink, R.M. 1982. Patterns of genic and morphologic variation among sparrows in the genera *Zonotrichia, Melospiza, Junco,* and *Passerella.* Auk 99: 632–649.

Zink, R.M. 1986. Patterns and evolutionary significance of geographic variation in the *shistacea* group of the fox sparrows *(Passerella iliaca).* Ornithol Mongr. No. 40.

Zink, R.M., and G.F. Barrowclough. 1984. Allozymes and song dialects: reassessment. Evolution 38: 444–448.

Zink, R.M., and J.V. Remsen Jr. 1986. Evolutionary process and patterns of geographical variation in birds. Current Ornithol. 4: 1–69.

Ziswiler, V., and D.S. Farner. 1972. Digestion and the digestive system. *In* Avian Biology, Vol. 2, D.S. Farner, J.R. King, and K.C. Parkes, Eds.: 343–430. New York: Academic Press.

Zusi, R.L. 1984. A functional and evolutionary analysis of rhynchokinesis in birds. Smithson. Contrib. Zool. No. 385.

Zusi, R.L. 1986. Birds: Charadriiformes (plovers, sandpipers, gulls, terns, auks). *In* Encyclopaedia Britannica (15th ed.) Macropaedia, Vol. 15: 54–63. Chicago: Encyclopaedia Britannica, Inc.

Zusi, R.L., and G.D. Bentz. 1982. Variation of a muscle in hummingbirds and swifts and its systematic implications. Proc. Biol. Soc. Wash. 95: 412–420.

Zusi, R.L., and G.D. Bentz. 1984. Myology of the Purple-throated Carib *(Eulampis jugularis)* and other hummingbirds (Aves: Trochilidae). Smithson. Contrib. Zool. No. 385.

Zusi, R.L., and D. Bridge. 1981. On the slit pupil of the Black Skimmer *(Rynchops niger).* J. Field Ornithol. 52: 338–340.

Zusi, R.L., and J.R. Jehl. 1970. The systematic relationships of *Aechmorhynchus, Prosobonia,* and *Phegornis* (Charadriiformes; Charadrii). Auk 87: 760–780.

Index

Frank B. Gill is curator and chairman of the Department of Ornithology at The Academy of Natural Sciences, Philadelphia, and adjunct professor of biology at The University of Pennsylvania. He received a B.S. and a Ph.D. in zoology from The University of Michigan. He has twice served as vice president of the American Ornithologists' Union and is an elected member of the International Ornithological Committee. In 1988, he received the Linnaean Society's Eugene Eisenmann Medal for excellence in ornithology and encouragement of the amateur. Dr. Gill's research interests include the feeding behavior and ecology of hummingbirds and sunbirds, the evolution of chickadees, and the evolution of new species of birds. He has participated in expeditions to South America, Africa, and the Indian Ocean, and he has written more than 80 research articles, reviews, and popular articles. He initiated the VIREO collection of bird photographs at The Academy of Natural Sciences and works extensively with both professional and amateur ornithological organizations.

James E. Coe received an A.B. from Harvard University and an M.F.A. from Parsons School of Design. He has contributed illustrations to *Field Guide to the Birds of New Guinea* (Princeton University Press) and to the *Audubon Society Master Guide to Birds* (Knopf). He illustrated *Birds of North America: West* (Macmillan) and is currently writing and illustrating two regional guides for beginner birdwatchers (Golden Press). He has exhibited his bird paintings at the Leigh Yawkey Woodson Art Museum (Wisconsin), the Cornell Laboratory of Ornithology, and the Museum of Comparative Zoology at Harvard University.

Guy Tudor's wildlife illustrations have appeared in many books and periodicals. His recent work includes serving as principal illustrator for *Manual of Neotropical Birds,* volume I (University of Chicago Press), *Complete Guide to North American Wildlife* (Harper and Row), and *A Guide to the Birds of Columbia* (Princeton University Press). He is principal illustrator and author of the field notes for *A Guide to the Birds of Venezuela* (Princeton University Press), and is coauthor (with Robert S. Ridgely) and illustrator of the four-volume *The Birds of South America* (University of Texas Press and World Wildlife Fund). Mr. Tudor's bird paintings have appeared in exhibits at the Field Museum of Natural History in Chicago, the Leigh Yawkey Woodson Art Museum (Wisconsin), the National Collection of Fine Arts, and the Princeton University Library. He is an elective member of the American Ornithologists' Union and of the American Birding Association.